Astrochemistry at high resolution

Space Telescope Science Institute, Baltimore, USA and online
31 May - 2 June 2023

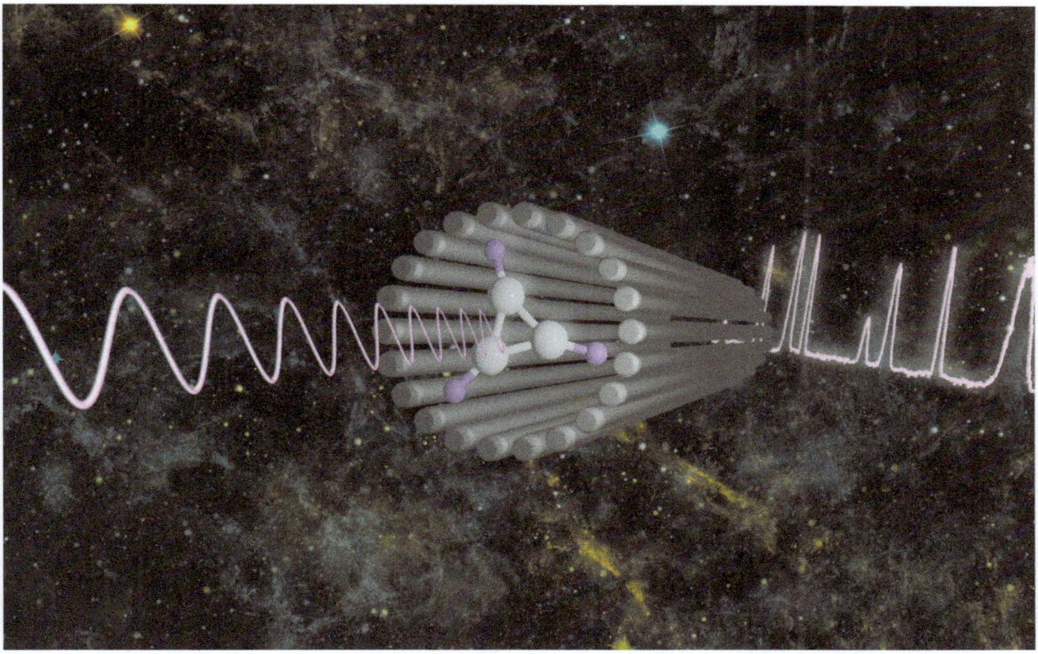

FARADAY DISCUSSIONS

Volume 245, 2023

ROYAL SOCIETY OF CHEMISTRY

The **Faraday Community for Physical Chemistry** of the Royal Society of Chemistry, previously the Faraday Society, was founded in 1903 to promote the study of sciences lying between chemistry, physics and biology.

Faraday Discussions (Print ISSN 1359-6640, Electronic ISSN 1364-5498) is published 8 times a year by the Royal Society of Chemistry, Thomas Graham House, Science Park, Milton Road, Cambridge, UK CB4 0WF.

Volume 245 ISBN 978-1-83767-092-5

2023 annual subscription price: print+electronic £1223 US $2154; electronic only £1165, US $2051.
Customers in Canada will be subject to a surcharge to cover GST. Customers in the EU subscribing to the electronic version only will be charged VAT.

All orders, with cheques made payable to the Royal Society of Chemistry, should be sent to the Royal Society of Chemistry Order Department, Royal Society of Chemistry, Thomas Graham House, Science Park, Milton Road, Cambridge, CB4 0WF, UK
Tel +44 (0)1223 432398; E-mail **orders@rsc.org**

If you take an institutional subscription to any Royal Society of Chemistry journal you are entitled to free, site-wide web access to that journal. You can arrange access via Internet Protocol (IP) address at **www.rsc.org/ip**

Customers should make payments by cheque in sterling payable on a UK clearing bank or in US dollars payable on a US clearing bank.

Printed in the UK

MIX
Paper from responsible sources
FSC® C013604
www.fsc.org

ROYAL SOCIETY OF CHEMISTRY

Faraday Discussions

Faraday Discussions are unique international discussion meetings that focus on rapidly developing areas of chemistry and its interfaces with other scientific disciplines.

Information for Authors

Astrochemistry at High Resolution

Faraday Discussions

www.rsc.org/faraday_d

A General Discussion on Astrochemistry at High Resolution was held in Baltimore, USA and online on the 31st May, 1st June and 2nd June 2023.

The Royal Society of Chemistry is the world's leading chemistry community. Through our high impact journals and publications we connect the world with the chemical sciences and invest the profits back into the chemistry community.

CONTENTS

ISSN 1359-6640; ISBN 978-1-83767-092-5

Cover
See Asvany *et al.*, *Faraday Discuss.*, 2023, **245**, 298–308.

Spectrum of a cyclic hydrocarbon ion measured using an ion trap is shown in the context of space.

Image reproduced with permission of Divita Gupta from Asvany *et al.*, *Faraday Discuss.*, 2023, **245**, 298–308.

PREFACE

INTRODUCTORY LECTURE

PAPERS AND DISCUSSIONS

CONCLUDING REMARKS

ADDITIONAL INFORMATION

Faraday Discussions

ROYAL SOCIETY OF CHEMISTRY

EDITORIAL

Astrochemistry at high resolution: preface

Martin R. S. McCoustra

DOI: 10.1039/d3fd90035e

The journey that brought us to the Space Telescope Science Institute in a warm and humid Baltimore in the early summer of 2023 began nearly seven years ago in 2016 with some preliminary discussions between members of the RSC Astrophysical Chemistry Group and the ACS Astrochemistry Sub-division committees at the Autumn ACS Meeting in Philadelphia. We recognised that our window on the Universe was rapidly expanding as new observational platforms were being developed. With ALMA (the Atacama Large Millimeter Array) in operation, JWST (the James Webb Space Telescope) then just about to fly, E-ELT (the European Extremely Large Telescope) breaking ground and several other resources well advanced in their planning, the chemically controlled cosmos is becoming even more open to our gaze. We were entering the age of *High Resolution Astrochemistry*. High resolution *spectroscopically*, as these new platforms offer unique insight into chemically rich environments at higher spectral resolution and sensitivity than many previous platforms for studying both the astrophysical gas and solid states. High resolution *spatially*, as these platforms offer unsurpassed chemical imaging capabilities in nearby astronomical objects at every evolutionary stage.

Astrochemical research truly requires a multidisciplinary approach, bringing together researchers from astronomy, and the gas phase, solid state and surface physics and chemistry communities of a range of flavours. Previous *Faraday Discussions* in this young discipline have explored aspects of the gas phase chemistry in astronomical environments (*Faraday Discussions* 109 and 133); have addressed the multitudinous roles of dust and ice found in large abundances in star and planet forming environments (*Faraday Discussion* Number 168); and have revealed the chemical complexities of planetary science (*Faraday Discussion* Number 147) in our own astronomical backyard and more widely.

We agreed that it was timely that a *Faraday Discussion* should seek to address the wide-ranging roles of chemistry in astronomy in the light of the wealth of observational detail becoming available in this era of *High Resolution Astrochemistry*. The discussion should explore the role of chemical spectroscopy in identifying chemical species and probing their environments; should attempt to understand the complex combination of gas phase and surface and solid-state chemistry that drives the chemical evolution of our Galaxy and others; explore

Institute of Chemical Sciences, Heriot-Watt University, Edinburgh, EH14 4AS, UK. E-mail: m.r.s.mccoustra@hw.ac.uk

how this chemical evolution is intimately involved in controlling the process of star and planet formation; and how chemistry brings us to the edge of biology. The scientific programme would bring together leading computational and experimental scientists working to unlock the secrets of astrochemistry with astronomers engaged in observing and understanding star and planet formation; and potentially chemists and biologists seeking to understand the first tentative steps toward life on our own planet and others!

Such a discussion meeting would be very timely. Existing and new observational platforms are delivering and will deliver increasingly detailed information on the chemical composition of our Galaxy and beyond. We are beginning to reveal in intimate detail the progress of cold and dense gas and dust through to stars and planets and onwards. We may even have the tools to detect the signatures of life on planets in nearby star systems. This is naught but stamp collecting unless we can understand what we see! The key role played by astrochemistry in this story is thoroughly accepted. Indeed, over the last two decades more than 25 million euros have been invested in developing structured astrochemical research programmes across Europe at national and international level. One can point to the Molecular Universe and LASSIE (Laboratory Astrochemical Surface Science in Europe) training networks in Framework 7; and to the EuroPAH training network in Horizon 2020 which played a powerful coordinating role in driving cross-Europe collaboration in astrochemistry by providing the necessary manpower to pursue astrochemical research objectives funded in part by national investment. These, of course, parallel in time the multi-billion euro investment in new observational platforms which are now beginning to deliver on their astronomical promise. A *Faraday Discussion* timed to coincide with the first tranche of observations from JWST and building on the insights provided by ALMA will attract a substantial audience and significant attention worldwide. It will also provide an ideal opportunity to discuss the future direction of such research by identifying the still unanswered questions – many of which can only be derived from new observational data.

Little did we appreciate in 2018 when this proposal was accepted by the Royal Society of Chemistry that it would be 5 years before we met in Baltimore. Delays in launching JWST and the COVID pandemic ensured that this discussion was a moveable feast. However, we finally made it! I hope that the readers of this Discussion volume will feel the real excitement that those of us in attendance, both physically on site and remotely *via* internet connections, felt during this discussion in being part of the rising tide of astrochemical research prompted by the observational revolution that we have seen with JWST.

Martin R. S. McCoustra
Chair on behalf of the Scientific Organising Committee
Ian R. Sims, Paola Caselli, Susanna Widicus Weaver, Anthony J. H. M. Meijer, Nick Walker, and Neill Reid

Faraday Discussions

PAPER

Spiers Memorial Lecture: Astrochemistry at high resolution

Cecilia Ceccarelli

Received 22nd May 2023, Accepted 25th May 2023
DOI: 10.1039/d3fd00106g

Astrochemistry is the science that studies the chemistry in the Universe, namely the combination of two fields: astronomy and chemistry. It started about fifty years ago and it has progressed in leaps and bounds, often triggered by the advent of new telescopes. From the collection of new interstellar molecule detections, astrochemistry has evolved more and more in the quest to understand how they are formed and thrive in the harsh conditions of the interstellar medium. Collaboration between astronomers and chemists has never been more necessary than today, when new powerful astronomical facilities provide us with ever sharper images of the regions where interstellar molecules are present. This review focuses on the special case of interstellar complex organic molecules (iCOMs), one the most debated astrochemical fields and where the astronomers–chemists collaboration and synergy is indispensable. The review will go through the various phases of the formation of planetary system similar to the solar system, providing the most recent observational picture at each step. The current scenarios of the iCOMs formation will be laid down and the critical chemical processes and quantities involved in each of them will be discussed. The major goal of this review is not only to present the progress but, more importantly, to highlight the many areas of uncertainty. A few specific cases will be discussed to give practical examples of why the huge challenge that represents the formation of iCOMs can only be won if chemists and astronomers work together.

1 Introduction

Why do we need a *Faraday Discussion* on astrochemistry at high resolution? The images reported in Fig. 1 say more than a thousand words: high resolution observations at different wavelengths obtained in the last decade have added so many details that astrochemistry now faces new challenges, adding to the old ones, to explain them. The figure shows only one of the numerous examples that could have been given. It shows the famous solar-type protostar NGC1333 IRAS4, a protobinary system composed of two objects, A1 and A2, separated by about $1''.8$ in the sky, equivalent to about 550 au. The figure shows methanol emission maps obtained with three different interferometers: SMA in the submillimeter

Univ. Grenoble Alpes, CNRS, IPAG, 38000 Grenoble, France. E-mail: Cecilia.Ceccarelli@univ-grenoble-alpes.fr

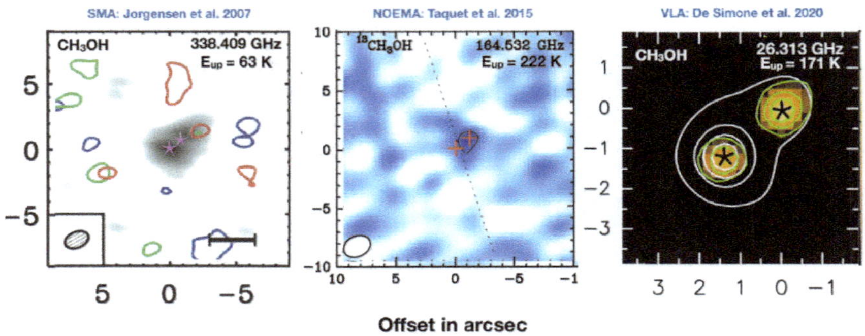

Fig. 1 Images of the methanol line emission towards the solar-type protostar NGC1333 IRAS4A, obtained with three interferometers at different wavelengths. *Left panel*: SMA map in the submillimeter range (the gray image is the continuum emission, whereas the green contours show the CH_3OH emission at the systemic velocity, and the red and blue contours at red-shifted and blue-shifted velocities, respectively).[1] *Central panel*: NOEMA map in the millimeter range (the image and contours show the CH_3OH emission).[2] *Right panel*: VLA map in the radio range (the color image is the continuum emission whereas the white contours show the CH_3OH emission).[3] IRAS4A is a binary system whose two objects, A1 and A2, are indicated by stars (left panel), crosses (central panel) and asterisks (right panel).

(submm),[1] NOEMA in the millimeter (mm),[2] and VLA in the radio range.[3] In the submm image methanol is not detected at the systemic velocity of the source and only marginally at red-shifted ones; in the mm it is detected towards A2 only; in the radio, methanol is clearly detected in both objects, A1 and A2. There are various conclusions from these images, the two most important of which are: both the spatial resolution and the wavelength of observations matter. Based on the first image, Jorgensen and collaborators concluded that methanol emission is associated with the outflow emanating from A2;[1] based on the second image, Taquet and collaborators and, later, López-Sepulcre and collaborators, concluded that only A2 possesses a hot corino;[2,4] finally, De Simone and collaborators found that, actually, both A1 and A2 possess hot corinos with similar abundance[3] (we will discuss later in this review the reason for these differences). Evidently, where methanol is present is crucial information for any theory of its formation and destruction, *i.e.*, for astrochemical theories. Needless to say, this applies to any of the almost 300 molecules detected so far in the interstellar medium (ISM).

In this review, I will try to lay down the observational scenario of the early stages of the solar-type star formation, focusing in particular on the class of species called interstellar complex organic molecules, here called interstellar complex organic molecules (iCOMs),[5,6] and the challenges that they represent to our current knowledge. Since their formation and destruction strongly depend on the formation of the ices that envelope the interstellar grains during the star and planet formation, the review will also briefly cover this aspect. Of course, this is only one part of the astrochemistry field. I chose to restrict this review because trying to discuss everything would end in discussing nothing in sufficient detail, so I decided to focus on something precise, choosing one of the currently most active fields of astrochemistry. Even with such a restriction, this review will be far from exhaustive so I will try to emphasize the most recent advances and challenges on the subject. I forward the interested reader to other recent reviews on the subject of astrochemistry during solar-type star and planet formation, where also different aspects and opinions are reported.[7–12]

Since this *Faraday Discussion* conference gathers together chemists and astronomers, I will try to have an interdisciplinary language understandable by the two communities, which may seem simplistic to either of the two communities at times, but I hope that in this way we understand each other on all the points discussed in this review.

Following this approach, this review will first summarise the overall framework of the solar-type formation process, with its major components and their physical and chemical characteristics (Section 2). I will then review the major observational characteristics of iCOMs during the solar-type star formation and in its different objects, with emphasis on recent high spatial resolution observations (Section 3). Third, I will discuss the major chemical processes occurring at each step and the critical quantities that regulate them, with the aim to emphasize what we do not know yet more than what we do (Section 4). In the subsequent section, I will provide some specific examples, with the goal of indicating what we can extract by comparing astronomical observations with the present astrochemical knowledge (Section 5). Finally, I will list the major conclusions and the take-home messages (Section 6).

2 From a molecular cloud to a planetary system

The process that leads the diffuse matter of a molecular cloud to form a solar-type star and its planetary formation, is long and complicated. At present, we have an overview of what we think happens, but several, even critical, passages are still obscure, as we will see in this review. That said, from a chemical point of view and for the context of this *Faraday Discussion*, four major phases can be identified, also corresponding to major changes and events in the physical processes and structure of the matter. Please note that previous reviews[7–12] exist on the same argument by several authors and with more details, depending on the review focus, so that here I just briefly emphasize the points important to understand the chemistry. The four phases identified here are schematically shown in Fig. 2, while the physical parameters of the objects of each phase are summarised in Fig. 3. In the following sub-sections, I briefly describe them.

2.1 Phase 1: prestellar core phase

The story starts from a dense clump of material in a molecular cloud of the Galaxy. Under the gravitational force, which is contrasted by the hydrostatic pressure, the presence of magnetic fields and the turbulence in the gas, matter slowly accumulates towards the clump center. The clumps that have enough mass to eventually overcome all forces acting against the gravitation are called prestellar cores (PSC).

Typically, PSC have a radius of $\sim$5000 au, a centrally condensed density structure, with the innermost part ($\leq$1000 au in radius) flattened at a density of 10^6–10^7 cm^{-3}, as summarised in Fig. 3. The temperature, in contrast, decreases towards the center: it is 10–20 K at the border of the PSC and decreases to 6–8 K at its center. The large density (with respect to the 10^4 cm^{-3} parental molecular

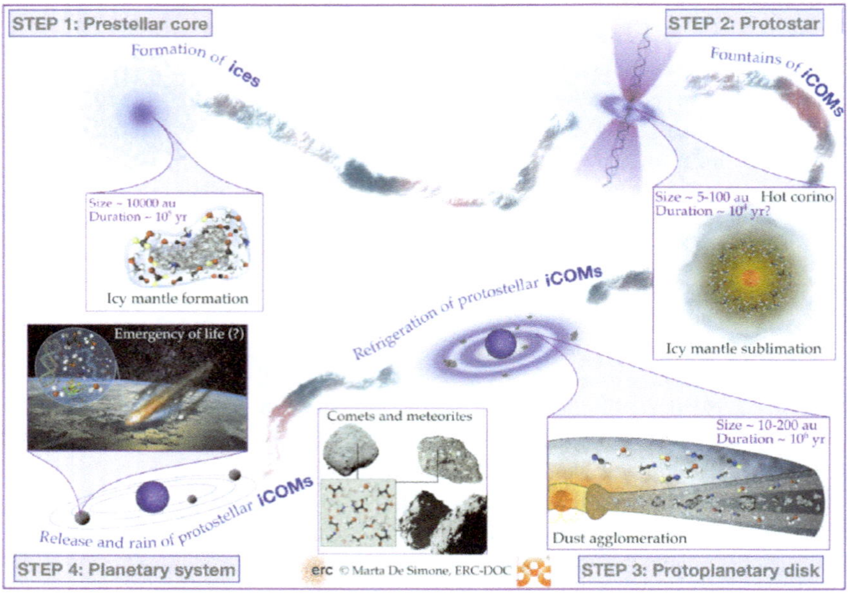

Fig. 2 Sketch of the four major steps involved in the formation of a solar-type planetary system and its iCOMs (courtesy of M. De Simone), as described in the text. Step 1: In the cold ($\leq$10 K) and dense ($\geq 10^5$ cm^{-3}) prestellar cores, ices are formed on the surface of the interstellar grains. Step 2: In the hot corinos of and shocks surrounding, young protostars, iCOMs are observed in large quantities. Step 3: In protoplanetary disks, the hot corino iCOMs are frozen into ices enveloping the grains. Step 4: The dust grains coagulate into pebbles and larger rocks that will eventually become planets, asteroids and comets; some of the ices fabricated in steps 1 and 2 are in this way transmitted to the final planetary system.

cloud) and low temperature across the PSC have very important "chemical" consequences. First, the formation of thick icy mantles around the dust grains; second, the disappearance in the inner 1000 au region of gaseous CO and, more generally, any species containing elements heavier than H and D; third, an increase of the molecular deuteration of all species formed in this stage, either in the gas or on the grain surfaces.

2.2 Phase 2: protostellar phase

Once the gravitational collapse starts, a protostar is formed, consisting of one or more central objects (the future stars), their circumstellar disks (with radii of $\leq$100 au), a possible circumbinary disk (radius $\sim$100–1000 au) and an extended envelope (radius $\sim$300–5000 au). Since the gravitational energy of the infalling material is ultimately released at the center of the protostar, the temperature increases from 10–20 K at the envelope outer border to $\geq$100 K in the central (typically) $\leq$100 au region. Likewise, the density increases towards the center.

While infalling, matter also rotates and, thus, to eventually accrete into the future star(s), it has to loose its angular momentum. This happens *via* the creation of the above-mentioned disks and the ejection of a substantial fraction of material, through what are known as jets and molecular outflows. On the other

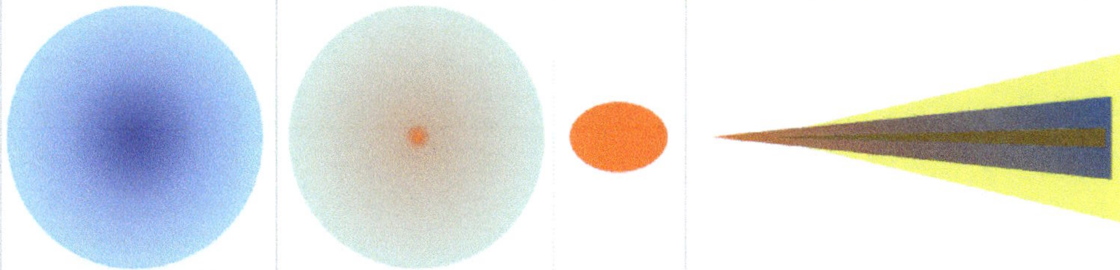

Component/ Parameter	Prestellar cores		Protostars			Protoplanetary disks			
	central zone	outer border	hot corino	outer envelope	molecular shocks	inner midplane	outer midplane	cold midplane	atmosphere
Life time [yr]	~10^5		10^4–10^5		≤2000−3000	~10^6			
Radius [au]	1000	3000−5000	≤100	1000-5000		≤10	≥10	10−100	10−100
Density [cm^{-3}]	10^6-10^7	10^5–10^4	≥10^7	10^6-10^4	10^5-10^8	≥10^{11}	10^5-10^{11}	≥10^6	≤10^4
Dust temperature [K]	6−8	10−20	≥100	50−10	10−20	≥100	20−≤100	≤20	20−100
Gas temperature [K]	6−8	10−20	50−≥100	50−10	50−2000	≥100	≥20	≤20	≥100
Chemical properties	heavy species depletion; super-deuteration	molecules & iCOM with low abundance	molecules & iCOM with high abundance	molecules & iCOM with low abundance	molecules & iCOM with high abundance	molecules & iCOM with high abundance	molecules & iCOM with low abundance	heavy species depletion; super-deuteration	UV-illuminated; ions and atoms
Most abundant C- & O- species		CO	H_2O, CO, O	CO	H_2O, CO, O	H_2O, CO_2, C_2H_2	CO		C^+, O
Most abundant ions	H_3^+, H_2D^+, HD_2^+	H_3^+, HCO^+, He^+	H_3O^+, H_3^+, HCO^+, He^+	H_3^+, HCO^+, He^+	H_3O^+, H_3^+, HCO^+, He^+	H_3O^+, H_3^+, HCO^+, He^+	H_3^+, HCO^+, He^+	H_3^+, H_2D^+, HD_2^+	C^+, S^+
Ionisation fraction	≤10^{-8}	$10^{-8}-10^{-7}$	≤10^{-9}	$10^{-8}-10^{-7}$	~$10^{-7}-10^{-6}$	≤10^{-8}	≤10^{-7}	≤10^{-8}	~10^{-4}

Fig. 3 Physical and chemical properties of the objects of the first three phases of the solar-type star and planet formation, as described in the text (Section 2): prestellar cores, protostars and protoplanetary disks. The values in the table are rough estimates to provide orders of magnitude rather than precise estimates, which depend on the specific object of study.

hand, the accretion of the matter from the rotating-infalling envelope towards the center likely proceeds in two major steps: trough large scale (~100–1000 au) streamers of material crashing onto the disk(s) and from the disk(s) onto the central objects. Both the ejected and the streaming material create shocks when they encounter the quiescent one, in the envelope and disk(s).

From a chemical point of view, the envelope can be divided in roughly three zones. Starting from the border and going inward they are: (i) the outer border, which is similar to PSC (as it is its evolution); (ii) the lukewarm envelope, where some species such as CO and CH_4 are desorbed from the grain mantles and reappear in the gas-phase; (iii) the hot envelope, where the dust temperature exceeds ~100 K and the entire (or most of the) mantle is sublimated, releasing all the frozen species in the gas-phase. This region is called the hot corino if it contains iCOMs. Please note that Fig. 3 only summarizes the physical and chemical parameters of the hot corino and the outer envelope, for simplicity.

In shocked regions, the gas temperature and density increase with respect to the quiescent gas, by a factor that depends on the velocity of the shock. Typically, temperature increases up to a few hundred K and density to 10^5–10^8 cm^{-3} in molecular shocks around solar-type protostars. Despite the dust remaining cold (at about 10–20 K, as the shock is inefficient to heat it), frozen species are released into the gas-phase, either because of the sputtering or mechanical destruction of the grain mantles.

2.3 Phase 3: protoplanetary disk phase

With time, the protostellar envelope dissipates and the circumstellar disk, now also called protoplanetary disk, appears surrounding an older protostar, a pre-main-sequence (PMS) star. The disk is exposed to the photons of the central PMS star and the interstellar radiation, which has consequences on the disk structure and chemical composition. First, the disk is flared, that is, its thickness increases with the radius. The disk density in the midplane increases inwardly, reaching $\geq 10^{11}$ cm^{-3} in the inner 10 au zone, while it decreases going horizontally away from the midplane. Likewise, the temperature increases going inward and upward.

From a chemical point of view, the disk can be divided in three zones: (i) the atmosphere, directly exposed to the UV, is similar to the PDRs (photo-dissociation regions) that surround the molecular clouds where atoms and ions dominate; (ii) the molecular layer, where several molecules are in the gas-phase, similar to the protostellar lukewarm envelope, but where UV photons may still have a role; (iii) the midplane, where chemistry is sort of similar to that in the protostellar envelopes: heavy-element bearing species are mostly frozen-onto the grain mantles in the outer zone (similar to the outer envelope), molecules are in the gas-phase in the intermediate zone (as in the lukewarm envelope), and mantles entirely sublimate in the inner midplane (as in the hot inner envelope). Note, however, that the icy mantles in the midplane also contain the new molecules formed in the prestellar and protostellar phases, even though it is not clear in what fraction. *Mutatis mutandis*, the chemistry in the circumstellar and circumbinary disks present in the protostellar phase is similar to that described here, except that these disks are surrounded by the envelope and, therefore, are less exposed to the UV photons.

That said, two important differences exist between the protostellar envelope and protoplanetary disk chemistry. First, the protoplanetary disks are thinner than the envelopes, and their chemistry is impacted more by the presence of UV photons, which are scattered across the molecular layer although not enough to photo-dissociate molecules. Second, the sizes of the dust grains hugely evolve with time: at the beginning the average grain radius is approximately ~0.1 μm as in molecular clouds, while at the end of the protoplanetary disk evolution, dust grains become planets, asteroids and comets, as described in the next phase. In between, roughly speaking, grains in the midplane coagulate and migrate inward and outward, whereas grains out of the plane fragment and migrate vertically.

2.4 Phase 4: planetary system formation and heritage

As said, grains in the disk midplane coagulate and grow until they become the seeds of giant gaseous planets, rocky planets, asteroids and comets. This evolution is not completely understood, but obviously it happens. Currently, we think that it occurs *via* four major steps: (i) interstellar grain coagulation into pebbles of ~10 cm; (ii) coagulation of pebbles into planetesimals of ~1 km; (iii) formation and migration of giant planets; (iv) formation of rocky planets, asteroids and comets.

From a chemical point of view and in the context of this *Faraday Discussion*, two major processes are important. First, the coagulation of the interstellar grains enveloped by the icy mantles formed during the previous three phases means that some of these ices are trapped in the interior of the pebbles and planetesimals, and so are protected. They constitute the heritage of the earliest phases of the planetary system formation. For example, water may be brought to the nascent planet by the rocky bricks that constitute it, as in the case of Earth.[13–15] More in general, the trapped ices may be conserved almost intact in the interior of the most pristine bodies of the planetary system, as in the case of solar system comets.

Second, the composition of the formed giant planets depend on their position in the disk, as the major carriers of the elements (*e.g.* H_2O, CO and CO_2 for O, or N_2 and N_3 for N, *etc.*) are frozen in different zones of the disk midplane. In principle, this can be used to pinpoint where a solar and/or extra-solar giant planet has formed in the disk, despite it now being at a different position because of the planet migration.[11,16] In practice, this is probably more difficult to obtain for the various uncertainties involved in the whole process,[17] some of which regard chemical properties, principally the binding energies of the various species (see below).

3 Observations of iCOMs and ices in solar-type star forming regions

iCOMs are here defined as species with at least six atoms of which at least one carbon and another heavy-element, such as O, N, S and so on. iCOMs represent a special class of the almost 300 detected interstellar species, because of their potentiality of being precursors of prebiotic molecules. For this reason, since their first detection in the 70s they have aroused curiosity and interest, and, at the same time, have started a decade long quarrel on how iCOMs are synthesised in the harsh conditions of the ISM. After all these years, there is still no consensus. This is also the reason why this review focuses on iCOMs.

In the following, I will summarise the observations towards the objects that represent different evolutionary stages or important components of the solar-like star formation process, emphasizing, when relevant, the high spatial resolution observations.

3.1 The cool chemistry in prestellar cores

3.1.1 The super-deuteration phenomenon. Before introducing the observed iCOMs in the PSC, it is important to emphasize a major characteristics of the chemistry in these objects, the so-called super-deuteration phenomenon. In PSC, the D- over H-bearing abundance ratio is much enhanced with respect to the elemental D/H ratio, which is a number that depends on the baryon to photons ratio in the Universe and it is fixed, as D was only formed at the Big Bang. Various observations constrain the elemental D/H equal to 1.6×10^{-5}.[18] If D and H were statistically distributed in molecules, then the D- over H-bearing abundance ratio of a molecule would be $\sim 10^{-5}$, the D_2 over H_2 ratio $\sim 10^{-10}$ and the D_3 over H_3 ratio $\sim 10^{-15}$. In contrast, the observed D_2CO/H_2CO abundance ratio in PSC is 0.01–0.1,[19] namely up to eight orders of magnitude larger than the elemental D/H.

Nowadays, we think we have understood reasonably well why this happens. In cold molecular gas, D is largely locked into HD. The most important way to "extract" D from HD and pass it to other species is *via* the reaction of HD with H_3^+, the most abundant ion (which is created by a chain of reactions started by the interaction of cosmic-rays with H_2). Once every three times H_2D^+ is formed. Of course, H_2D^+ reacts back with H_2 and, in the absence of peculiar conditions, the H_2D^+/H_3^+ abundance ratio is equal to 1/3 the elemental D/H. However, given the different zero point energy ($\sim$230 K) of H_2D^+ and H_3^+, the $H_2D^+ + H_2$ reaction, in cold enough gas, is much slower than the $H_3^+ +$ HD one. This causes the H_2D^+/H_3^+ abundance ratio to be enhanced with respect to the elemental D/H. Then, the species formed from H_2D^+ and H_3^+ will inherit the D/H enhancement. This is the case for H_2CO and its D-isotopologues (see later).

We have already seen that the temperature is very low in PSC, which is the first reason why the observed D_2CO/H_2CO abundance ratio in PSC is high. In addition, given the PSC low temperatures and large densities, CO, the most abundant gaseous neutral species in molecular clouds, freezes onto the dust grain surfaces disappearing from the gas-phase. Since CO is a major route of destruction of H_3^+, its disappearance from the gas-phase makes the destruction route of H_3^+ towards H_2D^+ increase, further enhancing the H_2D^+/H_3^+ abundance ratio.

In summary, any large molecular deuteration has origin in the prestellar phase. As we will see later in more detail, some iCOMs (and other molecules) in warm environments, such as hot corinos and shocks, are observed to have high D/H ratios. Since the temperature in these objects is too high for the deuteration to be a present day product, their high D/H ratio is indeed inherited, in one way or another, from the PSC phase.

3.1.2 iCOMs in the skin of prestellar cores. Just over a decade ago, two articles shook the implicit assumption that iCOMs can only form in warm environments. Bacmann *et al.*[20] and Cernicharo *et al.*[21] detected several iCOMs in PSC: methanol, acetaldehyde, methyl formate, dimethyl ether, and methyl cyanide.

Additional works have shown that iCOMs are also present in the cold envelopes of protostars.[22,23] Following up those first discoveries, more work showed the almost ubiquitous presence of methanol and acetaldehyde, the two easiest iCOMs to detect, in PSC.[24–28]

While now there is no doubt on the presence of iCOMs in PSC, it is important to pin down where their line emission comes from, both for correctly estimating their abundance and understanding their formation mechanism. Vastel and collaborators, based on a non-LTE radiative transfer analysis, argued that the bulk of the methanol line emission in L1544, the most studied PSC, originates in the cold ($\sim$10 K) and less dense ($\sim 2 \times 10^4$ cm^{-3}) gas at the border ($\sim$8000 au from the center) of the condensation.[24] Methanol line emission maps towards the same source confirmed that it does not come from the core itself but rather from the outer part of the L1544 condensation, less dense and exposed to weak UV illumination.[25,29] This analysis has been done for methanol only, but it is more than likely that the other iCOMs also originate in the skin of the PSC, a hypothesis strengthened by the observations of other iCOMs in the methanol peak of L1544.[26]

The location of the detected iCOMs means that it is very difficult to estimate their absolute abundance, as the H_2 column density is usually evaluated from the dust continuum emission, which only gives the integral towards the line of sight. Thus, if iCOMs are located in only a part of the line of sight probed by the dust emission, the derived abundance is a lower limit. Therefore, the most reliable quantity is the abundance ratio with respect to another species which emits in the same region. Usually, methanol, which is the most abundant iCOM, is taken as reference. Then, the survey of methanol and acetaldehyde carried out by Scibelli and Shirley,[28] of more than two dozen PSC in the Taurus Molecular Cloud, the largest PSC sample so far, indicates that [CH_3OH]/[CH_3CHO] abundance ratio is about 10, within a factor of 5 (Fig. 4) (see also ref. 30). The measurements of L1689B estimate an abundance of methyl formate similar to that of acetaldehyde and about three times larger than dimethyl formate.[20] As an indication, in L1544,

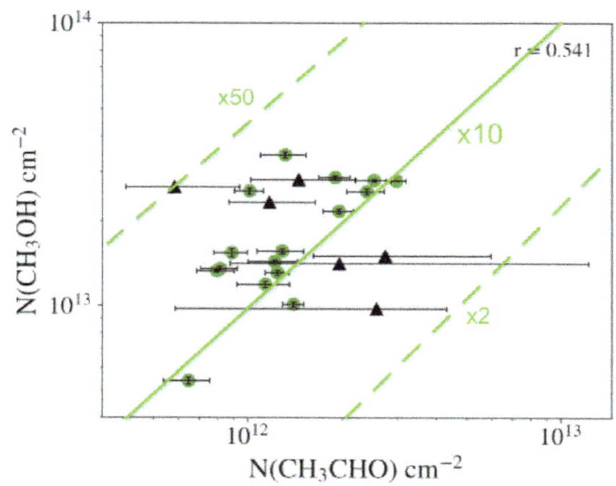

Fig. 4 Column density of methanol (*y*-axis) and acetaldehyde (*x*-axis) measured in pre-stellar cores (black circles with their error bars).[28] The green lines show the ratio [CH_3OH]/[CH_3CHO] equal to 10 (solid line) and multiplied and divided by a factor of 5 (dashed lines).

Table 1 List of the iCOMs detected in prestellar cores and at least three solar-type hot corinos, and their average measured abundance with respect to methanol

		[X]/[CH_3OH]	
Name	Formula	Prestellar cores	Hot corinos
Methanol	CH_3OH	1	1
Acetaldehyde	CH_3CHO	~0.1	~0.1
Ethanol	CH_3CH_2OH	—	0.01–0.1
Glycolaldehyde	$HCO(CH_2)OH$	—	0.001–0.01
Methyl formate	$HCOOCH_3$	~0.01	0.01–0.3
Dimethyl ether	CH_3OCH_3	~0.003	0.003–0.1
Formamide	NH_2CHO	—	~0.01
Methyl cyanide	CH_3CN	0.001–0.1	0.003–0.03

Vastel and collaborators suggest a methanol abundance (with respect to H_2) of ~6 $\times 10^{-9}$.[24] Table 1 summarises the estimated abundances, with respect to methanol, of the iCOMs so far detected in PSC.

Table 2 List of known class 0/I hot corinos and their major properties: distance of the parental cloud, luminosity, class and the reference reporting the first detection

Source	Cloud (distance) (pc)	L_{bol} ($L_\odot$)	Class	First detection references
IRAS 19347 + 0727 (B335)	Barnard 335 (~100)	~1	0	31
IRAS16293-2422 A	ρ-Ophiuchus (~140)	~20	0	32–34
IRAS16293-2422 B	ρ-Ophiuchus (~140)	~20	0	32–34
L1551-IRS5	Taurus (~140)	~35	I	35
BHR71-IRS1	Coalsack (~200)	~13	0	36
IRAS 18148-0440 (L483)	Aquila Rift (~200)	~13	0	37
Barnard 1b-S	Perseus (~300)	~0.3	0	38
Barnard 1-a	Perseus (~300)	~1	I	39
Barnard 1-c	Perseus (~300)	~1	I	36
NGC1333 IRAS 1A	Perseus (~300)	~3	I	36
NGC1333 IRAS 4A1	Perseus (~300)	~6	0	40
NGC1333 IRAS 4A2	Perseus (~300)	~6	0	40
NGC1333 IRAS 4B	Perseus (~300)	~6	0	41
NGC1333 IRAS 2A1	Perseus (~300)	~16	0	41
NGC1333 SVS13-A	Perseus (~300)	~32	I	42 and 43
L1455 IRS 1	Perseus (~300)	Unknown	I	36
IC 348 MMS A	Perseus (~300)	Unknown	I	36
Per-emb 26	Perseus (~300)	Unknown	I	36
HH 212	Orion (~400)	~9	0	44 and 45
HOPS-108	Orion (~400)	~38	0	46
MMS6	Orion (~400)	~	0	47
CSO33-b-a	Orion (~400)	≤23	0	48
SIMBA-a	Orion (~400)	~2	0	48
FIR6c-a	Orion (~400)	~8	0	48
MMS9-a	Orion (~400)	~9	0	48
MMS5	Orion (~400)	~16	0	48
Ser-emb 1	Serpens (~440)	~4	0	49
Ser-emb 17	Serpens (~440)	~4	I	50
Cep E-B	Cepheus (~800)	~75	0	51

3.2 Chemical richness of the hot corinos in young protostars

3.2.1 The census of known hot corinos and their iCOM inventory. The first objects where iCOMs were detected are the so-called hot cores, compact ($\leq$0.1 pc), hot (100–300 K) and dense ($\geq 10^7$ cm^{-3}) regions in massive star forming regions.[52,53] Much later, iCOMs were detected in solar-type forming regions, towards the so-called "hot corinos", to distinguish them by their more massive cousins.[32,54,55] For a long time, only a handful hot corinos were known. The most recent mm interferometers (namely NOEMA and ALMA) have allowed large surveys and, at present, about two dozen hot corinos are known,[36,39,48,56] as listed in Table 2. The majority of known hot corinos belong to low-mass star forming regions, with a few exceptions in the Orion molecular cloud,[44–46] the massive star forming region closest to the Sun.

By far the best studied hot corinos are those in the protobinary system of IRAS26293-2422, where just over a dozen iCOMs have been detected.[32–34,54,57–61] In contrast, just a few iCOMs have been detected in other hot corinos. Table 1 summarises the iCOMs detected in at least three hot corinos and their abundance relative to methanol.

Recent observations have also taught us that the detection of hot corinos and iCOMs can be dramatically hampered by the wavelength used for the observations and, therefore, the picture that we may derive from them. The figure shown at the beginning of this review (Fig. 1) illustrates this problem. Methanol towards the south source, IRAS4 A1, is only detected when observed in the radio range. In addition, the intensity of the methanol lines is very similar to that emitted by the north source, IRAS4 A2, the only one of the two detected in the mm, suggesting that the two hot corinos have likely similar iCOMs abundances. Therefore, restricting the observations to the mm range would have given a very different scenario: two very close siblings would have very different chemistry, a puzzle and challenge for astrochemical models…which indeed it is not. The reason for the above behavior is the absorption of the dust mixed with the gas in front of the hot corinos, which are embedded in the envelopes. Since the dust absorption efficiency decreases with increasing wavelength, it blocks the hot corino line emission in the submm and mm range, whereas it becomes transparent in the radio range. Therefore, it can be dangerous to limit the observations to the submm and mm wavelength ranges, because they may give a wrong picture and wrong iCOM abundances.

However, the dust absorption can also have a beneficial effect: to reveal the spatial stratification of the iCOM in a hot corino, even in the absence of the spatial resolution that would be required to resolve the line emission from different iCOMs. This is what has been observed so far in at least two objects, SVS13A and HH212-mm. The effect and principle is illustrated in Fig. 5.

3.2.2 Ices and deuteration of iCOMs. Despite the gas being warm in hot corinos, it is in these objects that doubly and even triply deuterated molecules were first discovered.[63,64] The detection of triply deuterated methanol, with $CD_3OH/CH_3OH \sim 1\%$, is particularly spectacular, as it shows that molecular deuteration can be different from the elemental D/H by up to 13 orders of magnitude.[65] These observations instantaneously gave us important information: these highly deuterated molecules can not be a present day, gas-phase product, they must be heritage of the cold prestellar phase. This has a straightforward

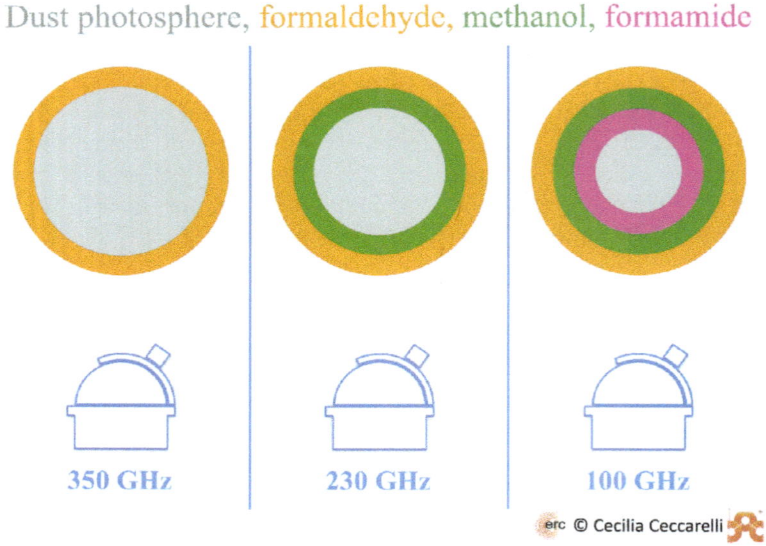

Fig. 5 Sketch of the combined effect of dust optical depth and iCOM binding energy (BE) on the iCOM detection towards the NGC1333-SVS13A hot corino over a 45 au scale.[62] Observations at 350 GHz (left panel) can only detect formaldehyde, whose BE allows it to be present over almost the entire envelope. Observations at 230 GHz (middle panel), where the dust optical depth is smaller than at 350 GHz, can also detect methanol, whose BE is larger than that of formaldehyde. Finally, observations at 100 GHz (right panel) can penetrate the most through the envelope, because the dust optical depth is the lowest and, therefore, detect formamide whose BE is the largest.

consequence: the chemistry in the hot corinos, and more generally in the warm objects of solar-type star forming regions, including shocked ones, is very much affected by the composition of the ices during the prestellar core phase.

Until very recent, only "simple" deuterated molecules were observed, with the most complex being methanol. These molecules are either formed in the gas phase directly *via* reactions with H_2D^+, such as DCO^+ or N_2D^+, or indirectly. Even in species formed on the grain surfaces, such as water and methanol, the deuteration originates from the enhanced H_2D^+/H_3^+ abundance ratio, which regulates the D/H atomic abundance in the gas and, hence, landing on the grain surfaces (see Section 4.2.2).

With the advent of the high sensitivity of ALMA and the very successful survey PILS,[59] deuterated iCOMs have also been detected: formamide, acetaldehyde, dimethyl ether, methyl formate, glycolaldehyde, ethanol and methyl cyanide.[43,66–71] The detection and abundance determination of deuterated iCOMs is important because it has great diagnostic power to discriminate the route of formation of the species, if predictions can be obtained for the different routes of formation, as will be explained in Section 5. In general, organics show a deuterium fractionation of about 1–10%, larger than that measured in water ($\leq$1%) towards the same sources, likely reflecting the history of the ice formation: water first, iCOMs later.[8]

Another application of the fact that prestellar ices are injected into the gas-phase is that their abundance can be used to constrain the prestellar core ice formation.[72] It is important to emphasize that we actually do not observe ices in the prestellar cores, because, as the ice observations are from absorption in the IR wavelength, we would need a strong IR source behind the prestellar core. While JWST will provide a huge amount of data on the ices in molecular clouds, it will not be able to sample the central zones of the prestellar cores, where the ices that govern the hot corino chemistry are formed. A first study towards the IRAS4a and IRAS4B hot corinos of two major ice components, methanol and ammonia, injected into and observed in the gas-phase, seems promising.[72] It suggests that the prestellar ice formation of these two objects was abruptly interrupted by the clash of an expanding bubble, coming from a distant and ancient supernova explosion, against the molecular cloud to which IRAS4a and IRAS4B belong. However, in order to extract this information, the observations have been compared with astrochemical model predictions which may or may not be reliable enough for the clash suggestion to be confirmed.

3.3 Shocks during star and planet formation

As mentioned in Section 2, the accretion process during the protostellar phase is accompanied by supersonically ejected or infalling material that crashes against quiescent matter, causing the so-called molecular shocks, where numerous and abundant molecules are observed. The shocks from outflowing matter have been known and studied for decades, and are called protostellar outflow shocks, while the discovery of streamers of material falling into the circumbinary or circumstellar disks and creating shocked regions is more recent (see also the contribution by Bianchi *et al.* in this volume).[73–75] At the moment of writing this review, very little is observationally known about the chemistry of the shocks caused by the infalling streamers, so in the following, I will focus on those caused by the protostellar outflows.

Despite the phenomenon of the outflows being known since the 80s and dozens of protostellar outflows studied in CO, in only two cases have iCOMs other than methanol been detected so far: the shocked region called L1157-B1 and the shocks caused by the outflows emanating from the IRAS4A binary system (mentioned in the Introduction). In both shocked regions, the "usual suspect" iCOMs have been detected: methanol, acetaldehyde, methyl formate, dimethyl ether, glycolaldehyde and methyl cyanide.

Fig. 6 shows an illustrative and important example, that of the deuterated formaldehyde, acetaldehyde and formamide line emission maps observed towards L1157-B1. The emission of deuterated formaldehyde clearly shows the regions where grain mantle species, previously formed in the cold phase, have been recently injected into the gas-phase. Acetaldehyde emission is overlapped, signalling a fresh release from the grain mantle too, whereas that of formamide is slightly but definitively displaced towards the south, where the dynamics of the region indicate an earlier passage of the shock. This example is illustrative because it shows that high-resolution maps of iCOMs provide the history of the shock passage and the evolution of chemistry: in particular, unless the ices are substantially different in the two regions where acetaldehyde and formamide are seen, the displacement means that formamide is a late product formed in the gas-phase from and after the injection of the ice components. It is important because this information provides an additional and precious constraint to the astrochemical models, in addition to the measured abundance and abundance ratios: how

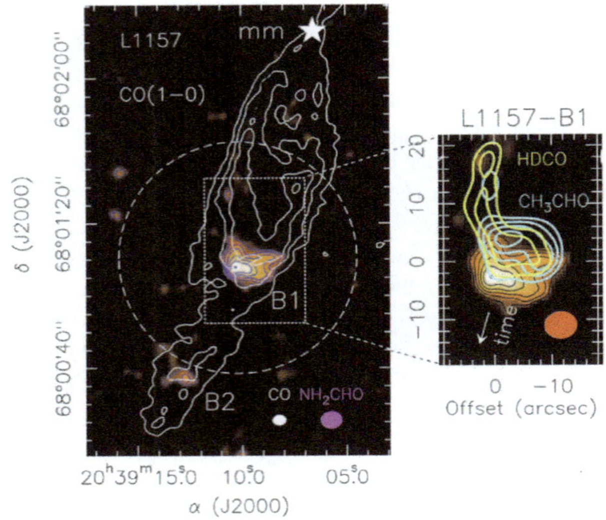

Fig. 6 Images of deuterated formaldehyde, acetaldehyde and formamide towards the shocked region of L1157-B1, in the southern lobe emanating from L1157-mm. *Left panel*: map of CO (1–0) line emission (white contours)[76] that shows the overall morphology of the outflow, where several clumpy cavities are excavated by the precessing jet, the brightest of which is B1. The color map shows the formamide line emission.[44] The dashed circle indicates the primary beam of the NH_2CHO image, namely the field of the observations. The magenta and white ellipses depict the synthesised beams of the CO (white) and formamide (magenta) observations. *Right panel*: map of the line emission of HDCO (yellow contours),[77] acetaldehyde (cyano contours)[44] and formamide (color image).[44] The region where HDCO emits shows where the grain mantles were recently sputtered by a shock. The region where HDCO does not emit but molecular emission is still visible is a region where a shock hit earlier. Therefore, the displacement between acetaldehyde (region coincident with the HDCO emitting one) and formamide (south region with no HDCO and CH_3CHO emission) is due to the time lapse between the two shocks.

species evolve with the time. We will see later, in Section 5.2, how this information can help to constrain the formation route of formamide in this source.

3.4 The hidden chemistry in protoplanetary disks

Observations of molecules in protoplanetary disks has been a decades long and painful hunt, where astronomers have struggled even to observe the simplest species. This is because protoplanetary disks are small in size and the molecular emission originates in the molecular layer described in Section 2.3, which has a relatively low column density. The advent of ALMA has greatly improved the situation providing the first detection of iCOMs in the outer regions of the disks and the first images of snow lines.[35,78–81] However, even methanol turns out to be a species difficult to detect in protoplanetary disks,[82] maybe because of the large dust opacity that absorbs much of its emission[83] or the relatively low temperature.[84] Fortunately, the new observations of JWST are already giving plenty of information on the inner regions of the disk,[85–87] an appetizer of which is also reported in this volume (*e.g.*, van Dishoeck *et al.* and Kamp *et al.*).

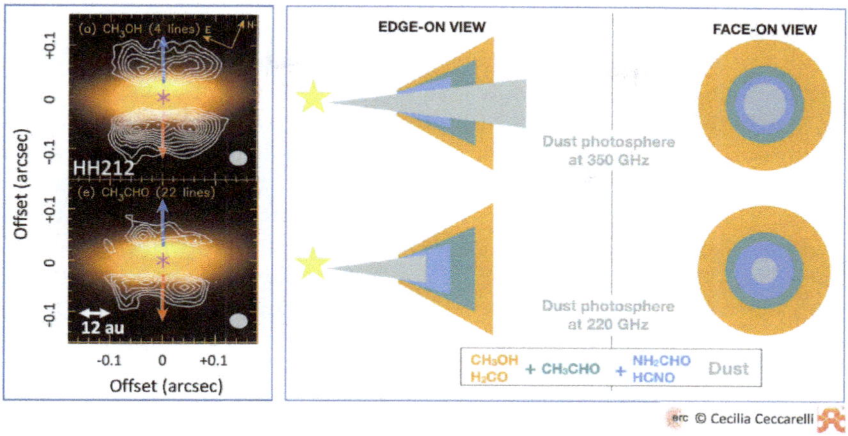

Fig. 7 iCOMs in the disk surrounding HH212-mm. *Left panel*: images of the methanol (top) and acetaldehyde (bottom) line emissions (white contours) overlapping the dust continuum emission (in color).[90] Note the lack of molecular line emission on the plane of the disk, which is seen almost edge-on, because of the absorption of the dust, giving it a hamburger-like shape. *Right panel*: sketch of the line emission from iCOMs and other species detected toward the disk: methanol, formaldehyde, acetaldehyde and fulminic acid. The regions where each species is detected at ~350 GHz (upper panels) and ~220 GHz (lower panels) are shown with the disk edge-on (left panels) and face-on (right panels) along with the dust photosphere. The molecular line emission depends on the combination of the wavelength at which each species is observed, and its binding energy. For example, methanol is only in the upper atmosphere if observed at 350 GHz, whereas it appears over the whole disk if observed at 220 GHz.

So far, the disk with the largest inventory of detected iCOMs is HH212-mm, a very young disk in the Orion region.[45,88–90] Once again, the usual iCOMs suspects are detected: methanol, acetaldehyde, methyl formate, formamide and methyl cyanide. This is another interesting case where observations at different wavelengths reveal the stratification of iCOMs in this disk due to the difference in the species binding energies[91] (Fig. 7). Actually, observations at different wavelengths could even be used to infer the succession of the binding energy of different species.

Another very promising disk is the one in the FuOri star V883. This source has recently undergone a burst in its luminosity. The burst was first announced in 1888 (but not measured) and, later, a bolometric luminosity of ~400 $L_\odot$ was measured.[92–94] The luminosity has been decreasing in the past years to a present value of 218 $L_\odot$.[95,96] The interesting point is that the sudden increase of the luminosity causes the heating of the dust at a larger distance from the center than in the quiescent status. In the region where the temperature exceeds that of water ice sublimation (~120 K, which correspond to a radius of about 320 au in the case of V388), iCOMs trapped in the ice are injected into the gas-phase, like in the hot corinos (Section 3.2), and become observable along with water.[97–101] Not surprisingly, the detection of methanol, acetaldehyde, methyl formate and methyl cyanide have been reported.[98,99]

3.5 Interstellar ices

While the discovery of ices enveloping the interstellar dust grains dates back to the 70s, the revolution in this field started with the advent of the first spatial telescope, the ESA satellite ISO (Infrared Space Observatory), which provided spectroscopic observations in the infrared (IR) of hundreds of sources. Two other space-borne telescopes followed, Spitzer and AKARI, to complete that first revolution. These observations have provided the inventory of the most abundant species present in the ices: H_2O, CO, CO_2, NH_3, CH_4 and CH_3OH. Other species, such as H_2CO, are less abundant and not always present. The review by Boogert and collaborators[102] offers an excellent view of what has been learned so far.

Important in the context of this review is when ices start to appear and with what composition. Water is the first solid species appearing already in molecular clouds with a visual extinction of ~3 mag, meaning at a cloud depth of ~1.6 mag (as the cloud is UV irradiated on both sides). Solid CO_2 is also detected in molecular clouds starting from a visual extinction of ~3 mag, with an abundance with respect to H_2O of ~15–40%. Solid CO appears at ~6 mag with an abundance of ~10–70%, followed by CH_3OH at ~18 mag with an abundance from $\lesssim$1 up to ~12%. Ammonia and methane are not detected towards molecular clouds with upper limits of ~5%.

In denser protostars, the abundance of solid CO_2 and CO (relative to H_2O) is similar to that measured in molecular clouds, whereas that of solid methanol is up to two-times larger, and methane and ammonia is only about ~1–10%. In practice, when considering the uncertainties, only methanol seems to be more abundant in the line of sight of dense envelopes of protostars than in those of molecular clouds.

Today we are at the verge of a second revolution thanks to JWST, which has unprecedented sensitivity and spectral resolution in the IR. A first work by McClure and collaborators[103] has provided a very appealing appetizer of what awaits. These authors presented observations towards two field stars, probing ices in the solar-type star forming region Chameleon I, in front of them. In addition to the most abundant and known species, the obtained spectra suggest the presence of less abundant species, notably ethanol, even though the detection is not confirmed. In addition, the high JWST sensitivity has allowed the observations of many more sources in the line of sight, which will provide a relatively high resolution map of the ice composition across the studied molecular cloud, the first ever. In the same vein, JWST observations by Yang and collaborators toward the envelope of a young protostar,[104] show the presence of the usual ice components and spectroscopic features that could be attributed to ethanol and, maybe, acetaldehyde.

4 Chemical processes and quantities

As we all know, having good and reliable observations is necessary but not sufficient for progressing our knowledge, namely in understanding what happens and why. For that, we need models, in our specific case astrochemical models. One can think of models as boxes that contain processes and quantities, which make a theory, and work them out into predictions that are, finally, compared

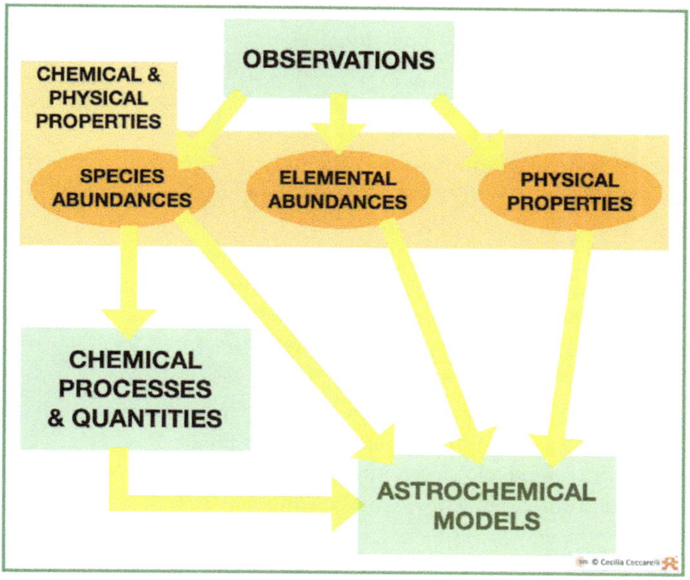

Fig. 8 Sketch of the input of astrochemical models. Astronomical observations provide an object's chemical and physical properties, namely the detected species and their abundance, the elemental abundances and physical parameters (density, temperature and so on). The observed species provide the input to ascertain the chemical processes and quantities. Finally, all information is combined to create the astrochemical models.

with observations. When the model predictions are able to reproduce the observations, the model and, thus, the theory are validated.

In general, astrochemical models need two classes of information: one relative to the physical and chemical properties of the object to be modeled and the second to the chemical processes occurring in the ISM conditions and the quantities associated with them (Fig. 8). In the next two sections, I briefly review how the object's physical and chemical properties are generally extracted from the observations and the chemistry of iCOMs, respectively.

4.1 Inputs of the astrochemical models

4.1.1 Physical and chemical properties of the astronomical object. Any astrochemical model needs the physical structure of the studied source as input, namely the gas density and temperature structures, the flux of UV photons and cosmic-rays, the dynamical behavior and so on. In addition, the measured species abundances constitute the constraints of the model predictions. These properties are evaluated from astronomical observations, more or less directly. For example, the species column density, gas density and temperature structures can be derived from the radiative transfer modeling of the observed lines, in some cases. However, these derivations have intrinsic not negligible uncertainties for at least three reasons.

The first one is that in order to have a reliable radiative transfer model one needs the so-called collisional coefficients for the studied species. These coefficients are usually obtained *via* quantum mechanical (QM) computations, because experiments of the state-to-state excitation are extremely difficult and have been obtained for very few species.[105–109] In the case of iCOMs, only the methanol and methyl cyanide have available collisional coefficients: methanol for collisions with H_2 (ortho and para) for the first 256 levels and temperature between 10 and 200 K;,[110] methyl cyanide for collisions with He only, between 20 and 140 K and for $J \leq 25$.[111]

The second source of uncertainty is the derivation of the species abundance, which requires knowledge of the H_2 column density of the same region emitting the species lines. As already mentioned in Section 3, deriving the H_2 column density is extremely difficult if not impossible, because the H_2 first excited level is at ~500 K and, therefore, the ground transition, by definition, only probes ≳50–100 K gas. In addition, as the transition is at 28 μm, observations can only be obtained with a space born telescope. JWST will provide new data in this respect, but, for the moment, the H_2 column density is mostly indirectly derived from the continuum observations (which have themselves uncertainties in the conversion from the dust to the gas column density, about a factor 3).

Finally, a third source of uncertainty is the inevitable approximation of the 3D structure of the studied object, because we only have 2D observations. Often, for an astronomer "the cow is a sphere", to say that a simple (*e.g.* spherical) symmetry is often assumed (for disks it would be cylinder symmetry, a slab for a cloud and so on). This is important to keep in mind because the estimates of the chemical abundances from observations have an uncertainty of, at least, a factor of three to be optimistic. Abundance ratios could be more reliable, if we have strong reason to believe that the two species originate exactly in the same region, which may not be the case even when the observations show them in the same line of sight (see, *e.g.*, the discussion in Section 3.2).

4.1.2 Elemental abundances: oxygen and carbon examples. Crucial to any astrochemical model are the elemental abundances. This is a long standing problem which various authors and methods have tried to solve.[112] Here, I do not have the ambition to review this huge and still debated field. I will just summarise briefly the major uncertainties in the elemental abundances, in particular, those of oxygen and carbon, the most abundant elements after hydrogen and helium.

Both carbon and oxygen are contained in the refractory component of the interstellar grains and in their volatile one. Oxygen is contained in silicates, while carbon is contained in so-called grains of carbonaceous material: a very difficult to characterize and quantify component of interstellar dust.[113]

In dense molecular clouds in the vicinity of the Sun, the sum of the abundances of gaseous and frozen CO, frozen H_2O, CO_2, CH_4 and CH_3O, and the oxygen trapped in the silicates should give a total equal to the elemental abundance measured in the Sun. Unfortunately, this is not the case and about one third of oxygen is "missing", *i.e.* we do not know where it is.[114,115] Please note that CO is measured to be the most abundant C- and O-bearing gaseous molecule in dense molecular clouds, with stringent measurements and upper limits of the H_2O and O_2 molecules, less abundant by at least a factor of 1000. However, we do not have straightforward measurements of atomic oxygen in cold molecular clouds, because of the difficulty of observing its ground line at 63 μm with adequate spectral resolution. In practice, very few observations exist and most were obtained by ISO. The ISO observations gave puzzling results, as they suggested that a substantial fraction of atomic oxygen could be gaseous.[116–118] More recent observations obtained with the airborne telescope SOFIA toward a couple of massive sources, confirm the presence of a substantial fraction of atomic oxygen ($\sim 10^{-5}$ with respect to H-nuclei) associated with molecular gas, but not enough to account for the missing elemental oxygen.[119,120] Therefore, the missing oxygen could be in refractory material difficult to observe, as O-rich carbonates or oxidized metals.[115]

Carbon elemental abundance is also difficult to constrain, because of the difficulty of quantifying the amount of carbon locked into the dust grains.[115]

Needless to say but better to do it: the choice of the oxygen and carbon elemental abundance has a huge impact on the predicted abundances of iCOMs in any model.

4.2 iCOMs chemistry

The second class of information necessary to any astrochemical model regards the chemical processes and quantities: they are the precious engine of the models. This section presents an overview of what is known so far, what is not completely clear and, finally, what is still missing in our understanding of the formation of iCOMs.

4.2.1 Overview and current scenarios. As mentioned at the beginning of Section 1, there is not yet a full consensus in the community on how iCOMs are formed, where and when. Roughly speaking, two alternative scenarios exist, shown schematically in Fig. 9. Importantly, both scenarios agree on the very first step: the formation on the grain surfaces of hydrogenated and oxygenated species from atoms or simple molecules formed in the gas-phase. Water, ices most abundant component, is thought to be formed from the hydrogenation of atomic oxygen, for example. Likewise, methanol, an important precursor of other iCOMs, is also believed to be formed by the hydrogenation of CO landing on the grains (see Section 4.2.2).

However, the two scenarios diverge when other species are considered, especially iCOMs. Briefly, to give an illustrative example, let me consider the case of the iCOM formation in hot corinos. In the so-called "Gas + grain chemistry", after the mantle formation, the frozen species are injected into the gas-phase when the dust grain temperature exceeds the water ice sublimation, $\sim$100 K. Once in the gas-phase, these species undergo a series of reactions that form iCOMs. In the other scenario, the "Only grain chemistry", after their formation, mantles are processed by photolysis and radiolysis creating radicals that remain trapped in the ices. With the evolution of the protostar, the dust temperature gradually increases. When it reaches a temperature where the trapped radicals can migrate, the latter can encounter other radicals and combine into iCOMs, also radicals can combine with closed-shell species. Also, reactions not involving the thermal migration of radicals could also take place, most of them ending up in radical combinations of iCOMs. Finally, when the dust temperature reaches the ice sublimation, the iCOMs, synthesised on the grain surfaces, are released into the gas-phase where they are observed by submm/mm/radio telescopes.

In the following sections, I briefly review the status of our understanding of some of the processes involved in the iCOM formation along with the quantities that characterise them, following the scheme in Fig. 9.

4.2.2 The formation of ices. The chemical composition of the ices turns out to be of fundamental importance for the successive chemical evolution. As

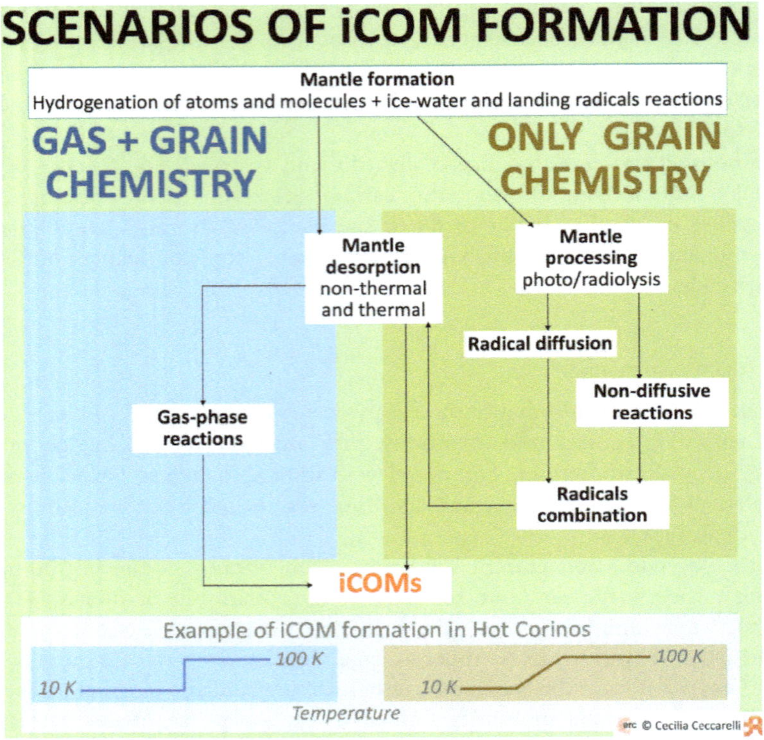

Fig. 9 Sketch of the two major scenarios invoked in the literature for the formation of iCOMs: "Gas + grain chemistry" and "Only grain chemistry". Both predict the formation of icy mantles enveloping the interstellar dust grains, prevalently during the cold molecular cloud and prestellar core stages. The mantle constituents are the result of the hydrogenation of atoms and simple molecules, such as CO, and the oxidation. Water and methanol, for example, are formed in this first step.[121–123] In addition, after the first layers of water-ice are formed, radicals landing on the grain surfaces can react with the ice-water molecules. The iCOM ethanol, for example, can be formed in this way.[124] After this first step, the two scenarios differ as follows. *Left blue panel*: In the "Gas + grain chemistry" scenario, the mantle components are either partially or completely injected into the gas-phase by thermal and non-thermal desorption processes, where they undergo gas-phase reactions that form iCOMs. This can occur either *via* mantle thermal desorption in hot corinos and inner protoplanetary disks, or *via* mantle non-thermal desorption in prestellar cores, outer protoplanetary disks and molecular outflows. *Right brown panel*: In the "Only grain chemistry" scenario, the icy mantles are processed by UV (photolysis) and CR (radiolysis) irradiation while being formed, and the radicals formed by the two processes remain trapped in the ice. When the dust temperature increases, radicals on the mantles become mobile and diffuse and, when they meet or meet some closed-shell species, they combine forming iCOMs. Alternatively, non-diffusive processes can also combine radicals into iCOMs. Likewise, some radicals are also predicted to meet and react in non-diffusive processes during the cold phase.

discussed in Section 3.5, the bulk of the interstellar ice is formed in the molecular cloud, when the density is no larger than $\sim 10^4$ cm^{-3}. Only methanol seems to have further formation in the denser protostellar phase, which is in the protostellar envelopes. One has to keep in mind that the measurements of the ice abundances are obtained *via* absorption observations: by definition, hence, they give the average along the line of sight. If a smaller region in the line has much larger/lower abundance, absorption observations cannot say. Even if we do not have so far, observations of ices towards PSCs (for the absence of illuminating IR sources behind), the values in the protostellar envelopes probably are a good proxy of those in PSCs, with the caveats above.

In the following, we briefly summarise the process of formation of the firmly detected O-, C- and N-bearing ice components.

Water. We think we have a good handle on the process forming water on the ices – the hydrogenation of atomic and molecular oxygen, landing on the grain surfaces – based on experiments[121] and quantum mechanics (QM) computations.[125] This also agrees with water being formed very early in molecular clouds (from atomic rather than molecular oxygen).

CO_2. The formation of CO_2 has been studied in laboratory experiments and the community agree that it mainly occurs *via* the CO + OH reaction, which forms the HOCO radical (assumed to be excited) that eventually becomes CO_2.[126] There are no explicit QM computations that have simulated the process yet.

CO. Obviously, solid CO is due to CO formed in the gas-phase freezing onto the grain surfaces (which also implies that seeing a species on the ice mantles does not imply that it is formed there).

Methanol. Since the early 2000s, laboratory experiments have shown that methanol is formed on the water ice surfaces *via* addition of hydrogen atoms to frozen CO [122] (see also the review by Tsuge and Watanabe[123]). The idea is that the two steps with an activation barrier, CO + H and H_2CO + H, are overcome by the H-atom tunnelling through them. Theoretical computations have been published later and confirmed the experimental results.[127,128] The form of the barriers (height and width) and the consequent rate coefficient of methanol formation, differ in the studies by Rimola and collaborators[127] and Song and Kästner,[128] as the first refers to Eley–Rideal reactions while the second one to Langmuir–Hinshelwood. Unfortunately, most astrochemical models adopt a rectangular barrier, with the height and width derived from approximating gas-phase computations,[129–132] predicting methanol formation rate coefficients more than four orders of magnitude smaller than the ones computed by Song and Kästner,[128] for example.

Ammonia. Solid ammonia is formed by the hydrogenation of atomic nitrogen, a process that has recently been studied *via* QM computations.[133] Later, ammonia formed in the gas-phase can also freeze-out onto the grain surfaces and enrich it further.[72,134] The fact that ammonia is actually detected towards protostellar envelopes may suggest that this second route is not negligible.

Methane. For a long time it was believed that methane is formed by successive hydrogenation of atomic carbon, as also shown in a recent laboratory experiment,[135] or by reactions with surface H_2.[136] However, recent QM studies are unanimous in showing that atomic carbon reacts with the water molecules of the ices and can not form methane.[137–139] Likely, in the ISM, the H addition chain starts rather from gaseous CH landing on the ice surfaces. The methane example also shows the importance of having QM computations that describe the micro-processes occurring on the grain surfaces, to be able to include the correct processes.

Ethanol. Recent QM computations predict that ethanol can be formed on the grain surfaces by the CCH, formed in the gas-phase, which instantaneously reacts with one water molecule of the ice to form vinyl alcohol in a barrierless reaction, which can then be hydrogenated into ethanol.[124]

4.2.3 Binding energies. A crucial parameter in the interpretation of the observed iCOMs is their binding energy (BE). The BE of the radicals that can potentially form iCOMs on the grain surfaces is crucial both for their presence/absence on the grain surfaces and for their diffusion, assuming that the diffusion barrier is proportional to the BE (see Section 4.2.5). The BE of iCOMs and their precursors, if the former are synthesised in the gas-phase, it is crucial to understand in what environment iCOMs can be observed.

Minissale and collaborators have recently published a comprehensive review of the BE of about 20 species (including H, O and C atoms).[140] Probably needless to say, the list does not contain any radicals. Also, only three iCOMs appear in the list: methanol, methyl cyanide and formamide. More recently, Ferrero and collaborators and Molpeceres and collaborators reported a joint experimental and theoretical study of the BE of acetaldehyde.[51,141]

An alternative means to obtain the BE of the species is *via* QM computations. Actually, this is probably the only way to obtain the BE of radicals. In the last few years, Ugliengo, Rimola and collaborators have carried out various QM computational studies and provided the BE of almost four dozen species, including radicals, iCOMs and S-bearing species.[142,143] These BEs have been obtained using models for the ice containing 64 water molecules. In the same vein, Bovolenta, Bovino and collaborators have carried out QM computations of the BE of almost two dozen species on 22 water molecules of ice, among which are two iCOMs – methanol and formic acid.[144,145] All these recent theoretical studies have highlighted an obvious point – adsorbed species do not have a single BE but

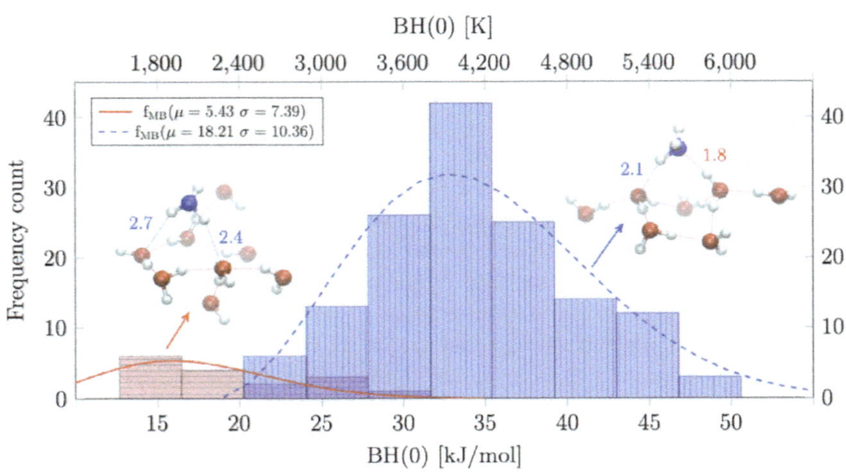

Fig. 10 Distribution of the binding energy, which is the binding enthalpy corrected for the zero point energy BH(0), of ammonia adsorbed on amorphous water ice as derived by quantum mechanical computations.[147] In these computations, the icy grain model constitutes 250 water molecules. The distribution can be fitted with two Gaussian curves, which reflect the two possible H-bonds that the ammonia can form with water: the stronger N···H(–OH) and the weaker N–H···O(H_2), where the ammonia is respectively H-bond acceptor and H-bond donor, as shown in the figure.

a distribution of values, which depend on the species (where and in what orientation they land on the ice) and the adsorbate (*e.g.* crystalline or amorphous water ice, or CO-rich ice). However, this is overlooked by models with only one exception.[146]

In this respect, Tinacci and collaborators have obtained the most extensive QM study and complete distribution of BE, obtained for ammonia and water over a large ice grain of 250 water molecules.[147,148] The example of the BE distribution of ammonia is reported in Fig. 10. The distribution is described by two Gaussian functions, which reflect the two possible H-bonds that the ammonia can form with water: the stronger the N···H(–OH) and the weaker N–H···O(H_2), where the ammonia is respectively H-bond acceptor and H-bond donor.

Finally, the other parameter entering in the astrochemical models to compute the sublimation temperature of a species, is the so-called pre-exponential factor. Usually, this is taken as a fixed number, $\sim 10^{12}$ s^{-1}, but several studies have now emphasized the necessity to have this number coupled with the BE of the species.[140,141] In other words, the $\sim 10^{12}$ s^{-1} may be very wrong for heavy molecules. For example, for acetaldehyde the pre-exponential factor is $\sim 10^{18}$ s^{-1}. In this case, using $\sim 10^{12}$ s^{-1} would lead to a sublimation temperature of about 10–20 K different from the correct one.[12]

4.2.4 Gas-phase reactions and networks. The processes at the base of the gas-phase reactions occurring in the ISM have been studied for decades. Here we just remind readers that the two pillars of the characterization of gas-phase reactions are: (i) laboratory experiments (*e.g.* the CRESU and cross-beam apparatus[149–152]) and (ii) quantum mechanical (QM) computations.

Present day gas-phase reaction networks contain more than 8000 reactions (more than double if D-isotopologues are taken into account) that involve about 500 (or more) species. Two major gas-phase networks are publicly available and used by several astrochemical models: KIDA†[153] and UMIST.‡[154] First, I would like to acknowledge the huge and heroic effort carried out by our colleagues that have first built the reaction networks on which the two above mentioned databases are based on, especially Eric Herbst and Tom Millar. That said, the KIDA and UMIST database networks suffer some problems that cannot be ignored, and which I here identify as reactions that (i) "should not be there because they are wrong" and (ii) "are missing".

Reactions that should not be there. As indicated in the two databases, the large majority ≥80% of the included reactions has never been studied in laboratory experiments or by theoretical computations. Even in the cases of reactions that have been investigated, the experimental conditions rarely reproduce those typical of solar-type star forming regions, either regarding the temperature or the density. In the absence of laboratory data, rate coefficients and their temperature dependence are mainly estimated with some chemical intuition or by drawing analogies with similarly known processes. When the data are available outside the temperature or density range of relevance, they are used as such or are extrapolated. However, both approaches can be seriously wrong for various reasons, discussed elsewhere.[12]

† https://kida.astrochem-tools.org/

‡ http://udfa.ajmarkwick.net/

One obvious error in writing down schemes of unknown reactions from similar arguments is that the formation enthalpy of the products may be larger than that of reactants, especially when unstable species, such as ions or radicals, or other uncommon species are involved. In this context, Tinacci and collaborators[155] carried out systematic work to "clean" the gas-phase reaction network GRETOBAPE,§ which is based on the KIDA network, from the most obvious source of error: the presence of endothermic reactions not recognized as such. Using the enthalpy of formation derived by electronic structure calculations,[156,157] these authors found that about 5% of the reactions in the original network resulted in endothermic reactions erroneously reported as barrierless or with too low activation energy. While on the one hand it is reassuring to know that the networks are not hugely wrong, one should not underestimate the impact of this 5% of erroneous reactions. For example, Tinacci *et al.* found that the overall silicon chemistry is hugely impacted by this.[155]

Other possible errors are the presence of reaction activation barriers which are either over- or under-estimated, wrong rate coefficients and wrong products. In these cases, dedicated experiments and/or QM computations are needed, which is extremely time consuming. The example of formamide that will be discussed later (Section 5) illustrates the difficulty of correcting the networks for relatively small, but still important in the interstellar context, activation barriers. Another example is provided by acetaldehyde, also discussed later (Section 5): about half of the acetaldehyde formation reactions listed in the KIDA database have either the wrong rate coefficient or products.[158]

Reactions that are missing. Despite the large dimensions of the available gas-phase reaction networks, there are no doubts that some important, and maybe also crucial, reactions are missing.

One obvious class of missing reactions are those in which neutrals react with the most abundant ions in molecular gas, namely HCO^+, H_3^+, He^+ and H^+. Tinacci and colleagues reported that, out of 238 neutral species present in the GRETOBAPE network, 111 of them do not have reactions with at least one of the above-mentioned ions, and 10 not even with one ion.[155]

Finally, it is more than probable that important gas-phase reactions are missing. Some of them may even be present in the literature but just overlooked. One example is the formation of methyl formate from dimethyl ether which was only introduced in 2015 (ref. 159), although the chain of reactions were measured and computed since 2005. Another possibility is just that reactions were not thought of, like the case of the formation of glycolaldehyde, for which no gas-phase formation routes were present in the database or literature: *ad hoc* new QM computations showed that there exists at least one formation route of glycolaldehyde in the gas-phase.[160]

As said, both laboratory experiments and QM computations are very time consuming, so it is important that chemists and astronomers work together. It is possible that artificial intelligence (machine learning) techniques could help to reduce the burden. A recent work by Stancil and collaborators on the application of an artificial neural network to compute collisional excitation rate coefficients[161] is very promising and perhaps a similar approach could be used for reactions.

§ https://aco-itn.oapd.inaf.it/aco-public-datasets/theoretical-chemistry-calculations

Another approach could also be the one recently applied by Komatsu and Suzuki,[162] to automatically search for paths of formation of cytosine.

4.2.5 Grain-surface processes, reactions and networks. The situation for the grain surface chemistry is even more complex than that of the gas-phase reactions, because the problem is more complex both to simulate in the terrestrial laboratories and to simulate with QM computations. For one thing, we do not really know the structure of the interstellar ices. Are those reproduced in the laboratory experiments reliably similar to them? Same for the simulated ices: what models are the best to simulate the real interstellar ices? One thing is certain: the nature of the ice does matter in several of the processes involved in the formation of iCOMs. To my best knowledge, the largest ice grain so far produced by QM computations contains about 1000 water molecules.[163] It is built by adding one water molecule at a time. The underlying assumption is that at each time step, the molecule has the time to find the position/orientation with the minimum energy, which may be the reality in the ISM. That said, so far only computations of some species' binding energies have been carried out using this ice grain (see Section 4.2.3), and none to simulate any of the other processes that may occur on the grains shown in Fig. 9 (diffusion, reactivity…). In the future, it will be necessary to simulate them on similar large ice grains in order to better understand what happens on the interstellar ices.

In Fig. 9, (at least) four processes are involved in the formation of iCOMs on the grain surfaces. They are briefly reviewed in the following.

Mantle processing. Once the "first generation" ice is formed (H_2O, CO, CO_2, NH_3, CH_3OH…), it is processed because of the UV illumination (photolysis) and/or cosmic-ray (radiolysis) irradiation forming radicals that remain trapped in the ice. To correctly describe the process and introduce it into astrochemical models, one needs to know what radicals are formed and at what rate. In the case of UV photolysis, models rely on the gas-phase reactions for introducing the products. While this may be correct, the possibility that the created radicals instantaneously react with the ice-water molecules and become something else, is mostly overlooked by models so far (see the cases of CN and CCH [124,164]). For the radical production rates, models again use the ones from the relevant gas-phase reactions, sometimes applying adjusting factors.[165] Unfortunately, while several experiments focused on the interstellar analogues photolysis, it is very difficult, if not impossible, for them to describe the microphysics process and, thus, estimate the radical production rates. One obvious reason is that the created radicals instantaneously react on the irradiated ices and, hence, experiments can only measure the final products and yields. Similar arguments apply to the cosmic-ray induced radiolysis.[166,167]

Radical diffusion. As said, it is difficult to experimentally measure the diffusion rate of radicals, because, by nature, they react fast. Experiments exist on a limited number of closed-shell molecules, such as CO, CO_2, N_2 and CH_4,[169–173] but none on radicals. To the best of my knowledge, QM computations exist only for the H atom[174] and not for any radical.

Usually, astrochemical models treat the diffusion as described by an Arrhenius equation with two parameters, the pre-exponential factor ν_{dif}, and the diffusion energy barrier E_{dif}, where E_{dif} is assumed to be a fraction of the binding energy. The fraction E_{dif}/BE, often assumed a constant value irrespective of the adsorbed species, probably depends on the species and can vary between 0.2 and 0.7, depending on the species.[172] In addition, as discussed in Section 4.2.3, also the binding energy is a poorly known parameter for radicals.

As a last comment with respect to the astrochemical treatment of the radical diffusion, even assuming that ν_{dif} and E_{dif} are known, it is important to emphasize that they would only refer to the diffusion on the ice surface and not through its bulk. A beautiful theoretical work by Ghesquière and collaborators[175] has shown how the mobility of species trapped in amorphous ice dramatically depends on the ice temperature. The radicals trapped in the bulk can only migrate when the ice undergoes morphology restructuring before it sublimates.

Non-diffusive reactions. Very recently, inspired by a series of laboratory experiments carried out by Theulè, Fedoseev, Ioppolo and collaborators,[176–178] Herbst and Garrod suggested surface chemistry based on non-diffusive processes.[179–181] The basic idea is that reactions on the grain surfaces can also occur efficiently enough because of: (i) direct reaction from a landing species with a species (radical if the landing species is a closed-shell molecule and *vice versa*) on the ice, *i.e.* Eley–Rideal mechanism; (ii) the product of a reaction is a radical close to another radical causing another reaction (and so on); (iii) reaction triggered by an excited product of a previous reaction.[130] As said, the aim of these models is to reproduce what some laboratory experiments obtain, but, unfortunately, there are no QM computations available for a detailed description of the associated processes.

Radical–radical combination. The vast majority of present astrochemical models follows the scheme introduced by Garrod and Herbst in 2006 (ref. 182 and 183) and assumes that, when radicals encounter grain surfaces, they combine in a barrierless way, as in a Lego game, and form iCOMs, drawing from the analogy with gas-phase reactions involving two radicals. However, while many experiments using UV-illumination have shown this possibility, theoretical QM computations warn that this is not always the case. First, barriers can be present

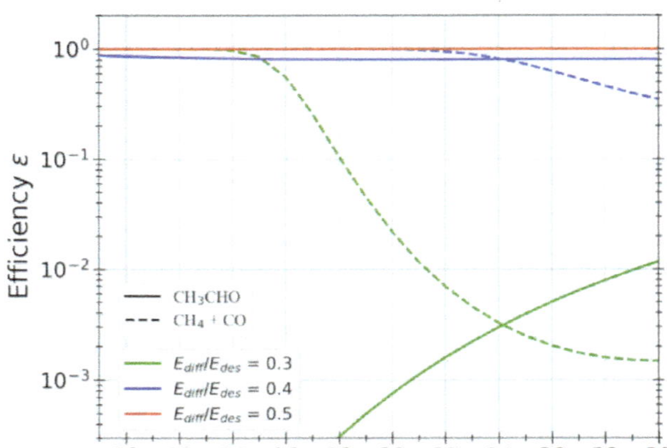

Fig. 11 QM computations of the efficiency of the reaction CH_3 + HCO leading to either CH_3CHO (solid lines) or CO + CH_4 (dashed lines) as a function of the temperature of the water ice.[168] Three different values for E_{dif}/BE ratios were considered in these computations: 0.3 (green), 0.4 (blue) and 0.5 (red). Depending on this value, which is practically unknown, acetaldehyde could easily form on the icy surfaces or not form at all.

because the radical has to break the bonds that keep it attached to the ice. Second, the direct combination of the two radicals can be in competition with the H abstraction from any of the radicals. For example, acetaldehyde, a system that has been extensively studied, will not be easily formed by the combination of CH_3 and HCO, as one can naively think (the formation of CH_4 and CO is a strong competitor).[168,184,185] Fig. 11 shows the computed efficiency of acetaldehyde formation on the grain surfaces as a function of the temperature reached by the warming ice before CH_3 sublimates (HCO has about two-times larger binding energy). The efficiency is given for three assumed values of the diffusion-over-binding energy and it goes from unity, for values larger than 0.4, and crashes when it is 0.3, showing the dramatic dependence of the acetaldehyde formation efficiency on a parameter that is practically unknown (see above).

Finally, it is important to emphasize that, in practice, each case is peculiar and requires a dedicated study. At present about two dozen systems have been studied on the water surfaces[186] and much less on CO-rich ices.[187]

Only one public database exists so far with a list of grain-surface reactions, posted on the KIDA website. The list is based on Garrod and collaborators' reaction network[129,183] and it is presented in a work by Ruaud and collaborators.[130]

5 Formation of iCOMs: observations against predictions

In this last part, I will discuss a few specific cases, trying to provide ideas rather than a precise comparison with specific models, which – as I hope I convince the

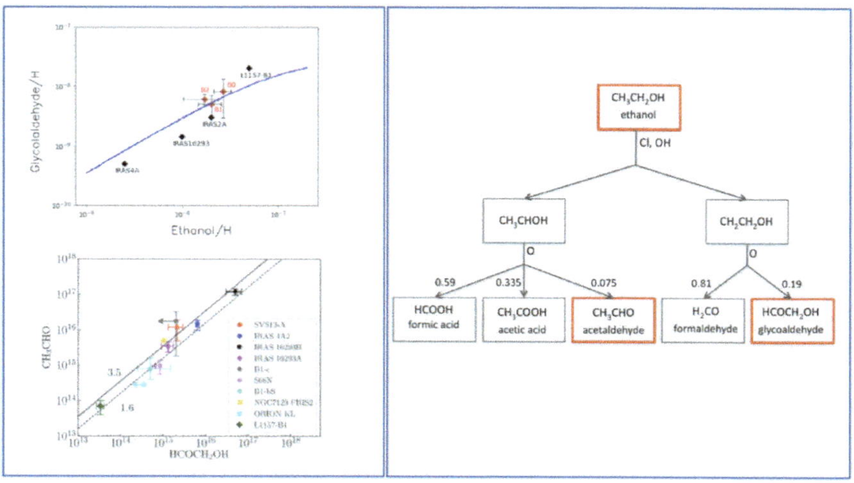

Fig. 12 Gas-phase formation route of glycolaldehyde and acetaldehyde from ethanol. *Left panel*: observed correlations between the measured abundances of glycolaldehyde and ethanol (upper panel)[160] and column densities of acetaldehyde and glycolaldehyde (bottom panel).[158] The upper panel shows the value derived for the shocked region L1157-B1: the black diamonds show the values obtained with the single-dish IRAM-30m observations[188] while the red diamonds show the values obtained in the three blobs in which the B1 region splits when observed with the interferometer NOEMA (Robuschi *et al.* in prep). *Right panel*: the proposed gas-phase chain of reactions that form acetaldehyde and glycolaldehyde starting from ethanol,[160] which would be the grain-surface product.[124]

reader – depend on many processes and quantities on which we do not have a full control, in my opinion.

5.1 Ethanol, glycolaldehyde and acetaldehyde

Observations towards warm objects (hot cores, hot corinos and molecular shocks) show that there exists a correlation between the measured ethanol and glycolaldehyde abundances.[36,188,189] In addition, the measured abundance of glycolaldehyde also correlates with acetaldehyde, as shown in Fig. 12. A simple explanation, very often assumed by several authors, is that the three species are all grain-surface chemistry products, basically following the scheme "Only grain chemistry" described in Section 4.2.5. Often, the combination of CH_3 and HCO radicals are invoked for acetaldehyde formation, but as discussed in Section 4.2.5 is far from the consolidated route of acetaldehyde formation.

However, there is also the possibility that the three species are chemically linked and not all of them are a grain-surface chemistry product. Skouteris, Balucani and collaborators proposed that glycolaldehyde is actually a gas-phase product, a daughter of ethanol, following the chain of reactions shown in Fig. 12 (right panel). The same chain of reactions would also form acetaldehyde. Following up on the possibility that ethanol, now perhaps observed in the solid form by JWST, is synthesised on the grain surfaces by the CCH + H_2O reaction (followed by an H addition),[124] glycolaldehyde and acetaldehyde could be siblings and both gas-phase products. The QM computations of the branching ratios rate coefficients of the glycolaldehyde formation would produce model predictions that fit very well with the observations (Fig. 12 (ref. 160)). In the same vein, the acetaldehyde over glycolaldehyde abundance ratio predicted by the QM computations branching ratios fit very well with the observed one (Fig. 12 (ref. 158)).

In addition, the measured deuteration of glycolaldehyde and acetaldehyde, and their isomers, are also in good agreement with those predicted by QM computations.[158,190] Unfortunately, no predictions of the deuteration for these two molecules exist in the literature. Therefore, while the gas-phase origin hypothesis has passed the possible tests so far available, the grain-surface one can not be excluded for lack of similar precise predictions from QM computations.

5.2 Formamide

This has been one of the most debated cases in the context of iCOM formation (see, for example, the exhaustive review by López-Sepulcre and collaborators[191]). In 2013, Kahane and collaborators suggested the possibility that NH_2CHO is formed in the gas-phase *via* the reaction of NH_2 with formaldehyde.[192] Two years later, Barone and collaborators presented QM computations showing that the reaction has an embedded barrier.[193] However, later QM computations by Song and Kästner[194] argued that the barrier is actually larger than that predicted by Barone *et al.* and that, consequently, the reaction is inefficient at the ISM temperatures. The height of the barrier was successively revised by Skouteris, Vazart and collaborators, who found a larger value than that of Barone *et al.*, but one still small enough for the reaction to be fast in warm gas ($\sim$100 K).[195,196] Specifically, Skouteris *et al.* predicted a rate constant at 100 K equal to 1.2×10^{-13} $cm^3\ s^{-1}$ and, using this value, were capable of reproducing the observations of formamide towards L1157-B1, as reported in Section 3.3.

Yet, the saga did not stop here. A recent experiment on the $NH_2 + H_2CO$ reaction carried out by Douglas, Heard and collaborators at 34 K with a CRESU apparatus, failed to detect formamide.[197] These authors put an upper limit on the rate constant equal to 6×10^{-12} cm^3 s^{-1}, namely 2.5 times larger than the predicted one by Skouteris *et al.* at the same temperature. Douglas *et al.* complemented the experimental results with new QM computations, whose accuracy is essentially the same as of those by Song and Kästner,[194] and insisted on the presence of an non-embedded activation barrier, which would lead to a much lower rate constant at temperatures $\leq$200 K.

The presence/absence of the barrier is linked to how the pre-reactant complex (PRC) is computationally treated and whether the zero-point-energy (ZPE) should be added or not to the transition state (TS) towards the formation of formamide. Barone, Skouteris and collaborators[193,195,196] claimed that the ZPE, computed with the standard methods adopted by the other authors,[194,197] should not be added to the PRC TS energy because the PRC has very loose modes which would not substantially alter the TS energy height. In contrast, Song, Kästner, Douglas, Heard and collaborators[194,197] think the opposite and they consider the ZPE of the PRC computed in the harmonic approximation. Since the various approximations used for the computations of the standard TS very likely do not apply to the PRC, at this stage, it is impossible to affirm with certainty who is right and whether the ZPE of the PRC may lead to an activation barrier for the $H_2CO + NH_2$. New calculations with more adapted methods should be employed, or a lower upper limit in experimental works should be obtained. In summary, unfortunately the new experimental work by Douglas *et al.*[197] does not bring new meaningful constraints when compared to the QM predictions of Barone, Skouteris and collaborators.

Astronomical observations could be used to distinguish whether the dominant route of formation of formamide is the gas-phase or grain-surface. The analysis of the displacement of acetaldehyde and formamide in L1157-B1, shown in Fig. 6, would be in favor of the gas-phase route for formamide.[44] In addition, as for glycolaldehyde and acetaldehyde, it is possible to compare the theoretical predictions, based on QM computations of the gas-phase reaction $NH_2 + H_2CO$, on the deuteration of formamide and its isomers,[196] with the observations towards the IRAS16293-2422 A2 hot corino observed by the PILS team.[67] In this case, the predictions agree extremely well with the observed values. As for the case of glycolaldehyde and acetaldehyde, the compatibility of the predicted abundances obtained assuming the gas-phase reaction route with the observations available in the literature, are necessary but insufficient.

The formamide formation on the grain surfaces has also been a saga. The observed correlation between the abundance of formamide with that of HCNO was immediately seen as proof that NH_2HCO is the result of HNCO hydrogenation.[198,199] This hypothesis was ruled out by the experiment by Noble, Theule and collaborators.[200] Successive similar experiments showed that there is indeed recycling between NH_2HCO and HNCO, with the H atom acquired and lost on each side, with a result in favor of H abstraction from NH_2HCO (and not the other way around).[201] Other possibilities evoked in the literature consider the NH_2 + HCO recombination on the grain surfaces[129] and the formation from the CN landing on the ice surface and reacting with a water molecule.[164]

6 Conclusions

The progress in astrochemistry and, in particular, the study of the formation of interstellar complex organic molecules (iCOMs) has proceeded at a vertiginous speed on all fronts: observations, laboratory experiments, quantum mechanics computations and modeling. It is difficult to consider everything and cite everybody in a review of a few pages, I did my best but I am conscious that my opinions enter in my writing. That said, I hope that this review provides a correct view of the passionate and fascinating discussions around the subject, as well as the many limits to our knowledge.

Progress towards a final theory on how iCOMs form and thrive during the formation of a planetary system similar to our solar system can only come from a very tight collaboration between astronomers and chemists. This is not only a curiosity, it may also be key for progressing our knowledge of the emergence of life on Earth, and maybe in other systems. After all, when one considers the major ingredients (elements) necessary for terrestrial life, they are an exact match of the most abundant elements at the moment our solar system formed: hydrogen, oxygen, carbon, nitrogen, sulphur and so on, literally reflecting the sequence of initial elemental abundances of the solar nebula. It may be a cause or a consequence: yet, life used what was available, in terms of elements. Notice, however, that once the Earth formed, the abundance of elements around for life to start was completely changed, with the most abundant ones no longer being hydrogen, oxygen, carbon and so on. It seems to me that life likely bears the chemical imprint of the first phases of solar system formation.

Finally, no doubt life used the laws that regulate the Universe, namely the physics and chemistry that we know. Whether the iCOMs formed during the infancy of the solar system had a role or not, this first chapter of our history has to be written in as much detail as possible.

Conflicts of interest

There are no conflicts to declare.

Acknowledgements

I would like to warmly thank my chemist friends and collaborators Nadia Balucani, Piero Ugliengo, Albert Rimola: without their help and teaching, this review would not have existed. I also thank my astronomer friends and collaborators, especially Claudio Codella, Ana López-Sepulcre, Charlotte Vastel and Bertrand Lefloch. A final very deep thank goes to the young researchers with whom I had the luck to work, especially Eleonora Bianchi, Mathilde Bouvier, Marta De Simone, Joan Enrique-Romero, Cecile Favre, Lorenzo Tinacci and Fanny Vazart. I acknowledge the funding within the European Union's Horizon 2020 research and innovation programme from the European Research Council (ERC) for the project "The Dawn of Organic Chemistry" (DOC), grant agreement No. 741002, and from the Marie Sklodowska-Curie for the project "Astro-Chemical Origins" (ACO), grant agreement No. 811312.

Notes and references

1 J. K. Jørgensen, T. L. Bourke, P. C. Myers, J. Di Francesco, E. F. van Dishoeck, C.-F. Lee, N. Ohashi, F. L. Schöier, S. Takakuwa, D. J. Wilner and Q. Zhang, *Astrophys. J.*, 2007, **659**, 479–498.

2 V. Taquet, A. López-Sepulcre, C. Ceccarelli, R. Neri, C. Kahane and S. B. Charnley, *Astrophys. J.*, 2015, **804**, 81.

3 M. De Simone, C. Ceccarelli, C. Codella, B. E. Svoboda, C. Chandler, M. Bouvier, S. Yamamoto, N. Sakai, P. Caselli, C. Favre, L. Loinard, B. Lefloch, H. B. Liu, A. López-Sepulcre, J. E. Pineda, V. Taquet and L. Testi, *Astrophys. J.*, 2020, **896**, L3.

4 A. López-Sepulcre, N. Sakai, R. Neri, M. Imai, Y. Oya, C. Ceccarelli, A. E. Higuchi, Y. Aikawa, S. Bottinelli, E. Caux, T. Hirota, C. Kahane, B. Lefloch, C. Vastel, Y. Watanabe and S. Yamamoto, *Astron. Astrophys.*, 2017, **606**, A121.

5 C. Ceccarelli, P. Caselli, F. Fontani, R. Neri, A. López-Sepulcre, C. Codella, S. Feng, I. Jiménez-Serra, B. Lefloch, J. E. Pineda, C. Vastel, F. Alves, R. Bachiller, N. Balucani, E. Bianchi, L. Bizzocchi, S. Bottinelli, E. Caux, A. Chacón-Tanarro, R. Choudhury, A. Coutens, F. Dulieu, C. Favre, P. Hily-Blant, J. Holdship, C. Kahane, A. Jaber Al-Edhari, J. Laas, J. Ospina, Y. Oya, L. Podio, A. Pon, A. Punanova, D. Quenard, A. Rimola, N. Sakai, I. R. Sims, S. Spezzano, V. Taquet, L. Testi, P. Theulé, P. Ugliengo, A. I. Vasyunin, S. Viti, L. Wiesenfeld and S. Yamamoto, *Astrophys. J.*, 2017, **850**, 176.

6 E. Herbst and E. F. van Dishoeck, *Annu. Rev. Astron. Astrophys.*, 2009, **47**, 427–480.

7 P. Caselli and C. Ceccarelli, *Astron. Astrophys. Rev.*, 2012, **20**, 56.

8 C. Ceccarelli, P. Caselli, D. Bockelée-Morvan, O. Mousis, S. Pizzarello, F. Robert and D. Semenov, *Protostars and Planets VI*, 2014, pp. 859–882.

9 K. I. Öberg, *Chem. Rev.*, 2016, **116**, 9631.

10 J. K. Jørgensen, A. Belloche and R. T. Garrod, *Annu. Rev. Astron. Astrophys.*, 2020, **58**, 727–778.

11 K. I. Öberg and E. A. Bergin, *Phys. Rep.*, 2021, **893**, 1–48.

12 C. Ceccarelli, C. Codella, N. Balucani, D. Bockelée-Morvan, E. Herbst, C. Vastel, P. Caselli, C. Favre, B. Lefloch and K. Öberg, arXiv, 2022, preprint, arXiv:2206.13270.

13 A. Morbidelli, J. Chambers, J. I. Lunine, J. M. Petit, F. Robert, G. B. Valsecchi and K. E. Cyr, *Elements*, 2022, **18**, 155.

14 L. Piani, Y. Marrocchi, T. Rigaudier, L. G. Vacher, D. Thomassin and B. Marty, *Science*, 2020, **369**, 1110–1113.

15 C. Ceccarelli and F. Du, *Elements*, 2022, **18**, 155.

16 F. Ober, S. Wolf, A. L. Uribe and H. H. Klahr, *Astron. Astrophys.*, 2015, **579**, A105.

17 P. Mollière, T. Molyarova, B. Bitsch, T. Henning, A. Schneider, L. Kreidberg, C. Eistrup, R. Burn, E. Nasedkin, D. Semenov, C. Mordasini, M. Schlecker, K. R. Schwarz, S. Lacour, M. Nowak and M. Schulik, *Astrophys. J.*, 2022, **934**, 74.

18 J. L. Linsky, B. T. Draine, H. W. Moos, E. B. Jenkins, B. E. Wood, C. Oliveira, W. P. Blair, S. D. Friedman, C. Gry, D. Knauth, J. W. Kruk, S. Lacour,

N. Lehner, S. Redfield, J. M. Shull, G. Sonneborn and G. M. Williger, *Astrophys. J.*, 2006, **647**, 1106–1124.

19 A. Bacmann, B. Lefloch, C. Ceccarelli, J. Steinacker, A. Castets and L. Loinard, *Astrophys. J.*, 2003, **585**, L55–L58.

20 A. Bacmann, V. Taquet, A. Faure, C. Kahane and C. Ceccarelli, *Astron. Astrophys.*, 2012, **541**, L12.

21 J. Cernicharo, N. Marcelino, E. Roueff, M. Gerin, A. Jiménez-Escobar and G. M. Muñoz Caro, *Astrophys. J.*, 2012, **759**, L43.

22 K. I. Öberg, S. Bottinelli, J. K. Jørgensen and E. F. van Dishoeck, *Astrophys. J.*, 2010, **716**, 825–834.

23 A. A. Jaber, C. Ceccarelli, C. Kahane and E. Caux, *Astrophys. J.*, 2014, **791**, 29.

24 C. Vastel, C. Ceccarelli, B. Lefloch and R. Bachiller, *Astrophys. J.*, 2014, **795**, L2.

25 L. Bizzocchi, P. Caselli, S. Spezzano and E. Leonardo, *Astron. Astrophys.*, 2014, **569**, A27.

26 I. Jiménez-Serra, A. I. Vasyunin, P. Caselli, N. Marcelino, N. Billot, S. Viti, L. Testi, C. Vastel, B. Lefloch and R. Bachiller, *Astrophys. J.*, 2016, **830**, L6.

27 C. Vastel, J. C. Loison, V. Wakelam and B. Lefloch, *Astron. Astrophys.*, 2019, **625**, A91.

28 S. Scibelli and Y. Shirley, *Astrophys. J.*, 2020, **891**, 73.

29 S. Spezzano, L. Bizzocchi, P. Caselli, J. Harju and S. Brünken, *Astron. Astrophys.*, 2016, **592**, L11.

30 C. Zhou, C. Vastel, J. Montillaud, C. Ceccarelli, K. Demyk, J. Harju, M. Juvela, I. Ristorcelli and T. Liu, *Astron.Astrophys.*, 2022, **658**, A131.

31 M. Imai, N. Sakai, Y. Oya, A. López-Sepulcre, Y. Watanabe, C. Ceccarelli, B. Lefloch, E. Caux, C. Vastel, C. Kahane, T. Sakai, T. Hirota, Y. Aikawa and S. Yamamoto, *Astrophys. J.*, 2016, **830**, L37.

32 S. Cazaux, A. G. G. M. Tielens, C. Ceccarelli, A. Castets, V. Wakelam, E. Caux, B. Parise and D. Teyssier, *Astrophys. J.*, 2003, **593**, L51–L55.

33 S. Bottinelli, C. Ceccarelli, R. Neri, J. P. Williams, E. Caux, S. Cazaux, B. Lefloch, S. Maret and A. G. G. M. Tielens, *Astrophys. J.*, 2004, **617**, L69–L72.

34 Y.-J. Kuan, H.-C. Huang, S. B. Charnley, N. Hirano, S. Takakuwa, D. J. Wilner, S.-Y. Liu, N. Ohashi, T. L. Bourke, C. Qi and Q. Zhang, *Astrophys. J.*, 2004, **616**, L27–L30.

35 E. Bianchi, C. J. Chandler, C. Ceccarelli, C. Codella, N. Sakai, A. López-Sepulcre, L. T. Maud, G. Moellenbrock, B. Svoboda, Y. Watanabe, T. Sakai, F. Ménard, Y. Aikawa, F. Alves, N. Balucani, M. Bouvier, P. Caselli, E. Caux, S. Charnley, S. Choudhury, M. De Simone, F. Dulieu, A. Durán, L. Evans, C. Favre, D. Fedele, S. Feng, F. Fontani, L. Francis, T. Hama, T. Hanawa, E. Herbst, T. Hirota, M. Imai, A. Isella, I. Jiménez-Serra, D. Johnstone, C. Kahane, B. Lefloch, L. Loinard, M. J. Maureira, S. Mercimek, A. Miotello, S. Mori, R. Nakatani, H. Nomura, Y. Oba, S. Ohashi, Y. Okoda, J. Ospina-Zamudio, Y. Oya, J. Pineda, L. Podio, A. Rimola, D. S. Cox, Y. Shirley, V. Taquet, L. Testi, C. Vastel, S. Viti, N. Watanabe, A. Witzel, C. Xue, Y. Zhang, B. Zhao and S. Yamamoto, *Mon. Not. R. Astron. Soc.: Lett.*, 2020, **498**, L87–L92.

36 Y.-L. Yang, N. Sakai, Y. Zhang, N. M. Murillo, Z. E. Zhang, A. E. Higuchi, S. Zeng, A. López-Sepulcre, S. Yamamoto, B. Lefloch, M. Bouvier, C. Ceccarelli, T. Hirota, M. Imai, Y. Oya, T. Sakai and Y. Watanabe, *Astrophys. J.*, 2021, **910**, 20.

37 Y. Oya, N. Sakai, Y. Watanabe, A. E. Higuchi, T. Hirota, A. López-Sepulcre, T. Sakai, Y. Aikawa, C. Ceccarelli, B. Lefloch, E. Caux, C. Vastel, C. Kahane and S. Yamamoto, *Astrophys. J.*, 2017, **837**, 174.

38 N. Marcelino, M. Gerin, J. Cernicharo, A. Fuente, H. A. Wootten, E. Chapillon, J. Pety, D. C. Lis, E. Roueff, B. Commerçon and A. Ciardi, *Astron. Astrophys.*, 2018, **620**, A80.

39 K. I. Öberg, T. Lauck and D. Graninger, *Astrophys. J.*, 2014, **788**, 68.

40 S. Bottinelli, C. Ceccarelli, B. Lefloch, J. P. Williams, A. Castets, E. Caux, S. Cazaux, S. Maret, B. Parise and A. G. G. M. Tielens, *Astrophys. J.*, 2004, **615**, 354–358.

41 S. Bottinelli, C. Ceccarelli, J. P. Williams and B. Lefloch, *Astron. Astrophys.*, 2007, **463**, 601–610.

42 E. Bianchi, C. Codella, C. Ceccarelli, F. Vazart, R. Bachiller, N. Balucani, M. Bouvier, M. De Simone, J. Enrique-Romero, C. Kahane, B. Lefloch, A. López-Sepulcre, J. Ospina-Zamudio, L. Podio and V. Taquet, *Mon. Not. R. Astron. Soc.*, 2019, **483**, 1850–1861.

43 E. Bianchi, C. Ceccarelli, C. Codella, A. López-Sepulcre, S. Yamamoto, N. Balucani, P. Caselli, L. Podio, R. Neri, R. Bachiller, C. Favre, F. Fontani, B. Lefloch, N. Sakai and D. Segura-Cox, *Astron. Astrophys.*, 2022, **662**, A103.

44 C. Codella, C. Ceccarelli, P. Caselli, N. Balucani, V. Barone, F. Fontani, B. Lefloch, L. Podio, S. Viti, S. Feng, R. Bachiller, E. Bianchi, F. Dulieu, I. Jiménez-Serra, J. Holdship, R. Neri, J. E. Pineda, A. Pon, I. Sims, S. Spezzano, A. I. Vasyunin, F. Alves, L. Bizzocchi, S. Bottinelli, E. Caux, A. Chacón-Tanarro, R. Choudhury, A. Coutens, C. Favre, P. Hily-Blant, C. Kahane, A. Jaber Al-Edhari, J. Laas, A. López-Sepulcre, J. Ospina, Y. Oya, A. Punanova, C. Puzzarini, D. Quenard, A. Rimola, N. Sakai, D. Skouteris, V. Taquet, L. Testi, P. Theulé, P. Ugliengo, C. Vastel, F. Vazart, L. Wiesenfeld and S. Yamamoto, *Astron. Astrophys.*, 2017, **605**, L3.

45 C.-F. Lee, Z.-Y. Li, P. T. P. Ho, N. Hirano, Q. Zhang and H. Shang, *Astrophys. J.*, 2017, **843**, 27.

46 L. Chahine, A. López-Sepulcre, R. Neri, C. Ceccarelli, S. Mercimek, C. Codella, M. Bouvier, E. Bianchi, C. Favre, L. Podio, F. O. Alves, N. Sakai and S. Yamamoto, *Astron. Astrophys.*, 2022, **657**, A78.

47 S.-Y. Hsu, S.-Y. Liu, T. Liu, D. Sahu, N. Hirano, C.-F. Lee, K. Tatematsu, G. Kim, M. Juvela, P. Sanhueza, J. He, D. Johnstone, S.-L. Qin, L. Bronfman, H.-R. V. Chen, S. Dutta, D. J. Eden, K.-S. Jhan, K.-T. Kim, Y.-J. Kuan, W. Kwon, C. W. Lee, J.-E. Lee, A. Moraghan, M. G. Rawlings, H. Shang, A. Soam, M. A. Thompson, A. Traficante, Y. Wu, Y.-L. Yang and Q. Zhang, *Astrophys. J.*, 2020, **898**, 107.

48 M. Bouvier, C. Ceccarelli, A. López-Sepulcre, N. Sakai, S. Yamamoto and Y.-L. Yang, *Astrophys. J.*, 2022, **929**, 10.

49 R. Martín-Doménech, J. B. Bergner, K. I. Öberg and J. K. Jørgensen, *Astrophys. J.*, 2019, **880**, 130.

50 J. B. Bergner, R. Martín-Doménech, K. I. Öberg, J. K. Jørgensen, E. Artur de la Villarmois and C. Brinch, *ACS Earth Space Chem.*, 2019, **3**, 1564–1575.

51 G. Molpeceres, J. Kästner, V. J. Herrero, R. J. Peláez and B. Maté, *Astron. Astrophys.*, 2022, **664**, A169.

52 R. H. Rubin, G. W. Swenson Jr, R. C. Benson, H. L. Tigelaar and W. H. Flygare, *Astrophys. J.*, 1971, **169**, L39.

53 G. A. Blake, E. C. Sutton, C. R. Masson and T. G. Phillips, *Astrophys. J.*, 1987, **315**, 621.
54 C. Ceccarelli, L. Loinard, A. Castets, A. Faure and B. Lefloch, *Astron. Astrophys.*, 2000, **362**, 1122–1126.
55 C. Ceccarelli, P. Caselli, E. Herbst, A. G. G. M. Tielens and E. Caux, *Protostars and Planets V*, 2007, pp. 47–62.
56 A. Belloche, A. J. Maury, S. Maret, S. Anderl, A. Bacmann, P. André, S. Bontemps, S. Cabrit, C. Codella, M. Gaudel, F. Gueth, C. Lefèvre, B. Lefloch, L. Podio and L. Testi, *Astron. Astrophys.*, 2020, **635**, A198.
57 C. Ceccarelli, L. Loinard, A. Castets, A. G. G. M. Tielens and E. Caux, *Astron. Astrophys.*, 2000, **357**, L9–L12.
58 J. K. Jørgensen, C. Favre, S. E. Bisschop, T. L. Bourke, E. F. van Dishoeck and M. Schmalzl, *Astrophys. J.*, 2012, **757**, L4.
59 J. K. Jørgensen, M. H. D. van der Wiel, A. Coutens, J. M. Lykke, H. S. P. Müller, E. F. van Dishoeck, H. Calcutt, P. Bjerkeli, T. L. Bourke, M. N. Drozdovskaya, C. Favre, E. C. Fayolle, R. T. Garrod, S. K. Jacobsen, K. I. Öberg, M. V. Persson and S. F. Wampfler, *Astron. Astrophys.*, 2016, **595**, A117.
60 S. Manigand, J. K. Jørgensen, H. Calcutt, H. S. P. Müller, N. F. W. Ligterink, A. Coutens, M. N. Drozdovskaya, E. F. van Dishoeck and S. F. Wampfler, *Astron. Astrophys.*, 2020, **635**, A48.
61 S. Manigand, A. Coutens, J. C. Loison, V. Wakelam, H. Calcutt, H. S. P. Müller, J. K. Jørgensen, V. Taquet, S. F. Wampfler, T. L. Bourke, B. M. Kulterer, E. F. van Dishoeck, M. N. Drozdovskaya and N. F. W. Ligterink, *Astron. Astrophys.*, 2021, **645**, A53.
62 E. Bianchi, A. López-Sepulcre, C. Ceccarelli, C. Codella, L. Podio, M. Bouvier and J. Enrique-Romero, *Astrophys. J.*, 2022, **928**, L3.
63 C. Ceccarelli, A. Castets, L. Loinard, E. Caux and A. G. G. M. Tielens, *Astron. Astrophys.*, 1998, **338**, L43–L46.
64 B. Parise, C. Ceccarelli, A. G. G. M. Tielens, A. Castets, E. Caux, B. Lefloch and S. Maret, *Astron. Astrophys.*, 2006, **453**, 949–958.
65 B. Parise, A. Castets, E. Herbst, E. Caux, C. Ceccarelli, I. Mukhopadhyay and A. G. G. M. Tielens, *Astron. Astrophys.*, 2004, **416**, 159–163.
66 C. Richard, L. Margulès, E. Caux, C. Kahane, C. Ceccarelli, J. C. Guillemin, R. A. Motiyenko, C. Vastel and P. Groner, *Astron. Astrophys.*, 2013, **552**, A117.
67 A. Coutens, J. K. Jørgensen, M. H. D. van der Wiel, H. S. P. Müller, J. M. Lykke, P. Bjerkeli, T. L. Bourke, H. Calcutt, M. N. Drozdovskaya, C. Favre, E. C. Fayolle, R. T. Garrod, S. K. Jacobsen, N. F. W. Ligterink, K. I. Öberg, M. V. Persson, E. F. van Dishoeck and S. F. Wampfler, *Astron. Astrophys.*, 2016, **590**, L6.
68 J. K. Jørgensen, H. S. P. Müller, H. Calcutt, A. Coutens, M. N. Drozdovskaya, K. I. Öberg, M. V. Persson, V. Taquet, E. F. van Dishoeck and S. F. Wampfler, *Astron. Astrophys.*, 2018, **620**, A170.
69 M. V. Persson, J. K. Jørgensen, H. S. P. Müller, A. Coutens, E. F. van Dishoeck, V. Taquet, H. Calcutt, M. H. D. van der Wiel, T. L. Bourke and S. F. Wampfler, *Astron. Astrophys.*, 2018, **610**, A54.
70 S. Manigand, H. Calcutt, J. K. Jørgensen, V. Taquet, H. S. P. Müller, A. Coutens, S. F. Wampfler, N. F. W. Ligterink, M. N. Drozdovskaya, L. E. Kristensen, M. H. D. van der Wiel and T. L. Bourke, *Astron. Astrophys.*, 2019, **623**, A69.

71 C. Richard, J. K. Jørgensen, L. Margulès, R. A. Motiyenko, J. C. Guillemin and P. Groner, *Astron. Astrophys.*, 2021, **651**, A120.

72 M. De Simone, C. Ceccarelli, C. Codella, B. E. Svoboda, C. J. Chandler, M. Bouvier, S. Yamamoto, N. Sakai, Y.-L. Yang, P. Caselli, B. Lefloch, H. B. Liu, A. López-Sepulcre, L. Loinard, J. E. Pineda and L. Testi, *Astrophys. J.*, 2022, **935**, L14.

73 R. Bachiller, S. Liechti, C. M. Walmsley and F. Colomer, *Astron. Astrophys.*, 1995, **295**, L51.

74 R. Bachiller and M. Pérez Gutiérrez, *Astrophys. J.*, 1997, **487**, L93–L96.

75 C. Vastel, F. Alves, C. Ceccarelli, M. Bouvier, I. Jiménez-Serra, T. Sakai, P. Caselli, L. Evans, F. Fontani, R. Le Gal, C. J. Chandler, B. Svoboda, L. Maud, C. Codella, N. Sakai, A. López-Sepulcre, G. Moellenbrock, Y. Aikawa, N. Balucani, E. Bianchi, G. Busquet, E. Caux, S. Charnley, N. Cuello, M. De Simone, F. Dulieu, A. Durân, D. Fedele, S. Feng, L. Francis, T. Hama, T. Hanawa, E. Herbst, T. Hirota, M. Imai, A. Isella, D. Johnstone, B. Lefloch, L. Loinard, M. Maureira, N. M. Murillo, S. Mercimek, S. Mori, F. Menard, A. Miotello, R. Nakatani, H. Nomura, Y. Oba, S. Ohashi, Y. Okoda, J. Ospina-Zamudio, Y. Oya, J. E. Pineda, L. Podio, A. Rimola, D. Segura Cox, Y. Shirley, L. Testi, S. Viti, N. Watanabe, Y. Watanabe, A. Witzel, C. Xue, Y. Zhang, B. Zhao and S. Yamamoto, *Astron. Astrophys.*, 2022, **664**, A171.

76 F. Gueth, S. Guilloteau and R. Bachiller, *Astron. Astrophys.*, 1996, **307**, 891–897.

77 F. Fontani, C. Codella, C. Ceccarelli, B. Lefloch, S. Viti and M. Benedettini, *Astrophys. J.*, 2014, **788**, L43.

78 K. I. Öberg, V. V. Guzmán, K. Furuya, C. Qi, Y. Aikawa, S. M. Andrews, R. Loomis and D. J. Wilner, *Nature*, 2015, **520**, 198–201.

79 C. Favre, D. Fedele, D. Semenov, S. Parfenov, C. Codella, C. Ceccarelli, E. A. Bergin, E. Chapillon, L. Testi, F. Hersant, B. Lefloch, F. Fontani, G. A. Blake, L. I. Cleeves, C. Qi, K. R. Schwarz and V. Taquet, *Astrophys. J.*, 2018, **862**, L2.

80 K. I. Öberg, V. V. Guzmán, C. Walsh, Y. Aikawa, E. A. Bergin, C. J. Law, R. A. Loomis, F. Alarcón, S. M. Andrews, J. Bae, J. B. Bergner, Y. Boehler, A. S. Booth, A. D. Bosman, J. K. Calahan, G. Cataldi, L. I. Cleeves, I. Czekala, K. Furuya, J. Huang, J. D. Ilee, N. T. Kurtovic, R. Le Gal, Y. Liu, F. Long, F. Ménard, H. Nomura, L. M. Pérez, C. Qi, K. R. Schwarz, A. Sierra, R. Teague, T. Tsukagoshi, Y. Yamato, M. L. R. van't Hoff, A. R. Waggoner, D. J. Wilner and K. Zhang, *Astrophys. J. Suppl.*, 2021, **257**, 1.

81 J. D. Ilee, C. Walsh, A. S. Booth, Y. Aikawa, S. M. Andrews, J. Bae, E. A. Bergin, J. B. Bergner, A. D. Bosman, G. Cataldi, L. I. Cleeves, I. Czekala, V. V. Guzmán, J. Huang, C. J. Law, R. Le Gal, R. A. Loomis, F. Ménard, H. Nomura, K. I. Öberg, C. Qi, K. R. Schwarz, R. Teague, T. Tsukagoshi, D. J. Wilner, Y. Yamato and K. Zhang, *Astrophys. J. Suppl.*, 2021, **257**, 9.

82 C. Walsh, R. A. Loomis, K. I. Öberg, M. Kama, M. L. R. van't Hoff, T. J. Millar, Y. Aikawa, E. Herbst, S. L. Widicus Weaver and H. Nomura, *Astrophys. J.*, 2016, **823**, L10.

83 P. Nazari, B. Tabone, G. P. Rosotti, M. L. van Gelder, R. Meshaka and E. F. van Dishoeck, *Astron. Astrophys.*, 2022, **663**, A58.

84 L. Podio, A. Garufi, C. Codella, D. Fedele, E. Bianchi, F. Bacciotti, C. Ceccarelli, C. Favre, S. Mercimek, K. Rygl and L. Testi, *Astron. Astrophys.*, 2020, **642**, L7.

85 L. Keyte, M. Kama, A. S. Booth, E. A. Bergin, L. I. Cleeves, E. F. van Dishoeck, M. N. Drozdovskaya, K. Furuya, J. Rawlings, O. Shorttle and C. Walsh, *Nat. Astron.*, 2023, **7**, 684.

86 S. L. Grant, E. F. van Dishoeck, B. Tabone, D. Gasman, T. Henning, I. Kamp, M. Güdel, P.-O. Lagage, G. Bettoni, G. Perotti, V. Christiaens, M. Samland, A. M. Arabhavi, I. Argyriou, A. Abergel, O. Absil, D. Barrado, A. Boccaletti, J. Bouwman, A. C. o. Garatti, V. Geers, A. M. Glauser, R. Guadarrama, H. Jang, J. Kanwar, F. Lahuis, M. Morales-Calderón, M. Mueller, C. Nehmé, G. Olofsson, E. Pantin, N. Pawellek, T. P. Ray, D. Rodgers-Lee, S. Scheithauer, J. Schreiber, K. Schwarz, M. Temmink, B. Vandenbussche, M. Vlasblom, L. B. F. M. Waters, G. Wright, L. Colina, T. R. Greve, K. Justannont and G. Östlin, *Astrophys. J.*, 2023, **947**, L6.

87 A. Banzatti, K. M. Pontoppidan, J. Pére Chávez, C. Salyk, L. Diehl, S. Bruderer, G. J. Herczeg, A. Carmona, I. Pascucci, S. Brittain, S. Jensen, S. Grant, E. F. van Dishoeck, I. Kamp, A. D. Bosman, K. I. Öberg, G. A. Blake, M. R. Meyer, E. Gaidos, A. Boogert, J. T. Rayner and C. Wheeler, *Astron. J.*, 2023, **165**, 72.

88 C.-F. Lee, P. T. P. Ho, Z.-Y. Li, N. Hirano, Q. Zhang and H. Shang, *Nat. Astron.*, 2017, **1**, 0152.

89 C. Codella, E. Bianchi, B. Tabone, C. F. Lee, S. Cabrit, C. Ceccarelli, L. Podio, F. Bacciotti, R. Bachiller, E. Chapillon, F. Gueth, A. Gusdorf, B. Lefloch, S. Leurini, G. Pineau des Forêts, K. L. J. Rygl and M. Tafalla, *Astron. Astrophys.*, 2018, **617**, A10.

90 C.-F. Lee, C. Codella, Z.-Y. Li and S.-Y. Liu, *Astrophys. J.*, 2019, **876**, 63.

91 C.-F. Lee, C. Codella, C. Ceccarelli and A. López-Sepulcre, *Astrophys. J.*, 2022, **937**, 10.

92 E. C. Pickering, *Astron. Nachr.*, 1890, **124**, 113.

93 M. S. Connelley and B. Reipurth, *Astrophys. J.*, 2018, **861**, 145.

94 K. M. Strom and S. E. Strom, *Astrophys. J.*, 1993, **412**, L63.

95 E. Furlan, W. J. Fischer, B. Ali, A. M. Stutz, T. Stanke, J. J. Tobin, S. T. Megeath, M. Osorio, L. Hartmann, N. Calvet, C. A. Poteet, J. Booker, P. Manoj, D. M. Watson and L. Allen, *Astrophys. J., Suppl. Ser.*, 2016, **224**, 5.

96 W. J. Fischer, E. Safron and S. T. Megeath, *American Astronomical Society Meeting Abstracts #228*, 2016, p. 308.01.

97 L. A. Cieza, S. Casassus, J. Tobin, S. P. Bos, J. P. Williams, S. Perez, Z. Zhu, C. Caceres, H. Canovas, M. M. Dunham, A. Hales, J. L. Prieto, D. A. Principe, M. R. Schreiber, D. Ruiz-Rodriguez and A. Zurlo, *Nature*, 2016, **535**, 258–261.

98 M. L. R. van't Hoff, J. J. Tobin, L. Trapman, D. Harsono, P. D. Sheehan, W. J. Fischer, S. T. Megeath and E. F. van Dishoeck, *Astrophys. J.*, 2018, **864**, L23.

99 J.-E. Lee, S. Lee, G. Baek, Y. Aikawa, L. Cieza, S.-Y. Yoon, G. Herczeg, D. Johnstone and S. Casassus, *Nat. Astron.*, 2019, **3**, 314–319.

100 M. Leemker, M. L. R. van't Hoff, L. Trapman, M. L. van Gelder, M. R. Hogerheijde, D. Ruíz-Rodríguez and E. F. van Dishoeck, *Astron. Astrophys.*, 2021, **646**, A3.

101 J. J. Tobin, M. L. R. van't Hoff, M. Leemker, E. F. van Dishoeck, T. Paneque-Carreño, K. Furuya, D. Harsono, M. V. Persson, L. I. Cleeves, P. D. Sheehan and L. Cieza, *Nature*, 2023, **615**, 227–230.
102 A. C. A. Boogert, P. A. Gerakines and D. C. B. Whittet, *Annu. Rev. Astron. Astrophys.*, 2015, **53**, 541–581.
103 M. K. McClure, W. R. M. Rocha, K. M. Pontoppidan, N. Crouzet, L. E. U. Chu, E. Dartois, T. Lamberts, J. A. Noble, Y. J. Pendleton, G. Perotti, D. Qasim, M. G. Rachid, Z. L. Smith, F. Sun, T. L. Beck, A. C. A. Boogert, W. A. Brown, P. Caselli, S. B. Charnley, H. M. Cuppen, H. Dickinson, M. N. Drozdovskaya, E. Egami, J. Erkal, H. Fraser, R. T. Garrod, D. Harsono, S. Ioppolo, I. Jiménez-Serra, M. Jin, J. K. Jørgensen, L. E. Kristensen, D. C. Lis, M. R. S. McCoustra, B. A. McGuire, G. J. Melnick, K. I. Ã-berg, M. E. Palumbo, T. Shimonishi, J. A. Sturm, E. F. van Dishoeck and H. Linnartz, *Nat. Astron.*, 2023, **7**, 431–443.
104 Y.-L. Yang, J. D. Green, K. M. Pontoppidan, J. B. Bergner, L. I. Cleeves, I. Evans, J. Neal, R. T. Garrod, M. Jin, C. H. Kim, J. Kim, J.-E. Lee, N. Sakai, C. N. Shingledecker, B. Shope, J. J. Tobin and E. F. van Dishoeck, *Astrophys. J.*, 2022, **941**, L13.
105 G. Sarma, C.-H. Yang, A. K. Saha, D. H. Parker and L. Wiesenfeld, *J. Chem. Phys.*, 2013, **138**, 024314.
106 S. Chefdeville, T. Stoecklin, C. Naulin, P. Jankowski, K. Szalewicz, A. Faure, M. Costes and A. Bergeat, *Astrophys. J., Lett.*, 2015, **799**, L9.
107 T. Stoecklin, A. Faure, P. Jankowski, S. Chefdeville, A. Bergeat, C. Naulin, S. B. Morales and M. Costes, *Phys. Chem. Chem. Phys.*, 2017, **19**, 189–195.
108 A. Bergeat, S. B. Morales, C. Naulin, L. Wiesenfeld and A. Faure, *Phys. Rev. Lett.*, 2020, **125**, 143402.
109 Q. Shuai, T. de Jongh, M. Besemer, A. van der Avoird, G. C. Groenenboom and S. Y. T. van de Meerakker, *J. Chem. Phys.*, 2020, **153**, 244302.
110 D. R. Flower, G. Pineau des Forêts and D. Rabli, *Mon. Not. R. Astron. Soc.*, 2010, **409**, 29–37.
111 S. Green, D. L. Cochrane and D. G. Truhlar, *J. Chem. Phys.*, 1986, **84**, 3865–3869.
112 A. Fuente, D. G. Navarro, P. Caselli, M. Gerin, C. Kramer, E. Roueff, T. Alonso-Albi, R. Bachiller, S. Cazaux, B. Commercon, R. Friesen, S. García-Burillo, B. M. Giuliano, J. R. Goicoechea, P. Gratier, A. Hacar, I. Jiménez-Serra, J. Kirk, V. Lattanzi, J. C. Loison, J. Malinen, N. Marcelino, R. Martín-Doménech, G. Muñoz-Caro, J. Pineda, M. Tafalla, B. Tercero, D. Ward-Thompson, S. P. Treviño-Morales, P. Riviére-Marichalar, O. Roncero, T. Vidal and M. Y. Ballester, *Astron. Astrophys.*, 2019, **624**, A105.
113 B. T. Draine, *Annu. Rev. Astron. Astrophys.*, 2003, **41**, 241–289.
114 C. A. Poteet, D. C. B. Whittet and B. T. Draine, *Astrophys. J.*, 2015, **801**, 110.
115 A. P. Jones and N. Ysard, *Astron. Astrophys.*, 2019, **627**, A38.
116 E. Caux, C. Ceccarelli, A. Castets, C. Vastel, R. Liseau, S. Molinari, B. Nisini, P. Saraceno and G. J. White, *Astron. Astrophys.*, 1999, **347**, L1–L4.
117 C. Vastel, E. Caux, C. Ceccarelli, A. Castets, C. Gry and J. P. Baluteau, *Astron. Astrophys.*, 2000, **357**, 994–1000.
118 D. C. Lis, J. Keene, T. G. Phillips, P. Schilke, M. W. Werner and J. Zmuidzinas, *Astrophys. J.*, 2001, **561**, 823–829.

119 P. F. Goldsmith, W. D. Langer, Y. Seo, J. Pineda, J. Stutzki, C. Guevara, R. Aladro and M. Justen, *Astrophys. J.*, 2021, **916**, 6.
120 D. C. Lis, P. F. Goldsmith, R. Güsten, P. Schilke, H. Wiesemeyer, Y. Seo and M. W. Werner, *Astron. Astrophys.*, 2023, **669**, L15.
121 F. Dulieu, L. Amiaud, E. Congiu, J. H. Fillion, E. Matar, A. Momeni, V. Pirronello and J. L. Lemaire, *Astron. Astrophys.*, 2010, **512**, A30.
122 N. Watanabe and A. Kouchi, *Astrophys. J.*, 2002, **571**, L173–L176.
123 M. Tsuge and N. Watanabe, *Proc. Jpn. Acad., Ser. B, Phys. Biol. Sci.*, 2023, **99**, 103–130.
124 J. Perrero, J. Enrique-Romero, B. Martínez-Bachs, C. Ceccarelli, N. Balucani, P. Ugliengo and A. Rimola, *ACS Earth Space Chem.*, 2022, **6**, 496–511.
125 G. Molpeceres, A. Rimola, C. Ceccarelli, J. Kästner, P. Ugliengo and B. Maté, *Mon. Not. R. Astron. Soc.*, 2019, **482**, 5389–5400.
126 F. Dulieu, E. Congiu, J. Noble, S. Baouche, H. Chaabouni, A. Moudens, M. Minissale and S. Cazaux, *Sci. Rep.*, 2013, **3**, 1338.
127 A. Rimola, V. Taquet, P. Ugliengo, N. Balucani and C. Ceccarelli, *Astron. Astrophys.*, 2014, **572**, A70.
128 L. Song and J. Kästner, *Astrophys. J.*, 2017, **850**, 118.
129 R. T. Garrod, *Astrophys. J.*, 2013, **765**, 60.
130 M. Ruaud, J. C. Loison, K. M. Hickson, P. Gratier, F. Hersant and V. Wakelam, *Mon. Not. R. Astron. Soc.*, 2015, **447**, 4004–4017.
131 Y. Aikawa, K. Furuya, S. Yamamoto and N. Sakai, *Astrophys. J.*, 2020, **897**, 110.
132 S. S. Jensen, S. Spezzano, P. Caselli, T. Grassi and T. Haugbølle, *arXiv*, 2023, preprint, arXiv:2305.05932.
133 S. Ferrero, S. Pantaleone, C. Ceccarelli, P. Ugliengo, M. Sodupe and A. Rimola, *Astrophys. J.*, 2023, **944**, 142.
134 R. Le Gal, P. Hily-Blant, A. Faure, G. Pineau des Forêts, C. Rist and S. Maret, *Astron. Astrophys.*, 2014, **562**, A83.
135 D. Qasim, G. Fedoseev, K. J. Chuang, J. He, S. Ioppolo, E. F. van Dishoeck and H. Linnartz, *Nat. Astron.*, 2020, **4**, 781–785.
136 T. Lamberts, G. Fedoseev, M. C. van Hemert, D. Qasim, K.-J. Chuang, J. C. Santos and H. Linnartz, *Astrophys. J.*, 2022, **928**, 48.
137 T. Shimonishi, N. Nakatani, K. Furuya and T. Hama, *Astrophys. J.*, 2018, **855**, 27.
138 D. Duflot, C. Toubin and M. Monnerville, *Front. Astron. Space Sci.*, 2021, **8**, 24.
139 G. Molpeceres, J. Kästner, G. Fedoseev, D. Qasim, R. Schömig, H. Linnartz and T. Lamberts, *J. Phys. Chem. Lett.*, 2021, **12**, 10854.
140 M. Minissale, Y. Aikawa, E. Bergin, M. Bertin, W. A. Brown, S. Cazaux, S. B. Charnley, A. Coutens, H. M. Cuppen, V. Guzman, H. Linnartz, M. R. S. McCoustra, A. Rimola, J. G. M. Schrauwen, C. Toubin, P. Ugliengo, N. Watanabe, V. Wakelam and F. Dulieu, *ACS Earth Space Chem.*, 2022, **6**, 597–630.
141 S. Ferrero, F. Grieco, A. S. Ibrahim Mohamed, F. Dulieu, A. Rimola, C. Ceccarelli, C. Nervi, M. Minissale and P. Ugliengo, *Mon. Not. R. Astron. Soc.*, 2022, **516**, 2586–2596.
142 S. Ferrero, L. Zamirri, C. Ceccarelli, A. Witzel, A. Rimola and P. Ugliengo, *Astrophys. J.*, 2020, **904**, 11.
143 J. Perrero, J. Enrique-Romero, S. Ferrero, C. Ceccarelli, L. Podio, C. Codella, A. Rimola and P. Ugliengo, *Astrophys. J.*, 2022, **938**, 158.
144 G. Bovolenta, S. Bovino, E. Vöhringer-Martinez, D. A. Saez, T. Grassi and S. Vogt-Geisse, *Mol. Astrophys.*, 2020, **21**, 100095.

145 G. M. Bovolenta, S. Vogt-Geisse, S. Bovino and T. Grassi, *Astrophys. J. Suppl.*, 2022, **262**, 17.
146 T. Grassi, S. Bovino, P. Caselli, G. Bovolenta, S. Vogt-Geisse and B. Ercolano, *Astron. Astrophys.*, 2020, **643**, A155.
147 L. Tinacci, A. Germain, S. Pantaleone, S. Ferrero, C. Ceccarelli and P. Ugliengo, *ACS Earth Space Chem.*, 2022, **6**, 1514–1526.
148 L. Tinacci, S. Ferrada-Chamorro, C. Ceccarelli, S. Pantaleone, D. Ascenzi, A. Maranzana, N. Balucani and P. Ugliengo, *arXiv*, 2023, preprint, arXiv:2302.14799.
149 Y. T. Lee and Y. R. Shen, *Phys. Today*, 1980, **33**, 52–59.
150 I. R. Sims and I. W. M. Smith, *Annu. Rev. Phys. Chem.*, 1995, **46**, 109–138.
151 P. Casavecchia, N. Balucani and G. G. Volpi, *Annu. Rev. Phys. Chem.*, 1999, **50**, 347–376.
152 A. Potapov, A. Canosa, E. Jiménez and B. Rowe, *Angew. Chem.*, 2017, **129**, 8742–8766.
153 V. Wakelam, J. C. Loison, E. Herbst, B. Pavone, A. Bergeat, K. Béroff, M. Chabot, A. Faure, D. Galli, W. D. Geppert, D. Gerlich, P. Gratier, N. Harada, K. M. Hickson, P. Honvault, S. J. Klippenstein, S. D. Le Picard, G. Nyman, M. Ruaud, S. Schlemmer, I. R. Sims, D. Talbi, J. Tennyson and R. Wester, *Astrophys. J., Suppl. Ser.*, 2015, **217**, 20.
154 D. McElroy, C. Walsh, A. J. Markwick, M. A. Cordiner, K. Smith and T. J. Millar, *Astron. Astrophys.*, 2013, **550**, A36.
155 L. Tinacci, S. Ferrada-Chamorro, C. Ceccarelli, S. Pantaleone, D. Ascenzi, A. Maranzana, N. Balucani and P. Ugliengo, *Astrophys. J.*, 2023, **266**, 38.
156 D. E. Woon and E. Herbst, *Astrophys. J. Suppl.*, 2009, **185**, 273–288.
157 L. Tinacci, S. Pantaleone, A. Maranzana, N. Balucani, C. Ceccarelli and P. Ugliengo, *Astrophys. J. Suppl.*, 2021, **256**, 35.
158 F. Vazart, C. Ceccarelli, N. Balucani, E. Bianchi and D. Skouteris, *Mon. Not. R. Astron. Soc.*, 2020, **499**, 5547–5561.
159 N. Balucani, C. Ceccarelli and V. Taquet, *Mon. Not. R. Astron. Soc.: Lett.*, 2015, **449**, L16–L20.
160 D. Skouteris, N. Balucani, C. Ceccarelli, F. Vazart, C. Puzzarini, V. Barone, C. Codella and B. Lefloch, *Astrophys. J.*, 2018, **854**, 135.
161 R. Wang, D. Mihalik and P. Stancil, *American Astronomical Society Meeting Abstracts*, 2021, p. 312.04.
162 Y. Komatsu and T. Suzuki, *ACS Earth Space Chem.*, 2022, **6**, 2491–2498.
163 A. Germain, L. Tinacci, S. Pantaleone, C. Ceccarelli and P. Ugliengo, *ACS Earth Space Chem.*, 2022, **6**, 1286–1298.
164 A. Rimola, D. Skouteris, N. Balucani, C. Ceccarelli, J. Enrique-Romero, V. Taquet and P. Ugliengo, *ACS Earth Space Chem.*, 2018, **2**, 720–734.
165 J. Kalvāns, *Mon. Not. R. Astron. Soc.*, 2018, **478**, 2753–2765.
166 C. N. Shingledecker, J. Tennis, R. L. Gal and E. Herbst, *Astrophys. J.*, 2018, **861**, 20.
167 A. Paulive, C. N. Shingledecker and E. Herbst, *Mon. Not. R. Astron. Soc.*, 2020, **500**, 3414.
168 J. Enrique-Romero, C. Ceccarelli, A. Rimola, D. Skouteris, N. Balucani and P. Ugliengo, *Astron. Astrophys.*, 2021, **655**, A9.
169 J. He, S. Emtiaz and G. Vidali, *Astrophys. J.*, 2018, **863**, 156.

170 A. Kouchi, K. Furuya, T. Hama, T. Chigai, T. Kozasa and N. Watanabe, *Astrophys. J.*, 2020, **891**, L22.
171 B. Maté, S. Cazaux, M. Á. Satorre, G. Molpeceres, J. Ortigoso, C. Millán and C. Santonja, *Astron. Astrophys.*, 2020, **643**, A163.
172 K. Furuya, T. Hama, Y. Oba, A. Kouchi, N. Watanabe and Y. Aikawa, *Astrophys. J.*, 2022, **933**, L16.
173 J. He, P. C. Pérez Rickert, T. Suhasaria, O. Sohier, T. Bäcker, D. Demertzi, G. Vidali and T. K. Henning, *arXiv*, 2023, preprint, arXiv:2302.08628.
174 B. Senevirathne, S. Andersson, F. Dulieu and G. Nyman, *Mol. Astrophys.*, 2017, **6**, 59–69.
175 P. Ghesquière, T. Mineva, D. Talbi, P. Theulé, J. A. Noble and T. Chiavassa, *Phys. Chem. Chem. Phys.*, 2015, **17**, 11455–11468.
176 P. Theulé, F. Duvernay, G. Danger, F. Borget, J. B. Bossa, V. Vinogradoff, F. Mispelaer and T. Chiavassa, *Adv. Space Res.*, 2013, **52**, 1567–1579.
177 G. Fedoseev, H. M. Cuppen, S. Ioppolo, T. Lamberts and H. Linnartz, *Mon. Not. R. Astron. Soc.*, 2015, **448**, 1288–1297.
178 S. Ioppolo, G. Fedoseev, K.-J. Chuang, H. M. Cuppen, A. R. Clements, M. Jin, R. T. Garrod, D. Qasim, V. Kofman, E. F. van Dishoeck and H. Linnartz, *Nat. Astron.*, 2020, **5**, 197.
179 Q. Chang and E. Herbst, *Astrophys. J.*, 2016, **819**, 145.
180 M. Jin and R. T. Garrod, *Astrophys. J., Suppl. Ser.*, 2020, **249**, 26.
181 R. T. Garrod, M. Jin, K. A. Matis, D. Jones, E. R. Willis and E. Herbst, *Astrophys. J. Suppl.*, 2022, **259**, 1.
182 R. T. Garrod and E. Herbst, *Astron. Astrophys.*, 2006, **457**, 927–936.
183 R. T. Garrod, S. L. Widicus Weaver and E. Herbst, *Astrophys. J.*, 2008, **682**, 283–302.
184 J. Enrique-Romero, A. Rimola, C. Ceccarelli and N. Balucani, *Mon. Not. R. Astron. Soc.: Lett.*, 2016, **459**, L6–L10.
185 J. Enrique-Romero, S. Álvarez-Barcia, F. J. Kolb, A. Rimola, C. Ceccarelli, N. Balucani, J. Meisner, P. Ugliengo, T. Lamberts and J. Kästner, *Mon. Not. R. Astron. Soc.*, 2020, **493**, 2523–2527.
186 J. Enrique-Romero, A. Rimola, C. Ceccarelli, P. Ugliengo, N. Balucani and D. Skouteris, *Astrophys. J. Suppl.*, 2022, **259**, 39.
187 T. Lamberts, M. N. Markmeyer, F. J. Kolb and J. Kästner, *ACS Earth Space Chem.*, 2019, **3**, 958–963.
188 B. Lefloch, C. Ceccarelli, C. Codella, C. Favre, L. Podio, C. Vastel, S. Viti and R. Bachiller, *Mon. Not. R. Astron. Soc.: Lett.*, 2017, **469**, L73–L77.
189 G. Baek, J.-E. Lee, T. Hirota, K.-T. Kim and M. K. Kim, *Astrophys. J.*, 2022, **939**, 84.
190 F. Vazart, C. Ceccarelli, N. Balucani and D. Skouteris, *Astrophys. J.*, 2022, **941**, 196.
191 A. López-Sepulcre, N. Balucani, C. Ceccarelli, C. Codella, F. Dulieu and P. Theulé, *ACS Earth Space Chem.*, 2019, **3**, 2122–2137.
192 C. Kahane, C. Ceccarelli, A. Faure and E. Caux, *Astrophys. J.*, 2013, **763**, L38.
193 V. Barone, C. Latouche, D. Skouteris, F. Vazart, N. Balucani, C. Ceccarelli and B. Lefloch, *Mon. Not. R. Astron. Soc.: Lett.*, 2015, **453**, L31–L35.
194 L. Song and J. Kästner, *Phys. Chem. Chem. Phys.*, 2016, **18**, 29278–29285.
195 F. Vazart, D. Calderini, C. Puzzarini, D. Skouteris and V. Barone, *J. Chem. Theory Comput.*, 2016, **12**, 5385–5397.

196 D. Skouteris, F. Vazart, C. Ceccarelli, N. Balucani, C. Puzzarini and V. Barone, *Mon. Not. R. Astron. Soc.*, 2017, **468**, L1–L5.
197 K. M. Douglas, D. I. Lucas, C. Walsh, N. A. West, M. A. Blitz and D. E. Heard, *Astrophys. J.*, 2022, **937**, L16.
198 E. Mendoza, B. Lefloch, A. López-Sepulcre, C. Ceccarelli, C. Codella, H. M. Boechat-Roberty and R. Bachiller, *Mon. Not. R. Astron. Soc.*, 2014, **445**, 151–161.
199 A. López-Sepulcre, A. A. Jaber, E. Mendoza, B. Lefloch, C. Ceccarelli, C. Vastel, R. Bachiller, J. Cernicharo, C. Codella, C. Kahane, M. Kama and M. Tafalla, *Mon. Not. R. Astron. Soc.*, 2015, **449**, 2438–2458.
200 J. A. Noble, P. Theule, E. Congiu, F. Dulieu, M. Bonnin, A. Bassas, F. Duvernay, G. Danger and T. Chiavassa, *Astron. Astrophys.*, 2015, **576**, A91.
201 K. A. Haupa, G. Tarczay and Y.-P. Lee, *74th International Symposium on Molecular Spectroscopy*, 2019.

Faraday Discussions

PAPER

The diverse chemistry of protoplanetary disks as revealed by JWST

Ewine F. van Dishoeck, *[ab] S. Grant,[b] B. Tabone,[c] M. van Gelder,[a] L. Francis,[a] L. Tychoniec,[d] G. Bettoni,[b] A. M. Arabhavi,[e] D. Gasman,[f] P. Nazari,[a] M. Vlasblom,[a] P. Kavanagh,[g] V. Christiaens,[h] P. Klaassen,[i] H. Beuther,[j] Th. Henning[j] and I. Kamp[e]

Received 18th January 2023, Accepted 16th March 2023

DOI: 10.1039/d3fd00010a

Early results from the James Webb Space Telescope-Mid-InfraRed Instrument (JWST-MIRI) guaranteed time programs on protostars (JOYS) and disks (MINDS) are presented. Thanks to the increased sensitivity, spectral and spatial resolution of the MIRI spectrometer, the chemical inventory of the planet-forming zones in disks can be investigated with unprecedented detail across stellar mass range and age. Here, data are presented for five disks, four around low-mass stars and one around a very young high-mass star. The mid-infrared spectra show some similarities but also significant diversity: some sources are rich in CO_2, others in H_2O or C_2H_2. In one disk around a very low-mass star, booming C_2H_2 emission provides evidence for a "soot" line at which carbon grains are eroded and sublimated, leading to a rich hydrocarbon chemistry in which even di-acetylene (C_4H_2) and benzene (C_6H_6) are detected. Together the data point to an active inner disk gas-phase chemistry that is closely linked to the physical structure (temperature, snowlines, presence of cavities and dust traps) of the entire disk and which may result in varying CO_2/H_2O abundances and high C/O ratios >1 in some cases. Ultimately, this diversity in disk chemistry will also be reflected in the diversity of the chemical composition of exoplanets.

[a]*Leiden Observatory, Leiden University, P. O. Box 9513, 2300 RA Leiden, The Netherlands. E-mail: ewine@strw.leidenuniv.nl*

[b]*Max-Planck Institut für Extraterrestrische Physik (MPE), Giessenbachstr. 1, 85748, Garching, Germany*

[c]*Université Paris-Saclay, CNRS, Institut d'Astrophysique Spatiale, 91405, Orsay, France*

[d]*European Southern Observatory, Karl-Schwarzschild-Strasse 2, 85748 Garching bei München, Germany*

[e]*Kapteyn Astronomical Institute, Rijksuniversiteit Groningen, P. O. Box 800, 9700 AV Groningen, The Netherlands*

[f]*Institute of Astronomy, KU Leuven, Celestijnenlaan 200D, 3001 Leuven, Belgium*

[g]*Dublin Institute for Advanced Studies, Astronomy & Astrophysics Section, 31 Fitzwilliam Place, Dublin 2, Ireland*

[h]*STAR Institute, Université de Liège, Allée du Six Août 19c, 4000 Liège, Belgium*

[i]*UK Astronomy Technology Centre, Royal Observatory Edinburgh, Blackford Hill, Edinburgh EH9 3HJ, UK*

[j]*Max-Planck-Institut für Astronomie (MPIA), Königstuhl 17, 69117 Heidelberg, Germany*

1 Introduction

Planets are built from gas and solids in the rotating disks around young stars. Their composition thus depends on the gas-grain chemistry that takes place in disks, which is known to change with position due to strong gradients in temperature, density and UV irradiation (*e.g.*, ref. 2–4). This chemistry may also change with evolutionary state: it is now well established that disks form early, already in the embedded state of star formation when the disk is still warmer[5] and surrounded by a larger scale envelope that can feed the disk with fresh material. Planetary cores are likely formed at these early stages as well.[6] The study of the chemistry of planet formation must therefore encompass the entire sequence of star formation, from cold collapsing clouds to warm circumstellar disks.

The James Webb Space Telescope (JWST) offers new opportunities to study the chemistry of the warm inner regions of disks, typically within a few au † radius from a solar-type star. This inner disk is the region in which most planets – both gas giants and terrestrial planets – are thought to form.[7,8] Compared with the Spitzer Space Telescope, which revealed that inner disks can have a very rich chemistry,[9–11] the JWST spectrometers have higher spectral resolving power ($R = \lambda/\Delta\lambda \approx 2000–4000$ *versus* 60–600) and more than two orders of magnitude higher sensitivity, combined with an order of magnitude higher spatial resolution. JWST also complements the Atacama Large Millimeter/submillimeter Array (ALMA) which resolves millimeter-sized dust emission in disks down to a few au but which probes chemistry primarily beyond $\sim$10 au (*e.g.*, ref. 12 and 13).

We present here some first results of the chemistry in the inner regions of disks based on JWST data from the medium resolution spectrometer (MRS)[14] that is part of the Mid-InfraRed Instrument (MIRI).[15,16] The data come from two guaranteed time programs: the JWST Observations of Young protoStars (JOYS) program (PIs: E. F. van Dishoeck and H. Beuther, program id: 1290), which targets two dozen young disks and their envelopes and outflows in the embedded stage of low- and high-mass star formation;[17] and the MIRI INfrared Disk Survey (MINDS) (PIs: Th. Henning and I. Kamp, program id: 1282), which observes about 50 disks around pre-main sequence stars across a range of stellar masses and ages.[18,19] Here, high mass refers to stars of O and B spectral type (typical masses 8 $M_\odot$ and higher) or protostars with total luminosities greater than a few $\times 10^3$ $L_\odot$. At the time of submission of this paper (January 2023), only one JOYS source has been observed—the high-mass protostar IRAS 23385 + 6052,[20] and data for a dozen MINDS sources have been taken, although none of them are disks that were found to be particularly line rich with Spitzer like AA Tau.[9] Here, we present results on small molecules, most notably CO_2, H_2O and hydrocarbons, as observed toward IRAS 23385 + 6052, and from disks around the T Tauri stars GW Lup, V1094 Sco, Sz 98 and the very low-mass star 2MASS-J16053215-1933159. Even within this limited sample observed so far, the data show the diversity of spectra and disk chemistry, not just among T Tauri disks but also across a range of stellar masses.

† 1 astronomical unit (au) = distance Sun–Earth = 1.5×10^{13} cm.

2 Observations and methods

2.1 JWST data reduction

The MIRI-MRS consists of four integral field units (IFUs) (also called "channels") covering different wavelength ranges within the 4.9–28.1 μm (355–2040 cm^{-1}) range. The field-of-view (FOV) varies between the four channels from $3.2'' \times 3.7''$ covered with 21 IFU slices of $0.18''$ at the shortest wavelengths, to $6.6'' \times 7.7''$ observed with 12 slices of $0.66''$ width at the longest wavelengths. Thus, the MIRI-MRS provides spectral images on arcsec scales but in this paper only point-source spectra centered on the disks are presented. Full wavelength coverage was obtained in three grating moves that are observed simultaneously for all four channels. The FASTR1 readout pattern was used with a 2-point dither pattern in JOYS and a 4-point dither pattern in MINDS.

IRAS 23385 + 6052 (luminosity $L = 3 \times 10^3$ $L_\odot$, distance $d = 4.9$ kpc,‡ [21] age few $\times$ 10^4 years, hereafter IRAS 23385) was observed as part of JOYS on August 22, 2022. This source actually consists of a small cluster of protostars, as revealed from previous millimeter[20,22] and near-infrared images,[23] with the most massive protostar estimated to be 9 $M_\odot$ from Keplerian rotation measurements of the surrounding gas.[20] Note that this source is too far north to be observable by ALMA. The total exposure time in each MRS grating setting was 200 s.

The spectra toward GW Lup ($M_* = 0.46$ $M_\odot$, $L = 0.33$ $L_\odot$,[24] $d = 155$ pc), V1094 Sco ($M_* \approx 0.9$ $M_\odot$, $L = 1.7$ $L_\odot$,[24] $d = 153$ pc), Sz 98 ($M_* \approx 0.74$ $M_\odot$, $L = 1.5$ $L_\odot$,[24] $d =$ 155 pc), and 2MASS-J16053215-1933159 ($M_* = 0.14$ $M_\odot$, $L = 0.04$ $L_\odot$,[25] $d = 152$ pc, hereafter J160532) were taken as part of the MINDS program on 2022 August 8 and August 1 for a total of ~2 h each including overheads (typically 800–1200 s per grating setting). The distances to these sources come from the Gaia DR3 catalogue,[26] whereas their ages are all typically a few Myr.[24,27]

The MIRI-MRS observations of all JOYS and MINDS sources were processed through the three reduction stages[28] using Pipeline version 1.8.4 of the JWST Science Calibration Pipeline43 and the CRDS context `jwst_1017.pmap`. Prior to processing of J160532, the background was removed using pair-wise dither subtraction. The default class Spec2 was applied, skipping the residual fringe correction. Instead, for GW Lup, V1094 Sco, and Sz 98, the alternative reference files as discussed in Gasman *et al.*[29] were applied, which correct the fringes sufficiently in Spec2. The data were then processed by the Spec3 class, which combines the calibrated data from the different dither observations into a final level3 spectral cube. The `outlier_detection` and `master_background` methods were skipped, since for both MINDS and JOYS this was done manually. Finally, a manual extraction of the spectra for each sub-band was performed, including aperture correction, with an aperture size of $2.5\lambda/D$ for the low-mass sources. The backgrounds of GW Lup, V1094 Sco, and Sz 98 were estimated using an annulus around the source, with an appropriate aperture correction accounting for the PSF signal present in the annulus.

The high-mass protostar IRAS 23385 turns out to be a binary source with a separation of ~$0.67''$ (~3280 au) at mid-infrared wavelengths. This means that the binary is resolved in Channel 1 and part of Channel 2, but that the

‡ 1 parsec = 206265 au = 3.26 lightyear = 3.086×10^{18} cm.

system is unresolved at longer wavelengths. The spectrum presented here was extracted using a circular aperture of 2.4″ diameter independent of wavelength that includes both sources. Background subtraction was done using the emission measured at off-source positions within the IFUs. This turned out to be more reliable than subtraction using the complementary dark field that was observed but that still contained strong extended emission. A residual fringe correction was applied at the spectrum level. More details for IRAS 23385 can be found in Beuther *et al.*,[30] and for the MINDS sources in Grant *et al.*[31] and Tabone *et al.*[1]

2.2 Methods

The mid-infrared wavelength range observed by MIRI-MRS covers the vibration–rotation bands of many molecules, from simple molecules like CO and CO_2 to more complex aliphatic and aromatic hydrocarbon molecules. For H_2O and OH, also pure rotational transitions between high-lying levels within the ground vibrational state are observable beyond 10 μm.

The continuum-subtracted spectra were fitted with slab models of the molecular emission that take optical depth effects and line overlap into account, following the methods used for analyzing Spitzer spectra.[10,32] The populations of the molecular levels are assumed to be characterized by a single excitation temperature, T_{ex}, which is taken to be equal to the gas temperature T assuming local thermodynamic equilibrium (LTE). Note that radiative pumping of the lines may also be significant in the inner disk,[33] in which case T_{ex} approaches the temperature of the radiation field, T_{rad}.

The line profile function is taken to be Gaussian with a full width at half maximum of $\Delta V = 4.7$ km s^{-1} ($\sigma = 2$ km s^{-1}), as in previous Spitzer studies. At high optical depth, the line shape becomes a flat-topped pseudo-Gaussian due to saturation at the line core. These models have only three free parameters: the line of sight column density N, the gas temperature T, and the emitting area given by πR^2. It is stressed that R does not necessarily correspond to a disk radius, but that the emission could come from a ring or any other region with an area equivalent to πR^2.

The molecular data, *i.e.*, energy levels, statistical weights, Einstein A coefficients and partition functions, were taken from the HITRAN 2020 database[34] and converted into LAMDA format.[35] For C_6H_6, molecular parameters based on the GEISA database were used.[36] Integrated line fluxes are computed, varying the emitting area and taking the known distances to the sources into account. The model spectrum is then convolved to the instrumental spectral resolving power[37] and the convolved model spectrum is resampled to have the same wavelength grid as the observed spectrum. Model grids were run for each molecule with T from 100 to 1500 K, N from 10^{14} to 10^{22} cm^{-2}, and emitting radius from 0.01 to 10 au for the low-mass sources. For the high-mass source, these ranges are 50–550 K, 10^{14}–10^{20} cm^{-2} and 0.01–10000 au. The best-fit N and T are determined using a χ^2 fit between the continuum subtracted data and the convolved and resampled model spectrum, using the best-fit emitting area for each N and T. The procedure is iterative, starting with fitting those molecules that have the most lines in a certain wavelength range that are overlapping with those of others, and ending with the weaker features.[1,31]

3 Results

3.1 Broad-band spectra

Fig. 1 presents part of the observed MIRI-MRS spectrum toward the high-mass binary protostar IRAS 23385.[30] Note that this source is more than a factor of 1000 weaker than the high-mass protostars studied with ISO,[38] only ~10 mJy, demonstrating the JWST sensitivity. As is typical for a deeply embedded young stellar object, the spectrum rises with wavelength due to the thermal emission from dust surrounding the binary protostar, which has a temperature gradient from warm to cold.[39] Superposed on this continuum are deep and broad absorption bands due to silicates and a variety of ices arising in the cold outer envelope.[40,41] A number of strong, narrow emission lines due to various atoms and H_2 are seen as well, associated with shocks due to the outflows from the protostars. Finally, PAH emission from a background cloud is seen at 11.3 and 8.6 μm;[42] the shorter wavelength PAH features are blocked by the high extinction and deep ice features in the envelope. Both the absorptions and spatially extended gas-phase lines will be described elsewhere.[30,43]

Fig. 2 compares the MIRI-MRS spectra of two low-mass sources, GW Lup and J160532. The GW Lup spectrum is typical of that of a disk around a young star whose protostellar envelope has already been dissipated: its spectral energy distribution drops at long wavelengths but it shows strong, broad emission bands due to amorphous and crystalline silicates[46–48] in the MIRI range.

In contrast to most disk sources including V1094 Sco and Sz 98, the J160532 spectrum is unusual: it shows no silicate emission feature but has two broad bumps centered at 7.7 and 13.7 μm as was already noted in low-resolution Spitzer data.[49,50] Tabone *et al.*[1] show that these bumps are due to the $\nu_4 + \nu_5$ and ν_5 bands of C_2H_2, respectively, which is present at very high column densities (~10^{21} cm^{-2}) and emitting at a temperature around 500 K within a small region of radius 0.033 au. In this optically thick component, the *Q*-branch is suppressed. The clearly

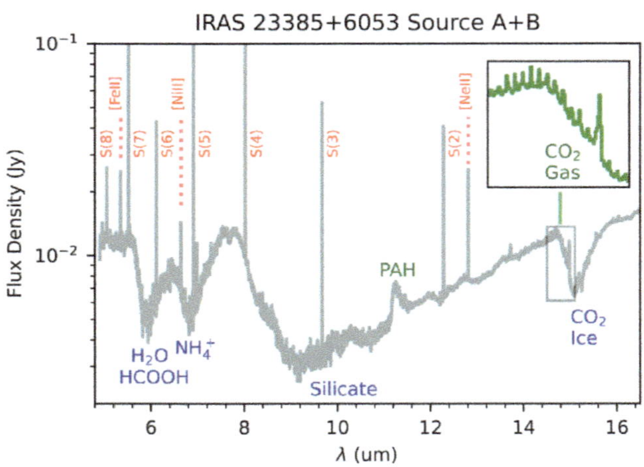

Fig. 1 Part of the MIRI-MRS spectrum toward the high-mass protostar IRAS 23385 + 6053 integrated over both sources. Blue labels refer to ice and solid-state absorption features, red to spatially extended narrow emission lines, and green to molecular emission with CO_2 centered on the sources and PAH emission extended.

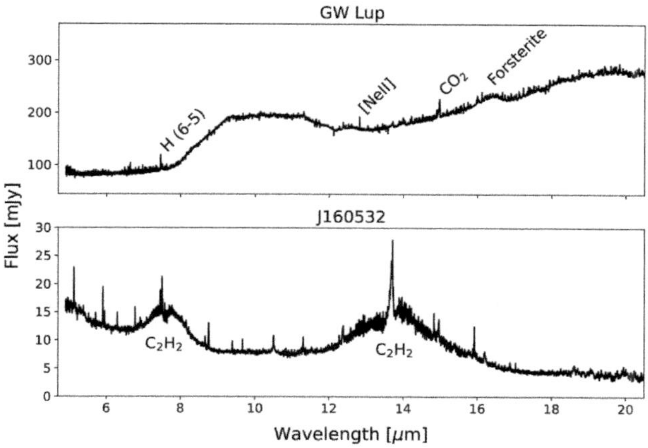

Fig. 2 MIRI-MRS 5–20 μm spectrum of the disk around the T Tauri star GW Lup[31] compared with that of the very low-mass star J160532 presented in ref. 1. The GW Lup spectrum shows amorphous silicate emission at 10 and 18 μm with superposed atomic and molecular lines as well as some crystalline forsterite emission. The broadband J160532 spectrum shows no silicate emission but two prominent bumps that can be ascribed to very abundant C_2H_2 emissions.[1]

visible *Q*-branch as well as other individual *R*- and *P*-lines (see below) require an additional optically thin C_2H_2 component with a 10^4 times lower column density.

None of these disk spectra shows emission from polycyclic aromatic hydrocarbons (PAHs). However, this lack of PAH features does not imply absence of these molecules: as shown by Geers *et al.*,[51] the UV radiation from stars with $T_{eff} <$ 4200 K (spectral type later than K6, assuming no significant excess UV radiation due to accretion) is too weak to produce detectable PAH bands.

3.2 Identified molecules

Of interest to this paper are the series of weak narrow lines with a line/continuum ratio of less than 10% on top of the continuum (see GW Lup spectrum in Fig. 2). Fig. 3 presents the continuum subtracted spectra in the 14.3–17.3 μm range for the five sources discussed here, with various emission features labelled. Fig. 4 presents the spectrum of J160532 at 13.3–16.1 μm including the prominent C_2H_2 *Q*-branch at 13.7 μm with weak HCN at 14.0 μm. The shorter wavelength parts of the spectra of the other sources showing C_2H_2 and HCN are included in the paper by Kamp *et al.*[18]

The most prominent band in Fig. 3 is the *Q*-branch of CO_2 at 15 μm. Its *R*- and *P*- branch lines are weaker but still visible, most notably in IRAS 23385. The *Q*-branches of the CO_2 hot bands and of the ^{13}C isotopolog of CO_2 are also detected in some sources, most notably GW Lup.[31] An irregular set of lines due to H_2O shows up most strongly in the Sz 98 disk, but these lines are also present in the GW Lup and V1094 Sco disks. In contrast, in the J160532 disk only a few weak H_2O lines are tentatively seen,[1] with none in IRAS 23385 in this wavelength range. The H_2 S(1) line at 17.0 μm and H I recombination lines are also seen in a few sources, as are some OH lines. Excitingly, bands of C_6H_6 and C_4H_2 are clearly detected in the J160532 disk (Fig. 4), and CH_4 emission is possibly found as well.[1] Many of these lines are seen with JWST for the first time in disks.

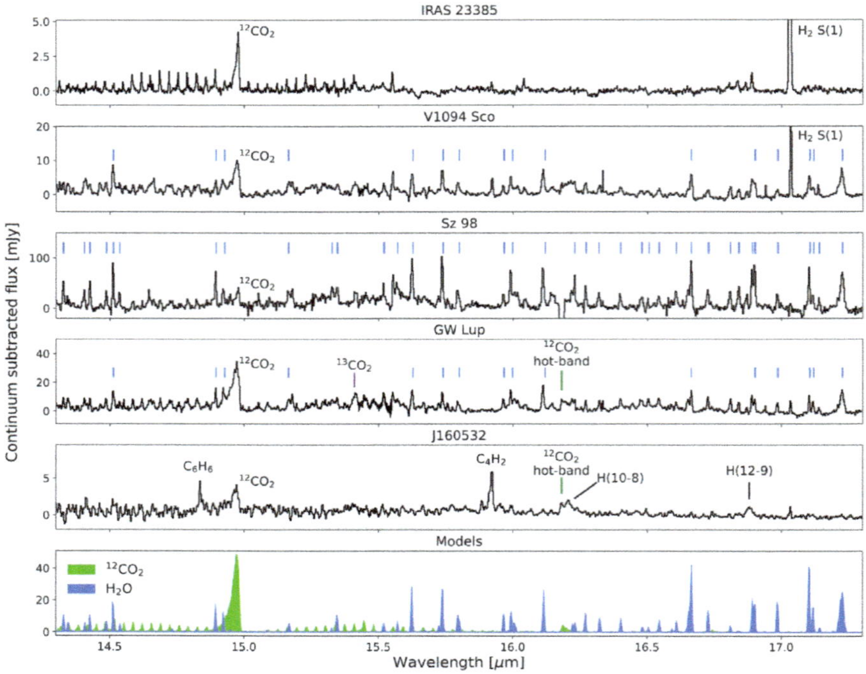

Fig. 3 Continuum-subtracted MIRI-MRS 14.3–17.3 μm spectra of the young high-mass protostar IRAS 23385,[44] and of the disks around the T Tauri stars V1094 Sco, Sz 98 (ref. 45) and GW Lup,[31] and the very low-mass star J160532.[1] The blue tickmarks in the V1094 Sco, Sz 98 and GW Lup panels indicate the locations of water lines. Note also the detected $^{13}CO_2$ and hot bands of $^{12}CO_2$ toward GW Lup.[31] The bottom panel shows the simulated LTE slab model spectra of CO_2 and H_2O at 600 K (arbitrary relative scaling).

Note that the bulk of the lines in this wavelength range can be fit with these molecules. There could be additional lower abundance molecules present, or isotopologs of identified molecules, whose weaker lines are blended with the

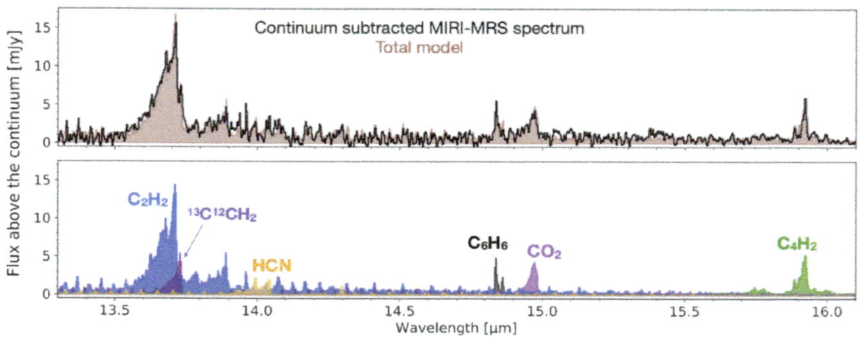

Fig. 4 Continuum-subtracted MIRI-MRS 13.3–16.1 μm spectrum of the very low-mass star J160532, showing prominent features of C_2H_2, C_4H_2 and C_6H_6. The ^{13}C isotopolog of C_2H_2 is also detected, as are HCN and CO_2 (see ref. 1 for details). The big bump at 13.7 μm (see Fig. 2) due to very high column density C_2H_2 component has been subtracted.

stronger lines that have not yet been identified at this stage of data processing. Also, line lists of various potentially interesting molecules are currently unavailable or incomplete at high temperatures, especially for hydrocarbons (see also paper by Kamp *et al.*[18]).

3.3 Inferred column densities and temperatures

Table 1 summarizes the inferred best-fitting column densities, temperatures and emitting areas of various molecules for two of the sources. An example of a typical χ^2 plot can be found in Fig. 5. As found in previous studies based on Spitzer data,[10,32] there is a wide range of fit parameters that reproduce the spectra well. Most notably, in the low column density regime (typically $<10^{17}$ cm^{-2} depending on molecule) where lines are optically thin, there is a complete degeneracy between column density and emitting radius. On the other hand, the latter can be determined from optically thick lines whose strength scales directly with emitting area. The temperature is often well constrained from the shape and position of the *Q*-branch feature, where lines with $\Delta J = 0$ pile up: the warmer the gas, the broader the feature. MIRI can resolve this *Q*-branch much better than before. An

Table 1 Best fit molecular model parameters[a]

	IRAS 23385			GW Lup		
Molecule	N (cm^{-2})	T (K)	R (au)	N (cm^{-2})	T (K)	R (au)
CO_2	1×10^{17}	150	35	2×10^{18}	400	0.11
H_2O	$<1 \times 10^{18}$	[150]	[35]	3×10^{18}	625	0.15
C_2H_2	7×10^{16}	210	8	4×10^{17}	500	0.05

[a] Brackets indicate parameters that were fixed in the fit.

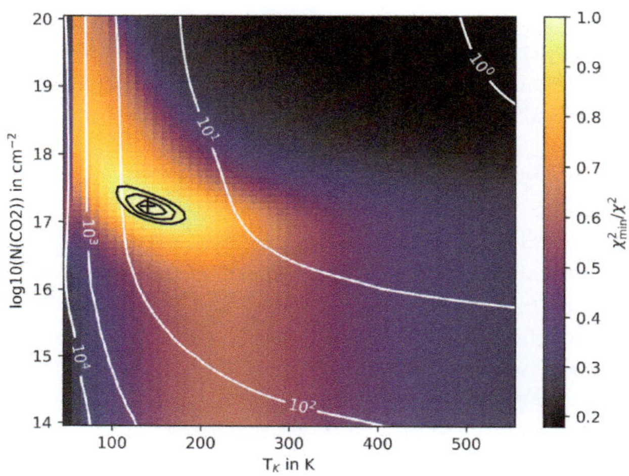

Fig. 5 Example of a χ^2 map to constrain temperature and column density, here for the case of CO_2 in IRAS 23385. The white lines indicate the best fitting emitting areas for those *N*, *T* combinations. The black lines indicate the 1σ, 2σ and 3σ contours.

Table 2 Column density ratios of various molecules[a]

Ratios	IRAS 23385	V1094 Sco	GW Lup	J160532
CO_2/H_2O	>0.1	0.004	0.69	>0.1
CO_2/C_2H_2	1.7	0.1	4.8	0.008

[a] Column density ratios do not reflect local abundance ratios. For J160532, the optically thick C_2H_2 component is used to compute the ratio.

excellent example of the latter is provided by the CO_2 *Q*-branch toward IRAS 23385, which is only consistent with a low temperature of 150 ± 50 K. The availability of both optically thick and thin lines with JWST as well as their ^{13}C isotopologs helps to resolve several of these degeneracies.

A few column density ratios are summarized in Table 2 to give an impression of the range of values, which will be discussed further below. For V1094 Sco, these column density ratios are obtained from a preliminary analysis assuming the same emitting radius of 0.13 au for all species. For more details, see ref. 1, 31 and 44.

4 Discussion

4.1 Diversity in spectra

Fig. 1–4 illustrate the diversity in mid-infrared spectra of embedded protostars and young stars with disks, and not just between these two classes of sources but even between disk-dominated sources themselves. High quality mid-infrared spectra are now possible for much fainter sources than before, down to the mJy continuum level, thanks to JWST. Focusing just on the molecular emission, one of the striking features of Fig. 3 is the difference in the strength of H_2O emission in the 14.3–17.3 μm region, with H_2O lines virtually absent in the spectra of the high mass binary IRAS 23385 and the very low-mass source J160532, but clearly present in those of GW Lup and V1094 Sco and dominant in Sz 98. Another obvious difference is the strength of the C_2H_2 band with respect to other features, with C_2H_2 and hydrocarbons dominating the spectrum of the very low-mass star J160532 but being much weaker toward other sources (see paper by Kamp *et al.*[18]). CO_2 is detected in all sources, most strongly relative to other features in the GW Lup disk.

How about nitrogen? One molecule that has not yet been detected in disks with MIRI is NH_3 whose ν_6 mode at 8.8 μm can now be observed at much higher spectral resolution with the MRS than was possible with Spitzer. The HCN *Q*-branch at 14.0 μm is seen in most sources but is now recognized to be heavily blended with both C_2H_2, CO_2 and/or H_2O lines (Fig. 4) resulting in an HCN contribution that is smaller than thought based on Spitzer data.[10,52] This means that an even lower fraction of nitrogen has been identified than before, suggesting that the bulk of the nitrogen in the inner disk may well be in the unobservable N_2 form.[53]

Where does the molecular emission originate? For the low-mass sources, the lines clearly come from the inner region of their disks, typically within 1 au, as is also evidenced by their small emitting areas (Table 1). Although the emission could in principle also come from a narrow ring further out in the disks, the high inferred temperatures also point to a location in the inner disk.

For high-mass protostars, the emitting area of only 35 au radius is also relatively small. Interestingly, for high-mass protostars these molecular bands are often seen in absorption rather than emission in spectra taken with the Infrared Space Observatory.[54–56] Those lines were thought to originate in the inner warm envelope where ices have sublimated ("hot core") or in shocked gas along the line of sight seen in absorption toward the strong mid-infrared continuum due to hot dust close to the protostars. However, more recent high spectral resolution ground-based and SOFIA-EXES data suggest an origin in the inner part of a disk around the massive young star.[57,58] Weak water absorption at 6 μm is seen toward IRAS 23385 but only towards source B of the binary.[44] Emission lines of CO_2, C_2H_2 and HCN at 13–15 μm have previously been seen only toward the Orion peak 1 and 2 shock positions away from the strong continuum,[59] and in the Cepheus A high-mass region.[60,61] Interestingly, in both cases moderate excitation temperatures of $\sim$200 K were found, lower than those seen for absorption bands, and similar to those derived here. For IRAS 23385, the molecular emission is clearly centered on the two protostars and is not detected in off-source shock positions.[44] Also, the lack of strong H_2O emission, universally observed to be associated with shocks,[62] argues against an outflow origin. Although we cannot exclude a shock origin at this stage, we assume here that the emission originates in the disk(s) around (at least one of) the protostars in the IRAS 23385 system.

4.2 Diversity in chemistry

Table 2 summarizes the column density ratios of the main molecules considered here in four of the sources. Uncertainties are up to an order of magnitude, based on the χ^2 plots. It should be stressed that these ratios should not be equated with abundance ratios since the emission of different molecules (or even of different bands of the same molecule) may originate from different regions or layers of the disk.[33,63] Moreover, the emission seen at mid-infrared wavelengths only probes the upper layers of the disk above the $\tau_{\text{mid-IR}} = 1$ contour where the dust continuum becomes optically thick.

In spite of these considerations, the similarities and differences in mid-infrared spectra can give some insight into the chemical processes at play in planet-forming zones of disks, especially since the differences in column density ratios are much larger than those of flux ratios of optically thick lines. As concluded from the appearance of the spectra, the J160532 inner disk is particularly rich in C_2H_2, whereas the V1094 Sco has the lowest CO_2/H_2O ratios, with Sz 98 likely even lower.[45] Apart from the larger size of the emitting region, the high-mass source IRAS 23385 does not stand out in terms of its abundance ratios compared with the orders of magnitude lower luminosity T Tauri sources.

GW Lup is part of a handful of sources that were found by Spitzer to have strong CO_2 emission but no emission from other molecules, the so-called "CO_2-only" sources.[10] Thanks to the detection of $^{12}CO_2$ hot bands, its *P*- and *R*- branch lines and the $^{13}CO_2$ *Q*-branch, the inferred CO_2 column density is two orders of magnitude higher (and emitting area smaller) than found in analyses of the Spitzer data, illustrating the importance of the additional weaker optically thin features now observed with MIRI-MRS.[31] H_2O and other molecules are now weakly detected as well with JWST, but its CO_2/H_2O column density ratio remains clearly higher than that of other low-mass sources.

4.3 Chemical processes

4.3.1 Disk models. The chemistry in disks is known to vary both radially and vertically owing to large gradients in temperature, density and UV irradiation. A detailed 2D (or axisymmetric 3D) model of the physical structure of a disk therefore needs to be constructed before the chemistry can be addressed. Such thermo-chemical models have been developed for disks around low-mass stars by various groups,[64–67] with some of them focused specifically on inner disk chemistry where high-temperature gas-phase chemistry dominates.[63,68–72] Snowlines, *i.e.*, the location in the disk where 50% of the molecule is in the gas and 50% in the ice, play an important role in setting the overall gas-phase chemical composition. Of particular interest is the C/O ratio in volatiles (*i.e.*, gas + ice), since ices naturally lock up more oxygen than carbon.[73] For CO, whose pure ice binding energy is $E/k \approx 855$ K,[74] its snowline lies around 20 K which is typically at tens of au in a disk around a low-mass star.[75] For H_2O, whose binding energy is much higher at $E/k \approx 5600$ K,[76] its snowline is at ~160 K at the high densities in disk midplanes, located at less than a few au.[77] The snowlines of other molecules like CO_2 generally lie in between those of CO and H_2O,[78] with the exception of those of large organic molecules. In particular, the "soot" line due to sublimation and erosion of refractory hydrocarbon material is thought to lie around 500 K in disks.[79]

If *R* listed in Table 1 is taken to be a disk radius, the emission observed here for low-mass stars originates from well inside the H_2O snowline. At these radii and temperatures, much of the oxygen that was locked up in ices should have returned to the gas phase.[73] The overall (gas + ices + refractory dust) C/O ratio in disks is expected to be similar to the stellar value, which in turn should be close to the solar value of C/O = 0.59 within the solar neighborhood.[80] For gas-phase chemistry outside the refractory dust sublimation radius, only the volatile (gas + ice) C/O ratio is relevant, which is generally taken to be somewhat lower than 0.6 in disk models. Its precise value does not matter much for the chemistry and mid-infrared line emission[63,72] as long as C/O <1.

It is important to note that any disk model reproducing mid-infrared lines needs to have a gas/dust ratio in the upper layers that is significantly higher than the standard interstellar medium value of 100, typically[33,63,81] of the order 10^3–10^4. Such high gas/dust ratios are naturally explained by grain growth and settling to the midplane.[82]

The physical and chemical structure of disks around high-mass O and B stars has not been modeled in as much detail as that of disks around A- and later-type stars.[71] An additional complication for such high-mass stars is that they are much younger and do not have an optically visible pre-main sequence stage. Although disks are commonly found to form in simulations of high-mass star formation,[83,84] they are still embedded in their natal envelopes. Moreover, accretion rates are still high, so heating of their inner disk midplane is dominated by viscous processes rather than passive irradiation.[85] Fig. 6 compares the 2D dust temperature structure for one quadrant of a disk around a low-mass star with that of a high-mass star,[86,87] obtained using the RADMC-3D code.[88] The gas temperature is closely coupled to that of the dust except in the upper surface layers. For high-mass disks, the midplane is actually warmer than the surface layers within the inner ~100 au [87] (depending on the accretion rate) giving rise to infrared absorption lines.[58] At larger radii, the disk switches to the usual structure of

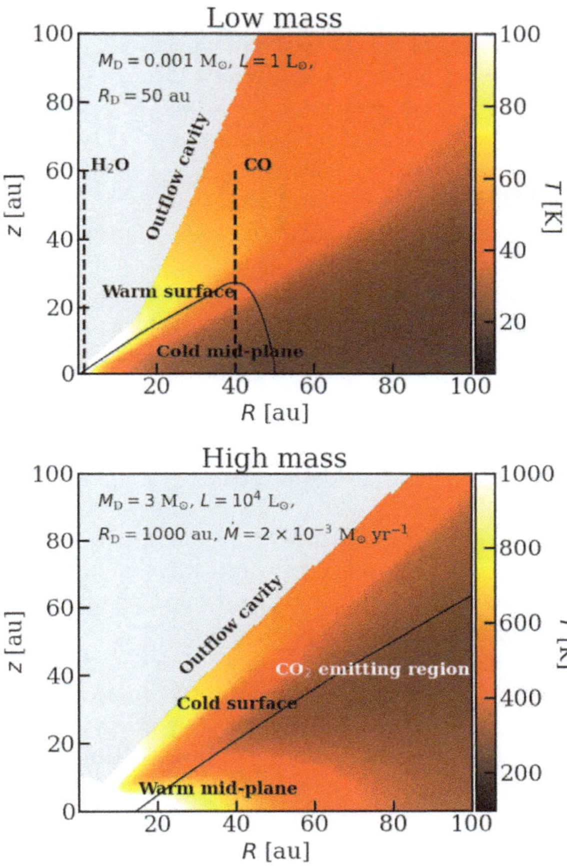

Fig. 6 Two dimensional map of the dust temperature structure of a disk around a low-mass star (top)[86] with that around a high-mass star (bottom).[87] The solid black lines show the approximate location of the disk, which is still embedded in an envelope for the high-mass source. Note that for the high-mass disk, the midplane temperature in the inner region (≲50–60 au) is larger than that at the disk surface due to accretional heating; no such heating is included for the low-mass source. Such a structure gives rise to mid-infrared absorption lines. At larger radii, the surface is warmer than the midplane due to irradiation resulting in mid-infrared emission lines (indicated in the CO_2 emitting region). Model parameters[86,87] are for low mass: $M_* = 0.5$ M$_\odot$, $L = 1$ L$_\odot$, $M_{disk} = 0.001$ M$_\odot$; high mass: $M_* = 30$ M$_\odot$, $L = 10^4$ L$_\odot$, $M_{disk} = 3$ M$_\odot$, $\dot{M} = 2 \times 10^{-3}$ M$_\odot$ per year. Both models use dust opacities appropriate for large grains.[86] For the low-mass disk, the H_2O snowline is at 1.5 au and the CO snowline at 40 au; for the high-mass disk, they are at 390 au and >20 000 au, respectively.

a vertically decreasing temperature structure from surface to midplane leading to mid-infrared emission lines. IRAS 23385 has a somewhat lower luminosity and accretion rate than assumed in the model in Fig. 6, suggesting that the switch moves well inside 100 au.

4.3.2 CO_2 *versus* H_2O: cavities and dust traps. Armed with the above background information on disk models, we now address the chemistry of the main species considered in this paper.

4.3.2.1 *Chemical processes relating CO_2 and H_2O.* The temperature structure of the inner disk plays a crucial role in setting the balance between the CO_2 and H_2O abundances.[71,89] Both molecules are formed through gas-phase reactions with the OH radical.[71,90] At temperatures of 100–250 K, the OH + CO → CO_2 + H reaction produces CO_2 with typical abundances of 10^{-7}–10^{-6} with respect to total H. At higher temperatures, the reaction OH + H_2 → H_2O + H, which has an energy barrier of 1740 K,[91] takes over and produces H_2O in favor of CO_2. Once H_2O becomes abundant, its column density becomes high enough ($\sim 10^{18}$ cm^{-2}) to become self-shielding[92] pushing the balance even further toward H_2O. Note that OH itself also needs moderately warm temperatures to form since the O + H_2 → OH + H reaction is endothermic by 900 K and has an energy barrier as well.[91] Both the formation of OH and H_2O are boosted by the presence of vibrationally excited H_2 where the vibrational energy can be used to overcome the energy barriers.

H_2O and CO_2 are both abundant components of ices in cold interstellar clouds and protostellar envelopes.[41] A fraction of these ices are likely preserved in the transition from envelope to disk, especially for the stronger bound molecules like H_2O,[93] as also illustrated by the similarity between interstellar and cometary ice abundances.[94,95] Inside their respective snowlines, the ice species sublimate and thus add to the gas-phase chemistry budget. Minor species intimately mixed with H_2O ice sublimate together with water at its snowline.[96]

Bosman *et al.*[89,98] find that H_2O mid-infrared emission dominates over CO_2 emission in the 13–17 μm range for low-mass disks, if H_2O self-shielding and efficient gas heating mechanisms are included. In particular, so-called (photo) chemical heating, in which the excess energy released by photodissociation and subsequent chemical reactions of the dissociation products is used to heat the gas,[99] can raise the temperatures in the upper layers significantly.[100]

4.3.2.2 *Role of small cavities.* Given the model predictions that H_2O emission should dominate mid-infrared spectra, which processes could then make CO_2 more prominent than H_2O in some disks but not others? One parameter that affects mid-infrared line fluxes is the amount of small grains in the upper layers. However, both CO_2 and H_2O fluxes are decreased by comparable amounts if the gas/dust ratio is decreased.[101] Another option would be to change the C/O ratio, but again both CO_2 and H_2O emission lines appear to be affected similarly.[63,72] Walsh *et al.*[71] show that for very low-mass stars, H_2O formation is indeed suppressed in the disk's surface layers due to their lower temperatures, consistent with the lowest-luminosity sources in our samples, GW Lup and J160532, showing the weakest H_2O lines. However, some higher luminosity sources, including IRAS 23385, also show weak water emission.

An alternative explanation would be to introduce a deep dust + gas cavity in the inner disk, caused for example by a companion. Its radius is thought to be in between that of the H_2O and CO_2 snowlines in order to remove most of the H_2O but not the CO_2 emission.[31,72,102] Fig. 7 presents an example of the 2D abundance structure of H_2O and CO_2 in a model[97] obtained using the DALI thermochemical code[66] for a disk with the same parameters as in Bosman *et al.*[89] Two cases are presented: a full disk and a disk with a deep dust + gas cavity out to 1 au, a location that is in between the two snowlines. The 70 and 200 K contours are also included to highlight the temperature range where CO_2 is efficiently produced.

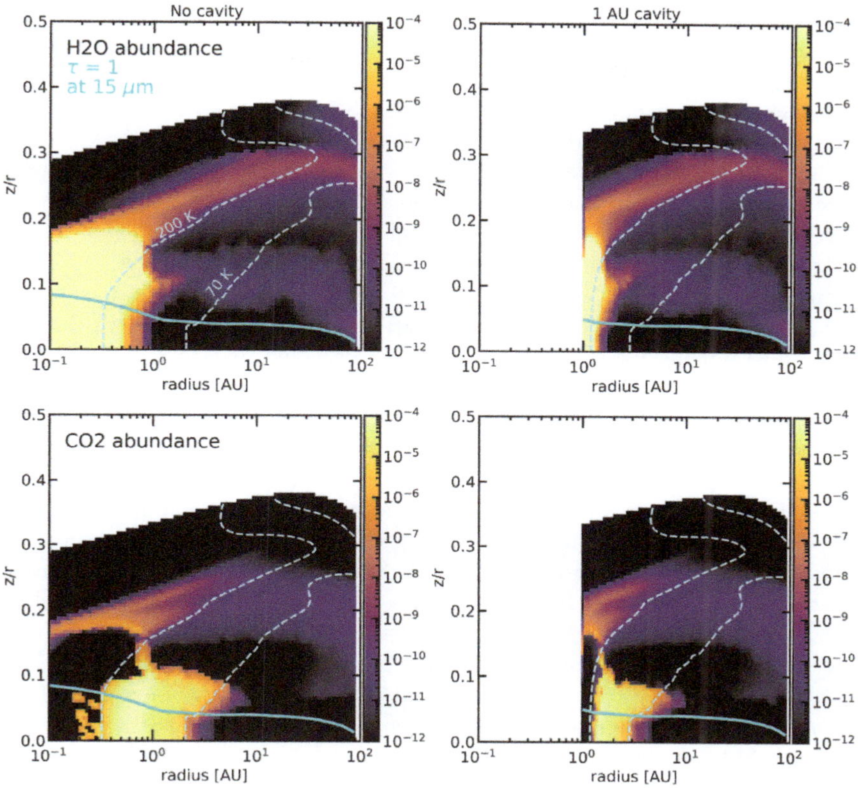

Fig. 7 Two dimensional map of the H_2O and CO_2 abundances of one quadrant of a disk around a low-mass star without a cavity (left) and with a 1 au cavity (right).[97] The dashed lines indicate the 70 and 200 K temperature contours; the blue line the $\tau_{\text{mid-IR}} = 1$ line at 15 μm. The cavity radius is located between the H_2O and CO_2 snowlines. The disk model parameters are taken from ref. 89.

If there is no cavity, it is seen that H_2O is highly abundant in the inner disk above the $\tau_{\text{mid-IR}} = 1$ (where the 15 μm dust continuum becomes optically thick) and would thus produce very strong H_2O lines. However, its column relative to CO_2 is strongly reduced in the model with a cavity (except at the very edge of the cavity). These models therefore provide a proof of concept for explaining those sources with high CO_2/H_2O column density ratios with disks that have small cavities, although further work suggests they can be larger, a few au.

For the GW Lup disk, the H_2O and CO_2 snowlines are expected to lie around 0.4 and 1.4 au, respectively,[103] so a qualitatively similar effect can be expected. Such cavities are too small to be resolved with ALMA but could be revealed in their spectral energy distributions or in spectrally resolved CO ro-vibrational line profiles.[104–106] With a few exceptions, these low-mass stars and their disks are generally too weak to image with infrared interferometers.[107]

Interestingly, the strong CO_2 emission and lack of H_2O in the high-mass source IRAS 23385 also fits into this scenario. The only difference is that there is not necessarily a cavity but rather a temperature inversion in the inner disk which would cause the H_2O-rich gas to be in absorption (most notably in its ν_2 band at 6 μm (ref. 44)) rather than emission. Fig. 6 shows that the temperature in the surface layers in the region where accretional heating no longer dominates (and where lines go back to being in emission), is around 200 K. Such a relatively cool temperature is consistent with that found for the CO_2 emission in this source (Table 1).

4.3.2.3 Role of dust traps. For low-mass sources, a related effect could be the existence of a dust trap triggered by a pressure bump due to a companion that created the cavity. If that dust trap lies between the H_2O and CO_2 snowlines, it locks up water-rich pebbles thereby suppressing H_2O. However, some gas enriched in carbon,[108–110] or in CO_2 due to ice sublimation inside its snowline, could still be transported through the trap to the inner disk.

A dust trap locking up the bulk of the H_2O ice would also lower the overall O/H abundance ratio in the inner disk.[111] The same holds for C/H, if CO, CH_4 and/or carbon-containing molecules are trapped. Some inner disks are indeed known to have low O/H and/or C/H abundances.[112–114] Disks that show spatially resolved dust traps with ALMA[75,115] are likely in this category, provided that the dust traps formed early in the disk's evolution, before the bulk of the grains have moved inwards.

In contrast, any disk in which drifting icy pebbles from the outer disk have reached the inner disk and crossed the H_2O snowline unimpeded should have O/H abundances that could be enhanced by 1–2 orders of magnitude.[116,117] Similarly, CO can be increased in the inner disk due to drifting icy grains.[118–120] Since these icy pebbles are also thought to contain considerable amounts of CO_2 ice, this should not change the relative ratios of the H_2O and CO_2 lines if both are continuously being supplied to the inner disk,[117] unless the extra CO_2 would remain hidden below the $\tau_{\text{mid-IR}} = 1$ line due to a lack of vertical mixing. Disks which show compact dust continuum emission with ALMA are good candidates to test this pebble drift scenario.

Fig. 8 illustrates these possibilities in a cartoon. It is clear that the combination of large samples of disks studied with both JWST mid-infrared spectroscopy and with ALMA at its highest spatial resolution, including disks that are small and currently unresolved with ALMA at moderate spatial resolution, can provide insight into the relative importance of these processes.

4.3.3 Rich hydrocarbon chemistry: "soot" line and C/O >1. The J160532 disk stands out in having unusually strong bands of C_2H_2 and other hydrocarbons including species that have not been identified in disks before the advent of JWST.[1] Equally interesting is the discovery[1] that the broad bands at 7.7 and 13.7 μm are due to very high column densities of hot C_2H_2.

What could be causing such abundant C_2H_2 in this disk? It is known that PAHs and carbonaceous grains can be destroyed by UV radiation, by chemical processes (*e.g.*, reactions with H, O or OH) or by grain sublimation under the conditions in the inner disk, with C_2H_2 as a main product.[121,122] When and where the carbon grain sublimation and erosion takes place depends on the type of carbonaceous material. Some laboratory data put the sublimation of amorphous carbon grains at 1200 K,[123] whereas other experiments suggest that thermal decomposition of refractory hydrocarbon material (also known as kerogen-like insoluble organic matter, IOM) occurs at lower temperatures, around 400–500 K depending on heating timescale.[79,124,125] This IOM material is thought to make up the bulk of the refractory carbon. Aliphatic and aromatic hydrocarbons, which contain of order 10% of total carbon, sublimate at somewhat lower temperatures of 300–400 K.[123]

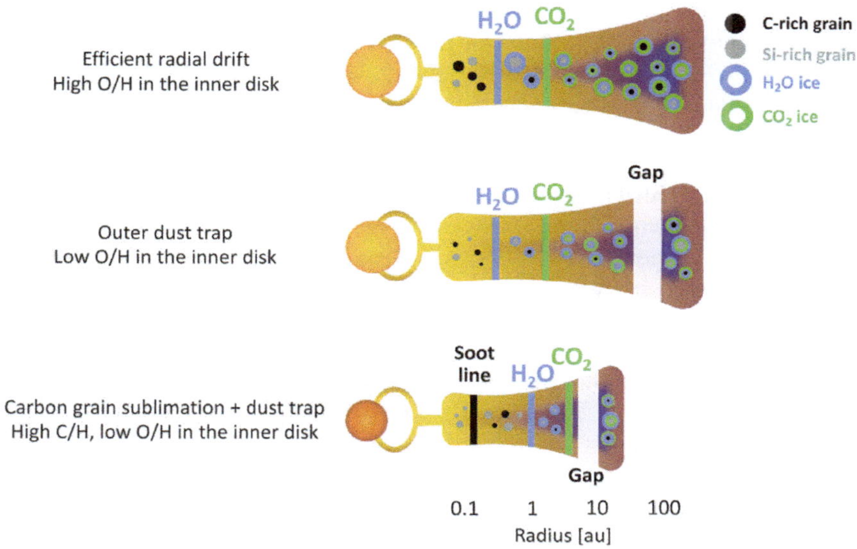

Fig. 8 Cartoons illustrating the cases of (top) enhanced H_2O and O/H in the inner disk due to icy pebble drift; (middle) reduced H_2O and O/H in the inner disk due to a dust trap outside the H_2O snowline. If this trap is still inside the CO_2 snowline, CO_2 could be enhanced; and (bottom) as (middle) but for a very low-mass star with enhanced C_2H_2 at the soot line.

Together, they form the so-called hydrocarbon "soot" line around 400–500 K, a term coined by Li *et al.*[79] to indicate the location where the destruction of hydrocarbon solids occurs. This temperature matches well with that found for the high abundance C_2H_2 gas component. One possibility is therefore that hydrocarbon grain destruction is being observed directly in this disk.

The destruction of carbon grains by thermal processing is thought to be irreversible. On the other hand, the conditions of warm dense gas with high C_2H_2 abundances are conducive to efficient formation of small aromatic molecules and PAHs up to temperatures where degradation starts to take over.[126,127] Indeed, the production of the smallest aromatic molecule, benzene, in the inner regions of disks as observed here was predicted by Woods & Willacy.[69,128] The J160532 and other similar disks around very low mass stars[129] are clearly excellent laboratories for studying such hydrocarbon chemistry in detail.

A related interesting question is how common the broad 7.7 and 13.7 μm features indicative of a "soot" line are in other sources. While disks around very low-mass stars are known to have strong C_2H_2 bands based on Spitzer data,[49,50,130] J160532 is the only such disk so far in which the broad bumps are clearly seen. What could make the J160532 disk special? One option is that we are able to look deep into the J160532 disk close to the midplane due to the lack of small grains, as evidenced by the large gas columns and strong H_2 lines.[1] In other disks, the midplane "soot" line could be hidden below the $\tau_{\text{mid-IR}} = 1$ line. Another option could be that we are witnessing the J160532 disk at a special time when destruction is taking place, perhaps triggered by a recent heating event due to accretion.

The J160532 disk presents another puzzle, namely that of water. The disk is clearly warm enough to sublimate any water ice from the grains and to trigger rapid formation of H_2O through the reactions of O and OH with H_2. In disk chemical models, such high abundances of C_2H_2 compared with H_2O (Table 2) are only found if the volatile C/O ratio of the gas in the inner disk is significantly increased compared with standard values of volatile C/O = 0.4–0.5. Specifically, C/O ratios >1 are needed to boost the hydrocarbon production and lock up the bulk of the volatile oxygen in CO and some CO_2.[52,63,72] A small amount of H_2O could still be produced in the upper layers of the disk where destruction of CO by UV photons or by He^+ frees up some atomic O that can react with H_2 to form H_2O. Walsh *et al.*[71] note that the atmospheres of disks around very low mass stars may naturally appear to be carbon rich (without the need for a high C/O ratio) because the unobservable O_2 rather than H_2O becomes the main gaseous oxygen carrier. However, the orders of magnitude higher C_2H_2 column and C_2H_2/CO_2 ratio observed for J160532 than found in these models, strongly points to a genuine carbon-rich atmosphere with C/O >1.

The most plausible mechanism to lead to C/O >1 and prevent oxygen from entering the inner disk would be to lock most of it up in water ice as part of pebbles and planetesimals in the outer disk beyond the water snowline, preventing them from migrating to the inner disk,[50,52] as illustrated in Fig. 8 (bottom).

4.3.4 Implications for planet formation. The JWST data provide detailed insight into the chemical composition of the gas in the upper layers of the disk. Planetesimals and planets are, however, formed in the disk midplanes which are not probed directly by mid-infrared data. Nevertheless, indirect information can be obtained if mid-plane material is vertically mixed upward. For example if icy pebbles drift across the water snowline and enhance the gaseous O/H ratio of the midplane gas, this can be mixed upward on comparatively short timescales (*e.g.*, ref. 120, 131 and 132) to result in enhanced OH and H_2O emission.

There is also downward gas transport: vertical transport of molecules can be followed by freeze-out onto the colder large grains in the midplane. If those grains become too large to be lifted back up, the disk surface composition becomes part of the planetesimal building material. This so-called "vertical cold finger effect" also limits the radial extent of the molecular emission: snow surfaces become vertical walls at the midplane snow radius rather than being curved with height in the disk, an effect that is well known for H_2O,[81,133,134] but likely also holds for other molecules. Finally, if a gap has opened in a disk, surface layer gas can be accreted directly onto forming giant planets through meridional flows,[135,136] thus leaving an imprint on the composition of their atmospheres.

Whether or not carbon-rich planets will be built in the J160532 disk is still an open question and depends also on the time of planet formation. A significant fraction of the carbon is in the gas phase and refractory carbon grains may even have been destroyed in this disk. If this carbon-rich gas were dispersed quickly from the system, only a small fraction of carbon would eventually be included in planets. This scenario has been proposed to explain why Earth is so carbon poor.[79,137]

5 Conclusions

The early JWST MIRI-MRS spectra presented here provide examples of how its increased spectral resolution and sensitivity provide new insights into the chemical composition of disks, from young to mature disks and from low- to high-mass stars. In particular, the detection of new species like C_4H_2 and C_6H_6 allows proposed chemical pathways to be tested. New lines of known species and of their isotopologs, with first detections of $^{13}CO_2$ and $^{13}C^{12}CH_2$, help to break degeneracies between model fit parameters so that their temperatures and column densities can be constrained more accurately. Indeed, orders of magnitude differences in inferred column densities have been found compared with earlier work. The lack of NH_3 suggests that most of the volatile nitrogen is in unobservable species.

The data highlight the diversity in mid-infrared spectroscopy of disks hinted at by Spitzer, from sources dominated by CO_2 with almost no H_2O, to those rich in H_2O to disks showing booming features of C_2H_2. There are also some similarities: the molecular emission observed in the high-mass protostellar source IRAS 23385, thought to originate from a massive young disk, indicates similar abundance ratios as those found in some CO_2-rich low-mass disks.

The diversity in chemistry and spectra is likely related to differences in the physical structures of the disks. Lines are brighter in disks around more luminous sources, which push the emitting area to larger radii and thus boost optically thick lines. The lack or weakness of lines of certain molecules, most notably H_2O, can be due to the very low luminosity of some sources in our sample. However, it is likely also linked to the presence of small cavities in the inner disk that cannot be resolved with ALMA and whose positions extend beyond the water snowline. A related possibility is the presence of dust traps outside of the respective snowlines of different molecules that can lock up certain elements, most notably oxygen, and prevent it from entering the inner disk.

The JWST data also provide evidence for a phenomenon that has been postulated but not seen prior to JWST: the destruction of hydrocarbon grains at the "soot" line boosting the C_2H_2 abundance by orders of magnitude[1] and raising the C/O ratio to >1.

There are a number of obvious next steps in this research. First, larger samples of disks have to be studied with JWST MIRI-MRS to search for more similarities and probe the limits of diversity, from young to old "debris" disks and across the stellar mass range. MRS spectra of line-rich sources will allow to hunt for other minor species, including NH_3. Also, adding NIRSPEC-IFU data for the same sources will be highly valuable to cover the CO ro-vibrational bands at 4.7 μm as well as the stretching bands of most of the species discussed here. Molecular data on vibration–rotation lines of many molecules, including their isotopologs, are however still incomplete and more laboratory work is needed.

Second, ALMA observations at the highest spatial resolution (down to a few au) are needed to search for small-scale substructures in the continuum that are indicative of dust traps. Their location with respect to different snowlines will help to understand which elements are locked in the outer disk and unable to reach the inner disk. In contrast, some disks may be dominated by drifting icy pebbles that enhance the inner disk in heavy elements. More work is also needed on high resolution studies of the physical structure of high-mass disks with ALMA.

Third, more sophisticated radiative transfer models are needed to analyse the spectra since the current column densities obtained from slab models are only a crude representation of actual abundances and their ratios. Retrieval models using a full 2D temperature and density disk structure but a simplified chemistry is one possibility.[138,139] Retrieval using full chemical models is computationally prohibitive,[140] but such models can then be used as inspiration for abundance distributions and the search for trends with physical parameters. In the longer run, full thermo-chemical disk models developed for individual sources can also serve as testbeds for these more simplified approaches. The distribution of small *versus* large dust and presence of substructures will be a key element in such models.

Finally, the consequences of the observed chemical diversity for the composition of any planets forming in these disks need to be investigated. Ultimately, these disk surveys need to be complemented by JWST surveys of the chemical composition of the atmospheres of mature planets to address the question of what sets their composition.

Author contributions

Investigation: Grant, Tabone, van Gelder, Francis, Beuther, Bettoni, Gasman, Arabhavi, Nazari, Vlasblom; writing – original draft: van Dishoeck; writing – review: all authors; software: Tabone, Grant, van Gelder, Francis, Arabhavi, Bettoni, Gasman, Klaassen, Christiaens; visualization: Grant, Nazari, Francis, Vlasblom; supervsision: van Dishoeck, Beuther, Kamp, Henning; funding: van Dishoeck, Kamp, Henning.

Conflicts of interest

There are no conflicts to declare.

Acknowledgements

The authors would like to thank the MIRI GTO JOYS and MINDS teams, as well as the entire MIRI European and US instrument team. Support from StScI is also appreciated. The following National and International Funding Agencies funded and supported the MIRI development: NASA; ESA; Belgian Science Policy Office (BELSPO); Centre Nationale d'Etudes Spatiales (CNES); Danish National Space Centre; Deutsches Zentrum fur Luftund Raumfahrt (DLR); Enterprise Ireland; Ministerio De Economiá y Competividad; Netherlands Research School for Astronomy (NOVA); Netherlands Organisation for Scientific Research (NWO); Science and Technology Facilities Council; Swiss Space Office; Swedish National Space Agency; and UK Space Agency. EvD acknowledges support from ERC Advanced grant 101019751 MOLDISK, the Danish National Research Foundation through the Center of Excellence "InterCat" (DNRF150), and DFG-grant 325594231, FOR 2634/2. BT is a Laureate of the Paris Region fellowship program, which is supported by the Ile-de-France Region and has received funding under the Marie Sklodowska-Curie grant agreement no. 945298. DG thanks the Research Foundation Flanders for co-financing the present research (grant number V435622N). TH is grateful for support from the ERC Advanced grant 832428 Origins; IK, MvG, LF, AA and EvD from TOP-1 grant 614.001.751 from the Dutch Research Council (NWO); and IK from H2020-MSCA-ITN- 2019, grant no. 860470 (CHAMELEON).

Notes and references

1 B. Tabone, *et al.*, *Nat. Astron*, 2023, DOI: 10.1038/s41550-023-01965-3.
2 K. I. Öberg and E. A. Bergin, *Phys. Rep.*, 2021, **893**, 1–48.
3 E. F. van Dishoeck and E. A. Bergin, *ExoFrontiers; Big Questions in Exoplanetary Science*, IOP, Bristol, UK, 2021, pp. 14–1.
4 A. Miotello, I. Kamp, T. Birnstiel, L. I. Cleeves and A. Kataoka, *arXiv*, 2022, e-prints, arXiv:2203.09818.
5 M. L. R. van't Hoff, D. Harsono, J. J. Tobin, A. D. Bosman, E. F. van Dishoeck, J. K. Jørgensen, A. Miotello, N. M. Murillo and C. Walsh, *Astrophys. J.*, 2020, **901**, 166.
6 Ł. Tychoniec, C. F. Manara, G. P. Rosotti, E. F. van Dishoeck, A. J. Cridland, T.-H. Hsieh, N. M. Murillo, D. Segura-Cox, S. E. van Terwisga and J. J. Tobin, *Astron. Astrophys.*, 2020, **640**, A19.
7 A. Morbidelli, J. I. Lunine, D. P. O'Brien, S. N. Raymond and K. J. Walsh, *Annu. Rev. Earth Planet. Sci.*, 2012, **40**, 251–275.
8 R. I. Dawson and J. A. Johnson, *Annu. Rev. Astron. Astrophys.*, 2018, **56**, 175–221.
9 J. S. Carr and J. R. Najita, *Science*, 2008, **319**, 1504.
10 C. Salyk, K. M. Pontoppidan, G. A. Blake, J. R. Najita and J. S. Carr, *Astrophys. J.*, 2011, **731**, 130.
11 K. M. Pontoppidan, C. Salyk, E. A. Bergin, S. Brittain, B. Marty, O. Mousis and K. L. Oberg, in *Protostars & Planets VI*, ed. Beuther H., Klessen R., Dullemond K. and Henning T., Tucson: Univ. Arizona Press, 2014, pp. 363–385.
12 L. I. Cleeves, *IAU Symposium*, 2018, vol. 332, pp. 57–68.
13 K. I. Öberg, V. V. Guzmán, C. Walsh, Y. Aikawa, E. A. Bergin, C. J. Law, R. A. Loomis, F. Alarcón, S. M. Andrews, J. Bae, J. B. Bergner, Y. Boehler, A. S. Booth, A. D. Bosman, J. K. Calahan, G. Cataldi, L. I. Cleeves, I. Czekala, K. Furuya, J. Huang, J. D. Ilee, N. T. Kurtovic, R. Le Gal, Y. Liu, F. Long, F. Ménard, H. Nomura, L. M. Pérez, C. Qi, K. R. Schwarz, A. Sierra, R. Teague, T. Tsukagoshi, Y. Yamato, M. L. R. van't Hoff, A. R. Waggoner, D. J. Wilner and K. Zhang, *Astrophys. J. Suppl.*, 2021, **257**, 1.
14 M. Wells, J. W. Pel, A. Glasse, G. S. Wright, G. Aitink-Kroes, R. Azzollini, S. Beard, B. R. Brandl, A. Gallie, V. C. Geers, A. M. Glauser, P. Hastings, T. Henning, R. Jager, K. Justtanont, B. Kruizinga, F. Lahuis, D. Lee, I. Martinez-Delgado, J. R. Martínez-Galarza, M. Meijers, J. E. Morrison, F. Müller, T. Nakos, B. O'Sullivan, A. Oudenhuysen, P. Parr-Burman, E. Pauwels, R. R. Rohloff, E. Schmalzl, J. Sykes, M. P. Thelen, E. F. van Dishoeck, B. Vandenbussche, L. B. Venema, H. Visser, L. B. F. M. Waters and D. Wright, *Publ. Astron. Soc. Pac.*, 2015, **127**, 646.
15 G. H. Rieke, G. S. Wright, T. Böker, J. Bouwman, L. Colina, A. Glasse, K. D. Gordon, T. P. Greene, M. Güdel, T. Henning, K. Justtanont, P. O. Lagage, M. E. Meixner, H. U. Nørgaard-Nielsen, T. P. Ray, M. E. Ressler, E. F. van Dishoeck and C. Waelkens, *Publ. Astron. Soc. Pac.*, 2015, **127**, 584.
16 G. Wright, *et al.*, *Publ. Astron. Soc. Pac.*, 2023, 048003.
17 Ewine F. van Dishoeck, *et al.*, *Astron. Astrophys.*, 2023, unpublished work.
18 I. Kamp, *et al.*, *Faraday Discuss.*, 2023, DOI: 10.1039/d3fd00013c.

19 Ewine F. van Dishoeck, *et al.*, *Astron. Astrophys.*, 2023, unpublished work.
20 R. Cesaroni, H. Beuther, A. Ahmadi, M. T. Beltrán, T. Csengeri, R. Galván-Madrid, C. Gieser, T. Henning, K. G. Johnston, P. D. Klaassen, R. Kuiper, S. Leurini, H. Linz, S. Longmore, S. L. Lumsden, L. T. Maud, L. Moscadelli, J. C. Mottram, A. Palau, T. Peters, R. E. Pudritz, Á. Sánchez-Monge, P. Schilke, D. Semenov, S. Suri, J. S. Urquhart, J. M. Winters, Q. Zhang and H. Zinnecker, *Astron. Astrophys.*, 2019, **627**, A68.
21 S. Molinari, L. Testi, J. Brand, R. Cesaroni and F. Palla, *Astrophys. J.*, 1998, **505**, L39–L42.
22 H. Beuther, J. C. Mottram, A. Ahmadi, F. Bosco, H. Linz, T. Henning, P. Klaassen, J. M. Winters, L. T. Maud, R. Kuiper, D. Semenov, C. Gieser, T. Peters, J. S. Urquhart, R. Pudritz, S. E. Ragan, S. Feng, E. Keto, S. Leurini, R. Cesaroni, M. Beltran, A. Palau, Á. Sánchez-Monge, R. Galvan-Madrid, Q. Zhang, P. Schilke, F. Wyrowski, K. G. Johnston, S. N. Longmore, S. Lumsden, M. Hoare, K. M. Menten and T. Csengeri, *Astron. Astrophys.*, 2018, **617**, A100.
23 F. Faustini, S. Molinari, L. Testi and J. Brand, *Astron. Astrophys.*, 2009, **503**, 801–816.
24 J. M. Alcalá, C. F. Manara, A. Natta, A. Frasca, L. Testi, B. Nisini, B. Stelzer, J. P. Williams, S. Antoniucci, K. Biazzo, E. Covino, M. Esposito, F. Getman and E. Rigliaco, *Astron. Astrophys.*, 2017, **600**, A20.
25 K. L. Luhman, K. A. Herrmann, E. E. Mamajek, T. L. Esplin and M. J. Pecaut, *Astron. J.*, 2018, **156**, 76.
26 Gaia Collaboration, A. Vallenari, A. G. A. Brown, T. Prusti, J. H. J. de Bruijne, F. Arenou, C. Babusiaux, M. Biermann, O. L. Creevey, C. Ducourant, D. W. Evans, L. Eyer, R. Guerra, A. Hutton, C. Jordi, S. A. Klioner, U. L. Lammers, L. Lindegren, X. Luri, F. Mignard, C. Panem, D. Pourbaix, S. Randich, P. Sartoretti, C. Soubiran, P. Tanga, N. A. Walton, C. A. L. Bailer-Jones, U. Bastian, R. Drimmel, F. Jansen, D. Katz, M. G. Lattanzi, F. van Leeuwen, J. Bakker, C. Cacciari, J. Castañeda, F. De Angeli, C. Fabricius, M. Fouesneau, Y. Frémat, L. Galluccio, A. Guerrier, U. Heiter, E. Masana, R. Messineo, N. Mowlavi, C. Nicolas, K. Nienartowicz, F. Pailler, P. Panuzzo, F. Riclet, W. Roux, G. M. Seabroke, R. Sordoørcit, F. Thévenin, G. Gracia-Abril, J. Portell, D. Teyssier, M. Altmann, R. Andrae, M. Audard, I. Bellas-Velidis, K. Benson, J. Berthier, R. Blomme, P. W. Burgess, D. Busonero, G. Busso, H. Cánovas, B. Carry, A. Cellino, N. Cheek, G. Clementini, Y. Damerdji, M. Davidson, P. de Teodoro, M. Nuñez Campos, L. Delchambre, A. Dell'Oro, P. Esquej, J. Fernández-Hernández, E. Fraile, D. Garabato, P. García-Lario, E. Gosset, R. Haigron, J. L. Halbwachs, N. C. Hambly, D. L. Harrison, J. Hernández, D. Hestroffer, S. T. Hodgkin, B. Holl, K. Janßen, G. Jevardat de Fombelle, S. Jordan, A. Krone-Martins, A. C. Lanzafame, W. Löffler, O. Marchal, P. M. Marrese, A. Moitinho, K. Muinonen, P. Osborne, E. Pancino, T. Pauwels, A. Recio-Blanco, C. Reylé, M. Riello, L. Rimoldini, T. Roegiers, J. Rybizki, L. M. Sarro, C. Siopis, M. Smith, A. Sozzetti, E. Utrilla, M. van Leeuwen, U. Abbas, P. Ábrahám, A. Abreu Aramburu, C. Aerts, J. J. Aguado, M. Ajaj, F. Aldea-Montero, G. Altavilla, M. A. Álvarez, J. Alves, F. Anders, R. I. Anderson, E. Anglada Varela, T. Antoja, D. Baines, S. G. Baker, L. Balaguer-Núñez, E. Balbinot, Z. Balog, C. Barache, D. Barbato, M. Barros,

M. A. Barstow, S. Bartolomé, J. L. Bassilana, N. Bauchet, U. Becciani, M. Bellazzini, A. Berihuete, M. Bernet, S. Bertone, L. Bianchi, A. Binnenfeld, S. Blanco-Cuaresma, A. Blazere, T. Boch, A. Bombrun, D. Bossini, S. Bouquillon, A. Bragaglia, L. Bramante, E. Breedt, A. Bressan, N. Brouillet, E. Brugaletta, B. Bucciarelli, A. Burlacu, A. G. Butkevich, R. Buzzi, E. Caffau, R. Cancelliere, T. Cantat-Gaudin, R. Carballo, T. Carlucci, M. I. Carnerero, J. M. Carrasco, L. Casamiquela, M. Castellani, A. Castro-Ginard, L. Chaoul, P. Charlot, L. Chemin, V. Chiaramida, A. Chiavassa, N. Chornay, G. Comoretto, G. Contursi, W. J. Cooper, T. Cornez, S. Cowell, F. Crifo, M. Cropper, M. Crosta, C. Crowley, C. Dafonte, A. Dapergolas, M. David, P. David, P. de Laverny, F. De Luise, R. De March, J. De Ridder, R. de Souza, A. de Torres, E. F. del Peloso, E. del Pozo, M. Delbo, A. Delgado, J. B. Delisle, C. Demouchy, T. E. Dharmawardena, P. Di Matteo, S. Diakite, C. Diener, E. Distefano, C. Dolding, B. Edvardsson, H. Enke, C. Fabre, M. Fabrizio, S. Faigler, G. Fedorets, P. Fernique, A. Fienga, F. Figueras, Y. Fournier, C. Fouron, F. Fragkoudi, M. Gai, A. Garcia-Gutierrez, M. Garcia-Reinaldos, M. García-Torres, A. Garofalo, A. Gavel, P. Gavras, E. Gerlach, R. Geyer, P. Giacobbe, G. Gilmore, S. Girona, G. Giuffrida, R. Gomel, A. Gomez, J. González-Núñez, I. González-Santamaría, J. J. González-Vidal, M. Granvik, P. Guillout, J. Guiraud, R. Gutiérrez-Sánchez, L. P. Guy, D. Hatzidimitriou, M. Hauser, M. Haywood, A. Helmer, A. Helmi, M. H. Sarmiento, S. L. Hidalgo, T. Hilger, N. Hładczuk, D. Hobbs, G. Holland, H. E. Huckle, K. Jardine, G. Jasniewicz, A. Jean-Antoine Piccolo, Ó. Jiménez-Arranz, A. Jorissen, J. Juaristi Campillo, F. Julbe, L. Karbevska, P. Kervella, S. Khanna, M. Kontizas, G. Kordopatis, A. J. Korn, Á. Kóspál, Z. Kostrzewa-Rutkowska, K. Kruszyńska, M. Kun, P. Laizeau, S. Lambert, A. F. Lanza, Y. Lasne, J. F. Le Campion, Y. Lebreton, T. Lebzelter, S. Leccia, N. Leclerc, I. Lecoeur-Taibi, S. Liao, E. L. Licata, H. E. P. Lindstrøm, T. A. Lister, E. Livanou, A. Lobel, A. Lorca, C. Loup, P. Madrero Pardo, A. Magdaleno Romeo, S. Managau, R. G. Mann, M. Manteiga, J. M. Marchant, M. Marconi, J. Marcos, M. M. S. Marcos Santos, D. Marín Pina, S. Marinoni, F. Marocco, D. J. Marshall, L. M. Polo, J. M. Martín-Fleitas, G. Marton, N. Mary, A. Masip, D. Massari, A. Mastrobuono-Battisti, T. Mazeh, P. J. McMillan, S. Messina, D. Michalik, N. R. Millar, A. Mints, D. Molina, R. Molinaro, L. Molnár, G. Monari, M. Monguió, P. Montegriffo, A. Montero, R. Mor, A. Mora, R. Morbidelli, T. Morel, D. Morris, T. Muraveva, C. P. Murphy, I. Musella, Z. Nagy, L. Noval, F. Ocaña, A. Ogden, C. Ordenovic, J. O. Osinde, C. Pagani, I. Pagano, L. Palaversa, P. A. Palicio, L. Pallas-Quintela, A. Panahi, S. Payne-Wardenaar, X. Peñalosa Esteller, A. Penttilä, B. Pichon, A. M. Piersimoni, F. X. Pineau, E. Plachy, G. Plum, E. Poggio, A. Prša, L. Pulone, E. Racero, S. Ragaini, M. Rainer, C. M. Raiteri, N. Rambaux, P. Ramos, M. Ramos-Lerate, P. Re Fiorentin, S. Regibo, P. J. Richards, C. Rios Diaz, V. Ripepi, A. Riva, H. W. Rix, G. Rixon, N. Robichon, A. C. Robin, C. Robin, M. Roelens, H. R. O. Rogues, L. Rohrbasser, M. Romero-Gómez, N. Rowell, F. Royer, D. Ruz Mieres, K. A. Rybicki, G. Sadowski, A. Sáez Núñez, A. Sagristà Sellés, J. Sahlmann, E. Salguero, N. Samaras, V. Sanchez Gimenez, N. Sanna, R. Santoveña, M. Sarasso, M. Schultheis, E. Sciacca, M. Segol, J. C. Segovia, D. Ségransan,

D. Semeux, S. Shahaf, H. I. Siddiqui, A. Siebert, L. Siltala, A. Silvelo, E. Slezak, I. Slezak, R. L. Smart, O. N. Snaith, E. Solano, F. Solitro, D. Souami, J. Souchay, A. Spagna, L. Spina, F. Spoto, I. A. Steele, H. Steidelmüller, C. A. Stephenson, M. Süveges, J. Surdej, L. Szabados, E. Szegedi-Elek, F. Taris, M. B. Taylo, R. Teixeira, L. Tolomei, N. Tonello, F. Torra, J. Torra, G. Torralba Elipe, M. Trabucchi, A. T. Tsounis, C. Turon, A. Ulla, N. Unger, M. V. Vaillant, E. van Dillen, W. van Reeven, O. Vanel, A. Vecchiato, Y. Viala, D. Vicente, S. Voutsinas, M. Weiler, T. Wevers, L. Wyrzykowski, A. Yoldas, P. Yvard, H. Zhao, J. Zorec, S. Zucker and T. Zwitter, *arXiv*, 2022, e-prints, arXiv:2208.00211.

27 N. Miret-Roig, P. A. B. Galli, J. Olivares, H. Bouy, J. Alves and D. Barrado, *Astron. Astrophys.*, 2022, **667**, A163.

28 H. Bushouse, J. Eisenhamer, N. Dencheva, J. Davies, P. Greenfield, J. Morrison, P. Hodge, B. Simon, D. Grumm, M. Droettboom, E. Slavich, M. Sosey, T. Pauly, T. Miller, R. Jedrzejewski, W. Hack, D. Davis, S. Crawford, D. Law, K. Gordon, M. Regan, M. Cara, K. MacDonald, L. Bradley, C. Shanahan and W. Jamieson, *spacetelescope/jwst: JWST 1.6.2*, Zenodo, 2022.

29 D. Gasman, *et al.*, *Astron. Astrophys.*, 2023, **673**, A102.

30 H. Beuther, *et al.*, *Astron. Astrophys.*, 2023, **673**, A121.

31 S. L. Grant, E. F. van Dishoeck, B. Tabone, D. Gasman, T. Henning, I. Kamp, M. Güdel, P.-O. Lagage, G. Bettoni, G. Perotti, V. Christiaens, M. Samland, A. M. Arabhavi, I. Argyriou, A. Abergel, O. Absil, D. Barrado, A. Boccaletti, J. Bouwman, A. C. o. Garatti, V. Geers, A. M. Glauser, R. Guadarrama, H. Jang, J. Kanwar, F. Lahuis, M. Morales-Calderón, M. Mueller, C. Nehmé, G. Olofsson, E. Pantin, N. Pawellek, T. P. Ray, D. Rodgers-Lee, S. Scheithauer, J. Schreiber, K. Schwarz, M. Temmink, B. Vandenbussche, M. Vlasblom, L. B. F. M. Waters, G. Wright, L. Colina, T. R. Greve, K. Justannont and G. Östlin, *arXiv*, 2022, e-prints, arXiv:2212.08047.

32 J. S. Carr and J. R. Najita, *Astrophys. J.*, 2011, **733**, 102.

33 S. Bruderer, D. Harsono and E. F. van Dishoeck, *Astron. Astrophys.*, 2015, **575**, A94.

34 I. Gordon, L. Rothman, R. Hargreaves, R. Hashemi, E. Karlovets, F. Skinner, E. Conway, C. Hill, R. Kochanov, Y. Tan, P. Wcisło, A. Finenko, K. Nelson, P. Bernath, M. Birk, V. Boudon, A. Campargue, K. Chance, A. Coustenis, B. Drouin, J. Flaud, R. Gamache, J. Hodges, D. Jacquemart, E. Mlawer, A. Nikitin, V. Perevalov, M. Rotger, J. Tennyson, G. Toon, H. Tran, V. Tyuterev, E. Adkins, A. Baker, A. Barbe, E. Canè, A. Császár, A. Dudaryonok, O. Egorov, A. Fleisher, H. Fleurbaey, A. Foltynowicz, T. Furtenbacher, J. Harrison, J. Hartmann, V. Horneman, X. Huang, T. Karman, J. Karns, S. Kassi, I. Kleiner, V. Kofman, F. Kwabia-Tchana, N. Lavrentieva, T. Lee, D. Long, A. Lukashevskaya, O. Lyulin, V. Makhnev, W. Matt, S. Massie, M. Melosso, S. Mikhailenko, D. Mondelain, H. Müller, O. Naumenko, A. Perrin, O. Polyansky, E. Raddaoui, P. Raston, Z. Reed, M. Rey, C. Richard, R. Tóbiás, I. Sadiek, D. Schwenke, E. Starikova, K. Sung, F. Tamassia, S. Tashkun, J. Vander Auwera, I. Vasilenko, A. Vigasin, G. Villanueva, B. Vispoel, G. Wagner, A. Yachmenev and S. Yurchenko, *J. Quant. Spectrosc. Radiat. Transfer*, 2022, **277**, 107949.

35 F. F. S. van der Tak, F. Lique, A. Faure, J. H. Black and E. F. van Dishoeck, *Atoms*, 2020, **8**, 15.

36 T. Delahaye, R. Armante, N. Scott, N. Jacquinet-Husson, A. Chédin, L. Crépeau, C. Crevoisier, V. Douet, A. Perrin, A. Barbe, V. Boudon, A. Campargue, L. Coudert, V. Ebert, J.-M. Flaud, R. Gamache, D. Jacquemart, A. Jolly, F. Kwabia Tchana, A. Kyuberis, G. Li, O. Lyulin, L. Manceron, S. Mikhailenko, N. Moazzen-Ahmadi, H. Müller, O. Naumenko, A. Nikitin, V. Perevalov, C. Richard, E. Starikova, S. Tashkun, V. Tyuterev, J. Vander Auwera, B. Vispoel, A. Yachmenev and S. Yurchenko, *J. Mol. Spectrosc.*, 2021, **380**, 111510.

37 A. Labiano, I. Argyriou, J. Álvarez-Márquez, A. Glasse, A. Glauser, P. Patapis, D. Law, B. R. Brandl, K. Justtanont, F. Lahuis, J. R. Martínez-Galarza, M. Mueller, A. Noriega-Crespo, P. Royer, B. Shaughnessy and B. Vandenbussche, *Astron. Astrophys.*, 2021, **656**, A57.

38 E. L. Gibb, D. C. B. Whittet, A. C. A. Boogert and A. G. G. M. Tielens, *Astrophys. J. Suppl.*, 2004, **151**, 35–73.

39 F. C. Adams, C. J. Lada and F. H. Shu, *Astrophys. J.*, 1987, **312**, 788.

40 E. L. Gibb, D. C. B. Whittet, W. A. Schutte, A. C. A. Boogert, J. E. Chiar, P. Ehrenfreund, P. A. Gerakines, J. V. Keane, A. G. G. M. Tielens, E. F. van Dishoeck and O. Kerkhof, *Astrophys. J.*, 2000, **536**, 347–356.

41 A. C. A. Boogert, P. A. Gerakines and D. C. B. Whittet, *Annu. Rev. Astron. Astrophys.*, 2015, **53**, 541–581.

42 S. Molinari, F. Faustini, L. Testi, S. Pezzuto, R. Cesaroni and J. Brand, *Astron. Astrophys.*, 2008, **487**, 1119–1128.

43 Ewine F. van Dishoeck, *et al.*, *Astron. Astrophys.*, 2023, submitted.

44 Ewine F. van Dishoeck, *et al.*, *Astron. Astrophys.*, 2023, unpublished work.

45 Ewine F. van Dishoeck, *et al.*, *Astron. Astrophys.*, 2023, submitted.

46 J. Kessler-Silacci, J.-C. Augereau, C. P. Dullemond, V. Geers, F. Lahuis, I. Evans, J. Neal, E. F. van Dishoeck, G. A. Blake, A. C. A. Boogert, J. Brown, J. K. Jørgensen, C. Knez and K. M. Pontoppidan, *Astrophys. J.*, 2006, **639**, 275–291.

47 E. Furlan, L. Hartmann, N. Calvet, P. D'Alessio, R. Franco-Hernández, W. J. Forrest, D. M. Watson, K. I. Uchida, B. Sargent, J. D. Green, L. D. Keller and T. L. Herter, *Astrophys. J. Suppl.*, 2006, **165**, 568–605.

48 T. Henning, *Annu. Rev. Astron. Astrophys.*, 2010, **48**, 21–46.

49 S. E. Dahm and J. M. Carpenter, *Astron. J.*, 2009, **137**, 4024–4045.

50 I. Pascucci, G. Herczeg, J. S. Carr and S. Bruderer, *Astrophys. J.*, 2013, **779**, 178.

51 V. C. Geers, J. C. Augereau, K. M. Pontoppidan, C. P. Dullemond, R. Visser, J. E. Kessler-Silacci, I. Evans, N. J. Evans Jr, F. van Dishoeck, G. A. Blake, A. C. A. Boogert, J. M. Brown, F. Lahuis and B. Merín, *Astron. Astrophys.*, 2006, **459**, 545–556.

52 J. R. Najita, J. S. Carr, K. M. Pontoppidan, C. Salyk, E. F. van Dishoeck and G. A. Blake, *Astrophys. J.*, 2013, **766**, 134.

53 K. M. Pontoppidan, C. Salyk, A. Banzatti, G. A. Blake, C. Walsh, J. H. Lacy and M. J. Richter, *Astrophys. J.*, 2019, **874**, 92.

54 F. P. Helmich, E. F. van Dishoeck, J. H. Black, T. de Graauw, D. A. Beintema, A. M. Heras, F. Lahuis, P. W. Morris and E. A. Valentijn, *Astron. Astrophys.*, 1996, **315**, L173–L176.

55 F. Lahuis and E. F. van Dishoeck, *Astron. Astrophys.*, 2000, **355**, 699–712.

56 A. M. S. Boonman, E. F. van Dishoeck, F. Lahuis and S. D. Doty, *Astron. Astrophys.*, 2003, **399**, 1063–1072.
57 C. Knez, J. H. Lacy, I. Evans, J. Neal, E. F. van Dishoeck and M. J. Richter, *Astrophys. J.*, 2009, **696**, 471–483.
58 A. G. Barr, A. Boogert, C. N. DeWitt, E. Montiel, M. J. Richter, J. H. Lacy, D. A. Neufeld, N. Indriolo, Y. Pendleton, J. Chiar and A. G. G. M. Tielens, *Astrophys. J.*, 2020, **900**, 104.
59 A. M. S. Boonman, E. F. van Dishoeck, F. Lahuis, S. D. Doty, C. M. Wright and D. Rosenthal, *Astron. Astrophys.*, 2003, **399**, 1047–1061.
60 P. Sonnentrucker, E. González-Alfonso, D. A. Neufeld, E. A. Bergin, G. J. Melnick, W. J. Forrest, J. L. Pipher and D. M. Watson, *Astrophys. J.*, 2006, **650**, L71–L74.
61 P. Sonnentrucker, E. González-Alfonso and D. A. Neufeld, *Astrophys. J.*, 2007, **671**, L37–L40.
62 E. F. van Dishoeck, L. E. Kristensen, J. C. Mottram, A. O. Benz, E. A. Bergin, P. Caselli, F. Herpin, M. R. Hogerheijde, D. Johnstone, R. Liseau, B. Nisini, M. Tafalla, F. F. S. van der Tak, F. Wyrowski, A. Baudry, M. Benedettini, P. Bjerkeli, G. A. Blake, J. Braine, S. Bruderer, S. Cabrit, J. Cernicharo, Y. Choi, A. Coutens, T. de Graauw, C. Dominik, D. Fedele, M. Fich, A. Fuente, K. Furuya, J. R. Goicoechea, D. Harsono, F. P. Helmich, G. J. Herczeg, T. Jacq, A. Karska, M. Kaufman, E. Keto, T. Lamberts, B. Larsson, S. Leurini, D. C. Lis, G. Melnick, D. Neufeld, L. Pagani, M. Persson, R. Shipman, V. Taquet, T. A. van Kempen, C. Walsh, S. F. Wampfler, U. Yıldız and WISH Team, *Astron. Astrophys.*, 2021, **648**, A24.
63 P. Woitke, M. Min, W. F. Thi, C. Roberts, A. Carmona, I. Kamp, F. Ménard and C. Pinte, *Astron. Astrophys.*, 2018, **618**, A57.
64 E. F. vanDishoeck, *Proc. Natl. Acad. Sci. U. S. A.*, 2006, **103**, 12249–12256.
65 P. Woitke, I. Kamp and W.-F. Thi, *Astron. Astrophys.*, 2009, **501**, 383–406.
66 S. Bruderer, E. F. van Dishoeck, S. D. Doty and G. J. Herczeg, *Astron. Astrophys.*, 2012, **541**, A91.
67 C. Walsh, H. Nomura, T. J. Millar and Y. Aikawa, *Astrophys. J.*, 2012, **747**, 114.
68 M. Agúndez, J. Cernicharo and J. R. Goicoechea, *Astron. Astrophys.*, 2008, **483**, 831–837.
69 P. M. Woods and K. Willacy, *Astrophys. J.*, 2009, **693**, 1360–1378.
70 A. E. Glassgold, R. Meijerink and J. R. Najita, *Astrophys. J.*, 2009, **701**, 142–153.
71 C. Walsh, H. Nomura and E. van Dishoeck, *Astron. Astrophys.*, 2015, **582**, A88.
72 D. E. Anderson, G. A. Blake, L. I. Cleeves, E. A. Bergin, K. Zhang, K. R. Schwarz, C. Salyk and A. D. Bosman, *Astrophys. J.*, 2021, **909**, 55.
73 K. I. Öberg, R. Murray-Clay and E. A. Bergin, *Astrophys. J.*, 2011, **743**, L16.
74 S. E. Bisschop, H. J. Fraser, K. I. Öberg, E. F. van Dishoeck and S. Schlemmer, *Astron. Astrophys.*, 2006, **449**, 1297–1309.
75 F. Long, P. Pinilla, G. J. Herczeg, D. Harsono, G. Dipierro, I. Pascucci, N. Hendler, M. Tazzari, E. Ragusa, C. Salyk, S. Edwards, G. Lodato, G. van de Plas, D. Johnstone, Y. Liu, Y. Boehler, S. Cabrit, C. F. Manara, F. Menard, G. D. Mulders, B. Nisini, W. J. Fischer, E. Rigliaco, A. Banzatti, H. Avenhaus and M. Gully-Santiago, *Astrophys. J.*, 2018, **869**, 17.
76 H. J. Fraser, M. P. Collings, M. R. S. McCoustra and D. A. Williams, *Mon. Not. R. Astron. Soc.*, 2001, **327**, 1165–1172.
77 G. D. Mulders, I. Pascucci and D. Apai, *Astrophys. J.*, 2015, **798**, 112.

78 M. Minissale, Y. Aikawa, E. Bergin, M. Bertin, W. A. Brown, S. Cazaux, S. B. Charnley, A. Coutens, H. M. Cuppen, V. Guzman, H. Linnartz, M. R. S. McCoustra, A. Rimola, J. G. M. Schrauwen, C. Toubin, P. Ugliengo, N. Watanabe, V. Wakelam and F. Dulieu, *ACS Earth Space Chem.*, 2022, **6**, 597–630.

79 J. Li, E. A. Bergin, G. A. Blake, F. J. Ciesla and M. M. Hirschmann, *Sci. Adv.*, 2021, 7, eabd3632.

80 M. Asplund, A. M. Amarsi and N. Grevesse, *Astron. Astrophys.*, 2021, **653**, A141.

81 R. Meijerink, K. M. Pontoppidan, G. A. Blake, D. R. Poelman and C. P. Dullemond, *Astrophys. J.*, 2009, **704**, 1471–1481.

82 A. J. Greenwood, I. Kamp, L. B. F. M. Waters, P. Woitke and W. F. Thi, *Astron. Astrophys.*, 2019, **626**, A6.

83 R. Kuiper, H. Klahr, H. Beuther and T. Henning, *Astrophys. J.*, 2011, **732**, 20.

84 B. Zhao, K. Tomida, P. Hennebelle, J. J. Tobin, A. Maury, T. Hirota, Á. Sánchez-Monge, R. Kuiper, A. Rosen, A. Bhandare, M. Padovani and Y.-N. Lee, *Space Sci. Rev.*, 2020, **216**, 43.

85 P. D'Alessio, J. Cantö, N. Calvet and S. Lizano, *Astrophys. J.*, 1998, **500**, 411–427.

86 P. Nazari, B. Tabone, G. P. Rosotti, M. L. van Gelder, R. Meshaka and E. F. van Dishoeck, *Astron. Astrophys.*, 2022, **663**, A58.

87 P. Nazari, B. Tabone and G. P. Rosotti, *arXiv*, 2022, e-prints, arXiv:2211.00126.

88 C. P. Dullemond, A. Juhasz, A. Pohl, F. Sereshti, R. Shetty, T. Peters, B. Commercon and M. Flock, *RADMC-3D: A multi-purpose radiative transfer tool*, Astrophysics Source Code Library, record ascl:1202.015, 2012.

89 A. D. Bosman, E. A. Bergin, J. K. Calahan and S. E. Duval, *Astrophys. J.*, 2022, **933**, L40.

90 E. F. van Dishoeck, E. Herbst and D. A. Neufeld, *Chem. Rev.*, 2013, **113**, 9043–9085.

91 D. Baulch, C. Cobos, R. Cox, C. Esser, P. Frank, T. Just, J. Kerr, M. Pilling, J. Troe, R. Walker and J. Warnatz, *J. Phys. Chem. Ref. Data*, 1992, **21**, 411–734.

92 T. Bethell and E. Bergin, *Science*, 2009, **326**, 1675.

93 R. Visser, E. F. van Dishoeck, S. D. Doty and C. P. Dullemond, *Astron. Astrophys.*, 2009, **495**, 881–897.

94 D. Bockelée-Morvan, D. C. Lis, J. E. Wink, D. Despois, J. Crovisier, R. Bachiller, D. J. Benford, N. Biver, P. Colom, J. K. Davies, E. Gérard, B. Germain, M. Houde, D. Mehringer, R. Moreno, G. Paubert, T. G. Phillips and H. Rauer, *Astron. Astrophys.*, 2000, **353**, 1101–1114.

95 M. N. Drozdovskaya, E. F. van Dishoeck, M. Rubin, J. K. Jørgensen and K. Altwegg, *Mon. Not. R. Astron. Soc.*, 2019, **490**, 50–79.

96 M. P. Collings, M. A. Anderson, R. Chen, J. W. Dever, S. Viti, D. A. Williams and M. R. S. McCoustra, *Mon. Not. R. Astron. Soc.*, 2004, **354**, 1133–1140.

97 M. Vlasblom, *et al.*, *Astron. Astrophys.*, 2023.

98 A. D. Bosman, E. A. Bergin, J. Calahan and S. E. Duval, *Astrophys. J.*, 2022, **930**, L26.

99 A. E. Glassgold and J. R. Najita, *Astrophys. J.*, 2015, **810**, 125.

100 M. Ádámkovics, A. E. Glassgold and J. R. Najita, *Astrophys. J.*, 2014, **786**, 135.

101 S. Antonellini, I. Kamp, P. Riviere-Marichalar, R. Meijerink, P. Woitke, W. F. Thi, M. Spaans, G. Aresu and E. Lee, *Astron. Astrophys.*, 2015, **582**, A105.

102 S. Antonellini, I. Kamp, F. Lahuis, P. Woitke, W. F. Thi, R. Meijerink, G. Aresu, M. Spaans, M. Güdel and A. Liebhart, *Astron. Astrophys.*, 2016, **585**, A61.

103 M. Temmink, *et al.*, *Astron. Astrophys.*, 2023.

104 C. Salyk, G. A. Blake, A. C. A. Boogert and J. M. Brown, *Astrophys. J.*, 2009, **699**, 330–347.

105 K. M. Pontoppidan, G. A. Blake, E. F. van Dishoeck, A. Smette, M. J. Ireland and J. Brown, *Astrophys. J.*, 2008, **684**, 1323–1329.

106 S. Antonellini, A. Banzatti, I. Kamp, W. F. Thi and P. Woitke, *Astron. Astrophys.*, 2020, **637**, A29.

107 GRAVITY Collaboration, K. Perraut, L. Labadie, J. Bouvier, F. Ménard, L. Klarmann, C. Dougados, M. Benisty, J. P. Berger, Y. I. Bouarour, W. Brandner, A. Caratti O Garatti, P. Caselli, P. T. de Zeeuw, R. Garcia-Lopez, T. Henning, J. Sanchez-Bermudez, A. Sousa, E. van Dishoeck, E. Alécian, A. Amorim, Y. Clénet, R. Davies, A. Drescher, G. Duvert, A. Eckart, F. Eisenhauer, N. M. Förster-Schreiber, P. Garcia, E. Gendron, R. Genzel, S. Gillessen, R. Grellmann, G. Heißel, S. Hippler, M. Horrobin, Z. Hubert, L. Jocou, P. Kervella, S. Lacour, V. Lapeyrère, J. B. Le Bouquin, P. Léna, D. Lutz, T. Ott, T. Paumard, G. Perrin, S. Scheithauer, J. Shangguan, T. Shimizu, J. Stadler, O. Straub, C. Straubmeier, E. Sturm, L. Tacconi, F. Vincent, S. von Fellenberg and F. Widmann, *Astron. Astrophys.*, 2021, **655**, A73.

108 A. Miotello, E. F. van Dishoeck, J. P. Williams, M. Ansdell, G. Guidi, M. Hogerheijde, C. F. Manara, M. Tazzari, L. Testi, N. van der Marel and S. van Terwisga, *Astron. Astrophys.*, 2017, **599**, A113.

109 L. I. Cleeves, R. A. Loomis, R. Teague, E. A. Bergin, D. J. Wilner, J. B. Bergner, G. A. Blake, J. K. Calahan, P. Cazzoletti, E. F. van Dishoeck, V. V. Guzmán, M. R. Hogerheijde, J. Huang, M. Kama, K. I. Öberg, C. Qi, J. Terwisscha van Scheltinga and C. Walsh, *Astrophys. J.*, 2021, **911**, 29.

110 A. D. Bosman, F. Alarcón, E. A. Bergin, K. Zhang, M. L. R. van't Hoff, K. I. Öberg, V. V. Guzmán, C. Walsh, Y. Aikawa, S. M. Andrews, J. B. Bergner, A. S. Booth, G. Cataldi, L. I. Cleeves, I. Czekala, K. Furuya, J. Huang, J. D. Ilee, C. J. Law, R. Le Gal, Y. Liu, F. Long, R. A. Loomis, F. Ménard, H. Nomura, C. Qi, K. R. Schwarz, R. Teague, T. Tsukagoshi, Y. Yamato and D. J. Wilner, *Astrophys. J. Suppl.*, 2021, **257**, 7.

111 A. Banzatti, I. Pascucci, A. D. Bosman, P. Pinilla, C. Salyk, G. J. Herczeg, K. M. Pontoppidan, I. Vazquez, A. Watkins, S. Krijt, N. Hendler and F. Long, *Astrophys. J.*, 2020, **903**, 124.

112 A. D. Bosman and A. Banzatti, *Astron. Astrophys.*, 2019, **632**, L10.

113 M. K. McClure, *Astron. Astrophys.*, 2019, **632**, A32.

114 J. A. Sturm, M. K. McClure, D. Harsono, S. Facchini, F. Long, M. Kama, E. A. Bergin and E. F. van Dishoeck, *Astron. Astrophys.*, 2022, **660**, A126.

115 S. M. Andrews, J. Huang, L. M. Pérez, A. Isella, C. P. Dullemond, N. T. Kurtovic, V. V. Guzmán, J. M. Carpenter, D. J. Wilner, S. Zhang, Z. Zhu, T. Birnstiel, X.-N. Bai, M. Benisty, A. M. Hughes, K. I. Öberg and L. Ricci, *Astrophys. J.*, 2018, **869**, L41.

116 R. A. Booth and J. D. Ilee, *Mon. Not. R. Astron. Soc.*, 2019, **487**, 3998–4011.

117 A. D. Bosman, A. G. G. M. Tielens and E. F. van Dishoeck, *Astron. Astrophys.*, 2018, **611**, A80.

118 A. S. Booth, C. Walsh, J. D. Ilee, S. Notsu, C. Qi, H. Nomura and E. Akiyama, *Astrophys. J.*, 2019, **882**, L31.
119 K. Zhang, E. A. Bergin, K. Schwarz, S. Krijt and F. Ciesla, *Astrophys. J.*, 2019, **883**, 98.
120 S. Krijt, A. D. Bosman, K. Zhang, K. R. Schwarz, F. J. Ciesla and E. A. Bergin, *Astrophys. J.*, 2020, **899**, 134.
121 M. E. Kress, A. G. G. M. Tielens and M. Frenklach, *Adv. Space Res.*, 2010, **46**, 44–49.
122 D. E. Anderson, E. A. Bergin, G. A. Blake, F. J. Ciesla, R. Visser and J.-E. Lee, *Astrophys. J.*, 2017, **845**, 13.
123 H.-P. Gail and M. Trieloff, *Astron. Astrophys.*, 2017, **606**, A16.
124 C. F. Chyba, P. J. Thomas, L. Brookshaw and C. Sagan, *Science*, 1990, **249**, 366–373.
125 Y. Kebukawa, S. Nakashima and M. E. Zolensky, *Meteoritics and Planetary Science Supplement*, 2010, **45**, 99–113.
126 M. Frenklach and E. D. Feigelson, *Astrophys. J.*, 1989, **341**, 372.
127 W. A. Morgan Jr, E. D. Feigelson, H. Wang and M. Frenklach, *Meteoritics*, 1991, **26**, 374.
128 P. M. Woods and K. Willacy, *Astrophys. J.*, 2007, **655**, L49–L52.
129 A. Arabhavi, *et al.*, *Astron. Astrophys.*, 2023.
130 I. Pascucci, D. Apai, K. Luhman, T. Henning, J. Bouwman, M. R. Meyer, F. Lahuis and A. Natta, *Astrophys. J.*, 2009, **696**, 143–159.
131 D. Semenov and D. Wiebe, *Astrophys. J. Suppl.*, 2011, **196**, 25.
132 P. Woitke, A. M. Arabhavi, I. Kamp and W. F. Thi, *Astron. Astrophys.*, 2022, **668**, A164.
133 S. M. Blevins, K. M. Pontoppidan, A. Banzatti, K. Zhang, J. R. Najita, J. S. Carr, C. Salyk and G. A. Blake, *Astrophys. J.*, 2016, **818**, 22.
134 A. D. Bosman and E. A. Bergin, *Astrophys. J.*, 2021, **918**, L10.
135 A. Morbidelli, J. Szulágyi, A. Crida, E. Lega, B. Bitsch, T. Tanigawa and K. Kanagawa, *Icarus*, 2014, **232**, 266–270.
136 R. Teague, J. Bae and E. A. Bergin, *Nature*, 2019, **574**, 378–381.
137 J.-E. Lee, E. A. Bergin and H. Nomura, *Astrophys. J.*, 2010, **710**, L21–L25.
138 A. M. Mandell, J. Bast, E. F. van Dishoeck, G. A. Blake, C. Salyk, M. J. Mumma and G. Villanueva, *Astrophys. J.*, 2012, **747**, 92.
139 A. D. Bosman, S. Bruderer and E. F. van Dishoeck, *Astron. Astrophys.*, 2017, **601**, A36.
140 P. Woitke, I. Kamp, S. Antonellini, F. Anthonioz, C. Baldovin-Saveedra, A. Carmona, O. Dionatos, C. Dominik, J. Greaves, M. Güdel, J. D. Ilee, A. Liebhardt, F. Menard, M. Min, C. Pinte, C. Rab, L. Rigon, W. F. Thi, N. Thureau and L. B. F. M. Waters, *Publ. Astron. Soc. Pac.*, 2019, **131**, 064301.

Faraday Discussions

ROYAL SOCIETY OF CHEMISTRY

PAPER

Chemical conditions on Hycean worlds†

Nikku Madhusudhan, *[a] Julianne I. Moses, [b] Frances Rigby [a] and Edouard Barrier [a]

Received 3rd April 2023, Accepted 2nd May 2023
DOI: 10.1039/d3fd00075c

Traditionally, the search for life on exoplanets has been predominantly focused on rocky exoplanets. The recently proposed Hycean worlds have the potential to significantly expand and accelerate the search for life elsewhere. Hycean worlds are a class of habitable sub-Neptunes with planet-wide oceans and H_2-rich atmospheres. Their broad range of possible sizes and temperatures lead to a wide habitable zone and high potential for discovery and atmospheric characterization using transit spectroscopy. Over a dozen candidate Hycean planets are already known to be transiting nearby M dwarfs, making them promising targets for atmospheric characterization with the James Webb Space Telescope (JWST). In this work, we investigate possible chemical conditions on a canonical Hycean world, focusing on (a) the present and primordial molecular composition of the atmosphere, and (b) the inventory of bioessential elements for the origin and sustenance of life in the ocean. Based on photochemical and kinetic modeling for a range of conditions, we discuss the possible chemical evolution and observable present-day composition of its atmosphere. In particular, for reduced primordial conditions the early atmospheric evolution passes through a phase that is rich in organic molecules that could provide important feedstock for prebiotic chemistry. We investigate avenues for delivering bioessential metals to the ocean, considering the challenging lack of weathering from a rocky surface and the ocean separated from the rocky core by a thick icy mantle. Based on ocean depths from internal structure modelling and elemental estimates for the early Earth's oceans, we estimate the requirements for bioessential metals in such a planet. We find that the requirements can be met for plausible assumptions about impact history and atmospheric sedimentation, and supplemented by other steady state sources. We discuss the observational prospects for atmospheric characterisation of Hycean worlds with JWST and future directions of this new paradigm in the search for life on exoplanets.

1 Introduction

The search for life elsewhere is the holy grail of exoplanet science. While the detection of an atmospheric signature of an Earth-like planet orbiting a sun-like

[a]*Institute of Astronomy, University of Cambridge, Cambridge, UK. E-mail: nmadhu@ast.cam.ac.uk*
[b]*Space Science Institute, Boulder, CO, USA*

† Electronic supplementary information (ESI) available. See DOI: https://doi.org/10.1039/d3fd00075c

star remains a difficult goal,[1–3] habitable-zone sub-Neptunes transiting nearby M dwarfs are more accessible; we refer to sub-Neptunes as planets with sizes between those of Earth and Neptune. Their bright and small host stars lead to larger planet–star size contrasts, making them more conducive to atmospheric characterisation using transit spectroscopy. Traditionally, the search for biosignatures in exoplanets has been focused on rocky exoplanets in the terrestrial habitable zone.[4–6] Like terrestrial planets in the solar system, such planets are expected to have predominantly rocky interior compositions but their atmospheres may be more diverse, ranging from heavy CO_2-rich or N_2-rich atmospheres to lighter H_2-rich atmospheres.[4,6,7] Recent transit surveys have discovered a number of super-Earths in the habitable zones of their host stars, notably around nearby M dwarfs, *e.g.* TRAPPIST-1,[8] LHS 1140,[9–11] and TOI-700.[12,13] Theoretical studies show that the James Webb Space Telescope (JWST) will have the capability to detect potential atmospheric biosignatures in such planets, albeit with significant investment of observing time.[14–16]

Recently, a new class of sub-Neptune exoplanets, Hycean worlds, has been proposed to significantly expand and accelerate the search for life elsewhere. Hycean worlds are temperate sub-Neptunes with ocean-covered surfaces underneath H_2-rich atmospheres.[17] The temperatures and pressures in their oceans allow for habitable conditions similar to those known to sustain life in Earth's oceans. Hycean planets offer significant advantages in the search for life compared to traditionally favoured habitable super-Earths with terrestrial-like atmospheres. Firstly, due to the possibility of a large water fraction such planets can be significantly larger in size for their mass compared to rocky planets, up to 2.6 $R_\oplus$ for 10 $M_\oplus$. Several recent studies have shown temperate sub-Neptunes in this size range to be capable of hosting surface liquid water and deep oceans for a range of atmospheric and interior conditions.[18–20] Secondly, the H_2-rich atmospheres allow for a much wider habitable-zone for Hycean worlds compared to the terrestrial habitable zone.[17] The potential of H_2-rich atmospheres to sustain surface habitable temperatures for large orbital separations has also been suggested previously in the context of rocky exoplanets.[21,22] Hycean planets at large orbital separations are also expected to be conducive for long-term habitability.[23]

Thanks to their physical properties Hycean worlds are more detectable and more accessible to atmospheric characterisation with current facilities compared to rocky habitable planets. The low mean molecular weight of their hydrogen-rich atmospheres and lower bulk gravity lead to larger atmospheric scale heights for Hycean planets relative to rocky planets of comparable mass. The extended atmospheres along with the large radii make them more favourable for transmission spectroscopy and searching for spectroscopic biomarkers.[17] The size range of Hycean planets ($\sim$1–2.6 $R_\oplus$) lies in the sub-Neptune regime ($\sim$1–4 $R_\oplus$), the radius range that dominates the currently known exoplanet population.[24,25] In particular, the Kepler[26] and TESS[27] missions have discovered tens of temperate sub-Neptunes orbiting nearby M dwarfs, which are excellent targets for transit spectroscopy. Over a dozen such planets have already been identified as candidate Hycean worlds with high potential for atmospheric follow-up.[17,28–31] Simulation studies show that such planets are indeed excellent candidates for detections of key volatile species and potential biomarkers in their atmospheres.[17,32–34,124]

Characterising the atmospheric composition is essential to identify a Hycean world. A degenerate set of internal structures can generally explain the observed mass, radius and equilibrium temperature of a candidate Hycean world.[18] As shown in Fig. 1, these solutions include rocky worlds with large H_2-rich envelopes or mini-Neptunes which contain both a rocky layer and an icy mantle along with a significant H_2-rich envelope, similar to the ice giants Uranus and Neptune in the solar system. Neither of these two scenarios would support a habitable surface, given the high temperature and pressure expected underneath the thick H_2 envelope. Hycean worlds would lie between these two extremes, with a thin H_2-rich atmosphere that allows for a liquid water layer underneath at habitable temperatures and pressures.[17] Recent studies have suggested using atmospheric compositions to infer the presence of surfaces in sub-Neptunes with H_2-rich atmospheres.[35–37] Thus, resolving the degeneracies in internal structure and reliably identifying a Hycean world would require an understanding of the expected atmospheric chemical processes and the observable compositions.

While Hycean worlds offer the required thermodynamic conditions for oceanic life their possible chemical conditions have not been explored in detail. A central question in determining the possibility of life on a planet is whether its environment has the potential for the origin and sustenance of life. The minimum ingredients required for life on Earth are known to be:[38,39] (a) an energy source, (b) liquid water, and (c) various bioessential elements. Hycean worlds clearly satisfy the requirements for the presence of an energy source and liquid water with the presence of the host star and an abundant source of water. However, as is common to all ocean worlds, including icy moons in the solar system,[40] the availability of chemical inventory accessible for seeding and sustaining life is a key challenge.[41–44] In particular, Hycean worlds, with their planet-wide oceans, hinder access to bioessential elements resulting from geochemical cycles involving weathering of rocky surfaces that would be natural for a rocky planet. The large water fraction possible on such planets, as for other ocean worlds and sub-Neptunes in general, can result in a thick layer of high-pressure ice disconnecting the ocean from the rocky core and limiting access to bioessential nutrients.[41,42,45,46] At the same time, other possible sources of nutrient delivery remain possible, including delivery through impacts,[39,46] extraterrestrial dust[39,46] and potential transport of nutrients from the inner core through a permeable ice layer.[47–51] In addition to bioessential elements, the availability of pre-biotic organic molecules also plays an essential role in the origin of life. Pre-biotic molecules such as HCN are thought to have been critical for abiogenesis in the

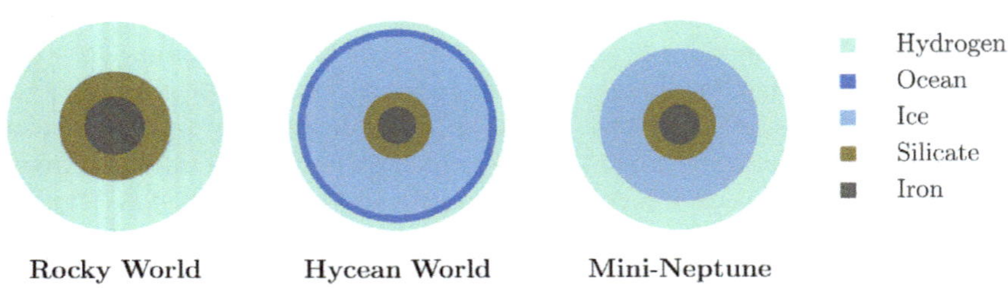

Fig. 1 Schematic of degeneracy between internal structures of Hycean worlds with those of other possible planet types with H_2-rich atmospheres.

Early Earth.[52] Whether such molecules can be delivered through extraterrestrial impactors or need to be created *in situ*, or both, remains an open debate.[53]

In the present work, we explore the chemical conditions possible on Hycean worlds. Using a canonical model of a Hycean world, we investigate the possible molecular inventory resulting from photochemical and kinetic processes in the atmosphere. We also explore sources of bioessential elements possible on such planets. In what follows, we describe our methods in Section 2, including the atmospheric structure, photochemistry, and internal structure models. We present our results and discuss the chemical conditions possible for life on Hycean worlds in Section 3. We discuss the observational prospects for identifying and characterising Hycean atmospheres with JWST in Section 4. We summarise our findings and discuss future directions in Section 5.

2 Methods

We explore the possible chemical conditions in a canonical Hycean planet focusing on the molecular composition of its atmosphere and the availability of bioessential elements in its ocean. We model the atmospheric pressure–temperature structure using a self-consistent radiative–convective equilibrium model, the atmospheric composition using a photochemical model, and the ocean properties using an internal structure model. Here we describe the model considerations.

2.1 Atmospheric modelling

We consider a plane-parallel H_2-rich atmosphere in radiative–convective equilibrium under Hycean conditions. We model the atmosphere using the GENESIS self-consistent modelling framework[55] adapted for sub-Neptune atmospheres.[17–19] This model self-consistently computes the temperature structure in radiative–convective equilibrium using the Rybicki scheme considering external irradiation, internal flux, possible scattering due to clouds/hazes, opacities from different chemical species (in chemical equilibrium or for a given composition), and day–night energy transport. The model computes line-by-line radiative transfer using the Feautrier method[56] and the discontinuous finite element method.[57] For given system parameters and chemical composition, the model self-consistently computes the temperature profile, flux profile, and the emergent spectrum of a planet.

We assume a canonical Hycean planet with the bulk properties of the habitable-zone sub-Neptune K2-18 b, which has been suggested as a candidate Hycean planet.[17,18] We adopt the system parameters reported by Benneke *et al.*,[58] with a planet mass of 8.63 $M_\oplus$ (ref. 59) and radius of 2.61 $R_\oplus$ (ref. 58) orbiting an M3V host star. The thermodynamic conditions on Hycean worlds can span a large range, with the ocean surface temperatures of 273–400 K and surface pressures of $1–10^3$ bar. For our canonical model, we consider a limiting case of a warm Hycean world in which the ocean surface is at a temperature (T) of $\sim$400 K and a pressure (P) of 100 bar. This choice is governed by the fact that most of the Hycean candidates currently identified are at zero-albedo equilibrium temperatures around 400 K. The bottom of the atmosphere is set at 100 bar, denoting the pressure at the ocean–atmosphere boundary, and the top at 10^{-6} bar. We derive the temperature profile with the required temperature and pressure at the bottom boundary by a combination of model parameters, which involve the incident irradiation, internal temperature, and scattering; an exploration of this parameter space is discussed in recent studies.[17–19]

We first consider our canonical planet to be irradiated with the maximal flux received by K2-18 b, *i.e.* at the sub-stellar point with no day–night energy redistribution (f_r), which corresponds to an irradiation temperature of 331 K. Alternately, assuming a nominal f_r of 0.35 this flux would correspond to an equivalent irradiation temperature of 369 K. We assume a nominal internal temperature of 30 K. The primary source of molecular opacity in this canonical model atmosphere is due to H_2O.[60,61] We also include H_2–H_2 and H_2–He collision-induced absorption.[62–65] We assume the H_2O abundance to be saturated in the lower atmosphere until a pressure of 0.1 bar at which point the H_2O abundance is assumed to be quenched higher up. We explore atmospheric pressure–temperature (P–T) profiles obtained using different levels of scattering in the atmosphere. Various sources of scattering can influence the P–T profile, including clouds and hazes of different compositions.[17–19] Here, we follow the approach of recent studies[17,19] and consider extinction due to hazes modelled as an enhanced Rayleigh scattering, with H_2 Rayleigh scattering enhanced by a multiplicative factor which is a free parameter in the model. We also consider P–T profiles from other recent studies[36,54] that assumed cloud/haze-free atmospheres for Hycean worlds. We discuss the specific P–T profiles used in this work in Section 3.1.

2.2 Chemical kinetics modeling

To investigate the possible atmospheric composition of our canonical Hycean world, we use the Caltech/JPL 1D KINETICS photochemical model[66–68] to solve the coupled continuity equations for the chemical production, loss, and vertical transport of atmospheric constituents. First, we consider models with a zero-flux boundary condition at the bottom of the atmosphere to investigate how the composition evolves with time chemically, when the atmospheric temperature at the ocean surface is too low to allow full chemical recycling of the photochemical products back into their original equilibrium constituents. Similar to previous investigations,[35,37] these toy models start with a thermochemical equilibrium composition for different assumptions about bulk elemental abundances, and then the composition is allowed to evolve with time under stellar UV irradiation, assuming zero flux boundary conditions for all species (*i.e.*, no atmospheric escape or external input at the top of the atmosphere, and no chemical exchange with or outgassing from the surface). The zero-flux assumption allows for a quick look at the timing and behavior of the transition from the assumed reduced initial atmosphere accreted from the protoplanetary disk—with H_2O, CH_4, and NH_3 containing the bulk of the oxygen, carbon, and nitrogen in this cool, H_2-rich atmosphere—to the CO_2, CO, H_2O, N_2 dominant end state (also still within a H_2-rich atmosphere) that occurs when the more photochemically fragile reduced forms of the elements are converted to molecules with stronger chemical bonds.[35,37] Of course, chemical exchange between the surface and atmosphere will occur on a real planet, with outgassing of dissolved CO_2 being particularly important for ocean worlds,[36,69] which we also consider as discussed below. Table 1 shows the cases reported in this work, with the zero-flux models corresponding to cases 1–6.

Table 1 Model assumptions and volume mixing ratios of prominent molecules at 1 mbar for the Hycean cases explored with the photochemical calculations in this work (see Section 2.2). The initial metallicity is assumed to be 100× solar. P_b is the pressure at the lower boundary of the atmosphere. The boundary conditions are those considered at the lower boundary. The first six cases consider zero-flux boundary conditions, *i.e.* an unreactive surface, whereas the last five cases consider boundary conditions as discussed in Section 2.2. Cases 7* and 8* compute chemistry at an age of 28 Myr in the early evolution of the planet, assuming reduced atmospheric boundary conditions representative of 100× solar metallicity. The remaining cases are evolved until 5 Gyr. The last five columns show the volume mixing ratios at 1 mbar of prominent CNO molecules (H_2O, CH_4, NH_3, CO_2 and CO) expected in H_2-rich atmospheres. The H_2O abundance is controlled by saturation at the lower boundary, that is mixing ratios of ~2×10^{-2} for the 100 bar cases, except when condensation occurs at higher altitudes (*e.g.*, PT2, PT3 and PT4 profiles), where the saturation minimum becomes ~1×10^{-2} for PT2, ~3×10^{-3} for PT3, and lower for PT4. We note that cooler temperature profiles can decrease the H_2O abundance further and affect the abundances of other molecules, especially decrease the CO_2 abundance

Case	*P–T* profile	P_b/bar	Star	Boundary condition	H_2O	CH_4	NH_3	CO_2	CO
1	PT1	100	Old	Zero flux	2.4×10^{-2}	1.1×10^{-2}	4.1×10^{-5}	1.3×10^{-2}	1.6×10^{-5}
2	PT2	100	Old	Zero flux	1.0×10^{-2}	6.6×10^{-3}	1.5×10^{-4}	4.5×10^{-3}	5.6×10^{-5}
3	PT3	100	Old	Zero flux	2.8×10^{-3}	2.5×10^{-3}	3.2×10^{-4}	3.5×10^{-4}	2.5×10^{-4}
4	PT1	1	Old	Zero flux	2.4×10^{-2}	6.8×10^{-6}	3.0×10^{-9}	4.1×10^{-2}	1.9×10^{-3}
5	PT3	1	Old	Zero flux	2.3×10^{-3}	8.2×10^{-5}	3.9×10^{-13}	4.2×10^{-2}	9.3×10^{-4}
6	PT4	1	Old	Zero flux	3.1×10^{-10}	7.8×10^{-5}	2.2×10^{-7}	9.4×10^{-15}	3.0×10^{-9}
7*	PT1	100	Young	Fixed H_2O, CH_4, NH_3	2.5×10^{-2}	5.0×10^{-2}	1.5×10^{-2}	8.5×10^{-5}	7.9×10^{-5}
8*	PT4	1	Young	Fixed H_2O, CH_4, NH_3	3.0×10^{-10}	4.9×10^{-2}	4.7×10^{-3}	1.8×10^{-19}	2.4×10^{-10}
9	PT1	100	Old	Fixed N_2, H_2O; $CO_2 = 1.4 \times 10^{-2}$	2.4×10^{-2}	1.2×10^{-4}	1.7×10^{-10}	1.3×10^{-2}	4.9×10^{-6}
10	PT1	100	Old	Fixed N_2, H_2O; $CO_2 = 1.0 \times 10^{-6}$	2.4×10^{-2}	3.5×10^{-6}	2.8×10^{-10}	1.0×10^{-6}	2.5×10^{-8}
11	PT4	1	Old	Fixed N_2, H_2O; $CO_2 = 1.0 \times 10^{-1}$	3.1×10^{-10}	5.5×10^{-8}	6.3×10^{-13}	1.0×10^{-1}	6.4×10^{-3}

These zero-flux models use the fully reversed reaction mechanism of Moses *et al.*[70] that includes 92 species containing the elements C, N, O, and H that interact with each other *via* ~1650 reactions; further model details are provided in Moses *et al.*[70] and Yu *et al.*[35] As is discussed by Hu *et al.*,[36] sulfur species are expected to be largely sequestered in the ocean on such planets, and volcanic outgassing is expected to be shut down by the overlying high pressures at the rock–ocean boundary.[69,71] The eddy diffusion coefficient K_{zz} in our model atmospheres is assumed to vary with pressure P according to $K_{zz} = 5.6 \times 10^4 P^{-0.5}$ cm^2 s^{-1}, with P in bar (see ref. 72), with K_{zz} capped at 10^{10} cm^2 s^{-1} in the upper atmosphere, and $K_{zz} = 10^6$ cm^2 s^{-1} at $P > 0.5$ bar. The stellar flux is taken from the HAZMAT database,[73] and we investigate results assuming both a young (45 Myr) and old (5 Gyr) M dwarf of 0.45 solar mass, with a medium EUV level for that age/mass; the younger star has a higher EUV flux. One change from Yu *et al.*[35] is that we consider H_2O condensation in these models, following procedures outlined in Moses *et al.*[74] Water vapor is saturated at the bottom (ocean) boundary. With some of the thermal profiles, H_2O will condense again higher up in the atmosphere if the H_2O vapor mixing ratio exceeds local saturation. Condensation/evaporation is assumed to be zero above the minimum in the saturation vapor mixing ratio near the tropopause (if one exists), such that the water vapor at higher altitudes never exceeds that minimum mixing ratio; *i.e.*, we assume condensed water droplets or ice particles are not carried to higher altitudes where they might re-evaporate. Model results are presented at various points in time as the composition evolves.

We use these zero-flux models to gain insight into the potential atmospheric composition at various stages in the planet's history. All the models pass initially through an organic-rich phase, whose lifetime depends on the atmospheric thermal structure, the pressure at the ocean surface, and the UV flux from the star. All models end up with at least partial conversion of the original CH_4 and NH_3 to CO_2, CO, N_2, and heavier organic molecules. We use the output from these simple models to define boundary conditions for more realistic photochemical models designed to explore the atmospheric composition of a Hycean planet with a surface ocean.

The atmospheric composition of these more realistic Hycean-world models will depend strongly on boundary conditions, which are not known *a priori* for these planets. Guided by our zero-flux model results and other discussions in the literature,[35–37,75–77] we examine an early organic-rich scenario where the atmosphere is assumed to start with a Neptune-like 100× solar metallicity composition in thermochemical equilibrium, with H_2O controlled by saturation at the bottom of the model. Mixing ratios of H_2O, NH_3, CH_4, and He are fixed at the lower boundary at these equilibrium (or saturated for H_2O) abundances. Fixing the mixing ratio in these models in combination with photochemical loss results in an upward flux of O, C, N that in steady state must end up being balanced by a loss of photochemical products through the oceanic lower boundary. Zero flux is assumed for all species at the top of the atmosphere. Based on several early Earth models and/or ocean-world models,[36,75–77] we have chosen deposition velocities ν_d at the lower boundary of zero for O_2, N_2, and several long-lived C_2H_x–C_4H_x hydrocarbons; $\nu_d = 10^{-8}$ cm s^{-1} for CO; $\nu_d = 10^{-5}$ cm s^{-1} for CO_2, C_2H_6, C_3H_8, C_4H_{10}, and stable hydrocarbons with 5 or more carbon atoms (precursors to organic hazes); $\nu_d = 10^{-3}$ cm s^{-1} for H_2CO, CH_3OH, H_2CCO, CH_3CHO, HCN, CH_3CN, HC_3N, CH_3NH_2, NO, HNCO; all other species are assumed to flow through the boundary at the maximum possible rate, given by K_{zz} divided by the atmospheric scale height. For our canonical Hycean world, the scale height at the surface is larger than that on Earth, and the maximum v_d is ~0.22 cm s^{-1}. We use the stellar flux from the younger star for this organic-rich "early-era" model. This model is used for cases 7 and 8 in Table 1.

We also consider "late-era" scenarios where the reduced parent species have already been depleted through conversion to more stable photochemical products, such that the mixing ratios of CO_2, N_2, H_2O, and He are fixed at the lower boundary. These cases are similar to the thin-atmosphere ocean-world models presented by Hu *et al.*,[36] where dissolved CO_2 is expected to be the dominant carbon phase in the ocean,[69] and the amount of atmospheric CO_2 is controlled by its equilibrium with the ocean, which in turn depends on the oceanic pH. Hu *et al.*[36] explored a range of possible fixed CO_2 mixing ratios in their model, and we also consider three cases: (1) a 100 bar, low-CO_2 case with a CO_2 partial pressure of 10^{-4} bar (*i.e.*, a mixing ratio of 10^{-6} for our 100 bar surface pressure), within the range of the lower bound calculated for a pH of 9–10 by Hu *et al.*,[36] (2) a 100 bar, intermediate CO_2 mixing ratio (along with N_2) that matches the mixing ratios at the lower boundary of our zero-flux model runs at the 5 Gyr time period (*i.e.*, 7.2×10^{-3} for N_2 and 1.37×10^{-2} for CO_2 in our warmest model), and (3) a 1 bar, high-CO_2 case (10% CO_2, 1% N_2) that aligns with the high-CO_2 case from Hu *et al.*[36] The H_2O mixing ratios in these models are fixed at saturation values at the lower boundary. The other boundary conditions are the same as in the organic-rich early-era model, except that CH_4 is assumed to have $v_d = 0$ and NH_3 diffuses through the bottom boundary at the maximum possible rate. These "late era" models use the stellar flux from the older M dwarf star and are run to steady state. These models are represented by cases 9–11 in Table 1. We discuss our results in Section 3.

2.3 Internal structure modelling

The internal structure is modelled following the approach used for Hycean candidates in previous studies.[18,20,78] For a given planet mass (M_p) and the mass fractions of the constituent layers (x_i), the planet radius (R_p) is obtained by solving the structure equations, *i.e.* of mass continuity and hydrostatic equilibrium. We assume four-layers – an H_2-rich atmosphere, an H_2O layer, a silicate outer core and an Fe inner core. For each layer, the equation of state (EOS) and temperature profile are specified, as described below. The structure equations, the adopted EOSs and the *P–T* profiles form a set of coupled differential equations which are solved using a fourth-order Runge–Kutta method.

In the H_2-rich atmosphere the pressures and temperatures required to allow habitable surface conditions are adequately low that the ideal gas EOS is sufficiently accurate. More generally, for the H_2-rich layer we use the EOS from Chabrier *et al.*[79] given the atmospheric *P–T* profile. The surface is taken to be at the ocean–atmosphere boundary, or H_2O–H_2/He boundary (HHB).[18] In the H_2O layer, we adopt an adiabatic *P–T* profile, with the initial pressure and temperature set by the HHB. The temperature-dependent EOS for H_2O is compiled from multiple sources – see previous studies[20,80] for a full description of these. The EOSs we adopt for the core[81] are in the form of a Birch–Murnaghan EOS[82] for $MgSiO_3$ perovskite[83] and Fe,[84] which are temperature independent. The core is assumed to be Earth-like in composition, with an Fe mass fraction of 0.33.

3 Results

The potential for life on a habitable-zone planet depends on a wide range of properties, including the thermodynamic and chemical conditions in both the atmosphere as well as the interior. It has been suggested previously that Hycean planets allow for a wide range of thermodynamic conditions in their atmospheres and oceans, similar to conditions known to be conducive for life in Earth's oceans.[17] Here we explore the possible chemical composition of a canonical Hycean atmosphere and the availability of bioessential elements in its ocean.

3.1 Atmospheric and interior structure

We consider a canonical Hycean planet and investigate the possible chemical inventory in its atmosphere for a range of physical conditions. The temperature structure of the atmosphere is arguably the central factor influencing the habitability at the ocean surface. A wide range of atmospheric temperature structures are possible in temperate sub-Neptunes, including Hycean worlds, depending on the stellar irradiation, internal temperature, atmospheric composition, and scattering due to clouds/hazes.[17–19,54,85–88] In particular, an adequate albedo due to clouds/hazes is required to maintain a liquid water surface for most candidate Hycean worlds.[17–19] An atmosphere without stong scattering due to clouds/hazes would generally render the surface too hot to maintain a habitable ocean.[19,54,85,86,88] The model considerations in this work are discussed in Section 2.1, based on the bulk properties of the candidate Hycean planet K2-18 b. We consider four *P*–*T* profiles for our models. First, we construct three profiles (PT1, PT2, and PT3) shown in Fig. 2, considering scattering due to hazes modelled with Rayleigh enhancement factors of 900, 1100 and 1300, respectively. These profiles

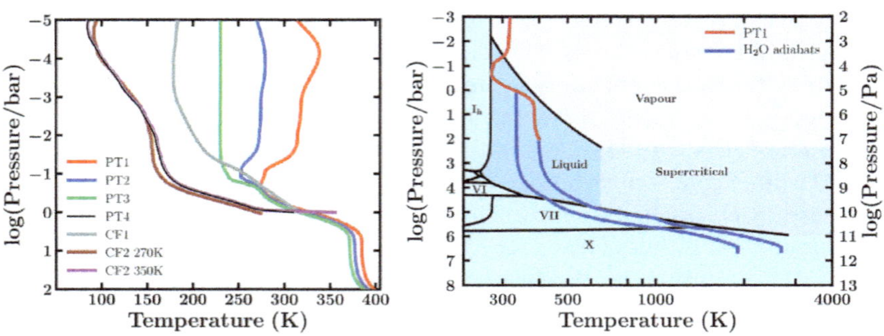

Fig. 2 Left: Model pressure–temperature (*P*–*T*) profiles of a canonical Hycean world considered in this study (see Section 3.1). The profiles PT1–PT3 consider scattering due to hazes modelled as enhanced Rayleigh scattering. CF1 and CF2 refer to *P*–*T* profiles from previous studies[36,54] assuming cloud/haze free (CF) model atmospheres for different conditions as discussed in Section 3.1. Two profiles are shown for CF2 (ref. 54) with different surface temperatures, and we consider a profile PT4 which has an intermediate surface temperature between the two CF2 profiles. Right: Internal temperature structure and the H_2O phase diagram. The black curves denote the different phase boundaries. The blue curves show H_2O adiabats starting at ocean surface pressures of 1 bar and 100 bar. The red curve shows the PT1 temperature profile of the canonical atmosphere from the left panel for reference.

represent a range of possible atmospheric temperatures but all have a maximum temperature of ~400 K and pressure of 100 bar at the ocean surface. As discussed in Section 2.1, our choice of this limiting case is driven by expectations for most of the candidate Hycean worlds currently known. This surface temperature is also close to the maximum ocean temperature of 395 K at which life is known to survive in Earth's oceans.[89,90] Any cooler Hycean worlds would be even more conducive to habitability, making the present case a conservative choice. We explore additional cases in which the PT1 and PT3 profiles are limited to the 1 bar pressure level, as shown in Table 1, representing shallower atmospheres with surfaces at a pressure of 1 bar and temperatures of ~310–330 K.

We also consider a *P–T* profile based on a cloud/haze-free (CF) model atmosphere of K2-18 b as pursued in recent works.[36,54] Hu *et al.*[36] consider an H_2-rich atmosphere of K2-18 b with Earth-like equivalent insolation, achieved assuming a Bond albedo of 0.3, but otherwise considering no clouds/hazes in the model. Innes *et al.*[54] report CF model *P–T* profiles for a K2-18 b-like planet with surface temperatures of 270 K and 350 K at 1 bar pressure, as shown in Fig. 2 and denoted by "CF2 270 K" and "CF2 350 K", respectively. These correspond to the inner edge of the CF Hycean habitable zone around an M dwarf similar to K2-18, which for a 1 bar atmosphere is at 0.28 au,[54] *i.e.* corresponding to the runaway greenhouse limit for a steam-dominated lower atmosphere. We note that for these profiles a significant part of the atmosphere lies below freezing temperatures, making the presence of clouds/hazes and, hence, a high albedo arguably inevitable and inherently inconsistent with the assumption of a CF atmosphere. Nevertheless, it is still instructive to investigate the effect of photochemistry in such cool Hycean atmospheres. We consider a variant of the CF2 350 K profile, denoted as PT4, restricting the surface temperature to 310 K, which is intermediate between the two CF2 profiles, avoids steam domination, and is close to the other profiles shown in Fig. 2 at 1 bar, enabling a comparative study. Overall, we consider four *P–T* profiles (PT1–PT4) for our photochemical models as shown in Table 1.

We also use the atmospheric temperature profiles to define the outer boundary condition for the internal structure model as pursued previously for K2-18 b.[18] The internal structure model is described in Section 2. The goal of the internal structure modelling is to estimate the range of ocean depths possible in our canonical Hycean world, which is then used to estimate the inventory of bioessential elements required in the ocean. For this purpose, we use the hottest of the four PT profiles (PT1), which leads to somewhat deeper oceans[20] and therefore requires a higher nutrient budget, *i.e.* our conservative case. We consider two different surface pressures for the ocean and estimate the range of ocean depths possible given the 1σ bounds on the mass and radius of the planet. For a 1 bar surface, we obtain a minimum ocean depth of 120 km, for an upper bound mass for K2-18 b of 9.98 $M_\oplus$.[58,59] From the atmospheric *P–T* profile, the surface temperature in this case is 328 K. Conversely, for the 100 bar case, we take the lower mass limit of 7.28 $M_\oplus$, obtaining a maximum ocean depth of 370 km. This is consistent with previous ocean depth estimates for the corresponding surface gravity.[20] In this case the surface lies at 398 K. In both cases, the total H_2O mass fraction in the interior is adjusted to minimise/maximise the ocean depth, with an upper limit of 90%.[17] The H_2O adiabats, showing the temperature structures in the H_2O mantle, for both cases are shown in Fig. 2. Therefore, based on our above estimates we nominally consider the ocean depths of our canonical Hycean world to range between 100–400 km, which we use in Section 3.4.

3.2 Atmospheric composition

Hycean worlds are characterised by their bulk atmospheric composition being dominated by H_2. However, the remaining molecular inventory of a Hycean atmosphere would depend on a number of factors, including the temperature structure, metallicity (or more generally, the bulk elemental abundances), surface pressure, the incident flux from the host star, potential atmospheric escape, and potential chemical fluxes to and from the ocean surface beneath the atmosphere. Many of these factors are unknown for Hycean worlds, particularly the initial conditions set by planet formation and the boundary conditions set by the ocean surface, which makes an *ab initio* modeling of their atmospheric chemistry challenging. Some recent studies have investigated the atmospheric chemistry of temperate sub-Neptunes with thin H_2-rich atmospheres for different assumptions about the surface and atmospheric parameters, which are also relevant for Hycean worlds.[35–37] Here we focus on our canonical Hycean atmosphere with temperature structures as discussed above and explore the possible atmospheric compositions for different assumptions about the initial and boundary conditions.

As discussed in Section 2.2, we pursue a two-step approach. We first explore models assuming a chemically inactive (zero-flux) lower boundary to assess the

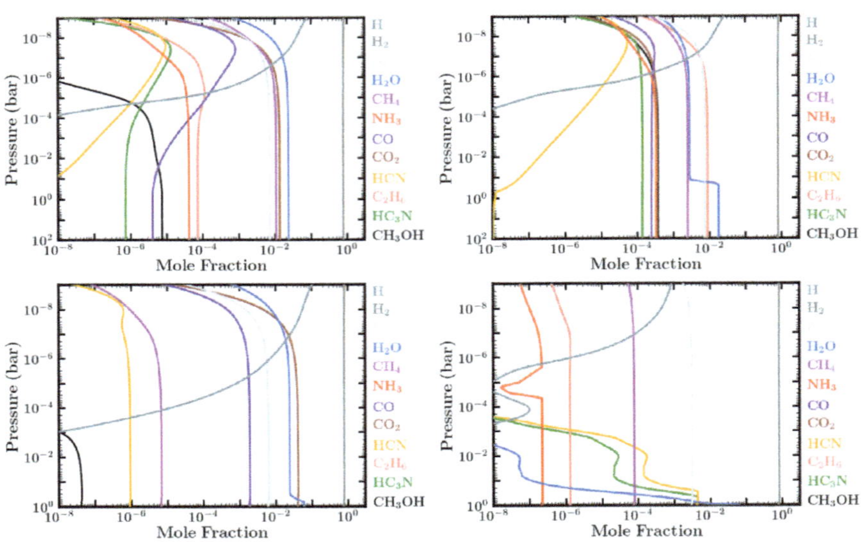

Fig. 3 Fully evolved (steady state) volume mixing-ratio profiles of prominent volatile species with zero-flux boundary conditions (inactive surface) for a 100× solar composition atmosphere and surface pressure of (top) 100 bar and (bottom) 1 bar: (top left) for the PT1 temperature profile in Fig. 2, corresponding to case 1 in Table 1; (top right) for the PT3 temperature profile, corresponding to case 3; (bottom left) for the PT1 temperature profile, corresponding to case 4; (bottom right) for the PT4 temperature profile, corresponding to case 6. Note that the sharp decline in H_2O, HCN, HC_3N, and NH_3 with altitude in case 6 is due to condensation.

sensitivity of the chemical processes to the assumed atmospheric properties, with no influence from the ocean, as pursued in recent work.[35,37] We then consider cases with fixed mixing ratio and deposition velocity boundary conditions for the ocean–atmosphere interface and assess their effect on the chemistry. We consider different types of boundary conditions based on expectations for a primordial atmosphere in thermochemical equilibrium or with significant CO_2 contribution from the ocean, as discussed in Section 2.2. We note that in this study we do not consider chemical fluxes resulting from potential life in the ocean which may significantly impact the atmospheric composition; we discuss this in Section 5.

The results for the photospheric (1 mbar) abundances of prominent molecules for the different model scenarios discussed in Section 2.2 are shown in Table 1. The vertical abundance profiles for some representative cases are shown in Fig. 3–5. Fig. 3 shows the abundance profiles for the models with zero-flux boundary conditions corresponding to cases 1, 3, 4 and 6 in Table 1. The abundances of the prominent molecules at 10 mbar and of other less abundant molecules are shown in the ESI.† In what follows, we present our findings on the prominent C, N, O molecules in the model atmospheres.

3.2.1 H_2 and H_2O. The most abundant molecules expected in a Hycean atmosphere are H_2, by definition, and H_2O, considering the ocean below. Even though H_2 itself does not have a strong spectral signature the H_2-rich nature of the atmosphere can be inferred *via* its low mean molecular weight, which in turn enables detections of other atmospheric constituents such as H_2O (*e.g.* ref. 18, 30 and 58). The H_2O abundance is governed by the temperature profile and atmospheric metallicity. For example, an atmospheric metallicity of 1× and 100× solar would imply an H_2O volume mixing ratio of ~0.1% and ~10%, respectively. If the abundance is above the local saturation vapor pressure, the excess H_2O can condense and precipitate out. Across our models with profiles PT1, PT2 and PT3, H_2O is abundant and is governed by saturation at the lower boundary. It is, therefore, generally expected that H_2O will be detectable in most Hycean atmospheres which are expected to be even warmer than K2-18 b.[17]

However, for the coolest Hycean worlds it is possible that the temperature at some point in the atmosphere is cold enough to lie within the ice sublimation

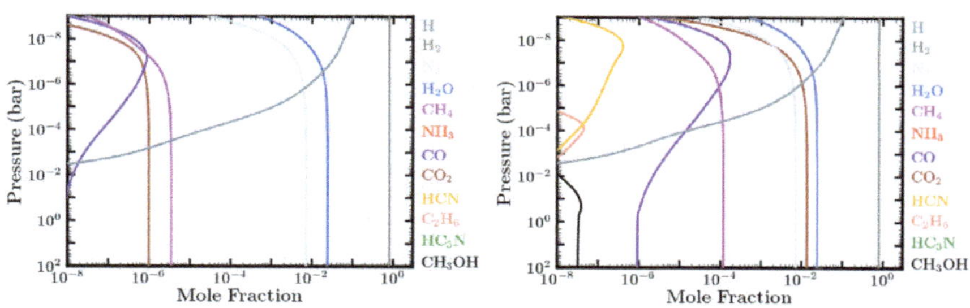

Fig. 4 Steady-state mixing-ratio profiles considering an atmosphere with lower boundary conditions of an active ocean surface with prescribed N_2, H_2O, and CO_2 mixing ratios and assuming the PT1 temperature profile in Fig. 2. Left: Abundances assuming a low CO_2 mixing ratio of 10^{-6} at the atmosphere–ocean boundary, corresponding to case 10 in Table 1. Right: Abundances assuming a high CO_2 mixing ratio of $\sim 1.4 \times 10^{-2}$ at the atmosphere–ocean boundary, corresponding to Case 9 in Table 1.

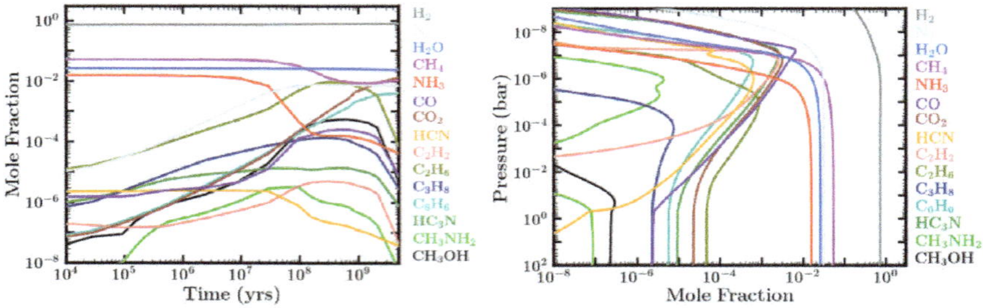

Fig. 5 Left: Evolution of atmospheric molecular composition at 0.01 bar for case 1 in Table 1. Right: Abundance profiles corresponding to Case 7 in Table 1, based on initial conditions of a reduced atmosphere in thermochemical equilibrium (similar to case 1) and fixed boundary conditions for H_2O, CH_4 and NH_3 (see Section 2.2). The abundances reflect the composition at 28 Myr, showing the availability of organic species early in the evolution.

regime causing H_2O to freeze out and further restricting the water vapor abundance at higher altitudes. This can be seen for our case 3 and case 6, using profiles PT3 and PT4, in Table 1 and in Fig. 3. This "cold trap" effect is seen both in the Earth's dry stratosphere and in the solar system giant planets where H_2O is negligible at the 0.1 bar level. In such cases it is possible that the H_2O may not be detectable in transmission spectra, which typically probe pressures below ~0.1 bar. Thus, whether H_2O is observable or not depends on the temperature structure of the atmosphere, which in turn depends on the various factors discussed above.

This potential cold-trapping effect is relevant to the ongoing debate regarding the observed transmission spectrum of the candidate Hycean planet K2-18 b, which has an insolation similar to that received by the Earth.[58,91] The transmission spectrum of the planet observed with HST has been explained by a degenerate set of solutions including H_2O and/or CH_4.[18,58,86,92] A non-detection or underabundance of H_2O in such an atmosphere could imply a cool and dry photosphere irrespective of the presence of an ocean underneath. This, in turn, could also imply the potential presence of H_2O clouds below the observable atmosphere. On the other hand, for hotter Hycean candidates an unambiguous detection of H_2O would be expected, as reported with the HST transmission spectrum of the planet TOI-270 d.[30] Thus, the detection or non-detection of H_2O in a Hycean atmosphere can place important constraints on the Bond albedo of the planet, which strongly dictates the temperature structure in the observable upper atmosphere. It is useful to note that besides the temperature profile, H_2O is relatively unaffected by other factors in a Hycean atmosphere, *e.g.* photochemistry does not significantly alter the H_2O abundance.

3.2.2 CH_4 and NH_3. In thick H_2-rich atmospheres, methane (CH_4) and ammonia (NH_3) are normally expected to be the dominant C and N bearing species under temperate conditions.[93,94,141] These molecules are ubiquitous in solar system giant planets. Both molecules are susceptible to photodissociation and other loss due to chemical kinetics in the observable atmosphere, but on giant planets are readily replenished due to their high abundances in the lower atmosphere, where higher temperatures maintain thermochemical equilibrium. However, the presence of a shallow surface boundary (whether ocean or solid) can significantly deplete the NH_3 and CH_4 abundances by curtailing their replenishment from the lower atmosphere.[35–37] In particular, we find that NH_3 is substantially depleted for surface pressures as high as 100 bar, even if no interaction with the surface occurs, as shown in Fig. 3 and Table 1. For comparison, for $100\times$ solar metallicity, the equilibrium abundance of NH_3 expected in our models is 1.5×10^{-2}. The lower the pressure at the surface, the greater the amount of NH_3 that is lost, as shown in Table 1 for cases 4–6. These findings are consistent with previous studies.[35,37]

NH_3 is also highly soluble in liquid water, thereby increasing its depletion potential due to dissolution in the Hycean ocean, which acts as a sink. This loss of NH_3 to the ocean is considered in our more realistic boundary condition models (see also ref. 36), where we find that the steady-state NH_3 mixing ratio becomes negligible; cases 9–11 in Table 1. Therefore, a substantial depletion or a lack of NH_3 may be considered as a characteristic signature of a Hycean atmosphere. We note, however, the possibility that the presence of life in a Hycean ocean may itself be a source of NH_3, as has been suggested previously for habitable rocky exoplanets with H_2-rich atmospheres.[95,96]

The abundance of CH_4 is less discriminating compared to NH_3. Even though CH_4 is also susceptible to photodissociation and other chemical loss, it is expected to be more abundant than NH_3 due to multiple factors. Firstly, considering cosmic abundances,[97] C is more than twice as abundant as N, and CH_4 is more thermochemically stable than NH_3 at the warmer temperatures near the ocean surfaces of Hycean planets. Secondly, in a H_2-rich atmosphere CH_4 is replenished by kinetic processes; that is, by photodissociation of other C bearing species such as CO and CO_2 in the upper atmosphere, and—more significantly—by recycling reactions deeper in the atmosphere, where higher pressures and temperatures allow photochemical products such as CO to be at least partially converted back to CH_4. We find significant CH_4 abundances for models evolved up to 5 Gyr for different surface pressures and metallicities, as shown in Table 1 and Fig. 3, consistent with previous studies.[35–37] We note that CH_4 is comparatively less abundant for lower metallicity and lower surface pressure, as also seen in previous work with zero-flux boundary conditions.[35]

Paradoxically, we find that warmer atmospheres in our studied range for 100 bar surface pressures retain the most CH_4, due to these aforementioned thermochemical recycling reactions being more efficient at higher temperatures. Atmospheres with ocean surfaces at lower pressures, on the other hand, lose more CH_4 as a result of missing out on these thermochemical reactions at greater depths, and they lose the methane on faster time scales. Depending on atmospheric temperatures, surface pressure, and planetary age, it is possible that CH_4 could be severely depleted and the C occupied primarily within CO_2 for atmospheres with high initial metallicities, shallow surfaces, older evolved compositions, or low carbon inventories (including low CO_2 dissolution scenarios, where CO_2 from the ocean is the only source of carbon). This latter scenario can be seen from our cases 10 and 11 in Table 1 and Fig. 4. Thus, a Hycean atmosphere might be expected to have observable amounts of CH_4 ($\gtrsim 10^{-4}$) under most conditions, but several factors can affect those expectations.

An important metric could be the relative abundance of CH_4 *vs.* NH_3. If both molecules are present in high abundances consistent with expectations from thermochemical equilibrium, a Hycean ocean may be confidently ruled out.

However, if NH_3 is absent or substantially depleted relative to CH_4 the presence of an ocean would be more likely.

3.2.3 Carbon oxides: CO and CO_2. In temperate and thick H_2-rich atmospheres, CO and CO_2 are generally expected through chemical disequilibrium processes and/or high metallicity. CO is expected through vertical mixing from a thick, deep atmosphere where high temperatures can thermochemically favor CO as the dominant C bearing species.[68] On the other hand, CO_2 is expected in the presence of high-metallicity. Both CO and CO_2 can also result from C based photochemistry in the presence of a high H_2O abundance. However, the presence of an ocean underneath a thin H_2-rich atmosphere, as in a Hycean planet, can significantly enhance the CO_2 abundance further due to release of dissolved CO_2 from the ocean.[36] Furthermore, over long timescales, as CH_4 is progressively photodissociated or otherwise chemically destroyed without replenishment from the lower atmosphere, CO_2 can become the dominant C bearing species in the atmosphere. We generally find a high CO_2 abundance across most of our cases, especially those with the warmer *P–T* profile (PT1) and 100× solar metallicity, as shown in Table 1 and Fig. 5. CO is generally less abundant than CO_2 with the exception of cases where the overall oxygen abundance is low, due to freeze out of H_2O in the atmosphere, as in case 6 for the coolest temperature profile we consider.

We note the important difference that the CO_2 abundance in our zero-flux boundary condition models is primarily due to chemical-kinetic processes in combination with initial metallicity assumptions, whereas that in cases 9, 10, & 11 and in the models of Hu *et al.*[36] is an input to the model based on the assumption for how much CO_2 is released from the ocean. In the Hu *et al.*[36] models, the abundance of atmospheric CO_2 is controlled by whatever fixed mixing ratio they adopt for CO_2 at the lower boundary, which they nominally assume to be between 4×10^{-4} and 0.1; their estimate for the lower-bound ranges between 5×10^{-5} and 7×10^{-4} bar partial pressure of CO_2. We also examine cases where the only carbon present in the atmosphere is that released through equilibrium with CO_2 dissolved in the ocean, with an assumed CO_2 mixing ratio at the ocean surface of 10^{-6} (*i.e.*, CO_2 partial pressure of 10^{-4} bar for our 100 bar atmosphere case), ~1.4%, or 10%. These correspond to cases 9–11 in Table 1 and Fig. 4.

As with the Hu *et al.*[36] study, we find that adopting fixed mixing-ratio lower-boundary conditions for CO_2, N_2, and H_2O in a canonical ocean model results in an atmosphere with much larger CO_2/CO ratios than for otherwise similar planets without ocean interaction. We agree that the CO_2/CO ratio could be a useful indicator to distinguish between high-metallicity deep atmospheres and planets with shallower atmospheres and liquid-water oceans. However, our high CO_2 ocean-release cases produce much less CH_4 than is derived for the Hu *et al.*[36] models, which end up with more CH_4 than CO_2, despite CO_2 being the only incoming source of carbon in the model.

While the presence of substantial atmospheric CO_2 and high CO_2/CO ratios may be strong indicators of a Hycean ocean, the absence of CO_2—while less likely on a warm Hycean planet—might also be less informative. For example, following the approach of Hu *et al.*[36] a low CO_2 abundance could indicate (a) a high efficiency of dissolution of CO_2 in the ocean and less availability in the atmosphere, or (b) more C sequestered in the core, making it less available in the ocean, or a generally low C content in the planetary interior as a whole. Secondly, the abundance of CO_2 in the atmosphere can also depend on the initial availability of CH_4 and H_2O in the atmosphere through the $CH_4 + H_2O \rightarrow CO_2$ conversion process (*i.e.*, assuming that the primordial atmosphere contained CH_4 not in immediate chemical equilibrium with the liquid ocean, such that the oceanic CO_2 does not control the entire atmospheric carbon budget). For cooler Hycean planets where H_2O can be frozen out in the observable atmosphere, CO_2 could also be underabundant from this process and hence less observable. Finally, it is also possible that microorganisms in a Hycean ocean could efficiently use dissolved CO_2 for biological processes and subsequently release other gases such as CH_4 or O_2. Microbial methanogenesis is known to be a significant source of CO_2 consumption in the Earth's oceans[98–100] and it is not inconceivable that life in a Hycean ocean may find efficient ways to do the same.

3.2.4 Other molecules. Most of our models contain moderate (typically ppb to ppm) levels of HCN, but because HCN is produced from photochemistry of both N_2 and NH_3 in the presence of other carbon-bearing molecules, it does not seem to be a good indicator of the presence/absence of an ocean. Tsai *et al.*[37] also emphasize that HCN is very sensitive to stellar properties and the strength of atmospheric mixing, so HCN becomes less reliable than NH_3 as an indicator of the presence of a surface. We do find that HCN is more depleted when a sink at the oceanic lower boundary is included, as in cases 9–11.

Similarly, CH_3OH becomes very abundant in some of our zero-flux boundary condition cases (*e.g.* cases 2 & 3 in Table 1 and Fig. 3) but is more depleted in our models with ocean interaction, due to dissolution in the ocean. In that way, CH_3OH can become a good discriminator of ocean *versus* solid surfaces, as suggested originally by Tsai *et al.*[37] Note also that the CH_3OH abundance in our zero-flux models has a high sensitivity to atmospheric temperatures, being less abundant in some atmospheres because it becomes a casualty of the more efficient conversion of CO into CH_4 at depth in warmer atmospheres (as in cases 1 and 4, see *e.g.* Fig. 3), or becomes a casualty of the low atmospheric oxygen abundance that results from H_2O freeze out in colder atmospheres (as in case 6, *e.g.*, Fig. 3).

With our cooler zero-flux models with a 100 bar surface, we find that C_2H_6 could also be an excellent indicator of a dry, unreactive planetary surface on a sub-Neptune planet. This can be seen in Fig. 3. If H_2O is moderately depleted through condensation/sublimation near a tropopause cold trap (*e.g.*, with our PT2 & PT3 profiles, cases 2 and 3), such that H_2O has a reduced abundance in the middle and upper atmosphere, conversion of CH_4 into CO and CO_2 becomes less effective than conversion of CH_4 into heavier hydrocarbons. In that situation, the overall photochemistry becomes more like that of Jupiter, except the hydrocarbons are not recycled efficiently back to CH_4 at depth. The eventual steady-state abundance of photochemically stable hydrocarbons such as C_2H_6 and other alkanes (*e.g.*, C_3H_8 and C_4H_{10}) becomes very large in cases with no surface loss, such that C_2H_6 actually contains the bulk of the carbon in the atmosphere. We are also finding a significant C_6H_6 abundance in our zero-flux models, but that result seems less robust, as we end the carbon chemistry at C_6H_6 and do not consider further loss of benzene to PAHs and other heavy organics. In our ocean models with fixed CO_2, N_2, and H_2O boundary conditions, we assume a moderate deposition velocity loss of C_2H_6 through the ocean boundary, and the steady-state C_2H_6 mixing ratio is less dramatic. Therefore, C_2H_6 should provide a better observational marker than

CH_3OH to distinguish ocean planets from solid-surface planets on sub-Neptunes that are cold enough for significant atmospheric water depletion. Another species of potential note is HC_3N, which becomes relatively abundant for some of our models.

3.3 Prebiotic molecular inventory

The molecular inventory of a habitable planet in the early stages of its evolution is expected to play an important role in its propensity for seeding life. On the Earth, key pre-biotic molecules such as HCN are thought to have played a pivotal role in abiogenesis.[52] Here, we investigate the molecular abundances over the lifetime of our canonical Hycean planet. At the outset, we find that most of the models considered undergo a phase that is rich in organic molecules early in the evolution, typically within the first ~1–300 Myr. The relative abundances of the different molecules depend on the boundary conditions and other factors discussed above. However, a common outcome in our zero-flux models is the prevalence of important organics such as CH_4, C_2H_6, C_6H_6, and nitriles over the early evolutionary history, as shown in Fig. 5, left panel.

We find that our models with fixed boundary conditions corresponding to an initially reduced atmosphere result in abundant organics starting very early in the evolution. We explored this in cases 7 and 8 in Table 1; the abundance profiles for case 7 are shown in Fig. 5, right panel. We find that hydrocarbons, nitriles, alcohols, and other organics can become extremely abundant in the first few million years of atmospheric evolution, if we assume the original starting composition were reduced (*e.g.*, CH_4, NH_3, and H_2O being the main carriers of C, N, and O, as would be expected for a cool H_2-rich accretion atmosphere in thermochemical equilibrium in the gaseous envelope, but not the ocean); see also ref. 35 and 37. High abundances of C_2H_6, C_3H_8, C_4H_{10}, C_6H_6, C_2H_2, HCN, H_3CN, CH_3NH_2, CH_3CN, and CH_3OH are particularly worth mentioning from our model results. While it may be unlikely that we manage to catch this early evolutionary stage in observations of Hycean planets, these photochemically produced molecules could provide pre-biotic molecules to the ocean. The abundance of HCN, an important prebiotic molecule on Earth,[52] is strongly correlated with the presence of NH_3 in the atmosphere. As NH_3 is photochemically destroyed over time, and is readily soluble in the ocean, most of the N is locked in N_2 thereby reducing the abundance of HCN at later times. However, other hydrocarbons with lesser or no dependence on N continue to be abundant. Heavier organic species are present but not particularly abundant in our final steady-state atmospheres with ocean interaction included, due to relatively low abundances of the primary parent molecule CH_4. However, if CH_4 continues to be abundant in the present-day atmosphere due to the reasons described above, a rich organic molecular inventory may continue as well.

3.4 Availability of bioessential elements

A central challenge to the habitability of Hycean worlds, as for all ocean worlds, is the availability of bioessential elements. Bioessential elements include the common volatile elements (CHNOPS) as well as essential metals such as Fe, Na, K, Mg, Mn, Ni, Ca, Cu, Zn, Co, and Mo,[39,40,46,101] which perform key functions in biochemical processes in terrestrial life. The challenge lies in the fact that ocean worlds would not have ready access to these elements which are typically acquired through interaction between the hydrological cycle and the silicate portion of the planet. This problem is particularly acute for planets with deep oceans where high-pressure ice at the bottom of the liquid water column can preclude access to elements from the rocky core. This challenge is well recognised both for volatile-rich sub-Neptunes[39,46] as well as for icy moons in the solar system.[40] For planets with H_2O-rich interiors and H_2-rich atmospheres, including Hycean worlds, H and O are naturally present in abundance while C and N are also found in large quantities in volatile ices (*e.g.* Rubin *et al.*[102]) and so could be present in the required abundances. However, this is not necessarily the case for P, S, and refractory metals. Here we assess possible sources of these elements for our canonical Hycean world.

3.4.1 Mass requirements. We first estimate the elemental inventory required for life in a canonical Hycean ocean assuming similar concentrations as those found in the Early Earth oceans.[101,103–105] We consider ocean depths of 100–400 km for a canonical Hycean world similar to K2-18 b, following Section 3.1, with $R_p = 2.61\ R_\oplus$. We note that here we are primarily concerned with the amount of nutrients required to seed life on the canonical Hycean as in the oceans of the Early Earth. Based on estimates of the elemental concentrations (in mol L^{-1}) in the Early Earth oceans and the volume of our canonical Hycean ocean, we calculate the total elemental mass requirements in the Hycean ocean such that elemental concentrations in these two oceans match.

We use Robbins *et al.*[101] estimates of Archean ocean metal concentrations. These originate from geochemical modelling by Saito *et al.*,[106] except for the Mo values which are adopted from Anbar and Knoll.[107] We take Jones *et al.*[105] and Crowe *et al.*[104] values for the P and S concentrations respectively. These estimates are obtained through a variety of methods. The Saito *et al.*[106] results come from geochemical modelling of the concentrations of various metal species under anoxic and iron-rich conditions representative of Archean oceans.[101,103,106,108,109] The modelling finds that the relative availability of various metals follows Fe > Ni,Mn,Co ≫ Zn,Cu. It is assumed that there is sufficient nutrient influx (*e.g.* from weathering) for the actual Earth ocean concentrations to match these modelling estimates. The P, S, and Mo values, in contrast, are taken from proxy sedimentary records. This can offer a more nuanced view of changes over time but can be limited by imperfect sorption models and post-deposition alteration of the sediments.[101,110,111]

A proportion of the elemental mass budget will be present in a precipitate instead of a dissolved form. Since the former will not be bioavailable, we adopt a dissolved fraction[104–107] to scale our Hycean mass budget for each element. This increases the required mass budget for each element. We note that we have not modelled self-consistent Hycean ocean chemistry or any kind of (geo)chemical cycle and so there is still significant uncertainty in this dissolved fraction. The Hycean ocean considered here would be warmer than the Earth ocean, meaning that the solubilities of various species would be different than on Earth. To reflect these unknowns, we introduce uncertainties of 1 dex in the primitive ocean concentrations, except for Mo for which we consider an increased uncertainty of 1.5 dex as its Archean abundance is relatively less accessible. The mass requirements are shown in Table 2 and Fig. 6.

Table 2 Estimates of element masses required to provide the necessary concentrations in our canonical Hycean ocean. C_{PO} is the estimated concentration of a given element in Earth's primeval ocean,[101] shown in log(moles/litre), and M_{req} is the median estimate of the element mass (in kg) required, assuming a median ocean depth of 200 km. The estimated range is shown in Fig. 6. f_{CC} refers to our adopted mass fraction of the element in carbonaceous chondrites, and F_{diss} is the solubility fraction of the element dissolved in Archean ocean conditions on Earth.[106,107] M_{imp} is the median estimate for the mass (in kg) of that element delivered from asteroid impacts

Element	C_{PO}	M_{req}	f_{CC}	F_{diss}	M_{imp}
Fe	−4.6	3.2×10^{18}	2.0×10^{-1}	3.0×10^{-1}	4.0×10^{17}
Ni	−8.2	8.5×10^{14}	1.3×10^{-2}	3.0×10^{-1}	2.6×10^{17}
Mn	−8.3	3.2×10^{14}	1.6×10^{-3}	6.0×10^{-1}	3.2×10^{16}
Co	−8.1	4.6×10^{14}	6.0×10^{-4}	7.0×10^{-1}	1.2×10^{16}
Mo	−9.0	2.1×10^{15}	1.5×10^{-6}	3.0×10^{-2}	3.0×10^{13}
Zn	−14.0	4.5×10^{15}	2.0×10^{-4}	1.0×10^{-7}	4.0×10^{15}
Cu	−22.0	4.4×10^{14}	1.0×10^{-4}	1.0×10^{-14}	2.0×10^{15}
P	−6.5	6.9×10^{15}	1.0×10^{-3}	9.8×10^{-1}	2.0×10^{16}
S	−5.0	5.7×10^{18}	1.0×10^{-2}	4.0×10^{-2}	2.0×10^{17}

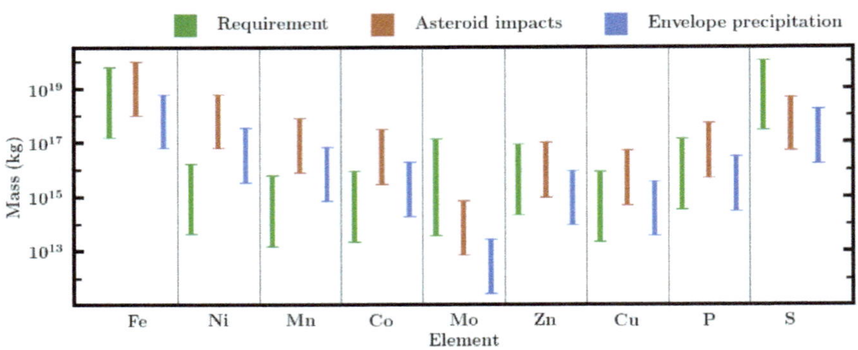

Fig. 6 Inventory of prominent bioessential elements. The estimated masses of the elements required for our canonical ocean are shown in green, along with the masses that can be available from impacts (brown) and from atmospheric settling (blue). See Section 3.4.

3.4.2 External delivery. One of the most common sources of elemental delivery in planetary environments is through external impacts by asteroids over the early history of the planet. Considering representative impactors on the Hadean Earth, we estimate the concentrations of nutrients that can be delivered to the Hycean ocean. We assume a composition of carbonaceous chondrites,[112] which are somewhat lower in metal content relative to other asteroids. For the impacting masses, we take the minimum estimate of post Moon-formation impacts,[113] itself derived from the crater size distribution on the Moon. This gives a total asteroid mass of 2×10^{21} kg which impacted the Earth, which we set as the total asteroid mass impacting the Hycean world. The mass distribution is weighted towards larger objects, $M(<m) \propto m^{0.5}$.[113] While the estimates for the total mass delivered to Earth in this Late Heavy Bombardment/Late Veneer scenario vary (as does the precise timing of these events, although this is not relevant to us), our value is on the lower end of usual estimates.[53,114]

Only a small fraction of the 2×10^{21} kg will end up dissolved in the Hycean oceans. Large impacts can travel through the ocean and deliver mass to the ice layers or even the core. Hydrodynamic modelling of asteroid impacts into Earth's oceans[115–117] suggests impactors will form impact craters if the ocean depth is less than 5–7 times the impactor diameter. If this factor applies to a 200 km deep Hycean ocean, only objects with diameters below 29 km would break apart completely in the ocean. Considering a chondritic asteroid density of 1600 kg m^{-3} [118] this would mean only objects lighter than $\sim 2 \times 10^{16}$ kg, and so only ∼0.5% of the total mass would be in impactors small enough to completely break apart in the ocean. Larger collisions that do reach a solid surface will throw up material back into the atmosphere, which will rain back down as small particles.[53,119] On the Earth, large collisions excavate material with a total mass nearly as large as the impacting mass,[120] significantly increasing the amount of mass deposited on the planet surface.

However, as little work on large impacts in deep oceans has been done, we are cautious about this additional amount of mass delivered to the oceans. We assume that between 0.25% and 25% of the total impact mass stays in the oceans, with a median of 1% or 2×10^{19} kg, and provides the required source of bioessential elements for seeding life. This fraction will change depending both on ocean depth and the precise physics of impact. The mass requirements and asteroid nutrient delivery rates are shown in Table 2 as well as Fig. 6. We find that in general asteroid impacts can plausibly deliver enough mass for the bioessential elements with the possible exception of molybdenum and sulphur.

Another avenue for nutrient delivery is through steady state accretion of extraterrestrial dust. The possibility of extraterrestrial dust being a sustained source of bioessential nutrients has been considered in the context of both the Earth[121] and sub-Neptunes.[46] For example, in the context of the Earth, the amount of Fe obtained through this source is estimated to be 4×10^6 kg per year,[122] compared to a global phytoplankton Fe assimilation of 7×10^8 kg per year.[123] So this flux would only be able to support a small biosphere by itself, and it would take of order 10^{12} years for it to raise the overall ocean Fe budget to the required $\sim 10^{18}$ kg. While it may help life continue once it has started, it may not be a substantial contributing factor to the start of life, unless the dust accretion rates are significantly higher at early times. On the other hand, depending on the system parameters the impactor and dust influx on Hycean worlds may be significantly higher than assumed here based on Earth values. Future work can investigate if the larger planet sizes of Hycean worlds and their closer proximity to their host stars, among other system properties, may enhance the mass influx rate.

3.4.3 Atmospheric condensation. Another important source of nutrients in a Hycean world would be the initial atmospheric composition. Depending on the formation scenario, the primordial hydrogen-rich atmosphere of a Hycean planet can contain a significant abundance of metals reflecting their proportions in the H-rich nebula. For example, a solar composition gas contains 0.13% by mass in Fe, 0.007% in Ni, *etc.*,[112] as shown in Table 3. However, this abundance is not a given considering that the planet could form outside the metal condensation fronts in the protoplanetary disk and accrete metal-poor gas. On the other hand, if the planet forms closer in or accretes significant solids along with the volatiles during formation it is conceivable that an equivalent amount of metals to a nebular metallicity gas, or even higher metallicity, may be accreted and

Table 3 Estimation of mass budget of selected elements contained in the atmosphere. X/H is the number fraction of the element relative to H in solar abundance.[112] f_{atm} is the atmospheric mass fraction in a given element for a solar composition atmosphere. M_{cond} is the median estimate mass (in kg) of the element delivered to the ocean; the estimated range is shown in Fig. 6

Element	X/H	f_{atm}	M_{req}	M_{cond}
Fe	3.0×10^{-5}	1.3×10^{-3}	3.2×10^{18}	6.6×10^{17}
Ni	1.6×10^{-6}	7.2×10^{-5}	8.5×10^{14}	3.7×10^{16}
Mn	3.3×10^{-7}	1.1×10^{-5}	3.2×10^{14}	7.1×10^{15}
Co	8.5×10^{-8}	4.2×10^{-6}	4.6×10^{14}	2.0×10^{15}
Mo	7.6×10^{-11}	5.4×10^{-9}	2.1×10^{15}	2.9×10^{12}
Zn	3.6×10^{-8}	1.7×10^{-6}	4.5×10^{15}	9.3×10^{14}
Cu	1.5×10^{-8}	7.2×10^{-7}	4.4×10^{15}	3.8×10^{14}
P	2.6×10^{-7}	5.9×10^{-6}	6.9×10^{15}	3.1×10^{15}
S	1.3×10^{-5}	3.2×10^{-4}	5.7×10^{18}	1.7×10^{16}

accessible in the volatile layer. If initially present in the atmosphere, these metals would eventually condense out given the temperate atmosphere of a Hycean world. We estimate whether such a scenario could meet the bioessential metal requirement in a Hycean ocean.

We consider a nominal atmospheric mass of our canonical Hycean planet to range between 10^{-6} and 10^{-4} M_p, informed by our internal structure model and previous estimates for K2-18 b.[18] Considering a solar composition for the atmosphere, we estimate the elemental budget of key metal species, as shown in Table 3. For P and S, which are not expected to condense out in the same fashion, we take a Neptune-like 100× solar metallicity atmosphere and assume that 1% of the atmospheric species end up in the ocean, very likely an underestimate.[36] We find that condensation in our primordial atmosphere provides comparable elemental inventory to the ocean requirements. Fig. 6 shows these estimates alongside the masses of required nutrients and possible asteroid impacts.

3.4.4 Convective transport from core. The lack of an ocean–rock boundary, preventing the dissolution of key nutrients and minerals, has been a longstanding objection to life both on ocean exoplanets and larger icy moons such as Ganymede.[42–44] However, evidence is building that even on planets with thick high-pressure ice layers, there may still be transport within the ice layer through convection.[47–49,51] Hernandez *et al.*[50] found that ice VII could transport up to 2.5% NaCl in its crystalline structure, although it is not clear what the situation for other nutrients would be. A number of unknowns prevent us from estimating the nutrient mass budget in the ocean possible through this mechanism. In particular, we expect ice X near the Hycean core boundary, and its behaviour is even less understood than ice VII. However, some form of nutrient transfer from the core to the ocean may be plausible.

4 Observational prospects with JWST

A central promise of Hycean planets is their accessibility to atmospheric characterisation with current facilities, particularly the JWST.[17] Their low gravities and low atmospheric mean molecular weight make Hycean worlds more conducive to atmospheric spectroscopy compared to rocky exoplanets of comparable mass and temperature. Several recent studies have demonstrated the feasibility of detecting key molecular signatures, including biomarkers, in candidate Hycean worlds using transit spectroscopy with JWST.[17,34,124] Our goal here is to assess (a) the spectral features of relevant chemical species observable with JWST transit spectroscopy, and (b) chemical diagnostics for identifying a Hycean planet.

4.1 Atmospheric spectroscopy with JWST

The high sensitivity and broad spectral range of JWST provide a promising avenue to detect key molecular species expected in Hycean atmospheres using transit spectroscopy. Here we assess the spectral features of the molecules predicted in our photochemical calculations, in Section 3, that may be observable with JWST. We model a transmission spectrum of K2-18 b using an adaptation of the AURA exoplanet atmospheric modeling and retrieval code.[125] We refer the reader to previous works[17,124] for details on the model setup using the AURA code and for the same planet bulk parameters as considered here. The model considers a plane parallel atmosphere at the day–night terminator of the planet in hydrostatic equilibrium and computes the transmission spectrum for a given temperature structure and composition. Besides the molecules considered in previous works using the AURA model,[17,124,125] we consider three additional molecules C_2H_6,[126–128] HC_3N,[126,127,129] and CH_3OH.[126,127,130,131] As seen in Section 3, the atmospheric composition can vary widely depending on a number of factors, including the temperature structure, atmospheric metallicity, ocean surface pressure, and ocean–atmosphere interactions. We consider an isothermal temperature structure at 300 K and uniform volume mixing ratios of 10^{-2} for H_2O, 10^{-4} for CH_4, and 10^{-5} for all other species. While these values lie within the wide range of abundances predicted by our chemical models, we note that some of the species may be significantly less or more abundant in a Hycean atmosphere than assumed here for illustration.

The model transmission spectrum along with the relative contributions of the molecules and synthetic JWST observations are shown in Fig. 7. Following

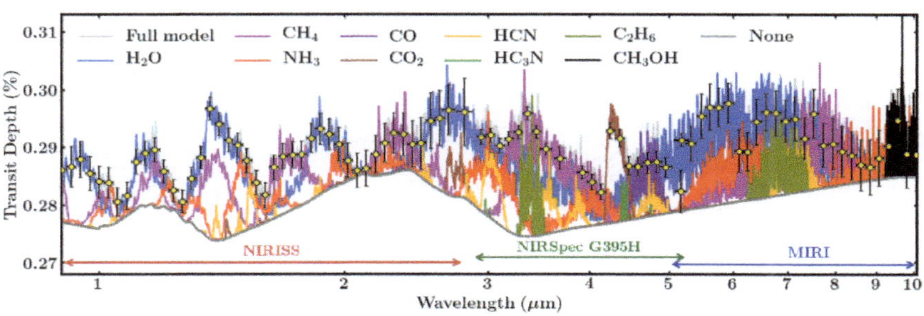

Fig. 7 Model transmission spectrum of our canonical Hycean world K2-18 b. The model spectrum with all the molecules included is shown in grey. The spectra with the individual contributions of the different molecular species are shown as denoted in the legend. The synthetic JWST observations corresponding to the full model are shown as the gold-centered black points with error bars binned to a resolution of $R = 40$. The observations include one transit with NIRISS SOSS, three transits with NIRSpec G395H and two transits with MIRI LRS, as discussed in Section 4.

previous work,[17] we assume 1 transit with JWST NIRISS[132] and 3 transits with NIRSpec G395H,[133,134] and an additional 2 transits with the MIRI[135] spectrograph to obtain a broad 1–10 μm spectral coverage. The simulated spectra were generated using Pandexo.[136] The resultant 6 transits of K2-18 b can be observed in ∼50 hours with JWST. As can be seen in Fig. 7, and consistent with previous studies,[17,36,37,124] the prominent molecules H_2O, CH_4, NH_3, CO_2 and CO have strong spectral features across the 1–10 μm range and are readily discernible with the expected JWST data quality if present at the atmospheric abundances assumed in the model. Conversely, if any of them are underabundant then robust upper-limits can be placed on their abundances. For example, simulated retrieval studies[17,124] have shown that the abundances of H_2O, CH_4, and NH_3 can be retrieved to precision better than 0.3 dex with JWST quality data. Among the minor species, HCN and C_2H_6 have the stronger features, whereas CH_3OH and HC_3N would be more challenging to detect, as previously reported for CH_3OH.[37]

4.2 Chemical diagnostics of a Hycean world

As discussed in Section 1, chemical characterisation of the atmosphere is essential to identify a Hycean world. The bulk properties of a candidate Hycean world, *i.e.* the mass, radius, and temperature, can generally be consistent with

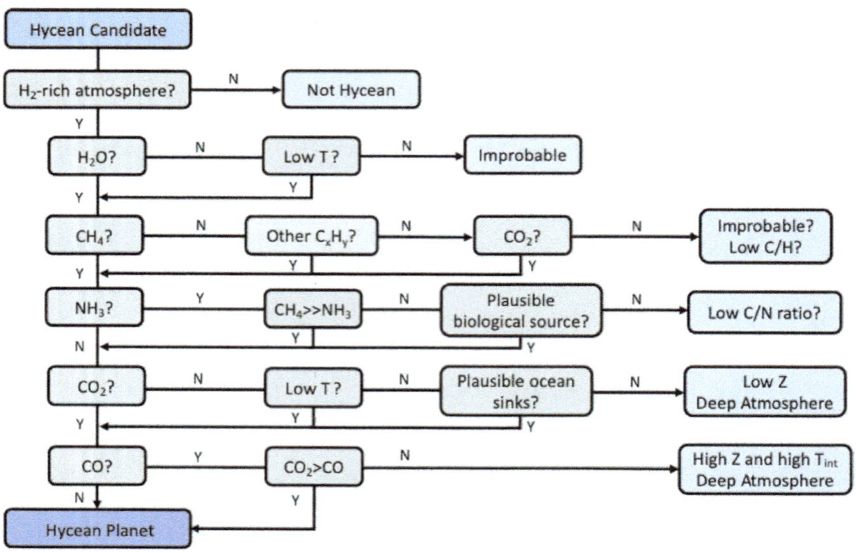

Fig. 8 Chemical diagnostics for Hycean atmospheres based on our photochemical models and motivated by recent works.[35–37] The relevance of each prominent molecule for the identification of a Hycean atmosphere is shown. For each molecule, the implications of its detection or non-detection are indicated, considering a minimal detectability threshold of 1 ppm abundance. Low T refers to cases where the atmospheric temperature is expected to be cold enough for H_2O condensation. C_xH_y refers to hydrocarbons. Z refers to atmospheric metallicity in a deep H_2-rich atmosphere, and T_{int} refers to the internal temperature. The pathways reflect the results of our photochemical calculations across the range of Hycean conditions considered and are generally consistent with previous works.[35–37] See Section 4.

a degenerate set of internal structures,[18] as shown in Fig. 1. Depending on the specific bulk properties, the possibilities could include (a) a Hycean world with a shallow H_2-rich atmosphere overlying a habitable ocean, (b) a sub-Neptune with a shallow H_2-rich atmosphere but with a solid surface underneath, or (c) a sub-Neptune with a deep H_2-rich atmosphere with temperatures and pressures at the bottom of the H_2 layer too high to be habitable. Several recent studies have investigated the possibility of atmospheric chemical signatures to identify the presence of solid or ocean surfaces in sub-Neptunes,[35–37] considering the possibility of a shallow H_2-rich atmosphere[18] in K2-18 b as a canonical example. Here we revisit this avenue based on the chemical calculations in the present work and motivated by previous works.[35–37] A flowchart of the chemical diagnostics of a Hycean world is shown in Fig. 8.

Given a candidate Hycean world,[17] the first requirement is to confirm the presence of an H_2-rich atmosphere using spectroscopic observations. An H_2-rich atmosphere is usually inferred indirectly through the detection of a spectral feature of one of the prominent molecules, *e.g.* H_2O, the spectral amplitude of which provides a constraint on the mean molecular mass of the atmosphere.[30,58] As discussed above, JWST will be able to robustly detect spectral features and constrain H_2-rich atmospheres for several nearby Hycean planets.

H_2O is naturally expected to be the dominant molecule observable in a Hycean atmosphere. However, if the temperature structure is too cool in the observable photosphere, H_2O can be depleted due to condensation, as in the Earth's stratosphere. As discussed in Section 3, for some of the *P–T* profiles of K2-18 b considered, H_2O can indeed be underabundant. On the other hand, warmer Hycean candidates would be expected to have discernible H_2O features.

The abundance of CH_4 in a Hycean atmosphere depends strongly on the boundary conditions. While CH_4 is susceptible to loss from photochemical processes, especially for low ocean surface pressures (*e.g.* 1 bar), it can still be present in significant abundances due to other production pathways. The underabundance of CH_4 would normally indicate the C present in other molecules such as CO_2 or, in the case of low-temperature Hyceans, higher-order hydrocarbons. The simultaneous depletion of all these molecules would be unlikely for a Hycean atmosphere, and may instead indicate a deep atmosphere with an unusually low C/H ratio.

A key predictor of a Hycean atmosphere is the NH_3 abundance. Our results confirm previous findings[36,37] that NH_3 is expected to be significantly underabundant in a Hycean atmosphere due to both its photochemical destruction in a shallow atmosphere and its high solubility in the ocean underneath. NH_3 is also expected to be significantly underabundant compared to CH_4, making a high CH_4/NH_3 ratio a useful diagnostic of a Hycean atmosphere. Conversely, a significant abundance of NH_3 in a Hycean candidate would need to be investigated for its potential biological origins[95] which are not accounted for in our models. In the absence of such a scenario, a low CH_4/NH_3 ratio may be indicative of a deep atmosphere with a low C/N ratio.

A significant abundance of CO_2 is another strong predictor of a Hycean atmosphere, consistent with previous studies.[36] However, we argue that the underabundance of CO_2 does not robustly rule out a Hycean candidate, as it can occur due to various factors, including a cold and dry atmosphere, low C content in the interior, high dissolution in the ocean and/or consumption by marine biota.[99,100] The lack of CO_2, in the presence of CH_4 and H_2O, would also be consistent with a deep atmosphere with a low metallicity.

Finally, while the CO abundance on its own is not a critical diagnostic, the CO_2/CO ratio is expected to be higher for a Hycean atmosphere compared to that expected in the absence of an ocean (see also ref. 36). Conversely, a CO/CO_2 ratio above unity would indicate a deep atmosphere with a high metallicity and high internal temperature as suggested previously.[36]

In summary, considering the major molecules, the key diagnostics of a candidate Hycean world would be (a) abundance of H_2O, CH_4, other hydrocarbons and/or CO_2, (b) underabundance of NH_3, particularly a high CH_4/NH_3 ratio, and (c) $CO_2/CO > 1$. Other molecules can also be abundant depending on the specific atmospheric conditions. HCN and CH_3OH are expected to be underabundant in Hycean atmospheres, as also found in previous studies.[37] In particular, HCN can be underabundant due to the absence of NH_3 and CH_3OH can be depleted due to dissolution in the ocean. We also find that other hydrocarbons such as C_2H_6 can be abundant for cool Hycean worlds where H_2O may be moderately depleted in the observable atmosphere through condensation; in this case CH_4 photodestruction leads primarily to other hydrocarbons rather than CO and CO_2.

Therefore, robustly establishing the presence of an ocean on a candidate Hycean world would require precise abundance estimates of multiple molecules discussed above and using their relative abundances to systematically constrain the key chemical pathways given the bulk properties of the planet in question. Previous work has also suggested other organic molecules as potentially observable biomarkers in Hycean worlds.[17] These include molecules such as methyl chloride, dimethyl sulphide, carbonyl sulphide, nitrous oxide or carbon sulphide, which are produced in relatively low quantities by life on Earth but can be promising biomarkers in H_2-rich atmospheres, depending on the biomass present and UV environment.[95]

5 Summary and discussion

Hycean worlds open a new paradigm in the search for life on exoplanets. Hycean planets are expected to be more detectable in transits and more amenable to atmospheric characterisation using transit spectroscopy than habitable rocky exoplanets. Previous work has shown that the thermodynamic conditions in the atmospheres and oceans of Hycean worlds can be similar to those known to be capable of hosting life in Earth's oceans.[17] In the present work, we investigate the chemical conditions possible on a canonical Hycean world, focusing on the present and prebiotic molecular inventory in the atmosphere and the availability of bioessential elements in the ocean. The observable atmospheric composition of a Hycean planet depends on a number of factors, including the temperature structure, atmospheric metallicity, ocean surface pressure, and ocean–atmosphere interactions. However, based on the results of several recent works[17,35–37] and the present study, the atmospheric compositions of Hycean worlds may be expected to include a range of prominent molecules that are detectable with modest investment of JWST time. In particular, the relative proportions of some of the key molecules, *e.g.* a high CH_4/NH_3 and/or a high CO_2/CO, could help distinguish a Hycean world from a non-habitable mini-Neptune or super-Earth with a thicker H_2-rich atmosphere.

5.1 Chemical inventory for life

We also investigate the question of whether a Hycean world would be conducive for the origin and sustenance of life. A number of factors affect the habitability of a planet, including its geophysical conditions, evolutionary history, architecture of the planetary system and astrophysical conditions.[137] From an ecological perspective, it is commonly agreed that life as we know on Earth has three essential requirements:[38,39] (a) energy sources, radiative and/or chemical, (b) liquid water, and (c) various bio-essential elements. In addition to these requirements, life also needs the right environmental conditions, *e.g.* temperature, pressure, pH, and UV radiation.

At the outset, by definition, the thermodynamic conditions (pressure and temperature) in a Hycean ocean would not be dissimilar to those experienced by microbial life in Earth's oceans.[17] The requirements of an energy source and presence of liquid water are also naturally met, more so than most solar system environments for which a liquid water surface is a key limitation.[39,40] The energy source includes both the radiant energy from the star as well as the chemical free energy from chemical reactions, *e.g.* between abundant H_2 and CO_2 in the ocean. In particular, as discussed above, the presence of abundant H_2 may be more conducive to processes such as methanogenesis that are prevalent in microbial communities in anoxic environments on Earth.[99,100] Previous studies have also discussed the feasibility of life in H_2-rich environments, as also seen on Earth.[95,138]

Considering atmospheric evolution starting from reduced primordial conditions the atmosphere undergoes a phase rich in organic molecules potentially conducive for prebiotic chemistry. In particular, several hydrocarbons and nitriles including C_2H_2, C_2H_6, C_3H_8, C_6H_6, HCN, and HC_3N become abundant (above $\sim$1 ppm) starting between 1–100 Myr with some remaining abundant for over a Gyr. We note however that the abundances of these molecules are strongly dependent on the initial conditions and boundary conditions describing the ocean–atmosphere interactions. Future studies can investigate the timescales on which the primordial ocean can equilibrate with a H_2-rich atmosphere[139,140] and how it would influence the primordial composition and chemical evolution of the atmosphere.

We investigated the availability of bioessential nutrients in a canonical Hycean ocean building on elemental estimates for oceans of the early Earth.[101,106] Life on Earth requires a range of bioessential elements, including key volatile elements (CHNOPS) and heavier metallic elements. While C, H, N, O and some heavier elements such as Na and K are naturally expected to be abundant in H_2-rich atmospheres,[46,93,94,141] the availability of P, S, and refractory metals is a bigger challenge. In particular, for ocean worlds, the presence of a thick icy mantle below a planet-wide ocean hinders ready access to weathering of rocky surfaces that is the dominant source of bioessential metals for life in Earth's oceans.[41,42,45,46]

Based on internal structure models for our canonical Hycean planet, we estimate the ocean depths to range between 100–400 km, consistent with previous studies for similar planets.[20,42] Assuming concentrations similar to those estimated for the early Earth's oceans[101,106] we estimate the mass requirements of P, S, and key bioessential refractory metals: Fe, Ni, Mn, Co, Mo, Zn and Cu. We find that the requirements can be reasonably met for plausible assumptions about the delivery of nutrients through asteroid impacts similar to the impact history of the Hadean Earth, considering a representative asteroid composition based on carbonaceous chondrites. We also consider nutrient availability in the planetary atmosphere/envelope accreted during formation assuming solar composition, consistent with the typical $\sim$1% dust mass fraction in protoplanetary and interstellar environments. The metals are present in condensates which sediment out of the relatively cold Hycean atmosphere post-formation, and can meet the abundance requirement of the Hycean ocean. Making conservative assumptions about the sequestration of P and S species in the Hycean ocean, we find that atmospheric species may also make a substantial contribution to these elemental budgets. Finally, extraterrestrial dust and potential convective transport of core material through the ice layer, may also provide supplemental steady-state sources of nutrients though unlikely to meet the entire primordial requirement. Overall, our results suggest that Hycean worlds provide adequate avenues to meet the chemical requirements of potential life similar to those estimated for the early Earth.

5.2 Future directions

Our initial investigation suggests that Hycean worlds have the potential to provide a promising environment for the origin and prevalence of life. However, the novelty of these worlds also open a wide range of open questions which affect their potential habitability. Here we outline some key directions for future research in this area.

• What are the atmospheric conditions at formation? In particular, a key question is about the timescale over which the primordial Hycean atmosphere equilibrates with the ocean underneath and what the resultant atmospheric composition could be. This sets the initial conditions for the later molecular evolution of the Hycean atmosphere, including the potential for prebiotic molecular chemistry.

• What are the boundary conditions at the ocean–atmosphere interface of a Hycean world? The presence of life in a Hycean ocean could add sources and sinks of prominent molecules which can significantly impact the observable chemistry. For example, methanogenesis as prevalent in anoxic environments on Earth can be a significant sink for CO_2 and source for additional CH_4 in a Hycean atmosphere.

• What is the possible impact history for a Hycean planet and how does the planetary system architecture influence it? Higher/lower impact fluxes or a different impactor size distribution would significantly change the bioessential element contributions. Furthermore, what happens when large bodies (10s or 100s km radius) land in very deep oceans? What percentage of the impact mass ends up in the hydrosphere?

• Can convection of the core material through the icy mantle in a Hycean interior contribute to the bioessential element budget in the ocean? The amount of material transport through an ice VII layer is uncertain, let alone through ice X. The element flux will depend on many other factors such as the heat flux from the core, the size of the ice layer, the presence of liquid water at the bottom of the ice layer, and a host of poorly known chemical and physical properties of high-pressure ices.

• How do (bio)(geo)chemical cycles operate on Hycean worlds? The lack of rocky surfaces makes an Earth-like cycle impossible. There could be some kind of equilibrium between precipitation onto the seafloor and these precipitates having their minerals leached back into the ocean.

• How do dynamical processes affect Hycean worlds? Accurate modeling of the atmospheric circulation is needed to properly establish the climates of Hycean planets, which in turn would have effects on the chemistry, ocean–atmosphere interactions and observable features of the atmospheres. Similarly, modeling of ocean dynamics is also needed to assess the homogeneity and turbulence of the ocean, which can have significant effects on nutrient profiles, sedimentation, and leaching rates, which in turn can affect biochemical processes.

Overall, in this work we find that the chemical conditions in Hycean atmospheres and nutrient availability in their oceans have the potential to meet the requirements for oceanic microbial life similar to those in the early Earth. As discussed above, several important questions remain and motivate new directions for further theoretical investigations on the potential habitability of Hycean worlds. Observationally, Hycean worlds show significant promise to expand and accelerate the search for signatures of extraterrestrial life with current and upcoming facilities within this decade.

Author contributions

NM conceived, planned and led the project and writing of the manuscript with contributions from JM, FR and EB. NM conducted the modelling of the atmospheric structure and JM conducted the photochemical modelling with inputs from NM. JM, NM and FR processed the outputs of the photochemical calculations. FR and NM conducted the interior modelling. EB and NM conducted the estimations of bioessential elements. NM and FR computed the atmospheric spectra and synthetic JWST observations.

Conflicts of interest

There are no conflicts to declare.

Acknowledgements

NM, FR and EB acknowledge support from UK STFC and UKRI towards the PhD studies of FR and EB. JM acknowledges support from the NASA Exoplanet Research Program grant 80NSSC23K0281. NM thanks Alexandra Turchyn, Mark Wyatt, Siddharth Gandhi, Anjali Piette, and Andrew Sellek for helpful discussions. EB thanks Skyla White and Richard Anslow for helpful discussions.

Notes and references

1 L. Arnold, *et al.*, *Astron. Astrophys.*, 2014, **564**, A58.
2 F. Rodler and M. López-Morales, *Astrophys. J.*, 2014, **781**, 54.
3 Y. K. Feng, *et al.*, *Astron. J.*, 2018, **155**, 200.

4 J. F. Kasting, D. P. Whitmire and R. T. Reynolds, *Icarus*, 1993, **101**, 108–128.
5 F. Selsis, *et al.*, *Astron. Astrophys.*, 2007, **476**, 1373–1387.
6 R. K. Kopparapu, *et al.*, *Astrophys. J.*, 2013, **765**, 131.
7 L. T. Elkins-Tanton and S. Seager, *Astrophys. J.*, 2008, **685**, 1237–1246.
8 M. Gillon, *et al.*, *Nature*, 2017, **542**, 456–460.
9 J. A. Dittmann, *et al.*, *Nature*, 2017, **544**, 333–336.
10 K. Ment, *et al.*, *Astron. J.*, 2019, **157**, 32.
11 J. Lillo-Box, *et al.*, *Astron. Astrophys.*, 2020, **642**, A121.
12 E. A. Gilbert, *et al.*, *Astron. J.*, 2020, **160**, 116.
13 J. E. Rodriguez, *et al.*, *Astron. J.*, 2020, **160**, 117.
14 J. K. Barstow and P. G. Irwin, *Mon. Not. R. Astron. Soc.: Lett.*, 2016, **461**, L92–L96.
15 I. A. G. Snellen, *et al.*, *Astron. J.*, 2017, **154**, 77.
16 J. Lustig-Yaeger, *et al.*, *Astron. J.*, 2019, **158**, 27.
17 N. Madhusudhan, A. A. A. Piette and S. Constantinou, *Astrophys. J.*, 2021, **918**, 1.
18 N. Madhusudhan, *et al.*, *Astrophys. J.*, 2020, **891**, L7.
19 A. A. A. Piette and N. Madhusudhan, *Astrophys. J.*, 2020, **904**, 154.
20 M. C. Nixon and N. Madhusudhan, *Mon. Not. R. Astron. Soc.*, 2021, **505**, 3414–3432.
21 D. J. Stevenson, *Nature*, 1999, **400**, 32.
22 R. Pierrehumbert and E. Gaidos, *Astrophys. J., Lett.*, 2011, **734**, L13.
23 M. Mol Lous, R. Helled and C. Mordasini, *Nat. Astron.*, 2022, **6**, 819–827.
24 F. Fressin, *et al.*, *Astrophys. J.*, 2013, **766**, 81.
25 B. J. Fulton and E. A. Petigura, *Astrophys. J.*, 2018, **156**, 264.
26 W. J. Borucki, *et al.*, *Science*, 2010, **327**, 977.
27 G. R. Ricker, *et al.*, *J. Astron. Telesc. Instrum. Syst.*, 2014, **1**, 014003.
28 A. Fukui, *et al.*, *Publ. Astron. Soc. Jpn.*, 2022, **74**, L1–L8.
29 K. Kawauchi, *et al.*, *Astron. Astrophys.*, 2022, **666**, A4.
30 T. Mikal-Evans, *et al.*, *Astron. J.*, 2023, **165**, 84.
31 C. Piaulet, *et al.*, *Nat. Astron.*, 2023, **7**, 206–222.
32 C. L. Phillips, *et al.*, *Astrophys. J.*, 2021, **923**, 144.
33 C. Phillips, *et al.*, *arXiv*, 2022, preprint, arXiv:2209.12919, DOI: 10.48550/arXiv.2209.12919.
34 M. Leung, *et al.*, *Astrophys. J.*, 2022, **938**, 6.
35 X. Yu, *et al.*, *Astrophys. J.*, 2021, **914**, 38.
36 R. Hu, *et al.*, *Astrophys. J., Lett.*, 2021, **921**, L8.
37 S.-M. Tsai, *et al.*, *Astrophys. J., Lett.*, 2021, **922**, L27.
38 C. P. McKay, *Proc. Natl. Acad. Sci. U. S. A.*, 2014, **111**, 12628–12633.
39 C. S. Cockell, *et al.*, *Astrobiology*, 2016, **16**, 89–117.
40 A. R. Hendrix, *et al.*, *Astrobiology*, 2019, **19**, 1–27.
41 H. Lammer, *et al.*, *Astron. Astrophys. Rev.*, 2009, **17**, 181–249.
42 L. Noack, *et al.*, *Icarus*, 2016, **277**, 215–236.
43 M. Lingam and A. Loeb, *Astron. J.*, 2018, **156**, 151.
44 B. Journaux, *et al.*, *Space Sci. Rev.*, 2020, **216**, 7.
45 S. Maruyama, *et al.*, *Geosci. Front.*, 2013, **4**, 141–165.
46 S. Seager, *et al.*, *Universe*, 2021, **7**, 172.
47 G. Choblet, *et al.*, *Icarus*, 2017, **285**, 252–262.
48 K. Kalousová, *et al.*, *Icarus*, 2018, **299**, 133–147.

49 K. Kalousová and C. Sotin, *Geophys. Res. Lett.*, 2018, **45**, 8096–8103.
50 J.-A. Hernandez, R. Caracas and S. Labrosse, *Nat. Commun.*, 2022, **13**, 3303.
51 L. Lebec, *et al.*, *Icarus*, 2023, **396**, 115494.
52 P. B. Rimmer, *et al.*, *Sci. Adv.*, 2018, **4**, eaar3302.
53 D. Parkos, *et al.*, *J. Geophys. Res.: Planets*, 2018, **123**, 892–909.
54 H. Innes, S.-M. Tsai and R. T. Pierrehumbert, *arXiv*, 2023, preprint, arXiv:2304.02698, DOI: 10.48550/arXiv.2304.02698.
55 S. Gandhi and N. Madhusudhan, *Mon. Not. R. Astron. Soc.*, 2017, **472**, 2334–2355.
56 I. Hubeny, *Mon. Not. R. Astron. Soc.*, 2017, **469**, 841–869.
57 J. I. Castor, P. G. Dykema and R. I. Klein, *Astrophys. J.*, 1992, **387**, 561.
58 B. Benneke, *et al.*, *Astrophys. J., Lett.*, 2019, **887**, L14.
59 R. Cloutier, *et al.*, *Astron. Astrophys.*, 2019, **621**, A49.
60 R. J. Barber, J. Tennyson, G. J. Harris and R. N. Tolchenov, *Mon. Not. R. Astron. Soc.*, 2006, **368**, 1087–1094.
61 L. Rothman, *et al.*, *J. Quant. Spectrosc. Radiat. Transfer*, 2010, **111**, 2139–2150.
62 J. Borysow, L. Frommhold and G. Birnbaum, *Astrophys. J.*, 1988, **326**, 509.
63 G. S. Orton, *et al.*, *Icarus*, 2007, **189**, 544–549.
64 M. Abel, L. Frommhold, X. Li and K. L. C. Hunt, *J. Phys. Chem. A*, 2011, **115**, 6805–6812.
65 C. Richard, *et al.*, *J. Quant. Spectrosc. Radiat. Transfer*, 2012, **113**, 1276–1285.
66 M. Allen, Y. L. Yung and J. W. Waters, *J. Geophys. Res.*, 1981, **86**, 3617–3627.
67 Y. L. Yung, M. Allen and J. P. Pinto, *Astrophys. J., Suppl. Ser.*, 1984, **55**, 465–506.
68 J. I. Moses, *et al.*, *Astrophys. J.*, 2011, **737**, 15.
69 E. S. Kite and E. B. Ford, *Astrophys. J.*, 2018, **864**, 75.
70 J. I. Moses, *et al.*, *Astrophys. J.*, 2013, **777**, 34.
71 E. S. Kite, M. Manga and E. Gaidos, *Astrophys. J.*, 2009, **700**, 1732–1749.
72 J. I. Moses, *et al.*, *Exp. Astron.*, 2022, **53**, 279–322.
73 S. Peacock, *et al.*, *Astrophys. J.*, 2020, **895**, 5.
74 J. I. Moses, *et al.*, *Icarus*, 2000, **145**, 166–202.
75 J. F. Kasting, *Origins Life Evol. Biosphere*, 1990, **20**, 199–231.
76 G. Arney, *et al.*, *Astrobiology*, 2016, **16**, 873–899.
77 S. Ranjan, *et al.*, *Astrophys. J.*, 2020, **896**, 148.
78 N. Madhusudhan, K. K. M. Lee and O. Mousis, *Astrophys. J., Lett.*, 2012, **759**, L40.
79 G. Chabrier, S. Mazevet and F. Soubiran, *Astrophys. J.*, 2019, **872**, 51.
80 S. W. Thomas and N. Madhusudhan, *Mon. Not. R. Astron. Soc.*, 2016, **458**, 1330–1344.
81 S. Seager, *et al.*, *Astrophys. J.*, 2007, **669**, 1279–1297.
82 F. Birch, *J. Geophys. Res.*, 1952, **57**, 227–286.
83 B. B. Karki, *et al.*, *Phys. Rev. B: Condens. Matter Mater. Phys.*, 2000, **62**, 14750–14756.
84 T. J. Ahrens, *Mineral Physics & Crystallography: A Handbook of Physical Constants*, American Geophysical Union, Washington, DC, 2000.
85 M. Scheucher, F. Wunderlich, J. L. Grenfell, M. Godolt, F. Schreier, D. Kappel, R. Haus, K. Herbst and H. Rauer, *Astrophys. J.*, 2020, **898**, 44.
86 D. Blain, B. Charnay and B. Bézard, *Astron. Astrophys.*, 2021, **646**, A15.

87 B. Charnay, D. Blain, B. Béard, J. Leconte, M. Turbet and A. Falco, *Astron. Astrophys.*, 2021, **646**, A171.
88 R. T. Pierrehumbert, *Astrophys. J.*, 2023, **944**, 20.
89 L. J. Rothschild and R. L. Mancinelli, *Nature*, 2001, **409**, 1092–1101.
90 N. Merino, *et al.*, *Front. Microbiol.*, 2019, **10**, 780.
91 K. K. Hardegree-Ullman, *et al.*, *Astrophys. J., Suppl. Ser.*, 2020, **247**, 28.
92 A. Tsiaras, *et al.*, *Nat. Astron.*, 2019, **3**, 1086–1091.
93 K. Lodders and B. Fegley, *Icarus*, 2002, **155**, 393–424.
94 J. I. Moses, *et al.*, *Astrophys. J.*, 2013, **777**, 34.
95 S. Seager, *et al.*, *Astrophys. J.*, 2013, **777**, 95.
96 S. Ranjan, *et al.*, *Astrophys. J.*, 2022, **930**, 131.
97 M. Asplund, *et al.*, *Annu. Rev. Astron. Astrophys.*, 2009, **47**, 481–522.
98 T. M. Hoehler, *et al.*, *FEMS Microbiol. Ecol.*, 2001, **38**, 33–41.
99 W. Bains, S. Seager and A. Zsom, *Life*, 2014, **4**, 716–744.
100 B. A. Berghuis, *et al.*, *Proc. Natl. Acad. Sci. U. S. A.*, 2019, **116**, 5037–5044.
101 L. J. Robbins, *et al.*, *Earth-Sci. Rev.*, 2016, **163**, 323–348.
102 M. Rubin, *et al.*, *Mon. Not. R. Astron. Soc.*, 2019, **489**, 594–607.
103 A. D. Anbar, *Science*, 2008, **322**, 1481–1483.
104 S. A. Crowe, *et al.*, *Science*, 2014, **346**, 735–739.
105 C. Jones, *et al.*, *Geology*, 2015, **43**, 135–138.
106 M. A. Saito, D. M. Sigman and F. M. Morel, *Inorg. Chim. Acta*, 2003, **356**, 308–318.
107 A. D. Anbar and A. H. Knoll, *Science*, 2002, **297**, 1137–1142.
108 E. D. Swanner, *et al.*, *Earth-Sci. Rev.*, 2020, **211**, 103430.
109 N. J. Tosca, *et al.*, *Geol. Soc. Am. Bull.*, 2016, **128**, 511–530.
110 L. J. Robbins, *et al.*, *GeoBiology*, 2013, **11**, 295–306.
111 I. Mukherjee and R. R. Large, *Geology*, 2020, **48**, 1018–1022.
112 K. Lodders, *Oxford Research Encyclopedia of Planetary Science*, 2020.
113 K. J. Zahnle, *et al.*, *Planet. Sci. J.*, 2020, **1**, 11.
114 D. Nesvorný, F. V. Roig, D. Vokrouhlický, W. F. Bottke, S. Marchi, A. Morbidelli and R. Deienno, *Icarus*, 2023, **399**, 115545.
115 G. Gisler, R. Weaver and M. Gittings, *Pure Appl. Geophys.*, 2011, **168**, 1187–1198.
116 T. Davison and G. S. Collins, *Meteorit. Planet. Sci.*, 2007, **42**, 1915–1927.
117 M. Nishizawa, *et al.*, *J. Geophys. Res.: Planets*, 2020, **125**, e06291.
118 B. Carry, *Planet. Space Sci.*, 2012, **73**, 98–118.
119 H. Genda, R. Brasser and S. Mojzsis, *Earth Planet. Sci. Lett.*, 2017, **480**, 25–32.
120 B. C. Johnson and H. J. Melosh, *Icarus*, 2012, **217**, 416–430.
121 P. W. Reiners and A. V. Turchyn, *Geology*, 2018, **46**, 863–866.
122 B. Peucker-Ehrenbrink, G. Ravizza and G. Winckler, *Elements*, 2016, **12**, 191–196.
123 I. Y. Fung, *et al.*, *Global Biogeochem. Cycles*, 2000, **14**, 281–295.
124 S. Constantinou and N. Madhusudhan, *Mon. Not. R. Astron. Soc.*, 2022, **514**, 2073–2091.
125 A. Pinhas, *et al.*, *Mon. Not. R. Astron. Soc.*, 2018, **480**, 5314–5331.
126 I. E. Gordon, *et al.*, *J. Quant. Spectrosc. Radiat. Transfer*, 2022, **277**, 107949.
127 F. Lattanzi, C. D. Lauro and J. V. Auwera, *J. Mol. Spectrosc.*, 2011, **267**, 71–79.
128 J. J. Harrison, N. D. C. Allen and P. F. Bernath, *J. Quant. Spectrosc. Radiat. Transfer*, 2010, **111**, 357–363.

129 A. Jolly, Y. Benilan and A. Fayt, *J. Mol. Spectrosc.*, 2007, **242**, 46–54.
130 L. Xu, *et al.*, *J. Mol. Spectrosc.*, 2004, **228**, 453–470.
131 H. Müller, *et al.*, *Astron. Astrophys.*, 2001, **370**, L49–L52.
132 R. Doyon, *et al.*, *Space Telescopes and Instrumentation 2012: Optical, Infrared, and Millimeter Wave*, 2012, p. 84422R.
133 P. Ferruit, *et al.*, *Space Telescopes and Instrumentation 2012: Optical, Infrared, and Millimeter Wave*, 2012, p. 84422O.
134 S. M. Birkmann, *et al.*, *Space Telescopes and Instrumentation 2014: Optical, Infrared, and Millimeter Wave*, 2014, p. 914308.
135 G. H. Rieke, *et al.*, *Publ. Astron. Soc. Pac.*, 2015, **127**, 584.
136 N. E. Batalha, *et al.*, *Publ. Astron. Soc. Pac.*, 2017, **129**, 064501.
137 V. S. Meadows and R. K. Barnes, *Handbook of Exoplanets*, Springer International Publishing AG, 2018, p. 57.
138 S. Seager, *et al.*, *Nat. Astron.*, 2020, **4**, 802–806.
139 E. S. Kite, *et al.*, *Astrophys. J.*, 2020, **891**, 111.
140 W. Misener, H. E. Schlichting and E. D. Young, *Mon. Not. R. Astron. Soc.*, 2023, **524**, 981–992.
141 N. Madhusudhan, *et al.*, *Space Sci. Rev.*, 2016, **205**, 285–348.

Faraday Discussions

ROYAL SOCIETY OF CHEMISTRY

PAPER

The chemical inventory of the inner regions of planet-forming disks – the JWST/MINDS program†

Inga Kamp, *[a] Thomas Henning, [b] Aditya M. Arabhavi, [a] Giulio Bettoni, [g] Valentin Christiaens, [d] Danny Gasman, [e] Sierra L. Grant, [g] Maria Morales-Calderón,[f] Benoît Tabone, [c] Alain Abergel,[c] Olivier Absil, [d] Ioannis Argyriou, [e] David Barrado, [f] Anthony Boccaletti,[h] Jeroen Bouwman, [b] Alessio Caratti o Garatti, [ij] Ewine F. van Dishoeck, [gk] Vincent Geers, [l] Adrian M. Glauser, [m] Manuel Güdel, [bmn] Rodrigo Guadarrama,[n] Hyerin Jang, [o] Jayatee Kanwar, [ap] Pierre-Olivier Lagage,[q] Fred Lahuis,[r] Michael Mueller, [a] Cyrine Nehmé,[s] Göran Olofsson, [t] Eric Pantin,[u] Nicole Pawellek, [nw] Giulia Perotti, [b] Tom P. Ray, [j] Donna Rodgers-Lee, [j] Matthias Samland, [b] Silvia Scheithauer, [b] Jürgen Schreiber,[b] Kamber Schwarz, [b] Milou Temmink, [k] Bart Vandenbussche, [e] Marissa Vlasblom, [k] Christoffel Waelkens,[e] L. B. F. M. Waters [ov] and Gillian Wright[l]

Received 22nd January 2023, Accepted 3rd April 2023
DOI: 10.1039/d3fd00013c

The understanding of planet formation has changed recently, embracing the new idea of pebble accretion. This means that the influx of pebbles from the outer regions of planet-forming disks to their inner zones could determine the composition of planets and their atmospheres. The solid and molecular components delivered to the planet-forming region can be best characterized by mid-infrared spectroscopy. With Spitzer low-resolution ($R = 100$, 600) spectroscopy, this approach was limited to the detection of abundant molecules, such as H_2O, C_2H_2, HCN and CO_2. This contribution will present the first results of the MINDS (MIRI mid-INfrared Disk Survey, PI:Th Henning) project. Due do the sensitivity and spectral resolution provided by the James Webb Space Telescope (JWST), we now have a unique tool to obtain the full inventory of chemistry in the inner disks of solar-type stars and brown dwarfs, including also less-abundant hydrocarbons and isotopologues. The Integral Field Unit (IFU) capabilities will enable at the same time spatial studies of the continuum and line emission in extended sources such as debris disks, the flying saucer and also the search for mid-IR signatures of forming planets in systems such as PDS 70. These JWST observations are complementary to ALMA and NOEMA observations of outer-disk chemistry; together these datasets will provide an integral view of the processes occurring during the planet-formation phase.

1 Introduction

Much of the exoplanet population studied so far resides inside 10 au from their host star. The composition of these planets should carry traces of the chemical composition of the inner regions of planet-forming disks. This can manifest itself in the bulk C/O ratio of gas-giant planet atmospheres,[1–4] but also affect the bulk interior composition of terrestrial planets (*e.g.* the sulphur content)[5] and the delivery of water to them.[6,7]

Planet-forming disks are expected to be layered in their chemical content, with the surface layers being ionized/atomic and deeper layers being molecular.[8,9] More recently, thermochemical models also showed that the spatial distribution of molecules comes in layers, with OH being closest to the surface and CO, H_2O, CO_2, HCN and C_2H_2 residing ever closer to the midplane[10] (Fig. 1). In addition, theory and observations have shown that dust grains can radially migrate in the disk if they reach sizes that allow them to dynamically decouple from the gas.[11,12] As a consequence, volatile ices carried along and sublimating at the respective ice-line locations could alter the elemental composition of the gas in the disk.[13] The inner-disk regions are generally highly optically thick (unless some processes have removed material and carved gaps/holes[14]), preventing us from probing down to the midplane. However, turbulent-mixing timescales are less than 10^4

[a]*Kapteyn Astronomical Institute, University of Groningen, PO Box 800, 9700 AV Groningen, The Netherlands. E-mail: kamp@astro.rug.nl*

[b]*Max-Planck-Institut für Astronomie (MPIA), Königstuhl 17, 69117 Heidelberg, Germany*

[c]*Université Paris-Saclay, CNRS, Institut d'Astrophysique Spatiale, 91405, Orsay, France*

[d]*STAR Institute, Université de Liège, Allée du Six Août 19c, 4000 Liège, Belgium*

[e]*Institute of Astronomy, KU Leuven, Celestijnenlaan 200D, 3001 Leuven, Belgium*

[f]*Centro de Astrobiología (CAB), CSIC-INTA, ESAC Campus, Camino Bajo del Castillo s/n, 28692 Villanueva de la Cañada, Madrid, Spain*

[g]*Max-Planck Institut für Extraterrestrische Physik (MPE), Giessenbachstr. 1, 85748, Garching, Germany*

[h]*LESIA, Université PSL, CNRS, Sorbonne Université, Université de Paris, 5 Place Jules Janssen, 92195 Meudon, France*

[i]*INAF – Osservatorio Astronomico di Capodimonte, Salita Moiariello 16, 80131 Napoli, Italy*

[j]*Dublin Institute for Advanced Studies, 31 Fitzwilliam Place, D02 XF86 Dublin, Ireland*

[k]*Leiden Observatory, Leiden University, 2300 RA Leiden, The Netherlands*

[l]*UK Astronomy Technology Centre, Royal Observatory Edinburgh, Blackford Hill, Edinburgh EH9 3HJ, UK*

[m]*ETH Zürich, Institute for Particle Physics and Astrophysics, Wolfgang-Pauli-Str. 27, 8093 Zürich, Switzerland*

[n]*Dept. of Astrophysics, University of Vienna, Türkenschanzstr 17, A-1180 Vienna, Austria*

[o]*Department of Astrophysics, IMAPP, Radboud University, PO Box 9010, 6500 GL Nijmegen, The Netherlands*

[p]*Space Research Institute, Austrian Academy of Sciences, Schmiedlstr. 6, A-8042 Graz, Austria*

[q]*Université Paris-Saclay, Université de Paris, CEA, CNRS, AIM, F-91191 Gif-sur-Yvette, France*

[r]*SRON Netherlands Institute for Space Research, PO Box 800, 9700 AV, Groningen, The Netherlands*

[s]*CEA, DSM, Irfu, Service d'Astrophysique – Laboratoire AIM, France*

[t]*Department of Astronomy, Stockholm University, AlbaNova University Center, 10691 Stockholm, Sweden*

[u]*IRFU, DAp Département D'Astrophysique CE Saclay, Gif-sur-Yvette, France*

[v]*SRON Netherlands Institute for Space Research, Niels Bohrweg 4, NL-2333 CA Leiden, The Netherlands*

[w]*Konkoly Observatory, Research Centre for Astronomy and Earth Sciences, Eötvös Loránd Research Network (ELKH), Konkoly-Thege Miklós út 15-17, H-1121 Budapest, Hungary*

† The MIRI mid-INfrared Disk Survey (MINDS) is part of the JWST/MIRI GTO time and led by Thomas Henning and Inga Kamp.

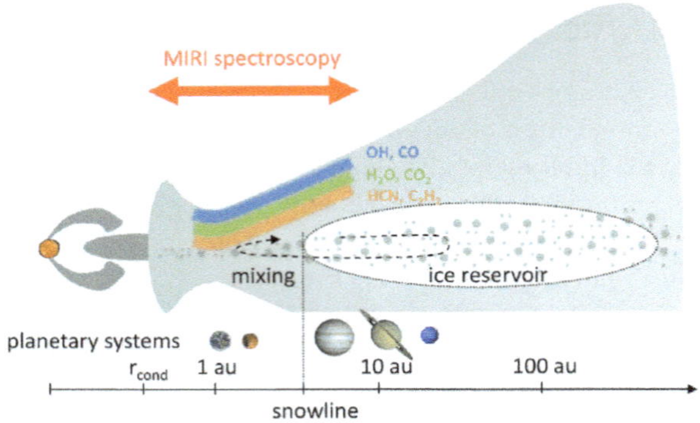

Fig. 1 Sketch of a typical disk around a T Tauri star. Thermochemical disk models suggest a layered molecular structure with OH and CO at the top, followed by H_2O and CO_2. HCN and C_2H_2 peak in abundance deeper in the disk. Theory and observations suggest the possibility of radial transport of icy grains, which can affect the inner-disk gas composition. The Mid-InfraRed Instrument (MIRI) on board the James Webb Space Telescope probes the inner ($\lesssim$10 au) warm disk surface.

years,[15] thus ensuring that any change of element abundance due to radial transport in the midplane should quickly spread vertically in the disk.

The mid-infrared wavelength range is key in probing the inner ~10 au of planet-forming disks. The gas in there is typically warm (several 100–1000 K) and we expect to see a mix of highly excited rotational lines and ro-vibrational ones. The Spitzer Space Telescope has frequently detected abundant molecules such as water, OH, CO_2, HCN and C_2H_2 in disks around solar-type stars, and less frequently around Herbig Ae stars.[16–19] Still, due to the low spectral resolution of $R \sim 600$, Spitzer saw only the tip of the iceberg; less abundant molecules (minor species, *e.g.* CH_4 and NH_3 with features <10 μm, and other hydrocarbons such as C_4H_2) remained undetected. Yet these minor species play a major role in the synthesis pathways of complex organic molecules and are key to understanding the types of chemistry occurring while planet formation is in full steam (*e.g.* C- or O-chemistry, combustion chemistry). Also, the quantitative interpretation of the molecular bands in the Spitzer data was strongly hampered by line blending of the various molecules. The recently launched James Webb Space Telescope (JWST) has the spectral resolution ($R \sim 1500$–3500) and the sensitivity to disentangle major species blends and to carry out a deep search for minor species.

In the following, we will present some first results on the richness of the inner disks of T Tauri and low-mass stars from Guaranteed Time Observations (GTO) with the MIRI on board of the James Webb Space Telescope.

2 The MINDS survey

The MIRI mid-INfrared Disk Survey (MINDS) is part of the JWST/MIRI GTO time and led by Thomas Henning and Inga Kamp (PID 1282). We use the medium-resolution spectrograph (MRS) of the Mid-InfraRed Instrument (MIRI),[20,21] which covers simultaneously the wavelength range 4.9 to 28.6 μm with a spectral resolution of $R \sim 1500$–3500.

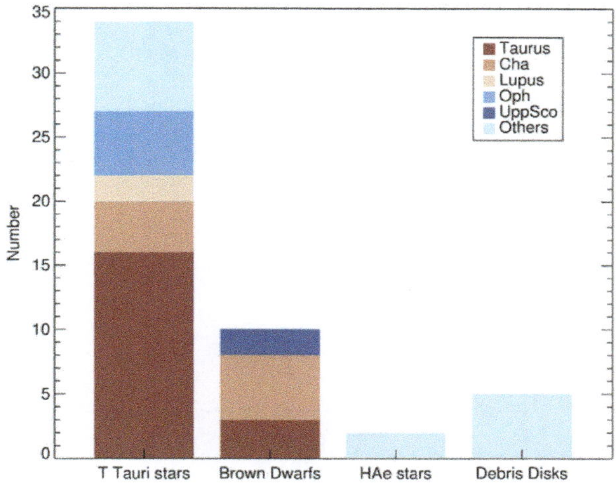

Fig. 2 Distribution of the 51 MINDS sources across planet-forming disk type and star-forming region.

2.1 The sample

We selected our sample to contain young gas-rich class II disks in nearby star-forming regions (Herbig Ae stars, T Tauri stars and very low mass stars/brown dwarfs). Brown dwarfs are sub-stellar objects that are not massive enough to sustain stable hydrogen fusion. The boundary between very low mass stars (VLMS) and brown dwarfs (BDs) is not well defined; some authors use color criteria[22,23] or spectral type (later than M6 for young BDs).[24] In addition, we selected a handful of young gas-rich debris disks, where ALMA has established the presence of CO gas. We mostly selected targets in the Taurus, Ophiuchus, Chamaeleon and Lupus star-forming regions largely based on the presence of molecular emission known from previous Spitzer observations and using ground-based CO ro-vibrational line profiles to make sure we look at disk emission. Fig. 2 shows the distribution of our targets across the various star-forming regions and object types; our main focus is on the Taurus star-forming region. We also included in our sample PDS 70, a young planet-forming disk, which is known to host two gas-giant protoplanets,[25,26] and TW Hya, one of the closest gas-rich planet-forming disks (d = 60 pc). Both systems have the potential to reveal intriguing mid-IR signatures related to forming planets.

2.2 The MIRI/MRS survey

The total observing time of our survey is ~120 h. We base our exposure time estimates on previous mid-IR Spitzer fluxes and the goal to obtain a S/N of 300–500 to maximize detection rates of minor species.[27] Due to the instrument sensitivity being a strong function of wavelengths and the Spectral Energy Distribution (SED) of our targets being non-flat, we will mostly achieve this S/N at shorter wavelengths. Since all our targets are relatively bright sources with well-known coordinates and proper motions (Gaia DR3 [28]), we omit a dedicated target acquisition (TA) with the exception of the brown dwarfs where TA is necessary. For the spatially extended sources in our sample, we schedule

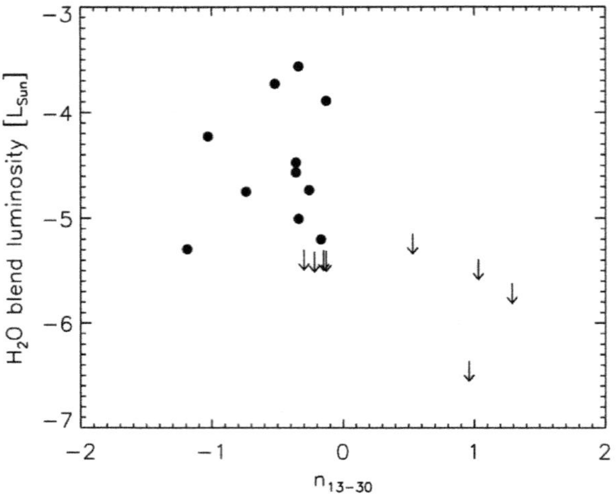

Fig. 3 Water luminosity *versus* SED slope between 13 and 30 μm for MINDS T Tauri disks that have high-resolution Spitzer spectra ($R \sim 600$). Arrows are upper limits. The higher spectral resolution and sensitivity of JWST MIRI/MRS allows detection of much fainter water fluxes in the MINDS program.

a dedicated background observation just before the science exposure with the same exposure time. For the science exposures, we use a 4-point dither strategy.

2.3 Data reduction

Our data reduction flow leverages both the JWST pipeline (v1.8.4) and routines from the VIP package.[29] From the JWST pipeline, we use (i) the class Detector1 to process uncalibrated raw data files using CRDS context jwst_1019.pmap and default parameters; (ii) Spec2 with default parameters, but skipping background subtraction, and using dedicated reference files for photometric and fringe flat calibrations as in Gasman *et al.*[40]; and (iii) Spec3 with default parameters, apart from the master_background and outlier_detection steps which are turned off. We leverage the four-point dither strategy to obtain a first guess of the background signal as the minimum of each quadruplet of dithered images (on the detector), then refined this estimate using a median-filter, which both smoothed the background map and removed residual star signals. This estimated background map is subtracted just before Spec3. Bad pixel correction is performed both before background estimation and after Spec3 using dedicated VIP routines. Spectrum extraction is performed manually through aperture photometry in 2.5 λ/D apertures in the final spectral frames, after identifying manually the star centroid. We use correction factors to compensate for the missing flux in the apertures (as in Argyriou *et al.*[40]), and filter out spikes affecting individual spaxels when integrating the signal in each aperture.

3 Small dust grains and gas in the inner disk

We have little direct imaging information on the detailed gas and dust structure of the inner 10 au of disks around low-mass stars ($M^* \lesssim 1\ M_\odot$). Much of what we know originates from the kinematic analysis of line profiles such as the CO ro-vibrational lines,[30,31] spectroscopy of dust emission features[32] or interferometry.[33] The Spitzer Space Telescope has studied samples of T Tauri disks down to the brown dwarfs regime both for their molecular content[17,19,34–37] and their dust mineralogy and grain sizes.[32,34,38] The inner disk surfaces are generally populated by micron-sized grains and disk models suggest that the 10 and 20 μm silicate feature and the molecular lines originate from a similar radial region.

Fig. 3 shows the Spitzer water luminosity[39] *vs.* SED slope n_{13-30} (MSc thesis Christian Lemmens) for a subset of our 19 T Tauri MINDS targets. Like previous work,[39] we see no obvious correlation across our sample; the only interesting point is that disks with large cavities ($n_{13-30} > 0.5$) have no water detection. It remains to be seen whether this was a sensitivity limit. Given the typical MINDS integration times, we measure noise levels of ~2–12 mJy around 17 μm,[40] corresponding to line luminosities of 5×10^{-8} to 3×10^{-7} $L_{\odot}$. Theoretical models predict for example water lines to be weaker in disks with inner cavities.[41,42]

Overall, we have not yet fully understood where the large diversity of molecular line flux strength originates from; in several cases, line fluxes themselves correlate well, *e.g.* water and CO ro-vib,[36] water and HCN.[19] It is unclear whether this is because these molecules are co-spatial radially and vertically. Disk models can often reproduce such correlations (*e.g.* Antonellini *et al.*[43]). Some molecules are not emitting in local thermodynamic equilibrium (LTE); OH is, for example, non-thermally excited through water photodissociation;[44] CO can be pumped by fluorescence or IR radiation;[30,45] and HCN can be pumped by IR radiation.[46] Also, the C_2H_2 fluxes do not correlate with water fluxes.[36] Even in T Tauri disks, the strength of the C_2H_2 ro-vibrational band is difficult to reproduce with simple gas-phase chemistry.[47,48] Higher abundances and column densities of this molecule occur in disk surfaces if the H/H_2 transition resides closer to the surface and temperatures are higher;[49] this can be achieved by, *e.g.*, considering only small grains instead of a wide grain size distribution up to mm-sizes and settling.[50,51]

There is still much to learn about the chemical composition of the inner 10 au and the Spitzer spectral resolution was a major limiting factor also in the analysis of the spectra. The blending of lines from different species was difficult to correct without large assumptions and the retrieval suffered from large degeneracies in column densities and temperatures. However, for understanding planetary diversity we need to obtain a quantitative inventory of the inner-disk molecular reservoir. The retrieval process will remain the most challenging part of the analysis. Hence, as in times of Spitzer data, we will start simple using 0D slab models and discuss towards the end pathways forward.

4 Slab model retrieval

This is the most simple retrieval method that can be applied to molecular emission spectra. It has been widely used on Spitzer data[18] ‡ and using radial temperature gradients (1D) also to near-IR ground based spectroscopy.[30] While the method is fast and powerful, it has not yet been assessed how the quantities of

‡ slabspec – https://doi.org/10.5281/zenodo.4037306

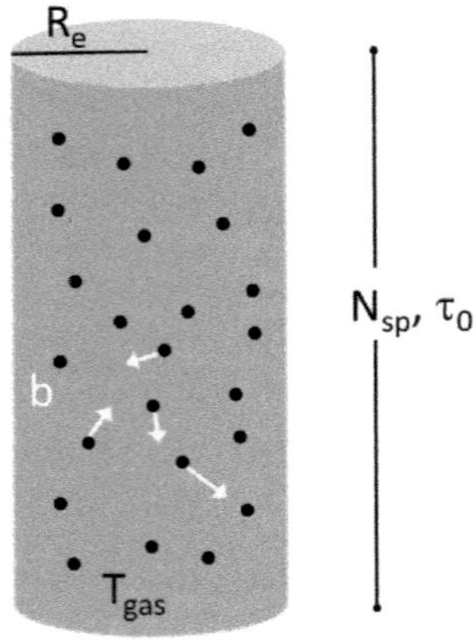

Fig. 4 Schematic of a 0D slab model.

temperature, column density and emitting area that we retrieve compare to the actual underlying disk properties. The subsequent paragraphs illustrate this and discuss remaining issues and future challenges for retrieval methods.

4.1 Slab model fluxes

The most simple 0D slab models calculate the line flux F_0 at the center of the line λ_0 for a selected number of line transitions of a single molecule (species sp). The slab of gas has a constant temperature T_{gas} and density $n_{<H>}$,§ and variable length, hence column density N_{sp} (Fig. 4). Within the slab, the abundance of the species ε_{sp} as well as the potential collision partners (electrons, protons, H, *o*-/*p*-H_2, and He) are free parameters, but constants. The velocity at which gas particles move b is the sum of their thermal velocity v_{th} and a turbulent component v_{turb}

$$b = \sqrt{{v_{th}}^2 + {v_{turb}}^2}. \tag{1}$$

The data characterizing a single line transition of a species sp are the lower and upper energy levels E_l, E_u, the respective statistical weights g_l, g_u and the Einstein A coefficient A_{ul}. The level populations are denoted as b_l and b_u and a superscript * denotes local thermodynamic equilibrium (LTE). The total line intensity emerging I_0 is then calculated using either a simple RT solution or from escape probability using

$$I_0 = \frac{h\nu A_{ul}}{4\pi} N_{\langle H \rangle} \varepsilon_{sp} b_u \beta_{esc}(\tau_0). \tag{2}$$

The Planck constant h and all other quantities are usually used in cgs units. The escape probability β_{esc} treatment depends on the geometry assumed[52] and the line center optical depth $\tau_0 = \tau(\lambda_0)$

§ This is the total hydrogen number density in the gas, so $n(H) + 2n(H_2)$.

$$\tau_0 = \frac{A_{\mathrm{ul}}\lambda_0{}^3}{8\pi b} N_{\mathrm{sp}} \left(b_{\mathrm{l}} \frac{g_{\mathrm{u}}}{g_{\mathrm{l}}} - b_{\mathrm{u}} \right). \tag{3}$$

For optically thin lines, the intensity scales directly with the column density of the species. However, as the line become optically thick in the center, the line broadening ensures that the flux keeps growing, albeit not linear anymore. The line photons do not then escape at the line center, but in the wings. In the most simple case, we assume a Gaussian line profile and high optical depth leads to a flat-topped quasi-rectangular profile.

The line intensity is converted into a flux assuming a specific emitting area $\pi R_e 2$ and the distance d to the source. For a specific turbulence, we can precalculate a grid of slab models for a range of temperatures and column densities and this can be reused for all spectra.

The level populations can be calculated either from LTE or from statistical equilibrium (SE). In the latter case, we need collisional cross sections for all transitions and potential collision partners. The availability of such data is often the limiting factor. Hence, we revert back to LTE for many of the molecules in the mid-IR spectral region. Deviations from LTE will occur if the gas densities are much smaller than the critical density of the line and/or if the molecules are subject to strong radiation fields (*e.g.* UV pumping for CO ro-vibrational lines,[45] and IR pumping by dust thermal emission for HCN[46]) and/or if the excitation occurs through chemical pathways (*e.g.* oxygen through photodissociation of OH,[53] and OH through photodissociation of water[44]). It is interesting to note that radiative pumping can drive molecules also into LTE, but the excitation temperature then corresponds to that of the radiation field and not to the gas temperature. Calculating level populations b_u with respect to the total population of the species in LTE requires knowing the partition functions $Z(T,P)$, where

$$b_{\mathrm{u}} = \frac{n_{\mathrm{sp,u}}}{n_{\mathrm{sp,tot}}} = \frac{g_{\mathrm{u}}}{Z(T,P)} \exp^{-E_{\mathrm{u}}/kT} \tag{4}$$

with E_u being the upper-level energy and k the Boltzmann constant. For many molecules these are provided in the form of tables covering a large range of temperature and pressures. In the mid-IR we have to drop the simplified approach of using only a restricted number of levels to calculate the partition functions, an approach often used when dealing with rotational lines at sub-mm wavelengths and very cold temperatures ($\lesssim$100 K).

We take the molecular data and partition functions mostly from three databases: the Leiden Atomic and Molecular Database (LAMDA[54,55]), the HITRAN database,[56] and the GEISA database.[57] The GEISA database lacks Einstein A coefficients and those are provided in selected cases by Agnes Perrin (private communication). We also note that several molecules have data only for a single or few ro-vibrational bands. Since the typical temperatures in the inner disks are a few 100 up to 1000 K, our analysis is severely hampered by this. We will come back to it later.

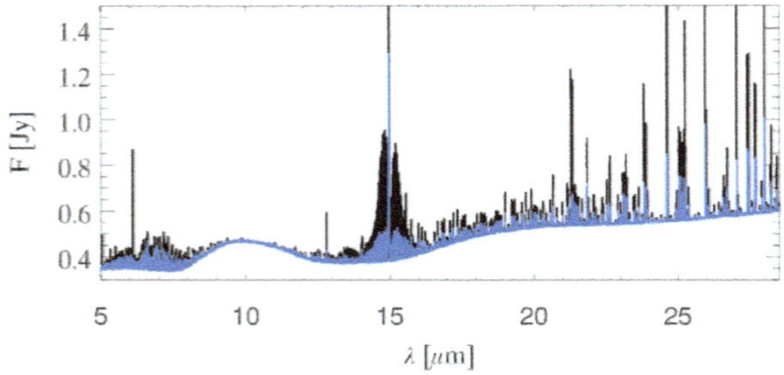

Fig. 5 Synthetic mid-IR spectrum for AA Tau based on the DIANA model[60] using a gas-to-dust ratio of 1000. The black line is the high-resolution synthetic model spectrum, and the blue one is convolved to $R = 3000$ with noise added (SNR = 300).

4.2 Slab model spectra and fitting procedure

The slab models outlined above provide total line fluxes in a certain wavelength range per molecule. For comparison with actual data, we need to convolve these fluxes with the appropriate spectral resolution of the instrument, in this case $R \sim 3000$ (JWST MIRI/MRS) using a very high spectral resolution wavelength grid. We then sample the spectra using the wavelength points of the observed spectra, thus reproducing the same oversampling as the instrument, and also facilitating the calculation of a formal (reduced) χ^2 in the fitting procedure.

Since the molecular emission of many molecules overlaps in the mid-IR spectra and the slab models can only deal with one molecule at a time, we define very restricted wavelength windows for the χ^2-minimalization in which we can be sure that only the molecule under investigation contributes. Often these are regions around the strong Q-branches of the molecules. We use the pre-

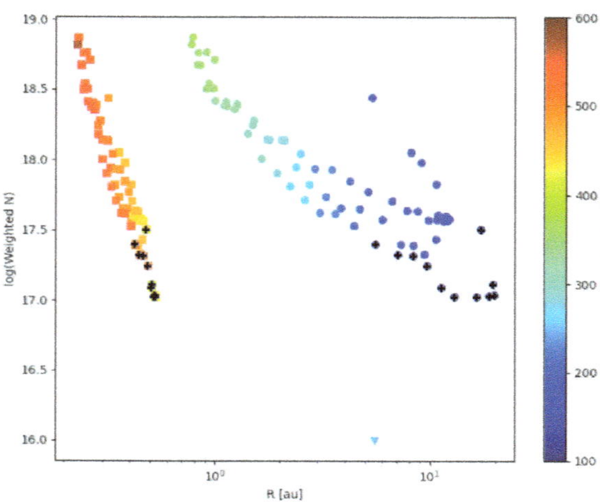

Fig. 6 Metadata (minimum, maximum gas temperatures, flux-weighted column densities, and inner/outer radius of the radial emitting region) for a representative sample of strong CO_2 lines extracted from the AA Tau thermochemical disk model. Black crosses indicate the strongest Q-branch lines. The triangle shows the retrieval result.

calculated grid of slab models (T,N) together with a range of emitting radii R_e and calculate the emergent slab model spectra F^{slab}. The goodness of each slab model for fitting the observations F^{obs} is evaluated using the χ^2 framework

$$\chi^2 = \frac{1}{n}\sum_{i=1,n}\frac{\left(F_i^{obs} - F_i^{slab}\right)}{\sigma^2}. \quad (5)$$

Here we assume a constant flux error across all chosen wavelength windows σ.

4.3 Test against 2D thermochemical models

The above described slab models have been widely used in the analysis of mid-IR spectra. However, it is not clear how the quantities (T, N and R_e) retrieved relate to the actual properties of the gas in the disk surface. To make a start, we decided to apply the same retrieval technique on a synthetic spectrum generated using a 2D thermochemical disk model. For that purpose, we use the codes ProDiMo[58,59] and FLiTs.[10] We take a prominent source AA Tau, which has a 2D RT disk model fitted to its large multi-wavelength photometry from the DIANA project.[60,61] We calculate the disk chemistry and gas heating/cooling self-consistently using a gas-to-dust ratio of 1000; this is required to ensure that line fluxes are of the same order as those observed by Spitzer.[10,62] We use the selection of molecules and molecular data from ref. 10 even though that means using a slightly outdated HITRAN version (2009 instead of 2020). We produce synthetic mid-IR spectra from 5–28.5 μm at a nominal spectral resolution of 100 000 using FLiTs. The spectra are then subsequently convolved to the lower resolution of MIRI ($R = 3000$), oversampled by a factor of two and we add noise with a typical S/N ratio of 300 (Fig. 5). We produce these spectra using either all selected molecules simultaneously, or just a single molecule at a time. This is to ensure that we can approach the understanding by stepwise increasing the complexity. Also, the line separation within the ro-vibrational band of CO_2 is large enough so that opacity overlap of individual lines is negligible. In the following, we will show the first results using pure CO_2 spectra.

Fig. 6 shows the first results from analysing the emitting conditions in the AA Tau thermochemical model and comparing them to the results of the slab retrieval. We analyse here only the strongest CO_2 lines from the model, but ensure that we have a coverage of P-, Q- and R-branch lines. For this representative sample of lines, we estimate for each line its emitting region by using the 15 and 85% radial and vertical contributions to the total line flux. We delimit the region at the lower end by the local dust continuum optical depth 1. Over this emitting region, we extract certain metadata for comparison to the retrieval results: the minimum and maximum gas temperatures, flux-weighted column densities, the inner/outer radius of the radial emitting region, and mass-averaged densities of potential collision partners, such as electrons, H, and H_2. It is interesting to note that the emitting region for the different individual CO_2 lines in the band span quite a range of radial emitting areas, with the strongest Q-branch lines originating the furthest out and thus also showing the lowest temperatures. It is important to note that the retrieval method is very biased towards the Q-branch when applying the χ^2-fitting approach outlined above. Even more puzzling is that the column density found in the retrieval is an order of magnitude lower than what the thermochemical models show. This is only a starting point and more in-depth work is underway, including the effects of opacity-overlap within molecular bands and also more elaborate fitting procedures/codes, such as CLIcK.[63]

The main warning to take away so far is to not use column density ratios from slab model retrieval to obtain abundances unless it is clear that the two molecules are co-spatial.

4.4 Degeneracies

The higher spectral resolution of JWST MIRI/MRS compared to Spitzer/IRS allows us to spectrally resolve the shape of the molecular Q-branches. This is a key element to break the degeneracies between column densities and temperatures in slab models. In several cases,[40,64] the JWST data is now better fitted with a high optical depth (high column density) slab model. For optically thin emission, large degeneracies remain between the column density and the emitting area.

4.5 Need for additional molecular data

The disks around BDs are very rich in hydrocarbons. Only limited data is available here *via* the HITRAN and GEISA databases. We lack especially the ^{13}C isotopologues of hydrocarbons, and the hot bands of several hydrocarbons (*e.g.*

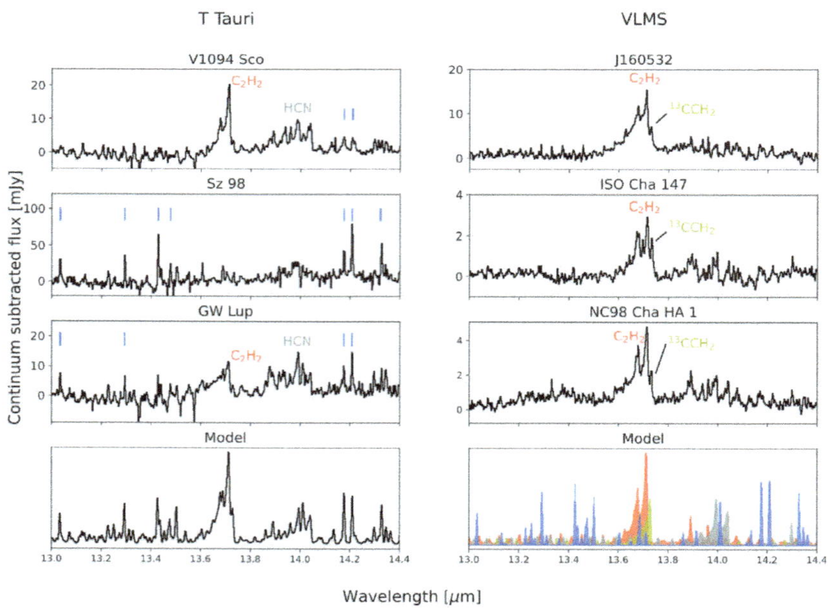

Fig. 7 Left: MIRI/MRS spectra of the T Tauri stars V1094 Sco, Sz 98 and GW Lup. Right: MIRI/MRS spectra of the VLMS and BDs J160532, ISO Cha 147 and NC98 Cha Hα1 (NC98 Cha HA1). The lowest panel on each side shows LTE slab models (T = 500 K, $N = 10^{16}$ cm^{-2}, arbitrary emitting area to match the observed spectra) of the main molecules detected in this wavelength range, C_2H_2 (red), $^{13}CCH_2$ (gold), HCN (grey) and water (blue – vertical tick marks). The lower left panel shows them co-added, while the lower right panel shows them individually with separate colors (see above).

benzene), but also more hydrocarbons in general. The GEISA database lacks the Einstein A coefficients, which are crucial for proper modelling of emission spectra from slabs/disks. We also need fits for partition functions as a function of T and P, covering the typical parameter space of inner disks (T = 100–5000 K, and P = 10^{-6}–1 mbar).

5 MINDS results

We focus here on two types of disks; those around T Tauri stars, and around very-low-mass stars (VLMS) or brown dwarfs (BDs). Their chemistry has already been found to be vastly different with Spitzer. The VLMS have higher C_2H_2/HCN and HCN/H_2O column density ratios than the T Tauri disks.[34,35] Clearly, the higher spectral resolution and sensitivity of JWST MIRI/MRS allow us to study these differences in much more detail, hopefully unraveling the cause of this profound dichotomy.

5.1 Disks around T Tauri stars

We show here a selected wavelength range centered on the two molecules C_2H_2 and HCN, including also a few stronger water lines (13–14.3 μm). Fig. 7 (left panel) shows the diversity in relative feature strength among three of our T Tauri sources: V1094 Sco (K6, d = 154 pc, and $L_* = 1.15\ L_\odot$ [65]), Sz 98 (K7, d = 156.22 pc, and $L_* = 1.53\ L_\odot$ [66]), and GW Lup (M1.5, d = 155 pc, and $L_* = 0.33\ L_\odot$ [67,68]).

GW Lup has been previously found to be a CO_2-rich source[17] and Grant *et al.*[40] recently reported the first detection of $^{13}CO_2$ and water in this source. The other two sources, Sz 98 and V1094 Sco, have only low-resolution Spitzer spectra taken and thus no previous molecule identifications in the mid-IR. Sz 98 is the strongest water source in this sample of three. Also, all three molecules, C_2H_2, water and HCN, are clearly detected in V1094 Sco; however, the molecular flux ratio of C_2H_2 to HCN is higher than in GW Lup, and the water lines are weaker.

Thus, the MIRI/MRS spectra show a pronounced diversity in feature ratio between C_2H_2 and HCN, but also in the relative strength of the water-line fluxes (indicated in blue in Fig. 7). The slab models in the lowest left panel of Fig. 7 are co-added to illustrate the richness in substructure in the HCN ro-vibrational wavelength range; with the higher spectra resolution of MIRI/MRS *versus* Spitzer, we can clearly resolve the structure and see the individual contributions of C_2H_2 and water lines in there. The question whether the diversity seen in the three T Tauri spectra is related to disk physical structure (gas + dust gaps, flaring, heating, and radial transport), different dust properties (settling, sizes, and gaps), or some other process remains open and will be discussed in Section 6.

5.2 Disks around VLMS and brown dwarfs

Similar to the T Tauri stars, we show here also three disks around VLMS/BDs: 2MASS-J16053215-1933159 (J160532, M4.75, d = 152 pc, and $L_* = 0.04\ L_\odot$ [37,69]), ISO Cha 147 (M5.75, d = 200 pc, and $L_* = 0.01\ L_\odot$ [34,70,71]), and NC98 Cha Hα1 (M7.5, d = 200 pc, and $L_* = 0.015\ L_\odot$ [71,72]).

From Fig. 7, we see that the VLMS/BD disks show structurally very strong C_2H_2 emission as already found from Spitzer spectra.[34,37] However, with the higher spectral resolution of MIRI/MRS, we now also detect in all three disks the isotopologue $^{13}CCH_2$.[64] This extra Q-branch on the red shoulder of the main isotopologue (see lowest right panel in Fig. 7, red and gold) is very pronounced in these disks compared to the T Tauri disks. Hence, they likely have much larger acetylene column densities or an unusually high isotope ratio $^{13}C/^{12}C$. In fact, the LTE slab model fitted to the MIRI spectrum of J160532 indicates an acetylene column density of $N = 3.2 \times 10^{20}$ cm^{-2}, making the emission highly optically thick and leading to a quasi-continuum.[64] Due to the overlap of these isotopologues, the slab model fitting should ideally be done for both simultaneously, and likely opacity overlap not only plays a role for within a molecule, but possibly also among isotopologues. More importantly, we are currently limited by the availability of molecular data for higher excited bands of the rare isotopologue.

In the disk around J160532, we also detected two new hydrocarbons, C_4H_2 and C_6H_6 ($N \sim 10^{17}$ cm^{-2}); there is also a tentative detection of CH_4. The emitting radius (circular emitting area πR^2) of these hydrocarbon molecules is very small within ~0.07 au. If this were to correspond to the physical radius in the disk from which the molecules emit, it would lie within the snowline expected of a typical BD disk.[84]

What is also striking is the absence of strong warm water emission in these observed disk spectra. The features at 14.2 μm, which are clearly attributed to water in the T Tauri disks, originate from C_2H_2 in these objects (see lowest right panel in Fig. 7 for the overlap of those species).

6 Discussion

6.1 Diversity in T Tauri disks: observations *versus* models

The diversity in T Tauri disk spectra is very intriguing. Many individual parameter studies have been carried out earlier and we summarize here the key findings in the light of how they affect mid-IR spectra specifically.

The gas-to-dust mass ratio in the disk surface directly impacts the strength of the molecular emission lines.[41,62] For disks with a nominal gas-to-dust mass ratio of 100, the dust continuum limits the molecular-line emitting region vertically; hence many disk models adopted a value of 1000 to match observed line fluxes (see also Fig. 5). The more the disk surface flares, the more radiation it intercepts and the warmer the gas becomes. This also boosts the strength of the molecular emission lines, radially extending the emitting area of mid-IR molecular lines.[41,48] The dust properties in the inner-disk surface are strongly affected by settling, radial drift, and growth, all of which can lower the local dust opacity. Simple parameter studies in radiation thermochemical models have shown that molecular lines become stronger if the local dust opacity is lowered.[41]

Connecting the basic understanding from those studies to disk evolutionary processes is a next step. A simplified dust evolution following the two-population approach[73] naturally lowers the dust opacity in the inner disk and together with settling this produces high gas-to-dust mass ratios in the disk surface;[74] all of these effects lead to a strengthening of mid-IR spectra in the presence of dust evolution. Ice transport along with radially drifting dust grains is shown to be very efficient and strongly affects the mid-IR spectra.[75] Coupling dust growth (pebble formation), dynamics and chemistry leads to an increasing complexity in predicting inner-disk composition;[13,76] the detailed composition will also depend on the presence of disk substructure (pressure bumps that keep icy pebbles from

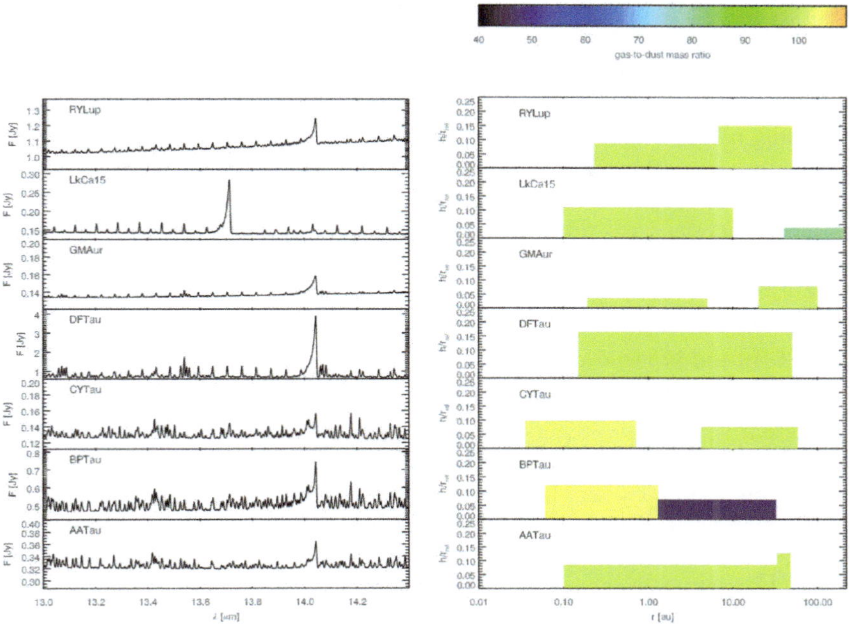

Fig. 8 (Left) Mid-infrared spectra predicted from a selected set of T Tauri DIANA disk models. (Right) Schematic structure for those selected DIANA models, showing the radial disk structure, the dimensionless scale height (scale height divided by the reference radius for the inner/outer zone, respectively) and the gas-to-dust mass ratio in color.

drifting[77]), the cold-finger effect (freezing out of ices enhancing growth and settling rates[62,78]), and details of ice condensation/diffusion/sublimation processes.[79,80]

Apart from the above advances in our understanding of how dust opacity, chemistry and disk evolution are intertwined, there are also more intricate effects directly related to molecules and line radiative transfer. The self-shielding of molecules and/or mutual shielding is a key example of that. Recent work[81–83] shows that this has a profound effect on the warm inner-disk chemistry since water self-shielding prevents the O from being unlocked as OH, inhibiting formation of CO_2. This affects the vertical location of the abundant warm reservoirs of water and CO_2 and hence also the relative strength of the CO_2-to-H_2O emission features.

Instead of parametrized standard disk models that are often used in building up understanding of the coupling of dynamics, physics, chemistry, and radiative transfer, we can also approach the interpretation from the perspective of observational diversity in disk structure. The DIANA project[58–61] has determined the physical parameters of 14 disks using multi-wavelength observational data and a coherent radiation thermochemical disk-modeling approach including dust and gas. Fig. 8 shows mid-IR spectra computed for a subset of disk models from this project. We took the models and calculated mid-IR spectra using the same selection for the molecules as used in ref. 10. We show here only the results using the simplified escape probability treatment for the line fluxes. This sample covers spectral types K5 to M1, so a range similar to the presented observed three sources. The fitting approach allowed the use of two zones that are either directly connected or separated by a gap (for details of the modeling approach and the fitting strategy, see ref. 58 and 60). We show here the same wavelength range as for the MIRI/MRS observations and also convolved the simulated spectra with a resolution of $R = 3000$. It is very intriguing to see that these models also cover a diversity of molecular feature ratios. Note, however, that these models were not tailored to fit any specific mid-IR line fluxes. The only point to be made is that the detailed inner-disk structure, *i.e.*, location of gaps, variation in scale height, and gas-to-dust mass ratio, affects the various molecular features in very different ways.

6.2 VLMS and brown-dwarf disks

The fact that these stars have much lower luminosity compared to their T Tauri counterparts is to some extent offset by the dust sublimation radius being closer to the star as well. So, to some extent, the disk surface should be self-similar in temperatures and chemistry, except that the warm region is spatially less extended and thus emitting areas should be much smaller compared to T Tauri disks.[51,84] There seems to also be no indication that the turbulence works differently across this mass range.[85] Herschel observations found that the disks around BDs may have a lower disk-to-star mass fraction than T Tauris. Various studies also show that the dynamical processes occurring in disks around VLMS and BDs can be different from their more massive counterparts. For example, dust evolution has been shown to proceed much faster in these lower-mass objects.[86]

The spectra of the disks around VLMS and BDs suggest that we are witnessing the carbon chemistry that proceeds in environments with a C/O ratio larger than one. The formation pathways for hydrocarbons in disks, especially C_2H_2, have been recently revisited (Kanwar *et al.* submitted).[40] The surface reservoir of this molecule is driven by chemical pathways that require H, H_2 and C to be abundant. C unlocking occurs *via* UV or X-ray dissociation of CO. Then there are neutral–neutral and ion–molecule pathways to drive the formation of C_2H_2. A high C/O ratio should allow for a much more radially and vertically extended hydrocarbon reservoir. The impact of this element ratio has been explored only for typical T Tauri disks[10,37,42,49] and shown to affect water, CO_2, HCN and C_2H_2 fluxes; the latter two increasing as C/O is pushed to values larger than 1. HCN increases gradually with an increasing C/O ratio, while C_2H_2 remains constant and starts increasing for C/O > 1.

The extremely high column densities of C_2H_2 observed in J160532 could be directly linked to the absence of silicate features in that disk.[64] If the absence of a clear silicate emission feature is due to the average grain size in the line-emitting region being larger than a few μm, then this implies a low opacity, and hence allows us to 'see' deep into the disk, and probe the full gas column almost down to the midplane. This difference in dust content in disks around VLMS and BDs could be linked to the faster dust radial migration discussed above.

6.3 Retrieval strategies

The limitations of using 0D slab models for retrieval are quite clear: single molecule, constant temperature, and density. From radiation thermochemical disk models, we know that the molecular line-emitting regions have radial and vertical density and temperature gradients. Also, the different molecules are not necessarily co-spatial. Hence, an approach that allows for a radial temperature gradient like CLIcK[63] is already a large improvement. The strong C_2H_2 band in the BD J160532 that requires a combination of optically thick and thin slab models to fit could indicate that we are in fact seeing that the emission is not confined to a narrow region in the disk (as assumed in a 0D slab model).

Even if sticking with 0D slab models, instead of a sequential fitting from the molecules with strongest flux contribution down to the ones with least flux contribution, simultaneous fitting of all molecules within an Markov Chain Monte Carlo (MCMC) approach should be developed. The largest issue with slab models pertaining to single molecules could be opacity overlap for isotopologues.

A full radiation thermochemical disk modeling is beyond reach for retrieval.[60] However, disk models fitted to a large range of observables of individual sources provide a great laboratory to enable deeper physical understanding and context, even to the 0D slab model results.

7 Outlook

JWST MIRI/MRS has tremendous potential to revolutionize our understanding of the chemistry in the inner 10 au of planet-forming disks. From the results shown above, we are left with some key questions: is the dichotomy between T Tauri and BD disk chemical composition a real dichotomy or is there a continuous spectrum? If there is a dichotomy, where is the break in SpType? What causes the difference in chemistry? Why would the C/O ratio be different in disks around VLMS and brown dwarfs? Are the hydrocarbons that we detect in disks around VLMS and BDs efficiently formed (combustion chemistry) from the bottom up, or do we witness the breaking down of large PAHs/carbonaceous dust?

We urgently need complete molecular data (including Einstein A coefficients and hot bands) for hydrocarbons, but also isotopologues.

Author contributions

Investigation: Arabhavi, Argyriou, Bettoni, Christiaens, Gasman, Grant, Guadarrama, Jang, Kanwar, Morales-Calderon, Pawellek, Perotti, Rodgers-Lee, Samland, Schreiber, Scheithauer, Schwarz, Tabone, Temmink and Vlasblom; funding acquisition: Henning, Kamp, van Dishoeck and Güdel; supervision: Abergel, Barrado, Bouwman, Boccaletti, Caratti o Caratti, van Dishoeck, Geers, Glauser, Güdel, Henning, Lagage, Lahuis, Müller, Nehmé, Olofsson, Pantin, Ray, Vandenbussche, Waelkens, Waters and Wright; software: Arabhavi, Grant, Kamp, Schwarz, Tabone and Waters; writing – original draft: Kamp, and Christiaens; writing – review & editing: Henning and all authors.

Conflicts of interest

There are no conflicts to declare.

Acknowledgements

The following National and International Funding Agencies funded and supported the MIRI development: NASA; ESA; the Belgian Science Policy Office

(BELSPO); Centre Nationale d'Etudes Spatiales (CNES); the Danish National Space Centre; Deutsches Zentrum fur Luftund Raumfahrt (DLR); Enterprise Ireland; Ministerio De Economía y Competividad; the Netherlands Research School for Astronomy (NOVA); the Netherlands Organisation for Scientific Research (NWO); the Science and Technology Facilities Council; the Swiss Space Office; the Swedish National Space Agency; and the UK Space Agency. I. K. acknowledges support from the H2020-MSCA-ITN-2019 grant no. 860470 (CHAMELEON). I. K., A. M. A., and E. v. D. acknowledge support from grant TOP-1614.001.751 from the Dutch Research Council (NWO). I. K. and J. K. acknowledge funding from H2020-MSCA-ITN-2019, grant no. 860470 (CHAMELEON). A. C. G. has been supported by PRIN-INAF MAIN-STREAM 2017 "Protoplanetary disks seen through the eyes of new-generation instruments" and from PRIN-INAF 2019 "Spectroscopically tracing the disk dispersal evolution (STRADE)". G. B. thanks the Deutsche Forschungsgemeinschaft (DFG) – grant 138 325594231, FOR 2634/2. E. v. D. acknowledges support from the ERC grant 101019751 MOLDISK and the Danish National Research Foundation through the Center of Excellence "InterCat" (DNRF150). D. G. would like to thank the Research Foundation Flanders for co-financing the present research (grant number V435622N). T. H. and K. S. acknowledge support from the European Research Council under the Horizon 2020 Framework Program *via* the ERC Advanced Grant Origins 832428. B. T. is a Laureate of the Paris Region fellowship program, which is supported by the Ile-de-France Region and has received funding under the Horizon 2020 innovation framework program and Marie Sklodowska-Curie grant agreement No. 945298. O. A. and V. C. acknowledge funding from the Belgian F. R. S.-FNRS. I. A. and D. G. thank the European Space Agency (ESA) and the Belgian Federal Science Policy Office (BELSPO) for their support in the framework of the PRODEX Programme. T. P. R. acknowledges support from ERC grant 743029 EASY. D. R. L. acknowledges support from Science Foundation Ireland (grant number 21/PATH-S/9339). D. B. and M. M. C. have been funded by Spanish MCIN/AEI/10.13039/501100011033 grants PID2019-107061GB-C61 and no. MDM-2017-0737.

Notes and references

1 C. Mordasini, R. van Boekel, P. Mollière, T. Henning and B. Benneke, The Imprint of Exoplanet Formation History on Observable Present-day Spectra of Hot Jupiters, *Astrophys. J.*, 2016, **832**, 41.

2 A. J. Cridland, E. F. van Dishoeck, M. Alessi and R. E. Pudritz, Connecting planet formation and astrochemistry. C/Os and N/Os of warm giant planets and Jupiter analogues, *Astron. Astrophys.*, 2020, **642**, A229.

3 A. D. Schneider and B. Bitsch, How drifting and evaporating pebbles shape giant planets. I. Heavy element content and atmospheric C/O, *Astron. Astrophys.*, 2021, **654**, A71.

4 P. Mollière, T. Molyarova, B. Bitsch, T. Henning, A. Schneider, L. Kreidberg, C. Eistrup, R. Burn, E. Nasedkin, D. Semenov, C. Mordasini, M. Schlecker, K. R. Schwarz, S. Lacour, M. Nowak and M. Schulik, Interpreting the Atmospheric Composition of Exoplanets: Sensitivity to Planet Formation Assumptions, *Astrophys. J.*, 2022, **934**, 74.

5 D. M. Jorge, I. E. E. Kamp, L. B. F. M. Waters, P. Woitke and R. J. Spaargaren, Forming Planets Around Stars With Non-Solar Elemental Composition, *Astron. Astrophys.*, 2022, **660**, A85.

6 A. Izidoro, K. de Souza Torres, O. C. Winter and N. Haghighipour, A Compound Model for the Origin of Earth's Water, *Astrophys. J.*, 2013, **767**, 54.

7 F. J. Ciesla, G. D. Mulders, I. Pascucci and D. Apai, Volatile Delivery to Planets from Water-rich Planetesimals around Low Mass Stars, *Astrophys. J.*, 2015, **804**, 9.

8 E. A. Bergin, Y. Aikawa, G. A. Blake and E. F. van Dishoeck, The Chemical Evolution of Protoplanetary Disks, in *Protostars and Planets V*, 2007, pp. 751–766.

9 T. Henning and D. Semenov, Chemistry in Protoplanetary Disks, *Chem. Rev.*, 2013, **113**, 9016–9042.

10 P. Woitke, M. Min, W.-F. Thi, C. Roberts, A. Carmona, I. Kamp, F. Ménard and C. Pinte, Modelling mid-infrared molecular emission lines from T Tauri stars, *Astron. Astrophys.*, 2018, **618**, A57.

11 T. Birnstiel, M. Fang and A. Johansen, Dust Evolution and the Formation of Planetesimals, *Space Sci. Rev.*, 2016, **205**, 41–75.

12 S. M. Andrews, Observations of Protoplanetary Disk Structures, *Annu. Rev. Astron. Astrophys.*, 2020, **58**, 483–528.

13 S. Krijt, A. D. Bosman, K. Zhang, K. R. Schwarz, F. J. Ciesla and E. A. Bergin, CO Depletion in Protoplanetary Disks: A Unified Picture Combining Physical Sequestration and Chemical Processing, *Astrophys. J.*, 2020, **899**, 134.

14 C. Espaillat, J. Muzerolle, J. Najita, S. Andrews, Z. Zhu, N. Calvet, S. Kraus, J. Hashimoto, A. Kraus and P. D'Alessio, *Protostars and Planets VI*, 2014, p. 497.

15 D. Semenov and D. Wiebe, Chemical Evolution of Turbulent Protoplanetary Disks and the Solar Nebula, *ApJS*, 2011, **196**, 25.

16 J. S. Carr and J. R. Najita, Organic Molecules and Water in the Planet Formation Region of Young Circumstellar Disks, *Science*, 2008, **319**, 1504.

17 K. M. Pontoppidan, C. Salyk, G. A. Blake, R. Meijerink, J. S. Carr and J. Najita, A Spitzer Survey of Mid-infrared Molecular Emission from Protoplanetary Disks. I. Detection Rates, *Astrophys. J.*, 2010, **720**, 887–903.

18 C. Salyk, K. M. Pontoppidan, G. A. Blake, J. R. Najita and J. S. Carr, A Spitzer Survey of Mid-infrared Molecular Emission from Protoplanetary Disks. II. Correlations and Local Thermal Equilibrium Models, *Astrophys. J.*, 2011, **731**, 130.

19 J. R. Najita, J. S. Carr, K. M. Pontoppidan, C. Salyk, E. F. van Dishoeck and G. A. Blake, The HCN-Water Ratio in the Planet Formation Region of Disks, *Astrophys. J.*, 2013, **766**, 134.

20 G. H. Rieke, M. E. Ressler, J. E. Morrison, L. Bergeron, P. Bouchet, M. García-Marín, T. P. Greene, M. W. Regan, K. G. Sukhatme and H. Walker, The Mid-Infrared Instrument for the James Webb Space Telescope, VII: The MIRI Detectors, *Publ. Astron. Soc. Pac.*, 2015, **127**, 665.

21 G. S. Wright, D. Wright, G. B. Goodson, G. H. Rieke, G. Aitink-Kroes, J. Amiaux, A. Aricha-Yanguas, R. Azzollini, K. Banks, D. Barrado-Navascues, T. Belenguer-Davila, J. A. D. L. Bloemmart, P. Bouchet, B. R. Brandl, L. Colina, Ö. Detre, E. Diaz-Catala, P. Eccleston, S. D. Friedman, M. García-Marín, M. Güdel, A. Glasse, A. M. Glauser, T. P. Greene, U. Groezinger, T. Grundy, P. Hastings, T. Henning, R. Hofferbert, F. Hunter, N. C. Jessen,

K. Justtanont, A. R. Karnik, M. A. Khorrami, O. Krause, A. Labiano, P. O. Lagage, U. Langer, D. Lemke, T. Lim, J. Lorenzo-Alvarez, E. Mazy, N. McGowan, M. E. Meixner, N. Morris, J. E. Morrison, F. Müller, H. U. N. rgaard-Nielson, G. Olofsson, B. O'Sullivan, J. W. Pel, K. Penanen, M. B. Petach, J. P. Pye, T. P. Ray, E. Renotte, I. Renouf, M. E. Ressler, P. Samara-Ratna, S. Scheithauer, A. Schneider, B. Shaughnessy, T. Stevenson, K. Sukhatme, B. Swinyard, J. Sykes, J. Thatcher, T. Tikkanen, E. F. van Dishoeck, C. Waelkens, H. Walker, M. Wells and A. Zhender, The Mid-Infrared Instrument for the James Webb Space TelescopeII: Design and Build, *Publ. Astron. Soc. Pac.*, 2015, **127**, 595.

22 C. Alves de Oliveira, E. Moraux, J. Bouvier, G. Duchêne, H. Bouy, T. Maschberger and P. Hudelot , Spectroscopy of brown dwarf candidates in IC 348 and the determination of its substellar IMF down to planetary masses, *Astron. Astrophys.*, 2013, **549**, A123.

23 S.-Y. Tang, W.-P. Chen, P.-S. Chiang, J. Jose, G. J. Herczeg and B. Goldman, Characterization of Stellar and Substellar Members in the Coma Berenices Star Cluster, *Astrophys. J.*, 2018, **862**(2), 106.

24 K. L. Luhman, The Formation and Early Evolution of Low-Mass Stars and Brown Dwarfs, *Annu. Rev. Astron. Astrophys.*, 2012, **50**, 65–106.

25 M. Keppler, M. Benisty, A. Müller, T. Henning, R. van Boekel, F. Cantalloube, C. Ginski, R. G. van Holstein, A. L. Maire, A. Pohl, M. Samland, H. Avenhaus, J. L. Baudino, A. Boccaletti, J. de Boer, M. Bonnefoy, G. Chauvin, S. Desidera, M. Langlois, C. Lazzoni, G. D. Marleau, C. Mordasini, N. Pawellek, T. Stolker, A. Vigan, A. Zurlo, T. Birnstiel, W. Brandner, M. Feldt, M. Flock, J. Girard, R. Gratton, J. Hagelberg, A. Isella, M. Janson, A. Juhasz, J. Kemmer, Q. Kral, A. M. Lagrange, R. Launhardt, A. Matter, F. Ménard, J. Milli, P. Mollière, J. Olofsson, L. Pérez, P. Pinilla, C. Pinte, S. P. Quanz, T. Schmidt, S. Udry, Z. Wahhaj, J. P. Williams, E. Buenzli, M. Cudel, C. Dominik, R. Galicher, M. Kasper, J. Lannier, D. Mesa, D. Mouillet, S. Peretti, C. Perrot, G. Salter, E. Sissa, F. Wildi, L. Abe, J. Antichi, J. C. Augereau, A. Baruffolo, P. Baudoz, A. Bazzon, J. L. Beuzit, P. Blanchard, S. S. Brems, T. Buey, V. De Caprio, M. Carbillet, M. Carle, E. Cascone, A. Cheetham, R. Claudi, A. Costille, A. Delboulbé, K. Dohlen, D. Fantinel, P. Feautrier, T. Fusco, E. Giro, L. Gluck, C. Gry, N. Hubin, E. Hugot, M. Jaquet, D. Le Mignant, M. Llored, F. Madec, Y. Magnard, P. Martinez, D. Maurel, M. Meyer, O. Möller-Nilsson, T. Moulin, L. Mugnier, A. Origné, A. Pavlov, D. Perret, C. Petit, J. Pragt, P. Puget, P. Rabou, J. Ramos, F. Rigal, S. Rochat, R. Roelfsema, G. Rousset, A. Roux, B. Salasnich, J. F. Sauvage, A. Sevin, C. Soenke, E. Stadler, M. Suarez, M. Turatto and L. Weber, Discovery of a planetary-mass companion within the gap of the transition disk around PDS 70, *Astron. Astrophys.*, 2018, **617**, A44.

26 S. Y. Haffert, A. J. Bohn, J. de Boer, I. A. G. Snellen, J. Brinchmann, J. H. Girard, C. U. Keller and R. Bacon, Two accreting protoplanets around the young star PDS 70, *Nat. Astron.*, 2019, **3**, 749–754.

27 A. D. Bosman, S. Bruderer and E. F. van Dishoeck, CO_2 infrared emission as a diagnostic of planet-forming regions of disks, *Astron. Astrophys.*, 2017, **601**, A36.

28 A. Vallenari, A. G. A. Brown, T. Prusti, J. H. J. de Bruijne, F. Arenou, C. Babusiaux, M. Biermann, O. L. Creevey, C. Ducourant, D. W. Evans,

L. Eyer, R. Guerra, A. Hutton, C. Jordi, S. A. Klioner, U. L. Lammers, L. Lindegren, X. Luri, F. Mignard, C. Panem, D. Pourbaix, S. Randich, P. Sartoretti, C. Soubiran, P. Tanga, N. A. Walton, C. A. L. Bailer-Jones, U. Bastian, R. Drimmel, F. Jansen, D. Katz, M. G. Lattanzi, F. van Leeuwen, J. Bakker, C. Cacciari, J. Castañeda, F. De Angeli, C. Fabricius, M. Fouesneau, Y. Frémat, L. Galluccio, A. Guerrier, U. Heiter, E. Masana, R. Messineo, N. Mowlavi, C. Nicolas, K. Nienartowicz, F. Pailler, P. Panuzzo, F. Riclet, W. Roux, G. M. Seabroke, R. Sordoørcit, F. Thévenin, G. Gracia-Abril, J. Portell, D. Teyssier, M. Altmann, R. Andrae, M. Audard, I. Bellas-Velidis, K. Benson, J. Berthier, R. Blomme, P. W. Burgess, D. Busonero, G. Busso, H. Cánovas, B. Carry, A. Cellino, N. Cheek, G. Clementini, Y. Damerdji, M. Davidson, P. de Teodoro, M. Nuñez Campos, L. Delchambre, A. Dell'Oro, P. Esquej, J. Fernández-Hernández, E. Fraile, D. Garabato, P. García-Lario, E. Gosset, R. Haigron, J. L. Halbwachs, N. C. Hambly, D. L. Harrison, J. Hernández, D. Hestroffer, S. T. Hodgkin, B. Holl, K. Janßen, G. Jevardat de Fombelle, S. Jordan, A. Krone-Martins, A. C. Lanzafame, W. Löffler, O. Marchal, P. M. Marrese, A. Moitinho, K. Muinonen, P. Osborne, E. Pancino, T. Pauwels, A. Recio-Blanco, C. Reylé, M. Riello, L. Rimoldini, T. Roegiers, J. Rybizki, L. M. Sarro, C. Siopis, M. Smith, A. Sozzetti, E. Utrilla, M. van Leeuwen, U. Abbas, P. Ábrahám, A. Abreu Aramburu, C. Aerts, J. J. Aguado, M. Ajaj, F. Aldea-Montero, G. Altavilla, M. A. Álvarez, J. Alves, F. Anders, R. I. Anderson, E. Anglada Varela, T. Antoja, D. Baines, S. G. Baker, L. Balaguer-Núñez, E. Balbinot, Z. Balog, C. Barache, D. Barbato, M. Barros, M. A. Barstow, S. Bartolomé, J. L. Bassilana, N. Bauchet, U. Becciani, M. Bellazzini, A. Berihuete, M. Bernet, S. Bertone, L. Bianchi, A. Binnenfeld, S. Blanco-Cuaresma, A. Blazere, T. Boch, A. Bombrun, D. Bossini, S. Bouquillon, A. Bragaglia, L. Bramante, E. Breedt, A. Bressan, N. Brouillet, E. Brugaletta, B. Bucciarelli, A. Burlacu, A. G. Butkevich, R. Buzzi, E. Caffau, R. Cancelliere, T. Cantat-Gaudin, R. Carballo, T. Carlucci, M. I. Carnerero, J. M. Carrasco, L. Casamiquela, M. Castellani, A. Castro-Ginard, L. Chaoul, P. Charlot, L. Chemin, V. Chiaramida, A. Chiavassa, N. Chornay, G. Comoretto, G. Contursi, W. J. Cooper, T. Cornez, S. Cowell, F. Crifo, M. Cropper, M. Crosta, C. Crowley, C. Dafonte, A. Dapergolas, M. David, P. David, P. de Laverny, F. De Luise, R. De March, J. De Ridder, R. de Souza, A. de Torres, E. F. del Peloso, E. del Pozo, M. Delbo, A. Delgado, J. B. Delisle, C. Demouchy, T. E. Dharmawardena, P. Di Matteo, S. Diakite, C. Diener, E. Distefano, C. Dolding, B. Edvardsson, H. Enke, C. Fabre, M. Fabrizio, S. Faigler, G. Fedorets, P. Fernique, A. Fienga, F. Figueras, Y. Fournier, C. Fouron, F. Fragkoudi, M. Gai, A. Garcia-Gutierrez, M. Garcia-Reinaldos, M. García-Torres, A. Garofalo, A. Gavel, P. Gavras, E. Gerlach, R. Geyer, P. Giacobbe, G. Gilmore, S. Girona, G. Giuffrida, R. Gomel, A. Gomez, J. González-Núñez, I. González-Santamaría, J. J. González-Vidal, M. Granvik, P. Guillout, J. Guiraud, R. Gutiérrez-Sánchez, L. P. Guy, D. Hatzidimitriou, M. Hauser, M. Haywood, A. Helmer, A. Helmi, M. H. Sarmiento, S. L. Hidalgo, T. Hilger, N. Hładczuk, D. Hobbs, G. Holland, H. E. Huckle, K. Jardine, G. Jasniewicz, A. Jean-Antoine Piccolo, Ó. Jiménez-Arranz, A. Jorissen, J. Juaristi Campillo, F. Julbe, L. Karbevska, P. Kervella, S. Khanna, M. Kontizas, G. Kordopatis, A. J. Korn, Á. Kóspál, Z. Kostrzewa-

Rutkowska, K. Kruszyńska, M. Kun, P. Laizeau, S. Lambert, A. F. Lanza, Y. Lasne, J. F. Le Campion, Y. Lebreton, T. Lebzelter, S. Leccia, N. Leclerc, I. Lecoeur-Taibi, S. Liao, E. L. Licata, H. E. P. Lindstrøm, T. A. Lister, E. Livanou, A. Lobel, A. Lorca, C. Loup, P. Madrero Pardo, A. Magdaleno Romeo, S. Managau, R. G. Mann, M. Manteiga, J. M. Marchant, M. Marconi, J. Marcos, M. M. S. Marcos Santos, D. Marín Pina, S. Marinoni, F. Marocco, D. J. Marshall, L. M. Polo, J. M. Martín-Fleitas, G. Marton, N. Mary, A. Masip, D. Massari, A. Mastrobuono-Battisti, T. Mazeh, P. J. McMillan, S. Messina, D. Michalik, N. R. Millar, A. Mints, D. Molina, R. Molinaro, L. Molnár, G. Monari, M. Monguió, P. Montegriffo, A. Montero, R. Mor, A. Mora, R. Morbidelli, T. Morel, D. Morris, T. Muraveva, C. P. Murphy, I. Musella, Z. Nagy, L. Noval, F. Ocaña, A. Ogden, C. Ordenovic, J. O. Osinde, C. Pagani, I. Pagano, L. Palaversa, P. A. Palicio, L. Pallas-Quintela, A. Panahi, S. Payne-Wardenaar, X. Peñalosa Esteller, A. Penttilä, B. Pichon, A. M. Piersimoni, F. X. Pineau, E. Plachy, G. Plum, E. Poggio, A. Prša, L. Pulone, E. Racero, S. Ragaini, M. Rainer, C. M. Raiteri, N. Rambaux, P. Ramos, M. Ramos-Lerate, P. Re Fiorentin, S. Regibo, P. J. Richards, C. Rios Diaz, V. Ripepi, A. Riva, H. W. Rix, G. Rixon, N. Robichon, A. C. Robin, C. Robin, M. Roelens, H. R. O. Rogues, L. Rohrbasser, M. Romero-Gómez, N. Rowell, F. Royer, D. Ruz Mieres, K. A. Rybicki, G. Sadowski, A. Sáez Núñez, A. Sagristà Sellés, J. Sahlmann, E. Salguero, N. Samaras, V. Sanchez Gimenez, N. Sanna, R. Santoveña, M. Sarasso, M. Schultheis, E. Sciacca, M. Segol, J. C. Segovia, D. Ségransan, D. Semeux, S. Shahaf, H. I. Siddiqui, A. Siebert, L. Siltala, A. Silvelo, E. Slezak, I. Slezak, R. L. Smart, O. N. Snaith, E. Solano, F. Solitro, D. Souami, J. Souchay, A. Spagna, L. Spina, F. Spoto, I. A. Steele, H. Steidelmüller, C. A. Stephenson, M. Süveges, J. Surdej, L. Szabados, E. Szegedi-Elek, F. Taris, M. B. Taylo, R. Teixeira, L. Tolomei, N. Tonello, F. Torra, J. Torra, G. Torralba Elipe, M. Trabucchi, A. T. Tsounis, C. Turon, A. Ulla, N. Unger, M. V. Vaillant, E. van Dillen, W. van Reeven, O. Vanel, A. Vecchiato, Y. Viala, D. Vicente, S. Voutsinas, M. Weiler, T. Wevers, L. Wyrzykowski, A. Yoldas, P. Yvard, H. Zhao, J. Zorec, S. Zucker and T. Zwitter, Gaia Data Release 3: Summary of the content and survey properties, *arXiv* e-prints, 2022, arXiv:2208.00211v1 [astro-ph.GA].

29 C. A. Gomez Gonzalez, O. Wertz, O. Absil, V. Christiaens, D. Defrère, D. Mawet, J. Milli, P.-A. Absil, M. Van Droogenbroeck, F. Cantalloube, P. M. Hinz, A. J. Skemer, M. Karlsson and J. Surdej, VIP: Vortex Image Processing Package for High-contrast Direct Imaging, *Astron. J.*, 2017, **154**, 7.

30 S. D. Brittain, J. R. Najita and J. S. Carr, Tracing the Inner Edge of the Disk Around HD 100546 with Rovibrational CO Emission Lines, *Astrophys. J.*, 2009, **702**, 85–99.

31 A. Banzatti, K. M. Pontoppidan, C. Salyk, G. J. Herczeg, E. F. van Dishoeck and G. A. Blake, The Depletion of Water During Dispersal of Planet-forming Disk Regions, *Astrophys. J.*, 2017, **834**, 152.

32 J. Olofsson, J. C. Augereau, E. F. van Dishoeck, B. Merín, F. Lahuis, J. Kessler-Silacci, C. P. Dullemond, I. Oliveira, G. A. Blake, A. C. A. Boogert, J. M. Brown, I. Evans, N. J. V. Geers, C. Knez, J. L. Monin and K. Pontoppidan, C2D Spitzer-IRS spectra of disks around T Tauri stars. IV. Crystalline silicates, *Astron. Astrophys.*, 2009, **507**, 327–345.

33 F. Anthonioz, F. Ménard, C. Pinte, J. B. Le Bouquin, M. Benisty, W. F. Thi, O. Absil, G. Duchêne, J. C. Augereau, J. P. Berger, S. Casassus, G. Duvert, B. Lazareff, F. Malbet, R. Millan-Gabet, M. R. Schreiber, W. Traub and G. Zins, The VLTI/PIONIER near-infrared interferometric survey of southern T Tauri stars. I. First results, *Astron. Astrophys.*, 2015, **574**, A41.
34 I. Pascucci, D. Apai, K. Luhman, T. Henning, J. Bouwman, M. R. Meyer, F. Lahuis and A. Natta, The Different Evolution of Gas and Dust in Disks around Sun-Like and Cool Stars, *Astrophys. J.*, 2009, **696**, 143–159.
35 J. S. Carr and J. R. Najita, Organic Molecules and Water in the Inner Disks of T Tauri Stars, *Astrophys. J.*, 2011, **733**, 102.
36 C. Salyk, G. A. Blake, A. C. A. Boogert and J. M. Brown, CO Rovibrational Emission as a Probe of Inner Disk Structure, *Astrophys. J.*, 2011, **743**, 112.
37 I. Pascucci, G. Herczeg, J. S. Carr and S. Bruderer, The Atomic and Molecular Content of Disks around Very Low-mass Stars and Brown Dwarfs, *Astrophys. J.*, 2013, **779**, 178.
38 J. Olofsson, J. C. Augereau, E. F. van Dishoeck, B. Merín, N. Grosso, F. Ménard, G. A. Blake and J. L. Monin, C2D Spitzer-IRS spectra of disks around T Tauri stars. V. Spectral decomposition, *Astron. Astrophys.*, 2010, **520**, A39.
39 A. Banzatti, I. Pascucci, A. D. Bosman, P. Pinilla, C. Salyk, G. J. Herczeg, K. M. Pontoppidan, I. Vazquez, A. Watkins, S. Krijt, N. Hendler and F. Long, Hints for Icy Pebble Migration Feeding an Oxygen-rich Chemistry in the Inner Planet-forming Region of Disks, *Astrophys. J.*, 2020, **903**, 124.
40 S. L. Grant, E. F. van Dishoeck, B. Tabone, D. Gasman, T. Henning, I. Kamp, M. Güdel, P.-O. Lagage, G. Bettoni, G. Perotti, V. Christiaens, M. Samland, A. M. Arabhavi, I. Argyriou, A. Abergel, O. Absil, D. Barrado, A. Boccaletti, J. Bouwman, A. C. o. Garatti, V. Geers, A. M. Glauser, R. Guadarrama, H. Jang, J. Kanwar, F. Lahuis, M. Morales-Calderón, M. Mueller, C. Nehmé, G. Olofsson, E. Pantin, N. Pawellek, T. P. Ray, D. Rodgers-Lee, S. Scheithauer, J. Schreiber, K. Schwarz, M. Temmink, B. Vandenbussche, M. Vlasblom, L. B. F. M. Waters, G. Wright, L. Colina, T. R. Greve, K. Justannont and G. Östlin, MINDS. The detection of 13CO2 with JWST-MIRI indicates abundant CO2 in a protoplanetary disk, *arXiv* e-prints, 2022, arXiv:2212.08047 [astro-ph.SR].
41 S. Antonellini, I. Kamp, P. Riviere-Marichalar, R. Meijerink, P. Woitke, W.-F. Thi, M. Spaans, G. Aresu and G. Lee, Understanding the water emission in the mid- and far-IR from protoplanetary disks around T Tauri stars, *Astron. Astrophys.*, 2015, **582**, A105.
42 D. E. Anderson, G. A. Blake, L. I. Cleeves, E. A. Bergin, K. Zhang, K. R. Schwarz, C. Salyk and A. D. Bosman, Observing Carbon and Oxygen Carriers in Protoplanetary Disks at Mid-infrared Wavelengths, *Astrophys. J.*, 2021, **909**, 55.
43 S. Antonellini, I. Kamp and L. B. F. M. Waters, Mid-infrared blends and continuum signatures of dust drift and accretion in protoplanetary disks, *Astron. Astrophys.*, 2023, **672**, A92.
44 B. Tabone, M. C. van Hemert, E. F. van Dishoeck and J. H. Black, OH mid-infrared emission as a diagnostic of H_2O UV photodissociation. I. Model and application to the HH 211 shock, *Astron. Astrophys.*, 2021, **650**, A192.
45 W. F. Thi, I. Kamp, P. Woitke, G. van der Plas, R. Bertelsen and L. Wiesenfeld, Radiation thermo-chemical models of protoplanetary discs. IV. Modelling CO

ro-vibrational emission from Herbig Ae discs, *Astron. Astrophys.*, 2013, **551**, A49.

46 S. Bruderer, D. Harsono and E. F. van Dishoeck, Ro-vibrational excitation of an organic molecule (HCN) in protoplanetary disks, *Astron. Astrophys.*, 2015, **575**, A94.

47 K. Willacy and P. M. Woods, Deuterium Chemistry in Protoplanetary Disks. II. The Inner 30 AU, *Astrophys. J.*, 2009, **703**, 479–499.

48 A. J. Greenwood, I. Kamp, L. B. F. M. Waters, P. Woitke and W. F. Thi, The infrared line-emitting regions of T Tauri protoplanetary disks, *Astron. Astrophys.*, 2019, **631**, A81.

49 J. R. Najita, M. Ádámkovics and A. E. Glassgold, Formation of Organic Molecules and Water in Warm Disk Atmospheres, *Astrophys. J.*, 2011, **743**, 147.

50 M. Agúndez, J. Cernicharo and J. R. Goicoechea, Formation of simple organic molecules in inner T Tauri disks, *Astron. Astrophys.*, 2008, **483**, 831–837.

51 C. Walsh, H. Nomura and E. van Dishoeck, The molecular composition of the planet-forming regions of protoplanetary disks across the luminosity regime, *Astron. Astrophys.*, 2015, **582**, A88.

52 P. Woitke, I. Kamp and W.-F. Thi, Radiation thermo-chemical models of protoplanetary disks. I. Hydrostatic disk structure and inner rim, *Astron. Astrophys.*, 2009, **501**, 383–406.

53 H. Störzer and D. Hollenbach, Optical Line Emission from Photodissociation Regions, *Astrophys. J.*, 2000, **539**, 751–759.

54 F. L. Schöier, F. F. S. van der Tak, E. F. van Dishoeck and J. H. Black, An atomic and molecular database for analysis of submillimetre line observations, *Astron. Astrophys.*, 2005, **432**, 369–379.

55 F. F. S. van der Tak, F. Lique, A. Faure, J. H. Black and E. F. van Dishoeck, The Leiden Atomic and Molecular Database (LAMDA): Current Status, Recent Updates, and Future Plans, *Atoms*, 2020, **8**, 15.

56 I. E. Gordon, L. S. Rothman, R. J. Hargreaves, R. Hashemi, E. V. Karlovets, F. M. Skinner, E. K. Conway, C. Hill, R. V. Kochanov, Y. Tan, P. Wcisło, A. A. Finenko, K. Nelson, P. F. Bernath, M. Birk, V. Boudon, A. Campargue, K. V. Chance, A. Coustenis, B. J. Drouin, J. M. Flaud, R. R. Gamache, J. T. Hodges, D. Jacquemart, E. J. Mlawer, A. V. Nikitin, V. I. Perevalov, M. Rotger, J. Tennyson, G. C. Toon, H. Tran, V. G. Tyuterev, E. M. Adkins, A. Baker, A. Barbe, E. Canè, A. G. Császár, A. Dudaryonok, O. Egorov, A. J. Fleisher, H. Fleurbaey, A. Foltynowicz, T. Furtenbacher, J. J. Harrison, J. M. Hartmann, V. M. Horneman, X. Huang, T. Karman, J. Karns, S. Kassi, I. Kleiner, V. Kofman, F. Kwabia-Tchana, N. N. Lavrentieva, T. J. Lee, D. A. Long, A. A. Lukashevskaya, O. M. Lyulin, V. Y. Makhnev, W. Matt, S. T. Massie, M. Melosso, S. N. Mikhailenko, D. Mondelain, H. S. P. Müller, O. V. Naumenko, A. Perrin, O. L. Polyansky, E. Raddaoui, P. L. Raston, Z. D. Reed, M. Rey, C. Richard, R. Tóbiás, I. Sadiek, D. W. Schwenke, E. Starikova, K. Sung, F. Tamassia, S. A. Tashkun, J. Vander Auwera, I. A. Vasilenko, A. A. Vigasin, G. L. Villanueva, B. Vispoel, G. Wagner, A. Yachmenev and S. N. Yurchenko, The HITRAN2020 molecular spectroscopic database, *J. Quant. Spectrosc. Radiat. Transfer*, 2022, **277**, 107949.

57 T. Delahaye, R. Armante, N. A. Scott, N. Jacquinet-Husson, A. Chédin, L. Crépeau, C. Crevoisier, V. Douet, A. Perrin, A. Barbe, V. Boudon, A. Campargue, L. H. Coudert, V. Ebert, J. M. Flaud, R. R. Gamache,

D. Jacquemart, A. Jolly, F. Kwabia Tchana, A. Kyuberis, G. Li, O. M. Lyulin, L. Manceron, S. Mikhailenko, N. Moazzen-Ahmadi, H. S. P. Müller, O. V. Naumenko, A. Nikitin, V. I. Perevalov, C. Richard, E. Starikova, S. A. Tashkun, V. G. Tyuterev, J. Vander Auwera, B. Vispoel, A. Yachmenev and S. Yurchenko, The 2020 edition of the GEISA spectroscopic database, *J. Mol. Spectrosc.*, 2021, **380**, 111510.

58 P. Woitke, M. Min, C. Pinte, W.-F. Thi, I. Kamp, C. Rab, F. Anthonioz, S. Antonellini, C. Baldovin-Saavedra, A. Carmona, C. Dominik, O. Dionatos, J. Greaves, M. Güdel, J. D. Ilee, A. Liebhart, F. Ménard, L. Rigon, L. B. F. M. Waters, G. Aresu, R. Meijerink and M. Spaans, Consistent dust and gas models for protoplanetary disks. I. Disk shape, dust settling, opacities, and PAHs, *Astron. Astrophys.*, 2016, **586**, A103.

59 I. Kamp, W.-F. Thi, P. Woitke, C. Rab, S. Bouma and F. Ménard, Consistent dust and gas models for protoplanetary disks. II. Chemical networks and rates, *Astron. Astrophys.*, 2017, **607**, A41.

60 P. Woitke, I. Kamp, S. Antonellini, F. Anthonioz, C. Baldovin-Saveedra, A. Carmona, O. Dionatos, C. Dominik, J. Greaves, M. Güdel, J. D. Ilee, A. Liebhardt, F. Menard, M. Min, C. Pinte, C. Rab, L. Rigon, W. F. Thi, N. Thureau and L. B. F. M. Waters, Consistent Dust and Gas Models for Protoplanetary Disks. III. Models for Selected Objects from the FP7 DIANA Project, *Publ. Astron. Soc. Pac.*, 2019, **131**, 064301.

61 O. Dionatos, P. Woitke, M. Güdel, P. Degroote, A. Liebhart, F. Anthonioz, S. Antonellini, C. Baldovin-Saavedra, A. Carmona, C. Dominik, J. Greaves, J. D. Ilee, I. Kamp, F. Ménard, M. Min, C. Pinte, C. Rab, L. Rigon, W. F. Thi and L. B. F. M. Waters, Consistent dust and gas models for protoplanetary disks. IV. A panchromatic view of protoplanetary disks, *Astron. Astrophys.*, 2019, **625**, A66.

62 R. Meijerink, K. M. Pontoppidan, G. A. Blake, D. R. Poelman and C. P. Dullemond, Radiative Transfer Models of Mid-Infrared H_2O Lines in the Planet-Forming Region of Circumstellar Disks, *Astrophys. J.*, 2009, **704**, 1471–1481.

63 Y. Liu, I. Pascucci and T. Henning, CLIcK: a Continuum and Line fItting Kit for circumstellar disks, *Astron. Astrophys.*, 2019, **623**, A106.

64 B. Tabone, G. Bettoni, E. F. van Dishoeck, A. M. Arabhavi, S. Grant, D. Gasman, Th. Henning, I. Kamp, M. Güdel, P. O. Lagage, T. Ray, B. Vandenbussche, A. Abergel, O. Absil, I. Argyriou, D. Barrado, A. Boccaletti, J. Bouwman, A. Caratti o Garatti, V. Geers, A. M. Glauser, K. Justannont, F. Lahuis, M. Mueller, C. Nehmé, G. Olofsson, E. Pantin, S. Scheithauer, C. Waelkens, L. B. F. M. Waters, J. H. Black, V. Christiaens, R. Guadarrama, M. Morales-Calderón, H. Jang, J. Kanwar, N. Pawellek, G. Perotti, A. Perrin, D. Rodgers-Lee, M. Samland, J. Schreiber, K. Schwarz, L. Colina, G. Östlin and G. Wright, A rich hydrocarbon chemistry and high C to O ratio in the inner disk around a very low-mass star, *Nat. Astron.*, 2023, DOI: 10.1038/s41550-023-01965-3.

65 E. Sanchis, L. Testi, A. Natta, C. F. Manara, B. Ercolano, T. Preibisch, T. Henning, S. Facchini, A. Miotello, I. de Gregorio-Monsalvo, C. Lopez, K. Mužić, I. Pascucci, A. Santamaría-Miranda, A. Scholz, M. Tazzari, S. van Terwisga and J. P. Williams, Demographics of disks around young very low-mass stars and brown dwarfs in Lupus, *Astron. Astrophys.*, 2020, **633**, A114.

66 S. E. van Terwisga, E. F. van Dishoeck, P. Cazzoletti, S. Facchini, L. Trapman, J. P. Williams, C. F. Manara, A. Miotello, N. van der Marel, M. Ansdell, M. R. Hogerheijde, M. Tazzari and L. Testi, The ALMA Lupus protoplanetary disk survey: evidence for compact gas disks and molecular rings from CN, *Astron. Astrophys.*, 2019, **623**, A150.

67 J. M. Alcalá, C. F. Manara, A. Natta, A. Frasca, L. Testi, B. Nisini, B. Stelzer, J. P. Williams, S. Antoniucci, K. Biazzo, E. Covino, M. Esposito, F. Getman and E. Rigliaco, X-shooter spectroscopy of young stellar objects in Lupus. Accretion properties of class II and transitional objects, *Astron. Astrophys.*, 2017, **600**, A20.

68 S. M. Andrews, J. Huang, L. M. Pérez, A. Isella, C. P. Dullemond, N. T. Kurtovic, V. V. Guzmán, J. M. Carpenter, D. J. Wilner, S. Zhang, Z. Zhu, T. Birnstiel, X.-N. Bai, M. Benisty, A. M. Hughes, K. I. Öberg and L. Ricci, The Disk Substructures at High Angular Resolution Project (DSHARP). I. Motivation, Sample, Calibration, and Overview, *Astrophys. J.*, 2018, **869**, L41.

69 K. L. Luhman, K. A. Herrmann, E. E. Mamajek, T. L. Esplin and M. J. Pecaut, New Young Stars and Brown Dwarfs in the Upper Scorpius Association, *Astron. J.*, 2018, **156**, 76.

70 C. F. Manara, C. Mordasini, L. Testi, J. P. Williams, A. Miotello, G. Lodato and A. Emsenhuber, Constraining disk evolution prescriptions of planet population synthesis models with observed disk masses and accretion rates, *Astron. Astrophys.*, 2019, **631**, L2.

71 C. F. Manara, L. Testi, G. J. Herczeg, I. Pascucci, J. M. Alcalá, A. Natta, S. Antoniucci, D. Fedele, G. D. Mulders, T. Henning, S. Mohanty, T. Prusti and E. Rigliaco, X-shooter study of accretion in Chamaeleon I. II. A steeper increase of accretion with stellar mass for very low-mass stars?, *Astron. Astrophys.*, 2017, **604**, A127.

72 K. L. Luhman, The Stellar Population of the Chamaeleon I Star-forming Region, *ApJS*, 2007, **173**, 104–136.

73 T. Birnstiel, H. Klahr and B. Ercolano, A simple model for the evolution of the dust population in protoplanetary disks, *Astron. Astrophys.*, 2012, **539**, A148.

74 A. J. Greenwood, I. Kamp, L. B. F. M. Waters, P. Woitke and W. F. Thi, Effects of dust evolution on protoplanetary disks in the mid-infrared, *Astron. Astrophys.*, 2019, **626**, A6.

75 A. D. Bosman, A. G. G. M. Tielens and E. F. van Dishoeck, Efficiency of radial transport of ices in protoplanetary disks probed with infrared observations: the case of CO_2, *Astron. Astrophys.*, 2018, **611**, A80.

76 C. Eistrup and T. Henning, Chemical evolution in ices on drifting, planet-forming pebbles, *Astron. Astrophys.*, 2022, **667**, A160.

77 P. Pinilla, T. Birnstiel, L. Ricci, C. P. Dullemond, A. L. Uribe, L. Testi and A. Natta, Trapping dust particles in the outer regions of protoplanetary disks, *Astron. Astrophys.*, 2012, **538**, A114.

78 S. Krijt, F. J. Ciesla and E. A. Bergin, Tracing Water Vapor and Ice During Dust Growth, *Astrophys. J.*, 2016, **833**, 285.

79 H. M. Cuppen, C. Walsh, T. Lamberts, D. Semenov, R. T. Garrod, E. M. Penteado and S. Ioppolo, Grain Surface Models and Data for Astrochemistry, *Space Sci. Rev.*, 2017, **212**, 1–58.

80 D. Powell, P. Gao, R. Murray-Clay and X. Zhang, Depletion of gaseous CO in protoplanetary disks by surface-energy-regulated ice formation, *Nat. Astron.*, 2022, **6**, 1147–1155.

81 T. Bethell and E. Bergin, Formation and Survival of Water Vapor in the Terrestrial Planet-Forming Region, *Science*, 2009, **326**, 1675.

82 A. D. Bosman, E. A. Bergin, J. Calahan and S. E. Duval, Water UV-shielding in the Terrestrial Planet-forming Zone: Implications from Water Emission, *Astrophys. J.*, 2022, **930**, L26.

83 A. D. Bosman, E. A. Bergin, J. K. Calahan and S. E. Duval, Water UV-shielding in the Terrestrial Planet-forming Zone: Implications for Carbon Dioxide Emission, *Astrophys. J.*, 2022, **933**, L40.

84 A. J. Greenwood, I. Kamp, L. B. F. M. Waters, P. Woitke, W. F. Thi, Ch. Rab, G. Aresu and M. Spaans, Thermochemical modelling of brown dwarf discs, *Astrophys. J.*, 2017, **601**, A44.

85 G. D. Mulders and C. Dominik, Probing the turbulent mixing strength in protoplanetary disks across the stellar mass range: no significant variations, *Astron. Astrophys.*, 2012, **539**, A9.

86 P. Pinilla, L. H. Quiroga-Nuñez, M. Benisty, A. Natta, L. Ricci, T. Henning, G. van der Plas, T. Birnstiel, L. Testi and K. Ward-Duong, Millimeter Spectral Indices and Dust Trapping By Planets in Brown Dwarf Disks, *Astrophys. J.*, 2017, **846**, 70.

Faraday Discussions

ROYAL SOCIETY OF CHEMISTRY

PAPER

New interstellar laboratories in the molecular ring†

Olivia H. Wilkins ‡* and Geoffrey A. Blake

Received 9th January 2023, Accepted 6th February 2023
DOI: 10.1039/d3fd00003f

Much of what is known about chemistry in star-forming regions comes from observations of nearby (d < 500 pc) low-mass protostars. For chemistry in high-mass star-forming regions, several more distant ($d \sim$ 2–8 kpc), exceptionally bright molecular sources have also been the subjects of repeated observations but with concomitantly poorer linear spatial resolution. Facilities such as ALMA and JWST, however, now provide the means for observing distant sources at dramatically higher spatial resolution and sensitivity. We used the modest resolving power of the Atacama Compact Array, a dedicated subset of ALMA antennas, to carry out a pilot survey of 11 giant molecular clouds selected from the Bolocam Galactic Plane Survey [Battisti & Heyer, *Astrophys. J.*, 2014, **780**, 173] within the so-called molecular ring between about 4 and 8 kpc from the galactic center. Within our observed sample, molecular emission regions—most of which correspond to at least one (candidate) young stellar object—exhibit a range of chemical complexity and diversity. Furthermore, nine target giant molecular clouds contain well-fit methanol emission, giving us a first look at the spatial chemical variability within the objects at relatively high (compared to past observations) resolutions of ~5″. This work lays the foundation for future high angular resolution studies of gas-phase chemistry with the full ALMA.

1 Introduction

Much of what is known about chemistry in star-forming regions comes from observations of nearby low-mass protostars, which have been found to host a wide range of molecules.[1–4] Such observations have shown that low-mass stars form in the coldest ($T \sim 10$ K) and densest ($n \sim 10^4$ cm^{-3}) cores. Orion KL, the closest massive star-forming region to the solar system at a distance of ~400 pc, is often considered the archetype of high-mass star-formation,[5] and several more distant, exceptionally bright molecular sources such as Sgr B2(N) ($d \sim 8$ kpc), W33A ($d \sim 4$

California Institute of Technology, 1200 East California Boulevard, Pasadena, CA 91125, USA. E-mail: olivia.h.wilkins@outlook.com

† Electronic supplementary information (ESI) available. See DOI: https://doi.org/10.1039/d3fd00003f

‡ Present address: NASA Postdoctoral Program Fellow, NASA Goddard Space Flight Center, 8800 Greenbelt Road, Greenbelt MD 20771, USA. E-mail: olivia.h.wilkins@nasa.gov

kpc), and W3(H_2O) ($d \sim 2$ kpc) have also been the subjects of repeated observations[6–8] but with concomitantly poorer linear spatial resolution.

The so-called molecular ring, which falls at radii of 4–8 kpc from the galactic center, has been mostly absent from astrochemical observations thus far despite containing the bulk of molecular matter in the Milky Way. Specifically, the surface density of giant HII regions along the galactic equator is highest between 4 and 8 kpc and peaks at ~6 kpc.[9] Within this region of the galaxy, an estimated 85% of the gas is molecular.[10] This suggests that much of our understanding of interstellar chemistry neglects the regions of the Milky Way where most of the molecular chemistry proceeds. Thus, observations of such regions are imperative to answer questions about whether and how chemistry varies at different galactic radii. Moreover, including molecular-ring laboratories in the current astrochemical sample will push the field toward larger, and thus more statistically meaningful, sample sizes.

We present the pilot results of an unbiased molecular line survey carried out toward a sample of 11 giant molecular clouds (GMCs) in the molecular ring. These observations were carried out with the Atacama Compact Array (ACA), a dedicated subset of Atacama Large Millimeter/submillimeter Array (ALMA) antennas, to efficiently identify molecule-rich gas associated with potential young stellar objects (YSOs) and YSO candidates in anticipation of future high angular resolution studies of gas-phase chemistry with the full ALMA. In this pilot survey, six 1 GHz-wide spectral windows were used to get a first look at the chemical stockrooms in the targeted GMCs.

The sample is presented in Section 2, and the observations and data reduction are presented in Section 3. Section 4 presents an overview of the molecular line data and, where possible, derived methanol column densities and rotation temperatures. The chemical implications of these results are discussed in Section 5, where the results are also compared to previous literature. Finally, Section 6 summarizes the findings of this pilot survey.

2 Source sample

The 11 sources observed in this study were selected from a sample of 437 GMCs with gas masses reported by Battisti and Heyer.[11] The GMCs were detected by the Bolocam Galactic Plane Survey (BGPS) of 1.1 mm dust continuum emission, and gas masses were derived from optically thin $^{13}CO\, J = 1 - 0$ emission from the Five College Radio Astronomy Observatory (FCRAO) Galactic Plane Surveys.

The *Spitzer* c2d (Cores to Disks) survey confirmed a linear relationship between a cloud's mass surface density and the star formation rate therein, and further that star-forming regions are clustered about the densest cores.[12] As such, when building our sample, possible target GMCs were first screened for mass surface density

$$\Sigma_{\mathrm{GMC}} = \frac{M_{\mathrm{GMC}}}{\pi {R_{\mathrm{GMC}}}^2} \tag{1}$$

where M_{GMC} is the mass and R_{GMC} is the effective radius of the cloud. A mass surface density of 200 $M_\odot$ pc^{-2} approximately corresponds to the 10^3 cm^{-3} density threshold for "dense" molecular clouds (for $R_{\mathrm{GMC}} \sim 5$ PC) and was therefore used as the cut-off for selecting targets.

We then considered GMCs with radii ≤ 2.0″ such that they could be observed with the ACA in a single session (*i.e.*, with ≤150 pointings) and selected only those GMCs that fall within the molecular ring based on their celestial coordinates. This left us with a sample of 11 molecular-ring GMCs at kinematic distances of $d \sim 4$ kpc or $d \sim 11$ kpc. The properties of the observed targets are given in Table 1. For simplicity, we refer to these objects by their BGPS IDs throughout this paper; their official names are listed in column (2) of Table 1 for reference.

3 Observations

Observations of the 11 GMCs were carried out during Cycle 6 (Project Code: 2018.1.01259, P.I.: Wilkins) using the ACA, which is a subset of antennas from ALMA and uses up to twelve 7 meter antennas. For each source, there were two local oscillator (LO) settings in ALMA Band 6: the "lower" LO setting centered at ~226 GHz and the "upper" centered at ~251 GHz. These LO settings were chosen to target tracer and otherwise astrochemically-interesting compounds and also included a continuum window in each. Continuum windows had bandwidths of 2 GHz with 2048 channels. The passband was examined for excess line emission, and line-free channels across all spectral windows were used to measure the continuum flux. The ACA has an angular resolution of 5.5″–6.2″ in the lower LO

Table 1 Properties of the 11 targeted GMCs

BGPS ID[a]	Name[a]	α_{J2000}[a]	δ_{J2000}[a]	v_{LSR}[a] (km s^{-1})	d_{kin}[a,b] (kpc)	R_{GMC}[a] (pc)	Σ_{GMC}[c] ($M_\odot$ pc^{-2})
3053	G023.368 − 00.290	18^{h}34^{m}53.0^s	−08°37′43″	78.1	10.67 ± 2.50	5.17	352.50
3474	G025.227 + 00.289	18^{h}36^{m}15.1^s	−06°42′50″	47.1	3.16 ± 0.73	1.61	233.32
4449	G030.536 + 00.021	18^{h}46^{m}53.2^s	−02°07′22″	47.8	3.03 ± 0.14	1.53	258.36
5623	G030.952 − 00.389	18^{h}57^{m}05.5^s	+02°06′44″	55.3	$10.49^{+1.32}_{-1.23}$	5.11	308.41
6029	G040.622 − 00.139	19^{h}06^{m}01.3^s	+06°47′01″	32.8	11.01 ± 2.79	4.94	341.74
6082	G041.741 + 00.095	19^{h}07^{m}15.1^s	+07°53′09″	13.7	11.91 ± 5.85	2.63	271.51
6112	G043.079 − 00.005	19^{h}10^{m}05.7^s	+09°01′06″	14.2	11.65 ± 1.61	3.64	394.00
6120	G043.237 − 00.047	19^{h}10^{m}33.0^s	+09°08′12″	7.5	12.04 ± 2.72	5.28	682.78
6299	G049.070 − 00.350	19^{h}22^{m}46.3^s	+14°09′25″	66.0	5.57 ± 1.25	1.66	323.44
6310	G049.170 − 00.208	19^{h}22^{m}26.3^s	+14°19′26″	63.1	5.56 ± 0.86	2.67	589.39
6318	G049.255 − 00.412	19^{h}23^{m}21.0^s	+14°17′54″	66.3	5.55 ± 1.06	1.48	508.62

[a] Values taken from Battisti and Heyer.[11] [b] The uncertainties for d_{kin} were not reported by Battisti and Heyer but were calculated for use in this work following the procedure given in the ESI. [c] Calculated using eqn (1).

Table 2 ACA image properties

BGPS ID	$\sigma_{RMS,cont}$ (mJy per beam)	θ_{syn}	Field of view
3053	1.7	5.9″ × 3.8″	210″ × 210″
3474	1.6	6.3″ × 4.0″	216″ × 216″
4449	1.6	6.7″ × 3.9″	210″ × 210″
5623	2.2	9.4″ × 4.1″	210″ × 210″
6029	1.9	7.1″ × 4.2″	180″ × 180″
6082	1.4	6.4″ × 3.8″	145″ × 145″
6112	1.2	7.5″ × 4.7″	163″ × 163″
6120	1.4	7.2″ × 4.7″	186″ × 186″
6299	1.8	7.4″ × 4.4″	165″ × 165″
6310	1.2	6.3″ × 4.7″	254″ × 254″
6318	1.5	8.2″ × 2.5″	145″ × 145″

and 5.2″–5.9″ in the upper LO setting. The mosaic fields of view were chosen based on GMC angular sizes reported by Battisti and Heyer.[11]

Calibration was completed using standard CASA (version 5.4.0-70) calibration pipeline scripts. The images presented here were primary-beam-corrected and created with continuum emission estimated from line-free channels and subtracted using the **tclean** algorithm with robust weighting and a Briggs parameter of 0.5 for deconvolution. The image noise-levels and synthesized beams are presented in Table 2. In general, the continuum noise level is ≤2.2 mJy per beam. A detailed description of the observations—including calibrators, number of antennas, and dates—for each source can be found in the ESI.†

4 Results

4.1 Spectral analysis

ACA spectra were inspected to compile a preliminary census of molecular compounds present in each source. Molecular transitions were identified using the Cologne Database for Molecular Spectroscopy (CDMS)[13–15] entries included in the Splatalogue§ database for astronomical spectroscopy. Table 3 lists all identified transitions along with the GMC(s) where transitions were detected. Fig. 1 shows the "lower" ALMA Band 6 spectra for BGPS 6120, which is one of the most chemically rich targets in our sample. A variety of molecular compounds are detected across the sample, including carbon monoxide (CO, $C^{18}O$), complex (having ≥6 atoms) organic compounds (CH_3OH, CH_3CCH), ionized species ($H^{13}CO^+$, $HC^{18}O^+$, HCS^+), nitrogen-bearing species (HC_3N), oxygen-bearing species (H_2CO, $H_2{}^{13}CO$), low gas density tracers (c-C_3H_2), and compounds formed in outflows (SO, SO_2, SiO).

As seen in Table 3, BGPS 6029 and BGPS 6120 are especially molecule-rich in this limited spectral line survey. BGPS 6318 is particularly molecule-poor, containing detectable CO and $C^{18}O$ emission only. For most compounds, only one line was present in our data. Multiple CH_3OH lines are detected in most sources. Several GMCs (BGPS 6029 and BGPS 6120) contain two lines of SO.

§ https://splatalogue.online

Table 3 Molecules with at least one transition detected toward the target GMCs

Molecule	Transition	Frequency (GHz)	BGPS ID				
			3053	3474	4449	5623	6029
CO	2–1, $v = 0$	230.538	X	X	X	X	X
$C^{18}O$	2–1, $v = 0$	219.560	X	X	X	X	X
SO	6_5–5_4	219.949	X	X	X	X	X
SiO	6–5	260.518	—	—	—	X	X
SO_2	$5_{(2,4)}$–$4_{(1,3)}$	241.616	—	—	—	—	X
	$4_{(3,1)}$–$4_{(2,2)}$	255.553	—	—	—	—	X
HDO	$2_{(1,1)}$–$2_{(1,2)}$	241.562	—	—	—	—	X
	$5_{(3,3)}$–$5_{(2,4)}$	256.247	—	—	—	—	X
HCS^+	6–5	256.027	—	—	—	—	X
$H^{13}CO^+$	3–2	260.255	X	X	X	X	X
$HC^{18}O^+$	3–2	255.479	—	—	—	—	X
H_2CO	$3_{(0,3)}$–$2_{(0,2)}$	218.222	X	X	X	X	X
	$3_{(2,2)}$–$2_{(2,1)}$	218.476	X	X	X	X	X
$H_2^{13}CO$	$3_{(1,2)}$–$2_{(1,1)}$	219.909	—	X	—	—	X
HNCO	$10_{(0,10)}$–$9_{(0,9)}$	219.798	—	X	—	—	X
HC_3N	24–23	218.325	—	X	—	—	X
c-C_3H_2	$6_{(1,5)}$–$5_{(0,5)}$	217.822	X	X	X	X	X
	$5_{(1,4)}$–$4_{(2,3)}$	217.940	—	X	X	X	X
CH_3OH	$4_{(-2,3)}$–$3_{(-1,2)}$ E	218.440	X	X	X	X	X
	$8_{(1,8)}$–$7_{(-0,7)}$ E	229.759	—	X	X	X	X
	$3_{(2,1)}$–$4_{(1,4)}$ E	230.027	—	—	—	X	X
	$5_{(-0,5)}$–$4_{(-0,4)}$ E	241.700	X	X	X	X	X
	$5_{(1,5)}$–$4_{(1,4)}$ E	241.767	X	X	X	X	X
	$5_{(0,5)}$–$4_{(0,4)}$ A	241.791	X	X	X	X	X
	$5_{(4,2)}$–$5_{(4,1)}$ A	241.807	—	—	—	—	X
	$5_{(4,2)}$–$5_{(4,1)}$ E	241.813	—	—	—	—	X

Table 3 (*Contd.*)

Molecule	Transition	Frequency (GHz)	BGPS ID				
			3053	3474	4449	5623	6029
	$5_{(3,2)}$–$4_{(3,1)}$ A	241.833	—	—	—	—	X
	$5_{(2,4)}$–$4_{(2,3)}$ A	241.842	—	—	—	—	X
	$5_{(3,2)}$–$4_{(3,1)}$ E	241.852	—	—	—	—	X
	$5_{(-1,4)}$–$4_{(-1,3)}$ E	241.879	X	X	—	X	X
	$5_{(2,3)}$–$4_{(2,2)}$ A	241.888	—	—	—	—	X
	$5_{(-2,4)}$–$4_{(-2,3)}$ E	241.905	X	X	X	X	X
CH_3CCH	15_3–14_3	256.293	—	—	—	—	X
	15_2–14_2	256.317	—	—	—	—	X
	15_1–14_1	256.332	—	—	X	—	X
	15_0–14_0	256.337	—	—	X	—	X

Molecule	Transition	Frequency (GHz)	BGPS ID					
			6082	6112	6120	6299	6310	6318
CO	2–1, $v = 0$	230.538	X	X	X	X	X	X
$C^{18}O$	2–1, $v = 0$	219.560	X	X	X	X	X	X
SO	6_5–5_4	219.949	X	X	X	X	X	—
SiO	6–5	260.518	—	X	X	—	—	—
SO_2	$5_{(2,4)}$–$4_{(1,3)}$	241.616	—	—	X	—	—	—
	$4_{(3,1)}$–$4_{(2,2)}$	255.553	—	—	X	—	—	—
HDO	$2_{(1,1)}$–$2_{(1,2)}$	241.562	—	—	—	—	—	—
	$5_{(3,3)}$–$5_{(2,4)}$	256.247	—	—	X	—	—	—
HCS^+	6–5	256.027	—	—	X	—	—	—
$H^{13}CO^+$	3–2	260.255	X	X	X	X	X	—
$HC^{18}O^+$	3–2	255.479	—	—	X	—	—	—

Table 3 (*Contd.*)

Molecule	Transition	Frequency (GHz)	BGPS ID					
			6082	6112	6120	6299	6310	6318
H_2CO	$3_{(0,3)}-2_{(0,2)}$	218.222	X	X	X	X	X	—
	$3_{(2,2)}-2_{(2,1)}$	218.476	X	X	X	X	X	—
$H_2^{13}CO$	$3_{(1,2)}-2_{(1,1)}$	219.909	—	—	X	—	—	—
HNCO	$10_{(0,10)}-9_{(0,9)}$	219.798	—	—	X	—	—	—
HC_3N	24–23	218.325	—	—	—	—	—	—
c-C_3H_2	$6_{(1,5)}-5_{(0,5)}$	217.822	—	—	X	—	—	—
	$5_{(1,4)}-4_{(2,3)}$	217.940	—	—	—	—	—	—
CH_3OH	$4_{(-2,3)}-3_{(-1,2)}$ E	218.440	X	X	X	X	X	—
	$8_{(1,8)}-7_{(-0,7)}$ E	229.759	—	X	X	X	—	—
	$3_{(2,1)}-4_{(1,4)}$ E	230.027	—	—	X	—	—	—
	$5_{(-0,5)}-4_{(-0,4)}$ E	241.700	—	X	X	X	X	—
	$5_{(1,5)}-4_{(1,4)}$ E	241.767	X	X	X	X	X	—
	$5_{(0,5)}-4_{(0,4)}$ A	241.791	X	X	X	X	X	—
	$5_{(4,2)}-5_{(4,1)}$ A	241.807	—	—	—	—	—	—
	$5_{(4,2)}-5_{(4,1)}$ E	241.813	—	—	—	—	—	—
	$5_{(3,2)}-4_{(3,1)}$ A	241.833	—	—	X	—	—	—
	$5_{(2,4)}-4_{(2,3)}$ A	241.842	—	—	X	—	—	—
	$5_{(3,2)}-4_{(3,1)}$ E	241.852	—	—	—	—	—	—
	$5_{(-1,4)}-4_{(-1,3)}$ E	241.879	—	X	X	—	—	—
	$5_{(2,3)}-4_{(2,2)}$ A	241.888	—	—	X	—	—	—
	$5_{(-2,4)}-4_{(-2,3)}$ E	241.905	X	X	X	X	X	—
CH_3CCH	15_3-14_3	256.293	—	—	X	—	—	—
	15_2-14_2	256.317	—	—	X	—	—	—
	15_1-14_1	256.332	—	—	X	—	—	—
	15_0-14_0	256.337	—	—	X	—	—	—

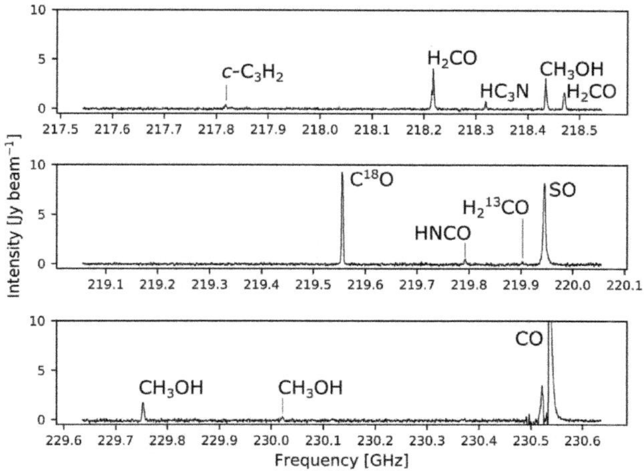

Fig. 1 ACA "lower" Band 6 spectra for BGPS 6120.

All sources except BGPS 6318 contain detectable CH_3OH and $C^{18}O$ emission concentrated around regions of 1.2 mm continuum emission. The targets (other than BGPS 6318) also all contain H_2CO, $H^{13}CO^+$, and SO. In this work, we focus on CH_3OH because multiple CH_3OH transitions are present in our data, allowing us to get a preliminary look at the gas temperature *via* spectral line fitting. Furthermore, CH_3OH is a hallmark of hot molecular cores,[2] signalling places where there may be new organic chemistry laboratories in the molecular ring.

Fig. 2 shows the regions where we see both $C^{18}O$ (a tracer of dense regions, such as cores)[16] and CH_3OH (commonly used as a tracer of other complex organics).[2,17] Each panel is labeled by its BGPS ID and an ACA# ID. For GMCs with multiple millimeter-emitting regions shown, the # was assigned by decreasing peak continuum flux. These regions are listed in Table 4.

4.2 CH_3OH rotation temperatures and column densities

Fig. 3–11 show the derived rotation temperature T_{rot} and column density N_{tot} maps for CH_3OH toward nine of the 11 target GMCs. Each map shows the entire ACA field of view (listed in Table 2) for each target. As mentioned in Section 4.1, BGPS 6318 did not contain CH_3OH emission. BGPS 6082 is also not represented in the figures because the lines in those spectra were too noisy to be fit successfully. Percent propagated uncertainty maps are presented in the ESI† and show that most fitted objects have fit uncertainties of $\leq$10% near the co-spatial continuum emission peaks.

The rotation temperature and column density maps were derived using a pixel-by-pixel fit of the spectra extracted within a single synthesized beam centered on each pixel. This follows the same procedure as Wilkins *et al.*[18] and is briefly described here. To avoid erroneous fits, only lines with peak amplitudes of $\geq 3\sigma_{RMS}$ were considered. Rotation temperature T_{rot} [K] and column density N_{tot} [cm^{-2}], along with line width and velocity shift, were determined simultaneously by least-squares fitting with LMFIT[19] (version 0.9.15) to model the spectra assuming optically thin lines at local thermodynamic equilibrium (LTE);

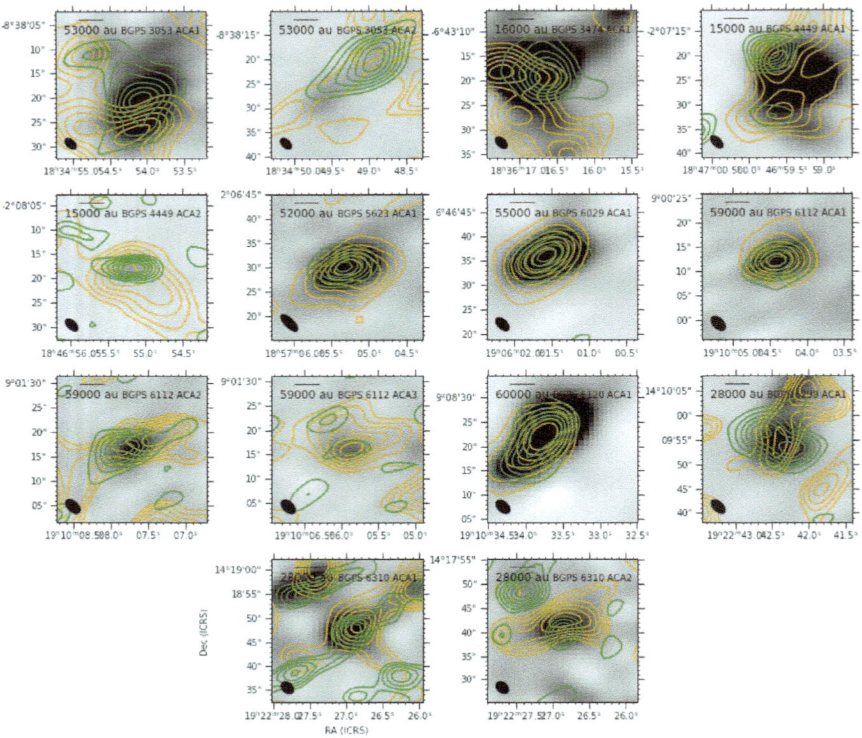

Fig. 2 Contours showing the distribution of $C^{18}O$ (orange) and CH_3OH (green) emission, represented by the integrated intensities of the 219.560 and 218.440 GHz lines, respectively, in mm sources. The greyscale images show the Band 6 (~240 GHz) continuum. The black ellipse in the bottom left of each panel shows the synthesized beam size.

commentary on the robustness of the LTE assumption is given in Section 4.3. The spectra were modeled using eqn (2) adapted from Remijan *et al.*[20]

$$N_{\text{tot}} = 2.04 \times 10^{20} C_\tau \frac{1}{\Omega_{\text{S}}(\Omega_{\text{S}} + \Omega_{\text{B}})} \frac{\int I_\nu \text{d}\nu}{\theta_{\text{a}}\theta_{\text{b}}} \frac{Q(T_{\text{rot}})\text{e}^{E_{\text{u}}/T_{\text{rot}}}}{\nu^3 S_{ij}\mu^2} \ \text{cm}^{-2} \tag{2}$$

where the $\int I_\nu \text{d}\nu$ is the integrated intensity of the line [Jy per beam km s^{-1}], the optical depth correction factor $C_\tau = \tau/(1 - \text{e}^{-\tau})$ is assumed to be unity (since we assume the optical depth τ is small), Ω_{S} and Ω_{B} are the solid angles of the source and beam respectively, θ_{a} and θ_{b} are the synthesized beam sizes [″], Q is the partition function, E_{u} is the upper-state energy level [K], ν is the rest frequency of the transition [GHz], and $S_{ij}\mu^2$ is the product of the transition line strength and the square of the electric dipole moment [Debye^2]. The lines fit in this analysis are given in Table 5.

We assume that the source fills the beam at all positions and that $\Omega \approx \pi\theta^2$ since θ is small (~5″), with the caveat that, because the objects targeted are overall not well characterized in the existing literature, this may not be a robust assumption; if there are objects in the sample where the source does not fill the beam, emission from those objects may be diluted in our beam and our column

Table 4 ALMA Band 6 (245 GHz, 1.2 mm) continuum emission regions with both $C^{18}O$ and CH_3OH emission

Source	α_{J2000}	δ_{J2000}	S_{peak} (mJy per beam)	τ	C_τ
BGPS 3053 ACA1	18h34m54.10s	−08°38′24.1″	67.3	0.06	1.03
BGPS 3053 ACA2	18h34m49.66s	−08°38′28.9″	18.1	0.04	1.02
BGPS 3474 ACA1	18h36m16.76s	−06°43′18.5″	206.9	0.34	1.18
BGPS 4449 ACA1[a] (N)	18h46m59.39s	−02°07′25.1″	273.6	0.04	1.02
BGPS 4449 ACA1[a] (S)				0.06	1.03
BGPS 4449 ACA2	18h46m55.18s	−02°08′17.7″	44.2	0.04	1.02
BGPS 5623 ACA1	18h57m05.29s	+02°06′30.1″	121.2	0.13	1.07
BGPS 6029 ACA1	19h06m01.54s	+06°46′35.2″	386.6	0.04	1.02
BGPS 6112 ACA1	19h10m04.40s	+09°01′12.3″	68.3	0.08	1.04
BGPS 6112 ACA2	19h10m07.72s	+09°01′16.5″	32.0	0.05	1.02
BGPS 6112 ACA3	19h10m05.92s	+09°01′16.0″	19.9	0.02	1.01
BGPS 6120 ACA1	19h10m33.69s	+09°08′23.3″	509.1	0.03	1.02
BGPS 6299 ACA1	19h22m42.43s	+14°09′53.1″	43.3	0.18	1.10
BGPS 6310 ACA1	19h22m26.99s	+14°18′47.3″	60.5	0.16	1.08
BGPS 6310 ACA2	19h22m26.87s	+14°17′40.2″	26.7	0.08	1.04

[a] BGPS 4449 ACA1 contains two distinct CH_3OH peaks, labeled N (north) and S (south). The optical depth τ values are listed for each CH_3OH peak. The coordinates and peak flux given are for the entire ACA1 continuum emission region.

density calculations may be lower limits. However, since the work described here is part of a pilot survey, the goal is to identify regions for follow-up high-resolution observations.

As seen in Fig. 3–11, the derived rotation temperature is less than 20 K for most objects. Under the assumptions of LTE, we can approximate the kinetic temperature of the objects by the rotation temperature. At the angular resolution of the ACA, we resolve the structure within the target sample GMCs on linear scales of $\sim10^4$ to 10^5 au. These large scales correspond to the spatial scales of dense clumps but are too large to resolve individual protostars; as such, these low temperatures are expected because warmer protostellar gas will be diluted by colder gas in the envelope.

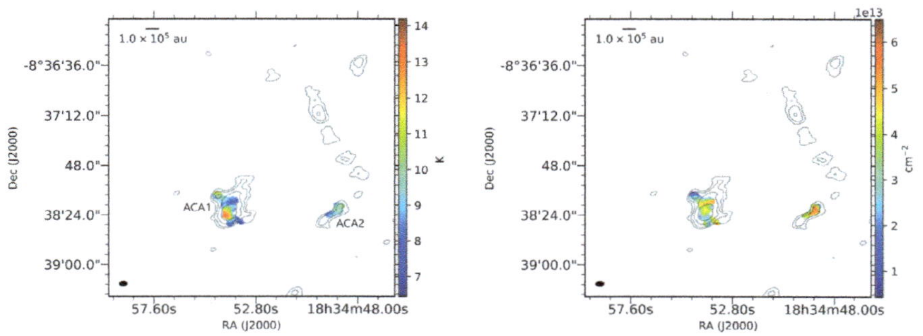

Fig. 3 Derived rotation temperature (left) and column density (right) for CH_3OH in BGPS 3053.

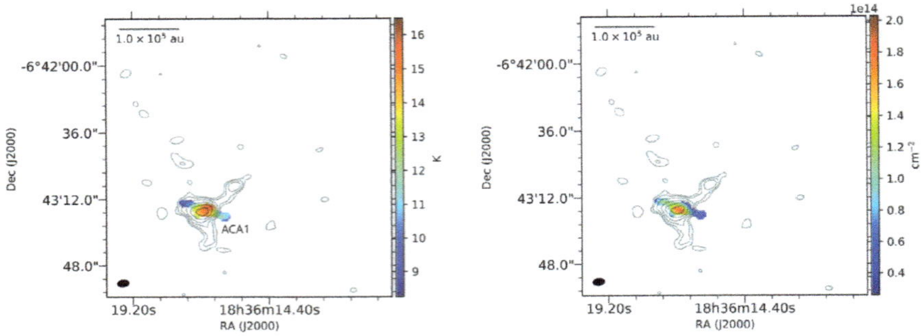

Fig. 4 Derived rotation temperature (left) and column density (right) for CH_3OH in BGPS 3474.

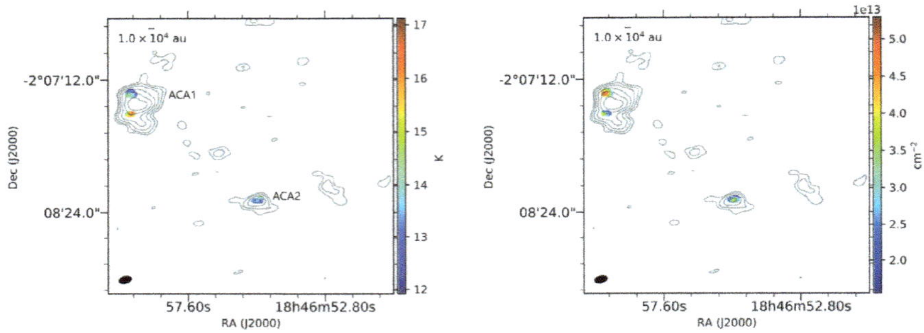

Fig. 5 Derived rotation temperature (left) and column density (right) for CH_3OH in BGPS 4449.

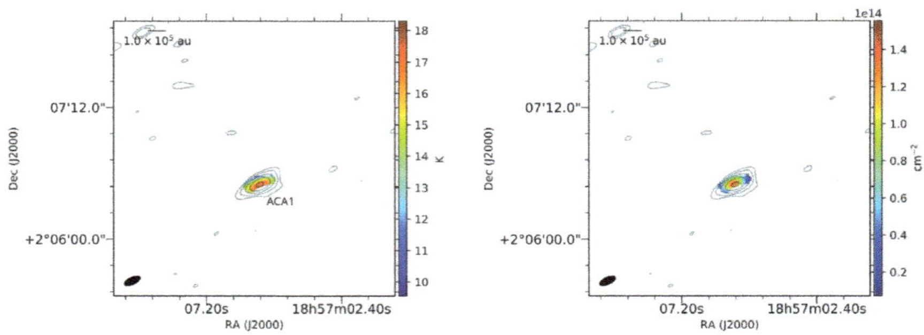

Fig. 6 Derived rotation temperature (left) and column density (right) for CH_3OH in BGPS 5623.

Two sources—BGPS 6029 (Fig. 7) and BGPS 6120 (Fig. 9)—have higher temperatures of up to 40 K and 35 K, respectively. These are still sub-hot core temperatures, but are warmer than the $\sim$10 K of the interstellar gas in which the seeds of infant stars are planted. This may indicate that the objects in these sources are more evolved and contain cores that are undergoing warm-up. As seen

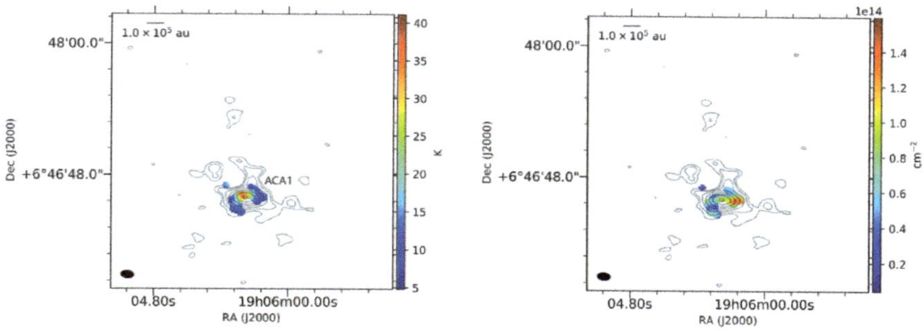

Fig. 7 Derived rotation temperature (left) and column density (right) for CH_3OH in BGPS 6029.

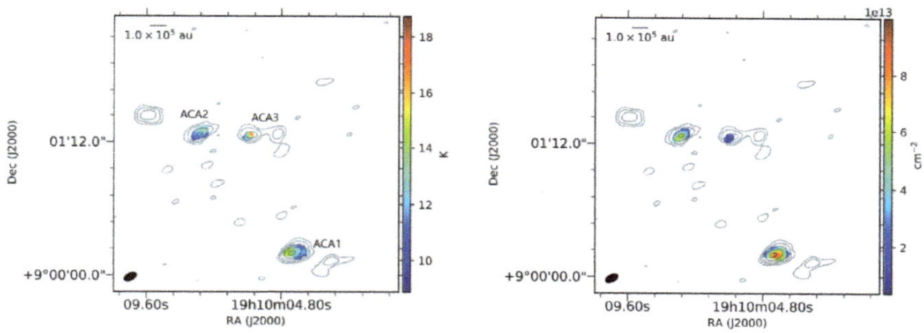

Fig. 8 Derived rotation temperature (left) and column density (right) for CH_3OH in BGPS 6112.

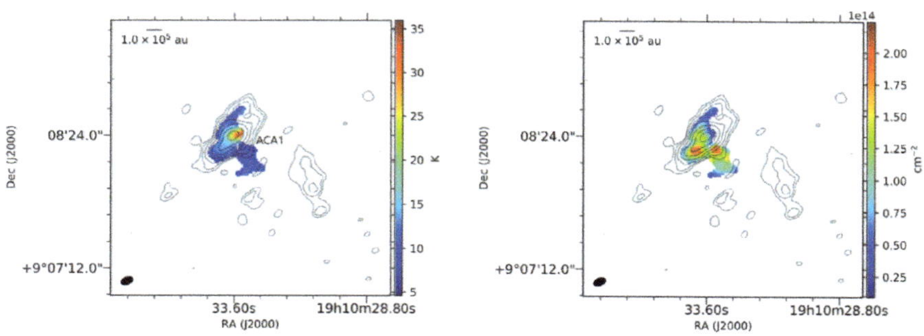

Fig. 9 Derived rotation temperature (left) and column density (right) for CH_3OH in BGPS 6120.

in Table 3, BGPS 6029 and BGPS 6120 are also the most chemically rich in our sample, which is a further sign that these regions may be more evolved than other regions identified in the sample.

CH_3OH column densities N_{tot} were derived to be on the order of 10^{13} to 10^{14} cm^{-2}, which is similar to abundances calculated for prestellar cores.[21] For the

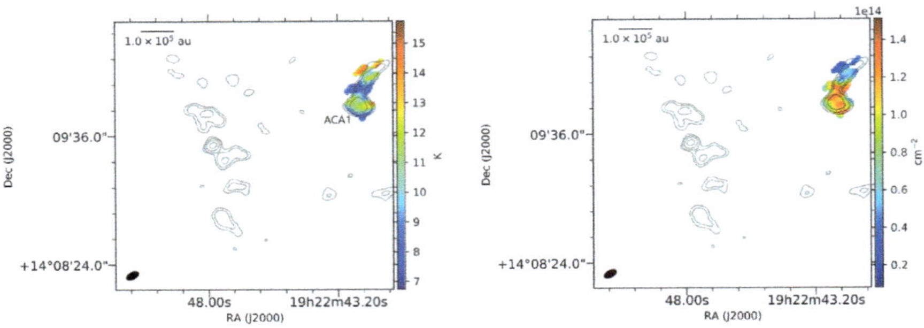

Fig. 10 Derived rotation temperature (left) and column density (right) for CH_3OH in BGPS 6299.

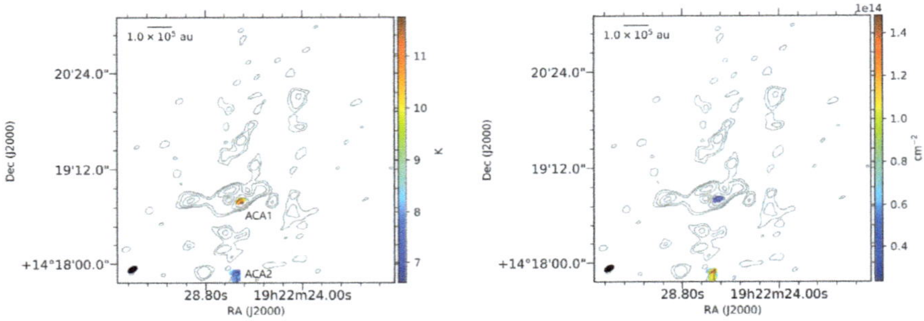

Fig. 11 Derived rotation temperature (left) and column density (right) for CH_3OH in BGPS 6310.

Table 5 CH_3OH lines fit to derive column density and rotation temperature

Transition	ν (GHz)	E_u (K)	g_u	A_{ul} (s^{-1})	$S_{ij}\mu^2$ (Debye^2)
$5_{(-0,5)}$–$4_{(-0,4)}$, E	241.7002	47.93	44	6.03×10^{-5}	16.16
$5_{(1,5)}$–$4_{(1,4)}$, E	241.7672	40.39	44	5.81×10^{-5}	15.53
$5_{(0,5)}$–$4_{(0,4)}$, A	241.7914	34.82	44	6.04×10^{-5}	16.17
$5_{(-1,4)}$–$4_{(-1,3)}$, E	241.8790	55.87	44	5.96×10^{-5}	15.92
$5_{(-2,4)}$–$4_{(-2,3)}$, E	241.9046	57.07	44	5.03×10^{-5}	13.43

most part, the sources exhibit a single peak in column density that is co-spatial with the millimeter continuum emission peak. This is the case for BGPS 3474 ACA1 (Fig. 4), BGPS 4449 ACA2 (Fig. 5), BGPS 5623 ACA1 (Fig. 6), BGPS 6112 ACA1, BGPS 6112 ACA2, BGPS 6112 ACA3 (Fig. 8), BGPS 6310 ACA1, and BGPS 6310 ACA2 (Fig. 11). The remaining sources—BGPS 4449 ACA1 (Fig. 5), BGPS 6029 ACA1 (Fig. 7), BGPS6120 ACA1 (Fig. 9), and BGPS 6299 ACA1 (Fig. 10)—have more peculiar $N_{tot}(CH_3OH)$ profiles.

BGPS 4449 ACA1 (Fig. 5) shows two fittable CH_3OH emission regions, indicating that this object may contain not one but two molecular cores. Ellsworth-Bowers *et al.*[22] reported a dense core in this region, and Saral *et al.*[23] identified this location as the site of a YSO and YSO candidates. In our analysis, the two candidate cores have different temperature and column density profiles. The northern source is slightly colder (13–14 K) but with a higher peak CH_3OH column density of $\sim 5.0 \times 10^{13}$ cm^{-2}. The southern source has temperature approaching 17 K but column densities about a factor of two lower than the northern source.

The peak column density ($\sim 1.4 \times 10^{14}$ cm^{-2}) in BGPS 6029 ACA1 is off-center to the west¶ of the continuum emission peak (Fig. 7). This is particularly curious because the maximum temperature ($\sim$40 K) in that region is nonetheless co-spatial with the continuum peak. The peaks in the measured parameters have uncertainties of $\lesssim$10% (see the ESI†), so this profile is not likely the result of poor spectral line fits. Instead, we may be seeing the overall temperature of a cluster of multiple objects early in stellar evolution. For instance, BGPS 6029 ACA1 reportedly corresponds to dense cores[24,25] and a YSO.[26,27] Lu *et al.*[25] report three distinct cores with radii of $\sim$0.03 pc ($\sim$6000 au) from ammonia (NH_3) observations with the Very Large Array (VLA). Fontani *et al.*[28] observed BGPS 6029 ACA1 as part of an astrochemical study to understand the nature of deuterium chemistry in high-mass star-forming regions. However, Fontani *et al.*[29] could not find deuterated methanol, which they used as a tracer of high-mass protostellar objects in which molecules formed through grain-surface chemistry are sputtered from the icy grain mantles, and concluded that the region is too evolved for grain-surface chemistry to be observed.

Like BGPS 4449 ACA1, the millimeter continuum region BGPS 6120 ACA1 (Fig. 9) contains multiple methanol column density peaks. There are two column density peaks of $\sim 2.0 \times 10^{14}$ cm^{-2} to the southeast and southwest of the continuum emission peak. There is another local peak with $N_{tot} \sim 1.5 \times 10^{14}$ cm^{-2} just to the east of the continuum peak. This agrees with past large surveys that reveal structures related to the early stages of star formation. For example, Morales *et al.*[30] characterized a partially embedded cluster about 66″ across in BGPS 6120. There are multiple YSOs and YSO candidates reported there as well.[23,31] Saral *et al.*[23] specifically report five YSO candidates. Our map of CH_3OH column density supports 2–3 of these being molecular cores.

BGPS 6299 ACA1 (Fig. 10) is perhaps the most abnormal fitted region in our data. It appears to be a complex comprising multiple millimeter continuum peaks. There are two regions with rotation temperatures of $\sim$12–15 K separated by a band of 7–8 K. Surprisingly, the cold band coincides with a region where $N_{tot}(CH_3OH) \sim 1.4 \times 10^{14}$ cm^{-2} that stretches south. North of the cold band, $N_{tot} \sim 6 \times 10^{13}$ cm^{-2}. The northern aspect of BGPS 6299 ACA1 has higher uncertainties of $\sim$20% whereas the cold band and the region south of it are fit well with uncertainties of $\leq$5%. The higher uncertainty to the north may be because the emission falls at the edge of the ACA's field of view and is thus noisier. In the south, the high-density region is broken up into several smaller regions, which may be indicative of a young stellar cluster. This is consistent with the existing literature, which reports five YSOs or YSO candidates in the vicinity of BGPS 6299 ACA1.[32,33] This region has also been reported to contain a dense core with a 1.00 probability of star formation as well as a H_2O maser.[26,32]

¶ In astronomical maps, north is up, south is down, east is right, and west is left.

In summary, across the nine GMCs containing fittable CH_3OH emission, there are ≥15 apparent molecular cores with CH_3OH column densities of 10^{13} to 10^{14} cm^{-2}. In general, derived rotation temperatures are ≤20 K, but a couple of regions have higher temperatures of ∼40 K.

4.3 Optical depth and LTE assumptions

The analysis in the previous section assumes that all observed CH_3OH emission is optically thin. However, CH_3OH is notoriously optically thick toward protostars.[8,34] Given the colder temperatures detected here, optically thin CH_3OH may be a more valid assumption, as is the case toward low-mass YSOs and cold prestellar cores.[4,21]

We tested this assumption by comparing the fitted parameters (N_{tot}, T_{rot}) to the CH_3OH spectrum for a pixel near the center of each emission region. The optical depth τ of a line can be calculated as

$$\tau = \frac{g_u}{\Delta V}\frac{N_{tot}}{Q(T_{rot})}\frac{A_{ul}c^3}{8\pi\nu^3}e^{-E_u/T_{rot}}\left(e^{h\nu/kT_{rot}} - 1\right) \quad (3)$$

where g_u is the degeneracy of the upper state, ΔV is the fitted line width, A_{ul} is the Einstein A coefficient, h is the Planck constant, and k is the Boltzmann constant. Other terms are the same as in eqn (2). Optical depth was calculated for each of the five lines listed in Table 5. From this, the optical depth correction factor $C_\tau = \tau/(1 - e^{-\tau})$ was calculated. Both τ and C_τ were averaged for each target and are summarized in Table 4.

Most sources in our sample have $\tau \ll 1$ or $C_\tau \sim 1$, confirming that the emission is indeed optically thin. A few sources, notably BGPS 3474 ACA1 ($C_\tau = 1.18$) and BGPS 6299 ACA1 ($C_\tau = 1.10$), have higher optical depth factors, suggesting CH_3OH may be optically thick toward the millimeter emission peaks, which would in turn result in the column densities being lower limits. Nonetheless, the assumption that the analyzed lines are optically thin is sufficient to achieve the present work's goal of gaining a preliminary look at the chemistry in previously unexplored interstellar laboratories.

Another consideration is the degree to which excitation, specifically rotation, temperatures T_{rot} reflect the kinetic temperatures T_{kin} in the targeted regions. This assumption typically holds in dense star-forming regions where $n_H \sim 10^5$ to 10^6 cm^{-3}. However, because the regions in our sample are not yet well characterized, making such an assumption is difficult. As discussed in Section 2, the target GMCs were screened to have mass surface densities ≥200 $M_\odot$ pc^{-2}, which approximately corresponds to $n_H \geq 10^3$ cm^{-3}. The mass surface density threshold is averaged over the entire field of view observed with the ACA, meaning that the continuum emission regions in which we fit the CH_3OH spectra are even denser. Many of these objects are associated with known YSOs or YSO candidates in other, large surveys,[27,32,33] suggesting a high density assumption is reasonable. Again, we stress that the purpose of this work is to survey the target GMCs for follow-up high-angular resolution studies, so the assumptions made here are sufficient.

4.4 SiO distribution

Silicon monoxide (SiO) is a reliable tracer of shock chemistry, particularly that associated with protostellar outflows.[35] SiO is produced in the gas phase from Si-bearing material sputtered off dust grains, from either the grain mantles or cores.[36] The rate of SiO production has been further shown to be influenced by the sputtering of other simple molecules, such as CO.[37]

As indicated in Table 3, SiO was detected in only a few sources in our sample: BGPS 5623, BGPS 6029, BGPS 6112, and BGPS 6120. The distribution of SiO in these GMCs is illustrated in Fig. 12. Of these, BGPS 6029 and BGPS 6120 were identified as being the most chemically rich, as well as the warmest, sources in the sample, which may indicate that these sources are more evolved than the others in the sample. Furthermore, BGPS 6029 ACA1 and BGPS 6120 ACA1 have been previously identified as active star-forming regions.[23,26,27] Thus, the presence of SiO emission, which has significantly enhanced abundances in protostellar outflows,[37–39] is to be expected.

Compared to other sources in our sample, BGPS 5623 ACA1 and BGPS 6112 ACA1 are not exceptional in terms of molecular diversity (Table 3) and temperature profile (Fig. 6 and 8, respectively). However, the presence of detectable SiO emission overlapping with the organic material traced by CH_3OH, as seen in the left column of Fig. 12, supports previous reports of YSOs in these regions.[31,32,40,41] This may also indicate that the YSOs therein are more evolved than other sources in the sample. However, the absence of a SiO detection toward other sources does not definitively preclude its presence.

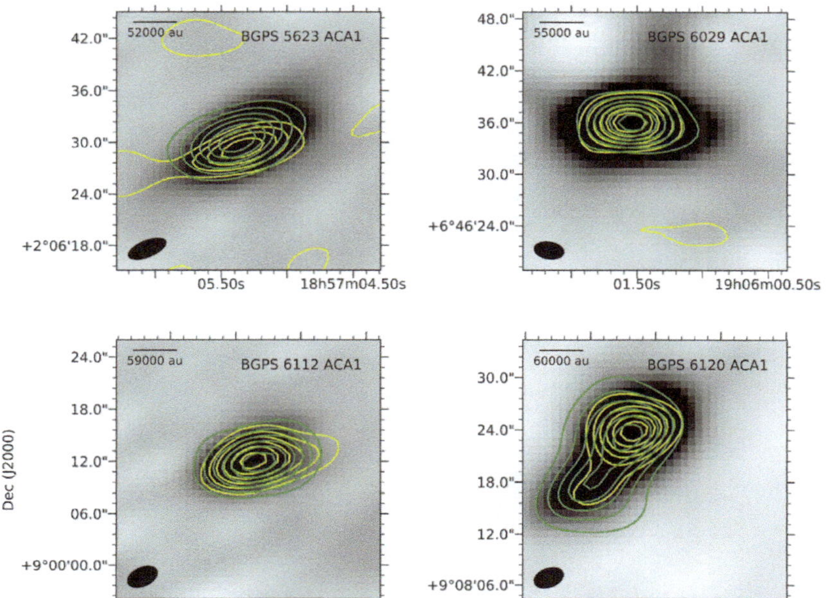

Fig. 12 Distribution of SiO showed by yellow contours compared to CH_3OH emission (green contours), represented by the integrated intensities of the 260.518 and 218.440 GHz lines, respectively. The underlying greyscale images show the Band 6 (~240 GHz) continuum. The black ellipse in the bottom left of each panel shows the synthesized beam size.

4.5 c-C_3H_2 distribution

Cyclopropenylidene (c-C_3H_2) has been suggested as a tracer of cold material in molecular clouds.[42] Observations of this molecule in L1544 showed that not only does it trace colder gas than CH_3OH, it originates in the outer, less dense layers of prestellar cores.[43]

We detected c-C_3H_2 that is cospatial with CH_3OH in BGPS 3053, BGPS 3474, BGPS 4449, BGPS 5623, and BGPS 6029. Fig. 13 shows the distribution of c-C_3H_2 (cyan contours) in these GMCs, with the CH_3OH distribution (green contours) shown for comparison. In three sources—BGPS 3474 ACA1, BGPS 5623 ACA1, and BGPS 6029 ACA1—the c-C_3H_2 emission peaks coincide with the CH_3OH peaks. In BGPS 3474 ACA1, the distribution of c-C_3H_2 is slightly less extended than CH_3OH, particularly toward the northern and southern edges of the 1.2 mm continuum emission region. In BGPS 5623 ACA1, the c-C_3H_2 emission peak is slightly west of the CH_3OH peak, but the distributions of these two molecules otherwise have similar spatial profiles. In BGPS 6029 ACA1, however, the c-C_3H_2 emission is more

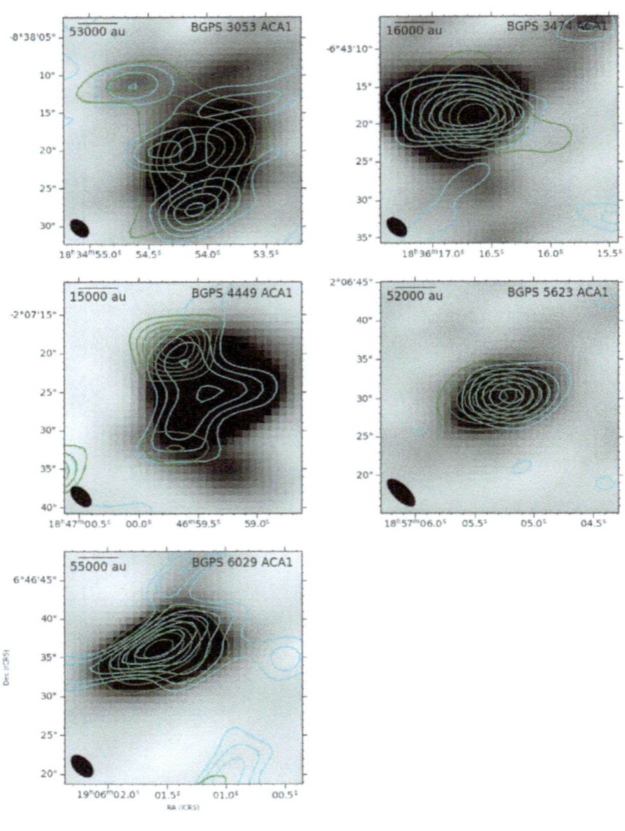

Fig. 13 Distribution of c-C_3H_2 showed by cyan contours compared to CH_3OH emission (green contours), represented by the integrated intensity (moment 0) maps of the 217.822 and 218.440 GHz lines, respectively. The underlying greyscale images show the Band 6 (~240 GHz) continuum. The black ellipse in the bottom left of each panel shows the synthesized beam size.

extended than that of CH_3OH, stretching toward the north and west. In these directions, there are also c-C_3H_2 emission peaks cospatial with continuum emission; these regions could perhaps correspond to prestellar cores in which CH_3OH has not yet formed or has not yet been liberated from the dust grains.

The c-C_3H_2 and CH_3OH profiles are much different in BGPS 3053 ACA1 and BGPS 4449 ACA1. There are at least two CH_3OH emission peaks in BGPS 3053 ACA1, including one centered on the strongest 1.2 mm continuum emission region. The second CH_3OH peak lies to the northeast and coincides with a continuum emission region that is somewhat separated from the larger complex. In this source, c-C_3H_2 has at least four peak emission regions. Only one is cospatial with a CH_3OH emission peak, specifically that to the northeast. The remaining three are distributed across the main continuum region, with the strongest peak lying to the south of the CH_3OH peak.

Similarly, the c-C_3H_2 and CH_3OH distributions do not have overlapping profiles in BGPS 4449 ACA1. In this source, the CH_3OH emission has distinct north and south peaks, whereas the c-C_3H_2 emission is more extended and spans the entire BGPS 4449 ACA1 1.2 mm continuum region. This molecular emission may thus trace the colder gas enveloping the two molecular cores traced by CH_3OH.

BGPS 6299 also contains c-C_3H_2 emission, but that emission is not cospatial with ACA1. Rather, the c-C_3H_2 in this source traces much of the continuum emission seen in the eastern half of Fig. 10 where CH_3OH was not detected. As such, the c-C_3H_2 emission in BGPS 6299 is not discussed here.

5 Discussion

The maps presented in Section 4 reveal a number of previously unexplored astrochemical laboratories in the molecular ring. Except for BGPS 6029 (in which NH_3 was previously observed[25]), the objects targeted in this pilot survey have not, as far as we know, been previously observed in terms of molecular emission. From this first look at the spectral line emission in these molecular-ring GMCs, we see a rich display of chemistry, including so-called complex organics, ionized species, nitrogen-bearing species, and tracers of protostellar outflows.

5.1 Possible snapshots of protostellar evolution

From the chemical inventories and integrated intensity and fitted parameter maps, we see that there may be objects at different stages of stellar evolution across different GMCs in the sample (*i.e.*, the warmer and more chemically-rich BGPS 6029 and BGPS 6120 in contrast with the colder and relatively molecule-poor BGPS 6310). For GMCs in which multiple YSOs and YSO candidates have been identified, there is evidence of a range of YSO stages within the individual clouds. Specifically, BGPS 6120 ACA1 contains multiple CH_3OH column density peaks that span a range of rotation temperatures from <10 K to ∼30 K (Fig. 9). Furthermore, the SiO emission in this region (Fig. 12) is less extended than that of CH_3OH and is concentrated near where the rotation temperature in ACA1 is the warmest, which may indicate more evolved material to the northwest of the ACA1 continuum region and less evolved material to the southeast.

The detection of SiO in BGPS 5623 ACA1 and BGPS 6112 ACA1 is also interesting. The other two targets in our sample with detectable SiO emission are also the most molecule-rich, but BGPS 5623 and BGPS 6112 each have about half the number of detected molecular species as BGPS 6029 and BGPS 6120 in this survey. This, along with the derived rotation temperatures for these sources, is characteristic of many of the other observed sample targets. It is possible, then, that the combination of SiO emission and cold temperatures in BGPS 5623 ACA1 and BGPS 6112 ACA1 means these objects are in an evolutionary state between those of BGPS 6029 ACA1 and BGPS 6120 ACA1 and of the CH_3OH-containing objects in which SiO was not detected. If this is the case, the sample provides several snapshots along the evolutionary sequence of YSOs in molecular-ring star-forming regions.

Furthermore, while BGPS 6112 ACA1 has detectable SiO emission, its ACA2 and ACA3 counterparts do not. This may be because BGPS 6112 ACA2 and ACA3 are, respectively, not as evolved as and more evolved than BGPS 6112 ACA1. ACA2 is slightly colder than the other two BGPS 6112 molecular emission regions (Fig. 9), and its classification as a Class I YSO candidate[23] may mean that it is too early in its evolution to be emitting detectable molecular outflows traced by SiO emission. Conversely, ACA3 is slightly warmer than ACA1 and has been classified as a Class II YSO;[23] as such, ACA3 may be nearing the end of its protostellar evolution.

Other evolutionary clues may come from c-C_3H_2 emission. The extended c-C_3H_2 emission and absence of detectable SiO emission in the northeast of BGPS 3053 ACA1 and between BGPS 4449 ACA1(N) and ACA1(S) may indicate early-stage star formation, perhaps at the prestellar stage, is present in these regions. Similarly, while BGPS 6029 ACA1 contains SiO emission, it is confined to the strongest 1.2 mm continuum emission region; the c-C_3H_2 emission is much more extended, with two cold regions to the north and west extending from the warmer CH_3OH region, suggesting this source could be a site to study how chemistry varies with temperature within a single object.

5.2 Comparison to other sources

Curiously, even the warmest regions (based on fitted CH_3OH emission) in our sample are at most ~40 K, well below typical rotation temperatures of ≥100 K for CH_3OH in hot cores such as AFGL 2591 ($d \sim 3.3$ kpc),[44] NGC 7538 IRS1 ($d \sim 2.6$ kpc),[45] and the Orion KL Hot Core ($d \sim 400$ pc).[8,18] Furthermore, CH_3OH desorbs in laboratory ice experiments at temperatures of 120–130 K.[46] Low CH_3OH rotation temperatures of <20 K have been reported toward low-mass protostars in nearby clouds, with investigators noting rotation temperatures do not necessarily reflect kinetic temperatures.[4,47] As noted in Section 4.3, the GMCs in the present sample are not yet well-characterized, making it difficult to determine whether the kinetic temperatures of the CH_3OH gas are, in fact, much higher than the derived rotation temperatures.

Methanol has been detected in prestellar cores, for example, with cold kinetic temperatures of ~10 K in Taurus Molecular Cloud,[21] thus we cannot completely rule out cold kinetic temperatures in our sample of molecular-ring GMCs. If the gas is indeed cold and the derived rotation temperatures of <40 K are not the result of beam dilution or discrepancies between excitation and kinetic temperature, a mechanism other than thermal desorption of CH_3OH following formation on icy dust grain mantles is required. One possibility is low-temperature sputtering of CH_3OH off grains by, for example, CO_2, models of which reproduced observed gas-phase CH_3OH abundances in Taurus Molecular Cloud within a factor of 2.5.[48] In addition, grain–grain collisions resulting from turbulence have been offered as an explanation for gas-phase organics in cold regions.[49] X-ray photodesorption of CH_3OH at 15 K in disks has also been demonstrated experimentally.[50] In BGPS 5623 ACA1, BGPS 6029 ACA1, BGPS 6112 ACA1, and BGPS 6120 ACA1, CH_3OH may have been liberated by shocks, which are traced by SiO.

As with temperature, the derived column densities on the order of $\sim10^{13}$ to 10^{14} cm^{-2} are much lower than those previously detected in hot cores. For example, Bisschop *et al.*[8] observed a sample of seven hot cores at distances between 1.0 and 7.7 kpc with the James Clerk Maxwell Telescope and reported CH_3OH column densities on the order of $\sim10^{16}$ to 10^{18} cm^{-2}. Similarly, IRAM 30 m observations of Orion KL by Menten *et al.*[51] found CH_3OH column densities of $\sim10^{17}$ cm^{-2}.

The derived column densities of our sample are more similar to those of 10^{13} to 10^{14} cm^{-2} calculated for prestellar cores in the Taurus Molecular Cloud.[21] However, as shown in Table 6, all of the CH_3OH emission regions for which column density maps were derived correspond to YSOs or at least YSO candidates, suggesting that these regions are at a more evolved protostellar phase. Instead, it is likely that, as perhaps with rotation temperature, the relatively low derived column densities are the result of beam dilution. As mentioned in Section 4.2, we assumed that the sources in this pilot survey fill the beam. Because these sources are not yet well-characterized, it is difficult to estimate the sizes of individual sources, especially in more complex sources like BGPS 6120 ACA1. However, it is likely that few sources fill the beam, especially in observations of more distant ($d \sim 11$ kpc) GMCs.

If we assume, for example, that the massive YSO corresponding to BGPS 5623 ACA1 (Table 6) has a radius of at most 4000 au (assuming a size similar to the Orion KL hot core, which is 10″ at a distance of ~400 pc (ref. 34))—or ~0.4″ at BGPS 5623's distance ~10.49 kpc—we can estimate that the source fills roughly 0.4% of the ACA synthesized beam. Furthermore, if beam dilution is the cause of

Table 6 Objects corresponding to previously identified YSOs

Identifier	Corresponding YSO(s)
BGPS 3053 ACA1	Embedded protostar[57]
BGPS 3474 ACA1	High-mass,[58] flat-spectrum YSO[59]
BGPS 4449 ACA1	High-mass YSO[33]
BGPS 5623 ACA1	High-mass YSO[40,41]
BGPS 6029 ACA1	High-mass YSO[26,27]
BGPS 6112 ACA1	YSO[31,32]
BGPS 6112 ACA2	Class I YSO candidate[23]
BGPS 6112 ACA3	Class II YSO[23]
BGPS 6120 ACA1	5 YSO candidates[23]
BGPS 6299 ACA1	2 YSO candidates[33]
BGPS 6310 ACA1	YSO-bearing clump[31]

the low rotation temperatures calculated for our sample, the partition function (Q in eqn (2)) would be about an order of magnitude greater for temperatures closer to ~150 K. In other words, we expect that the column densities are underestimated by at least three orders of magnitude. If this is indeed the case for all sources, the true beam-dilution-corrected CH_3OH column density is likely closer to $\sim 10^{16}$ to 10^{17} cm^{-2}. Such values would be on the lower end of, but still on par with, CH_3OH column densities derived for other hot cores, including W33A, a molecular-ring hot core with a CH_3OH column density of 2.0×10^{17} cm^{-2}.[8]

For relatively close GMCs ($d \sim$ 3–5 kpc), beam dilution is expected to be considerably less. If we again assume the YSO corresponding to BGPS 4449 ACA1 has a radius of 4000 au, we can assume that it has an angular size of ~1.3″ at BGPS 4449's distance of 3.03 kpc, which in turn suggests it fills roughly 6.5% of the ACA synthesized beam. Coupled with a higher rotation temperature of ~150 K, the true beam-dilution-corrected CH_3OH column densities for relatively nearby GMCs is likely closer to the $\sim 10^{15}$ to 10^{16} cm^{-2}, which is indeed lower than that other high-mass star-forming regions in the existing literature. This invites the question of whether star-forming regions in our molecular-ring sample have lower column densities than previously studied (massive) YSOs and suggests that the profile of "typical" high-mass YSOs in which $N_{tot}(CH_3OH) \sim 10^{16}$ cm^{-2} is not necessarily a robust representative of the early stages of massive star formation throughout the galaxy. High-angular resolution observations with the full ALMA are thus needed to determine whether the molecular-ring YSOs are indeed less exceptional in terms of molecular abundance or if the CH_3OH column density profiles from ACA observations suffer from beam dilution.

One of the big questions in astrochemistry is whether the chemical inventories and processes observed are universal across the galaxy. As discussed in Section 1, most of our understanding of chemical processes in star-forming regions comes from the local ISM. However, whether chemistry in the molecular ring is different remains to be seen despite there being some key differences between star-forming regions in the molecular ring *versus* the local ISM. One difference is that the typical astrochemical laboratories—including the Taurus Molecular Cloud and Perseus Molecular Cloud—are low-mass star-forming regions. However, high-mass stars, such as those that form in GMCs perhaps like the ones observed in this pilot study, can severely affect the formation, and thus accompanying chemistry, of neighboring YSOs.[52]

Even if the chemistry in the molecular ring is ultimately the same as what is observed in the local ISM, our current understanding of chemical evolution in the galaxy is limited by small sample sizes. This is especially true for the high-mass star-forming regions, our knowledge of which is observationally biased toward the shortest and rarest evolutionary stages located up to $d \sim$ 3 kpc.[53] Therefore, broader surveys are necessary to fully investigate the physical and chemical evolutionary sequences of such stars in a statistically meaningful way.

5.3 Outlook

Dedicated high-resolution follow-up investigations of sources in our sample would not only constrain the nature of these objects but could help astrochemists better understand chemistry in the molecular ring generally. The number of molecular-ring objects that have been studied in the context of astrochemistry is limited. The massive star-forming region W33A ($d \sim$ 4 kpc) is one such region that falls within the molecular ring at a radius of 4.2 kpc from the galactic center based on its celestial coordinates ($\alpha_{J2000} = 18^h14^m39.4^s$ $\delta_{J2000} = -17°52'01''$). For example, McCall *et al.*[54] reported on gas-phase H_3+, Dartois *et al.*[55] observed CH_3OH ice, and Gibb *et al.*[56] presented the overall ice inventory of W33A. With the resolving power of new facilities, studying these more distant molecular-ring regions, even if not exceptionally molecule-rich, will yield even more astrochemical findings.

Fortunately, the chemical emission for many of the objects presented here corresponds to previously identified YSOs and YSO candidates. Some of these were already discussed in Section 4.2. A comprehensive list, based on a thorough literature search using the SIMBAD database, is given in Table 6. The fact that many of the CH_3OH emission regions presented here correspond to objects that have been, at least preliminarily, characterized in terms of their physical aspects and evolutionary stage emphasizes the potential for these objects to provide new insights about the chemistry alongside (massive) star formation.

6 Conclusions

We present a pilot survey of 11 GMCs selected for their high mass surface density and location in the so-called molecular ring:

1. We identify at least 15 molecular cores with both $C^{18}O$ and fittable CH_3OH emission across 9 (of 11) target GMCs.
2. Rotation temperature and column density maps were derived for regions containing fittable CH_3OH emission. In general, these regions, many of which correspond to previously identified YSOs or YSO candidates, have CH_3OH rotation temperatures of ≤20 K and column densities on the order of 10^{13} to 10^{14} cm^{-2}. Two GMCs—BGPS 6029 and 6120—have peak derived rotation temperatures of 35–40 K.
3. Silicon- and sulphur-bearing species, possible evidence of shocks propagated by outflows, are found in 10/11 GMCs targeted in this survey. Specifically, SiO is detected in 4/11 GMCs, SO in 10/11, and SO_2 in 2/11. SiO emission in BGPS 5623 and BGPS 6112 may indicate that these clouds contain sources at a different evolutionary stage than other sources in the sample with similar chemical inventories.
4. Some sources harbor multiple YSOs or YSO candidates, some of which are not definitively distinguished in the existing literature. BGPS 4449 ACA1 contains two distinct CH_3OH peaks (not previously reported, to the best of our knowledge), and BGPS 6120 ACA1 contains two or three distinct CH_3OH peaks (and is the site of five YSO candidates, based on previous reports). BGPS 6299 ACA1 is also the site of multiple YSO candidates, according to the existing literature, and it seems to have multiple CH_3OH column density peaks. These sources are especially interesting because they may be targets to study variability in chemical environments within the same physical environment (*i.e.*, a single clump).

The work presented here lays the foundation for follow-up high-angular-resolution studies of the gas-phase chemistry in the molecular ring with ALMA. The variety of tracers identified in the spectra of these sources indicates that these objects contain a rich chemical inventory in which different chemical environments—from shocks propagated by outflows (*via* SiO) to cold, more diffuse gas (*via* *c*-C_3H_2)—can be investigated. Furthermore, investigations of ices in these objects (*e.g.*, with JWST), will help to constrain the coupling between gas and grain chemistry.

Author contributions

O. H. W. contributed to the conceptualization, investigation, methodology, visualization, and writing the original manuscript draft. G. A. B. contributed to conceptualization, funding acquisition, methodology, and reviewing the prepared manuscript.

Conflicts of interest

There are no conflicts to declare.

Acknowledgements

This paper makes use of the following ALMA data: ADS/JAO.ALMA#2018.1.01259. ALMA is a partnership of ESO (representing its member states), NSF (USA), and NINS (Japan), together with NRC (Canada), MOST and ASIAA (Tawain), and KASI (Republic of Korea), in cooperation with the Republic of Chile. The Joint ALMA Observatory is operated by ESO, AUI/NRAO, and NAOJ. The National Radio Astronomy Observatory (NRAO) is a facility of the National Science Foundation (NSF) operated under Associated Universities, Inc. (AUI). This research has made use of the SIMBAD database, operated at CDS, Strasbourg, France,[60] and APLpy, an open-source plotting package for Python.[61,62] During much of this work, O. H. W. was supported by an NSF Graduate Research Fellowship under Grant No. DGE-1144469 and by an ARCS Los Angeles Founder Chapter scholarship. G. A. B. gratefully acknowledges support from the NSF AAG (AST-1514918) and NASA Astrobiology (NNX15AT33A) and Exoplanet Research (XRP, NNX16AB48G) programs. This work benefited from discussions with Susanna Widicus Weaver. O. H. W thanks Sarah Wood, Melissa Hoffman, and the NRAO North American ALMA Science Center (NAASC) for their data reduction guidance.

References

1 E. F. van Dishoeck and G. A. Blake, *Annu. Rev. Astron. Astrophys.*, 1998, **36**, 317–368.

2 E. Herbst and E. F. van Dishoeck, *Annu. Rev. Astron. Astrophys.*, 2009, **47**, 427–480.

3 K. I. Öberg, N. van der Marel, L. E. Kristensen and E. F. van Dishoeck, *Astrophys. J.*, 2011, **740**, 14.

4 K. I. Öberg, T. Lauck and D. Graninger, *Astrophys. J.*, 2014, **788**, 68.

5 G. A. Blake, E. C. Sutton, C. R. Masson and T. G. Phillips, *Astrophys. J.*, 1987, **315**, 621.

6 F. F. S. van der Tak, E. F. van Dishoeck and P. Caselli, *Astron. Astrophys.*, 2000, **361**, 327–339.

7 F. F. S. van der Tak, E. F. van Dishoeck, I. Evans, J. Neal and G. A. Blake, *Astrophys. J.*, 2000, **537**, 283–303.

8 S. E. Bisschop, J. K. Jørgensen, E. F. van Dishoeck and E. B. M. de Wachter, *Astron. Astrophys.*, 2007, **465**, 913–929.

9 N. Z. Scoville and P. M. Solomon, *Astrophys. J., Lett.*, 1975, **199**, L105–L109.

10 D. B. Sanders, P. M. Solomon and N. Z. Scoville, *Astrophys. J.*, 1984, **276**, 182–203.

11 A. J. Battisti and M. H. Heyer, *Astrophys. J.*, 2014, **780**, 173.

12 I. Evans, J. Neal, M. M. Dunham, J. K. Jørgensen, M. L. Enoch, B. Merín, E. F. van Dishoeck, J. M. Alcalá, P. C. Myers, K. R. Stapelfeldt, T. L. Huard, L. E. Allen, P. M. Harvey, T. van Kempen, G. A. Blake, D. W. Koerner, L. G. Mundy, D. L. Padgett and A. I. Sargent, *Astrophys. J., Suppl. Ser.*, 2009, **181**, 321–350.

13 H. S. P. Müller, S. Thorwirth, D. A. Roth and G. Winnewisser, *Astron. Astrophys.*, 2001, **370**, L49–L52.

14 H. S. P. Müller, F. Schlöder, J. Stutzki and G. Winnewisser, *J. Mol. Struct.*, 2005, **742**, 215–227.

15 C. P. Endres, S. Schlemmer, P. Schilke, J. Stutzki and H. S. P. Müller, *J. Mol. Spectrosc.*, 2016, **327**, 95–104.

16 Z. Lin, Y. Sun, Y. Xu, J. Yang and Y. Li, *Astrophys. J., Suppl. Ser.*, 2021, **252**, 20.

17 K. I. Öberg, A. C. A. Boogert, K. M. Pontoppidan, S. van den Broek, E. F. van Dishoeck, S. Bottinelli, G. A. Blake and N. J. Evans, *Astrophys. J.*, 2011, **740**, 109.

18 O. H. Wilkins, P. B. Carroll and G. A. Blake, *Astrophys. J.*, 2022, **924**, 4.

19 M. Newville, R. Otten, A. Nelson, A. Ingargiola, T. Stensitzki, D. Allan, A. Fox, F. Carter, Michał, D. Pustakhod, Y. Ram, Glenn, C. Deil, Stuermer, A. Beelen, O. Frost, N. Zobrist, G. Pasquevich, A. L. R. Hansen, T. Spillane, S. Caldwell, A. Polloreno, A. Hannum, J. Zimmermann, J. Borreguero, J. Fraine, deep 42-thought, B. F. Maier, B. Gamari and A. Almarza, *lmfit/lmfit-py 0.9.15*, 2019, DOI: 10.5281/zenodo.3567247.

20 A. Remijan, L. E. Snyder, D. N. Friedel, S. Y. Liu and R. Y. Shah, *Astrophys. J.*, 2003, **590**, 314–332.

21 S. Scibelli and Y. Shirley, *Astrophys. J.*, 2020, **891**, 73.

22 T. P. Ellsworth-Bowers, J. Glenn, A. Riley, E. Rosolowsky, A. Ginsburg, I. Evans, J. Neal, J. Bally, C. Battersby, Y. L. Shirley and M. Merello, *Astrophys. J.*, 2015, **805**, 157.

23 G. Saral, J. L. Hora, S. E. Willis, X. P. Koenig, R. A. Gutermuth and A. T. Saygac, *Astrophys. J.*, 2015, **813**, 25.

24 H. Beuther, P. Schilke, K. M. Menten, F. Motte, T. K. Sridharan and F. Wyrowski, *Astrophys. J.*, 2002, **566**, 945–965.

25 X. Lu, Q. Zhang, H. B. Liu, J. Wang and Q. Gu, *Astrophys. J.*, 2014, **790**, 84.

26 B. E. Svoboda, Y. L. Shirley, C. Battersby, E. W. Rosolowsky, A. G. Ginsburg, T. P. Ellsworth-Bowers, M. R. Pestalozzi, M. K. Dunham, I. Evans, J. Neal, J. Bally and J. Glenn, *Astrophys. J.*, 2016, **822**, 59.

27 M. B. Areal, S. Paron, M. Celis Peña and M. E. Ortega, *Astron. Astrophys.*, 2018, **612**, A117.

28 F. Fontani, T. Sakai, K. Furuya, N. Sakai, Y. Aikawa and S. Yamamoto, *Mon. Not. R. Astron. Soc.*, 2014, **440**, 448–456.

29 F. Fontani, G. Busquet, A. Palau, P. Caselli, Á. Sánchez-Monge, J. C. Tan and M. Audard, *Astron. Astrophys.*, 2015, **575**, A87.

30 E. F. E. Morales, F. Wyrowski, F. Schuller and K. M. Menten, *Astron. Astrophys.*, 2013, **560**, A76.

31 J. S. Urquhart, C. König, A. Giannetti, S. Leurini, T. J. T. Moore, D. J. Eden, T. Pillai, M. A. Thompson, C. Braiding, M. G. Burton, T. Csengeri, J. T. Dempsey, C. Figura, D. Froebrich, K. M. Menten, F. Schuller, M. D. Smith and F. Wyrowski, *Mon. Not. R. Astron. Soc.*, 2018, **473**, 1059–1102.

32 J. S. Urquhart, T. Csengeri, F. Wyrowski, F. Schuller, S. Bontemps, L. Bronfman, K. M. Menten, C. M. Walmsley, Y. Contreras, H. Beuther, M. Wienen and H. Linz, *Astron. Astrophys.*, 2014, **568**, A41.

33 G. Saral, J. L. Hora, M. Audard, X. P. Koenig, J. R. Martínez-Galarza, F. Motte, Q. Nguyen-Luong, A. T. Saygac and H. A. Smith, *Astrophys. J.*, 2017, **839**, 108.

34 N. R. Crockett, E. A. Bergin, J. L. Neill, C. Favre, P. Schilke, D. C. Lis, T. A. Bell, G. Blake, J. Cernicharo, M. Emprechtinger, G. B. Esplugues, H. Gupta, M. Kleshcheva, S. Lord, N. Marcelino, B. A. McGuire, J. Pearson, T. G. Phillips, R. Plume, F. van der Tak, B. Tercero and S. Yu, *Astrophys. J.*, 2014, **787**, 112.

35 J. Martin-Pintado, R. Bachiller and A. Fuente, *Astron. Astrophys.*, 1992, **254**, 315–326.

36 P. Schilke, C. M. Walmsley, G. Pineau des Forets and D. R. Flower, *Astron. Astrophys.*, 1997, **321**, 293–304.

37 I. Jiménez-Serra, P. Caselli, J. Martín-Pintado and T. W. Hartquist, *Astron. Astrophys.*, 2008, **482**, 549–559.

38 I. Jiménez-Serra, J. Martín-Pintado, A. Rodríguez-Franco and S. Martín, *Astrophys. J., Lett.*, 2005, **627**, L121–L124.

39 I. Jiménez-Serra, J. Martín-Pintado, P. Caselli, S. Viti and A. Rodríguez-Franco, *Astrophys. J.*, 2009, **695**, 149–155.

40 J. M. Rathborne, J. M. Jackson and R. Simon, *Astrophys. J.*, 2006, **641**, 389–405.

41 Q. Nguyen Luong, F. Motte, M. Hennemann, T. Hill, K. L. J. Rygl, N. Schneider, S. Bontemps, A. Men'shchikov, P. André, N. Peretto, L. D. Anderson, D. Arzoumanian, L. Deharveng, P. Didelon, J. di Francesco, M. J. Griffin, J. M. Kirk, V. Könyves, P. G. Martin, A. Maury, V. Minier, S. Molinari, M. Pestalozzi, S. Pezzuto, M. Reid, H. Roussel, M. Sauvage, F. Schuller, L. Testi, D. Ward-Thompson, G. J. White and A. Zavagno, *Astron. Astrophys.*, 2011, **535**, A76.

42 J. M. Vrtilek, C. A. Gottlieb and P. Thaddeus, *Astrophys. J.*, 1987, **314**, 716.

43 Y. Lin, S. Spezzano, O. Sipilä, A. Vasyunin and P. Caselli, *Astron. Astrophys.*, 2022, **665**, A131.

44 K. G. Johnston, D. S. Shepherd, T. P. Robitaille and K. Wood, *Astron. Astrophys.*, 2013, **551**, A43.

45 L. Moscadelli, M. J. Reid, K. M. Menten, A. Brunthaler, X. W. Zheng and Y. Xu, *Astrophys. J.*, 2009, **693**, 406–412.

46 K. I. Öberg, R. T. Garrod, E. F. van Dishoeck and H. Linnartz, *Astron. Astrophys.*, 2009, **504**, 891–913.

47 D. M. Graninger, O. H. Wilkins and K. I. Öberg, *Astrophys. J.*, 2016, **819**, 140.

48 A. Paulive, J. T. Carder and E. Herbst, *Mon. Not. R. Astron. Soc.*, 2022, **516**, 4097–4110.

49 J. Kalvāns and K. Silsbee, *Mon. Not. R. Astron. Soc.*, 2022, **515**, 785–794.

50 R. Basalgète, R. Dupuy, G. Féraud, C. Romanzin, L. Philippe, X. Michaut, J. Michoud, L. Amiaud, A. Lafosse, J. H. Fillion and M. Bertin, *Astron. Astrophys.*, 2021, **647**, A35.

51 K. M. Menten, C. M. Walmsley, C. Henkel and T. L. Wilson, *Astron. Astrophys.*, 1988, **198**, 253–266.

52 C. J. Lada and E. A. Lada, *Annu. Rev. Astron. Astrophys.*, 2003, **41**, 57–115.

53 F. Motte, S. Bontemps and F. Louvet, *Annu. Rev. Astron. Astrophys.*, 2018, **56**, 41–82.

54 B. J. McCall, T. R. Geballe, K. H. Hinkle and T. Oka, *Astrophys. J.*, 1999, **522**, 338–348.

55 E. Dartois, W. Schutte, T. R. Geballe, K. Demyk, P. Ehrenfreund and L. D'Hendecourt, *Astron. Astrophys.*, 1999, **342**, L32–L35.

56 E. L. Gibb, D. C. B. Whittet, W. A. Schutte, A. C. A. Boogert, J. E. Chiar, P. Ehrenfreund, P. A. Gerakines, J. V. Keane, A. G. G. M. Tielens, E. F. van Dishoeck and O. Kerkhof, *Astrophys. J.*, 2000, **536**, 347–356.

57 S. E. Ragan, E. A. Bergin and R. A. Gutermuth, *Astrophys. J.*, 2009, **698**, 324–349.

58 M. Wienen, F. Wyrowski, F. Schuller, K. M. Menten, C. M. Walmsley, L. Bronfman and F. Motte, *Astron. Astrophys.*, 2012, **544**, A146.

59 T. Baug, L. K. Dewangan, D. K. Ojha and J. P. Ninan, *Astrophys. J.*, 2016, **833**, 85.

60 M. Wenger, F. Ochsenbein, D. Egret, P. Dubois, F. Bonnarel, S. Borde, F. Genova, G. Jasniewicz, S. Laloë, S. Lesteven and R. Monier, *Astron. Astrophys., Suppl. Ser.*, 2000, **143**, 9–22.

61 T. Robitaille and E. Bressert, *APLpy: Astronomical Plotting Library in Python*, Astrophysics Source Code Library, 2012.

62 T. Robitaille, *APLpy v2.0: The Astronomical Plotting Library in Python*, 2019, DOI: 10.5281/zenodo.2567476.

Faraday Discussions

PAPER

Streamers feeding the SVS13-A protobinary system: astrochemistry reveals accretion shocks?

Eleonora Bianchi, [*a] Ana López-Sepulcre, [bc] Cecilia Ceccarelli, [b] Claudio Codella, [db] Linda Podio, [d] Mathilde Bouvier, [e] Joan Enrique-Romero, [f] Rafael Bachiller [g] and Bertrand Lefloch [bh]

Received 27th January 2023, Accepted 16th March 2023

DOI: 10.1039/d3fd00018d

We report Atacama Large Millimeter/submillimeter Array (ALMA) high-angular resolution (~50 au) observations of the binary system SVS13-A. More specifically, we analyse deuterated water (HDO) and sulfur dioxide (SO_2) emission. The molecular emission is associated with both the components of the binary system, VLA4A and VLA4B. The spatial distribution is compared to that of formamide (NH_2CHO), previously analysed in the system. Deuterated water shows an additional emitting component spatially coincident with the dust-accretion streamer, at a distance ≥120 au from the protostars, and at blue-shifted velocities (>3 km s^{-1} from the systemic velocities). We investigate the origin of the molecular emission in the streamer, in light of thermal sublimation temperatures calculated using updated binding energy (BE) distributions. We propose that the observed emission is produced by an accretion shock at the interface between the accretion streamer and the disk of VLA4A. Thermal desorption is not completely excluded in case the source is actively experiencing an accretion burst.

[a]*Excellence Cluster ORIGINS, Boltzmannstraße 2, Garching bei München, 85748, Germany. E-mail: eleonora.bianchi@origins-cluster.de*

[b]*Univ. Grenoble Alpes, CNRS, IPAG, 38000 Grenoble, France*

[c]*Institut de Radioastronomie Millimétrique, 38406 Saint-Martin d' Hères, France*

[d]*INAF, Osservatorio Astrofisico di Arcetri, Largo E. Fermi 5, Firenze, I-50125, Italy*

[e]*Leiden Observatory, Leiden University, P.O. Box 9513, 2300 RA Leiden, The Netherlands*

[f]*Leiden Institute of Chemistry, Gorlaeus Laboratories, Leiden University, P.O. Box 9502, 2300 RA Leiden, The Netherlands*

[g]*Observatorio Astronómico Nacional (OAN-IGN), Alfonso XII 3, 28014 Madrid, Spain*

[h]*Laboratoire d'astrophysique de Bordeaux, Univ. Bordeaux, CNRS, B18N, Allée Geoffroy Saint-Hilaire, 33615 Pessac, France*

1 Introduction

Our understanding of how a star and a planetary system form has evolved substantially in recent years. On the one hand, observations have revealed the presence of substructures, such as gaps and rings, in very young protostellar disks (age $\sim 10^5$ years).[2–4] On the other hand, evolved Class II disks (age > 10^6 years) do not contain enough dust mass to explain the observed exoplanet population, in contrast to early Class 0 and I disks, which are more massive.[5] These indications suggest that planetesimal formation has started already in the early phases of protostellar disks. In this respect, it is of paramount importance to investigate the properties of young Class 0 (age $\sim 10^4$ years) and Class I disks (age $\sim 10^5$ years) in order to define the initial conditions for planet formation. Since young disks are typically deeply embedded in the parent envelope, the characterisation of their physical and chemical properties is not trivial from an observation point of view. If simple parameters, such as the disk dust and gas mass, are already difficult to measure (see *e.g.* ref. 6 and 7), their chemical composition is even less explored. Nevertheless, the molecular complexity present in the disk will be inherited from the forming planets, at least in the outer regions (see *e.g.* ref. 8 and 9 and references therein). Shedding light on the chemical composition of young disks is the only way to investigate the initial chemical budget available for planets when they start to form. Because of an important envelope contribution, the standard gas tracers (*e.g.* CO isotopologues) fail in the observation of the inner disk regions of young protostars. Other tracers, such as interstellar complex organic molecules (iCOMs; molecules with 6 or more atoms and based on carbon; see *e.g.* ref. 9 and 10) have been demonstrated to be more effective in studying the complex processes happening in the disk. However, their employment requires, at the same time, sensitive observations and a better understanding of the chemical/physical processes that release the molecules we observe in the gas. In this context, a strong synergy between astronomical observations, quantum chemistry calculations and laboratory experiments is required. In this paper, we present new observations of the protobinary system SVS13-A performed using the Atacama Large Millimeter/submillimeter Array (ALMA). In section 2 we present the source background and in section 3 we describe the ALMA observations. The main results of our analysis are presented in section 4. Section 5 discusses the possible origin of the detected molecular emission, using the most up-to-date binding-energy distributions. The main conclusions are reported in section 6.

2 The protostellar binary system SVS13-A

SVS13-A is a perfect target on which to perform astrochemical studies and tackle some of the questions above. It is very well studied, and is located in Perseus, at a distance of ~299 pc (*Gaia*[11]). The system has a bolometric luminosity of 58.8 $L_\odot$,[12] and it is at the origin of the famous Herbig–Haro (HH) chain 7–11,[13] a molecular jet and a large-scale outflow.[14,15] SVS13-A is composed of a close binary system with a separation of 0.3″.[16] The proper motion of the two components, VLA4A and VLA4B, has been investigated for almost 30 years.[17,18] A third companion, called SVS13A2, is detected at a distance of ~5″.[19,20] The continuum emission at high-angular resolution shows two circumstellar disks and circumbinary material distributed in spiral arms, which suggest accretion streamers (see Fig. 1).[1,18,20] More specifically, a bright arc is observed from the position of VLA4A extending towards the south-east. Two other arcs are extending from VLA4B towards the east and west sides of VLA4A. In addition, we detect three elongated structures that extend from the northern edge of the continuum associated with VLA4A and B towards the north. For simplicity, later in the text we refer to the central arc-like structure as the northern streamer. In order to verify that these structures are not artifacts introduced by the cleaning process, we test different beam tapering. All the elongated arc-like structures are still present when using a larger circular beam. We conclude that the dust emission is tracing accretion streamers extending towards the north and the south and feeding VLA4A and VLA4B.

The optical source position measured using *Gaia* is very close to VLA4B, even though there is a non-negligible discrepancy suggesting that the protostar is very embedded and the optical emission is due to light scattered by the circumbinary material.[18] The source driving the molecular jet, as well as a microjet traced by [Fe II], has also been identified for VLA4B.[15,21] When observed at mm wavelengths, VLA4B is brighter than VLA4A by a factor of 3.8 at 7 mm and 3.7 at 9 mm,[16] while the flux becomes comparable at longer wavelengths.[5,20] The situation reverses at 1.2 mm where VLA4A is brighter by a factor of 1.5.[1] The analysis of spectral energy distribution (SEDs) reveals the presence of free–free emission for both VLA4A and VLA4B, consistent with the presence of radio jets.[18] The different spectral indices measured toward the sources, as well as the change in intensities, suggest that the dust at 1.2 mm is optically thick in VLA4A and very optically thick in VLA4B.[5,18] In this

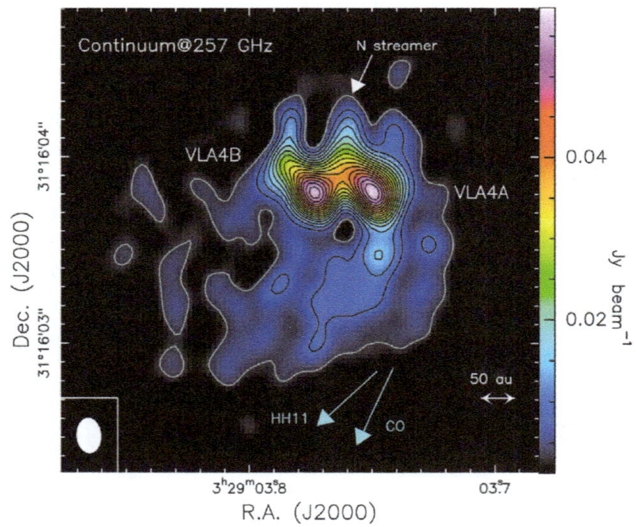

Fig. 1 Continuum emission as observed by ALMA at 1.2 mm (color scale and contours).[1] The first contours (gray) and steps (black) are 8σ, corresponding to 3.35 mJy per beam. The beam size is $0''19 \times 0''13$ (PA = +5°). The binary system is composed of VLA4A (right) and VLA4B (left). The coordinates are $\alpha_{J2000} = 03^h\ 29^m\ 3.75^s$, $\delta_{J2000} = +31°\ 16'\ 03.81''$ and $\alpha_{J2000} = 03^h\ 29^m\ 3.773^s$, $\delta_{J2000} = +31°\ 16'\ 03.81''$ for VLA4A and VLA4B, respectively. The blue arrows indicate the directions toward HH11 (133°) as well as the main axis of the H_2/CO jet (155°), as inferred by Lefèvre *et al.*[15] The white arrow indicates the position of the northern streamer discussed in the text.

context, VLA4B is interpreted as the primary and more massive component of the binary system, while VLA4A would be the less evolved secondary component. Taking into account the high uncertainties related to the free–free contamination, the disk masses have been estimated to be $\sim$0.2 $M_\odot$ for VLA4B and $\sim$0.08 $M_\odot$ for VLA4A,[5] confirming this picture. Interestingly, SVS13-A has also been deeply investigated in the NIR and found to be in a phase of active accretion.[12] The quick rise of magnitude in the K-band, followed by a slow decline, is reminiscent of FU Ori sources, while the small bursts amplitude and the line-rich spectra resemble EX Ori sources.[21]

2.1 The prototype Class I hot corinos

SVS13-A has been found to host hot corino chemistry. More specifically, young Class 0 protostars possess an inner region, typically $\leq$100 au, where the gas temperature is high enough ($\geq$100 K) to desorb the icy mantles of dust grains and release complex molecules into the gas phase, which can in turn react in the gas phase and form more complex molecular species. The hot corino phenomenon was first observed in Class 0 protostars (*e.g.* IRAS16293-2422 (ref. 22)) and only recently it was also discovered in more evolved Class I objects. In this respect, SVS13-A provided the first evidence that Class I objects can be as chemically rich as young Class 0 sources.[23] More specifically, water and iCOM emission have been detected from a compact (<100 au) and hot (>100 K) region[23–25] towards the source. Since then, the chemistry of the system has been deeply investigated thanks to several observational Large Programs including ASAI (Astrochemical Surveys At IRAM 30 m [26]), CALYPSO (Continuum and Lines in Young ProtoStellar Objects[27]) with the Plateau de Bure interferometer, SOLIS (Seed Of Life In Space[28]) with IRAM-NOEMA and PEACHES (Perseus ALMA Chemistry Survey[29]). These projects identified more than 100 emission lines emitted by iCOMs, including those of CH_3OH and its isotopologues, CH_3CHO, $HCOOCH_3$, CH_3OCH_3, $HCOCH_2OH$, NH_2CHO and CH_3CH_2OH. The broad spectral coverage of these surveys allowed a detailed multi-line analysis, which results in an accurate determination of the gas properties. In particular, rotational temperatures and column density of each molecule have been determined and compared to those measured in other objects. The relative iCOM abundances are in good agreement with measurements both in Class 0 protostars and in the 67P/Churyumov–Gerasimenko comet, suggesting to some extent inheritance of complexity from the early stages.[23] ALMA high-angular resolution observations ($\sim$50 au) shed new light on the distribution of iCOMs across the binary system. In particular, a striking chemical segregation was observed. Some molecular species, such as methanol and dimethyl ether, are detected in both VLA4A and VLA4B, while others, *e.g.* formamide and ethylene glycol, are detected only in VLA4A.[1,18] Interestingly, the line emission appears to be elongated towards the north (up to $\sim$150 au), where one of the accretion streamers in the dust is present (see Fig. 1). The origin of such a molecular line distribution needs to be clarified using different molecular tracers. For this purpose, we report here the analysis of sulfur dioxide (SO_2) and deuterated water (HDO) in the SVS13-A system.

3 Observations

The observations presented here were acquired using the Atacama Large Millimeter/submillimeter Array (ALMA) on 8 September 2019 during the project

Table 1 List of transitions and line properties (on the T_{MB} scale) of the HDO, SO_2 and NH_2CHO emission. The columns report the transitions and their frequency ν (GHz), the upper-level energy E_{up} (K), the line strength $S\mu^2$ (D^2), the velocity-integrated line intensity I_{int} (mK km s^{-1}) and the intensity ratio

					I_{int}^{b} (mK km s^{-1})		Intensity ratio
Species	Transition	ν^a (GHz)	E_{up}^{a} (K)	$S\mu^{2a}$ (D^2)	VLA4A	VLA4B	VLA4A/VLA4B
HDO	2(1,1)–2(1,2)	241.561550	95	0.4	349 (70)	177 (36)	2.0 (0.6)
SO_2^{c}	5(2,4)–4(1,3)	241.615797	24	5.7	≥126	≥175	—
NH_2CHO^{d}	12(1,12)–11(1,11)	243.521044	79	156	267 (59)	40 (11)	7 (2)

^{a} Frequencies and spectroscopic parameters have been taken from ref. 31 for HDO, and retrieved from the Jet Propulsion Laboratory molecular database (https://spec.jpl.nasa.gov/).[32] For NH_2CHO, spectroscopic parameters are taken from ref. 34 and retrieved from the Cologne Database for Molecular spectroscopy (https://cdms.astro.uni-koeln.de/).[33] ^{b} Errors on the integrated intensity include 20% calibration uncertainty. ^{c} Contaminated by the CH_2DOH ($4_{3,1}$–$5_{1,5}$)e1 line at 241.61928840 GHz, at −4.3 km s^{-1} from the SO_2 rest line. ^{d} From ref. 1.

(2018.1.01461.S, see also ref. 1). The telescope was in the C43-6 configuration, with baselines between 43 m and 5.9 km. The center of the observed field is $\alpha_{J2000} = 03^h\ 29^m\ 3.8^s$, $\delta_{J2000} = +31°\ 16'\ 03.8''$. The bandpass and flux calibrator was the quasar J0510 + 1800, while the phase calibrator was J0336 + 3218. Data calibration was performed using the standard ALMA calibration pipeline in CASA.[30] Successively, we used the IRAM-GILDAS† package to determine line-free continuum channels, perform phase self-calibration and apply the solution both to the continuum and the spectral cubes. The final calibration uncertainty on the absolute flux is 20%.

4 Results

We detect and analyse the HDO 2(1,1)–2(1,2) transition at 241.561550 GHz with an upper-level energy (E_{up}) of 95 K. In the same spectral window we also detect the SO_2 5(2,4)–4(1,3) transition at 241.615797 GHz with E_{up} of 24 K. The line spectroscopic parameters are listed in Table 1. The rms (root mean square) noise is typically 4.8 mJy per beam km s^{-1} using spectral channels of 122 kHz (0.15 km s^{-1}). The synthesized beam is 0.21″ × 0.14″ for both the line cubes. The analysis is supported by the NH_2CHO 12(1,12)–11(1,11) transition at 243.521044 GHz, previously published in ref. 1.

Fig. 2 shows the spatial distribution of the HDO (left panel), NH_2CHO (middle panel) and SO_2 (right panel) lines with respect to the dust continuum emission. In particular, the line emission is integrated over the whole velocity interval and reported in a color scale superposed to the continuum emission shown as white contours. As already noticed in ref. 1, the NH_2CHO emission is more compact and centered in VLA4A, while HDO and SO_2 have a broader emission, extending also towards VLA4B. This finding confirms the chemical segregation reported in the

† https://www.iram.fr/IRAMFR/GILDAS.

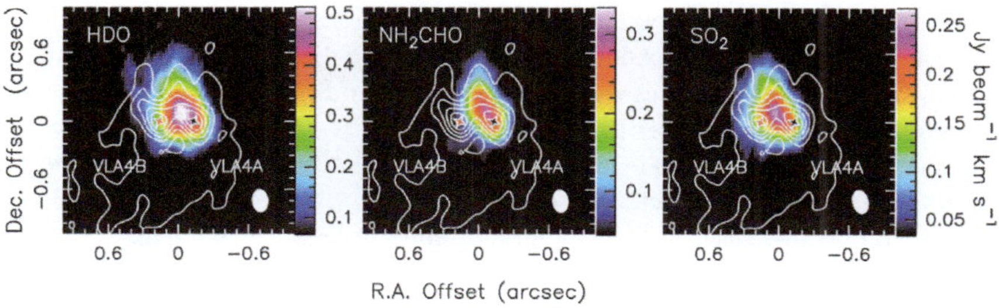

Fig. 2 Velocity-integrated emission (moment 0) of the HDO 2(1,1)–2(1,2) (left panel), NH_2CHO 12(1,12)–11(1,11) (middle panel) and SO_2 5(2,4)–4(1,3) (right panel) lines, in a colour scale, superposed to the dust emission shown as white contours. The synthesised beams are shown in white in the lower right corner of each panel. The black and red stars indicate the VLA4A and VLA4B positions, respectively.

binary system using iCOMs.[1,18] Moreover, HDO and SO_2 emissions show an elongation toward the north, as already observed for iCOMs.

In order to further investigate the lines' radial and vertical distributions, we plot in Fig. 3 the intensity [Jy] profiles of SO_2, HDO and NH_2CHO extracted from the moment 0 maps (see Fig. 2) along a horizontal axis connecting the two protostars (left panel) and a vertical axis between the protostars, crossing the northern streamer (right panel). The *x*-axis reports the offset in astronomical units with respect to the position of VLA4A. The vertical dashed lines indicate the positions of the two protostars. The radial profiles show that HDO and SO_2 are extended in a region covering both protostars with a peak of emission in between. Despite the difference in the peak intensity, there are no substantial differences between the spatial distributions of HDO and SO_2. Conversely, NH_2CHO has

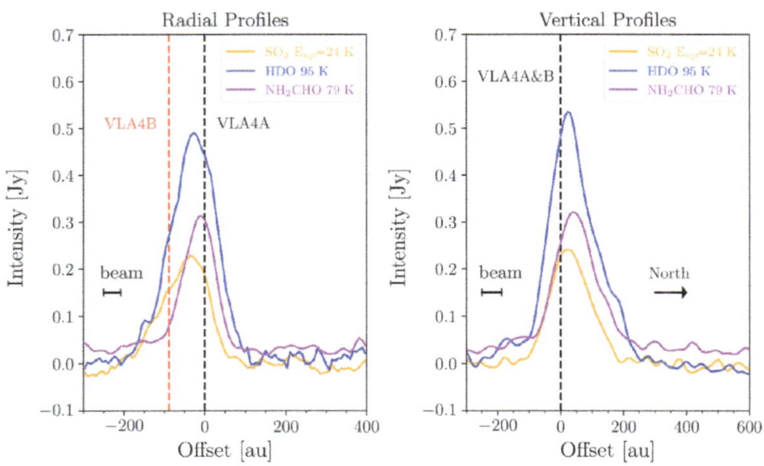

Fig. 3 Left panel: Intensity profile of the HDO line at 241.56155 GHz (blue) and the SO_2 line at 241.615797 GHz (orange) extracted along a line connecting VLA4A with VLA4B. The two vertical dashed lines show the positions of VLA4A (black) and VLA4B (red). Right panel: Intensity profile of the same lines extracted along a vertical line crossing the northern streamer. The vertical dashed line shows the position of VLA4A and VLA4B.

a peak of the emission closer to VLA4A.[1] In addition, the vertical profiles show that all the molecules are extended toward the north direction at a larger distance than the dust streamer, located at ~0.4″ (corresponding to 120 au at the source distance). The radial and vertical profiles do not show a trend with the upper level energy of the different lines, confirming that the molecular distributions do not strongly depend on the excitation conditions.[1]

We report in Table 1 the integrated line intensities extracted from the continuum peak positions of VLA4A and VLA4B, respectively, and their intensity ratios. The intensity ratio for HDO is 2.0 (0.6), suggesting that water is distributed in both the protostars, even if it is more abundant in VLA4A. Similar intensity ratios were derived for iCOMs such as CH_3OH and CH_3CHO.[1] On the other hand, the intensity ratio measured for the NH_2CHO line is 7 (2), confirming the presence of formamide predominantly towards VLA4A. We also report the intensity of SO_2

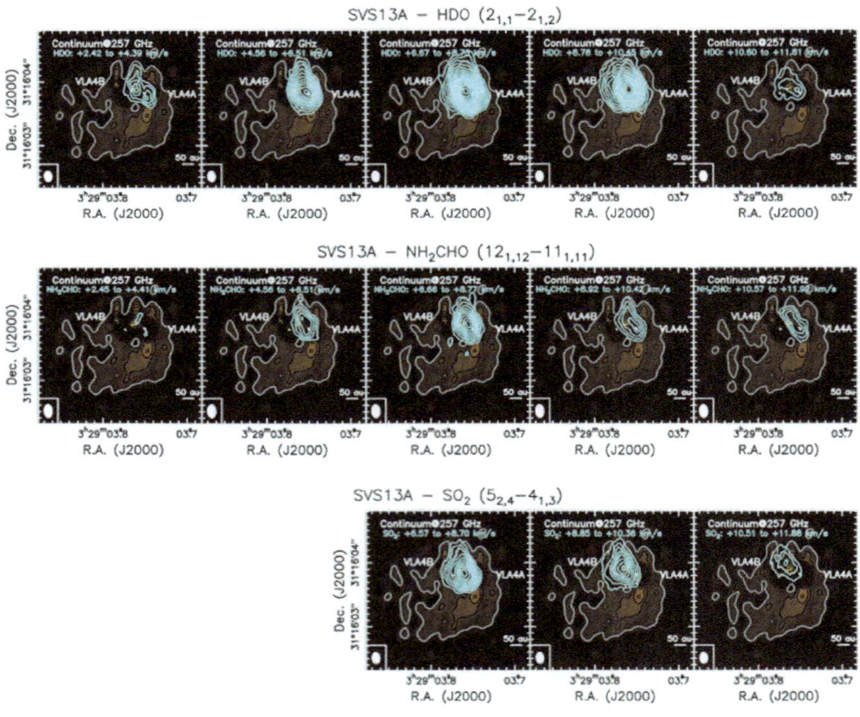

Fig. 4 Upper panels: Channel map of the HDO($2_{1,1}$–$2_{1,2}$) emission (cyan contours) overlapped on the 1.2 mm continuum map of SVS13-A VLA4A and VLA4B (see Fig. 1). Five velocity intervals have been considered (reported in the top-left corner of each panel). The first contours and steps correspond to 5σ (20 mJy per beam km s^{-1} for all the channels but the most blue- and red-shifted ones, with 16 mJy per beam km s^{-1}), and 2σ, respectively. The systemic velocities are +7.7 km s^{-1} (VLA4A) and +8.5 km s^{-1} (VLA4B). The synthesised beam (bottom-left of each panel) is 207 × 136 mas (PA = +5°). Middle panels: Same as the upper panels, for NH_2CHO ($12_{1,12}$–$11_{1,11}$). First contours and steps as for HDO. Lower panels: Same as the upper panels, for SO_2 ($5_{2,4}$–$4_{1,3}$). First contours and steps as for HDO. In this case, only three velocity intervals can be drawn because of the contamination due to the CH_2DOH($4_{3,1}$–$5_{1,5}$)e1 line at 241.61928840 GHz, *i.e.* at −4.3 km s^{-1} from the SO_2 rest line (see Table 1).

in VLA4A and VLA4B. The ratio is not reported in this case since the blue-shifted part of the line is contaminated and the line intensity extracted for VLA4A is a lower limit. The HDO line intensity of the brighter source, VLA4A, is consistent with the rotational temperature of 150–260 K and a column density of 4×10^{17} cm^{-2}, derived using IRAM-30 m data considering the emission from both protostars.[24] The NH_2CHO rotational temperatures and column densities derived in ref. 1 are $T_{rot} = 140$ K and $N_{tot} = 5.9(0.9) \times 10^{15}$ cm^{-2} in VLA4A and $T_{rot} = 170$ K and $N_{tot} = 1.4(0.3) \times 10^{15}$ cm^{-2} in VLA4B.

Fig. 4 shows the spatial distributions of HDO (upper panels), NH_2CHO (middle panels) and SO_2 (lower panels) with respect to the dust continuum emission. More specifically, the line emission is integrated in five different velocity channels and showed with cyan contours superposed to the continuum emission in a color scale. The NH_2CHO emission in all the velocity intervals traces only VLA4A and it is elongated toward the northern streamer. For HDO, the blue-shifted emission in the velocity intervals 2.4–4.4 km s^{-1} and 4.6–6.5 km s^{-1} shows the same distribution. In the velocity intervals 6.7–8.7 km s^{-1}, 8.8–10.5 km s^{-1} and 10.6–11.8 km s^{-1} the line emission is associated with both protostars, peaking in between. The central panels correspond to the systemic velocities of VLA4A and VLA4B, which are +7.7 km s^{-1} and +8.5 km s^{-1}, respectively, and have been determined from methanol emission.[1] SO_2 emission is also associated with both protostars in the velocity intervals 6.6–8.7 km s^{-1}, 8.9–10.4 km s^{-1}, and 10.5–11.9 km s^{-1}. Even if the blue-shifted part of the line (velocities lower than +6.5 km s^{-1}) is contaminated by the $CH_2DOH(4_{3,1}-5_{1,5})$e1 line at 241.61928840 GHz (see Table 1), the SO_2 emission in the other available channels shows a spatial distribution similar to that of HDO. The overall line distribution confirms what was previously reported in ref. 1, but the HDO emission highlights an additional structure (i) with a peak spatially shifted with respect to the protostar position, and (ii) at velocities blue-shifted by ≥3 km s^{-1}. This component is perfectly aligned with the northern streamer traced by the dust, thus revealing molecular gas associated with the streamer itself. Interestingly, in the same blue-shifted velocity range, NH_2CHO shows hints of this component even if the S/N is lower than that of HDO (see Fig. 3).

5 Discussion

5.1 Binding energies and sublimation temperatures

As we will discuss in more detail in the next subsection, the presence of deuterated water, SO_2 and formamide in the gas phase provides new insights into their origin.

First, deuterated water is prevalently formed in the cold prestellar phase on grain surfaces, in the so-called mantle (see *e.g.* ref. 35). Therefore, HDO remains frozen unless it is released into the gas phase by some process. In general, two processes can possibly inject HDO into the gas phase (*e.g.* see the discussion in ref. 9): the thermal sublimation from warm dust grains or a (non-thermal) shock. Other possible mechanisms, such as chemical desorption or photo-desorption, would not be efficient here. Concerning formamide, there has been some discussion on its formation route. In principle, it can be synthesised on grain surfaces (see *e.g.* ref. 36) or in the gas phase[37] from reactants (H_2CO and NH_2) that are formed on grain surfaces, or both. If formed on grain surfaces, formamide would need to be released into the gas phase by either thermal sublimation or a (non-thermal) shock, as HDO. The same consideration applies if formamide is formed in the gas phase, because the two reactants H_2CO and NH_2 would need to be in the gaseous form. Finally, SO_2 is formed in the gas phase, but sulfur is observed to be depleted, by about a factor of 20, in cold molecular clouds,[38] so it is often taken as a tracer of warm and shocked regions. In summary, the presence of HDO, SO_2 and NH_2CHO implies that either they (or their precursors) are thermally sublimated or they are injected from the frozen mantles into the gas phase by a shock.

In order to distinguish which of the two hypotheses is correct, one has to evaluate the thermal sublimation temperature of deuterated water, SO_2 and formamide. To this end, we used the half-life time t_{sub} of the frozen species:

$$t_{\text{sub}} = \ln(2)/k_{\text{des}} \tag{1}$$

The rate of desorption k_{des} is given by the usual equation:

$$k_{\text{des}} = \nu_{\text{des}} \exp(-\text{BE}/T) \tag{2}$$

where BE is the binding energy, given in units of temperature, and ν_{des} is the pre-exponential factor, which depends on the adsorbate and surface (see *e.g.* ref. 39 and 40).

Table 2 lists the BEs of HDO, SO_2 and NH_2CHO over an amorphous water surface, as computed in ref. 41 and 42 on a ice model of 60 water molecules. Note that the HDO BE was assumed to be equal to that of H_2O, an assumption demonstrated to be valid by new computations by Tinacci *et al.* (submitted) on a large (200 molecules) water-ice grain. The table also reports the pre-exponential factors ν of H_2O and NH_2CHO, computed in ref. 39 following the formalism in ref. 43 (see Table 5 of ref. 39). The pre-exponential factor of SO_2 was estimated by us using the formalism in ref. 43.

Fig. 5 shows the derived sublimation temperatures of deuterated water and formamide, for a time interval between 10^3 and 10^5 years. We do not show the case of SO_2 for clarity reasons, since its sublimation temperature would lie below that of HDO. The figure shows three sublimation temperature curves for each species, corresponding to the average, minimum and maximum BEs computed in ref. 41. Assuming that, in reality, the BE has a distribution of values approximately described by a Gaussian function (see *e.g.* ref. 45), the curve corresponding to the average would represent where most of the relevant molecules would thermally sublimate. While HDO being in the gaseous form is, in principle (see below), compatible with thermal sublimation, the same is not

Table 2 Binding energies (BEs) and pre-exponential factors (ν) used for deriving the sublimation temperatures shown in Fig. 5. The BEs are from ref. 41 and 42. The pre-exponential factors are from ref. 39 (see text). Note that the second column reports minimum and maximum values of BE computed on a amorphous water surface model of 60 molecules. The third column reports the pre-exponential factors computed using the method in ref. 43, also described and discussed in ref. 39 and 40

Species	BE [K]	ν [s^{-1}]
HDO	3605–6111	5×10^{15}
SO_2	2105–5827	2×10^{16}
NH_2CHO	5793–10 960	4×10^{18}

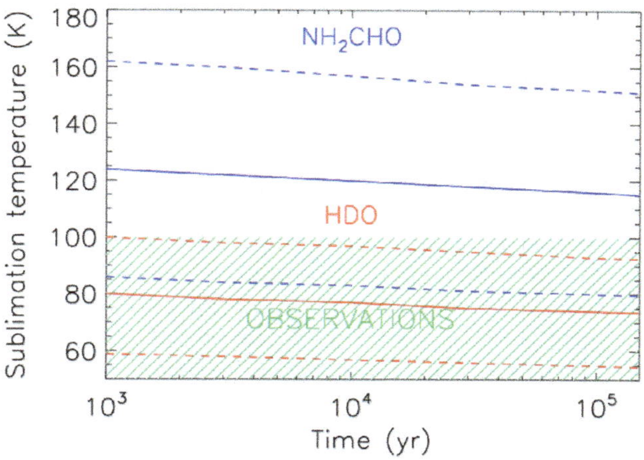

Fig. 5 Sublimation temperatures of deuterated water (HDO; red lines) and formamide (NH_2CHO; blue lines). The three curves for each species correspond to the average (solid line), minimum and maximum (dashed lines) BEs computed in ref. 41 and reported in Table 2 (see text). Note that for HDO we used the same BE as H_2O, as demonstrated to be valid by Tinacci *et al.* (submitted). The green area shows the grain temperature derived in ref. 44 assuming a bolometric luminosity of 55 $L_\odot$ and a distance from the center of 120 au (see text).

true for formamide. The implications of this fact will be discussed in the following subsection.

5.2 Accretion streamers feeding the disks

The bolometric luminosity (L_{bol}) of VLA4A and VLA4B is not measured singularly due to angular resolution limitations at infrared wavelengths. The total bolometric luminosity of VLA4A + VLA4B is estimated to be in the range 45–55 $L_\odot$,[20,46] at the source distance of 299 pc. The dust streamer traced by the continuum emission extends at a distance larger than 0.4″ (120 au). If we consider a spherical envelope-like distribution, and conservatively a bolometric luminosity of 55 $L_\odot$, the temperature at this distance from the protostars is lower than 100 K (see *e.g.* ref. 47). The presence of formamide at this position, considering the information on the binding energies and the corresponding sublimation temperatures (see section 5.1 and Fig. 5), suggests that formamide is not thermally desorbed. In the case of HDO and SO_2, the binding energies are lower, so their presence cannot exclude thermal desorption. If the continuum emission around the protostars originates from an inclined disk (instead of a spherical envelope), as suggested by the observed geometry,[18] the snowlines of NH_2CHO, HDO, and SO_2, would be even closer to the source. Therefore, the detection of molecules at a distance of ≥120 au would exclude the thermal evaporation scenario.

However, episodic accretion is a common process in young stellar objects. More specifically, observations performed using the Very Large Telescope indicate that SVS13-A is in a phase of active accretion, deriving a mass accretion rate of a few 10^{-6} $M_\odot$ per year[12] (see also ref. 48). The presence of an accretion burst is in principle expected to move the snowlines outwards and inject into the gas phase the molecules stored in the dust mantles, as in the case of the Fu Ori V833 disk.[49] However, at near-IR wavelengths VLA4A is not detected,[21] suggesting a negligible contribution to the bolometric luminosity, which will instead be dominated by VLA4B.[12] This is consistent with VLA4B being the primary and more massive component of the binary system, while VLA4A would be the less massive secondary component. The process of differential accretion in a binary system has been studied using numerical simulations and the secondary component is expected to accrete most of the gas, except in the case of very eccentric orbits (see *e.g.* ref. 50 and 51). In this scenario, VLA4A would be the component actively accreting, and the associated accretion burst may be at the origin of the detected molecular emission associated with VLA4A. In order to definitely exclude the possibility that thermal sublimation is responsible for the chemical enrichment at distances larger than 120 au, it is crucial to have an estimate of the individual VLA4A and VLA4B luminosities.

Alternatively, the emission may be produced by an accretion shock, sputtering the grains, occurring when the accretion streamer impacts the disk of VLA4A. This would also justify the fact that the molecular emission shows emission at both blue-shifted and red-shifted velocities in the streamer (see Fig. 4), which may be due to the velocity dispersion caused by the shock. Indeed, a similar scenario has been proposed to explain SO and SO_2 emission, where molecular streamers impact onto the disks in Class I/II sources, observed by the ALMA-DOT project.[52] Also, in these sources we observe both blue-shifted and red-shifted SO and SO_2 emission at the impact region.

Recently, the possible presence of accretion shocks or self-gravitating substructures (see *e.g.* ref. 53) has been invoked to explain local temperature enhancements in the circumbinary material of the Class 0 source IRAS16293-2422.[54] However, in the case of SVS13-A the circumstellar disks around VLA4A and VLA4B, as well as the circumbinary material, are expected to be stable against local gravitational instabilities, as found in ref. 18.

In summary, a major result of the presented observations is the detection of molecular gas in the accretion streamer feeding the system SVS13-A, on a scale of ~100–200 au Interestingly, an accretion streamer connecting the source to the outer envelope, up to ~700 au, has been recently imaged in DCN using the IRAM NOEMA interferometer.[55] The detected DCN streamer is associated with the southern dusty streamer, even though some emission is also present in the northern region (see Fig. C.1 in ref. 55). However, the limited angular resolution of the NOEMA observations (synthesized beam of $1.2'' \times 0.7''$) prevents a proper comparison with the ALMA continuum emission. Finally, the present observations confirm that deuterated water and iCOMs, including formamide, are as effective as classical tracers, such as SO and SO_2, in revealing and investigating accretion processes in young embedded Class 0/I disks.

6 Conclusions

We report new ALMA observations of HDO and SO_2 at an ~50 au angular resolution in the protobinary system SVS13-A. The line spatial distribution and the radial and vertical profiles indicate that HDO and SO_2 are associated with both components of the binary system, VLA4A and VLA4B. We compare the spatial distribution with that of NH_2CHO, which instead mainly traces the VLA4A component, confirming a chemical differentiation around the two protostars. In addition, the channel maps of HDO emission reveal a further component at blue-shifted velocities (+2.42–4.39 km s^{-1}) spatially coincident with the dust-accretion streamer. The streamer traced by HDO is located at a distance ≥120 au from the VLA4A and VLA4B protostars. Formamide is also detected at the same spatial position. In order to investigate the origin of the molecular emission in the streamer, we evaluate the thermal sublimation temperatures of deuterated water, SO_2 and formamide, using updated BE distributions. We find that a thermal sublimation origin of the molecular emission is excluded, unless the VLA4A source is currently experiencing an accretion burst. Alternatively, the emission morphology is consistent with an accretion shock, produced by an accretion streamer impacting the VLA4A disk, as previously found in more evolved protoplanetary disks. Finally, water has a key role in the planet formation process and it is relevant to investigate its abundance across the different evolutionary stages. The spatial distribution of HDO in SVS13-A demonstrates that in Class I disks water is not necessarily confined in the inner 10 au, as suggested by previous ALMA and NOEMA $H_2{}^{18}O$ observations in 5 Class I disks.[56] In this respect, accretion processes are a powerful tool to investigate the amount of water present in the young disks that will be eventually inherited from the forming planetary systems.

Conflicts of interest

There are no conflicts to declare.

Acknowledgements

This project has received funding from: (1) the Deutsche Forschungsgemeinschaft (DFG, German Research Foundation) under Germany's Excellence Strategy – EXC 2094 – 390783311; (2) the European Research Council (ERC) under the European Union's Horizon 2020 Research and Innovation Program, for the Project "The Dawn of Organic Chemistry" (DOC), grant agreement No 741002; (3) the PRIN-INAF 2016 The Cradle of Life – GENESIS-SKA (GEneral coNditions in Early planetary Systems for the rISe of life with SKA); and (4) the European Union's Horizon 2020 Research and Innovation Programs under projects "Astro-Chemistry Origins" (ACO), Grant No. 811312. This paper makes use of the following ALMA data: ADS/JAO.ALMA#2018.1.01461.S. ALMA is a partnership of the ESO (representing its member states), NSF (USA) and NINS (Japan), together with the NRC (Canada), MOST and ASIAA (Taiwan), and the KASI (Republic of Korea), in cooperation with the Republic of Chile. The Joint ALMA Observatory is operated by the ESO, AUI/NRAO and NAOJ. Most of the computations presented in this paper were performed using the GRICAD infrastructure (https://gricad.univ-grenoble-alpes.fr), which is partly supported by the Equip@Meso project (reference ANR-10-EQPX-29-01) of the programme Investissements d'Avenir supervised by the Agence Nationale pour la Recherche. Software: Astropy[57,58] and Matplotlib.[59]

Notes and references

1 E. Bianchi, A. López-Sepulcre, C. Ceccarelli, C. Codella, L. Podio, M. Bouvier and J. Enrique-Romero, *Astrophys. J., Lett.*, 2022, **928**, L3.

2 ALMA Partnership, C. L. Brogan, L. M. Pérez, T. R. Hunter, W. R. F. Dent, A. S. Hales, R. E. Hills, S. Corder, E. B. Fomalont, C. Vlahakis, Y. Asaki, D. Barkats, A. Hirota, J. A. Hodge, C. M. V. Impellizzeri, R. Kneissl, E. Liuzzo, R. Lucas, N. Marcelino, S. Matsushita, K. Nakanishi, N. Phillips, A. M. S. Richards, I. Toledo, R. Aladro, D. Broguiere, J. R. Cortes, P. C. Cortes, D. Espada, F. Galarza, D. Garcia-Appadoo, L. Guzman-Ramirez, E. M. Humphreys, T. Jung, S. Kameno, R. A. Laing, S. Leon, G. Marconi, A. Mignano, B. Nikolic, L.-A. Nyman, M. Radiszcz, A. Remijan, J. A. Rodón, T. Sawada, S. Takahashi, R. P. J. Tilanus, B. Vila Vilaro, L. C. Watson, T. Wiklind, E. Akiyama, E. Chapillon, I. de Gregorio-Monsalvo, J. Di Francesco, F. Gueth, A. Kawamura, C.-F. Lee, Q. Nguyen Luong, J. Mangum, V. Pietu, P. Sanhueza, K. Saigo, S. Takakuwa, C. Ubach, T. van Kempen, A. Wootten, A. Castro-Carrizo, H. Francke, J. Gallardo, J. Garcia, S. Gonzalez, T. Hill, T. Kaminski, Y. Kurono, H.-Y. Liu, C. Lopez, F. Morales, K. Plarre, G. Schieven, L. Testi, L. Videla, E. Villard, P. Andreani, J. E. Hibbard and K. Tatematsu, *Astrophys. J., Lett.*, 2015, **808**, L3.

3 P. D. Sheehan, J. J. Tobin, S. Federman, S. T. Megeath and L. W. Looney, *Astrophys. J.*, 2020, **902**, 141.

4 D. M. Segura-Cox, A. Schmiedeke, J. E. Pineda, I. W. Stephens, M. Fernández-López, L. W. Looney, P. Caselli, Z.-Y. Li, L. G. Mundy, W. Kwon and R. J. Harris, *Nature*, 2020, **586**, 228–231.

5 Ł. Tychoniec, J. J. Tobin, A. Karska, C. Chandler, M. M. Dunham, R. J. Harris, K. M. Kratter, Z.-Y. Li, L. W. Looney, C. Melis, L. M. Pérez, S. I. Sadavoy, D. Segura-Cox and E. F. van Dishoeck, *Astrophys. J., Suppl. Ser.*, 2018, **238**, 19.

6 A. Miotello, I. Kamp, T. Birnstiel, L. I. Cleeves and A. Kataoka, *arXiv*, 2022, arXiv:2203.09818, DOI: 10.48550/arXiv.2203.09818.

7 P. D. Sheehan, J. J. Tobin, L. W. Looney and S. T. Megeath, *Astrophys. J.*, 2022, **929**, 76.

8 K. I. Öberg and E. A. Bergin, *Phys. Rep.*, 2021, **893**, 1–48.

9 C. Ceccarelli, C. Codella, N. Balucani, D. Bockelée-Morvan, E. Herbst, C. Vastel, P. Caselli, C. Favre, B. Lefloch and K. Öberg, *arXiv*, 2022, arXiv:2206.13270, DOI: 10.48550/arXiv.2206.13270.

10 E. Herbst and E. F. van Dishoeck, *Annu. Rev. Astron. Astrophys.*, 2009, **47**, 427–480.

11 C. Zucker, E. F. Schlafly, J. S. Speagle, G. M. Green, S. K. N. Portillo, D. P. Finkbeiner and A. A. Goodman, *Astrophys. J.*, 2018, **869**, 83.

12 E. Fiorellino, C. F. Manara, B. Nisini, S. Ramsay, S. Antoniucci, T. Giannini, K. Biazzo, J. Alcalà and D. Fedele, *Astron. Astrophys.*, 2021, **650**, A43.

13 B. Reipurth, R. Chini, E. Krugel, E. Kreysa and A. Sievers, *Astron. Astrophys.*, 1993, **273**, 221–238.

14 B. Lefloch, A. Castets, J. Cernicharo, W. D. Langer and R. Zylka, *Astron. Astrophys.*, 1998, **334**, 269–279.

15 C. Lefèvre, S. Cabrit, A. J. Maury, F. Gueth, B. Tabone, L. Podio, A. Belloche, C. Codella, S. Maret, S. Anderl, P. André and P. Hennebelle, *Astron. Astrophys.*, 2017, **604**, L1.
16 G. Anglada, L. F. Rodríguez and J. M. Torrelles, *Astrophys. J., Lett.*, 2000, **542**, L123–L126.
17 C. Carrasco-González, G. Anglada, L. F. Rodríguez, J. M. Torrelles and M. Osorio, *Astron. J.*, 2008, **136**, 2238–2243.
18 A. K. Diaz-Rodriguez, G. Anglada, G. Blázquez-Calero, M. Osorio, J. F. Gómez, G. A. Fuller, R. Estalella, J. M. Torrelles, S. Cabrit, L. F. Rodríguez, C. Lefèvre, E. Macías, C. Carrasco-González, L. A. Zapata, I. de Gregorio-Monsalvo and P. T. P. Ho, *Astrophys. J.*, 2022, **930**, 91.
19 L. F. Rodríguez, G. Anglada and S. Curiel, *Astrophys. J., Lett.*, 1997, **480**, L125–L128.
20 J. J. Tobin, L. W. Looney, Z.-Y. Li, C. J. Chandler, M. M. Dunham, D. Segura-Cox, S. I. Sadavoy, C. Melis, R. J. Harris, K. Kratter and L. Perez, *Astrophys. J.*, 2016, **818**, 73.
21 K. W. Hodapp and R. Chini, *Astrophys. J.*, 2014, **794**, 169.
22 S. Cazaux, A. G. G. M. Tielens, C. Ceccarelli, A. Castets, V. Wakelam, E. Caux, B. Parise and D. Teyssier, *Astrophys. J., Lett.*, 2003, **593**, L51–L55.
23 E. Bianchi, C. Codella, C. Ceccarelli, F. Vazart, R. Bachiller, N. Balucani, M. Bouvier, M. De Simone, J. Enrique-Romero, C. Kahane, B. Lefloch, A. López-Sepulcre, J. Ospina-Zamudio, L. Podio and V. Taquet, *Mon. Not. R. Astron. Soc.*, 2019, **483**, 1850–1861.
24 C. Codella, C. Ceccarelli, E. Bianchi, L. Podio, R. Bachiller, B. Lefloch, F. Fontani, V. Taquet and L. Testi, *Mon. Not. R. Astron. Soc.*, 2016, **462**, L75–L79.
25 M. De Simone, C. Codella, L. Testi, A. Belloche, A. J. Maury, S. Anderl, P. André, S. Maret and L. Podio, *Astron. Astrophys.*, 2017, **599**, A121.
26 B. Lefloch, R. Bachiller, C. Ceccarelli, J. Cernicharo, C. Codella, A. Fuente, C. Kahane, A. López-Sepulcre, M. Tafalla, C. Vastel, E. Caux, M. González-García, E. Bianchi, A. Gómez-Ruiz, J. Holdship, E. Mendoza, J. Ospina-Zamudio, L. Podio, D. Quénard, E. Roueff, N. Sakai, S. Viti, S. Yamamoto, K. Yoshida, C. Favre, T. Monfredini, H. M. Quitián-Lara, N. Marcelino, H. M. Boechat-Roberty and S. Cabrit, *Mon. Not. R. Astron. Soc.*, 2018, **477**, 4792–4809.
27 A. Belloche, A. J. Maury, S. Maret, S. Anderl, A. Bacmann, P. André, S. Bontemps, S. Cabrit, C. Codella, M. Gaudel, F. Gueth, C. Lefèvre, B. Lefloch, L. Podio and L. Testi, *Astron. Astrophys.*, 2020, **635**, A198.
28 C. Ceccarelli, P. Caselli, F. Fontani, R. Neri, A. López-Sepulcre, C. Codella, S. Feng, I. Jiménez-Serra, B. Lefloch, J. E. Pineda, C. Vastel, F. Alves, R. Bachiller, N. Balucani, E. Bianchi, L. Bizzocchi, S. Bottinelli, E. Caux, A. Chacón-Tanarro, R. Choudhury, A. Coutens, F. Dulieu, C. Favre, P. Hily-Blant, J. Holdship, C. Kahane, A. Jaber Al-Edhari, J. Laas, J. Ospina, Y. Oya, L. Podio, A. Pon, A. Punanova, D. Quenard, A. Rimola, N. Sakai, I. R. Sims, S. Spezzano, V. Taquet, L. Testi, P. Theulé, P. Ugliengo, A. I. Vasyunin, S. Viti, L. Wiesenfeld and S. Yamamoto, *Astrophys. J.*, 2017, **850**, 176.
29 Y.-L. Yang, N. Sakai, Y. Zhang, N. M. Murillo, Z. E. Zhang, A. E. Higuchi, S. Zeng, A. López-Sepulcre, S. Yamamoto, B. Lefloch, M. Bouvier, C. Ceccarelli, T. Hirota, M. Imai, Y. Oya, T. Sakai and Y. Watanabe, *Astrophys. J.*, 2021, **910**, 20.

30 J. P. McMullin, B. Waters, D. Schiebel, W. Young and K. Golap, in *CASA Architecture and Applications*, ed. R. A. Shaw, F. Hill and D. J. Bell, 2007, vol. 376, p. 127.
31 J. K. Messer, F. C. De Lucia and P. Helminger, *J. Mol. Spectrosc.*, 1984, **105**, 139–155.
32 H. M. Pickett, R. L. Poynter, E. A. Cohen, M. L. Delitsky, J. C. Pearson and H. S. P. Müller, *J. Quant. Spectrosc. Radiat. Transfer*, 1998, **60**, 883–890.
33 H. S. P. Müller, F. Schlöder, J. Stutzki and G. Winnewisser, *J. Mol. Struct.*, 2005, **742**, 215–227.
34 S. G. Kukolich and A. C. Nelson, *Chem. Phys. Lett.*, 1971, **11**, 383–384.
35 C. Ceccarelli, P. Caselli, D. Bockelée-Morvan, O. Mousis, S. Pizzarello, F. Robert and D. Semenov, *Protostars and Planets VI*, 2014, pp. 859–882.
36 R. T. Garrod, S. L. Widicus Weaver and E. Herbst, *Astrophys. J.*, 2008, **682**, 283–302.
37 D. Skouteris, F. Vazart, C. Ceccarelli, N. Balucani, C. Puzzarini and V. Barone, *Mon. Not. R. Astron. Soc.*, 2017, **468**, L1–L5.
38 A. Fuente, P. Rivière-Marichalar, L. Beitia-Antero, P. Caselli, V. Wakelam, G. Esplugues, M. Rodríguez-Baras, D. Navarro-Almaida, M. Gerin, C. Kramer, R. Bachiller, J. R. Goicoechea, I. Jiménez-Serra, J. C. Loison, A. Ivlev, R. Martín-Doménech, S. Spezzano, O. Roncero, G. Muñoz-Caro, S. Cazaux and N. Marcelino, *arXiv*, 2022, arXiv:2212.03742, DOI: 10.48550/arXiv.2212.03742.
39 M. Minissale, Y. Aikawa, E. Bergin, M. Bertin, W. A. Brown, S. Cazaux, S. B. Charnley, A. Coutens, H. M. Cuppen, V. Guzman, H. Linnartz, M. R. S. McCoustra, A. Rimola, J. G. M. Schrauwen, C. Toubin, P. Ugliengo, N. Watanabe, V. Wakelam and F. Dulieu, *ACS Earth Space Chem.*, 2022, **6**, 597–630.
40 S. Ferrero, F. Grieco, A. S. Ibrahim Mohamed, F. Dulieu, A. Rimola, C. Ceccarelli, C. Nervi, M. Minissale and P. Ugliengo, *Mon. Not. R. Astron. Soc.*, 2022, **516**, 2586–2596.
41 S. Ferrero, L. Zamirri, C. Ceccarelli, A. Witzel, A. Rimola and P. Ugliengo, *Astrophys. J.*, 2020, **904**, 11.
42 J. Perrero, J. Enrique-Romero, S. Ferrero, C. Ceccarelli, L. Podio, C. Codella, A. Rimola and P. Ugliengo, *Astrophys. J.*, 2022, **938**, 158.
43 S. L. Tait, Z. Dohnálek, C. T. Campbell and B. D. Kay, *J. Chem. Phys.*, 2005, **122**, 164708.
44 E. Bianchi, C. Ceccarelli, C. Codella, A. López-Sepulcre, S. Yamamoto, N. Balucani, P. Caselli, L. Podio, R. Neri, R. Bachiller, C. Favre, F. Fontani, B. Lefloch, N. Sakai and D. Segura-Cox, *Astron. Astrophys.*, 2022, **662**, A103.
45 L. Tinacci, A. Germain, S. Pantaleone, S. Ferrero, C. Ceccarelli and P. Ugliengo, *ACS Earth Space Chem.*, 2022, **6**, 1514–1526.
46 L. Podio, B. Tabone, C. Codella, F. Gueth, A. Maury, S. Cabrit, B. Lefloch, S. Maret, A. Belloche, P. André, S. Anderl, M. Gaudel and L. Testi, *Astron. Astrophys.*, 2021, **648**, A45.
47 C. Ceccarelli, A. Castets, E. Caux, D. Hollenbach, L. Loinard, S. Molinari and A. G. G. M. Tielens, *Astron. Astrophys.*, 2000, **355**, 1129–1137.
48 T.-H. Hsieh, N. M. Murillo, A. Belloche, N. Hirano, C. Walsh, E. F. van Dishoeck, J. K. Jørgensen and S.-P. Lai, *Astrophys. J.*, 2019, **884**, 149.

49 J.-E. Lee, S. Lee, G. Baek, Y. Aikawa, L. Cieza, S.-Y. Yoon, G. Herczeg, D. Johnstone and S. Casassus, *Nat. Astron*, 2019, **3**, 314–319.

50 M. R. Bate and I. A. Bonnell, *Mon. Not. R. Astron. Soc.*, 1997, **285**, 33–48.

51 S. Ceppi, N. Cuello, G. Lodato, C. Clarke, C. Toci and D. J. Price, *Mon. Not. R. Astron. Soc.*, 2022, **514**, 906–919.

52 A. Garufi, L. Podio, C. Codella, D. Segura-Cox, M. Vander Donckt, S. Mercimek, F. Bacciotti, D. Fedele, M. Kasper, J. E. Pineda, E. Humphreys and L. Testi, *Astron. Astrophys.*, 2022, **658**, A104.

53 K. Kratter and G. Lodato, *Annu. Rev. Astron. Astrophys.*, 2016, **54**, 271–311.

54 M. J. Maureira, M. Gong, J. E. Pineda, H. B. Liu, K. Silsbee, P. Caselli, J. Zamponi, D. M. Segura-Cox and A. Schmiedeke, *Astrophys. J., Lett.*, 2022, **941**, L23.

55 T. H. Hsieh, D. M. Segura-Cox, J. E. Pineda, P. Caselli, L. Bouscasse, R. Neri, A. Lopez-Sepulcre, M. T. Valdivia-Mena, M. J. Maureira, T. Henning, G. V. Smirnov-Pinchukov, D. Semenov, T. Möller, N. Cunningham, A. Fuente, S. Marino, A. Dutrey, M. Tafalla, E. Chapillon, C. Ceccarelli and B. Zhao, *Astron. Astrophys.*, 2023, **669**, A137.

56 D. Harsono, M. V. Persson, A. Ramos, N. M. Murillo, L. T. Maud, M. R. Hogerheijde, A. D. Bosman, L. E. Kristensen, J. K. Jørgensen, E. A. Bergin, R. Visser, J. C. Mottram and E. F. van Dishoeck, *Astron. Astrophys.*, 2020, **636**, A26.

57 Astropy Collaboration, T. P. Robitaille, E. J. Tollerud, P. Greenfield, M. Droettboom, E. Bray, T. Aldcroft, M. Davis, A. Ginsburg, A. M. Price-Whelan, W. E. Kerzendorf, A. Conley, N. Crighton, K. Barbary, D. Muna, H. Ferguson, F. Grollier, M. M. Parikh, P. H. Nair, H. M. Unther, C. Deil, J. Woillez, S. Conseil, R. Kramer, J. E. H. Turner, L. Singer, R. Fox, B. A. Weaver, V. Zabalza, Z. I. Edwards, K. A. Bostroem, D. J. Burke, A. R. Casey, S. M. Crawford, N. Dencheva, J. Ely, T. Jenness, K. Labrie, P. L. Lim, F. Pierfederici, A. Pontzen, A. Ptak, B. Refsdal, M. Servillat and O. Streicher, *Astron. Astrophys.*, 2013, **558**, A33.

58 Astropy Collaboration, A. M. Price-Whelan, B. M. Sipőcz, H. M. Günther, P. L. Lim, S. M. Crawford, S. Conseil, D. L. Shupe, M. W. Craig, N. Dencheva, A. Ginsburg, J. T. VanderPlas, L. D. Bradley, D. Pérez-Suárez, M. de Val-Borro, T. L. Aldcroft, K. L. Cruz, T. P. Robitaille, E. J. Tollerud, C. Ardelean, T. Babej, Y. P. Bach, M. Bachetti, A. V. Bakanov, S. P. Bamford, G. Barentsen, P. Barmby, A. Baumbach, K. L. Berry, F. Biscani, M. Boquien, K. A. Bostroem, L. G. Bouma, G. B. Brammer, E. M. Bray, H. Breytenbach, H. Buddelmeijer, D. J. Burke, G. Calderone, J. L. Cano Rodríguez, M. Cara, J. V. M. Cardoso, S. Cheedella, Y. Copin, L. Corrales, D. Crichton, D. D'Avella, C. Deil, É. Depagne, J. P. Dietrich, A. Donath, M. Droettboom, N. Earl, T. Erben, S. Fabbro, L. A. Ferreira, T. Finethy, R. T. Fox, L. H. Garrison, S. L. J. Gibbons, D. A. Goldstein, R. Gommers, J. P. Greco, P. Greenfield, A. M. Groener, F. Grollier, A. Hagen, P. Hirst, D. Homeier, A. J. Horton, G. Hosseinzadeh, L. Hu, J. S. Hunkeler, Ž. Ivezić, A. Jain, T. Jenness, G. Kanarek, S. Kendrew, N. S. Kern, W. E. Kerzendorf, A. Khvalko, J. King, D. Kirkby, A. M. Kulkarni, A. Kumar, A. Lee, D. Lenz, S. P. Littlefair, Z. Ma, D. M. Macleod, M. Mastropietro, C. McCully, S. Montagnac, B. M. Morris, M. Mueller, S. J. Mumford, D. Muna, N. A. Murphy, S. Nelson, G. H. Nguyen, J. P. Ninan, M. Nöthe, S. Ogaz, S. Oh, J. K. Parejko, N. Parley, S. Pascual, R. Patil, A. A. Patil, A. L. Plunkett, J. X. Prochaska, T. Rastogi, V. Reddy Janga,

J. Sabater, P. Sakurikar, M. Seifert, L. E. Sherbert, H. Sherwood-Taylor, A. Y. Shih, J. Sick, M. T. Silbiger, S. Singanamalla, L. P. Singer, P. H. Sladen, K. A. Sooley, S. Sornarajah, O. Streicher, P. Teuben, S. W. Thomas, G. R. Tremblay, J. E. H. Turner, V. Terrón, M. H. van Kerkwijk, A. de la Vega, L. L. Watkins, B. A. Weaver, J. B. Whitmore, J. Woillez and V. Zabalza, *Astron. J.*, 2018, **156**, 123.

59 J. D. Hunter, *Comput. Sci. Eng.*, 2007, **9**, 90–95.

Faraday Discussions

ROYAL SOCIETY OF CHEMISTRY

PAPER

Tracing the chemical footprint of shocks in AGN-host and starburst galaxies with ALMA multi-line molecular studies

Ko-Yun Huang *[a] and Serena Viti [ab]

Received 11th January 2023, Accepted 17th March 2023
DOI: 10.1039/d3fd00007a

Multi-line molecular observations are an ideal tool for a systematic study of the physico-chemical processes in the Interstellar Medium (ISM), given the wide range of critical densities associated with different molecules and their transitions, and the dependencies of chemical reactions on the energy budget of the system. Recently high spatial resolution of typical shock tracers – SiO, HNCO, and CH_3OH – have been studied in the potentially shocked regions in two nearby galaxies: NGC 1068 (an AGN-host galaxy) (Huang *et al.*, *Astron. Astrophys.*, 2022, **666**, A102; Huang *et al.*, in prep.) and NGC 253 (a starburst galaxy) (K.-Y. Huang *et al.*, *arXiv*, 2023, preprint, arXiv:2303.12685, DOI: 10.48550/arXiv.2303.12685). This paper is dedicated to the comparative study of these two distinctively different galaxies, with the aim of determining the differences in their energetics and understanding large-scale shocks in different types of galaxies.

1 Introduction

The evolution of galaxies is impacted by many key physical and chemical processes in the interstellar medium (ISM) that are often associated with star-formation, Active Galactic Nuclei (AGN), large-scale outflows, and shocks. In this context, nearby galaxies that host starburst and AGN-dominated environments are prime laboratories for the investigation of these feedback mechanisms and their impact on the ISM.

An AGN is considered a manifestation of a mass-accreting supermassive black hole (SMBH) at the center of galaxies, which convert gravitational potential energy and kinetic energy into radiation and powers the jet and the associated multi-phase mass outflow. As AGNs are often identified by prominent point sources in the K band with an associated hard X-ray peak and a bright and compact radio continuum source, the gas chemistry in the nuclear regions of AGN-host galaxies

[a]*Leiden Observatory, Leiden University, PO Box 9513, NL-2300 RA Leiden, The Netherlands. E-mail: kyhuang@strw.leidenuniv.nl*

[b]*Department of Physics and Astronomy, University College London, Gower Street, London WC1E 6BT, UK*

is often considered to be X-ray dominated regions (XDRs). Moreover, AGN-dominated galaxies are often thought to have an enhanced cosmic ray ionization rate (CRIR) and radiation field compared to normal spiral galaxies such as the Milky Way. NGC 1068 is a nearby ($D = 14$ Mpc,[1] $1'' \sim 70$ pc) Seyfert 2 galaxy and is considered to be the archetype of a composite AGN-starburst system. The proximity of this galaxy makes it an ideal target for studying in detail the feedback of the AGN activity at its central $r \sim 200$ pc circumnuclear disc (CND). It also allows the simultaneous study of the starburst (SB) activities at the outer $r \sim 1$–1.5 kpc SB ring with the SB ring being spatially resolved from the AGN feedback from the central region of the galaxy. The molecular gas in the CND of NGC 1068 was shown to be outflowing[2] in multi-line CO, HCN, and HCO^+ observations.[2,3] The deprojected outflow velocity of NGC 1068 at the CND region was estimated ranging between 85–200 km s^{-1} with the increase in radial distance that runs between $r \sim 50$–200 pc.[3] The outflowing molecular gas in the CND of NGC 1068 is a manifestation of ongoing AGN feedback,[3] for the outflow is likely launched by the interaction between the molecular gas in the CND and both AGN ionized wind and the radio jet plasma.[3] This interaction has produced a large-scale molecular shock on spatial scales of up to $r = 400$ pc from the AGN and likely triggers rich shock chemistry signatures across the CND.[4]

Starburst galaxies are extremely luminous systems that are powered by bursts of massive star formation. Strong stellar feedback induced by high star-forming rates (SFR) can trigger multi-phase outflows that involve ionized, neutral, and molecular gas. As starburst activities inject a significant amount of energy into the ambient environment, starburst galaxies are also important targets to study the feedback mechanisms in the ISM. NGC 253 is a barred spiral galaxy that is almost edge-on with an inclination of 76°,[5] and is one of the nearest starburst systems ($D \sim 3.5 \pm 0.2$ Mpc (ref. 6)). NGC 253 is considered a prototype of nuclear starburst with an SFR of $\sim 2\ M_\odot\ yr^{-1}$ coming from its central molecular zone (CMZ),[7,8] which is half of its global SF activity. The CMZ of NGC 253 spans about 300×100 pc across.[9] A large-scale outflow in NGC 253 has been revealed by multi-wavelength observations: in X-rays,[10,11] Hα,[12] molecular emission,[13–16] and dust.[17] This large-scale outflow is thought to be driven by the galaxy's starburst activity,[18] for there are no signs of AGN influence[19,20] despite there being a bright compact radio source associated with the nucleus of the galaxy.[21] The deprojected outflow velocity of NGC 253 was estimated as ranging between 110–360 km s^{-1} which increases with distance above the plane up to $r \sim 1$ kpc.[12] The presence of shocks in NGC 253 has been suggested by the detection of two molecular shock tracers, HNCO and SiO,[22,23] as well as the detection of Class I CH_3OH masers,[24] and the enhanced fractional abundances of CO_2.[25]

Both silicon monoxide, SiO, and isocyanic acid, HNCO, are well-known shock tracers,[26–31] and have been observed in nearby galaxies.[22,23,32–39] The simultaneous detection of HNCO and SiO has been found to be very useful for the characterization of different types of shocks (*e.g.* fast *versus* slow shocks) in NGC 1068,[38,39] NGC 1097,[37] in the nearby weakly barred spiral galaxy IC 342 which hosts moderate starburst activities,[32,33] and in NGC 253.[23] A high abundance of silicon in the gas phase can only be explained by significant sputtering from the core of the dust grains by high-velocity ($v_s \geq 50$ km s^{-1}) shocks.[38] Once silicon is in the gas phase, it is expected to quickly react with molecular oxygen or a hydroxyl radical to form SiO.[41] An enhanced gas-phase SiO abundance could thus be a sensitive indicator of the heavily shocked regions. The formation of HNCO has been suggested to be mainly on the icy mantles of dust grains,[42] or possibly in the gas phase with subsequent freeze-out onto the dust grains when the temperatures are low.[43] Regardless of how it forms, icy mantles sputtering associated with low-velocity ($v_s \leq 20$ km s^{-1}) shocks can lead to an enhanced abundance of HNCO in the gas phase. The good correlation between HNCO and SiO revealed in Galactic dense molecular cores by Zinchenko *et al.*[28] hinted that both species trace shocks, although the absence of HNCO in the higher velocity wings observed in the SiO spectral profile also hinted to the fact that high-velocity shock conditions may suppress the HNCO abundance. A follow-up survey over sources towards the Galactic Center performed by Martín *et al.*[30] reveals that HNCO can however also be heavily destroyed by UV radiation in PDR regions (later also found in NGC 253 in Martín *et al.*[34]). On the other hand, chemical modelling of HNCO and SiO in NGC 1068 performed by Kelly *et al.*[38] confirmed that the HNCO abundance can be suppressed in high-velocity shocks due to the destruction of its precursor, the molecule NO. In fact, Kelly *et al.*[38] showed that HNCO can also be thermally desorbed from the surface of dust grains when the gas and dust remain coupled at higher gas densities ($n_{H_2} \geq 10^4$ cm^{-3}). Hence HNCO is a complex molecule to interpret and may not be a unique tracer of shock activity.

In our two recent works, we presented the observations of multiple transitions of HNCO and SiO observed with ALMA towards both the CND of NGC 1068 [39] and the CMZ of NGC 253 (Huang *et al.*,[40]). These two studies have shown clear signatures of shocks in these two different environments under the influence of shocks from potentially multiple origins.

In this paper we revisit these observational results and present an astro-chemical modeling study with the aim of comparing these two galaxies in HNCO and SiO. For completeness, we shall also include a brief comparison with the results from our most recent work (Huang *et al.*, in prep.) where we explore the shock properties as traced by CH_3OH in the CND of NGC 1068. The aim will be (1) to determine whether these three species (HNCO, SiO, and CH_3OH) are uniquely sensitive to certain type(s) of shocks in the two galaxies in consideration; (2) to see if we can get a handle on determining the shock origins in these regions; and finally (3) to explore the possibility of differentiating the energetics of these shocks (*e.g.* AGN-driven or starburst-driven) in different types of galaxies.

In Section 2 we briefly summarize the observations of HNCO and SiO in NGC 253 and NGC 1068. In Section 3 we describe our chemical modeling analysis of HNCO and SiO. Section 4 covers a brief discussion of CH_3OH. In Section 5 we discuss the shock properties that each species is tracing, their shock history and structure of the shocks. We also explore the potential in differentiating the shock chemistry in the AGN-dominated *versus* starburst environments, on the basis of these two galaxies – NGC 1068 and NGC 253. Our conclusions are summarized in Section 6.

2 Summary of previous observational results

In Huang *et al.* 2022[39] we examined the potential shock signatures traced by HNCO and SiO in the CND of NGC 1068 at a spatial resolution of ~56 pc, which is comparable to the size of a giant molecular cloud (GMC), using ALMA observations. Fig. 1 shows some of the line intensity maps from Huang *et al.*[39] from

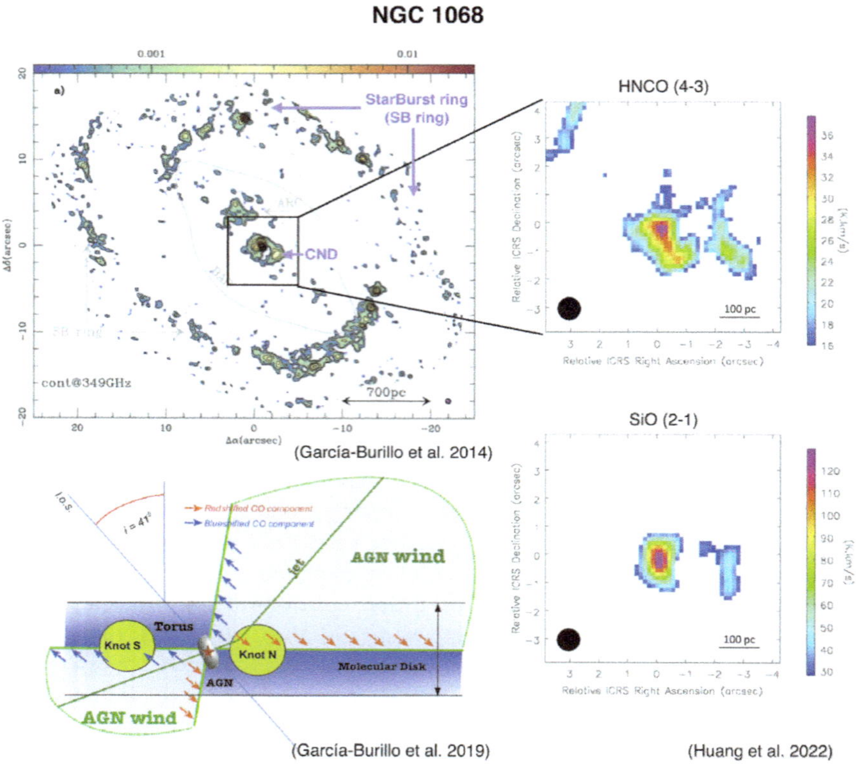

Fig. 1 The two example intensity maps, HNCO (4-3) and SiO (2-1), from the CND of NGC 1068 (two maps on the right, adapted from Huang *et al.*[39]) relative to the layout of the galaxy. *Top left:* dust continuum map of NGC 1068 including the SB ring, adapted from García-Burillo *et al.*;[2] *bottom left*: a schematic adapted from García-Burillo *et al.*,[3] illustrating the relative layout of the AGN wind, the molecular disk/torus, and the line of sight of the observer for NGC 1068.

HNCO and SiO, and their relevant layout with the outflow and the host galaxy itself shown in the dust continuum map adapted from García-Burillo *et al.*[2]

In Huang *et al.*,[40] the same pair of shock tracers, HNCO and SiO, were used to study the CMZ of NGC 253 at higher spatial resolution of ~28 pc using ALMA observations as part of the ALCHEMI large program. Fig. 2 shows examples of the line intensity maps from Huang *et al.*[40] of HNCO and SiO, and their relevant layout with the outflow and the host galaxy itself with maps adapted from Bolatto *et al.*[14]

We list below the main observational results and the remaining open questions from these two previous studies that are relevant to the current work:

(1) From both studies it was confirmed that HNCO and SiO are consistently tracing very different gas components. HNCO tends to trace cooler and denser gas components than those traced by SiO. The high gas temperatures (few hundreds K) traced by SiO signal the evidence of strong shock heating.

(2) In NGC 1068, the discussion of the chemical origin(s) of the two species in Huang *et al.* 2022[39] is based on the prior chemical modeling performed by Kelly

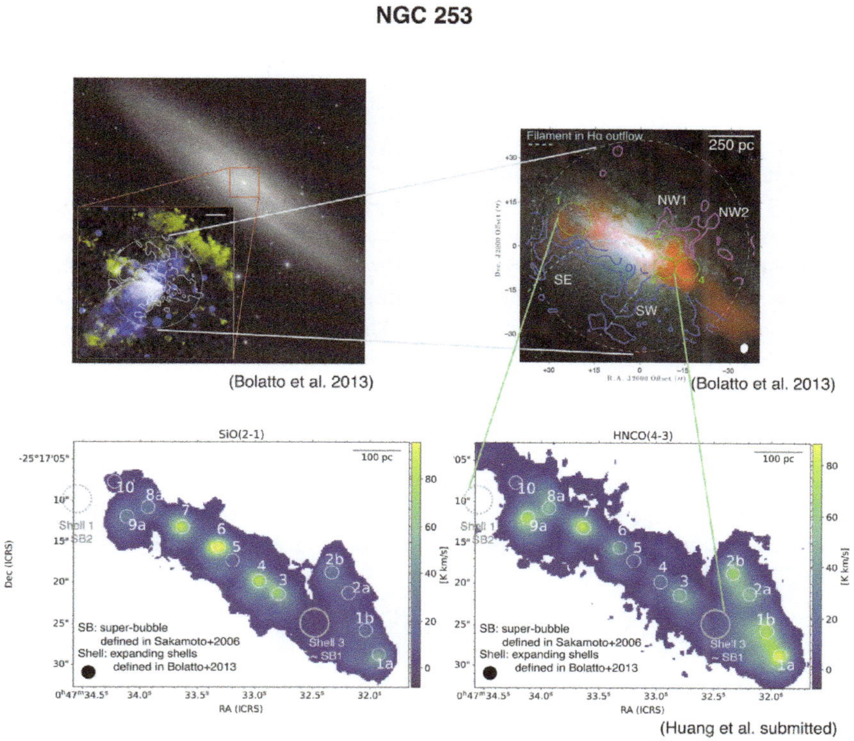

Fig. 2 The two example intensity maps, HNCO (4-3) and SiO (2-1), from the CMZ of NGC 253 (two maps on the bottom, adapted from Huang *et al.*[40]) relative to the layout of the galaxy. *Top panels*: maps illustrating the multi-phase galactic wind in NGC 253, adapted from Bolatto *et al.*[14] and reference therein.

et al. 2017[38] with only low CRIR assumed for the CND ($\zeta = \zeta_0$). In the latter, SiO was clearly associated with strong shocks, yet the chemical origin of HNCO could either be weak shocks or thermal sublimation in denser gas ($n_{H_2} \geq 10^4$ cm^{-3}) environment. The chemical modeling performed by Huang *et al.*[40] strongly supports the fast-shock origin of SiO and slow-shock origin of HNCO in the cases tailored for the high cosmic-ray ionization rate (CRIR) environment ($\zeta = 10^{3-5}\zeta_0$, $\zeta_0 = 1.3 \times 10^{-17}$ s^{-1} is the standard galactic CRIR) of NGC 253. Huang *et al.*[40] shows higher CRIR not only affects both HNCO and SiO abundances in shock chemistry, but makes the thermal sublimation an unviable route in enhancing HNCO abundance.

(3) With well-constrained gas density and gas temperature inferred from the RADEX modeling coupled with Bayesian inference processes that are informed with observational data, Huang *et al.*[40] have shown it to be possible to further infer the age since the last shock episode with each shock tracer. This is because the shock influence timescale is inversely proportional to the gas density. Gas components showing different gas temperatures suggest that they are at different post-shock cooling stages relative to this shock influence timescale.

(4) It remained unclear, however, whether the weak shock episode traced by HNCO and the strong shock episode evidenced by SiO – both in NGC 1068 and

NGC 253 – point to the same occurrence of shock. In other words, it is possible that the two types of shock episodes arise from events that are not associated in causality. However, if one associates the weak shock episode and the strong shock episode to be the same occurrence of shock event, Huang *et al.*[40] showed that there is a global trend in the age of shocks, which increases from the inner to the outer regions of the CMZ of NGC 253.

(5) The chemical origin of the observed HNCO at gas density $n_{H_2} \sim 10^3$ cm^{-3} remains unclear. Neither shock scenarios nor thermal sublimation can account for the HNCO abundance observed at this low gas density in the CMZ of NGC 253.

The summary of key properties concerning these two galaxies that will be used in the current work are listed in Table 1. The listed fractional molecular abundance is derived based on the observational data and the gas property inference *via* radiative transfer modeling presented by Huang *et al.* (2022)[39] and Huang *et al.*[40] We call this observation-based fractional abundance $X_{\text{species,obs}}$. In particular, from the gas property inference performed by both previous studies, both molecular gas density (n_{H_2}) and the molecular total column density per species (N_{HNCO} and N_{SiO}) can be obtained. These gas properties can lead to an estimate of an "observation-based" fractional molecular abundance by the following relation:

$$X_{\text{species,obs}} = \frac{N_{\text{species}}}{n_{H_2}\Delta z_{\text{cloud}}\eta_{\text{ff}}} \sim \frac{N_{\text{species}}}{n_{H_2}\theta_{\text{beam}}\eta_{\text{ff}}} \geq \frac{N_{\text{species}}}{n_{H_2} \times \theta_{\text{beam}} \times 1.0}, \tag{1}$$

where Δz_{cloud} refers to the line-of-sight dimension of the gas component in consideration, θ_{beam} stands for the beam size of the observation, and η_{ff} is the beam filling factor which is a parameter illustrating the source size relative to the beam size as defined by $\eta_{\text{ff}} = \frac{{\theta_S}^2}{{\theta_{\text{beam}}}^2 + {\theta_S}^2}$. As an approximation, we use the beam

Table 1 List of physical and chemical properties of the CND in NGC 1068 and the CMZ of NGC 253 based on previous observational studies[a]

	NGC 1068		NGC 253	
Property	CND	Ref.	CMZ	Ref.
Beam size [pc]	56	(a)	28	(b)
n_{H_2}[HNCO]	10^3 to 10^6	(a)	10^3–10^6	(b)
n_{H_2}[SiO]	10^3 to 10^4	(a)	10^2–10^4	(b)
$X_{\text{HNCO,obs}}$	$\geq(2.71 \times 10^{-12}$ to $1.21 \times 10^{-8})$		$\geq(2.65 \times 10^{-12}$ to $6.60 \times 10^{-8})$	
$X_{\text{SiO,obs}}$	$\geq(1.13 \times 10^{-10})$		$\geq(2.47 \times 10^{-8}$ to $1.61 \times 10^{-7})$	
ζ [ζ_0]	1.0–10.0	(c)	10^3–10^5	(d, e, f and g)
v_{outflow} [km s^{-1}]	85–200	(h)	110–360	(i)

[a] References: (a) Huang *et al.* 2022;[39] (b) Huang *et al.*[40] (submitted); (c) Scourfield *et al.* 2020[44] – CRIR inferred from multi-J CS emissions in the CND region of NGC 1068; (d) Holdship *et al.* 2021;[45] (e) Harada *et al.* 2021;[46] (f) Holdship *et al.* 2022;[47] (g) Behrens *et al.* 2022;[48] (h) García-Burillo *et al.* 2019;[3] (i) Westmoquette *et al.* 2011[12] – this estimate is up to radial distance of 1 kpc.

size of the observation (θ_{beam}) as the estimate of the line-of-sight dimension of the gas component (Δz_{cloud}), and in the final inequality we used the 1.0 as the upper limit for the η_{ff}.

3 Modeling shock chemistry

As mentioned earlier, for NGC 1068, Huang *et al.* (2022)[39] did not model the chemistry but based our conclusions on prior modeling work by Kelly *et al.*[38] Here, therefore, we present a chemical modeling analysis that covers the most comprehensive set of physical and chemical conditions which takes both galaxies – NGC 1068 and NGC 253 – into consideration.

The chemical modeling was performed with the open source time dependent gas–grain `UCLCHEM` code[49] (https://uclchem.github.io) in order to further explore the chemical origin of the observed HNCO and SiO emissions in these two distinctively different galaxies. `UCLCHEM` is a 3-phase gas–grain chemical modelling code that incorporates user-defined chemical networks to produce chemical abundances along user-defined physics modules that can simulate a variety of physical conditions.

We perform modeling cases that cover the physical conditions measured from the CND of NGC 1068 and the CMZ of NGC 253 – this includes a wide CRIR range ($\zeta = 1.0$–$10^5\zeta_0$, $\zeta_0 = 1.3 \times 10^{-17}$ s^{-1} is the standard galactic CRIR), slow ($v_s = 10$ km s^{-1}) *versus* fast ($v_s = 50$ km s^{-1}) shocks, and a wide range of molecular gas densities n_{H_2}. The chosen CRIR range is based on the existing CRIR measurements[44–48] in these two extragalactic environments. In particular, in the CND of NGC 1068, the inferred CRIR based on multi-transition CS observations ranges between $\zeta = 1.0$–$10.0\zeta_0$.[44] And in the CMZ of NGC 253, the measured CRIR based on multi-molecule, multi-transition observations is estimated to be $\sim\zeta = 10^{3-5}\zeta_0$,[45–48] which is a few orders of magnitude higher than NGC 1068. A brief summary of the parameter space explored in our chemical modeling is listed in Table 2.

`UCLCHEM v3.1` includes an improved sputtering module in the parameterized C-shock model following Jiménez-Serra *et al.*[29] as well as a 3-phase chemistry where chemistry is computed for the gas phase, the grain surfaces, and the bulk ice. These improvements, compared to the model used in Kelly *et al.*,[38] ensure a better treatment of the sputtering of refractory species, such as Si-bearing species, during the shock process. Aside from these technical differences, the work by Kelly *et al.* 2017[38] only covers the modeling with standard galactic CRIR (ζ_0).

Table 2 The parameter space explored in our chemical modelling. Note that $\zeta_0 = 1.3 \times 10^{-17}$ s^{-1} and $X_{Si,gas,\odot} = 4.07 \times 10^{-5}$

Variable	Grid
Gas density,[a] n_{H_2} [cm^{-3}]	[10^3, 10^4, 10^5, 10^6]
C-shock velocity, v_{shock} [km s^{-1}]	[10.0, 50.0]
CRIR ζ [ζ_0]	[1.0, 10.0, 10^3, 10^5]
Physical model	Shock or non-shocked scenario

[a] Gas density refers to the pre-shock gas density in the shock scenario, and the initial gas density in the non-shocked scenario.

In Fig. 3–6 we show the results of our modeling for four gas density setups ($n_{H_2} = 103$ to $n_{H_2} = 106$ cm^{-3}, ordered from Fig. 3 to 6) with two shock velocities ($v_s = 10$ km s^{-1} and $v_s = 50$ km s^{-1}) to represent the slow and fast shock scenarios, and these two shock scenarios are shown in the left (slow shock) and right (fast shock) panels respectively. In each row we show the modeling results using different CRIR from the top row (lowest CRIR) to bottom row (highest CRIR). In

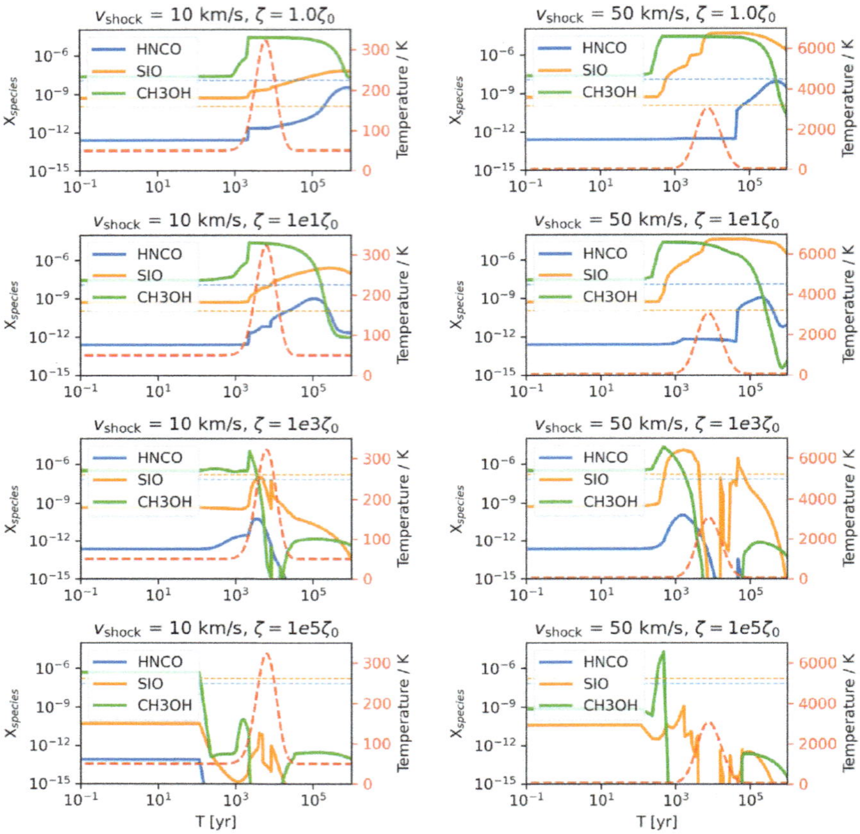

Fig. 3 Chemical abundances as a function of time for slow shock ($v_s = 10$ km s^{-1}) models (left panel) and fast shock ($v_s = 50$ km s^{-1}) models (right panel). The pre-shock gas density in the shock models and the gas density in non-shock models is 10^3 cm^{-3}. The red dashed curve represents the temperature profile, with the temperature scale on the vertical axis on the right, also in red. For the shock models, within each panel we present from top to bottom: [CRIR (ζ) = ζ_0], [CRIR (ζ) = $10\zeta_0$], [CRIR (ζ) = $10^3\zeta_0$], and [CRIR = $10^5\zeta_0$]. The dashed, colored horizontal lines indicate the lower limit of the species fractional abundances "measured" from our RADEX-Bayesian inference based on observational data and with an assumed hydrogen column density, for HNCO (blue) and SiO (orange) respectively. The fractional abundance values used are the minimum derived values among the studied GMC-sized regions in the two galaxies. In the top 4 figures the "measured" fractional abundances are cited from observations towards the CND of NGC 1068 which was found of lower CRIR with corresponding colors for each species when observed cases are available. On the other hand, the "measured" fractional abundances in the bottom 4 figures are cited from observations towards the CMZ of NGC 253, as found with higher CRIR.

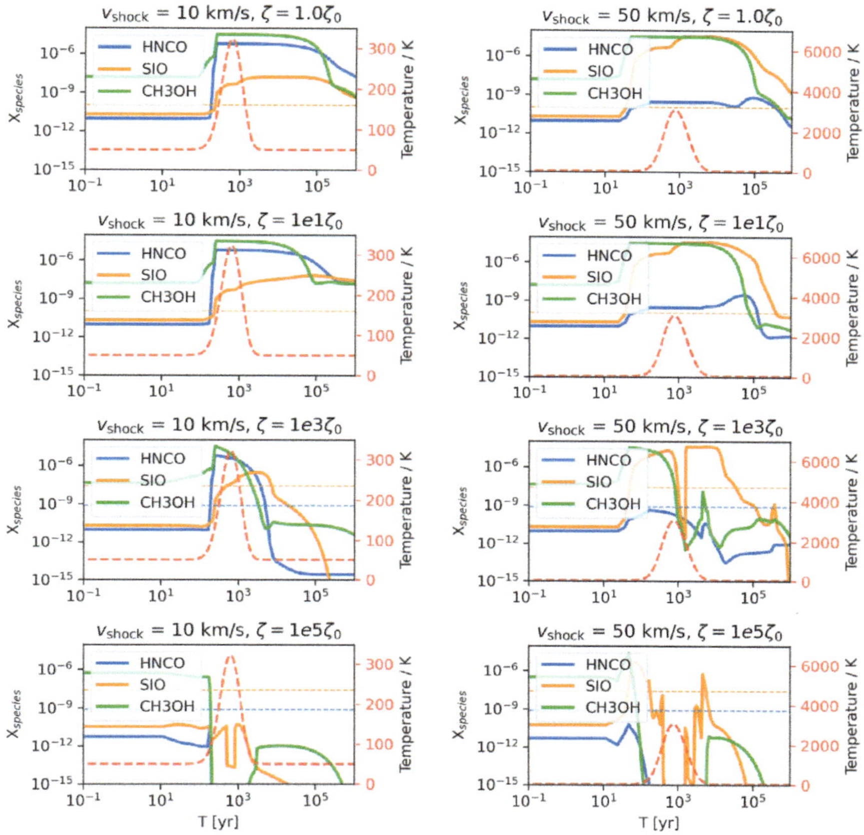

Fig. 4 As in Fig. 3 but for a pre-shock gas density of 10^4 cm^{-3}.

particular, in each figure (*e.g.* Fig. 3) the top two rows are models at lower CRIR for the CND of NGC 1068, and the bottom two rows are models at higher CRIR for the CMZ of NGC 253. Each Figure is associated with one specific gas density *e.g.* Fig. 3 at a gas density $n_{H_2} = 10^3$ cm^{-3}. In each figure we also overlay the minimum fractional abundances as constrained by the observations in NGC 1068 and NGC 253 respectively, $X_{species,obs}$, which was explained earlier in Section 2 with the corresponding gas density and CRIR.

First of all for the SiO abundance (orange solid curves), it is clear that fast-shock chemistry dominates over slow-shock chemistry. While slow shocks can also enhance the SiO abundance, the enhancement is still more than 2 orders of magnitude lower than the enhancement arising from fast shocks. The case for the HNCO abundance (blue solid curves) is reversed, where its enhancement is dominated by slow-shock chemistry. The only exception is in the least dense gas environment ($n_{H_2} = 10^3$ cm^{-3}, Fig. 3), where the enhancement of HNCO in none of the cases is sufficient to reach the minimum observed, $X_{HNCO,obs}$.

In general, high CRIR suppresses the abundances of all shock tracers of interest, especially the highest CRIR tested, $\zeta = 10^5\zeta_0$. In the denser gas environments ($n_{H_2} = 10^{5-6}$ cm^{-3}) such suppression seems less severe.

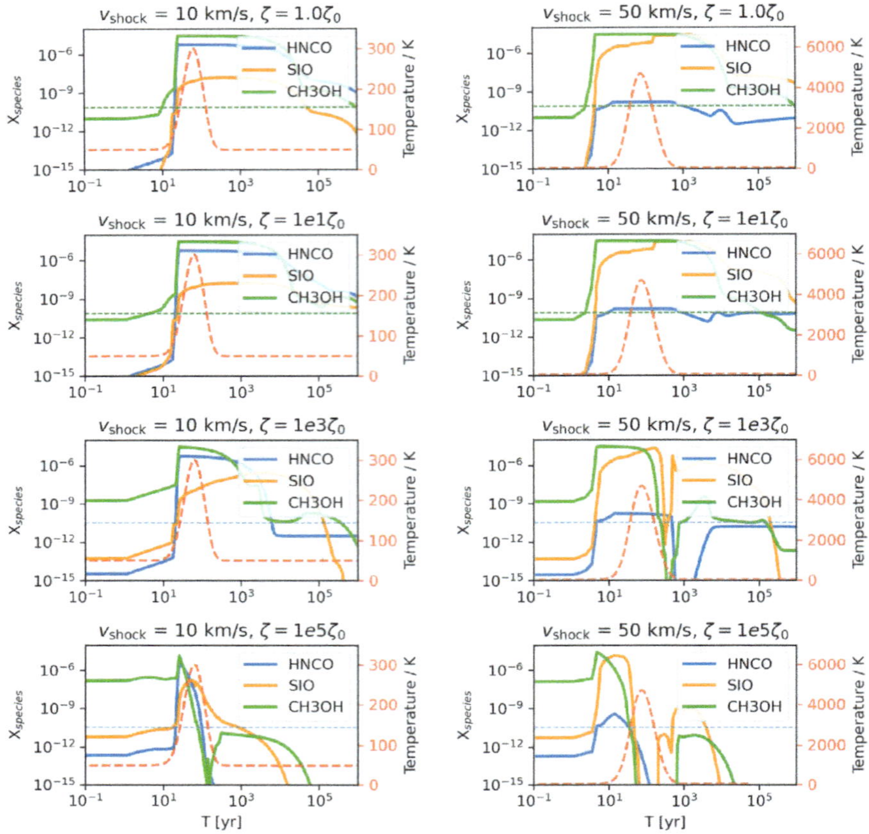

Fig. 5 As in Fig. 3 but for a pre-shock gas density of 10^5 cm^{-3}. The dashed, green horizontal lines in the top 4 figures indicate the lower limit of the species fractional abundances of methanol "measured" from our RADEX-Bayesian inference based on observational data and with an assumed hydrogen column density for the CND of NGC 1068.

4 Methanol in the CND of NGC 1068

Methanol (CH_3OH) can be formed efficiently in cold environments (12–20 K) through repeated hydrogenation of CO on interstellar ices[50–54] as well as *via* the radical-molecule H-atom abstraction reaction.[55] Sources of energy such as shocks can remove CH_3OH from the grain surface and enhance the gas-phase CH_3OH either through a sputtering process or sublimation. Although gas-phase formation routes of CH_3OH have also been proposed, it cannot account for the observed CH_3OH in general.[56,57] CH_3OH has been therefore also proposed as a good candidate tracer of shocks in the ISM, although it has also been detected in cold and quiescent environments such as pre-stellar cores[58,59] where alternative non-thermal desorption processes aside from shocks may be needed.

In Fig. 3–6 we also display the CH_3OH fractional abundances (in green solid curves) over slow *versus* fast shock chemistry. It is clear that both slow shock and fast shock are able to enhance the CH_3OH abundance to comparable levels. This

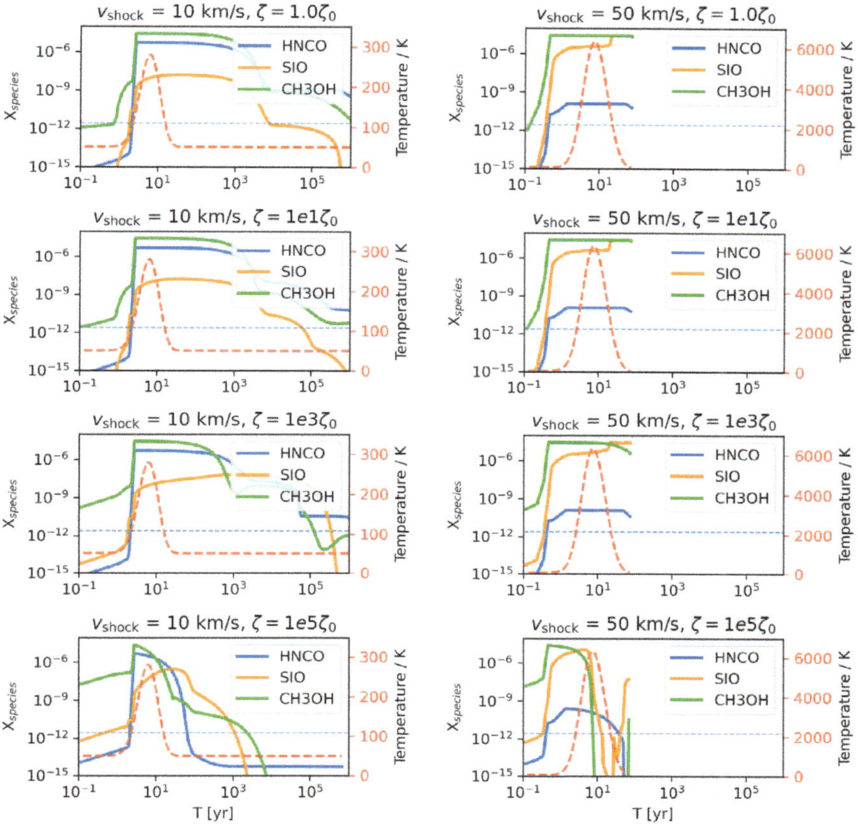

Fig. 6 As in Fig. 3 but for a pre-shock gas density of 10^6 cm^{-3}.

is consistent with prior modeling work,[49] where CH_3OH was shown enhanced for a large span of shock velocities. One of the key pieces of information to differentiate the shock origin – fast or slow shock – of the observed CH_3OH is the gas temperature. In the slow-shock scenario the gas temperature can only reach a few hundred Kelvin, while in the fast-shock scenario the gas temperature can reach up to a few thousand Kelvin. Note that, however, a measured low gas temperature does not necessarily mean only slow-shock origin, for it could arise from a gas component that experienced full post-shock cooling down from even a few thousand Kelvin.

Such degeneracy will be an interesting piece of knowledge when one wants to reconstruct the shock structure and shock history in the target field. As pointed out by Huang *et al.*,[40] it remains unclear whether the fast shock episode traced by SiO and the slow shock episode traced by HNCO are caused by the same shock occurrence. It is possible that they arise from the same shock events but belong to different parts of the gas structure, or they could be triggered by sporadic events that are not linked at all. As CH_3OH is enhanced to comparable levels in both fast shock and slow shock, how well matched in its spatial distribution and the gas properties it traces compared to the other shock-tracing species (*e.g.* HNCO and

SiO) will be critical to the overall understanding of the shock structure and history reconstruction.

For instance, our preliminary analysis of methanol data in the CND in NGC 1068 from Huang *et al.*, in prep. indicates that CH_3OH is tracing similarly high-density and cold (T < 50 K) gas components as HNCO in the CND regions. In terms of the chemical origin of the observed CH_3OH, it could arise from the post-shock cooling condition gas under the influence of both fast and slow shocks. One should therefore be careful in using these three species at once to portrait the shock structure in the field. The degeneracy and the great complexity are inevitable when we include more and more molecular tracers in consideration, but we want to emphasize that with careful analysis and interpretation it will at the same time provide enormous potential in fully reconstructing the shock structure and history with well-covered parameter space measured from observations.

5 Discussion

5.1 Spatial distribution of shocked gas in the two galaxies

Although observed with lower spatial resolution (~56 pc), the HNCO emission tends to be more spatially extended than SiO in the CND of NGC 1068 (*e.g.* Fig. 1), as pointed out by Huang *et al.*[39] It is different for the case of CMZ in NGC 253, where HNCO and SiO emission seems to be co-spatial in most transitions, see Fig. 2 for example.

There are a few possible reasons concerning such differences in the two systems:

• The spatial offset between HNCO and SiO identified in the CND of NGC 1068 (but not in the CMZ of NGC 253) could arise from the difference in the sensitivity levels of the two sets of observations. In other words it is possible that SiO in NGC 1068 is also extended throughout the CND rather than just localized in the East and West spots of the CND.

• As NGC 253 is more edge-on, it is possible that the difference is due to a difference in viewing angle.

• The less extended SiO emission in the CND of NGC 1068 could instead arise from chemical differential across the CND of NGC 1068. It is possible that SiO is highlighting the strongly shocked gas layer(s) and HNCO is mapping out more extended and weakly shocked region, potentially correlated to the geometry of the AGN-driven, large-scale outflow. Meanwhile the shocks induced in the CMZ of NGC 253 may originate from the mini-starburst events in each GMC, rather than from an ordered, large-scale outflow.

5.2 Weak shock tracers – degeneracy with non-shock chemistry

As we mentioned in earlier sections (Section 1 and Section 4), both HNCO and CH_3OH can also be enhanced *via* thermal sublimation of the ices induced by events other than shocks. In Fig. 7 we show a chemical model for a gas density $n_{H_2} = 10^5$ cm^{-3} in its chemical evolution of the gas being warmed by the presence of star forming processes (including outflows), and/or X-ray or cosmic rays instead of being shocked. This shows that both HNCO and CH_3OH abundances can be enhanced in a warm-gas environment without shocks particularly at low CRIR cases ($\zeta = 1.0$–$10.0\zeta_0$, top two panels). This degeneracy (shock chemistry or

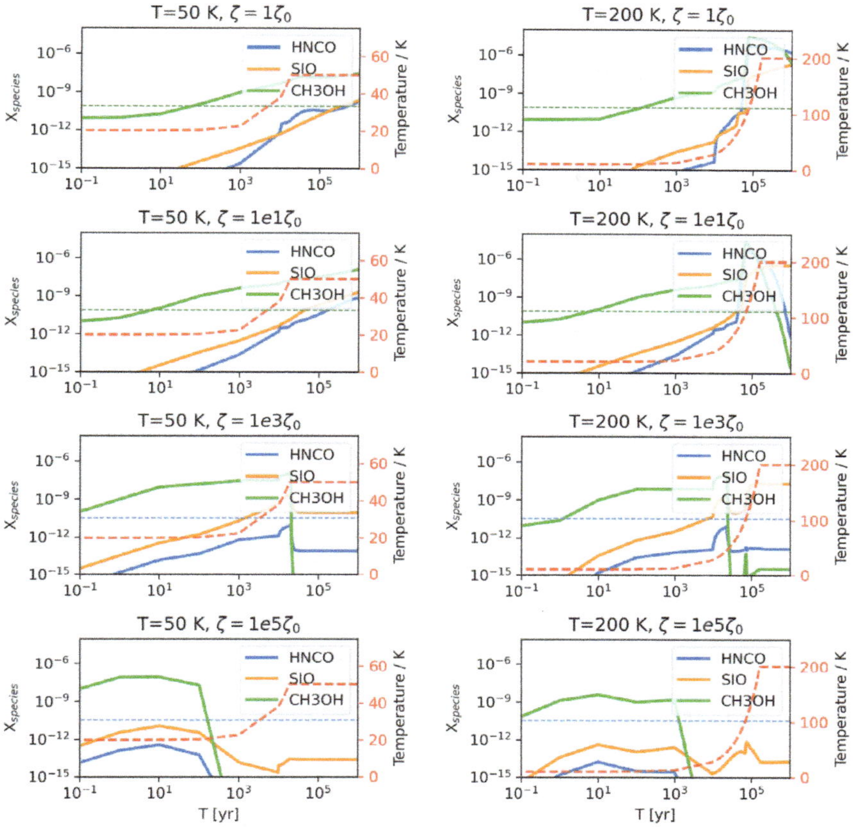

Fig. 7 As in Fig. 3 but for the non-shock models with gas density 10^5 cm^{-3}. The dashed, green horizontal lines in the top 4 figures indicate the lower limit of the species fractional abundances of methanol "measured" from our RADEX-Bayesian inference based on observational data and with an assumed hydrogen column density for the CND of NGC 1068.

thermal sublimation) will impact more the interpretation of the CND in NGC 1068, where lower (than in NGC 253) CRIR values were reported.[44]

5.3 Shock history reconstruction in both galaxies and the shock origin

In NGC 253 it remains unclear whether the fast shock episodes traced by SiO are associated with the slow shock episodes traced by HNCO; however if one associates the shock timescales traced by these two species there seems to be a trend of increasing shock ages from the inner to the outer region of the CMZ. Such trend could be related to the large-scale outflow that stems from the nuclear region due to starburst activities. On the other hand, it remains possible that the shock episodes traced by HNCO and SiO were triggered by sporadic, localized events which are only sensitive to their neighboring environment.

In the case of NGC 1068, however, due to the poorly constrained gas temperature, it is still difficult to gauge such shock timescales and make comparison across the CND region. Future observations of more transitions of both species are needed to properly constrain the gas temperature.

5.4 Shock chemistry in the two environments – NGC 1068 and NGC 253

5.4.1 Velocities of AGN-driven *vs.* starburst-driven outflows. As it is also often thought that AGN-driven outflows can be more powerful than those starburst-driven, we also inspected the outflow velocities from the two systems reported in the literature as listed in Table 1. It seems in fact that the outflow velocities are pretty comparable, and both velocities are high enough to trigger both fast and slow shocks in their sphere of influence. There is no fundamental difference from this aspect for the two galaxies in consideration.

5.4.2 Cosmic-ray ionization rate (CRIR). Generally the CRIR in the CMZ of NGC 253 is a few orders of magnitude higher than in the CND of NGC 1068. From Section 3 it was clear that the three shock-tracing species are all sensitive to CRIR, where high CRIR is suppressing the abundances of these species regardless of the gas density and shock velocity.

The assumed CRIR for the CND in NGC 1068 is based on the modeling conducted by Scourfield *et al.*[44] using multi-J CS observations. CS is often considered a dense gas tracer and could be more embedded in the gas clumps, therefore could be more shielded from radiation and cosmic rays. Indeed Scourfield *et al.*[44] showed that the gas densities traced by CS are generally at $n_{H_2} \sim 10^{5.5}$ to 10^7 cm^{-3}, which are much higher than the densities traced by the CRIR tracer adopted for the NGC 253 study, *e.g.* Holdship *et al.*[47]. A follow-up study of H_3O^+ and SO in NGC 1068 should be done to see if the derived CRIR may in fact be different from the CS study.[44]

6 Conclusions

We have revisited the shock chemistry of two nearby galaxies, the AGN-host galaxy NGC 1068 and the starburst galaxy NGC 253, by analysing three shock tracers: SiO, HNCO, and CH_3OH. We presented a chemical modelling study of these two distinctively different environments and explored the potential of using HNCO and SiO in determining the shock structure and history, as well as in probing the origin(s) of shocks in the nuclei of the two galaxies. We also briefly discussed the potential of using CH_3OH as an additional shock tracer. We briefly summarize here our main conclusions:

(1) SiO can only be enhanced by fast shocks, HNCO can only be enhanced by slow shocks, and CH_3OH can respond to the presence of both slow and fast shocks.

(2) Generally a high CRIR suppresses the chemical abundances of all three shock tracers, especially for the highest CRIR case explored, $\zeta = 10^5\zeta_0$.

(3) In the lower CRIR regimes ($\zeta = 1$ to $10\zeta_0$), the abundance of HNCO and CH_3OH can also be enhanced by thermal sublimation instead of shocks. This has to be taken into careful consideration when interpreting results especially from the CND of NGC 1068.

(4) There are more similarities than differences in the properties traced by HNCO and SiO in both systems. Follow-up studies, including the observations of more transitions of these species for NGC 1068, are needed to better constrain the gas properties of the shocked gas.

Author contributions

KYH contributed to the conceptualization, chemical modeling and analysis, and article text discussion. SV contributed to the conceptualization, analysis, and article text discussion.

Conflicts of interest

There are no conflicts to declare.

Acknowledgements

This work is supported by European Research Council (ERC) Advanced Grant MOPPEX 833460.

Notes and references

1 J. Bland-Hawthorn, J. F. Gallimore, L. J. Tacconi, E. Brinks, S. A. Baum, R. R. J. Antonucci and G. N. Cecil, *Astrophys. Space Sci.*, 1997, **248**, 9–19.
2 S. García-Burillo, F. Combes, A. Usero, S. Aalto, M. Krips, S. Viti, A. Alonso-Herrero, L. K. Hunt, E. Schinnerer, A. J. Baker, F. Boone, V. Casasola, L. Colina, F. Costagliola, A. Eckart, A. Fuente, C. Henkel, A. Labiano, S. Martín, I. Márquez, S. Muller, P. Planesas, C. Ramos Almeida, M. Spaans, L. J. Tacconi and P. P. van der Werf, *Astron. Astrophys.*, 2014, **567**, A125.
3 S. García-Burillo, F. Combes, C. Ramos Almeida, A. Usero, A. Alonso-Herrero, L. K. Hunt, D. Rouan, S. Aalto, M. Querejeta, S. Viti, P. P. van der Werf, H. Vives-Arias, A. Fuente, L. Colina, J. Martín-Pintado, C. Henkel, S. Martín, M. Krips, D. Gratadour, R. Neri and L. J. Tacconi, *Astron. Astrophys.*, 2019, **632**, A61.
4 S. Viti, S. García-Burillo, A. Fuente, L. K. Hunt, A. Usero, C. Henkel, A. Eckart, S. Martin, M. Spaans, S. Muller, F. Combes, M. Krips, E. Schinnerer, V. Casasola, F. Costagliola, I. Marquez, P. Planesas, P. P. van der Werf, S. Aalto, A. J. Baker, F. Boone and L. J. Tacconi, *Astron. Astrophys.*, 2014, **570**, A28.
5 A. McCormick, S. Veilleux and D. S. N. Rupke, *Astrophys. J.*, 2013, **774**, 126.
6 R. Rekola, M. G. Richer, M. L. McCall, M. J. Valtonen, J. K. Kotilainen and C. Flynn, *Mon. Not. R. Astron. Soc.*, 2005, **361**, 330–336.
7 A. K. Leroy, A. D. Bolatto, E. C. Ostriker, E. Rosolowsky, F. Walter, S. R. Warren, J. Donovan Meyer, J. Hodge, D. S. Meier, J. Ott, K. Sandstrom, A. Schruba, S. Veilleux and M. Zwaan, *Astrophys. J.*, 2015, **801**, 25.
8 G. J. Bendo, R. J. Beswick, M. J. D'Cruze, C. Dickinson, G. A. Fuller and T. W. B. Muxlow, *Mon. Not. R. Astron. Soc.: Lett.*, 2015, **450**, L80–L84.
9 K. Sakamoto, R.-Q. Mao, S. Matsushita, A. B. Peck, T. Sawada and M. C. Wiedner, *Astrophys. J.*, 2011, **735**, 19.
10 D. K. Strickland, T. M. Heckman, K. A. Weaver and M. Dahlem, *Astron. J.*, 2000, **120**, 2965–2974.
11 D. K. Strickland, T. M. Heckman, K. A. Weaver, C. G. Hoopes and M. Dahlem, *Astrophys. J.*, 2002, **568**, 689–716.
12 M. S. Westmoquette, L. J. Smith and I. Gallagher, *Mon. Not. R. Astron. Soc.*, 2011, **414**, 3719–3739.

13 B. E. Turner, *Astrophys. J.*, 1985, **299**, 312–333.
14 A. D. Bolatto, S. R. Warren, A. K. Leroy, F. Walter, S. Veilleux, E. C. Ostriker, J. Ott, M. Zwaan, D. B. Fisher, A. Weiss, E. Rosolowsky and J. Hodge, *Nature*, 2013, **499**, 450–453.
15 F. Walter, A. D. Bolatto, A. K. Leroy, S. Veilleux, S. R. Warren, J. Hodge, R. C. Levy, D. S. Meier, E. C. Ostriker, J. Ott, E. Rosolowsky, N. Scoville, A. Weiss, L. Zschaechner and M. Zwaan, *Astrophys. J.*, 2017, **835**, 265.
16 N. Krieger, A. D. Bolatto, F. Walter, A. K. Leroy, L. K. Zschaechner, D. S. Meier, J. Ott, A. Weiss, E. A. C. Mills, R. C. Levy, S. Veilleux and M. Gorski, *Astrophys. J.*, 2019, **881**, 43.
17 R. C. Levy, A. D. Bolatto, A. K. Leroy, M. C. Sormani, K. L. Emig, M. Gorski, L. Lenkić, E. A. C. Mills, E. Tarantino, P. Teuben, S. Veilleux and F. Walter, *Astrophys. J.*, 2022, **935**, 19.
18 P. J. McCarthy, W. van Breugel and T. Heckman, *Astron. J.*, 1987, **93**, 264.
19 F. Müller-Sánchez, O. González-Martín, J. A. Fernández-Ontiveros, J. A. Acosta-Pulido and M. A. Prieto, *Astrophys. J.*, 2010, **716**, 1166–1177.
20 B. D. Lehmer, D. R. Wik, A. E. Hornschemeier, A. Ptak, V. Antoniou, M. K. Argo, K. Bechtol, S. Boggs, F. E. Christensen, W. W. Craig, C. J. Hailey, F. A. Harrison, R. Krivonos, J. C. Leyder, T. J. Maccarone, D. Stern, T. Venters, A. Zezas and W. W. Zhang, *Astrophys. J.*, 2013, **771**, 134.
21 J. L. Turner and P. T. P. Ho, *Astrophys. J. Lett.*, 1985, **299**, L77–L81.
22 S. García-Burillo, J. Martín-Pintado, A. Fuente and R. Neri, *Astron. Astrophys.*, 2000, **355**, 499–511.
23 D. S. Meier, F. Walter, A. D. Bolatto, A. K. Leroy, J. Ott, E. Rosolowsky, S. Veilleux, S. R. Warren, A. Weiß, M. A. Zwaan and L. K. Zschaechner, *Astrophys. J.*, 2015, **801**, 63.
24 P. K. Humire, C. Henkel, A. Hernández-Gómez, S. Martín, J. Mangum, N. Harada, S. Muller, K. Sakamoto, K. Tanaka, Y. Yoshimura, K. Nakanishi, S. Mühle, R. Herrero-Illana, D. S. Meier, E. Caux, R. Aladro, R. Mauersberger, S. Viti, L. Colzi, V. M. Rivilla, M. Gorski, K. M. Menten, K. Y. Huang, S. Aalto, P. P. van der Werf and K. L. Emig, *Astron. Astrophys.*, 2022, **663**, A33.
25 N. Harada, S. Martín, J. G. Mangum, K. Sakamoto, S. Muller, V. M. Rivilla, C. Henkel, D. S. Meier, L. Colzi, M. Yamagishi, K. Tanaka, K. Nakanishi, R. Herrero-Illana, Y. Yoshimura, P. K. Humire, R. Aladro, P. P. van der Werf and K. L. Emig, *Astrophys. J.*, 2022, **938**, 80.
26 J. Martín-Pintado, P. de Vicente, A. Fuente and P. Planesas, *Astrophys. J. Lett.*, 1997, **482**, L45–L48.
27 S. Hüttemeister, G. Dahmen, R. Mauersberger, C. Henkel, T. L. Wilson and J. Martin-Pintado, *Astron. Astrophys.*, 1998, **334**, 646–658.
28 I. Zinchenko, C. Henkel and R. Q. Mao, *Astron. Astrophys.*, 2000, **361**, 1079–1094.
29 I. Jiménez-Serra, P. Caselli, J. Martín-Pintado and T. W. Hartquist, *Astron. Astrophys.*, 2008, **482**, 549–559.
30 S. Martín, M. A. Requena-Torres, J. Martín-Pintado and R. Mauersberger, *Astrophys. J.*, 2008, **678**, 245–254.
31 N. J. Rodríguez-Fernández, M. Tafalla, F. Gueth and R. Bachiller, *Astron. Astrophys.*, 2010, **516**, A98.
32 D. S. Meier and J. L. Turner, *Astrophys. J.*, 2005, **618**, 259–280.

33 A. Usero, S. García-Burillo, J. Martín-Pintado, A. Fuente and R. Neri, *Astron. Astrophys.*, 2006, **448**, 457–470.

34 S. Martín, J. Martín-Pintado and R. Mauersberger, *Astrophys. J.*, 2009, **694**, 610–617.

35 S. García-Burillo, A. Usero, A. Fuente, J. Martín-Pintado, F. Boone, S. Aalto, M. Krips, R. Neri, E. Schinnerer and L. J. Tacconi, *Astron. Astrophys.*, 2010, **519**, A2.

36 D. S. Meier and J. L. Turner, *Astrophys. J.*, 2012, **755**, 104.

37 S. Martín, K. Kohno, T. Izumi, M. Krips, D. S. Meier, R. Aladro, S. Matsushita, S. Takano, J. L. Turner, D. Espada, T. Nakajima, Y. Terashima, K. Fathi, P. Y. Hsieh, M. Imanishi, A. Lundgren, N. Nakai, E. Schinnerer, K. Sheth and T. Wiklind, *Astron. Astrophys.*, 2015, **573**, A116.

38 G. Kelly, S. Viti, S. García-Burillo, A. Fuente, A. Usero, M. Krips and R. Neri, *Astron. Astrophys.*, 2017, **597**, A11.

39 K. Y. Huang, S. Viti, J. Holdship, S. García-Burillo, K. Kohno, A. Taniguchi, S. Martín, R. Aladro, A. Fuente and M. Sánchez-García, *arXiv*, 2022, preprint, arXiv:2202.05005, DOI: 10.48550/arXiv.2202.05005.

40 K.-Y. Huang, S. Viti, J. Holdship, J. G. Magnum, S. Martín, N. Harada, S. Muller, K. Sakamoto, K. Tanaka, Y. Yoshimura, R. Herrero-Illana, D. S. Meier, E. Behrens, P. P. van der Werf, C. Henkel, S. García-Burillo, V. M. Rivilla, K. L. Emig, L. Colzi, P. K. Humire, R. Aladro and M. Bouvier, *arXiv*, 2023, preprint, arXiv:2303.12685, DOI: 10.48550/arXiv.2303.12685.

41 P. Schilke, C. M. Walmsley, G. Pineau des Forets and D. R. Flower, *Astron. Astrophys.*, 1997, **321**, 293–304.

42 G. Fedoseev, S. Ioppolo, D. Zhao, T. Lamberts and H. Linnartz, *Mon. Not. R. Astron. Soc.*, 2015, **446**, 439–448.

43 A. López-Sepulcre, A. A. Jaber, E. Mendoza, B. Lefloch, C. Ceccarelli, C. Vastel, R. Bachiller, J. Cernicharo, C. Codella, C. Kahane, M. Kama and M. Tafalla, *Mon. Not. R. Astron. Soc.*, 2015, **449**, 2438–2458.

44 M. Scourfield, S. Viti, S. García-Burillo, A. Saintonge, F. Combes, A. Fuente, C. Henkel, A. Alonso-Herrero, N. Harada, S. Takano, T. Nakajima, S. Martín, M. Krips, P. P. van der Werf, S. Aalto, A. Usero and K. Kohno, *Mon. Not. Roy. Astron. Soc.*, 2020, **496**, 5308–5329.

45 J. Holdship, S. Viti, S. Martín, N. Harada, J. Mangum, K. Sakamoto, S. Muller, K. Tanaka, Y. Yoshimura, K. Nakanishi, R. Herrero-Illana, S. Mühle, R. Aladro, L. Colzi, K. L. Emig, S. García-Burillo, C. Henkel, P. Humire, D. S. Meier, V. M. Rivilla and P. van der Werf, *Astron. Astrophys.*, 2021, **654**, A55.

46 N. Harada, S. Martín, J. G. Mangum, K. Sakamoto, S. Muller, K. Tanaka, K. Nakanishi, R. Herrero-Illana, Y. Yoshimura, S. Mühle, R. Aladro, L. Colzi, V. M. Rivilla, S. Aalto, E. Behrens, C. Henkel, J. Holdship, P. K. Humire, D. S. Meier, Y. Nishimura, P. P. van der Werf and S. Viti, *Astrophys. J.*, 2021, **923**, 24.

47 J. Holdship, J. G. Mangum, S. Viti, E. Behrens, N. Harada, S. Martín, K. Sakamoto, S. Muller, K. Tanaka, K. Nakanishi, R. Herrero-Illana, Y. Yoshimura, R. Aladro, L. Colzi, K. L. Emig, C. Henkel, Y. Nishimura, V. M. Rivilla and P. P. van der Werf, *arXiv*, 2022, preprint, arXiv:2204.03668, DOI: 10.48550/arXiv.2204.03668.

48 E. Behrens, J. G. Mangum, J. Holdship, S. Viti, N. Harada, S. Martin, K. Sakamoto, S. Muller, K. Tanaka, K. Nakanishi, R. Herrero-Illana,

Y. Yoshimura, R. Aladro, L. Colzi, K. L. Emig, C. Henkel, K.-Y. Huang, P. K. Humire, D. S. Meier and V. M. Rivilla, *arXiv*, 2022, preprint, arXiv:2209.06244, DOI: 10.48550/arXiv.2209.06244.
49 J. Holdship, S. Viti, I. Jiménez-Serra, A. Makrymallis and F. Priestley, *Astron. J.*, 2017, **154**, 38.
50 A. G. G. M. Tielens and W. Hagen, *Astron. Astrophys.*, 1982, **114**, 245–260.
51 S. B. Charnley, A. G. G. M. Tielens and T. J. Millar, *Astrophys. J. Lett.*, 1992, **399**, L71.
52 K. Hiraoka, N. Ohashi, Y. Kihara, K. Yamamoto, T. Sato and A. Yamashita, *Chem. Phys. Lett.*, 1994, **229**, 408–414.
53 N. Watanabe and A. Kouchi, *Astrophys. J. Lett.*, 2002, **571**, L173–L176.
54 G. W. Fuchs, H. M. Cuppen, S. Ioppolo, C. Romanzin, S. E. Bisschop, S. Andersson, E. F. van Dishoeck and H. Linnartz, *Astron. Astrophys.*, 2009, **505**, 629–639.
55 J. C. Santos, K.-J. Chuang, T. Lamberts, G. Fedoseev, S. Ioppolo and H. Linnartz, *Astrophys. J. Lett.*, 2022, **931**, L33.
56 W. D. Geppert, F. Hellberg, F. Österdahl, J. Semaniak, T. J. Millar, H. Roberts, R. D. Thomas, M. Hamberg, M. A. Ugglas, A. Ehlerding, V. Zhaunerchyk, M. Kaminska and M. Larsson, *Astrochemistry: Recent Successes and Current Challenges*, 2005, pp. 117–124.
57 R. Garrod, I. H. Park, P. Caselli and E. Herbst, *Faraday Discuss.*, 2006, **133**, 51.
58 L. Bizzocchi, P. Caselli, S. Spezzano and E. Leonardo, *Astron. Astrophys.*, 2014, **569**, A27.
59 C. Vastel, C. Ceccarelli, B. Lefloch and R. Bachiller, *Astrophys. J. Lett.*, 2014, **795**, L2.

Faraday Discussions

ROYAL SOCIETY OF CHEMISTRY

DISCUSSIONS

Observational astrochemistry in the age of ALMA, NOEMA, JWST and beyond!: general discussion

Eleonora Bianchi, Sandra Brünken, Cecilia Ceccarelli, Martin Cordiner, Athena Flint, Robin T. Garrod, Murthy S. Gudipati, Divita Gupta, Dwayne E. Heard, Eric Herbst, Ko-Yun Huang, Inga Kamp, Jialu Li, Nikku Madhusudhan, Rafael Martin Domenech, Martin R. S. McCoustra, Anthony Meijer, Stefanie Milam, Tom J. Millar, Karolina Plakitina, Ian Sims, Ewine F. van Dishoeck, Serena Viti, Nicholas Walker, Laurent Wiesenfeld and Olivia H. Wilkins

DOI: 10.1039/d3fd90024j

Rafael Martin Domenech opened the discussion of the introductory Spiers Memorial Lecture by Cecilia Ceccarelli: The formation of methanol in interstellar ices is a fundamental question, since methanol is usually seen as the smallest possible complex organic molecule (COM), and has long been thought to be the precursor for (or at least to participate in) the formation of other complex organic molecules (see, *e.g.*, ref. 1). Methanol has traditionally been thought to be formed through hydrogenation of CO (see, *e.g.*, ref. 2). However, there is now evidence that alternative formation pathways might be required to explain the presence of methanol in regions where hydrogenation of CO is not possible/efficient (see, for example, detection of methanol in interstellar ices in ref. 3). These alternative formation pathways would also contribute to the formation of other COMs, and should thus be taken into account when addressing COM formation in ices. Otherwise, we could be limited to H-atom addition reactions that might not represent the main COM formation pathway in some interstellar regions.

1 K. I. Öberg, R. T. Garrod, E. F. van Dishoeck and H. Linnartz, *Astron. Astrophys.*, 2009, **504**(3), 891.
2 N. Watanabe and A. Kouchi, *Astrophys. J.*, 2002, **567**, 651.
3 M. K. McClure, W. R. M. Rocha, K. M. Pontoppidan, N. Crouzet, L. E. U. Chu, E. Dartois, T. Lamberts, J. A. Noble, Y. J. Pendleton, G. Perotti, D. Qasim, M. G. Rachid, Z. L. Smith, F. Sun, T. L. Beck, A. C. A. Boogert, W. A. Brown, P. Caselli, S. B. Charnley, H. M. Cuppen, H. Dickinson, M. N. Drozdovskaya, E. Egami, J. Erkal, H. Fraser, R. T. Garrod, D. Harsono, S. Ioppolo, I. Jiménez-Serra, M. Jin, J. K. Jørgensen, L. E. Kristensen, D. C. Lis, M. R. S. McCoustra, B. A. McGuire, G. J. Melnick, K. I. Öberg, M. E. Palumbo, T. Shimonishi, J. A. Sturm, E. F. van Dishoeck and H. Linnartz, *Nat. Astron.*, 2023, **7**, 431.

Cecilia Ceccarelli answered: I fully agree with Martin's comment. We have to think of alternative ways of forming methanol and interstellar complex organic molecules. One recent development is indeed the realization that radicals landing on a water-rich ice do react with the water molecules to form complex molecules. In this way, gaseous carbon atoms (which are present when CO is not yet) landing on the water-ice react and form H_2CO, as shown by Molpeceres *et al.* (2021).[1] Likewise, Perrero *et al.* (2022)[2] have shown that landing CCH leads to the formation of ethanol, while Rimola *et al.* (2018)[3] have shown that CN leads to formamide. The reactions of radicals with the water molecules of the ice have a twofold consequence. On the one hand, they can indeed form complex molecules. On the other hand, they sort of hamper the possibility that the radical combinations, often assumed in astrochemical models, are efficient, because radicals do react with the water molecules (which surround them!) before being able to migrate and encounter other radicals. Thus, the radical reactions with water-ice molecules are potentially a game-changer. Likewise, reactions of radicals on the CO-rich ices have a similar effect, as shown by Ferrero *et al.* (2023),[4] for example. Many more studies are necessary on this aspect, and likely the most important role on this will be played by quantum mechanics computations, in my view.

1 G. Molpeceres, J. Kästner, G. Fedoseev, D. Qasim, R. Schömig, H. Linnartz and T. Lamberts, *J. Phys. Chem. Lett.*, 2021, **12**, 10854–10860.
2 J. Perrero, J. Enrique-Romero, B. Martínez-Bachs, C. Ceccarelli, N. Balucani, P. Ugliengo and A. Rimola, *ACS Earth Space Chem.*, 2022, **6**, 496–511.
3 A. Rimola, D. Skouteris, N. Balucani, C. Ceccarelli, J. Enrique-Romero, V. Taquet and P. Ugliengo, *ACS Earth Space Chem.*, 2018, **2**, 720–734.
4 S. Ferrero, S. Pantaleone, C. Ceccarelli, P. Ugliengo, M. Sodupe and A. Rimola, *Astrophys. J.*, 2023, **944**, 142.

Karolina Plakitina commented: Thank you for your great talk, Cecilia (https://doi.org/10.1039/d3fd00106g). I was wondering if you had the chance to examine the emission of higher transitions of CO in L1157-B1. If you did, I'm curious to know if it exhibits a velocity-integrated emission similar to that of HDCO.

Cecilia Ceccarelli answered: CO transitions with J_{up} between 5 and 16 were observed with the Herschel telescope by Lefloch *et al.*[1] The spatial resolution is (by far) not enough to make the comparison with the HDCO emission observed with the interferometer NOEMA. However, when considering the CO line profiles, Lefloch *et al.* found that there are three major velocity components emitting in the beam encompassing B1, of which one has the physical characteristics (temperature and density) of the cavity of B1, which is the region probed by the HDCO emission. Thus, in the paper by Codella *et al.*,[2] where we studied the HDCO, acetaldehyde and formamide emission, we used those parameters to model the emitting-gas properties and to run the chemical model, which concluded that formamide is likely a gas-phase product.

1 B. Lefloch, S. Cabrit, G. Busquet, C. Codella, C. Ceccarelli, J. Cernicharo, J. R. Pardo, M. Benedettini, D. C. Lis and B. Nisini, *Astrophys. J., Lett.*, 2012, **757**, L25.
2 C. Codella, C. Ceccarelli, P. Caselli, N. Balucani, V. Barone, F. Fontani, B. Lefloch, L. Podio, S. Viti, S. Feng, R. Bachiller, E. Bianchi, F. Dulieu, I. Jiménez-Serra, J. Holdship, R. Neri, J. E. Pineda, A. Pon, I. Sims, S. Spezzano, A. I. Vasyunin, F. Alves, L. Bizzocchi, S. Bottinelli, E. Caux, A. Chacón-Tanarro, R. Choudhury, A. Coutens, C. Favre, P. Hily-Blant, C. Kahane, A. Jaber Al-Edhari, J. Laas, A. López-Sepulcre, J. Ospina, Y. Oya, A. Punanova, C. Puzzarini, D. Quenard, A. Rimola, N. Sakai, D. Skouteris, V. Taquet, L. Testi, P. Theulé, P. Ugliengo, C. Vastel, F. Vazart, L. Wiesenfeld and S. Yamamoto, *Astron. Astrophys.*, 2017, **605**, L3.

Martin R. S. McCoustra opened the discussion of the paper by Inga Kamp: I will take Chair's privilege and ask the opening question in this discussion session, which has so nicely highlighted the capabilities of the James Webb Space Telescope (JWST). It is nice to note the observations of acetylene (C_2H_2), diacetylene (C_4H_2) and benzene (C_6H_6) in the IR data presented in this session. These may be linked through Diels–Alder chemistry. However, I note the presence of hydrogen cyanide (HCN). In the presence of C_4H_2, Diels–Alder chemistry can result in the formation of pyridine (C_5H_5N); the heterocyclic analogue of C_6H_6 with a significant dipole moment. Is there any evidence for this species? If not, what laboratory data would you need to search for it?

Inga Kamp responded: That is a very interesting suggestion; we will certainly "hunt" for pyridine in our JWST data if there would be mid-infrared spectroscopic data available for the molecule at high ($\sim$500 K) temperature. So far, we have searched for all molecules for which good line lists, Einstein A values and partition functions are readily available, most notably from HITRAN but also from GEISA and through collaboration with molecular spectroscopists. As a starting point for mere identification, 'a' mid-IR emission spectrum of this species would be good enough.

Nicholas Walker addressed all: Martin asked "Where is pyridine?!" In these conversations, we should also keep imidazole in mind. The dipole moment of pyridine is about 2.2 D whereas that of imidazole is higher, about 3.67 D. So should we also be asking "Where is imidazole?!" Rotational transition frequencies for the molecule are provided in the paper "Rotational spectroscopy of imidazole: improved rest frequencies for astrophysical searches."[1]

1 B. M. Giuliano, L. Bizzocchi, A. Pietropolli Charmet, B. E. Arenas, A. L. Steber, M. Schnell, P. Caselli, B. J. Harris, B. H. Pate, J.-C. Guillemin and A. Belloche, *Astron. Astrophys.*, 2019, **628**, A53.

Serena Viti opened the discussion of the paper by Ewine F. van Dishoeck: The destruction of carbon grains by thermal processing at a temperature of $\sim$500 K will change the carbon budget in the disk, and as a consequence the C/O ratio. Can you comment on the consequences of such changes?

Ewine F. van Dishoeck responded: Indeed, in the general interstellar medium (ISM) about half the available carbon is thought to be locked up in refractory carbon grains, leading to a C/O ratio in the gas of about 0.4–0.5. If this refractory carbon is all destroyed, the gaseous C/O ratio will double and comes very close to unity. If in addition some of the oxygen is depleted, for example in dust traps beyond the H_2O snowline, as we are arguing for the case of J160532 (https://doi.org/10.1039/d3fd00010a), the gaseous C/O ratio will exceed unity. This will completely change the gas-phase chemistry in favor of hydrocarbon molecules, as is also known for the case of C-rich *vs.* O-rich asymptotic giant branch (AGB) stars.

Serena Viti opened the discussion of the paper by Nikku Madhusudhan: The C/O ratio in planetary atmospheres is often a free input parameter in atmospheric models as this may differ from the initial value as found in the parent molecular clouds or in any case may even be different from cloud to cloud. Would variations in this ratio affect the diagnostic diagram in Fig. 8 of your paper (https://doi.org/10.1039/d3fd00075c)?

Nikku Madhusudhan replied: In our models, the C/O ratio of the primordial atmosphere is indeed a free input parameter. However, the final atmospheric composition is strongly affected by the boundary conditions at the surface–atmosphere interface. While the C/O ratio could influence the initial atmospheric composition to some extent, the steady-state atmospheric composition of a Hycean world that is several Gyr old would likely be dominated by the ocean–atmosphere interactions and the atmospheric temperature profile. Therefore, the diagnostics for a Hycean atmosphere are expected to be relatively less sensitive to the initial C/O ratio over a reasonable range expected for primordial H_2-rich planetary atmospheres, *e.g.*, ∼0.1–1. The C/O ratio has a stronger effect on atmospheric chemistry for gas giant planets, and is a more reliable metric to trace their formation histories, as they are expected to host primary atmospheres that are not significantly altered by surface–atmosphere interactions.

Robin T. Garrod remarked: In the paper, you indicate that the temperature structure in the models is perhaps the most important determinant of the chemical outcomes. I am wondering to what extent the influence of dust is considered in the models? Some extrasolar planets with small orbital radii are suggested to have capture rates of interplanetary dust grains that are perhaps several orders of magnitude higher than those of solar system planets. How would interplanetary dust capture in the upper atmosphere affect the temperature structure through cloud/haze formation, water nucleation and such effects?

Nikku Madhusudhan replied: The models don't explicitly include dust; the nature and abundance of dust possible in Hycean atmospheres are presently unknown. However, a general effect of dust in the form of cloud/haze is modelled using parametric prescriptions. For example, in the present study (https://doi.org/10.1039/d3fd00075c), we model haze as an enhanced Rayleigh scattering, as discussed in Section 2.1 of our paper. Varying the haze parameter affects the albedo due to scattering, which in turn affects the temperature structure, as shown in Section 3.1 and Fig. 2 of our paper. There are other models, especially of giant exoplanets, in the literature that involve detailed microphysical cloud/haze modelling, including the nucleation, condensation and growth processes. This is a future avenue to pursue for Hycean atmospheres.

Robin T. Garrod asked: How would the diurnal nature of a planet affect the physical and chemical outcomes? Are there dayside/nightside effects that could alter your results?

Nikku Madhusudhan replied: Most of the candidate Hycean planets currently known are expected to be tidally locked, with permanent dayside and nightside. The temperature structure and chemistry presented in this work are for average dayside atmospheres. The nightside atmospheres would naturally be expected to be non-irradiated and cooler, depending on the efficiency of the day–night energy redistribution, with corresponding effects on the chemistry, *e.g.* H_2O condensation is more likely on the nightside than dayside. While Hycean worlds are generally expected to host planet-wide oceans that are habitable, there are also planets where the dayside may be too hot to be habitable but the nightside could still be cool enough to be habitable. Such planets are referred to as 'Dark Hycean' worlds[1] and their atmospheric chemistry may be more inhomogeneous across the planet compared to regular Hycean worlds.

1 N. Madhusudhan, A. A. A. Piette and S. Constantinou, *Astrophys. J.*, 2021, **918**, 1, DOI: 10.3847/1538-4357/abfd9c.

Olivia H. Wilkins returned to the discussion of the papers by Ewine van Dishoeck and Inga Kamp: Since submitting the papers about the JWST Observations of Young protoStars (JOYS) and MIRI INfrared Disk Survey (MINDS) programs, additional targets have been observed. As of now, about what proportion of disks in these programs have been observed with JWST? With these additional observations, have any trends emerged that either support or challenge current models of disk chemistry? Or is it too soon to tell?

Inga Kamp responded: About 60% of the MINDS targets have been observed, but it is too early to tell whether there are any trends emerging, other than that there is a large diversity among disks and stars with otherwise similar parameters. Note also that the appearance of the spectra, *i.e.*, their line fluxes, do not directly represent column densities or abundances. To obtain those, one needs to do the detailed model analyses taking optical depths into account. The weak lines (hot bands, ^{13}C isotopologues) are telling us that column density ratios can be orders of magnitude larger than flux ratios, as demonstrated in the case of GW Lup by Grant *et al.*[1]

1 S. L. Grant, E. F. van Dishoeck, B. Tabone, D. Gasman, T. Henning, I. Kamp, M. Güdel, P.-O. Lagage, G. Bettoni, G. Perotti, V. Christiaens, M. Samland, A. M. Arabhavi, I. Argyriou, A. Abergel, O. Absil, D. Barrado, A. Boccaletti, J. Bouwman, A. Caratti o Garatti, V. Geers, A. M. Glauser, R. Guadarrama, H. Jang, J. Kanwar, F. Lahuis, M. Morales-Calderón, M. Mueller, C. Nehmé, G. Olofsson, E. Pantin, N. Pawellek, T. P. Ray, D. Rodgers-Lee, S. Scheithauer, J. Schreiber, K. Schwarz, M. Temmink, B. Vandenbussche, M. Vlasblom, L. B. F. M. Waters, G. Wright, L. Colina, T. R. Greve, K. Justannont and G. Östlin, *Astrophys. J., Lett.*, 2023, **947**, L6.

Laurent Wiesenfeld addressed Inga Kamp: Among the main simple molecules that have no permanent electric dipole moment, CO_2 is evidently prominent. However, most simple hydrocarbons have a high enough symmetry that preclude any permanent electric dipole. Among those are all naturally occurring hydrocarbons with one or two carbon atoms, with methane and acetylene most abundant (propyne and cyclo-C_3H_2 are the simplest hydrocarbons to have a permanent dipole). How advanced is the analysis of these molecules, and, especially, what is the situation with methane?

Inga Kamp responded: We have two disks around very low mass stars that are extremely rich in such small hydrocarbons. We detected C_2H_2 (and its isotopologue $^{13}C^{12}CH_2$), C_4H_2, C_6H_6 in the disk around the low mass star J16053215 (ref. 1). We also have a tentative detection of CH_4 in that same disk. and a clear detection in another disk.[2] In that other disk, we also detect many more small hydrocarbons. Again, we are primarily limited by the available molecular data – as a minimum we need an IR spectrum – to hunt for these molecules, their isotopologues and corresponding ions. We use at this stage simple slab model analysis to obtain temperatures, column densities and emitting areas of these molecules; this works only for those molecules for which we have the basic molecular data, such as transitions, Einstein A coefficients, partition functions. At the same time, we improve our chemical networks to include larger hydrocarbons, so that we can also model their formation/destruction pathways in high density, warm disk environments.[3]

1 B. Tabone, G. Bettoni, E. F. van Dishoeck, A. M. Arabhavi, S. Grant, D. Gasman, Th. Henning, I. Kamp, M. Güdel, P. O. Lagage, T. Ray, B. Vandenbussche, A. Abergel, O. Absil, I. Argyriou, D. Barrado, A. Boccaletti, J. Bouwman, A. Caratti o Garatti, V. Geers, A. M. Glauser, K. Justannont, F. Lahuis, M. Mueller, C. Nehmé, G. Olofsson, E. Pantin, S. Scheithauer, C. Waelkens, L. B. F. M. Waters, J. H. Black, V. Christiaens, R. Guadarrama, M. Morales-Calderón, H. Jang, J. Kanwar, N. Pawellek, G. Perotti, A. Perrin, D. Rodgers-Lee, M. Samland, J. Schreiber, K. Schwarz, L. Colina, G. Östlin and G. Wright, *Nat. Astron.*, 2023, 7, 805–814.
2 Arabhavi *et al.*, submitted.
3 Kanwar *et al.*, submitted.

Ian Sims addressed Ewine F. van Dishoeck and Inga Kamp: You have mentioned the need for more spectroscopic data, especially at high temperatures. High-temperature high-resolution spectroscopy is an active research field, and one of my colleagues at Rennes, Robert Georges, along with his collaborators, has pioneered a high enthalpy source that has enabled them to make considerable progress in this area.[1] For which species and up to what temperature would you most like to have such data (a wish list, if you like)?

1 R. Georges, J. Thiévin, A. Benidar, S. Carles, B. Amyay, M. Louviot, V. Boudon and J. Vander Auwera, *Rev. Sci. Instrum.*, 2019, **90**, 093103.

Ewine F. van Dishoeck answered: Thank you for this comment and reference, that is very encouraging. We are still working on such a "wish list" but certainly the ^{13}C isotopologues of C_2H_2, C_2H_4 and other hydrocarbons that are seen prominently in the JWST Mid-InfraRed Instrument (MIRI)/Medium Resolution Spectrometer (MRS) spectra are high on this list.

Inga Kamp added: Another example is C_3H_4 for which we have incomplete data only from the GEISA database. And we also lack the hotter bands for C_6H_6 that clearly show in the spectra. Apart from the spectroscopic data, we also need the corresponding partition functions for the relevant pressure/temperature regimes.

Murthy S. Gudipati asked: The hydrocarbon molecules (C_2H_2, HCN, CH_3CN, C_6H_6, *etc.*) that you have observed in disks are similar to what we see in Titan's atmosphere, except for CO_2. Is it possible that the inner-disk regions are similar to Titan's oxygen-depleted upper atmosphere that receives energetic photons that lead to the formation of organic molecules? If so, is the CO_2 observed in the same data from a different region in the line of sight?

Ewine F. van Dishoeck answered: This is an interesting suggestion and worth further exploration. CO_2 tends to have slightly larger emitting areas than some other molecules in some disks (*e.g.*, GW Lup) as also found in models of Agúndez *et al.*[1] It is too early to tell whether this is a general trend. In the hydrocarbon-rich disk J160532, CO_2 emission is consistent with coming from the same area as the hydrocarbons. Of course, this does not mean that the molecules are in the same location of the disk, either radially or vertically. One would need spectrally and/or spatially resolved data to better pinpoint their relative locations.

1 M. Agúndez, J. Cernicharo and J. R. Goicoechea, *Astron. Astrophys.*, 2008, **483**, 831.

Inga Kamp responded: We estimated the typical temperatures and pressures in the inner disks around these very-low-mass stars based on the column densities and temperatures derived from the JWST spectra through slab models. The temperatures are typically higher in these disks and the pressures overlap with the lowest pressure in Titan's atmosphere, or are even 2–3 orders of magnitude lower. So, these disks 'live' in a somewhat different part of the (T, P) parameters space. However, more work is required to understand the chemistry and some pathways could be similar to planetary atmospheres (including, *e.g.*, three-body reactions). In disk models with a steady-state chemistry, the CO_2 reservoir contributing to the mid-IR emission is often more radially extended; we need to analyse more objects to see if we find this also structurally back from analysing JWST spectra using slab models (*e.g.*, lower temperatures and larger emitting areas compared to hydrocarbons).

Murthy S. Gudipati returned to the discussion of the paper by Nikku Madhusudhan: The search for life on exoplanets is a "tall order" and we are limited by our remote-sensing as well as *in situ* observational abilities to search and potentially detect life several light-years away. It would be prudent to undertake a "search for potentially habitable exoplanets" as a first step that is more realistic.

Nikku Madhusudhan responded: I agree that establishing the potential for habitability of an exoplanet is arguably a more realistic first step than robustly detecting signs of life.

Murthy S. Gudipati said: We simulate exoplanet atmospheres in the laboratory, especially those of hot Jupiter-type planets reaching up to 1700 K. However, the issue is to have uniform temperature throughout the reaction cell, as the optical elements at the end require ambient temperatures, resulting in temperature gradients from 1700 K to 300 K within the cell. How important is to have rate constants for various reactions in the gas phase at one single temperature, instead of at several peak temperatures with temperature gradients (which can be used to retrieve temperature dependent rate constants with or without UV photochemistry)?

Nikku Madhusudhan replied: Temperature dependent rate constants are presently adequate for chemical models as pursued in this study (https://doi.org/10.1039/d3fd00075c). Furthermore, for most Hycean worlds, the atmospheres are expected to be at terrestrial-like atmospheric temperatures between $\sim$200–400 K, which may be more accessible for laboratory experiments compared to those of hot Jupiters.

Ewine F. van Dishoeck asked: Can you clarify the origin of the term Hycean worlds, and is there a difference with what we call "ocean worlds"?

Nikku Madhusudhan replied: The word "Hycean" is a combination of the words hydrogen and ocean. The term "Hycean worlds" refers to habitable planets with hydrogen-rich atmospheres and planet-wide oceans, as defined in our original work.[1]

The term "ocean worlds" typically refers to planets with substantial oceans and atmospheres that are dominated by H_2O or other heavy molecules, such as N_2, O_2 or CO_2, as in atmospheres of terrestrial planets, and may or may not be habitable. In the solar-system context, ocean worlds also refer to icy moons with sub-surface oceans.

1 N. Madhusudhan, A. A. A. Piette and S. Constantinou, *Astrophys. J.*, 2021, **918**, 1, DOI: 10.3847/1538-4357/abfd9c.

Stefanie Milam returned to the discussion of the paper by Ewine F. van Dishoeck: There was a recent press release on water (and no CO_2) observed towards a main belt comet.[1] Looking at it from a disk perspective, in this situation where there is water from the main belt region but no CO_2, what are the implications for disk chemistry and water distribution through the solar system?

1 https://www.nasa.gov/feature/goddard/2023/nasa-s-webb-finds-water-and-a-new-mystery-in-rare-main-belt-comet.

Ewine F. van Dishoeck responded: That is a very intriguing result! I haven't studied the paper yet, but from an ISM perspective one would expect CO_2 ice wherever there is H_2O ice. If you do not see CO_2 ice (but do see H_2O ice), perhaps the comet (or its building blocks) has been heated in the past to temperatures in between the CO_2 and H_2O sublimation temperatures. Or perhaps there has been segregation of the ices due to heating and/or UV. Some variations in CO_2/H_2O ratios across the comet have also been seen in comet 67P by Rosetta.[1] Looking at other molecules with a range of sublimation temperatures, such as CO and H_2CO, will be important to test various scenarios.

1 K. Altwegg, H. Balsiger and S. A. Fuselier, *Annu. Rev. Astron. Astrophys.*, 2019, **57**, 113 and references cited therein.

Dwayne E. Heard commented: The reactions of OH + CO producing CO_2, and the reaction of OH + H_2 producing H_2O, produce molecules whose abundances have been measured. What do you need to assume about the rovibrational distribution in the various modes in these products of the two reactions? When you have polyatomics it is complicated to know what this may be, therefore, do you need to make an assumption of thermodynamic equilibrium for the conversion of the observed astronomically measured absorption to an abundance? Can you please comment on this?

Ewine F. van Dishoeck answered: This is a good question: generally, the reaction networks do not consider state-to-state reactions, with the exception of reactions with vibrationally excited H_2 where a generic H_2* molecule is taken (see for example, ref. 1 and 2). Applications include the formation of OH and CN[3] and the formation of ions like CH^+, OH^+ and CO^+ that react on every collision and whose population distribution is therefore not thermalized. This was highlighted by John Black in a *Faraday Discussions* paper[4] and applied subsequently to ions (see, *e.g.*, ref. 5 and 6).

For most neutral molecules, inelastic collisions with H_2 are fast enough to thermalize the population distribution. Also, to retrieve column densities and abundances, the simple slab models assume that the population distribution can be characterized by a single excitation temperature, but that temperature does not have to be the same as the kinetic temperature. Even for cases of formation pumping, having a single excitation temperature is a good zeroth-order assumption if one is just interested in retrieving the column density.

There is one other case where state-to-state reactions need to be taken into account and that is highly relevant to JWST: the photodissociation of H_2O leading to OH in highly excited rotational states. It is well known from molecular physics experiments (*e.g.*, ref. 7) and theory (*e.g.*, ref. 8) that absorption into the B excited electronic state, for example by Lyman α photons, produces OH in very high excited rotational states up to $N = 47$ ($E_{\rm up} = 47\,000$ K!). This process has already been observed with the Spitzer Space Telescope in shocks and disks (*e.g.*, ref. 9 and 10) and is now also being seen even more prominently with JWST in the 9–12 micron range (*e.g.*, ref. 11). The same JWST spectra also reveal the O + H_2 → OH (J) + H formation pumping in the 14–25 micron range. See the models developed by Tabone *et al.*[12]

1 A. G. G. M. Tielens and D. Hollenbach, *Astrophys. J.*, 1985, **291**, 722.
2 S. Bruderer, E. F. van Dishoeck, S. D. Doty and G. J. Herczeg, *Astron. Astrophys.*, 2012, **541**, A91.
3 P. Cazzoletti, E. F. van Dishoeck, R. Visser, S. Facchini and S. Bruderer, *Astron. Astrophys.*, 2018, **609**, A93.
4 J. H. Black, *Faraday Discuss.*, 1998, **109**, 257.
5 P. Stäuber and S. Bruderer, *Astron. Astrophys.*, 2009, **505**, 195.
6 A. Zanchet, B. Godard, N. Bulut, O. Roncero, P. Halvick and J. Cernicharo, *Astrophys. J.*, 2013, **766**, 80.
7 S. A. Harich, D. W. H. Hwang, X. Yang, J. J. Lin, X. Yang and R. N. Dixon, *J. Chem. Phys.*, 2000, **113**, 10073.
8 R. van Harrevelt and M. C. van Hemert, *J. Chem. Phys.*, 2000, **112**, 5787.
9 A. Tappe, C. J. Lada, J. H. Black and A. A. Muench, *Astrophys. J.*, 2008, **680**, L117.
10 J. S. Carr and J. R. Najita, *Astrophys. J.*, 2014, **788**, 66.
11 D. Gasman, E. F. van Dishoeck, S. L. Grant, M. Temmink, B. Tabone, T. Henning, I. Kamp, M. Güdel, P.-O. Lagage, G. Perotti, V. Christiaens, M. Samland, A. M. Arabhavi, I. Argyriou, A. Abergel, O. Absil, D. Barrado, A. Boccaletti, J. Bouwman, A. Caratti o Garatti, V. Geers, A. M. Glauser, R. Guadarrama, H. Jang, J. Kanwar, F. Lahuis, M. Morales-Calderón, M. Mueller, C. Nehmé, G. Olofsson, E. Pantin, N. Pawellek, T. P. Ray, D. Rodgers-Lee, S. Scheithauer, J. Schreiber, K. Schwarz, B. Vandenbussche, M. Vlasblom, L. B. F. M. Waters, G. Wright, L. Colina, T. R. Greve and Göran Östlin, *Astron. Astrophys.*, 2023, submitted.
12 B. Tabone, B. Godard, G. Pineau des Forêts, S. Cabrit and E. F. van Dishoeck, *Astron. Astrophys.*, 2020, **636**, A60.

Dwayne E. Heard said: You mention the importance of vibrationally excited hydrogen in order to form the hydroxyl radical, OH, in order to assist in overcoming energy barriers to some reactions (https://doi.org/10.1039/d3fd00010a).

Can you please outline the evidence for vibrationally excited hydrogen in these regions?

Ewine F. van Dishoeck responded: Indeed, both the O + H_2 and OH + H_2 reactions are endothermic and/or have energy barriers that need to be overcome, either by having a higher gas temperature or having H_2 in excited states, or both. In several of our sources (*e.g.*, J160532), we are directly detecting H_2 in excited rotational levels up to at least $J = 7$ ($E_{up} = 7200$ K), enough to overcome these barriers. Pure rotational lines within $\nu = 1$ are also observed in some sources. The H_2 $\nu = 1$–0 lines lie in the NIRSpec wavelength range at 2 micron; those MINDS data have not yet been taken. There are some ground-based detections of the $\nu =$ 1–0 S(1) line, dating back to Bary *et al.*,[1] although one has to be careful to assess that these lines are indeed coming from the disk surface rather than a disk wind or jet. The high spatial resolution of NIRSpec (0.2″) will help to distinguish disks from jets.

1 J. S. Bary, D. A. Weintraub and J. H. Kastner, *Astrophys. J.*, 2003, **586**, 1136.

Martin Cordiner addressed Inga Kamp and Ewine F. van Dishoeck: On the topic of linking JWST observations of protoplanetary disks to planet formation, do we know enough about the physical and thermal structures of disks, as well as the chemical mixing processes, to be able to infer the mid-plane abundances based on observations of higher vertical layers?

Ewine F. van Dishoeck responded: Linking the observed abundances of molecules in the surface layers of disks to those in the midplane is indeed a very important issue that the disk modeling community is working hard to characterize. For example, the detection of $^{13}CO_2$ is thought to be one way of quantifying the CO_2 abundance deeper in the disk including possible enhancement by sublimating icy pebbles in the midplane.[1,2] In the case of J160532, we seem to be fortunate to be able to look deep into the disk, close to the midplane, as also evidenced by the strong H_2 lines in this spectrum.

1 A. D. Bosman, S. Bruderer and Ewine F. van Dishoeck, *Astron. Astrophys.*, 2017, **601**, A36.
2 A. D. Bosman, E. A. Bergin, J. K. Calahan and S. E. Duval, *Astrophys. J., Lett.*, 2022, **933**, L40.

Inga Kamp added: This is indeed a very valid concern. I would say that observationally we do not yet know enough about these inner disks to claim that. In terms of theory, there are many studies investigating radial and vertical mixing and transport processes and diffusion processes, and how they affect the elemental composition in disks. In my view, we will not always get abundances of molecules directly from JWST MIRI/MRS spectra as this would make the implicit assumption that all molecules reside radially/vertically in the same region. Disk models show clearly that we cannot expect that to be true in general. In my view, JWST will also help us a lot to understand better the processes occurring in the inner disks and to quantify them. This includes also testing our current understanding of chemistry, thermal disk structures, vertical layering of disks, disk winds, *etc.*

Jialu Li addressed all: Analysis of spectra at MIRI's resolution ($R = 3500$) might be hampered by the blending of individual transitions into broad line features, optical depth effects, and line-depth reduction due to partial coverage of the source. The most "prominent" lines are usually optically thick and specifically suffer from the degeneracy problem between the column density and the partial coverage. Depending on how the partial coverage is taken care of, the derived column density can differ by up to two orders of magnitude (for example, comparing Infrared Space Observatory Short Wavelength Spectrometer (ISO-SWS), $R \sim 1500$, and Stratospheric Observatory for Infrared Astronomy/Echelon Cross Echelle Spectrograph (SOFIA/EXES), $R \sim 50\,000$, studies on the same massive protostar).

Jialu Li returned to the discussion of the paper by Ewine F. van Dishoeck: Do you have any chance of finding absorption lines among JWST observations toward massive protostars other than IRAS23385?

Ewine F. van Dishoeck answered: Indeed, we have several more high-mass protostars in our JOYS program that will be observed by this fall. Of course, JWST does not provide the beautiful spectrally resolved deep absorption bands seen with instruments like the Very Large Telescope-CRyogenic high-resolution InfraRed Echelle Spectrograph (VLT-CRIRES(+)) or SOFIA-EXES due to its lower spectral resolving power. The information that we can derive from these absorption data will therefore be more limited than what you are able to conclude with your data.

Murthy S. Gudipati returned to the discussion of the paper by Inga Kamp: Do we have any observational constraints on radial material transportation/mixing/turbulence in protoplanetary disks? What are their length scales (1 au or 10 au?) from the protostar. This is very important to determine how much thermal cycling ice grains/gas undergo as well as chemical transformations that exclusively occur in the protoplanetary phase.

Inga Kamp replied: Yes, we do have some constraints both from dust and gas observations. The mm dust grains are settled in a very thin disk, thus indicating very low turbulent motions stirring them up vertically (*e.g.*, ref. 1). Sub-mm line kinematics with the Atacama Large Millimeter/submillimeter Array (ALMA) routinely suggest subsonic turbulence (*e.g.*, ref. 2 for a review on disks around Herbig stars). Radial transport of dust has been shown in a couple of cases, where the outer radius of mm dust as observed with ALMA is much smaller than that of CO gas (*e.g.*, ref. 3 and 4). How much ice mixing occurs vertically, radially and on which length scales is mostly informed by theory (*e.g.*, ref. 5). ALMA observations of the CO isotopologues in two disks, TW Hya and IM Lup, show that the location of the rare isotopologues ^{13}CO and $C^{18}O$ agrees well with the expected location of the ice surface in the outer disk (*e.g.*, ref. 6 and 7).

1 C. Pinte, W. R. F. Dent, F. Ménard, A. Hales, T. Hill, P. Cortes and I. de Gregorio-Monsalvo, *Astrophys. J.*, 2016, **816**, 25.
2 S. D. Brittain, I. Kamp, G. Meeus, R. D. Oudmaijer and L. B. F. M. Waters, *Space Sci. Rev.*, 2023, **219**, 7.

3 S. Facchini, E. F. van Dishoeck, C. F. Manara, M. Tazzari, L. Maud, P. Cazzoletti, G. Rosotti, N. van der Marel, P. Pinilla and C. J. Clarke, *Astron. Astrophys.*, 2019, **626**, L2.
4 L. Trapman, S. Facchini, M. R. Hogerheijde, E. F. van Dishoeck and S. Bruderer, *Astron. Astrophys.*, 2019, **629**, A79.
5 (*a*) S. Krijt and F. J. Ciesla, *Astrophys. J.*, 2016, **822**, 111; (*b*) S. Krijt, K. R. Schwarz, E. A. Bergin and F. J. Ciesla, *Astrophys. J.*, 2018, **864**, 78.
6 K. R. Schwarz, E. A. Bergin, L. Ilsedore Cleeves, G. A. Blake, K. Zhang, K. I. Öberg, E. F. van Dishoeck and C. Qi, *Astrophys. J.*, 2016, **823**, 91.
7 C. Pinte, F. Ménard, G. Duchêne, T. Hill, W. R. F. Dent, P. Woitke, S. Maret, G. van der Plas, A. Hales, I. Kamp, W. F. Thi, I. de Gregorio-Monsalvo, C. Rab, S. P. Quanz, H. Avenhaus, A. Carmona and S. Casassus, *Astron. Astrophys.*, 2018, **609**, A47.

Murthy S. Gudipati asked: During the protoplanetary phase (in protoplanetary disks), have there been observations that show bidirectional material transportation (radial or vertical)? If yes, what are the driving forces?

Inga Kamp answered: Yes, there are observations (*e.g.,* emission lines) showing gas accretion onto the star (so inward moving material) as well as winds from the inner disk transporting material outward/away from the disk (velocity resolved line profiles). Recent work points towards magneto-hydrodynamical winds, *e.g.*, ref. 1 and 2. Dust grains move inward due to gas drag forces, but in the presence of gas pressure bumps, this transport can be stalled for certain grain sizes. Diffusion of course acts in both directions. Dust grains also vertically settle towards the disk midplane due to the vertical component of stellar gravity. Winds can drag along very small dust grains and thus lift them vertically upward.

1 M. N. Simon, I. Pascucci, S. Edwards, W. Feng, U. Gorti, D. Hollenbach, E. Rigliaco and J. T. Keane, *Astrophys. J.*, 2016, **831**, 169.
2 J. Bae, A. Isella, Z. Zhu, R. Martin, S. Okuzumi and S. Suriano, *arXiv*, 2022, preprint, arXiv:2210.13314 [astro-ph.EP], DOI: 10.48550/arXiv.2210.13314.

Murthy S. Gudipati returned to the discussion of the paper by Nikku Madhusudhan: Do the models for Hycean worlds include the water–rock interactions that are required for their potential habitability?

Nikku Madhusudhan answered: The models currently do not include the water–rock interactions, because for most Hycean worlds the water layer is expected to be so deep that a thick mantle of high-pressure water-ice separates the rock in the interior from the ocean precluding direct interaction between the two. This is why we have investigated other possible avenues that can deliver bioessential elements to the ocean from external sources, such as asteroid impacts and atmospheric precipitation, as discussed in section 3.4 of the paper. We find that such sources can meet the bioessential metal requirements for habitability on Hycean worlds.

Murthy S. Gudipati queried: On Hycean planets that could potentially have liquid water oceans on the surface and a H_2-dominated atmosphere, how would photochemistry (from the parent star's photons) affect ocean–atmosphere interactions? What kinds of H_2 atmospheric pressure is expected in these environments and would stellar UV/VIS/IR photons make their way to the ocean?

Nikku Madhusudhan replied: The effect of photochemistry in a Hycean atmosphere would be interlinked with the ocean–atmosphere interactions. For example, photochemistry can deplete ammonia (NH_3), which is otherwise normally expected to be abundant in cool H_2-rich atmospheres. For a deep H_2-rich atmosphere, the depleted NH_3 in the upper atmosphere would be efficiently replenished by NH_3 from the lower atmosphere that is in thermochemical equilibrium. The presence of an ocean surface under a shallow H_2-rich atmosphere in a Hycean planet can cut off the thermochemical recycling of NH_3 and make it under-abundant in the observable upper atmosphere. Photochemistry can also act as a sink for some chemical species released from the ocean and in turn generate new photochemical byproducts in the atmosphere. The atmospheric pressure due to H_2 at the surface of the ocean is expected to range between $\sim$1–1000 bars, as originally defined for Hycean worlds.[1] It is possible for some of the incident UV/VIS/IR photons to reach the ocean surface depending on a number of factors, including the atmospheric composition, the nature and coverage of clouds/hazes, and surface pressure. Naturally, all other factors remaining the same, shallower atmospheres will have more stellar photons reaching the surface.

1 N. Madhusudhan, A. A. A. Piette and S. Constantinou, *Astrophys. J.*, 2021, **918**, 1, DOI: 10.3847/1538-4357/abfd9c.

Murthy S. Gudipati asked: On Hycean ocean worlds with H_2-dominant atmospheres, what are the sources of energy that are required for the potential habitability of these planets?

Nikku Madhusudhan replied: The primary source of energy is expected to be the irradiation from the host star, with some additional energy possible due to internal heat from within the planet.

Cecilia Ceccarelli returned to the discussion of the paper by Inga Kamp: How much does your model depend on basic chemistry, and how does that impact the parameters that you derive by comparing model predictions and observations? For example, how do you form methane? In most models it is assumed that methane is given by the C atom additions but the first step is impossible, as many studies have shown that C landing on a water-ice will interact with the water molecule and be not available for hydrogenation.

Inga Kamp answered: The ProDiMo disk models shown in the paper (https://doi.org/10.1039/d3fd00013c) contain gas-phase chemistry with an extended network of 235 species, including also adsorption and (thermal/non-thermal) desorption processes, but no further grain-surface chemistry.[1] In the models presented here, the methane formation is thus not proceeding on grain surfaces. Methane forms in the very inner disk at high densities and temperatures through gas-phase chemistry. The chemistry, physics and radiative transfer are intertwined in the models and hence our conclusions to some extent depend on how well we understand the chemistry. We often use the models to inform us how specific parameters affect the observables, like mid-IR line emission. This is likely a safer approach than to claim that a model that agrees well with observations is the ground truth.

1 I. Kamp, W.-F. Thi, P. Woitke, C. Rab, S. Bouma and F. Ménard, *Astron. Astrophys.*, 2017, **607**, A41.

Divita Gupta opened the discussion of the paper by Olivia H. Wilkins: You presented the rotational temperature plots for methanol (https://doi.org/10.1039/d3fd00003f). Is there any other molecule you could also do this for using your data? This could potentially help compare the chemistry/origin of these two species.

Olivia H. Wilkins responded: Thanks for this question. Calculating rotational temperatures and relative abundances of different molecules would be possible if future observations target multiple rotational transitions of a given molecule. However, in this pilot survey, most other molecules detected have only one or two transitions covered in our spectra, whereas we would need at least three lines to derive these parameters. The exception is CH_3CCH, which has four transitions detected in the spectra of both BGPS 6029 and BGPS 6120. Nevertheless, rotational temperature and column density plots were not derived for CH_3CCH in this study because they would not provide insight into the chemistry across a substantial part of the sample.

Eric Herbst opened the discussion of the paper by Ko-Yun Huang: Although there are many studied galaxies, both near and quite distant, it often appears that galactic scientists spend much of their time on the study of one well-known barred spiral galaxy, NGC 1068. Why is this galaxy so popular and studied in so much detail? Do the accretion disk or disks play a dominant role? What about the dominance of Active Galactic Nuclei (AGN) structures? Is this galaxy dominated by star bursts or photon-dominated regions? What allows unsaturated carbon-chain molecules to form?

Ko-Yun Huang answered: Indeed NGC 1068 is of particular interest for many communities for several reasons: (1) it hosts both AGN at its center and starburst ring at its arms; (2) its proximity ($\sim$10–14 Mpc) makes it interesting as, with state-of-the-art observations, we are able to spatially resolve the AGN-dominated regions from the starburst-dominated regions; (3) due to AGN feedback from ionized winds and jets, a massive molecular outflow has been found with ALMA observations (*e.g.*, ref. 1) – the interaction with the ambient material has produced large-scale shocks at its circumnuclear disk up to 400 pc.

All of the above makes this galaxy particularly interesting for extragalactic astrochemists, especially in studying the rich shock chemistry from this source. In addition, the detection of carbon-chain molecules including C_2H and c-C_3H_2 suggests interesting chemistry potentially arising from photon-dominated region (PDR) or X-ray dominated region (XDR) chemistry, which is important for the study of both carbon-chain growth and the dissociation of PAHs.

1 S. García-Burillo, F. Combes, A. Usero, S. Aalto, M. Krips, S. Viti, A. Alonso-Herrero, L. K. Hunt, E. Schinnerer, A. J. Baker, F. Boone, V. Casasola, L. Colina, F. Costagliola, A. Eckart, A. Fuente, C. Henkel, A. Labiano, S. Martín, I. Márquez, S. Muller, P. Planesas, C. Ramos Almeida, M. Spaans, L. J. Tacconi and P. P. van der Werf, *Astron. Astrophys.*, 2014, **567**, A125.

Robin T. Garrod returned to the discussion of the paper by Olivia H. Wilkins: You investigate sources in a ring with a galactocentric distance between 4–8 kpc. In the graphical abstract image of your paper (https://doi.org/10.1039/d3fd00003f), there seemed to be a bimodal distribution with many sources situated either close to the inner edge or the outer edge of the ring. Given that metallicity is a function of galactocentric distance, is there a metallicity or gas-to-dust ratio influence on the results that could be discerned from the data? Do the inner- *vs.* outer-ring sources show any systematic differences in their chemistry?

Olivia H. Wilkins answered: The sources in this pilot survey fall between approximately 4.60 and 8.28 kpc from the galactic center (Table 1), and the appearance of a bimodal distribution comes from a split between near ($d < 6$ kpc) and far ($d > 10$ kpc) sources. From the present data, there are no clear systematic differences in the chemistry with respect to galactocentric distance. Both of the more molecule-rich sources, BGPS 6029 and BGPS 6120, fall toward the outer edge of the ring at 7.18 and 8.28 kpc from the galactic center, respectively, but many of the sources closer to the galactic center, including BGPS 3474 and BGPS 4449 at 5.31 and 5.61 kpc from the galactic center, respectively, are also quite molecule-rich. Measuring the gas-to-dust ratio was out of the scope of this pilot study, but seeing whether there is a connection between this or other parameters and galactocentric distances would certainly make for an interesting study.

Table 1 Distances to galactic center R_{gal} derived for targeted GMCs

BGPS ID	l (deg)	d_{kin} (kpc)	R_{gal} (kpc)[†]
3053	023.373	10.67	4.60
3474	025.229	3.16	5.31
4449	030.524	3.03	5.61
5623	035.456	10.49	6.11
6029	040.628	11.01	7.18
6082	041.747	11.91	7.98
6112	043.076	11.65	7.97
6120	043.233	12.04	8.28
6299	049.069	5.57	6.05
6310	049.178	5.56	6.06
6318	049.260	5.55	6.07

† R_{gal} is estimated from the law of cosines as $R_{gal} = \sqrt{R_0^2 + d_{kin}^2 - 2R_0 d_{kin} \cos l}$ where $R_0 \sim 8$ kpc is the distance between the solar system and the galactic center and l is galactic longitude.

Anthony Meijer returned to the discussion of the paper by Ko-Yun Huang: Is there any evidence in your models or observations for the formation of H_2NCO^+ (*i.e.*, protonated isocyanic acid), which is a potential intermediate in the formation of, *e.g.*, urea? (see, *e.g.*, ref. 1).

1 E. C. S. Slate, R. Barker, R. T. Euesden, M. R. Revels and A. J. H. M. Meijer, *Mon. Not. R. Astron. Soc.*, 2020, **497**(4), 5413.

Serena Viti responded: To my knowledge, our observational spectra of NGC 253 or NGC 1068 do not include H_2NCO^+, but we did not explicitly look for transitions of this species – in fact, the data for NGC 253 are from the ALMA spectral survey ALCHEMI so the transitions of this molecule may be detected. In the specific chemical network used for the SiO and HNCO models in the paper led by Huang (https://doi.org/10.1039/d3fd00007a), H_2NCO^+ was not included.

Serena Viti returned to the discussion of the paper by Olivia H. Wilkins: Many of the conclusions are derived based on the kinetic temperature estimate from methanol. The temperature derived was, however, rather low, especially for gas that is believed to arise from hot cores/young stellar objects (YSOs). This is, as indeed your paper states, a bit surprising, especially as the spatial resolution is probably not high enough to disentangle all the gas components in the beam. Could it be that (some of the) methanol is subthermally excited? Or could line blending be affecting the analysis? Or is the excitation temperature T_{ex} really the kinetic temperature T_{kin}? What is your opinion?

Olivia H. Wilkins responded: These are all excellent questions that I hope will be answered by future investigations of these sources. I think that at least some of the methanol is subthermally excited. Since most of the methanol emission overlaps with places where young stellar objects (YSOs) or YSO candidates have been identified, but not characterized, in previous large surveys, I would expect higher angular resolution observations to yield higher methanol rotation temperatures of >100 K. This is characteristic of methanol gas observed in high-mass cores (*e.g.*, ref. 1). In terms of line blending, all lines used to produce the temperature and column density maps were inspected to make sure there were no other strong lines overlapping. However, there may be some blending of lines from different components that emit within the same beam, but have only slightly different velocity shifts, making them fittable but perhaps slightly broadened if they contain multiple velocity components. I think the biggest culprit of the low temperatures, especially toward the continuum emission regions, is that the emission from hot gas is likely diluted by cooler gas. In previous work using these methods,[2] we calculated excitation temperature maps for methanol and methyl cyanide (acetonitrile) in Orion KL using two beam sizes and found that the temperature profile varied substantially. Favre *et al.*[3] also noticed this phenomenon but for methyl formate. If hot/warm methanol emission in the molecular-ring sources is much smaller than the synthesized beam and is surrounded by cold gas, the average temperature across the beam would be lower than the true peak temperature. However, to ascertain the cause of the low temperatures in the present study, additional observations at high spatial resolution are needed.

1 S. E. Bisschop, J. K. Jørgensen, E. F. van Dishoeck and E. B. M. de Wachter, *Astron. Astrophys.*, 2007, **465**, 913.
2 O. H. Wilkins, P. Brandon Carroll and G. A. Blake, *Astrophys. J.*, 2022, **924**, 4.
3 C. Favre, D. Despois, N. Brouillet, A. Baudry, F. Combes, M. Guélin, A. Wootten and G. Wlodarczak, *Astron. Astrophys.*, 2011, **532**, A32.

Serena Viti opened the general of the paper by Eleonora Bianchi: You make the point in the discussion that to determine whether the species in question are released to the gas from the ice *via* thermal sublimation or shocks, one has to accurately know the binding energies. Do you think binding energies are the main "key" in determining the amount that thermally sublimates or are there other factors to consider? I am thinking here of the results from TPD experiments over the years that have shown that some species have multiple peaks of desorption.

Eleonora Bianchi responded: The conclusion that NH_2CHO emission is the result of an accretion shock is suggested by the evidence that the emission extends at a distance of ~120 au from the central protostar. Considering a bolometric luminosity of 55 L_{solar}, the temperature at this distance is lower than the average sublimation temperature. However, considering the distribution of binding energies, thermal sublimation cannot be totally excluded for the lower range (as shown in Fig. 5 of our paper (https://doi.org/10.1039/d3fd00018d)). Since the desorption from the shock is caused by the sputtering of the dust-grain mantles, what matters is the binding energy of the formamide. The other desorption peaks observed by TPD experiments are due to the trapping of the species in the water-rich ices (*i.e.*, connected with the water binding energy (BE) when a piece of the mantle sublimates) and restructuring of the water-ice, which are processes mostly induced by dust heating, not occurring in shocks.

The major uncertainty is related to the bolometric luminosity measurements. Indeed, we lack the measurements of the bolometric luminosities of the single sources. However, there are measurements of the system bolometric luminosity at the same epoch of our observations[1] and they are consistent with this value.

1 E. Fiorellino, C. F. Manara, B. Nisini, S. Ramsay, S. Antoniucci, T. Giannini, K. Biazzo, J. Alcalà and D. Fedele, *Astron. Astrophys.*, 2021, **650**, 43.

Ewine F. van Dishoeck returned to the discussion of the paper by Olivia H. Wilkins: This comment concerns the discussion on local thermodynamic equilibrium (LTE) modeling. Generally, what is meant in practice by this term is that the level populations of the molecule can be characterized by a single excitation temperature, but that the excitation temperature does not have to be the same as the kinetic gas temperature. There can be subthermal excitation at low densities (as may well be the case on the larger scales that you are looking at), or the excitation temperature can reflect the radiation temperature. See Fig. 6 of ref. 1 which shows one illustrative example for the case of CH_3OH: for gas at $T = 100$ K and $n = 1 \times 10^7$ cm^{-3}, the rotation temperature is only 24 K; for $n = 1 \times 10^9$ cm^{-3}, it is 80 K. The derived column densities are similar.

1 D. Johnstone, A. M. S. Boonman and E. F. van Dishoeck, *Astron. Astrophys.*, 2003, **412**, 157.

Ewine F. van Dishoeck returned to the discussion of the paper by Eleonora Bianchi: The word "accretion" is prominently in the title of your paper (https://doi.org/10.1039/d3fd00018d) and also in the main text. How do you know that the gas in the streamer is accreting and not outflowing? Do you have velocity information? I appreciate that this is not easy to obtain observationally and can sometimes be confusing.

Eleonora Bianchi replied: We have the information on the velocity, having a spectral resolution of ~0.15 km s^{-1}. Understanding the kinematics is not easy in this case because the system is complex and the two protostars have different systemic velocities. Nevertheless, we observe in the HDO channel maps a blue-shifted peak at velocities larger than 3 km s^{-1} and the gas velocity increases approaching the protostars, which suggests infalling. In addition, HDO emission is co-spatial with the streamer traced by dust emission and the direction is different to that of the outflow, which is very well characterised in this source with both CO and Herbig–Haro objects. Finally, a streamer has been recently detected in the same source at larger scales by Hsieh *et al.*[1]

1 T.-H. Hsieh, D. M. Segura Cox, J. E. Pineda, P. Caselli, L. Bouscasse, R. Neri, A. López-Sepulcre, M. T. Valdivia-Mena, M. J. Maureira, T. Henning, G. V. Smirnov-Pinchukov, D. Semenov, Th. Möller, N. Cunningham, A. Fuente, S. Marino, A. Dutrey, M. Tafalla, E. Chapillon, C. Ceccarelli and B. Zhao, *Astron. Astrophys.*, 2023, **669**, 137.

Ewine F. van Dishoeck remarked: Nice work! Do you see a spatial difference between the three molecules depending on which line you image? Lines with different upper energy levels and/or Einstein *A* coefficients can sometimes show very different emission patterns.

Eleonora Bianchi answered: Thank you! In a previous paper,[1] we analysed the spatial distribution of different transitions of CH_3OH, CH_3CHO, CH_3OCH_3 and NH_2CHO. In order to investigate the chemical segregation, we compare the transitions with close upper-level energies (E_{up}) and we can exclude the possibility that the molecular distribution mainly depends on the excitation conditions. Line peak intensities between the selected transitions of different species are also comparable. In general, in this source we do not see dramatic differences between the morphology of lines of the same species with different upper energy levels and/or Einstein *A* coefficients. We see instead major differences between the morphology of different species.

1 E. Bianchi, A. López-Sepulcre, C. Ceccarelli, C. Codella, L. Podio, M. Bouvier and J. Enrique-Romero, *Astrophys. J., Lett.*, 2022, **928**, L3.

Sandra Brünken returned to the discussion of the paper by Olivia H. Wilkins: Earlier papers by Spezzano *et al.*,[1,2] and Lin *et al.*,[3] use the same molecular tracers (CH_3OH, C_3H_2) that you use in your study to decipher the chemical differentiation in cold starless cores. The ultimate goal is to use these tracers to link their abundance to large-scale physical changes in these cores. For example, a higher abundance of C_3H_2 is linked to a higher illumination from the interstellar radiation field, releasing more C atoms into the gas phase and forming more carbon-chain molecules, whereas methanol traces regions with less illumination, thus more CO, the precursor for methanol. Can you comment on the prospects to reproduce similar findings in the selection of high-mass star-forming regions (HMSFRs) you studied? Is the resolution capable to trace these regions, with observed differentiation on spatial scales of a few thousand AU (astronomical units) for the low-mass prestellar cores?

1 S. Spezzano, L. Bizzocchi, P. Caselli, J. Harju and S. Brünken, *Astron. Astrophys.*, 2017, **592**, L11.
2 S. Spezzano, P. Caselli, J. E. Pineda, L. Bizzocchi, D. Prudenzano and Z. Nagy, *Astron. Astrophys.*, 2020, **643**, A60.
3 Y. Lin, S. Spezzano, O. Sipilä, A. Vasyunin and P. Caselli, *Astron. Astrophys.*, 2022, **665**, A131.

Olivia H. Wilkins replied: Thank you for sharing these interesting insights. Unfortunately, there are not enough lines in the pilot survey data to derive abundances of c-C_3H_2, but future observations to target this molecule and additional methanol lines could provide insight into these questions. High-mass analogues to low-mass prestellar cores (FWHM $\sim$ 0.01–0.1 pc) remain elusive, and it has been proposed that high-mass stars may instead evolve from large (0.1–1 pc) gas reservoirs, such as starless massive dense cores or starless clumps (*e.g.*, ref. 1). The Atacama Compact Array (ACA) angular resolutions in this study correspond to linear resolutions of 10^4–10^5 au, so my guess would be that an analogous study to those by Spezzano *et al.*[2,3] and Lin *et al.*[4] would not be feasible for most sources unless the full array is used. However, since the FWHM estimates for massive dense cores is one or two orders of magnitude greater than those for low-mass prestellar cores, using c-C_3H_2 and CH_3OH abundances to trace large-scale physical changes with the ACA may be sufficient for 'near' sources in our sample, namely BGPS 3474 and BGPS 4449, where the linear resolution is around 10^4 au (the other three 'near' sources, BGPS 6299, 6310, and 6318 did not contain detectable c-C_3H_2 emission). Finally, just because c-C_3H_2 was not detected for some clouds in this survey does not mean that it is not present; rather, this molecule might be revealed in higher angular resolution observations if it exists on spatial scales smaller than those probed by the ACA. Regardless, I think the questions you asked would make for an interesting future observation project!

1 F. Motte, S. Bontemps and F. Louvet, *Annu. Rev. Astron. Astrophys.*, 2018, **56**, 41.
2 S. Spezzano, L. Bizzocchi, P. Caselli, J. Harju and S. Brünken, *Astron. Astrophys.*, 2017, **592**, L11.
3 S. Spezzano, P. Caselli, J. E. Pineda, L. Bizzocchi, D. Prudenzano and Z. Nagy, *Astron. Astrophys.*, 2020, **643**, A60.
4 Y. Lin, S. Spezzano, O. Sipilä, A. Vasyunin and P. Caselli, *Astron. Astrophys.*, 2022, **665**, A131.

Cecilia Ceccarelli commented: You derived low T_{ex} values from the methanol maps. I reckon that because the transitions' upper-level energy range is not very large, your T_{ex} estimates have relatively large errors. That said, the derived low T_{ex} may probe a really relatively cold and tenuous gas associated with the skin of the molecular cloud.

We found something analogous in a survey of protostars in the Orion Molecular Clouds 2 and 3,[1] where the single-dish methanol emission was dominated by the PDR cloud skin gas. To reach these conclusions we (i) compared the linewidth and shape of the methanol lines with those of c-C_2H_3, finding that they were very similar, and (ii) carried out a non-LTE analysis for both species lines and found a low kinetic temperature.

Since your sources are in the molecular ring, probably they are irradiated by large UV fields, so maybe your observations really probe relatively cool methanol.

1 M. Bouvier, A. López-Sepulcre, C. Ceccarelli, C. Kahane, M. Imai, N. Sakai, S. Yamamoto and P. J. Dagdigian, *Astron. Astrophys.*, 2020, **636**, A19.

Olivia H. Wilkins responded: Thank you for these perspectives! Certainly a challenge/limitation of conducting limited line surveys such as this study (https://doi.org/10.1039/d3fd00003f) is the use of small upper-state energy ranges, which may factor into the low derived temperatures. Nevertheless, I appreciate the ideas for how to test whether methanol might originate in the cool 'skin' of the clouds. One of the motivations of this work was that the chemistry of high-mass molecular-ring sources might be more affected by external physical parameters, such as strong UV fields, when compared to low-mass star-forming regions. I did a quick comparison of the linewidth and shape of the CH_3OH and c-C_3H_2 lines in a couple of sources. For BGPS 4449 ACA1(N), the line shapes look comparable, with the caveat that the c-C_3H_2 line (at ~217.8 GHz) is much weaker than the CH_3OH line (at ~218.4 GHz) in the same spectral window, making such visual inspection difficult. However, the approximate FWHM values for these c-C_3H_2 and CH_3OH lines are about 3.7 and 5.5 km s^{-1}, respectively. In this source, c-C_3H_2 is quite extended whereas CH_3OH is not, as seen in Fig. 13 of the manuscript. In BGPS 5623 ACA1, however, the c-C_3H_2 and CH_3OH emission are quite co-spatial. Furthermore, the 217.8 and 218.4 GHz lines have similar linewidths of about 4.1 km s^{-1}. A non-LTE analysis for both cases would be ideal to test whether the methanol emission in BGPS 4449 is embedded in ACA1 (and has a higher kinetic temperature), whereas that in BGPS 5623 is on the skin (with a lower kinetic temperature).

Tom J. Millar returned to the discussion of the paper by Ko-Yun Huang: You mentioned that the SiO abundance correlates well with shock velocity. Is this SiO being formed by the removal of an ice mantle or by destruction of the silicate grain core? If the latter, at what shock velocities does this occur?

Ko-Yun Huang responded: Indeed both slow shock and fast shock enhance the gas-phase SiO abundances for most cases, by releasing Si from the mantle, and grain core in the case of stronger shocks. However, SiO enhancement due to fast shock is at least 2–3 orders of magnitude higher than that by slow shock. The slow shock is indeed capable of removing the Si-bearing species from the ice mantles, but the amount is by no means comparable to the Si budget embedded in the grain core that can only be liberated by fast shock.

Tom J. Millar commented: You consider very high cosmic-ray ionisation rates, up to 10^5 times the standard rate, in your models. Since 10% of the cosmic-ray ionisations of H_2 lead to $H^+ + H$, *i.e.*, the break-up of the H_2 molecule and which is not easily reformed, do you see any decrease in the H_2 abundance in your models?

Ko-Yun Huang responded: The dissociation of the H_2 molecule into the atomic form of hydrogen induced by high cosmic ray ionization rate (CRIR), in particular for CRIR $= 10^3$–10^5 standard rate, does not appear to dominate. H_2 abundance does decrease (in the higher-density cases, $n_{H_2} = 10^5$–10^6 cm^{-3}), but it is not dominated by H_2 dissociation into H atoms (or protons).

Inga Kamp asked: Since you are simulating quite long timescales, including also the cool-down phase after the shock heating (dust sublimation), I was wondering if you also account for recondensation of the dust? I am not sure what the condensation timescale is for the conditions you simulate and if this can compete with the Si gas-phase chemistry.

Ko-Yun Huang answered: Our observation results from both NGC 1068 and NGC 253 suggested most post-shock gas temperatures are above 50 K in most cases – thus our modeling stops the post-shock cooling at 50 K. At this temperature (50 K), we expect negligible freeze-out to take place. In other words, with this set up, we expect the re-condensation timescale to be much longer than the modeled time duration.

Dwayne E. Heard returned to the discussion of the paper by Olivia H. Wilkins: In your table of elements that were seen, you do not list phosphorus. Was that a surprise that no phosphorus-containing species were observed? Are the spectral transitions not favourable? This really stood out to me. Would you not expect phosphorus to be formed in supernova explosions, for example, perhaps in the vicinity?

Olivia H. Wilkins responded: It was not a surprise that phosphorus-containing species were not observed in our survey. Such molecules remain elusive in the broad context of astrochemistry, with very few detected in inter- and circumstellar environments. Furthermore, there are relatively few phosphorus-bearing molecules included in the Cologne Database for Molecular Spectroscopy (CDMS),[1] for example. Phosphorus is in only about 2% of the nearly 1200 CDMS entries. Nitrogen, which is isoelectronic to phosphorus, is found in nearly 40% of entries. So even if there are phosphorus-bearing species in our sources, we might not be able to detect them. Unfortunately, we do not yet know enough about phosphorus chemistry in star-forming regions to know whether phosphorus-containing molecules should be expected to reside in molecular-ring sources.

1 http://cdms.de.

Athena Flint commented: In your surveys of these HMSFRs, you mention that one detection of note is that of c-C_3H_2. Interestingly, the lower end of the frequency range over which you observe overlaps with the survey range of Rojas-García *et al.*[1] where they observe a peak that may correspond to methylenecyclopropene (c-$C_3H_2CH_2$) at 217.429650 GHz. It is possible that c-$C_3H_2CH_2$ can form from c-C_3H_2 in regions warm enough to allow for traversal of the reaction barrier between c-C_3H_2 and the methyl radical.[2] Are there any peaks towards the lower end of your frequency range that may agree with the conclusions of Rojas-García *et al.* that you were initially hesitant to identify?

1 O. S. Rojas-García, A. I. Gómez-Ruiz, A. Palau, M. T. Orozco-Aguilera, M. Chavez Dagostino and S. E. Kurtz, *Astrophys. J., Suppl. Ser.*, 2022, **262**, 13.
2 A. R. Flint, B. N. Rogers and R. C. Fortenberry, *Mon. Not. R. Astron. Soc.*, 2023, **522**, 4215–4223.

Olivia H. Wilkins responded: Unfortunately, our range doesn't go quite low enough to cover the potential methylenecyclopropene line at 217.43 GHz to check whether this molecule can be detected in the targeted high-mass star-forming regions. The lower end of our spectral windows falls around 217.55 GHz, and with the Doppler effect, the methylenecyclopropene line would be shifted even further out of the survey frequency range.

Cecilia Ceccarelli returned to the discussion of the paper by Ko-Yun Huang: Following up on an earlier question, how can you be sure that SiO is due to shocks and not X-rays or cosmic rays?

Ko-Yun Huang responded: Shock strength (*e.g.*, no shock, weak shock, and strong shock) and cosmic ray ionization rates are two of the several parameters we explored in our chemical modeling grids. From our results, it is clear that strong shocks are dominant in enhancing the SiO abundance, and not ionization rates.

Cecilia Ceccarelli asked: Does your model include the X-rays or CR-induced SiO desorption from the grain mantles?

Ko-Yun Huang answered: Yes, our chemical models include cosmic ray-induced desorption.

Ewine F. van Dishoeck remarked: This is a follow-up to previous questions: do you know how molecular your shocks are, *i.e.*, is molecular hydrogen destroyed in the higher velocity shocks? In some cases, your abundances are seen to drop very dramatically; is it because the gas becomes atomic due to shock destruction or due to the high cosmic ray ionization rates destroying H_2 and CO (as found in the work of Thomas Bisbas *et al.*[1,2])?

1 T. G. Bisbas, P. P. Papadopoulos and S. Viti, *Astrophys. J.*, 2015, **803**, 37.
2. T. G. Bisbas, E. F. van Dishoeck, P. P. Papadopoulos, L. Szűcs, S. Bialy and Z.-Y. Zhang, *Astrophys. J.*, 2017, **839**, 90.

Ko-Yun Huang answered: We are modelling non-dissociative C-shocks. The shock velocities explored in our modeling went up to 60 km s^{-1}. Indeed, there is a sharp drop of all the shock tracers – HNCO, SiO and CH_3OH – and the destruction of these three species are dominated by high-CRIR-induced reactions. In a high CRIR environment, the gas is indeed more atomic, but the CRIR dissociation does not dominate the destruction of our molecules of interest. On the other hand, in the strong shock regime, during the shock passage, species such as CO and H_2 are dissociated (and react to form other molecules), from 1 to a few orders of magnitude drop in molecular abundances, depending on the species of discussion.

Conflicts of interest

There are no conflicts to declare.

Faraday Discussions

ROYAL SOCIETY OF CHEMISTRY

PAPER

Fingerprinting fragments of fragile interstellar molecules: dissociation chemistry of pyridine and benzonitrile revealed by infrared spectroscopy and theory†

Daniël B. Rap,[a] Aude Simon,[b] Kim Steenbakkers,[a] Johanna G. M. Schrauwen,[a] Britta Redlich[a] and Sandra Brünken*[a]

Received 25th January 2023, Accepted 22nd March 2023
DOI: 10.1039/d3fd00015j

The cationic fragmentation products in the dissociative ionization of pyridine and benzonitrile have been studied by infrared action spectroscopy in a cryogenic ion trap instrument at the Free-Electron Lasers for Infrared eXperiments (FELIX) Laboratory. A comparison of the experimental vibrational fingerprints of the dominant cationic fragments with those from quantum chemical calculations revealed a diversity of molecular fragment structures. The loss of HCN/HNC is shown to be the major fragmentation channel for both pyridine and benzonitrile. Using the determined structures of the cationic fragments, potential energy surfaces have been calculated to elucidate the nature of the neutral fragment partner. In the fragmentation chemistry of pyridine, multiple non-cyclic structures are formed, whereas the fragmentation of benzonitrile dominantly leads to the formation of cyclic structures. Among the fragments are linear cyano-(di)acetylene$^{\bullet+}$, methylene-cyclopropene$^{\bullet+}$ and *o*- and *m*-benzyne$^{\bullet+}$ structures, the latter possible building blocks in interstellar polycyclic aromatic hydrocarbon (PAH) formation chemistry. Molecular dynamics simulations using density functional based tight binding (MD/DFTB) were performed and used to benchmark and elucidate the different fragmentation pathways based on the experimentally determined structures. The implications of the difference in fragments observed for pyridine and benzonitrile are discussed in an astrochemical context.

[a]*Institute for Molecules and Materials, FELIX Laboratory, Radboud University, Toernooiveld 7, 6525 ED Nijmegen, The Netherlands. E-mail: sandra.bruenken@ru.nl*

[b]*Laboratoire de Chimie et Physique Quantiques (LCPQ), Fédération FeRMI, CNRS & Université Toulouse III – Paul Sabatier, 118 Route de Narbonne, 31062 Toulouse, France*

† Electronic supplementary information (ESI) available. See DOI: https://doi.org/10.1039/d3fd00015j

Introduction

Recently, various aromatic molecules have been detected in the cold interstellar medium (ISM) *via* their rotational transitions in the radio-regime. Among these detections are cyano-substituted aromatic molecules such as cyano-cyclopentadiene,[1,2] cyano-benzene,[3] cyano-indene[4] and cyano-naphthalene[5] isomers. The discovery of these species has shown that nitrogen can be a part of aromatic molecules and that (cyano)-substitution can be a proxy for the existence of analog pure carbon species.[6] On the other hand, alternative nitrogen-containing molecules such as pyridine and aniline, also possessing substantial dipole moments, have not been found in the ISM yet despite sensitive searches.[7,8] Surprisingly, even more complex nitrogen containing molecules such as functionalized pyridines and quinolines have been found in meteorite samples,[9,10] and have been viewed as possible building blocks of molecules that we consider as necessary for life. Their biological importance lies in the fact that they can be processed to molecules that are part of DNA and RNA similar to life on Earth. Possible so-called bottom-up formation processes towards these larger organic building blocks have been studied using a variety of experimental and theoretical methodologies.[11–17] However, other processes can counteract these growth mechanisms, *e.g.*, through fragmentation, and can play an important role in the complex interstellar chemistry. These processes can be induced by the presence of high energetic electrons from electromagnetic winds in environments such as Titan's atmosphere,[18] or cosmic rays[19] and ultraviolet (UV) photons[20] from nearby stars that can induce specific chemistry at the edges of molecular clouds close to photo-dissociation regions (PDRs).[21] Small carbon-containing molecules have been detected in these regions and are regarded as indirect evidence of top-down hydrocarbon chemistry.[22,23] Laboratory experiments have shown that both bottom-up and top-down processes can be at play simultaneously. Fragmentation of carbon cycles such as benzene[24,25] and naphthalene[26,27] using plasma sources, but also discharges of mixtures of small species such as CH_4 and N_2 (ref. 28) have shown that this approach is not always completely destructive, but that recombination chemistry of the fragments and other molecules plays an important role in the formation of larger species. This emphasizes how fragmentation and formation mechanisms are often intertwined.

A variety of studies has been investigating the fragmentation processes of astronomical relevant molecules in detail ranging from theoretical[29–31] to experimental studies using dissociative ionization and infrared spectroscopy[32–37] or imaging Photoelectron-Photoion-Coincidence (iPEPICO) spectroscopy.[38–40] By using a combination of techniques, some of these studies have provided structural information of the fragment ions.

The fragmentation processes of pyridine have been investigated previously using theoretical studies,[41] electron impact ionization,[42] VUV photoionization,[43] kinetic energy release distribution (KERDs) measurements[44,45] and collision-induced luminescence spectroscopy.[46] This has shown that the lowest energy fragmentation channel yields ions with the molecular formula $C_4H_4^{\bullet+}$ formed through loss of HCN and/or HNC, which remains the dominant cationic fragment over all ionizing energies.[42,47] A similar major cationic fragmentation channel was observed in the dissociation of benzonitrile, which has been studied using photodissociation in an ion-cyclotron resonance ion trap,[48] appearance energy determinations[49,50] and Rice–Ramsperger–Kassel–Marcus (RRKM) calculations.[51] These studies pointed toward the formation of the $C_6H_4^{\bullet+}$ fragment upon the loss of HCN and/or HNC. However, no direct structural assignments have been made for these and other cationic fragments of the astrochemically relevant pyridine and benzonitrile molecules to date.

To understand the chemistry in the various environments in space, it is important to have a detailed look at the fragmentation mechanisms of these small nitrogen-containing molecules that have already been detected (benzonitrile) and that are yet unobserved (pyridine). This detailed look should involve accurate structural knowledge of the fragments that, when detected using astronomical measurements, can serve as proxies for the presence of their either undetectable or former parent molecules.

Here we apply a multifaceted approach to investigate the fragmentation chemistry of the nitrogen-containing molecules pyridine and benzonitrile in detail. Combining dissociative ionization and cryogenic infrared pre-dissociation spectroscopy (IRPD) using the free-electron laser FELIX,[52,53] we measure the infrared fingerprints of the major ionic fragments. The molecular structures of the fragment products are assigned by comparison to quantum-chemical calculations and are used to guide the construction of potential energy surfaces (PESs) that explain the nature of the fragmentation reaction. Finally, we have performed molecular dynamics (MD) simulations with the electronic structure determined at the self-consistent-charge density functional based tight binding (SCC-DFTB)[54] level of theory to predict the observed fragmentation channels including the molecular structures. Such a combined approach has been successfully used to disentangle the dissociation mechanism of cationic PAHs such as anthracene and phenanthrene.[37] This work can guide experimental and astronomical studies in finding new astrochemical relevant species, and, furthermore, provides vibrational fingerprint spectra of dominant fragment ions as reference data for sensitive infrared astronomical searches, *e.g.*, with the James Webb Space Telescope (JWST).[55]

Results and discussion

Mass spectrometry of pyridine and benzonitrile fragmentation

The experimental and theoretical methods used in the present work are described in more detail in the ESI.† Only the results are reported and discussed in the rest of the manuscript. Main fragmentation channels were identified by measuring the number of ions of the different fragment masses (m/z) in the dissociative ionization of pyridine and benzonitrile at various electron impact energies, see Supplementary Methods in the ESI† for details of the experimental apparatus. An example breakdown curve for the dissociative ionization of benzonitrile is shown in Supplementary Fig. 1,† and it agrees within our uncertainties with previous electron impact ionization studies. The ionization energies of pyridine and benzonitrile are 9.34 eV (electron impact ionization)[56] and 9.73 eV (threshold electron detection),[57] respectively. At low electron impact energies of 15(2) eV, mainly the HCN/HNC fragment channel, with an appearance energy (AE) of 12.34 eV (ref. 58) and 12.54 eV (ref. 50) for pyridine and benzonitrile, respectively, is accessible and other fragment channels are present in minor abundance. This is demonstrated in the mass spectra for both pyridine and benzonitrile as shown in Fig. 1a and b, respectively. Upon higher electron energies (50 eV), we observe more contribution from other energetically higher-lying fragment channels. From the *m*/*z* values we can deduce the possible neutral loss fragments. We observe the loss of a hydrogen atom (H, −1 *m*/*z*, AE = 14 eV for pyridine[47]), acetylene (C_2H_2, −26 *m*/*z*, AE = 13.84 eV for pyridine[47]) and/or cyano radical $CN^{\bullet}$ (−26 *m*/*z*, AE = 13.52 eV for benzonitrile[58]), hydrogen cyanide and/or hydrogen isocyanide (HCN and/or HNC, −27 *m*/*z*, AE = 12.34 eV (ref. 58) and 12.54 eV (ref. 50) for pyridine and benzonitrile, respectively) and ethylene and/or dihydrogen cyanide (C_2H_4 and/or $H_2CN^{\bullet}$, −28 *m*/*z*, AE = 16.6 eV for pyridine[47]) for both parent molecules.

For both pyridine and benzonitrile, the most abundant fragment, *m*/*z* 52 and *m*/*z* 76, is formed upon the loss of either HCN or HNC. However, based on the observed mass difference alone, the exact mechanism cannot be deduced as structurally different neutral and cationic fragment molecules can be formed. Looking at the nature of the cationic fragment species can help to understand the fragmentation mechanism. In order to determine the molecular structure of the observed fragment ions, we have performed infrared pre-dissociation spectroscopy using the free-electron laser FELIX.[52] The fragments that were produced in the ion source were mass-selected and their infrared fingerprint were measured in the 550–2200 cm^{-1} range (see Supplementary Methods† for details of the experimental setup). Here, the HCN/HNC loss fragments were produced using electron impact energies of 15(2) eV; however, to get a sufficient number of ions to probe the minor fragmentation channels, an electron energy of 50(2) eV was used. Comparing the recorded fingerprint spectra to calculated infrared vibrational modes using density functional theory (DFT) allows assignment of molecular structures to the fragment ions, see Supplementary Methods† for details of these calculations. The major fragmentation channels and their accompanying molecular structures that have been identified in this way for both the dissociative ionization of pyridine and benzonitrile are discussed throughout the paper and are schematically shown in Fig. 2.

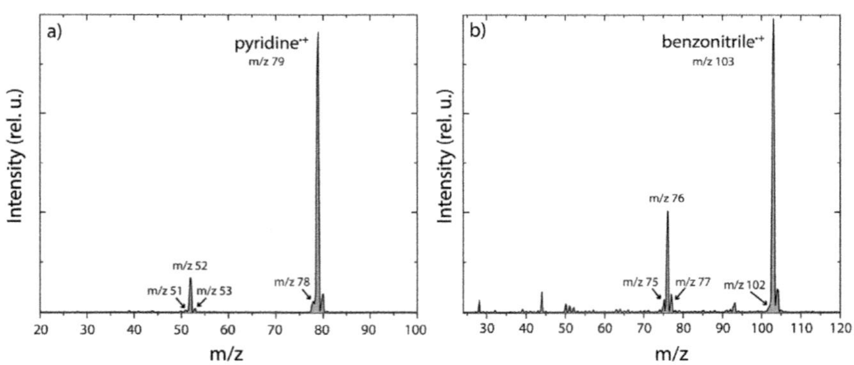

Fig. 1 Experimental mass spectra of the dissociative ionization of (a) pyridine and (b) benzonitrile at an electron impact energy of 15 eV. The fragments that are studied in this paper are labelled with their *m*/*z* value.

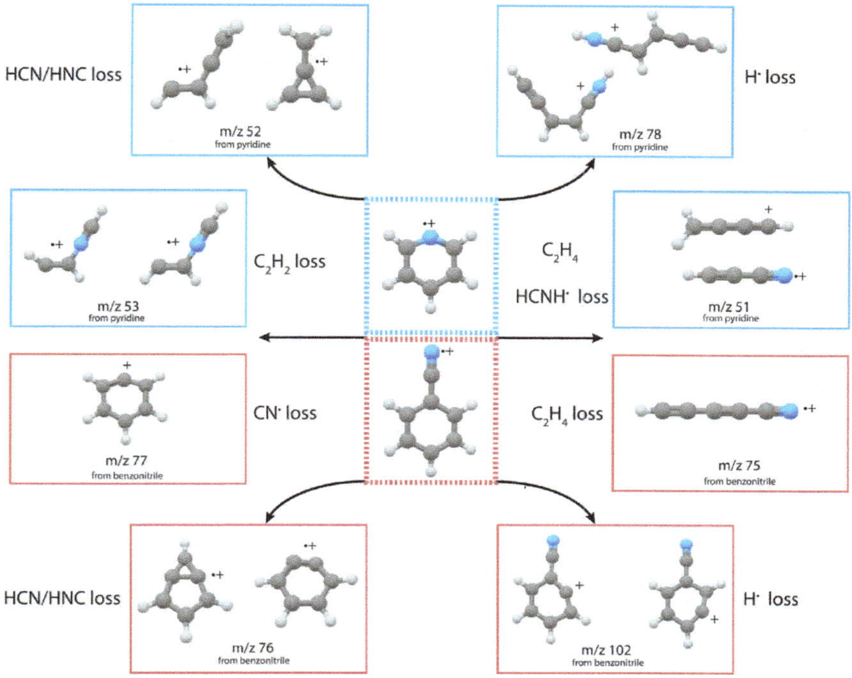

Fig. 2 Schematic overview of the discovered structures from the major fragmentation channels upon dissociative ionization of pyridine (blue boxes) and benzonitrile (red boxes).

Spectroscopy of pyridine$^+$ fragmentation products

Upon dissociative ionization of pyridine, we observe the formation of cationic fragments with *m*/*z* 51, 52, 53 and 78. For each of the fragment channels, the recorded infrared fingerprint spectrum is shown in Fig. 3 together with the calculated or previously measured infrared spectra and the structures of the assigned molecules. Multiple additional candidate structures have been evaluated and optimized, and their infrared spectra have been calculated as described in the Supplementary Methods and are shown in the ESI (Supplementary Fig. 3 and 7–9),† but do not show a sufficient match with the experimental data.

The major fragmentation channel (*m*/*z* 52) can be assigned to the lowest energy isomeric methylene-cyclopropene$^{\bullet+}$ structure containing a 3-membered ring. A minor contribution of a non-cyclic $CHCHCCH_2{}^{\bullet+}$ species, hinting to a ring-opening process of methylene-cyclopropene$^{\bullet+}$, is observed. Using a saturation depletion experiment, the abundance of the major isomer is estimated to be around 85–90%. The other isomer is measured to account for 15–20% (more details in Supplementary Fig. 5 and 6†). This non-cyclic structure is significantly higher in energy (130.7 kJ mol^{-1}) than the other isomers butatriene$^{\bullet+}$ (9.7 kJ mol^{-1}) or cyclobutadiene$^{\bullet+}$ (39.9 kJ mol^{-1}) (Supplementary Fig. 3†), with the latter being characterized previously by Ar-IRPD spectroscopy.[60] This hints to some excess energy in the ion that enables further ring opening, for which a barrier of 174 kJ mol^{-1} (1.8 eV) (at the MP2/6-311++G(d,p) level of theory) has

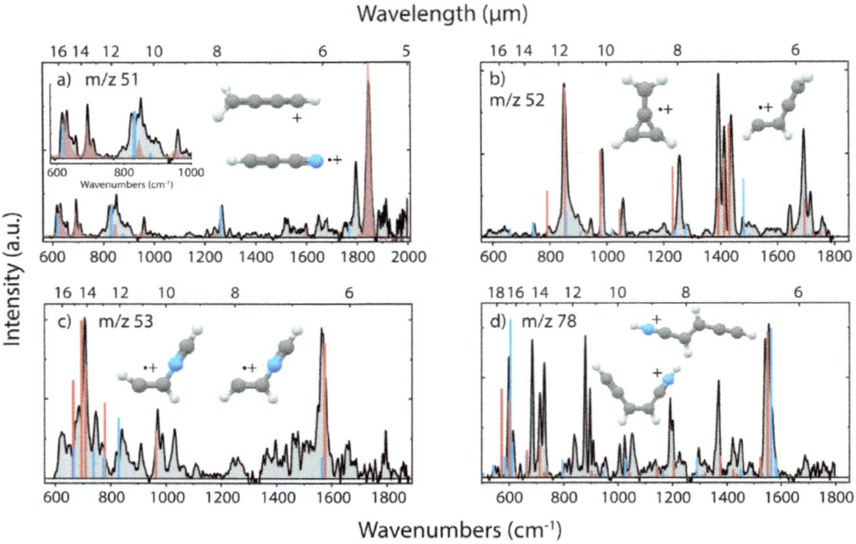

Fig. 3 Experimental infrared fingerprint spectra (grey) of (a) *m*/*z* 51, (b) *m*/*z* 52, (c) *m*/*z* 53 and (d) *m*/*z* 78 fragment channels of the dissociative ionization of pyridine. For each panel, the *x*-axis is showing wavenumbers in cm^{-1} (lower axis) and wavelength in μm (upper axis). An experimental reference spectrum of cyanoacetylene$^{\bullet+}$ ($HC_3N^{\bullet+}$) (red trace) from Steenbakkers *et al.*[59] is shown in panel (a). Calculated vibrational modes of the assigned structures are shown for (a) protonated diacetylene$^+$ (l-CH_2CCCH^+) (blue), (b) methylene-cyclopropene$^{\bullet+}$ (c-$C_3H_2(CH_2)^{\bullet+}$) (red), $CHCHCCH_2^{\bullet+}$ (blue), (c) protonated *trans*-isocyanoacetylene$^+$ (*trans*-$CHCHNCH^+$) (red), protonated *cis*-isocyanoacetylene$^+$ (*cis*-$CHCHNCH^+$) (blue) and (d) *cis*-$CHCCHCHCNH^+$ (red), *trans*-$CHCCHCHCNH^+$ (blue) as sticks. The infrared spectra have been calculated at the harmonic B3LYP-GD3/N07D level of theory. The anharmonic spectra of methylene-cyclopropene$^{\bullet+}$ (*m*/*z* 52) and *cis*/*trans* $CHCCHCHCNH^+$ (*m*/*z* 78) have been used to improve the match with the experimental spectrum due to the calculated strong combination and overtone modes. The spectrum of protonated diacetylene (*m*/*z* 51) has been calculated at the B2PLYPD3/aug-cc-pVTZ level of theory.

been calculated.[61] For the other fragmentation channels, we generally observe non-cyclic compounds.

The *m*/*z* 78 fragment can be assigned to the H-loss fragments from cationic pyridine where a hydrogen is migrated to the nitrogen atom and the aromatic ring has opened. The spectrum can be assigned to the *cis* and *trans* isomers that have similar energies, based on the double peak structure around 1550 cm^{-1}, related to the C=C double bond stretching vibration from each of the two isomers. Small frequency shifts between the calculated and experimental spectra are observed between 550 and 750 cm^{-1} due to anharmonicity effects and are hampering a clear assignment. To validate the structural assignment, the infrared spectrum in the N–H stretch region was measured using a table-top OPO/OPA system and showed two features, presenting both the NH and CH stretching modes, thereby confirming the presence of an N–H group in the molecule (Supplementary Fig. 10†). Both assigned species are 43.2 and 41.3 kJ mol^{-1} higher in energy than the global minimum structure, the cyclic 1-dehydro-pyridine$^{\bullet+}$ (Supplementary Fig. 7†), where a hydrogen has been removed on the C1 position. This demonstrates that under the experimental conditions (EI of 50 eV, single collision conditions) a hydrogen migration towards the nitrogen and subsequent ring opening is more probable than direct H-loss from the ring. Clear evidence of H-migration to the nitrogen atom before dissociation of pyridine molecules has been discussed previously by Wasowicz *et al.*[62,63]

To elucidate the structure of the *m*/*z* 53 fragment, many potential species with the molecular formula $C_3H_3N^+$ as discussed by Bera *et al.*[64] have been considered (Supplementary Fig. 8†). Two isomers, protonated *cis*- and *trans*-isocyanoacetylene$^+$, can be assigned and contain a diagnostic C=C double bond stretching mode around 1565 cm^{-1}, similar as for the H-loss species. A few experimental bands have not been assigned but can come from anharmonic contributions such as overtones and combination bands. The assignment of these N-containing species for *m*/*z* 53 confirms the neutral fragment to be C_2H_2 rather than the loss of CN, which would lead to the $C_4H_5^+$ ionic fragment, whose lowest energy isomer is the methyl-cyclopropyl cation, for which high-level vibrational frequency calculations exist.[65] The observed species are not the global minimum on the $C_3H_3N^+$ PES, which has a CCCN backbone geometry as calculated by Bera *et al.*,[64] but they have been proposed as one of the intermediates in the $C_2H_2^+$ + HCN ion–molecule reaction.[66]

The major fragment in the *m*/*z* 51 channel is assigned to cyanoacetylene$^{\bullet+}$ ($HC_3N^{\bullet+}$) where a neutral C_2H_4 fragment is formed. This open-shell linear ion exhibits a $^2\Pi$ electronic ground state and is therefore subjected to Renner–Teller vibronic coupling effects. The spectrum of this ion has been recently recorded in our group by means of Ne-tagging IRPD,[59] where it was produced from the vinyl cyanide neutral precursor, and we use it here as a reference spectrum to assign the fragment ion structure. We may see some contribution of isocyanoacetylene$^{\bullet+}$, based on the slightly shifted experimental band around 1795 cm^{-1} due to the CC stretching vibration of the CNC bond. This open-shell linear ion is also Renner–Teller affected and we have constructed an effective Hamiltonian for the bending vibrations using harmonic frequencies and spin–orbit parameters calculated using the RCCSD(T)-F12a/cc-pVTZ-F12 level of theory (Supplementary Fig. 9†). This allows the making of a comparison with similarly determined frequencies of the affected bending modes of cyanoacetylene$^{\bullet+}$ (Supplementary Fig. 9, see also Supplementary Methods†). All bending modes of isocyanoacetylene$^{\bullet+}$ are shifted to lower frequencies and do not fully explain the additional experimental bands in the range below 900 cm^{-1}. However, the other vibrational modes, both in the range lower than 900 cm^{-1} and the feature around 1265 cm^{-1}, can be assigned to protonated diacetylene which is formed by the loss of a species with molecular formula $H_2CN^{\bullet}$. The cationic fragment has been calculated to be the global minimum on the $C_4H_3^+$ PES, with an analogous structure to the assigned *m*/*z* 52 fragment, c-$C_3H_2(CH)^+$, lying 89 kJ mol^{-1} higher in energy (Supplementary Fig. 9†). Whereas the electronic spectrum of protonated diacetylene is known,[67,68] only a limited range of the infrared spectrum was covered in a neon matrix spectroscopy study.[67] This all leads to the conclusion that we observe the presence of both cyanoacetylene$^{\bullet+}$ and protonated diacetylene$^+$, but we cannot exclude some additional contribution of isocyanoacetylene$^{\bullet+}$.

Spectroscopy of benzonitrile$^+$ fragmentation products

In the dissociative ionization of benzonitrile, we observe the formation of *m*/*z* 75, 76, 77 and 102, corresponding to similar fragmentation channels as for pyridine, namely $C_2H_4/H_2CN^•$, HCN/HNC, $C_2H_2/CN^•$ and H-loss, respectively. The experimental infrared fingerprint spectra of these fragments are shown in Fig. 4, together with the calculated infrared spectra of the assigned molecular structures. The infrared frequencies of multiple additional potential structures were also calculated and are shown in Supplementary Fig. 13–16.†

Generally, we observe more cyclic structures across the fragment channels than for the pyridine fragmentation. The molecular structures of the most abundant fragment with *m*/*z* 76 can be assigned to a combination of two isomers, *i.e.*, the *ortho* (*o*) and a bicyclic *meta* (*m*)-benzyne$^{•+}$ species. The characteristic CC stretch of *o*-benzyne$^{•+}$ is calculated around 1740 cm^{-1} and is experimentally observed. The structure of *o*-benzyne$^{•+}$ shows a deformation of the ring making the structure non-planar, which has been previously observed by Kaiser *et al.*[70] using photoelectron spectroscopy. No saturation depletion measurements have been performed due to experimental difficulties, but the recorded relative depletion of the strong and unique 1740 cm^{-1} stretching band can be regarded as

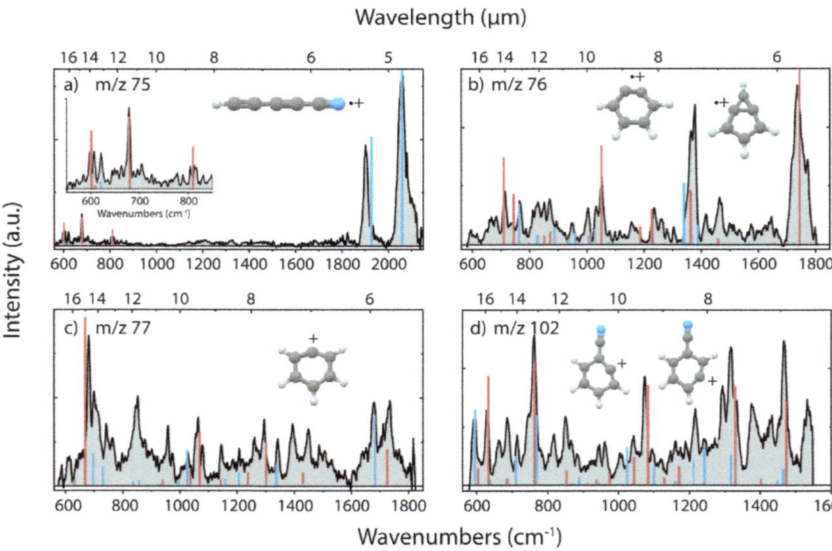

Fig. 4 Experimental infrared fingerprint spectra (grey) of (a) *m*/*z* 75, (b) *m*/*z* 76, (c) *m*/*z* 77 and (d) *m*/*z* 102 fragment channels of the dissociative ionization of benzonitrile. For each panel, the *x*-axis is showing wavenumbers in cm^{-1} (lower axis) and wavelength in μm (upper axis). Calculated vibrational modes are shown for (a) cyano-diacetylene$^{•+}$ ($HC_5N^{•+}$) (red/blue), (b) *o*-benzyne$^{•+}$ ($C_6H_4^{•+}$) (red), bicyclic *m*-benzyne$^{•+}$ ($C_6H_4^{•+}$) (blue), (c) phenyl$^+$ ($C_6H_5^+$) (red), ^{13}C-*o*-benzyne$^{•+}$ ($^{13}CC_5H_4^{•+}$) (blue) and (d) 2-dehydro ($C_7H_4N^+$) (blue) and 3-dehydro-benzonitrile$^+$ ($C_7H_4N^+$) (red) as sticks. The infrared spectra have been calculated at the harmonic B3LYP-GD3/N07D level of theory. The anharmonic infrared spectrum of bicyclic *m*-benzyne$^{•+}$ (*m*/*z* 76) has been calculated to improve the match with the experimental spectrum. The Renner–Teller affected bending modes of $HC_5N^{•+}$ (red, panel (a)) have been determined using an effective Hamiltonian approach and are combined with (calculated) harmonic stretching modes (blue) by Gans *et al.*[69]

a lower limit for the *o*-benzyne$^{\bullet+}$ abundance, indicating the abundance to be larger than 65%.

To confirm the assignment of the other isomer, we have measured the experimental infrared spectrum of the H_2 loss fragment of cationic benzene, as for neutral benzene dissociation it has been calculated that neutral (*o*/*m*/*p*)-benzyne isomers can be formed[71] (Supplementary Fig. 11†). Comparison with anharmonic infrared frequency calculations of both *m*-benzyne$^{\bullet+}$ and a more bicyclic *m*-benzyne$^{\bullet+}$ species shows clear evidence of the existence of the latter. Comparing these reference frequencies to the spectrum of *m*/*z* 76 from benzonitrile confirms the presence of this bicyclic *m*-benzyne$^{\bullet+}$ structure as the other minor fragment (Supplementary Fig. 12†). The most important difference between the two *meta* isomers is that their CC bond lengths differ between 1.968 Å and 1.422 Å (at the B3LYP-GD3/N07D level of theory) for the *m*-benzyne$^{\bullet+}$ and bicyclic *m*-benzyne$^{\bullet+}$, respectively, which changes the infrared spectrum significantly and makes them spectroscopically distinguishable. While for neutral *m*-benzyne the biradicaloid structure is favored to the closed shell bicyclic structure,[72] we observe that for the radical cation the bicyclic structure exists as a stable fragmentation product of benzonitrile.

Two dehydro-benzonitrile^{+} isomers are assigned as the H-loss fragments with *m*/*z* 102 indicating that direct H-loss processes take place without ring opening. This contrasts with the pyridine H-loss channel where two non-cyclic structures are observed. Based on the calculated vibrational modes, we observe predominantly the 2- and 3-dehydro isomers where the hydrogen is lost at the C2 and C3 carbon atom, respectively (Supplementary Fig. 14†). The 3- and 4-dehydro isomers are both similar in energy and the lowest for this channel, however, calculated modes of the latter do not contribute significantly to the experimental spectrum.

In the experimental spectrum of the *m*/*z* 77 fragment, $CN^{\bullet}$- or C_2H_2-loss, two features around 1680 and 1735 cm^{-1} are observed, indicating two structures with different CC stretching vibrations. The 1735 cm^{-1} band and many other features can be associated with the phenyl^{+} structure. This assignment is strengthened by the comparison with experimental infrared frequencies from Wiersma *et al.*[73] The result shows that we observe the loss of $CN^{\bullet}$, rather than the loss of C_2H_2. The other CC stretching mode around 1680 cm^{-1} is likely due to an interesting structure containing a CH group and a 5-membered ring that was observed at the B3LYP-GD3/N07D level of theory (Supplementary Fig. 15†). However, we are more confident to assign this mode to some contribution of the ^{13}C isotopologues of the abundant *o*-benzyne$^{\bullet+}$ fragment that overlaps in this mass channel. Interestingly, also for protonated benzonitrile^{+}, the major fragmentation channel, which is HCN/HNC loss, leads to the formation of phenyl^{+}.[74]

The smallest fragment with *m*/*z* 75 can be assigned to cyanodiacetylene$^{\bullet+}$ ($HC_5N^{\bullet+}$), a structure similar to cyanoacetylene$^{\bullet+}$ as the fragment from pyridine, where in both systems neutral C_2H_4 is lost. This open-shell linear ion is again Renner–Teller affected, but to our knowledge its vibrational spectrum has not been recorded previously. The assignment was therefore based on a combination of harmonic vibrational frequencies for the stretching modes and an effective Hamiltonian approach for the Renner–Teller affected bending modes. Both the harmonic frequencies and the Renner–Teller and spin–orbit parameters were taken from Gans *et al.*[69] (calculated at ic-MRCI+Q/AVTZ level of theory). The intensities here were calculated at the B3LYP-GD3/N07D level of theory. We see no trace of pure carbon species such as cyclic $C_6H_3^+$ which is calculated to be stable but difficult to observe experimentally[75] (Supplementary Fig. 16†).

Potential energy surfaces of HCN/HNC loss

Identifying the structure of the cationic fragments has revealed the formation of many cyclic and non-cyclic structures. The accompanying neutral fragment can be deduced from these structural assignments for some of the channels. However, the exact nature of the neutral counter fragment in the major loss channels leading to methylene-cyclopropene$^{\cdot+}$ (from pyridine) and *o*-benzyne$^{\cdot+}$/bicyclic *m*-benzyne$^{\cdot+}$ (from benzonitrile) cannot be directly deduced from the spectral assignment. Both the loss of HCN and HNC could occur, each leading to different fragmentation mechanisms. To study these different pathways, we have constructed the PESs using DFT calculations, as outlined in the Supplementary Methods,† guided by the experimentally obtained structures, leading to either the HCN or HNC neutral product for both pyridine (Fig. 5) and benzonitrile fragmentation (Fig. 6).

Due to the energy difference between HCN and HNC, HCN being lower by 59 kJ mol^{-1} (0.6 eV),[76,77] the total energy of the cationic fragment and the HNC neutral is higher compared to the HCN loss pathway. However, the potential existence of higher lying transition states on either pathway can influence the kinetics and thereby the observed fragment structures. The additional internal energy after the ionization process can be up to around 5(2) eV, which is sufficient

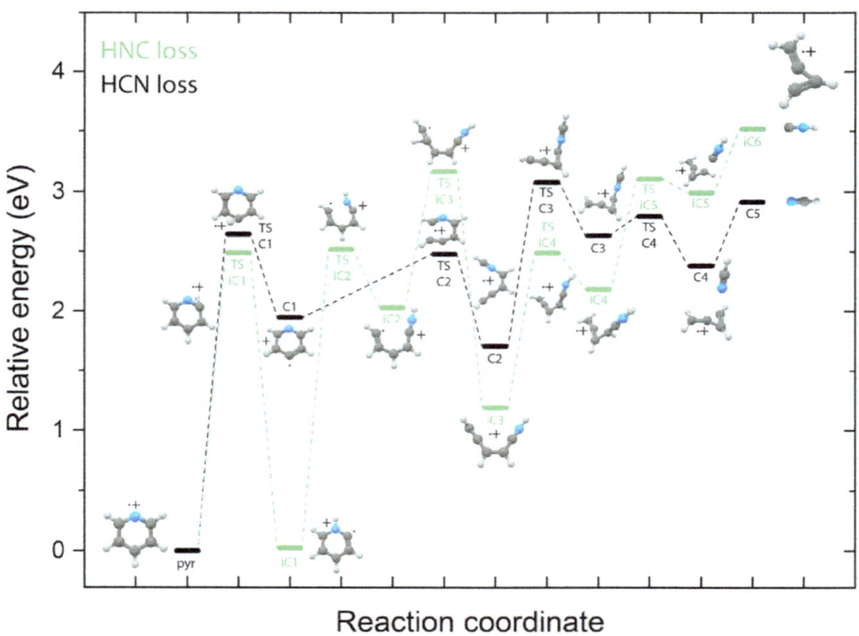

Fig. 5 Potential energy surface for the fragmentation to methylene-cyclopropene$^{\cdot+}$ from pyridine$^{\cdot+}$ leading to either HCN loss (C-pathway, black pathway) or HNC loss (iC-pathway, green pathway). The energies are calculated at the B3LYP-GD3/N07D level of theory and are corrected for the zero-point vibrational energy.

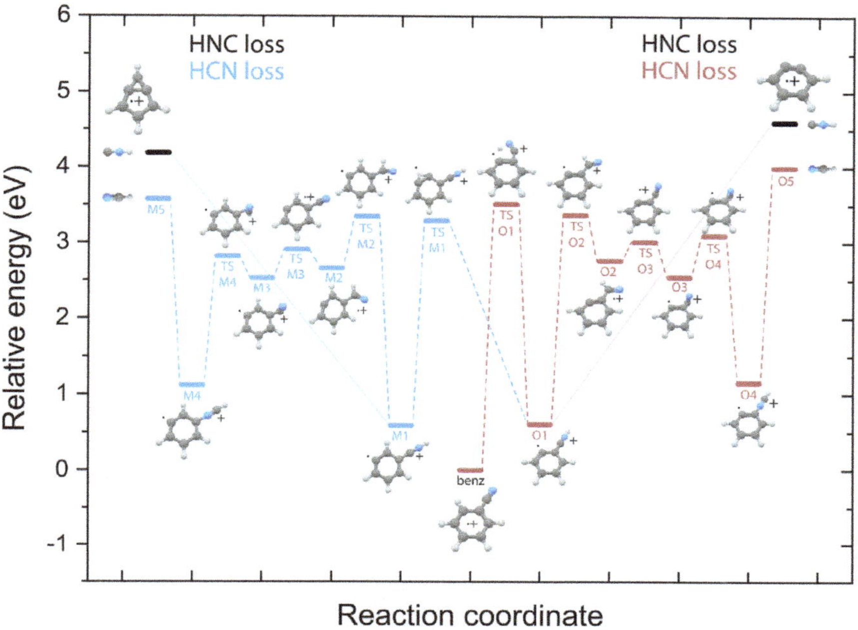

Fig. 6 Potential energy surface for the fragmentation to both *o*-benzyne$^{•+}$ (red pathway) and bicyclic *m*-benzyne$^{•+}$ (blue pathway) from benzonitrile$^{•+}$ leading to either HCN loss (O-pathway, red), (M-pathway, blue) or HNC loss (black energy levels, solid lines). The energies are calculated at the B3LYP-GD3/N07D level of theory and are corrected for the zero-point vibrational energy.

to surmount the calculated barriers in the HCN/HNC loss PESs of pyridine and benzonitrile. Two fragmentation pathways to methylene-cyclopropene$^{•+}$ that lead to the two neutral isomers have been found. Both pathways start with hydrogen migration either to the nitrogen (**TS iC1**) or onto the ring to create a CH_2 moiety (**TS C1**). This is followed by ring opening by **TS C2** and **TS iC2** to yield analogue structures **C2** and **iC3**. The HCN loss pathway has been described earlier by Yim *et al.*[41] showing a reverse barrier for formation (**TS C3**). On the other hand, the HNC loss pathway stays below the energy of the exit channel without any reverse barrier. Furthermore, on the HNC potential, we can locate two potential precursors, **iC2** or **iC3**, for the H-loss product, *cis*-CHCCHCHCNH$^+$, that we have observed experimentally. The relative energy of this H-loss product is calculated at 3.83 eV and can be formed upon the loss of a hydrogen atom from both the **iC2** and **iC3** intermediates. Similarly, using the experimental fingerprint of *m*/*z* 78, we can also exclude analogue H-loss fragments from the HCN pathway intermediate **C2**, iso *cis*-CHCCHCHCNH$^+$ (Supplementary Fig. 7†). For the dissociative ionization of aniline, it has been observed that the H and HNC loss channels are coupled *via* the same intermediate structures.[32] In that case, no direct H-loss was observed, but both H-migration and isomerization steps occurred prior to the loss of a hydrogen atom. Although the end products of the HNC loss pathway are energetically less favourable, the intermediates such as **iC1** and **iC3** are significantly lower in energy than their corresponding structures on the HCN pathway.

Together with the experimental detection of the H-loss product, *cis*-$HCCCHCHCNH^+$, which can exist on the same potential, it points towards the existence of the HNC pathway. This indicates that for pyridine, not the thermodynamically favored products (which is neutral HCN) are formed but the higher lying HNC isomer. However, we cannot exclude the HCN pathway as we do not have experimental evidence on the neutral structure.

Unlike on the PES for the pyridine fragmentation, the HNC loss pathways for benzonitrile do not require many isomerization steps. The fragmentation towards HNC starts with a hydrogen migration step to the nitrogen atom (**TS O1**), and in the case of the *m*-benzyne pathway an additional hydrogen migration on the ring (**TS M1**) is required, and is followed by the direct loss of HNC from the intermediates **O1** or **M1** leading to the *o*-benzyne$^{\bullet+}$ and bicyclic *m*-benzyne$^{\bullet+}$ products, respectively. These two distonic benzonitrile isomers (**O1**/**M1**) exist as reaction intermediates at relatively low energies. For protonated benzonitrile$^+$, where protonation has occurred on the CN group, Wincel *et al.*[78] has shown that upon high internal energies a similar entropically favored direct HNC loss is likely. In the case of the HCN loss pathways, for both cationic fragments the CN group needs to rotate in order to form neutral HCN. This requires multiple transition states, however all lying below the energy of the end products. The HNC loss pathway, where one isomerization step is required followed by the ejection of HNC would be a significantly faster process compared to the loss of HCN that requires more steps prior to dissociation and is expected to be slower due to steric effects, in particular the accessibility of the different atoms within the molecule that are involved in the isomerization process. The identified structures of the H-loss fragments of benzonitrile suggest a direct H-loss mechanism. In contrast to pyridine (and aniline[32]) no precursor for H-loss can therefore be observed in the PES.

Molecular dynamics simulations of fragmentation

We have tried to simulate the observed fragmentation chemistry of pyridine and benzonitrile by running extensive MD/DFTB simulations (see Supplementary Methods†) in order to estimate the capability of such an approach to accurately predict the fragment channels and their accompanying molecular structures.

First, both the fragmentation of the canonical pyridine$^{\bullet+}$ ion and its energetically low-lying alpha-distonic pyridine$^{\bullet+}$ were studied. A maximum contribution of 20% of the alpha-distonic pyridine$^{\bullet+}$ can be expected experimentally[79] which can contribute differently to the fragmentation chemistry. Similar dissociation pathways with the two sets of parameters were found and the results for the alpha-distonic$^{\bullet+}$ ions revealed a better agreement with the structures determined in Fig. 3. With the same energy input, the dissociation of the alpha-distonic pyridine$^{\bullet+}$ ions turned out to be more efficient than that of the canonical pyridine$^{\bullet+}$. The determined branching ratios (BRs) and kinetics for an internal energy of 6.6 eV are reported in Supplementary Fig. 17.†

As can be seen in Supplementary Fig. 17,† the loss of HCN/HNC as the major channel is similar to the experimental results (Fig. 1a). For the pyridine$^{\bullet+}$ fragmentation, both HCN and HNC losses were observed, the majority being the loss of HCN (83% HCN *vs.* 17% HNC with Set2 at 6.6 eV and 79% HCN *vs.* 21% HNC with Set1 at 6.74 eV. Set1 and Set2 refer to two sets of SCC-DFTB parameters, see the Supplementary Methods†). Major loss of HNC was observed from the alpha-distonic pyridine$^{•+}$ isomer (99.8% of HNC at 500 ps). Interestingly, isomerization of pyridine$^{•+}$ into the alpha-distonic pyridine$^{•+}$ isomer was found to occur for Set1 for which the energy difference was found to be in better agreement with the DFT results (see Supplementary Methods†). Two examples of reaction pathways starting from the canonical pyridine$^{•+}$ from the simulations are reported in Supplementary Fig. 19.† Both pathways involve H migration and ring opening to form $C_4H_4^{•+}$ in a non-cyclic form. The cyclic form was occasionally observed in the simulations, but the non-cyclic form was found to be the final one at 1 ns. The remaining internal energy for the majority of the $C_4H_4^{•+}$ fragments was determined to be between 2–2.5 eV. This is sufficient to both induce ring-opening processes, such as calculated for methylene-cyclopropene$^{•+}$ (barrier of 1.8 eV)[61] and further H migration processes (barrier of 1.8 eV) to form other linear species.

Loss of C_2H_4 (or $H_2CN^•$ in one simulation) was found to be a minor fragment channel. Interestingly, the two products with *m*/*z* 51 reported in Fig. 3a were found to be formed from the alpha-distonic pyridine$^{•+}$ (Fig. 7a and b). The simulation also predicts the formation of the isocyanoacetylene$^{•+}$ (Supplementary Fig. 20†), whose formation cannot be completely experimentally ruled out (Supplementary Fig. 9†).

The loss of C_2H_2 was also observed as a minor fragment channel for both pyridine$^{•+}$ and the alpha-distonic pyridine$^{•+}$ (Fig. 7f and Supplementary Fig. 21†). The formed *m*/*z* 53 fragment from pyridine$^{•+}$ contains a similar CCNC backbone as the structure observed experimentally (Fig. 3c), only differing by a H migration. The fragment structure resulting from the alpha-distonic pyridine$^{•+}$ (Supplementary Fig. 21†) is not identified in the experiment (Supplementary Fig. 8†), making it more probable that this experimentally observed fragmentation channel goes *via* the canonical pyridine$^{•+}$ rather than *via* H migration and the alpha-distonic pyridine$^{•+}$ isomer.

The loss of H was observed as a minor channel in all simulations. Both the direct loss of H, preserving the cyclic structure of the *m*/*z* 78 cationic fragment, and the formation of non-cyclic fragments resulting from H migrations and ring opening were observed (Fig. 7c). The simulated non-cyclic structure is observed experimentally as one of the isomers (Fig. 3d).

The experimentally observed structures are more consistent with an initial C–N breaking, favored in the case of alpha-distonic pyridine$^{•+}$, than with an initial C–C breaking, initiated by a H migration towards the carbon atom (Supplementary Fig. 20†). Most of the fragmentation pathways proceed through the formation of the alpha-distonic pyridine$^{•+}$ and subsequent dissociation, as was simulated for the fragments with *m*/*z* 78, 52 and 51. Stronger spectral signatures of the alpha-distonic pyridine$^{•+}$ isomer have been experimentally observed at higher electron energies,[79] which could mean that isomerization and subsequent fragmentation occurs *via* the distonic isomer at the electron energy of 50 eV where the lower abundant H, C_2H_2 and C_2H_4 loss channels have been probed.

Secondly, simulations of the dissociation of benzonitrile$^+$ showed that the major loss channel, similar to pyridine$^+$, is the loss of HCN/HCN (Fig. 8b). Both HCN and HNC losses were observed (Supplementary Fig. 22†). However, for Set1 for which the energy difference between benzonitrile and its alpha-distonic isomer is lower and more consistent with DFT results, HNC is the major product (see Supplementary Table 1†). HCN loss is major for Set2 where the

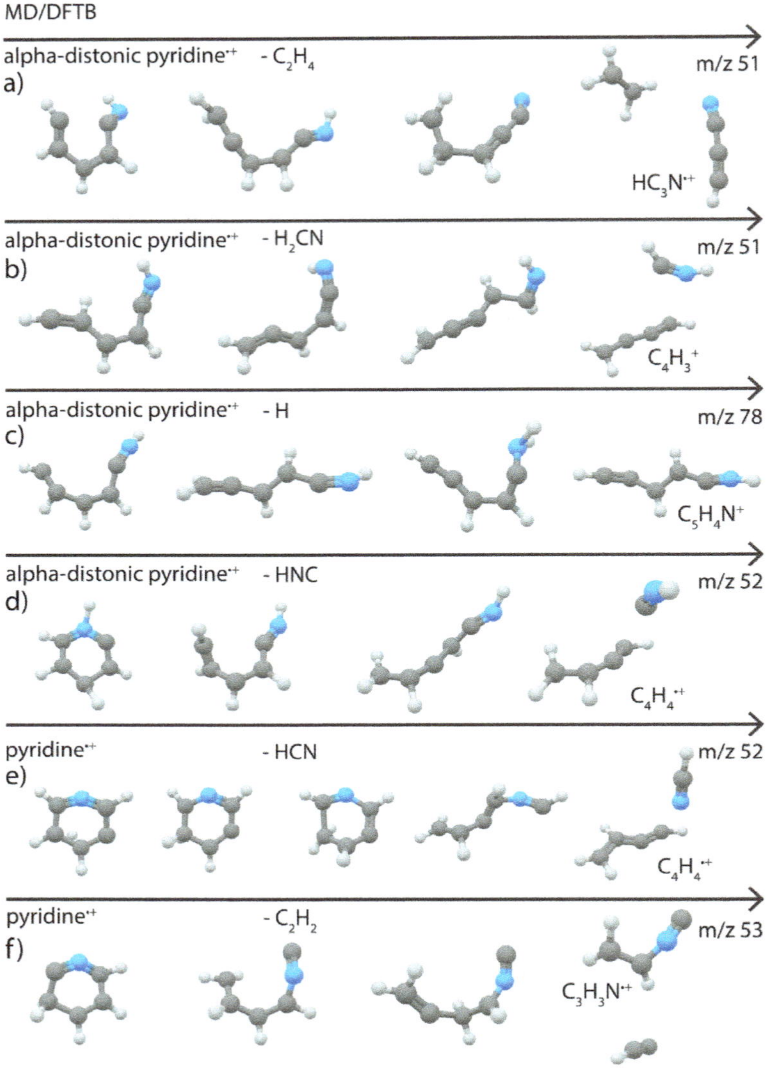

Fig. 7 Exemplary pathways for loss of (a) C_2H_4, (b) H_2CN, (c) H and (d) HNC loss from alpha distonic pyridine$^{\bullet+}$ and (e) HCN and (f) C_2H_2 loss from canonical pyridine$^{\bullet+}$ as observed in the MD/DFTB simulations at 6.74 eV internal energy of the respective cation using Set1 parameters. The arrows indicate time.

formation of the alpha-distonic isomer is energetically less favorable (see Supplementary Methods†). Both the cationic *m*- and *o*-benzyne$^{\bullet+}$ species were observed at the lowest energy for Set1, whereas only *o*-benzyne$^{\bullet+}$ was observed for Set2 (with one non-cyclic form). When the energy increases, non-cyclic forms of $C_6H_4^{\bullet+}$ were observed for both sets. Also, the $H^{\bullet}$ and $CN^{\bullet}$ direct losses were observed as minor channels (see Supplementary Table 1†) to lead to molecular ions with structures similar to those experimentally observed at *m*/*z* 102 and 77, respectively (Fig. 4d and c). Also, the loss of C_2H_4 was observed, leading to the

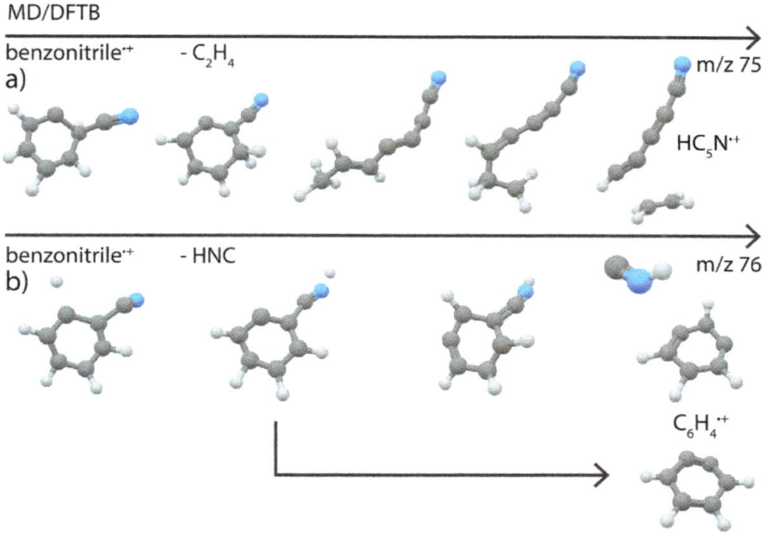

Fig. 8 Exemplary pathways for loss of (a) C_2H_4 and (b) HNC from canonical benzonitrile$^{\bullet+}$ as observed in the MD/DFTB simulations at 8.8 eV internal energy of the respective cation using Set1 parameters. The arrows indicate time.

cyanodiacetylene$^{\bullet+}$ structure that was discovered experimentally for *m*/*z* 75 (Fig. 4a) (exemplary pathway in Supplementary Fig. 8a†).

As for the case of pyridine, the pathways are driven by H migrations that lead to C–C or C–N weakening with opening and formation of a linear chain in the case of the loss of C_2H_4. A few other minor channels that were only abundant in very minor quantities in the experiments were seen in the simulations (Supplementary Table 1†).

This study shows that the MD/DFTB simulations provide fragment channels leading to specific products that are in most cases also experimentally observed. Slightly alternative fragments are also observed in the MD/DFTB simulations, for example structures that encountered different degrees of H-migration processes, which may be due to some uncertainties in the DFTB potential or to differences between the experimental conditions and those of the simulations, *e.g.*, the available internal energy after electron impact ionization. The MD/DFTB simulations can give insights into dissociation mechanisms and fragment structures that could be hardly guessed intuitively due to chemical complexity. The fragment structures can be further optimized with other commonly used DFT methods and their calculated infrared spectra or rotational constants can be compared with experimental data.

Astrophysical implications and conclusions

Although both pyridine and benzonitrile exhibit similar (partial) ionization cross sections,[80] their formed fragments are clearly distinct (*e.g.* cyclic/non-cyclic/ different *m*/*z*). Mapping the different fragmentation products of the two astrochemically relevant molecules pyridine and benzonitrile can help to understand the detection or non-detection in different astronomical environments. Similar to cyano-substituted aromatics that are proxies for their pure carbon forms, abundances of specific 'fragment' species in the investigated top-down processes could point towards the former existence of their parent molecules. For example, the small hydrocarbon ion l-C_3H^+ has been detected in the Horsehead PDR as evidence of top-down hydrocarbon chemistry, where it is expected to be a fragment from larger PAHs.[23] *Vice versa*, the knowledge of the most common fragment products can also help to construct reverse ion-molecule formation processes. For most of the HCN/HNC loss pathways that resulted in the $C_4H_4^{\bullet+}$ and $C_6H_4^{\bullet+}$ fragments for pyridine and benzonitrile, respectively, no reverse barriers have been found in the PESs. Only for the $C_4H_4^{\bullet+}$ + HCN fragment channel a small barrier of 15.9 kJ mol^{-1} (0.16 eV) is present that can slow down the direct reverse formation of pyridine$^{\bullet+}$ when both the cationic and neutral fragment collide reactively.

A good example is the rich chemistry in the cold molecular cloud TMC-1 where recently small hydrocarbons such as *o*-benzyne,[81] cyclopentadiene[82] and (cyano-substituted) PAHs[5,82,83] have been detected. This source was already known for the presence of many large neutral cyanopolyyne molecules such as HC_3N, also isocyanoacetylene HC_2NC (ref. 84) and HC_5N, with an odd number of carbons up to the largest found chain $HC_{11}N$ (ref. 85) and also some of their protonated forms, HC_3NH^+ and HC_5NH^+ (tentatively).[86] It is speculated that larger (linear) chains can undergo cyclization, which is thermodynamically favorable.[87] It is also discussed that the fragmentation products of larger cyclic species can lead to modest-sized molecules that can be important precursors for small aromatic molecules, but are not (accurately) considered in the models.[88] The interplay between these two bottom-up and top-down viewpoints is still not clear. From this study we can conclude a few points that can support this discussion.

Firstly, we can identify two cationic cyanopolyyne species, $HC_3N^{\bullet+}$ and $HC_5N^{\bullet+}$, as fragmentation products of pyridine and benzonitrile, respectively. When starting from species with more carbons, such as cyano-substituted PAHs, we can expect even larger cyanopolyyne species as fragments.

Secondly, the most abundant fragment from the dissociative ionization of benzonitrile is the *o*-benzyne$^{\bullet+}$ radical cation. Interestingly, the neutral form has recently been detected in TMC-1.[81] Besides neutral-radical bottom-up formation reactions, *o*-benzyne could also be formed from dissociative electron recombination of phenyl$^+$,[81] which we here identified as a fragment of benzonitrile. This (cationic) *o*-benzyne structure is not only an important fragment observed here, but also an essential precursor for PAH growth. This has already been demonstrated experimentally where the PAH indene was formed upon the reaction of *o*-benzyne and the allyl radical.[89] The cationic form formed here from benzonitrile can be regarded as even more reactive, either with H_2, possibly forming benzene, or with other small neutral hydrocarbons to form PAHs. In addition, for the H-loss channel of benzonitrile we observe no ring-opening but direct H removal which maintains the cyclic structure of the parent. Altogether, we form different fragments, *o*-benzyne$^{\bullet+}$, phenyl$^+$ and dehydro-benzonitrile$^+$ which are all likely to react with H_2 and/or $CN^{\bullet}$ radicals and/or undergo (dissociative) electron recombination to form molecules such as benzene[90] and sequentially benzonitrile.[91]

For pyridine, the H-loss and C_2H_2 loss fragments are non-cyclic species which are less likely to directly react back to the cyclic parent. They can react with H_2 to form more stable and more saturated carbon or nitrile species but, on the other hand, play a role as reactive intermediates in other bottom-up growth processes. One interesting example shown in this study is the fragmentation product with *m*/*z* 53 from pyridine, assigned to two protonated isocyanide acetylene$^+$ isomers. Interestingly, fragmentation studies of pyrimidine also showed the formation of these isomers containing the CCNC backbone.[92] Experimental and theoretical studies on the $C_3H_3N^+$ PES and ion–molecule reactions of these intermediates have shown that structures with this backbone can be important intermediates for reactions with HCN to form the diazine species pyrazine$^{\bullet+}$ and pyrimidine$^{\bullet+}$.[64,93] All these results combined show that fragmentation of pyridine can lead to potential higher energetic intermediate species that can further react to form species with multiple incorporated nitrogen atoms, thereby leading to even more complexity.

Thirdly, the same interplay between formation and destruction is important for even smaller species such as HCN and HNC. These are among the most abundant species in interstellar space but also with significantly varying relative abundances in different environments. The HNC/HCN ratio can, *e.g.*, differ between ~0.013 in star-forming regions[94] and 0.54–4.5 in dark clouds.[95] There are multiple competitive processes present and it is important, as Loison *et al.*[96] mentioned, to look for new sources and sinks of HCN and HNC in order to understand the different isomeric ratios. Destruction mechanisms are important in the dissociative electron recombination of $HCNH^+$ leading to both HCN and HNC.[97] Also, the different photo-dissociation rates of HNC and HCN can alter the ratio when UV interstellar radiation is present.[98] Moreover, the observationally derived abundances can be influenced by the different collisional excitation cross sections.[99] Analogous to the speculated H_2 formation from the photo-destruction of PAHs in PDRs,[36,100] also the HCN/HNC fragmentation pathway of nitrogen-containing (polycyclic) aromatic molecules could play an important role in those environments. Observations have shown that the HNC/HCN ratio decreases with a factor ~5 towards the star.[101] Also, the $CN^\bullet$ abundance increases towards the star, whereas the abundances of HCN and HNC decrease.

We can speculate that also (small) nitrogen containing molecules such as the ones discussed in this study, pyridine and benzonitrile, can add to the formation of both HCN and HNC. For both precursors we have observed that the dominant fragmentation pathways lead to the loss of HCN/HNC, demonstrating the vulnerability of the nitrogen containing aromatic molecules. This can also been seen if one compares the loss of H as the major channel in the fragmentation of benzene which has a higher appearance energy (AE = 13.7 eV)[102] compared to the loss of HCN/HNC for pyridine (AE = 12.34 eV).[58] Two different fragmentation pathways releasing either HCN or HNC as the neutral can be found for both molecules. If the neutral does not gain sufficient internal energy after the fragmentation, no further isomerization towards the thermodynamically stable HCN species is expected due to the significant barrier of 1.91 eV (184 kJ mol^{-1}).[76] Based on the experimentally determined structures, calculated PESs and the results of the MD/DFTB simulations, we tend to form the higher lying HNC isomer upon the fragmentation process. Also, when using a higher electron impact ionization energy of 50 eV, benzonitrile is fragmented to phenyl$^+$ and $CN^\bullet$ is formed, a process that can also be initiated by UV radiation from stars although the appearance energy of 13.6 eV is close to the H atom ionization energy. But to completely map the abundances of all the molecules correctly, one needs an astrochemical model that includes both bottom-up reactions and top-down processes. This study provides the structural information and branching ratios of the important top-down processes that can be used to construct such a model.

Overall, the experimental approach described here allows identification of the different fragmentation products by measuring their fingerprint infrared spectra to deduce the molecular structures. But in these spectra, more information is present, as they can be used as reference data for astronomical detections. The fingerprint infrared spectra allow to distinguish many different spectral features that are indicative for characteristic groups within the molecules. For example, the CC triple bond stretches of the cyanopolyyne$^{•+}$ structures are strong and informative features to look for in high-resolution astronomical infrared spectra such as those coming from JWST. Also, many of the discovered fragments possess a sufficient permanent dipole moment required for radio-astronomical observations. A particular interesting species is the most abundant fragment of pyridine, the cyclic methylene-cyclopropene$^{•+}$. This molecule is stabilized due to the aromatic character of the three-membered cyclopropenylium moiety.[61] In the atmosphere of Titan, a possible precursor of this species, c-$C_3H_3^+$, cyclopropenyl$^+$, has been detected by *in situ* mass-spectrometry.[103] Possible derivatives of this molecule such as methyl-cyclopropenyl$^+$ ($C_4H_5^+$) or the here identified methylene-cyclopropene$^{•+}$ ($C_4H_4^{•+}$) are discussed to exist as well. Methylene-cyclopropene$^{•+}$ in Titan's atmosphere can also be an important fragmentation proxy for the presence of pyridine, as protonated pyridine has also been found there using mass-spectrometry.[104]

Besides providing infrared reference spectra, and to structurally assign molecules formed in the dissociative ionization, this study also allowed us to benchmark MD-DFTB modeling to predict the structures of the fragments. The good agreement between the outcome of the simulations and the spectroscopic assignments encourages us to propose dissociation pathways for larger and more complex species such as the recently astronomically detected PAH 1-cyanonaphthalene.[5] Exploratory simulations for the dissociation of 1-cyanonaphthlene$^{•+}$ ($C_{11}H_7N^{•+}$) were performed for internal energies of 12.7 eV and 13.6 eV with Set1 parameters (720 simulations of 1 ns were run). At both energies, the major loss channel of HNC/HCN was observed with for instance 105 $C_{10}H_6^{•+}$ fragments at 1 ns at 12.7 eV, where HNC was found to be the major neutral loss channel with 100% HNC and no HCN at 12.7 eV, and 94% HNC and 6% HCN at 13.6 eV. At the higher energy of 13.6 eV, the loss of H, CN$^•$ and C_2H_2 appeared (Supplementary Table 2†). The structures of the potentially important cationic fragments of the HNC loss channel are shown in Fig. 9 (exemplary pathways in Supplementary Fig. 23†), showing both

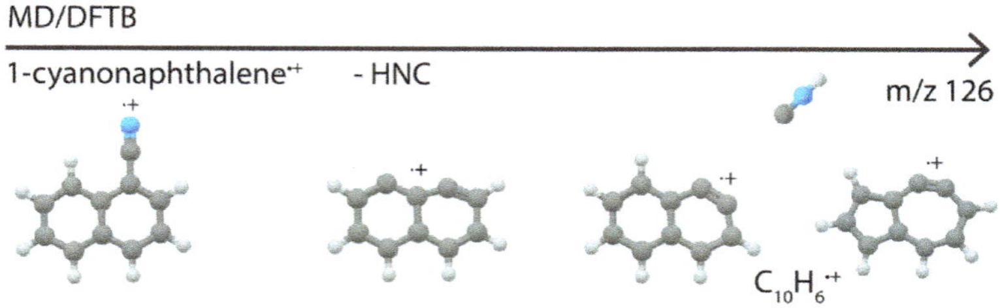

Fig. 9 Cationic fragment structures upon the loss of HNC from canonical 1-cyano-naphthalene$^{•+}$ as observed in the MD/DFTB simulations at 12.7 eV internal energy of the respective cation using Set1 parameters.

didehydronaphthalene (arynes) cation structures, as well as the formation of isomers containing five- and seven-membered rings. These MD/DFTB simulations provide a valuable basis to experimentally study the dissociative ionization of cyanonaphthalenes with the same multi-faceted approach as presented here for the monocyclic pyridine and benzonitrile species, and to investigate if similar structural motifs as observed here can be found in the dissociation of larger N-heterocyclic and cyano-substituted polycyclic aromatic hydrocarbons.

Author contributions

D. B. R.: conception and design of work, data collection, data analysis and interpretation, theoretical calculations, visualization, drafting and editing of article. A. S.: theoretical calculations, data interpretation, editing of article. K. S.: theoretical calculations, data interpretation, editing of article. J. G. M. S.: data collection, data analysis and interpretation, editing of article. B. R.: funding acquisition, project administration, supervision, data interpretation, editing of article. S. B.: conception and design of work, data collection, data interpretation, methodology, project administration, funding acquisition, supervision, drafting and editing of article. All authors discussed the manuscript.

Conflicts of interest

The authors declare no competing interests.

Acknowledgements

We gratefully acknowledge the support of Radboud University and of the Nederlandse Organisatie voor Wetenschappelijk Onderzoek (NWO), for providing the required beam time at the FELIX laboratory and the skilful assistance of the FELIX staff. This work was sponsored by NWO Exact and Natural Sciences for the use of supercomputer facilities at SURFsara in Amsterdam (NWO Rekentijd grant 2021.055). KS and JGMS are supported by the research program "HFML-FELIX: a Dutch Centre of Excellence for Science under Extreme Conditions" (with project number 184.035.011) of the research program "Nationale Roadmap Grootschalige Wetenschappelijke Infastructuur", which is financed by the Netherlands Organisation for Scientific Research (NWO). AS thanks the computing mesocenter CALMIP ("CALculs en Midi-Pyrénées", UMS CNRS 3667) for generous allocation of computing resources (project p17002). We thank the Cologne Laboratory Astrophysics group for providing the FELion ion trap instrument for the current experiments and the Cologne Center for Terahertz Spectroscopy funded by the Deutsche Forschungsgemeinschaft (DFG, grant SCHL 341/15-1) for supporting its operation. We thank Jerry Kamer for the helpful discussions on the benzonitrile dissociation pathways, and Sven Thorwirth for data on the m/z 53 ion from pyridine dissociation taken during his FELIX beamshift.

References

1 M. C. McCarthy, K. L. K. Lee, R. A. Loomis, A. M. Burkhardt, C. N. Shingledecker, S. B. Charnley, M. A. Cordiner, E. Herbst, S. Kalenskii,

E. R. Willis, C. Xue, A. J. Remijan and B. A. McGuire, *Nat. Astron.*, 2021, **5**, 176–180.
2 K. L. Kelvin Lee, P. B. Changala, R. A. Loomis, A. M. Burkhardt, C. Xue, M. A. Cordiner, S. B. Charnley, M. C. McCarthy and B. A. McGuire, *Astrophys. J., Lett.*, 2021, **910**, L2.
3 B. A. McGuire, A. M. Burkhardt, S. Kalenskii, C. N. Shingledecker, A. J. Remijan, E. Herbst and M. C. McCarthy, *Science*, 2018, **359**, 202–205.
4 M. L. Sita, P. B. Changala, C. Xue, A. M. Burkhardt, C. N. Shingledecker, K. L. Kelvin Lee, R. A. Loomis, E. Momjian, M. A. Siebert, D. Gupta, E. Herbst, A. J. Remijan, M. C. McCarthy, I. R. Cooke and B. A. McGuire, *Astrophys. J., Lett.*, 2022, **938**, L12.
5 B. A. McGuire, R. A. Loomis, A. M. Burkhardt, K. L. K. Lee, C. N. Shingledecker, S. B. Charnley, I. R. Cooke, M. A. Cordiner, E. Herbst, S. Kalenskii, M. A. Siebert, E. R. Willis, C. Xue, A. J. Remijan and M. C. McCarthy, *Science*, 2021, **371**, 1265–1269.
6 J. Cernicharo, M. Agúndez, R. I. Kaiser, C. Cabezas, B. Tercero, N. Marcelino, J. R. Pardo and P. de Vicente, *Astron. Astrophys.*, 2021, **655**, L1.
7 T. J. Barnum, M. A. Siebert, K. L. K. Lee, R. A. Loomis, P. B. Changala, S. B. Charnley, M. L. Sita, C. Xue, A. J. Remijan, A. M. Burkhardt, B. A. McGuire and I. R. Cooke, *J. Phys. Chem. A*, 2022, **126**, 2716–2728.
8 S. B. Charnley, Y. J. Kuan, H. C. Huang, O. Botta, H. M. Butner, N. Cox, D. Despois, P. Ehrenfreund, Z. Kisiel, Y. Y. Lee, A. J. Markwick, Z. Peeters and S. D. Rodgers, *Adv. Space Res.*, 2005, **36**, 137–145.
9 P. G. Stoks and A. W. Schwartz, *Geochim. Cosmochim. Acta*, 1982, **46**, 309–315.
10 Z. Martins, *Life*, 2018, **8**, 28.
11 D. S. N. Parker and R. I. Kaiser, *Chem. Soc. Rev.*, 2017, **46**, 452–463.
12 J. Bouwman, A. Bodi and P. Hemberger, *Phys. Chem. Chem. Phys.*, 2018, **20**, 29910–29917.
13 D. B. Rap, J. G. M. Schrauwen, A. N. Marimuthu, B. Redlich and S. Brünken, *Nat. Astron.*, 2022, **6**(9), 1059–1067.
14 D. S. N. Parker, F. Zhang, Y. S. Kim, R. I. Kaiser, A. Landera, V. V. Kislov, A. M. Mebel and A. G. G. M. Tielens, *Proc. Natl. Acad. Sci. U. S. A.*, 2012, **109**, 53–58.
15 L. Zhao, R. I. Kaiser, B. Xu, U. Ablikim, M. Ahmed, M. M. Evseev, E. K. Bashkirov, V. N. Azyazov and A. M. Mebel, *Nat. Astron.*, 2018, **2**, 973–979.
16 S. Doddipatla, G. R. Galimova, H. Wei, A. M. Thomas, C. He, Z. Yang, A. N. Morozov, C. N. Shingledecker, A. M. Mebel and R. I. Kaiser, *Sci. Adv.*, 2021, **7**, 1–12.
17 T. Yang, T. P. Troy, B. Xu, O. Kostko, M. Ahmed, A. M. Mebel and R. I. Kaiser, *Angew. Chem., Int. Ed.*, 2016, **55**, 14983–14987.
18 M. Galand, R. Yelle, J. Cui, J. E. Wahlund, V. Vuitton, A. Wellbrock and A. Coates, *J. Geophys. Res.: Space Phys.*, 2010, **115**, 7312.
19 R. Gredel, S. Lepp, A. Dalgarno and E. Herbst, *Astrophys. J.*, 1989, **347**, 289.
20 T. Allain, S. Leach and E. Sedlmayr, *Astron. Astrophys.*, 1996, **305**, 602–615.
21 D. J. Hollenbach and A. G. G. M. Tielens, *Rev. Mod. Phys.*, 1999, **71**, 173–230.
22 J. Pety, D. Teyssier, D. Fossé, M. Gerin, E. Roueff, A. Abergel, E. Habart and J. Cernicharo, *Astron. Astrophys.*, 2005, **435**, 885–899.
23 V. V. Guzmán, J. Pety, J. R. Goicoechea, M. Gerin, E. Roueff, P. Gratier and K. I. Öberg, *Astrophys. J., Lett.*, 2015, **800**, L33.

24 A. K. Lemmens, D. B. Rap, S. Brünken, W. J. Buma and A. M. Rijs, *Phys. Chem. Chem. Phys.*, 2022, **24**, 14816–14824.

25 M. C. McCarthy, K. L. K. Lee, P. B. Carroll, J. P. Porterfield, P. B. Changala, J. H. Thorpe and J. F. Stanton, *J. Phys. Chem. A*, 2020, **124**, 5170–5181.

26 A. K. Lemmens, D. B. Rap, J. M. M. Thunnissen, B. Willemsen and A. M. Rijs, *Nat. Commun.*, 2020, **11**, 269.

27 M. Alliati, D. Donaghy, X. Tu and J. W. Bradley, *J. Phys. Chem. A*, 2019, **123**, 2107–2113.

28 D. Dubois, N. Carrasco, L. Jovanovic, L. Vettier, T. Gautier and J. Westlake, *Icarus*, 2020, **338**, 113437.

29 A. Simon, M. Rapacioli, G. Rouaut, G. Trinquier and F. X. Gadéa, *Philos. Trans. R. Soc., A*, 2017, **375**, 20160195.

30 J. C. Choe, N. R. Cheong and S. M. Park, *Int. J. Mass Spectrom.*, 2009, **279**, 25–31.

31 K. Ramanathan, S. Arun, J. Bouwman, L. Avaldi, M. V. Vinitha, P. Bolognesi, R. Richter and U. R. Kadhane, *J. Chem. Phys.*, 2022, **157**, 064303.

32 D. B. Rap, T. J. H. H. van Boxtel, B. Redlich and S. Brünken, *J. Phys. Chem. A*, 2022, **126**, 2989–2997.

33 A. J. de Haas, J. Oomens and J. Bouwman, *Phys. Chem. Chem. Phys.*, 2017, **19**, 2974–2980.

34 G. Wenzel, A. Simon, S. Banhatti, P. Jusko, S. Schlemmer, S. Brünken and C. Joblin, *J. Mol. Spectrosc.*, 2022, **385**, 111620.

35 J. Bouwman, A. J. de Haas and J. Oomens, *Chem. Commun.*, 2016, **52**, 2636–2638.

36 S. Panchagnula, J. Bouwman, D. B. Rap, P. Castellanos, A. Candian, C. Mackie, S. Banhatti, S. Brünken, H. Linnartz and A. G. G. M. G. M. Tielens, *Phys. Chem. Chem. Phys.*, 2020, **22**, 21651–21663.

37 S. Banhatti, D. B. Rap, A. Simon, H. Leboucher, G. Wenzel, C. Joblin, B. Redlich, S. Schlemmer and S. Brünken, *Phys. Chem. Chem. Phys.*, 2022, **24**, 27343–27354.

38 B. West, C. Joblin, V. Blanchet, A. Bodi, B. Sztáray and P. M. Mayer, *J. Phys. Chem. A*, 2012, **116**, 10999–11007.

39 A. Candian, J. Bouwman, P. Hemberger, A. Bodi and A. G. G. M. Tielens, *Phys. Chem. Chem. Phys.*, 2018, **20**, 5399–5406.

40 U. R. Kadhane, M. V. Vinitha, K. Ramanathan, A. S., J. Bouwman, L. Avaldi, P. Bolognesi and R. Richter, *J. Chem. Phys.*, 2022, **156**, 244304.

41 M. K. Yim and J. C. Choe, *J. Phys. Chem. A*, 2011, **115**, 3087–3094.

42 C. Lifshitz, *J. Phys. Soc. Jpn.*, 1982, **86**, 1235.

43 J. N. Sousa, L. C. Ribeiro, M. S. Arruda, M. G. P. Homem, H. K. Tanaka, B. C. Credidio, R. R. T. Marinho, A. Medina, I. Prazeres, A. C. F. Santos and F. V. Prudente, *J. Phys. Chem. A*, 2019, **123**, 5164–5170.

44 P. Urbain, B. Leyh, F. Remacle and J. C. Lorquet, *Int. J. Mass Spectrom.*, 1999, **185–187**, 155–163.

45 E. Gridelet, R. Locht, A. J. Lorquet, J. C. Lorquet and B. Leyh, *J. Phys. Chem. A*, 2008, **112**, 10086–10095.

46 J. H. D. Eland, J. Berkowitz, H. Schulte and R. Frey, *Int. J. Mass Spectrom. Ion Phys.*, 1978, **28**, 297–311.

47 J. Momigny, J. Urbain and H. Wankenne, *Bull. Soc. R. Sci. Liege*, 1965, **34**, 337–353.

48 H. Y. So and R. C. Dunbar, *J. Am. Chem. Soc.*, 1988, **110**, 3080–3083.
49 M. A. Baldwin, *Org. Mass Spectrom.*, 1979, **14**, 601–605.
50 A. Maccoll and D. Mathur, *Org. Mass Spectrom.*, 1981, **16**, 261–263.
51 W. J. Chesnavich and M. T. Bowers, *J. Am. Chem. Soc.*, 1980, **102**, 3662.
52 D. Oepts, A. F. G. van der Meer and P. W. van Amersfoort, *Infrared Phys. Technol.*, 1995, **36**, 297–308.
53 P. Jusko, S. Brünken, O. Asvany, S. Thorwirth, A. Stoffels, L. Van Der Meer, G. Berden, B. Redlich, J. Oomens and S. Schlemmer, *Faraday Discuss.*, 2019, **217**, 172–202.
54 M. Elstner, D. Porezag, G. Jungnickel, J. Elsner, M. Haugk, T. Frauenheim, S. Suhai and G. Seifert, *Phys. Rev. B: Condens. Matter Mater. Phys.*, 1998, **58**, 7260–7268.
55 J. P. Gardner, J. C. Mather, M. Clampin, R. Doyon, M. A. Greenhouse, H. B. Hammel, J. B. Hutchings, P. Jakobsen, S. J. Lilly, K. S. Long, J. I. Lunine, M. J. McCaughrean, M. Mountain, J. Nella, G. H. Rieke, M. J. Rieke, H. W. Rix, E. P. Smith, G. Sonneborn, M. Stiavelli, H. S. Stockman, R. A. Windhorst and G. S. Wright, *Space Sci. Rev.*, 2006, **123**(4), 485–606.
56 M. Arimura and Y. Yoshikawa, *J. Mass Spectrom. Soc. Jpn.*, 1984, **32**, 375–380.
57 M. Araki, S. I. Sato and K. Kimura, *J. Phys. Chem.*, 1996, **100**, 10542–10546.
58 P. C. Burgers and J. L. Holmes, *Int. J. Mass Spectrom. Ion Processes*, 1984, **58**, 15–24.
59 K. Steenbakkers, A. N. Marimuthu, B. Redlich, G. C. Groenenboom and S. Brünken, *J. Chem. Phys.*, 2023, **158**, 084305.
60 R. A. Relph, J. C. Bopp, J. R. Roscioli and M. A. Johnson, *J. Chem. Phys.*, 2009, **131**, 114305.
61 K. S. Kharnaior, A. K. Chandra and R. H. Duncan Lyngdoh, *Comput. Theor. Chem.*, 2017, **1115**, 158–168.
62 T. J. Wasowicz and B. Pranszke, *J. Phys. Chem. A*, 2016, **120**, 964–971.
63 T. J. Wasowicz, I. Dabkowska, A. Kivimaki, M. Coreno and M. Zubek, *J. Phys. B: At., Mol. Opt. Phys.*, 2017, **50**, 015101.
64 P. P. Bera, T. J. Lee and H. F. Schaefer, *J. Chem. Phys.*, 2009, **131**, 074303.
65 C. Puzzarini, N. Tasinato, J. Bloino, L. Spada and V. Barone, *Phys. Chem. Chem. Phys.*, 2019, **21**, 3431–3439.
66 N. Takagi, K. Fukuzawa, Y. Osamura and H. F. Schaefer III, *Astrophys. J.*, 1999, **525**, 791.
67 A. Batalov, J. Fulara, I. Shnitko and J. P. Maier, *J. Phys. Chem. A*, 2006, **110**, 10404–10408.
68 K. J. Catani, G. Muller, P. Jusko, P. Theulé, E. J. Bieske and C. Jouvet, *J. Chem. Phys.*, 2017, **147**, 084302.
69 B. Gans, S. Boyé-Péronne and J. Liévin, *J. Chem. Phys.*, 2019, **150**, 244303.
70 D. Kaiser, E. Reusch, P. Hemberger, A. Bodi, E. Welz, B. Engels and I. Fischer, *Phys. Chem. Chem. Phys.*, 2018, **20**, 3988–3996.
71 V. V. Kislov, T. L. Nguyen, A. M. Mebel, S. H. Lin and S. C. Smith, *J. Chem. Phys.*, 2004, **120**, 7008.
72 W. Sander, M. Exner, M. Winkler, A. Balster, A. Hjerpe, E. Kraka and D. Cremer, *J. Am. Chem. Soc.*, 2002, **124**, 13072–13079.

73 S. D. Wiersma, A. Candian, J. M. Bakker, G. Berden, J. R. Eyler, J. Oomens, A. G. G. M. Tielens and A. Petrignani, *Phys. Chem.Chem. Phys.*, 2021, **23**, 4334–4343.
74 U. Jacovella, J. A. Noble, A. Guliani, C. S. Hansen, A. J. Trevitt, J. Mouzay, I. Couturier-Tamburelli, N. Pietri and L. Nahon, *Astron. Astrophys.*, 2022, **657**, A85.
75 R. Peverati, P. P. Bera, T. J. Lee and M. Head-Gordon, *J. Phys. Chem. A*, 2014, **118**, 10109–10116.
76 A. E. DePrince and D. A. Mazziotti, *J. Phys. Chem. B*, 2008, **112**, 16158–16162.
77 C.-F. Pau and W. J. Hehre, *J. Phys. Chem.*, 1982, **86**, 321–322.
78 H. Wincel, R. H. Fokkens and N. M. M. Nibbering, *J. Am. Soc. Mass Spectrom.*, 1990, **1**, 225–232.
79 D. B. Rap, A. N. Marimuthu, B. Redlich and S. Brünken, *J. Mol. Spectrosc.*, 2020, **373**, 111357.
80 D. Gupta, H. Choi, S. Singh, P. Modak, B. Antony, D. C. Kwon, M. Y. Song and J. S. Yoon, *J. Chem. Phys.*, 2019, **150**, 064313.
81 J. Cernicharo, M. Agundez, R. I. Kaiser, C. Cabezas, B. Tercero, N. M. J. R. Pardo and P. de Vicente, *Astron. Astrophys.*, 2021, **652**, 9.
82 J. Cernicharo, M. Agúndez, C. Cabezas, B. Tercero, N. Marcelino, J. R. Pardo and P. De Vicente, *Astron. Astrophys.*, 2021, **649**, L15.
83 A. M. Burkhardt, K. Long Kelvin Lee, P. Bryan Changala, C. N. Shingledecker, I. R. Cooke, R. A. Loomis, H. Wei, S. B. Charnley, E. Herbst, M. C. McCarthy and B. A. McGuire, *Astrophys. J., Lett.*, 2021, **913**, L18.
84 K. Kawaguchi, M. Ohishi, S.-I. Ishikawa, N. Kaifu, K. Kawaguchi, M. Ohishi, S.-I. Ishikawa and N. Kaifu, *Astrophys. J., Lett.*, 1992, **386**, L51.
85 R. A. Loomis, A. M. Burkhardt, C. N. Shingledecker, S. B. Charnley, M. A. Cordiner, E. Herbst, S. Kalenskii, K. L. K. Lee, E. R. Willis, C. Xue, A. J. Remijan, M. C. McCarthy and B. A. McGuire, *Nat. Astron.*, 2021, **5**(2), 188–196.
86 B. A. McGuire, *Astrophys. J., Suppl. Ser.*, 2022, **259**, 30.
87 R. A. Loomis, C. N. Shingledecker, G. Langston, B. A. Mcguire, N. M. Dollhopf, A. M. Burkhardt, J. Corby, S. T. Booth, P. B. Carroll, B. Turner and A. J. Remijan, *Mon. Not. R. Astron. Soc.*, 2016, **463**, 4175–4183.
88 A. M. Burkhardt, R. A. Loomis, C. N. Shingledecker, K. L. K. Lee, A. J. Remijan, M. C. McCarthy and B. A. McGuire, *Nat. Astron.*, 2021, **5**, 181–187.
89 M. N. McCabe, P. Hemberger, E. Reusch, A. Bodi and J. Bouwman, *J. Phys. Chem. Lett.*, 2020, **11**, 2859–2863.
90 V. Vuitton, R. V. Yelle and J. Cui, *J. Geophys. Res.: Planets*, 2008, **113**, 5007.
91 N. Balucani, O. Asvany, A. H. H. Chang, S. H. Lin, Y. T. Lee, R. I. Kaiser, H. F. Bettinger, P. V. R. Schleyer and H. F. Schaefer, *J. Chem. Phys.*, 1999, **111**, 7457.
92 K. Chatterjee and O. Dopfer, *Phys. Chem. Chem. Phys.*, 2020, **22**, 17275–17290.
93 A. M. Hamid, P. P. Bera, T. J. Lee, S. G. Aziz, A. O. Alyoubi and M. S. El-Shall, *J. Phys. Chem. Lett.*, 2014, **5**, 3392–3398.
94 P. Schilke, C. M. Walmsley, G. Pineau Des Forets, E. Roueff, D. R. Flower and S. Guilloteau, *Astron. Astrophys.*, 1992, **256**, 595–612.
95 T. Hirota, S. Yamamoto, H. Mikami and M. Ohishi, *Astrophys. J.*, 1998, **503**, 717–728.

96 J. C. Loison, V. Wakelam and K. M. Hickson, *Mon. Not. R. Astron. Soc.*, 2014, **443**, 398–410.
97 T. Amano, Z. Zelinger, T. Hirao, J. Takano and R. Toyoda, *J. Mol. Spectrosc.*, 2008, **251**, 252–255.
98 A. Aguado, O. Roncero, A. Zanchet, M. Agúndez and J. Cernicharo, *Astrophys. J.*, 2017, **838**, 33.
99 B. M. Hays, D. Gupta, T. Guillaume, O. Abdelkader Khedaoui, I. R. Cooke, F. Thibault, F. Lique and I. R. Sims, *Nat. Chem.*, 2022, **14**(7), 811–815.
100 H. Andrews, A. Candian and A. G. G. M. Tielens, *Astron. Astrophys.*, 2016, **595**, A23.
101 A. Fuente, J. Martin-Pintado, J. Cernicharo, R. Bachiller, A. Fuente, J. Martin-Pintado, J. Cernicharo and R. Bachiller, *Astron. Astrophys.*, 1993, **276**, 473–488.
102 H. M. Rosenstock, K. E. McCulloh and F. P. Lossing, *Int. J. Mass Spectrom. Ion Phys.*, 1977, **25**, 327–341.
103 A. Ali, E. C. Sittler, D. Chornay, B. R. Rowe and C. Puzzarini, *Planet. Space Sci.*, 2013, **87**, 96–105.
104 A. Ali, E. C. Sittler, D. Chornay, B. R. Rowe and C. Puzzarini, *Planet. Space Sci.*, 2015, **109–110**, 46–63.

Faraday Discussions

PAPER

Kinetics of CN ($v = 1$) reactions with butadiene isomers at low temperature by cw-cavity ring-down in a pulsed Laval flow with theoretical modelling of rates and entrance channel branching

Shameemah Thawoos,[a] Gregory E. Hall,[b] Carlo Cavallotti[c] and Arthur G. Suits *[a]

Received 6th February 2023, Accepted 16th March 2023
DOI: 10.1039/d3fd00029j

We present an experimental and theoretical investigation of the reaction of vibrationally excited CN ($v = 1$) with isomers of butadiene at low temperature. The experiments were conducted using the newly built apparatus, UF-CRDS, which couples near-infrared cw-cavity ring-down spectroscopy with a pulsed Laval flow. The well-matched hydrodynamic time and long ring-down time decays allow measurement of the kinetics of the reactions within a single trace of a ring-down decay, termed Simultaneous Kinetics and Ring-down (SKaR). The pulsed experiments were carried out using a Laval nozzle designed for the 70 K uniform flow with nitrogen as the carrier gas. The measured bimolecular rates for the reactions of CN ($v = 1$) with 1,3-butadiene and 1,2-butadiene are $(3.96 \pm 0.28) \times 10^{-10}$ and $(3.06 \pm 0.35) \times 10^{-10}$ cm^3 per molecule per s, respectively. The reaction rate measured for CN ($v = 1$) with the 1,3-butadiene isomer is in good agreement with the rate previously reported for the reaction with ground state CN ($v = 0$) under similar conditions. We report the rate of the reaction of CN ($v = 1$) with the 1,2-butadiene isomer here for the first time. The experimental results were interpreted with the aid of variable reaction-coordinate transition-state theory calculations to determine rates and branching of the addition channels based on a high-level multireference treatment of the potential energy surface. H-abstraction reaction rates were also theoretically determined. For the 1,2-butadiene system, theoretical estimates are then combined with literature values for the energy-dependent product yields from the initial adducts to predict overall temperature-dependent product branching. H loss giving 2-cyano-1,3-butadiene + H is the main product channel, exclusive of abstraction, at all energies, but methyl loss forming 1-cyano-prop-3-yne is 15% at low temperature growing to 35% at 500 K. Abstraction forming HCN and various radicals is important at 500 K and above. The astrochemical implications of these results are discussed.

[a]Department of Chemistry, University of Missouri, Columbia, MO 65211, USA. E-mail: Suitsa@missouri.edu
[b]Chemistry Division, Brookhaven National Laboratory, Upton, NY 11973, USA
[c]Department of Chemistry, Materials, and Chemical Engineering "G. Natta", Politecnico di Milano, Milano 20133, Italy

Introduction

The recent identification of benzonitrile, as well as the first branched alkyl species, isopropyl cyanide, in interstellar environments has propelled interest in understanding reactions leading to nitrogen-rich complex molecules at low temperatures.[1–8] The interstellar medium (ISM) and planetary atmospheres such as that of Titan are both rich in neutral molecules including cyano radicals (CN) and a variety of hydrocarbons. Reactions of CN with unsaturated hydrocarbons can be a key aspect of the growth of small molecules into nitrogen-containing complex organic molecules. Although the temperatures and densities are very different in these different environments, understanding pathways that contribute to such products at low temperature is essential to permit predictions of the temperature-dependent product branching, which will have a significant impact on the chemistry downstream, and is of great fundamental chemical interest as well.

The CRESU technique (a French acronym for reaction kinetics in uniform supersonic flows) has opened up experimental investigation of such reactions in laboratory settings at temperatures of interstellar molecular clouds and planetary atmospheres.[9–13] The CRESU technique creates a "wall-less" reactor at temperatures as low as 15 K by creating uniform supersonic expansion.[14–16] The uniform supersonic flow is achieved by expanding a gas from a high-pressure stagnation reservoir to a low-pressure reaction chamber through a convergent-divergent nozzle commonly known as a Laval nozzle.[9,17] Most of the CRESU work has been performed using laser-induced fluorescence (LIF) detection, which is very sensitive and well-suited to the flow environment. However, there is a rather limited range of species amenable to LIF detection. Recent work in our group and others has expanded the range of detection techniques for application in the supersonic flow environment, including chirped-pulse microwave spectroscopy, VUV photoionization, and VUV-LIF.[18–22] Another recent development from our group has been the implementation of cw-cavity ring-down spectroscopy in the uniform flow (UF-CRDS). As we have recently shown, the hydrodynamic time in the flow is well-matched to the ring-down decay, so that with this approach we can record the full kinetic trace on each 400 μs ring-down, a method termed Simultaneous Kinetics and Ring-down (SKaR).[23] Here we report rates of the reactions of vibrationally excited CN ($v = 1$) with butadiene isomers at 70 K using the UF-CRDS method with SKaR.[24] We combine this experimental investigation with high-level *ab initio* calculations and transition-state theory modeling to interpret the observations.

Theoretical investigations of low-temperature CN radical reactions and their experimental study through advances in spectroscopic detection methods coupled with CRESU[24–26] have provided a solid basis for considering these reactions as a major contributor to the formation of nitrogen-rich compounds in astrochemical environments.[25] Experimental and theoretical findings for reactions of ground-state CN with isomers 1,2-butadiene and 1,3-butadiene suggest the CN adds without a barrier to form long-lived C_5H_6N complexes leading to a range of possible products.[25,27] The results of crossed-molecular beam studies conducted under single collision conditions and supported by electronic-structure calculations suggest a nearly exclusive formation of the thermodynamically less favorable product isomer 1-cyano-1,3-butadiene with minor branching to pyridinyl radicals,[25] although it is not clear whether addition to the C2 carbon giving 2-cyano-1,3-butadiene was considered (*vide infra*). The rates for the reaction of 1,3-butadiene with CN ($\nu = 0$) have been reported for a wide range of temperatures ranging from 1200 K (ref. 26 and 28) to as low as 23 K.[25] The rates measured above room temperature by both Butterfield *et al.*[26] and Gardez *et al.*[28] show a negative temperature dependence, while the experiments conducted at temperatures lower than room temperature using the CRESU apparatus by Morales *et al.*[25] show a fast reaction with a mild positive temperature dependence and good agreement for the room-temperature rate. In contrast, no experimental measurement of the rate of reaction of CN with 1,2-butadiene has been reported to our knowledge. However, theoretical studies carried out by Jamal and Mebel.[27] for the reaction between CN and 1,2-butadiene predicts it can proceed *via* CN addition (to the C1, C2 or C3 carbon, or insertion into the C=C double bonds) or by H abstraction either directly or *via* a CN roaming mechanism. This can be followed by H loss mainly giving 2-cyano-1,3-butadiene, or by methyl loss giving either 1-cyano-prop-3-yne or cyanoallene, with the relative yields sensitive to the nature of the initial adduct. They suggested the dominant product from the CN reaction with 1,2-butadiene is 2-cyano-1,3-butadiene + H, which arises from the pathway leading to CN addition to the C2 carbon atom, but we will investigate this more closely below.

Experimental methods

The experiments were performed using the UF-CRDS apparatus.[24] In brief, the UF-CRDS apparatus is a home-built instrument that couples a highly sensitive near-infrared continuous-wave cavity ring-down spectrometer (cw-CRDS) with a pulsed Laval flow. The CRD spectrometer consists of two high-reflectivity plano-concave mirrors (reflectivity at 7070 cm^{-1} is 99.9965%) separated by 80 cm. A distributed feedback (DFB) diode laser centered at 7070 cm^{-1} is used as the laser source to probe the rotational levels of CN ($\nu = 1$). The laser output is fiber-coupled and most of it is sent through an acoustic optical modulator (AOM), which acts as a fast switch interrupting the laser input and allowing ring-down events. About 10% of the output is sent to a wavemeter that monitors the wavelength continuously. The output laser from the AOM is steered into the cavity using two steering mirrors and a mode-matching lens. The output cavity mirror is mounted on a piezoelectric transducer, which allows the cavity length modulation at a set frequency and amplitude. The transmitted light at the output high-reflectivity mirror of the cavity is focused on to a InGaAs photodiode. All timing, triggering, monitoring, and data storage are done using a multi-function DAQ board and a custom-built LabVIEW code.

The CRD spectrometer is mounted perpendicular to the pulsed Laval flow. The pulsed Laval flow is generated using a home-built, high-throughput piezoelectric stack valve[29] and a 3D printed Laval nozzle. The stack valve rapidly fills up a reservoir of approximately 20 cm^3, and the gas is allowed to expand through a Laval nozzle into a vacuum chamber that is pumped by a turbomolecular pump backed by a dry roots pump. The stack valve, the reservoir, and the Laval nozzle are mounted on a motorized translational stage that could be moved vertically and horizontally. This allows us to move the Laval nozzle with respect to the stationary CRDS and a pitot probe. The pulsed flow generated is characterized using a two-dimensional pitot map, as well as spectroscopic measurements conducted in the flow. In the present study, a nominally 70 K 3D printed Laval nozzle is used. The temperature measured in the isentropic core using the pitot probe was 68 $\pm$ 2 K, and the temperature retrieved using a Boltzmann distribution plot from the spectra recorded for acetylene overtone ($1_0^2 5_0^1$) was 66 $\pm$ 6 K.[24]

Previously reported timing and data acquisition were adopted based on the method introduced by Hippler *et al.*[30] and implemented successfully by many others.[31–34] However, this method was quite susceptible to environmental changes such as vibrations and temperature causing it to skip cavity resonances or drift with respect to the modulation frequency. This limited the throughput of the data acquisition. Thus, a revised version of the cavity modulation and data acquisition was implemented, which will be described in greater detail in a future publication. In brief, the triangular voltage modulation was replaced with sinusoidal voltage modulation. An independent servo feedback loop was implemented to adjust the DC bias for the cavity modulation drive to keep the modulation cycle centered on one transmission fringe, eliminating slow drift of fringe timing. The cavity is modulated at a fixed 50 Hz frequency. The 50 Hz sinusoidal signal and a phase-synchronized timing pulse at a selected, reduced repetition rate for the pulsed flow and excimer photolysis were generated continuously. The pulsed Laval flow is generated using the piezoelectric stack valve, which is held open for 3 ms. The phase for the synchronization pulse is chosen such that the next fringe after the trigger will occur near the central, most uniform portion of the gas pulse. The cavity transmission is monitored constantly by a photodiode. When the signal exceeds a threshold, a trigger pulse is sent to the control input of the radio frequency driver for the AOM, interrupting the injection of resonant laser light into the cavity and initiating a ring-down event. A 400 μs ring-down signal is recorded at 1 MHz sampling rate. The excimer laser is fired to initiate photochemistry at the same rate as the pulsed flow, at a selected time interval (23 μs) after the onset of the first ring-down acquisition following the piezo stack trigger. The delayed photolysis provides a pre-trigger ring-down signal for normalization, in addition to the reference ring-down signals, repeatedly measured during the intervals between successive gas pulses. The precise elapsed time between the stack trigger and the optical triggering at the beginning of the next ring-down fringe is measured and used to accumulate averages of the ring-down data in bins of 200 μs width. When analyzing the CN kinetics, the ring-down events are averaged for two time bins (400 μs) centered within the gas pulse.

High purity $\geq$ 99.0% 1,3-butadiene ($CH_2CHCHCH_2$) and 97% cyanogen bromide (BrCN) were purchased from Sigma-Aldrich, 98% 1,2-butadiene (CH_2-$CCHCH_3$) was purchased from ChemSampCo, Inc., and ultra-high purity nitrogen gas was purchased from Airgas. All the chemicals are used without further purification. BrCN was used as the source of vibrationally excited CN ($\nu = 1$) and the radicals were produced by photolyzing the flow mixture containing BrCN using an excimer laser operated at 248 nm. Photolysis of BrCN has been studied extensively and the literature suggests that the vibrational excitation is higher in 248 nm photodissociation of BrCN compared to that at 193 nm.[35–39] The 248 nm excimer was operated at an average power of 75 mJ per pulse and the laser fluence at the exit of the nozzle is approximately 133 mJ cm^{-2}. The excimer is aligned along the axis of the nozzle and counter-propagates to the flow. To maximize the overlap of the CRDS probing region and the photolysis volume, the rectangular shape of the excimer is rotated using a periscope and loosely focused vertically using a 2 m focal length cylindrical lens. This creates a nearly uniform column of CN ($\nu = 1$) that is rapidly rotationally cooled to the flow temperature and reacts with the diene that is co-seeded in the flow. The BrCN concentration is set such that the maximum estimated CN produced was 2.7×10^{11} molecule per cm^3. This reacting column of gas passes by the probe region where it is detected by CRDS. The rate of reaction is measured by monitoring the loss of CN ($\nu = 1$) by tuning the DFB diode laser at 7070.240 cm^{-1} ($j = 5.5$), R1 branch of the CN ($\nu = 1$) electronic transition $A^2\Pi$–$X^2\Sigma^+$. The entire course of the reaction is monitored during each ring-down pulse, and this is repeated and averaged over ~1200 gas pulses or ~5 minutes at each concentration of the co-reactant.

Experimental results

The measurement of rates of reaction is accomplished by a method described by Brown *et al.*,[23] referred to as SKaR (Simultaneous Kinetics and Ring-down). In the absence of the photolysis pulse and the intra-cavity absorption by CN radicals, the measured ring-down signal is a single-exponential decay with a time constant τ_0. When there is a time-dependent concentration of a resonant absorber inside the cavity, the instantaneous ring-down rate is accelerated relative to the empty cavity, leading to a faster, non-exponential decay that depends on the time-dependent concentration of the absorber, $[A](t)$ according to eqn (1).[23]

$$I(t) = I_0 \exp\left[-c\sigma\frac{L_a}{L}\int_0^t [A](t)\mathrm{d}t - \frac{t}{\tau_0}\right] \quad (1)$$

Here I_0 is the intensity at time zero, c is the speed of light, σ is the absorption cross section of the absorbing species, and $\frac{L_a}{L}$ is the ratio of the absorbing path length to the total cavity length. The dependence on the empty cavity ring-down is removed by computing the ratio of $I(t)$ to the empty-cavity ring-down signal:

$$\mathrm{Ratio}(t) = \frac{I(t)}{I_0\,\mathrm{e}^{-t/\tau_0}} = \exp\left[-c\sigma\frac{L_a}{L}\int_0^t [A](t)\mathrm{d}t\right] \quad (2)$$

The time-dependent concentration can then be formally obtained by differentiating the logarithm of the ratio:

$$[A](t) = \frac{-1}{c\sigma}\frac{L}{L_a}\frac{\mathrm{d}\ln[\mathrm{ratio}(t)]}{\mathrm{d}t} \quad (3)$$

No explicit functional form for the time dependence need be assumed to transform the measured ring-down signals to a kinetic trace with eqn (3), although ratio(t) becomes increasingly poorly determined as the ring-down decay proceeds, and the numerical derivative amplifies the noise, as discussed by Brown *et al.*[23] The time-dependent signal attributable to the absorption of CN ($\nu = 1$) in selected low rotational states is empirically found to follow a sequential growth and decay model of the form

$$[CN](t) = a(e^{-k_1 t} - e^{-k_2 t}) \tag{4}$$

where a is an adjustable amplitude, the rise rate k_2 qualitatively accounts for the rotational thermalization of the initially hot (and unobserved) rotational states of CN ($v = 1$) formed in the photolysis of BrCN,[38,40] and the fall-rate k_1 is nearly negligible in the absence of the hydrocarbon reactant, but increases linearly with added hydrocarbon.

Fig. 1A shows an example of the background-ring-down-corrected ratio(t) for CN ($v = 1$, and $J = 5.5$) in the presence of 1,3-butadiene in a 70 K flow. The inset to Fig. 1A shows the averaged foreground and background ring-down signals used to compute this ratio. In Fig. 1B, the blue trace is derived from the numerical derivative of the logarithm of the ratio. In practice, the fit is performed by adjusting parameters a, k_1 and k_2 in eqn (4), while minimizing the squared error in the integrated form of eqn (2), when compared directly to the experimentally determined ratio(t), thus avoiding the numerical differentiation step in the analysis. The red line fit in Fig. 1B is the rise–fall function from eqn (4) that produces the best fit to the ratio in a time window from the time of photolysis at $t = 23$ μs until $t = 180$ μs.

Measurements were performed with incremental additions of 1,3- or 1,2-butadiene, with reaction conditions as summarized in Table 1. Concentrations of the added hydrocarbons were small enough to avoid significant perturbation to the flow conditions, but high enough to ensure pseudo-first-order conditions, in large excess over the photolytic CN radical concentration.

Fig. 2A and B illustrate the variation of the rise–fall fits with changing concentrations of added 1,3-butadiene and 1,2-butadiene, respectively, in a 70 K flow. The rise rates are nearly independent of the added hydrocarbon, and the peak concentrations of CN ($v = 1$, and $J = 5.5$) occur about 20 μs after laser photolysis in all cases. The decay rates k_1 increase linearly with the added hydrocarbon concentration, as plotted in Fig. 2C and D, according to eqn (5):

$$k_1 = k_b[C_4H_6] + k_0 \tag{5}$$

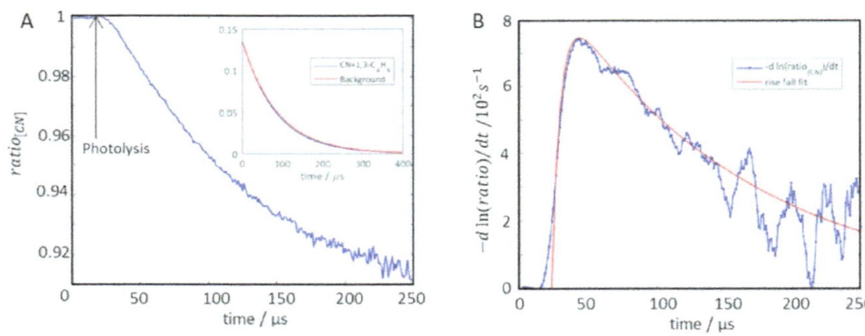

Fig. 1 (A) Background normalized ring-down decay. Inset: Background ring-down (red), ring-down with the absorber, CN ($v = 1$) in the presence of 1,3-butadiene. Arrow indicates when the photolysis laser is fired. (B) Negative derivative of the natural logarithm of the smoothed data from (A) (blue) and the rise–fall fit (red) where the time limits are set at 23 and 180 μs for the lower and upper limits, respectively.

Table 1 Summary of reaction flow conditions and rates obtained for the reaction CN (v = 1) + isomers of butadiene

Isomer	Gas	Flow density/10^{16} molecules per cm^3	Range of [C_4H_6]/10^{13} molecules per cm^3	No. of measurements	Rate/10^{-10} cm^3 per molecule per s
1,2-Butadiene	N_2	5.5	0.81–6.52	7	3.06 ± 0.35
1,3-Butadiene	N_2	5.5	1.42–7.07	7	3.96 ± 0.28

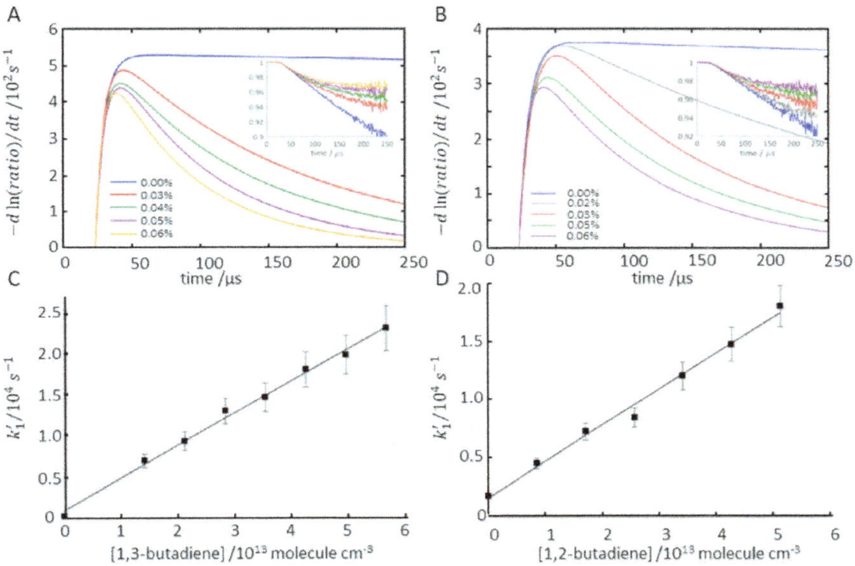

Fig. 2 Selected rise–fall fit for varying (A) 1,3-butadiene and (B) 1,2-butadiene densities. Insets: The respective background normalized ring-down ratio traces. The percent butadiene isomer added is the percent fraction of the flow density, 5.5 × 10^{16} molecules per cm^3. (C) and (D) are respective bimolecular rate plots for pseudo first-order rate *vs.* 1,3 and 1,2 butadiene isomer densities.

The intercept k_0 is the first-order rate of loss of CN ($\nu = 1$) in the absence of co-reactant, likely from side reactions or diffusion of CN ($\nu = 1$) from the probe region. We have observed a significant dependence of k_0 on excimer alignment along the axis of the flow. The slope from Fig. 2C gives a bimolecular rate $k_{1,3}$ for 1,3-butadiene of 3.96 ± 0.28 $\times 10^{-10}$ cm^3 per molecule per s at 70 K. The rate reported by Morales *et al.*[25] for the reaction of CN ($\nu = 0$) with 1,3-butadiene at 70 K is in good agreement with our present measurement, suggesting the low temperature rate of the reaction of CN with 1,3-butadiene has little if any dependence on the vibrational excitation of the CN radical. The rate measured for the reaction of CN ($\nu = 1$) with the 1,2-butadiene isomer ($k_{1,2}$) at 70 K is 3.06 ± 0.35 × 10^{-10} cm^3 per molecule per s (Fig. 2B and D). This is significantly slower than the corresponding rate for the 1,3 isomer. The uncertainty given is at the 95% confidence interval and is dominated by uncertainties in the fits to the first-order decay rates and estimated from the reproducibility of replicate measurements.

Theoretical methods

The rate constants of the entrance and H-abstraction channels of the reaction between CN, 1,2-butadiene and 1,3-butadiene were theoretically determined. Simulations were performed for CN ($\nu = 0$) as it was assumed that the CN vibrational excitation energy does not contribute to the system reactivity. This means that the population of the density of states at the transition state is considered to be adiabatic with respect to the CN vibrational excitation. This assumption is supported by experimental evidence, which shows that the rate constants for the reaction of CN ($\nu = 0$) and CN ($\nu = 1$) are in good agreement for 1,3-butadiene where both measurements have been made. Total rate constants were determined as the sum of contributions from central (two separate sites in the case of 1,2-butadiene, one for 1,3-butadiene) and terminal carbon atom additions, as well as from H-abstraction from all possible sites.

Addition reactions were investigated using variable reaction-coordinate transition-state theory (VRC-TST)[41] to determine reactive fluxes at short and long ranges for each examined reaction channel. The addition rate constant was then computed through a master equation (ME) simulation using the two-transition-state model,[42] which allows determination of a phenomenological rate constant for a system whose reactivity is controlled by the presence of two bottlenecks along the reaction pathway. VRC-TST simulations were performed using the partially automated protocol that is implemented in EStokTP.[43,44] In short-range simulations, pivot atoms are placed along the axis of the bond that is established between the reactants in the minimum-energy structure formed when the fragments are at a 2.4 Å separation and displaced with respect to the reactive centers by 0.001, 0.1, and 0.2 Bohr. Two pivot points are used for butadiene and one for CN, centered on the carbon atom. To restrict the stochastic sampling performed in VRC-TST simulations to the investigated addition site, repulsive potentials were placed on the competitive reaction sites. This is expected to give results equal to those that can be obtained using multifaceted dividing surfaces placing pivot points on each competitive reactive site.[41] Interaction potentials between the fragments were determined on CASPT2(5e,5o)/dz geometries at the CASPT2(11e,10o) level with energies computed with the aug-cc-pVTZ and aug-cc-pVQZ basis sets and extrapolated to the complete basis set limit. VRC-TST simulations were performed at the CASPT2(5e,5o)/dz level using correction potentials for geometry relaxation and high-level energy extrapolation.

Phenomenological rate constants for H-abstraction reactions were determined with the *ab initio* transition-state theory-based master equation approach (AITSTME,[45]), using the automated 1TS model implemented in EStokTP.[43] Geometries and Hessians at the saddle points were determined at the ωB97X-D/jun-cc-pVTZ level, and energies at the CCSD(T) level extrapolated to the complete basis set.[46] A 1D hindered rotor model was used for butadiene internal torsions.

Three different H-abstraction reaction channels were investigated for both 1,2-butadiene and 1,3-butadiene.

Density functional theory simulations were performed using Gaussian 09,[47] CASPT2 and CCSD(T) calculations with Molpro,[48] and ME simulations with MESS.[49]

Computational results

The rate of addition of CN to butadiene is controlled by two dynamic bottlenecks, located at long and short ranges. Short-range interactions control the reactive flux at high temperatures, while at the temperature of the present experiments, 70 K, long-range interactions are dominant over short-range interactions. The potential energy surfaces (PESs) for the addition and abstraction processes for CN ($\nu = 0$) + 1,3-butadiene and CN ($\nu = 0$) + 1,2-butadiene are shown in Fig. 3A and B, respectively.

The total rate constants were multiplied by correction factors of 0.9 and 0.8 to account for recrossing, which is not included in VRC-TST. Though there is some uncertainty in this parameter, which is often system dependent,[45] a good fit is usually obtained using recrossing coefficients of 0.8–0.9. Indeed, the experimental rates measured for 1,3-butadiene, shown in Fig. 4, are in excellent agreement when the recrossing coefficient is set to 0.8, while those for 1,2-butadiene measured in this work agree well when using a recrossing coefficient of 0.9, as shown in Fig. 5. It can also be seen that the total rate constant for the reaction between CN and 1,3-butadiene is well described by the theoretical model for the whole range of temperatures for which experimental data are available. The contributions of the different channels to the total rate constant reported in Fig. 4 and 5 highlight that the most reactive site for addition at low temperatures is the terminal site for 1,3-butadiene and the C2 site for 1,2-butadiene. This is consistent with the PESs of the two reactive systems, as the fastest channels are also the most exothermic. As can be seen, up to about 1000 K, the dominant reaction channel is addition, after which abstraction starts gaining relevance.

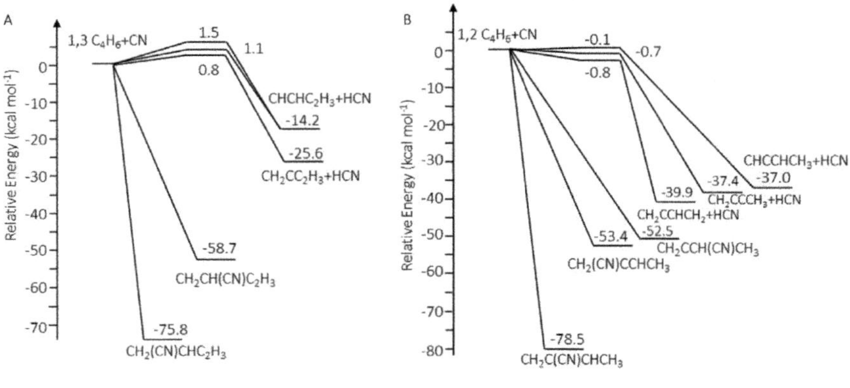

Fig. 3 Calculated PESs for the reaction between CN ($\nu = 0$) and 1,3-butadiene (A) and 1,2-butadiene (B). Energies are reported in kcal mol^{-1} and include zero-point energy corrections.

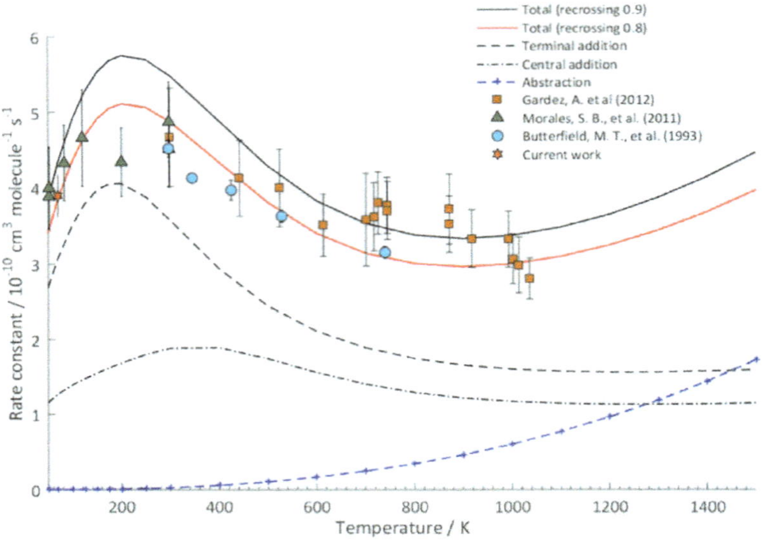

Fig. 4 Comparison between calculated and experimental data (ref. 25, 26, and 28 the present study) for the total rate constant for the reaction between 1,3-C_4H_6 and CN.

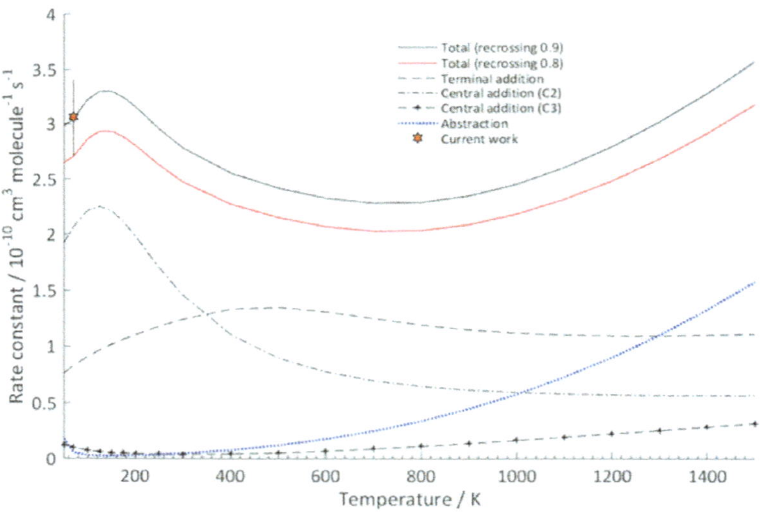

Fig. 5 Comparison between the calculated and experimental rate constant measured in the present work for the reaction between 1,2-C_4H_6 and CN.

Discussion

It is evident that the reactions of CN ($\nu = 1$) with both butadiene isomers at 70 K are quite fast and the reaction with the 1,3 isomer is significantly faster than the 1,2 isomer ($k_{1,3}/k_{1,2}$ is 1.29). Although the conjugated 1,3-butadiene is more than 50 kJ mol^{-1} lower in energy than 1,2 butadiene, it has two accessible terminal sites for addition that can produce conjugated products. These terminal sites are more reactive, and this can contribute to the long-range attraction important at low temperature. Both reactions presented here are believed to be candidates leading to the formation of pyridine, the nitrogen-containing aromatic analog to benzene. However, both quantum chemical calculations[27,50] and crossed molecular beam experiments[50] suggested that the only reaction that possibly could lead to pyridine is the reaction with the 1,3 isomer. However, the reaction with the 1,3 isomer mainly produced the thermodynamically less favorable linear 1-cyano-1,3-butadiene ($CH(CN)CHC_2H_3$) product[25] with branching to pyridine predicted to be less than 0.2% and likely much lower. This is an interesting contrast to the reaction of the isoelectronic CCH radical with 1,3-butadiene, which gives a significant yield of benzene. The difference is chiefly attributed to the much greater stability of the initial adduct in the CCH case.

As noted in the introduction, however, the prediction above assumed solely CN addition to the terminal carbon. Our results examining the entrance channel in detail show significant branching to central addition in 1,3-butadiene: about 25% at 70 K and over 50% at 300 K and above. The subsequent fate of the central addition intermediate is not clear, however, as that region of the PES and subsequent dynamics were not investigated by Morales *et al.*, nor by us. Earlier theoretical treatment by Sun *et al.*[51] suggested that the intermediate formed by C2 addition would isomerize to the C1 adduct, so perhaps this is why C2 addition was neglected. In any case, formation of pyridine from the C2 adduct seems even less likely than from the terminal addition adduct, so the implications for the low-pressure product branching, insofar as pyridine is concerned, are likely unchanged. In any case, this is an interesting subject for further investigation.

For the CN + 1,2-butadiene reaction, although there are no reported experimental studies, there is an analogous theoretical investigation conducted by Jamal and Mebel.[27] For this system, they have performed a detailed investigation of the energy-dependent branching starting from the intermediates formed from addition at the C1, C2, or C3 carbons or by insertion into the double bonds, but without knowing the weighting of the initial addition or abstraction. The product channels predicted by Jamal and Mebel include 1-cyano-prop-3-yne ($CH_2(CN)CCH$) + CH_3 (P1), 2-cyano-1,3-butadiene ($CH_2C(CN)C_2H_3$) + H (P2), and cyanoallene ($CH_2CCH(CN)$) + CH_3 (P3). We can combine our results for the branching of the initial addition step and abstraction with theirs for the decay from the various intermediates to predict the overall branching. The results are presented in Table 2.

Table 2 Temperature-dependent branching (%) for the indicated product channels (see text) combining the present entrance channel branching with calculations of product branching arising from C1, C2 or C3 addition from ref. 30. We take their 0 K value at 50 K and use 1 kcal mol^{-1} ≈ 500 K. H abstraction is summed over all H atom sites

T (K)	P1	P2	P3	Abstraction
50	15	78	1	6
500	35	60	1	4
1000	30	44	3	24
1500	21	31	4	44

At low temperature, CN addition at the C2 carbon is favored nearly 3:1 over the terminal site. According to Jamal and Mebel, C2 addition leads exclusively to the H-loss channel, P2, while addition at the terminal site branches 3:2 for P1 and P2, respectively. As a result, H loss is strongly favored at low temperature. As the temperature is increased, addition at the terminal carbon grows in importance, exceeding that at C2 at around 350 K. At 500 K, C1 is favored over C2 56:33 so that branching to P1 reaches 35%. Abstraction is still almost negligible. As the temperature is further increased, C1 addition continues to grow in importance, and abstraction also comes into play reaching 31% at 1000 K and 41% at 1500 K. H loss is always the dominant product of addition, but still, it is only 60% at 500 K and 31% at 1500 K. Addition at C3 is always minor but increases with temperature, reaching 9% at 1500 K. As this is the only pathway to cyanoallene (P3), this channel never exceeds 4%.

Although the present results are consistent with earlier conclusions that pyridine formation is unlikely in these reactions, the fast reaction rates and the dominance of CN-substituted unsaturated products formed implies the above reactions could be prominent elementary steps in systematic growth of more complex nitrogen-containing compounds detected in the ISM.[1–4,52,53] Although differing in density and collision frequency by many orders of magnitude, these same reactions are certainly important in the chemistry of planetary atmospheres, such as that of Titan, which possess rich chemistry driven by energetic processing of the dense N_2–CH_4 atmosphere.[54–56] Nitriles and unsaturated hydrocarbons are abundant, leading to the formation of aerosol haze layers that have been the subject of intense investigation for decades, while analogous hazes on exoplanets frustrate interrogation of their atmospheres.[57,58] Understanding the barrierless CN addition reactions studied here is part of the ongoing effort to unravel the detailed chemistry of these hazes. Another fascinating aspect of Titan chemistry is its "cryominerology", rocks comprising butadiene, HCN and other species formed as the hazes rain down on the surface.[59,60] Understanding formation of these extraordinary minerals will also benefit from this effort.

As noted in the introduction, isopropyl cyanide was the first branched-chain alkyl species detected in the ISM, following 5 years after detection of the normal species.[8] Modeling shows distinct formation pathways for each. The initial modeling ascribed iso-formation to the CN reaction with 2-propyl radicals in ice mantles, while *n*-propyl cyanide was formed by ethyl radical reaction with CH_2CN. The iso species is now attributed to methyl addition to CH_3CHCN, with the latter formed by CN reaction with ethylene in the ice.[61] In the newer models, the normal species is still formed largely by the same pathway, but CN + propene is also found to play a major role. However, the model restricts branching in that reaction exclusively to *n*-propyl cyanide, otherwise it is found difficult to account for the predominance of the normal isomer.[62] The nature of the initial addition site is thus critical in determining the branching downstream. As we have shown, the temperature dependence of the addition site and addition *vs.* abstraction can vary strongly with temperature.

Conclusions

We have studied the kinetics of the reaction of CN ($\nu = 1$) with 1,2- and 1,3-butadiene at 70 K in a pulsed Laval flow under pseudo-first order conditions. The density of the reactant CN was monitored using cw-CRDS in the near-IR using the SKaR technique. Both reactions were fast, but that with 1,3-butadiene was significantly faster: $(3.96 \pm 0.28) \times 10^{-10}$ and $(3.06 \pm 0.35) \times 10^{-10}$ cm^3 per molecule per s for 1,3- and 1,2-butadiene, respectively. The results were interpreted with the aid of VRC-TST calculations of the entrance channel branching based on multireference *ab initio* calculations of the initially formed adducts, which gave good agreement with the measured rates. The theoretical prediction of the initial adduct formation was combined with previous Rice–Ramsperger–Kassel–Marcus (RRKM) calculations of the subsequent product yields to predict overall product branching. At low temperature, H loss forming 2-cyano-1,3-butadiene is the dominant product as a result of addition at C2, but CH_3 + 1-cyano-prop-3-yne following addition at C1 is also significant and grows in importance with temperature. Addition at C3 is always minor and, as a result, the cyanoallene + CH_3 product channel yield is less than 4%. Abstraction contributes a few percent at low temperature, increasing to 44% at 1500 K.

Data availability

Master equation inputs for all the reactions studied, as well as all the details on stationary points and VRC-TST fluxes necessary to reproduce the simulations are available at https://doi.org/10.5281/zenodo.7695851.

Author contributions

AGS conceived the experiment. ST carried out all experiments, GEH conceived the data acquisition and analysis protocol, and GEH and ST implemented it. CC conceived and performed all theoretical calculations. ST and AGS wrote the paper with contributions from all authors.

Conflicts of interest

There are no conflicts to declare.

Acknowledgements

This work was supported by the NSF under award number CHE-1955239 and by the Italian MUR (PRIN 2020 – Grant 202082CE3T). CC acknowledges the CINECA award HP10BCGTXF, under the ISCRA initiative, for the availability of high-performance computing resources and support. The contributions of GEH, as a senior scientist emeritus at BNL, have not been directly supported by any specific BNL or DOE funded program.

Notes and references

1 B. A. McGuire, A. M. Burkhardt, S. Kalenskii, C. N. Shingledecker, A. J. Remijan, E. Herbst and M. C. McCarthy, *Science*, 2018, **359**, 202–205.

2 M. C. McCarthy, K. L. K. Lee, R. A. Loomis, A. M. Burkhardt, C. N. Shingledecker, S. B. Charnley, M. A. Cordiner, E. Herbst, S. Kalenskii and E. R. Willis, *Nat. Astron.*, 2021, **5**, 176–180.

3 B. A. McGuire, R. A. Loomis, A. M. Burkhardt, K. L. K. Lee, C. N. Shingledecker, S. B. Charnley, I. R. Cooke, M. A. Cordiner, E. Herbst and S. Kalenskii, *Science*, 2021, **371**, 1265–1269.

4 A. Belloche, R. Garrod, H. Müller, K. Menten, C. Comito and P. Schilke, *Astron. Astrophys.*, 2009, **499**, 215–232.

5 F. J. Lovas, J. Hollis, A. J. Remijan and P. Jewell, *Astrophys. J.*, 2006, **645**, L137.

6 J. E. Elsila, J. P. Dworkin, M. P. Bernstein, M. P. Martin and S. A. Sandford, *Astrophys. J.*, 2007, **660**, 911.

7 A. Arnau, I. Tunon, E. Silla and J. Andres, *J. Chem. Educ.*, 1990, **67**, 905.

8 A. Belloche, R. T. Garrod, H. S. Muller and K. M. Menten, *Science*, 2014, **345**, 1584–1587.

9 B. Rowe, G. Dupeyrat, J. Marquette and P. Gaucherel, *J. Chem. Phys.*, 1984, **80**, 4915–4921.

10 B. Rowe, G. Dupeyrat, J. Marquette, D. Smith, N. Adams and E. Ferguson, *J. Chem. Phys.*, 1984, **80**, 241–245.

11 B. Rowe and J. Marquette, *Int. J. Mass Spectrom. Ion Processes*, 1987, **80**, 239–254.

12 I. W. Smith, *Chem. Soc. Rev.*, 2008, **37**, 812–826.

13 M. Fournier, S. D. Le Picard and I. R. Sims, in *Cold Chemistry*, ed. O. Dulieu and A. Osterwalder, Royal Society of Chemistry, 2017, pp. 1–45.

14 I. Sims, J. L. Queffelec, A. Defrance, C. Rebrion-Rowe, D. Travers, P. Bocherel, B. Rowe and I. W. Smith, *J. Chem. Phys.*, 1994, **100**, 4229–4241.

15 I. Sims, I. Smith, D. Clary, P. Bocherel and B. Rowe, *J. Chem. Phys.*, 1994, **101**, 1748–1751.

16 C. Sleiman, G. El Dib, D. Talbi and A. Canosa, *ACS Earth Space Chem.*, 2018, **2**(10), 1047–1057.

17 D. B. Atkinson and M. A. Smith, *Rev. Sci. Instrum.*, 1995, **66**, 4434–4446.

18 J. M. Oldham, C. Abeysekera, B. Joalland, L. N. Zack, K. Prozument, I. R. Sims, G. B. Park, R. W. Field and A. G. Suits, *J. Chem. Phys.*, 2014, **141**, 154202.

19 C. Abeysekera, L. N. Zack, G. B. Park, B. Joalland, J. M. Oldham, K. Prozument, N. M. Ariyasingha, I. R. Sims, R. W. Field and A. G. Suits, *J. Chem. Phys.*, 2014, **141**, 214203.

20 O. Durif, M. Capron, J. P. Messinger, A. Benidar, L. Biennier, J. Bourgalais, A. Canosa, J. Courbe, G. A. Garcia and J.-F. Gil, *Rev. Sci. Instrum.*, 2021, **92**, 014102.

21 S. Soorkia, C.-L. Liu, J. D. Savee, S. J. Ferrell, S. R. Leone and K. R. Wilson, *Rev. Sci. Instrum.*, 2011, **82**, 124102.

22 D. Chastaing, S. D. Le Picard and I. R. Sims, *J. Chem. Phys.*, 2000, **112**, 8466–8469.

23 S. S. Brown, A. Ravishankara and H. Stark, *J. Phys. Chem. A*, 2000, **104**, 7044–7052.

24 N. Suas-David, S. Thawoos and A. G. Suits, *J. Chem. Phys.*, 2019, **151**, 244202.

25 S. B. Morales, C. J. Bennett, S. D. Le Picard, A. Canosa, I. R. Sims, B. Sun, P. Chen, A. H. Chang, V. V. Kislov, A. M. Mebel, X. Gu, F. Zhang, P. Maksyutenko and R. I. Kaiser, *Astrophys. J.*, 2011, **742**, 26.

26 M. T. Butterfield, T. Yu and M. Lin, *Chem. Phys.*, 1993, **169**, 129–134.

27 A. Jamal and A. M. Mebel, *J. Phys. Chem. A*, 2013, **117**, 741–755.

28 A. Gardez, G. Saidani, L. Biennier, R. Georges, E. Hugo, V. Chandrasekaran, V. Roussel, B. Rowe, K. Reddy and E. Arunan, *Int. J. Chem. Kinet.*, 2012, **44**, 753–766.

29 C. Abeysekera, B. Joalland, Y. Shi, A. Kamasah, J. M. Oldham and A. G. Suits, *Rev. Sci. Instrum.*, 2014, **85**, 116107.

30 M. Hippler and M. Quack, *Chem. Phys. Lett.*, 1999, **314**, 273–281.

31 P. Birza, T. Motylewski, D. Khoroshev, A. Chirokolava, H. Linnartz and J. Maier, *Chem. Phys.*, 2002, **283**, 119–124.

32 T. Motylewski and H. Linnartz, *Rev. Sci. Instrum.*, 1999, **70**, 1305–1312.

33 H. Verbraak, A. Ngai, S. Persijn, F. Harren and H. Linnartz, *Chem. Phys. Lett.*, 2007, **442**, 145–149.

34 J. Van Helden, R. Peverall, G. Ritchie, G. Berden and R. Engeln, *Cavity enhanced techniques using continuous wave lasers*, Wiley-Blackwell, West Sussex, UK, 2009.

35 J. A. Russell, I. A. McLaren, W. M. Jackson and J. B. Halpern, *J. Phys. Chem.*, 1987, **91**, 3248–3253.

36 C. Huang, W. Li, R. Silva and A. G. Suits, *Chem. Phys. Lett.*, 2006, **426**, 242–247.

37 W. Fisher, R. Eng, T. Carrington, C. Dugan, S. Filseth and C. Sadowski, *Chem. Phys.*, 1984, **89**, 457–471.

38 I. Nadler, H. Reisler and C. Wittig, *Chem. Phys. Lett.*, 1984, **103**, 451–457.

39 K. Kanda, S. Katsumata, T. Nagata, Y. Ozaki, T. Kondow, K. Kuchitsu, A. Hiraya and K. Shobatake, *Chem. Phys.*, 1993, **175**, 399–411.

40 S. Hay, F. Shokoohi, S. Callister and C. Wittig, *Chem. Phys. Lett.*, 1985, **118**, 6–11.

41 Y. Georgievskii and S. J. Klippenstein, *J. Phys. Chem. A*, 2003, **107**, 9776–9781.

42 E. E. Greenwald, S. W. North, Y. Georgievskii and S. J. Klippenstein, *J. Phys. Chem. A*, 2007, **111**, 5582–5592.

43 C. Cavallotti, M. Pelucchi, Y. Georgievskii and S. J. Klippenstein, *J. Chem. Theory Comput.*, 2019, **15**, 1122–1145.

44 C. Cavallotti, *Proc. Combust. Inst.*, 2023, **39**(1), 11–28.

45 S. J. Klippenstein and C. Cavallotti, in *Computer Aided Chemical Engineering*, Elsevier, 2019, vol. 45, pp. 115–167.

46 B. Hanamirian, A. Della Libera, L. Pratali Maffei and C. CCavallotti, *J. Phys. Chem. A*, 2023, **127**(5), 1314–1328.

47 M. J. Frisch, G. W. Trucks, H. B. Schlegel, G. E. Scuseria, M. A. Robb, J. R. Cheeseman, G. Scalmani, V. Barone, B. Mennucci and G. A. Petersson, *Gaussian 09, Rev. D.01*, Gaussian, Inc., Wallingford, CT, 2009.

48 H. J. Werner, P. J. Knowles, F. R. Manby, J. A. Black, K. Doll, A. Hesselmann, D. Kats, A. Kohn, T. Korona, D. A. Kreplin, Q. Ma, T. F. Miller 3rd, A. Mitrushchenkov, K. A. Peterson, I. Polyak, G. Rauhut and M. Sibaev, *J. Chem. Phys.*, 2020, **152**, 144107.

49 Y. Georgievskii, J. A. Miller, M. P. Burke and S. J. Klippenstein, *J. Phys. Chem. A*, 2013, **117**, 12146–12154.

50 S. B. Morales, S. D. Le Picard, A. Canosa and I. R. Sims, *Faraday Discuss.*, 2010, **147**, 155–171.

51 B. Sun, C. Huang, S. Chen, S. Chen, R. Kaiser and A. Chang, *J. Phys. Chem. A*, 2014, **118**, 7715–7724.

52 M. C. McCarthy and B. A. McGuire, *J. Phys. Chem. A*, 2021, **125**, 3231–3243.

53 K. K. Singh, P. Tandon, A. Misra, M. Yadav and A. Ahmad, *Int. J. Astrobiol.*, 2021, **20**, 62–72.
54 V. Vuitton, R. V. Yelle, S. J. Klippenstein, S. M. Hörst and P. Lavvas, *Icarus*, 2019, **324**, 120–197.
55 J. Loison, E. Hébrard, M. Dobrijevic, K. Hickson, F. Caralp, V. Hue, G. Gronoff, O. Venot and Y. Bénilan, *Icarus*, 2015, **247**, 218–247.
56 S. M. Hörst, *J. Geophys. Res.: Planets*, 2017, **122**, 432–482.
57 S. M. Hörst, C. He, N. K. Lewis, E. M.-R. Kempton, M. S. Marley, C. V. Morley, J. I. Moses, J. A. Valenti and V. Vuitton, *Nat. Astron.*, 2018, **2**, 303–306.
58 C. He, *et al.*, *Planet. Sci. J.*, 2020, **1**, 51.
59 C. Ennis, M. L. Cable, R. Hodyss and H. E. Maynard-Casely, *ACS Earth Space Chem.*, 2020, **4**, 1195–1200.
60 M. L. Cable, T. e. Runčevski, H. E. Maynard-Casely, T. H. Vu and R. Hodyss, *Acc. Chem. Res.*, 2021, **54**, 3050–3059.
61 R. Garrod, A. Belloche, H. Müller and K. Menten, *Astron. Astrophys.*, 2017, **601**, A48.
62 A. Belloche, R. Garrod, O. Zingsheim, H. Müller and K. Menten, *Astron. Astrophys.*, 2022, **662**, A110.

Faraday Discussions

ROYAL SOCIETY OF CHEMISTRY

PAPER

Experimental, theoretical, and astrochemical modelling investigation of the gas-phase reaction between the amidogen radical (NH_2) and acetaldehyde (CH_3CHO) at low temperatures†

Kevin M. Douglas, *[a] Lok Hin Desmond Li, [a] Catherine Walsh, [b] Julia H. Lehman, [c] Mark A. Blitz [ad] and Dwayne E. Heard *[a]

Received 20th February 2023, Accepted 30th March 2023
DOI: 10.1039/d3fd00046j

The first experimental study of the low-temperature kinetics of the gas-phase reaction of NH_2 with acetaldehyde (CH_3CHO) has been performed. Experiments were carried out using laser-flash photolysis and laser-induced fluorescence spectroscopy to create and monitor the temporal decay of NH_2 in the presence of CH_3CHO. Low temperatures relevant to the interstellar medium were achieved using a pulsed Laval nozzle expansion. Rate coefficients were measured over the temperature and pressure range of 29–107 K and 1.4–28.2 $\times$ 10^{16} molecules per cm^3, with the reaction exhibiting a negative temperature dependence and a positive pressure dependence. The yield of CH_3CO from the reaction has also been determined at 67.1 and 35.0 K, by observing OH produced from the reaction of CH_3CO with added O_2. *Ab initio* calculations of the potential energy surface (PES) were combined with Rice–Rampsberger–Kessel–Marcus (RRKM) calculations to predict rate coefficients and branching ratios over a broad range of temperatures and pressures. The calculated rate coefficients were shown to be sensitive to the calculated density of states of the stationary points, which in turn are sensitive to the inclusion of hindered rotor potentials for several of the vibrational frequencies. The experimentally determined rate coefficients and yields have been used to fit the calculated PES, from which low-pressure limiting rate coefficients relevant to the ISM were determined. These have been included in a single-point dark cloud astrochemical model, in which the reaction is shown to be a potential source of gas-phase CH_3CO radicals under dark cloud conditions.

[a]*School of Chemistry, University of Leeds, Leeds, LS2 9JT, UK. E-mail: k.m.douglas@leeds.ac.uk; d.e.heard@leeds.ac.uk*
[b]*School of Physics and Astronomy, University of Leeds, Leeds, LS2 9JT, UK*
[c]*School of Chemistry, University of Birmingham, Edgbaston, B15 2TT, UK*
[d]*National Centre for Atmospheric Science (NCAS), University of Leeds, Leeds, LS2 9JT, UK*

† Electronic supplementary information (ESI) available. See DOI: https://doi.org/10.1039/d3fd00046j

1. Introduction

An active field of astrochemical research remains the understanding of the reaction mechanisms for the formation of complex organic molecules (COMs), which are molecules observed in space containing carbon with at least 6 atoms.[1] Since the early 1960s, the development of ground based radio astronomy has led to a rapid rate of detection of COMs with a wide range of functional groups in the interstellar medium (ISM).[2] The detection of COMs at very high spatial resolution in cold objects suggests a gas-phase route to their formation rather than just *via* a pure surface-grain chemistry mechanism.[3] However, the mechanisms for forming many gas-phase COMs are still unknown, with both gas-phase and grain-surface (*via* non-thermal desorption) routes possible.

The formation of biomolecules such as amino acids from COMs under pre-biotic conditions has attracted much interest.[4] Amino acids such as glycine (NH_2CH_2COOH), which has been observed in the coma of a comet,[5] are building blocks for proteins, which are essential components of all living systems. The peptide bond (NH–C=O) the plays a key role in the linking of amino acids into peptide chains and proteins, and contains all the components necessary for the formation of nucleic polymers under prebiotic conditions.[6] However, only two molecules containing this bond have been observed in space, formamide (NH_2CHO),[7] and acetamide ($NH_2C(O)CH_3$).[8]

In a previous paper[9] we investigated the low-temperature kinetics of the gas-phase reaction between the amidogen radical (NH_2) and formaldehyde (CH_2O) as a possible source of interstellar formamide. In this paper, we have performed an experimental and theoretical study of the kinetics of the reaction of NH_2 with acetaldehyde, the next aldehyde in the homologous series after formaldehyde, for which there are three energetically favourable channels (ΔH calculated in this study, see Section 4):

$$NH_2 + CH_3CHO \rightarrow NH_3 + CH_3CO,\ \ \Delta H^\circ_{0\ K} = -72\ \left(kJ\ mol^{-1}\right) \qquad (R1a)$$

$$NH_2 + CH_3CHO \rightarrow NH_3 + CH_2CHO,\ \ \Delta H^\circ_{0\ K} = -43\ \left(kJ\ mol^{-1}\right) \qquad (R1b)$$

$$NH_2 + CH_3CHO \rightarrow NH_2COCH_3 + H,\ \ \Delta H^\circ_{0\ K} = -30\ \left(kJ\ mol^{-1}\right) \qquad (R1c)$$

Reaction (R1) therefore, is both a potential source of gas-phase organic radicals (CH_3CO and CH_2CHO *via* reactions (R1a) and (R1b), respectively), and of acetamide (NH_2COCH_3, *via* reaction (R1c)), the largest interstellar molecule with a peptide bond that has so far been observed in space.[8] We present measurements of rate coefficients for the reaction of NH_2 and CH_3CHO in the range 29–107 K, using a pulsed laser photolysis-laser induced fluorescence (PLP-LIF) technique coupled with a Laval nozzle to achieve the low temperatures relevant to the ISM. We also report an experimental measurement of the absolute branching yield for reaction (R1a) generating NH_3 + CH_3CO products at 35 K and 67 K. We also performed a theoretical investigation of the reaction using *ab initio* methods to calculate the potential energy surface, and used reaction rate theory to calculate both rate coefficients and product branching ratios over the temperature range 20–400 K. These kinetic data are then incorporated into a single point dark cloud astronomical model to assess the impact of the newly determined rate coefficients. A more detailed introduction to the low-temperature kinetics of gas-phase reactions, to the astrochemistry of acetamide, and to the related reaction between NH_2 and formaldehyde, is given in the ESI in Section S1.†

2. Methodology

2.1 Experimental study

The use of a Laval nozzle expansion coupled with a PLP-LIF technique has been employed by this group to study a range of low temperature neutral–neutral reactions, including reactions of OH with unsaturated hydrocarbons[10] and VOCs,[11–14] $^1CH_2$ (singlet methylene) with atmospheric gases and hydrocarbons,[15,16] CH with CH_2O,[17] and NH_2 with CH_2O.[9] In the current study, we employ the same technique to study the low temperature reaction of NH_2 with CH_3CHO. As the experimental apparatus employed in the current and previous studies has been discussed in detail elsewhere,[10–13] only a very brief overview is given here, with additional details given in the ESI in Section S2.†

The low temperatures employed in this study were obtained by the isentropic expansion of a gas mixture from a high pressure reservoir to a low pressure chamber through a Laval nozzle, producing a thermalized low temperature gas flow that is stable for several centimetres from the nozzle exit. A set of 4 nozzles were used during the experiments to achieve flow temperatures of between 29 and 107 K. The density and temperature profile of the flows were characterized by impact pressure measurements, and the temperature of several of the uniform flows was confirmed by rotationally resolved LIF spectroscopy.[15,17]

The gas mixture introduced to the reservoir consists primarily of an inert bath gas (Ar (99.9995%, BOC), He (99.9995%, BOC), or N_2 (99.9995%, BOC)), together with smaller amounts of the NH_2 precursor (NH_3 (99.98%, BOC)), the CH_3CHO co-reagent (Sigma-Aldrich, ≥99.0%), and CH_4 (99.995%, BOC). These reagent and bath gases were combined in the required quantities in a mixing manifold using calibrated mass flow controllers (MFCs; MKS Instruments). Both the bath gases and the CH_4 were introduced as pure gases, while the NH_3 was introduced as a dilute mixture of ~10% in Ar. The dilute NH_3 flow typically accounted for less than 0.2% of the total flow, resulting in ~0.02% NH_3 present in the gas expansions. The CH_3CHO co-reagent was introduced by entraining CH_3CHO vapour in a flow of the bath gas passing through a glass bubbler (a modified Dreschel bottle) containing CH_3CHO liquid. By measuring the pressure of the bath gas over the CH_3CHO, the concentration of CH_3CHO entrained in the gas flow was determined *via* its known vapour pressure,[18] and confirmed by UV absorption spectroscopy (see ESI Section S2†).

NH_2 radicals were generated from the PLP of NH_3 at 213 nm (Reaction (R2)) by the 5th harmonic of a Nd:YAG laser (Quantel Q-Smart 850), with a typical pulse energy of ~10 mJ. NH_2 radicals were observed by time-resolved LIF spectroscopy, probing the $A\,^2A_1$ (0,9,0) ← $X\,^2B_1$ (0,0,0) transition near 597.7 nm (ref. 19 and 20) using the output of a Nd:YAG pumped dye laser (a Quantel Q-smart 850 pumping a Sirah Cobra-Stretch). The non-resonant fluorescence at ~620 nm was collected *via* a series of lenses through an optical filter (Semrock Brightline interference filter, $\lambda_{max} = 620$ nm, fwhm = 14 nm), and observed by a temporally gated channel photomultiplier (CPM; PerkinElmer C1952P), mounted at 90° to both laser beams. The temporal evolution of the LIF signal was recorded by varying the time delay between the photolysis and probe lasers. A typical time-resolved LIF profile (Fig. 1) consisted of 165 delay steps and resulted from the average or between 5 and 15 individual delay scans.

$$NH_3 + \hbar\nu\ (213\ \text{nm}) \rightarrow NH_2 + H \tag{R2}$$

Photolysis of NH_3 produces both ground and vibrationally excited NH_2,[20,21] and as a consequence of this the profiles of NH_2 ($\nu = 0$) exhibit a growth resulting from the relaxation of vibrationally excited NH_2 (see Fig. 1). As kinetics measurements are limited by the dynamic time of the low temperature expansions, it was important that this relaxation was as efficient as possible, in order to maximise the time in which we could observe the loss of NH_2 ($\nu = 0$). As such, CH_4, which has been shown to efficiently relax NH_2 ($\nu > 0$),[9] was added to our gas flows.

Experiments looking at the dimerization of CH_3CHO at low temperatures were conducted using the same experimental apparatus as described above. CH_3CHO was observed by probing the $^3(n,\pi^*)$ $A\ ^3A'' \leftarrow X\ ^1A'$ transition at 308.0 nm,[22] using the frequency doubled output of a Nd:YAG pumped dye laser (same system as described above with a BBO doubling crystal). The non-resonant fluorescence at $\lambda > 390$ nm was discriminated using a long pass Perspex filter.

Some additional experiments were also carried out observing the production of OH radicals from the reaction of CH_3CO with O_2 (R4), in order to obtain the yield of CH_3CO produced from reaction (R1) (*via* channel (R1a)). As a part of these experiments, the temporal removal of OH by CH_3CHO, both with and without O_2 present, was also recorded, with the OH produced by the PLP of *tert*-butyl

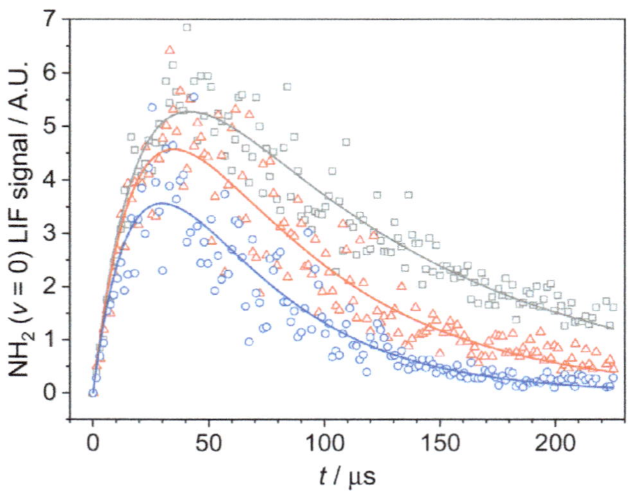

Fig. 1 NH_2 ($v = 0$) temporal traces collected at 52.2 K, a total Ar density of 6.2×10^{16} molecules per cm^3, $[NH_3] = 2.5 \times 10^{13}$ molecules per cm^3, $[CH_4] = 1.8 \times 10^{15}$ molecules per cm^3, and $[CH_3CHO]$ of 1.3, 2.8 and 5.1×10^{14} molecules per cm^3 (grey squares, red triangles, and blue circles respectively). Solid lines are the least squares fitting of a bi-exponential function to the traces from which k'_{obs} is obtained (see Section 3.1 for details).

hydrogen peroxide ($(CH_3)_3COOH$; *t*BuOOH). These experiments were all conducted using the same experimental apparatus as described above. The *t*BuOOH precursor was introduced in the same method as the CH_3CHO co-reagent, by entraining its vapour in a flow of bath gas passing through a bubbler containing *t*BuOOH liquid (Alfa-Aesar, 70% aqueous solution). The concentration of *t*BuOOH in our low temperature flows was estimated using its known vapour pressure at room temperature, and confirmed by UV absorption spectroscopy (see ESI Section S2†). OH radicals were observed by probing the $A^2\ \Sigma^+\ (v' = 0) \leftarrow X^2\ \Pi\ (v'' = 0)$ transition at ~307.9 nm, using the laser system as described above. The resonant fluorescence was collected using a gated photomultiplier tube, in which the scattered laser probe light was gated out, fitted with an interference filter centred at 308.5 nm (Barr Associates, fwhm = 5 nm).

2.2 Theoretical calculations

All *ab initio* electronic structure calculations were carried out using the Gaussian 09 programme.[23] Geometric structures of the stationary points involved (reactants, products, and intermediates including adducts, pre-reaction complexes (PRCs), and transition states (TSs)) were optimised at the M062X/aug-cc-pVTZ level of theory,[24–27] using the ultrafine integration grids in Gaussian. Rotational constants, harmonic vibrational frequencies, and zero-point energies (ZPEs) were obtained from the same level of theory. TSs were found to have only one imaginary vibrational frequency, while for the reactants, products and intermediates, all of the vibrational frequency values were positive. ZPEs obtained from the harmonic frequencies were corrected with a scaling factor of 0.956 for M062X/aug-cc-pVTZ.[28] Intrinsic reaction coordinate (IRC) calculations were performed for all the TSs identified to verify their linkage with the respective local minima on the PES. More accurate single-point energy values were obtained at the CCSD(T)/aug-cc-pVTZ level[29,30] upon the optimised structures. RRKM calculations were performed using the Master Equation Solver for the Multi-Energy Well Reactions (MESMER) program.[31]

3. Results

3.1 Kinetics

Typical NH_2 ($v = 0$) LIF temporal profiles produced following the PLP of NH_3 can be seen in Fig. 1. As can be seen from these profiles there is an initial growth in the NH_2 signal, with no instant signal observed. As discussed above, this growth in the NH_2 ($v = 0$) signal is due to relaxation of vibrationally excited NH_2 produced following the photolysis of NH_3. Although the growth of the NH_2 ($v = 0$) signal is not strictly a single exponential growth (due to the presence of several vibrationally excited states relaxing down), we were able to fit the NH_2 traces satisfactorily treating the growth as a single process:

$$NH_2\ (v \geq 1) + X \xrightarrow{k_{rel}} NH_2\ (v = 0) + X \quad (R3)$$

where X can be CH_4, the bath gas, NH_3, and the co-reagent R under investigation. In practice, relaxation by the bath gas (Ar, He, or N_2) is slow,[32,33] necessitating the addition of CH_4 to our flows to promote efficient vibrational relaxation. Together with the reaction for the removal of NH_2 ($v = 0$):

$$NH_2\ (v=0) + CH_3CHO \xrightarrow{k_1} \text{products} \quad (R1)$$

and as experiments were carried out under pseudo-first-order conditions (*i.e.* $[NH_2] \ll [CH_3CHO]$ and $[CH_4]$), the temporal evolution of the NH_2 ($v = 0$) LIF signal (which is proportional to the NH_2 ($v = 0$) concentration) is given by:

$$[NH_2\ (v=0)]_t = \left(\frac{k'_{rel}}{k'_{obs} - k'_{rel}}\right)[NH_2\ (v \geq 1)]_0\left(\exp^{-k'_{rel}t} - \exp^{-k'_{obs}t}\right) \quad (E1)$$

and

$$k'_{obs} = k_1[R] + k'_{loss} \quad (E2)$$

where k'_{rel} and k'_{obs} are the pseudo-first-order rate coefficients for the reactions producing and removing NH_2 ($v = 0$), $[NH_2\ (v \geq 1)]_0$ is the initial amount of NH_2 ($v \geq 1$) produced following photolysis of NH_3 that is subsequently relaxed down to NH_2 ($v = 0$), and k'_{loss} is the total rate coefficient for other loss processes of NH_2 ($v = 0$), such as diffusion out of the probe laser beam volume. Eqn (E1) was fitted to the NH_2 ($v = 0$) temporal profiles, and the parameters k'_{rel}, k'_{obs}, and $[NH_2\ (v \geq 1)]_0$ extracted. As can be seen from Fig. 1, the NH_2 traces are satisfactorily fit using a biexponential function, indicating the validity of treating the growth of the NH_2 ($v = 0$) signal as a single process. Plotting k'_{obs} *vs.* $[CH_3CHO]$ should then yield a straight line (eqn (E2)), with a gradient equal to the bimolecular rate constant, k_1, and intercept k'_{loss}. An example of such a bimolecular plot can be seen in Fig. 2. As can be seen from Fig. 2, in some of our bimolecular plots we observe a curvature in the k'_{obs} values at high $[CH_3CHO]$, which we attribute to the formation of CH_3CHO dimers in our low temperature flows. Depending on the reactivity of the dimer with NH_2 compared to the monomer, the bimolecular plots may curve

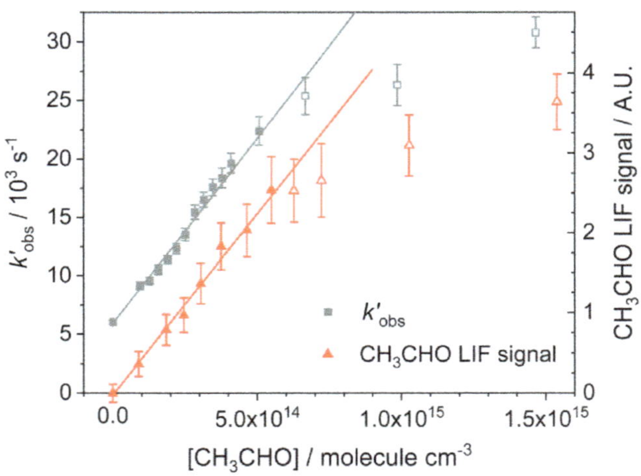

Fig. 2 Left axis: bimolecular plot of k'_{obs} *vs.* $[CH_3CHO]$ for the reaction between NH_2 + CH_3CHO (R1). Right axis: dimerization experiments observing CH_3CHO LIF signal *vs.* $[CH_3CHO]$. The deviation from linearity above approximately 5.5×10^{14} molecules per cm^3 CH_3CHO in both plots indicates significant dimer formation above this concentration. All data collected at 52.2 K and a total Ar density of 6.2×10^{16} molecules per cm^3.

upwards or downwards, or display no curvature if the dimer removes NH_2 at exactly twice the rate as the monomer. In order to help identify at which CH_3CHO concentration significant amounts of dimers form, experiments were carried out monitoring the relative concentration of CH_3CHO in the flow *via* laser-induced fluorescence as a function of the [CH_3CHO] added (see Fig. 2). In regions in which little or no dimerization is occurring, the amount of CH_3CHO monomer present in the flow will increase linearly with the [CH_3CHO] added; as such the CH_3CHO LIF signal will increase linearly (closed red triangles in Fig. 2). However, in regions in which CH_3CHO dimerization is occurring, the amount of CH_3CHO monomer present in the flows will actually be less than the [CH_3CHO] added, and as such the CH_3CHO LIF signal will begin to curve over with increasing [CH_3CHO] (open red triangles in Fig. 2). These dimerization experiments were carried out for each low temperature uniform flow employed in this study. When determining the bimolecular rate coefficient, k_1, from a bimolecular plot, only data in the linear region below the CH_3CHO concentration at which dimers formed were used (solid black squares in Fig. 2). As can be seen from Fig. 2, both the bimolecular plot and the dimerization plot exhibit the same shape with increasing [CH_3CHO] and are in good agreement on the concentration at which CH_3CHO dimers begin to form (above 6×10^{14} molecules per cm^3).

The bimolecular rate coefficients for the reaction of NH_2 with CH_3CHO (R1) determined in this study are presented in Table 1. The errors reported are the 1σ confidence intervals of the linear least squares fits of the bimolecular plots. A selection of these results recorded using He as a bath gas are shown in Fig. 3. Presenting only those rate coefficients collected in a particular bath gas and at the same approximate pressure, it can be seen that the reaction exhibits a clear negative temperature dependence over the temperature range investigated (see Fig. 3 top panel and Fig. S3†). By presenting the rate coefficients collected at the same approximate temperature (from the same nozzle) and again in the same bath gas, it can also be seen that the reaction exhibits a strong positive pressure dependence over the pressure range investigated (see Fig. 3 bottom panel and Fig. S4†). Both the negative T dependence and the positive P dependence are indicative of reactions with a pre-reaction complex on their potential energy surface (PES; see Fig. 7). There have been no previous studies of the reaction between NH_2 and CH_3CHO at low temperatures, and only one study at higher temperatures, over the temperature range 297–543 K, conducted using an isothermal discharge flow system.[34] These high temperature rate coefficients are compared with our theoretically determined rate coefficients in Section 4, and in Fig. S13.†

3.2 NH_2 ($\nu > 0$) removal

The strong pressure dependence of reaction (R1), which at certain temperatures and with certain bath gases exhibited an almost 1 : 1 relationship between rate coefficient and pressure, was slightly unexpected, and indicated that the reaction is firmly within the low-pressure region over the pressures investigated (1.4–28.2 $\times 10^{16}$ molecules per cm^3). In contrast, the reaction between OH and CH_3CHO, which has also been studied at low temperatures, does not exhibit any pressure dependence over the temperature and pressure range of 21–106 K and 1–20 $\times 10^{16}$ molecules per cm^3, respectively.[35] One method for estimating the high-

Table 1 Bimolecular rate coefficients for the reaction between NH_2 + CH_3CHO together with relevant experimental conditions

Nozzle	Gas	~T^a/K	T^b/K	Density^b/10^{16} molecules per cm^3	$k_1{}^c$/10^{-11} cm^3 per molecule per s
1	Ar	31	34.2 ± 3.0	2.1 ± 0.3	3.77 ± 0.61
			30.3 ± 2.7	3.4 ± 0.4	5.82 ± 0.45
			28.7 ± 2.6	6.2 ± 0.8	10.7 ± 1.9
	He	37	42.2 ± 5.4	2.6 ± 0.5	1.74 ± 0.50
			37.2 ± 3.1	4.6 ± 0.6	3.56 ± 0.54
			35.0 ± 2.8	6.1 ± 0.7	3.69 ± 0.53
			34.3 ± 3.4	8.8 ± 1.3	4.72 ± 0.91
	N_2	66	70.0 ± 2.7	1.4 ± 0.1	0.25 ± 0.27
			67.1 ± 2.1	2.5 ± 0.2	0.84 ± 0.38
			67.1 ± 2.1	2.5 ± 0.2	1.53 ± 0.33
			64.6 ± 2.0	4.5 ± 0.3	0.92 ± 0.39
			62.9 ± 2.4	6.4 ± 0.6	2.80 ± 0.31
2	Ar	45	52.2 ± 2.9	6.2 ± 0.5	3.17 ± 0.09
			47.0 ± 2.8	11.2 ± 1.0	4.37 ± 0.41
			41.6 ± 1.6	28.2 ± 1.6	9.88 ± 0.90
			39.3 ± 2.7	16.5 ± 1.7	8.46 ± 0.37
	He	51	55.4 ± 3.8	10.0 ± 1.0	1.76 ± 0.10
			55.4 ± 3.8	10.0 ± 1.0	1.56 ± 0.09
			50.9 ± 1.9	17.9 ± 1.0	3.48 ± 0.25
			47.6 ± 1.5	23.5 ± 1.2	3.89 ± 0.38
	N_2	83	89.9 ± 2.5	4.2 ± 0.3	0.72 ± 0.10
			84.4 ± 2.8	7.4 ± 0.6	0.57 ± 0.22
			80.5 ± 2.9	13.0 ± 1.2	2.57 ± 0.31
			78.4 ± 2.9	17.7 ± 1.6	1.43 ± 0.42
3	Ar	45	47.9 ± 4.3	6.2 ± 0.8	4.89 ± 0.11
			44.4 ± 4.3	11.2 ± 1.6	9.95 ± 0.91
			43.3 ± 4.3	15.9 ± 2.3	11.6 ± 0.7
	He	52	54.6 ± 4.0	7.4 ± 0.8	1.4 ± 0.12
			51.8 ± 3.7	10.1 ± 1.1	2.3 ± 0.12
			49.6 ± 3.9	14 ± 1.6	3.13 ± 0.18
	N_2	89	92.9 ± 6.6	5.3 ± 0.9	0.17 ± 0.05
			87.4 ± 6.4	9.1 ± 1.7	0.92 ± 0.18
			85.6 ± 6.0	12.7 ± 2.2	1.14 ± 0.18
4	Ar	55	59.4 ± 7.4	8.8 ± 1.6	3.7 ± 0.21
			54.2 ± 7.9	15.0 ± 3.2	6.28 ± 0.59
			52.4 ± 7.8	20.5 ± 4.5	9.69 ± 0.61
	He	62	63.2 ± 6.2	6.8 ± 0.9	0.80 ± 0.14
			62.9 ± 5.7	13.5 ± 1.9	0.73 ± 0.14
			60.0 ± 6.2	18.6 ± 2.9	1.61 ± 0.22
	N_2	101	106.7 ± 8.6	7.6 ± 7.6	0.42 ± 0.04
			99.5 ± 9.8	12.7 ± 3.1	0.21 ± 0.03
			97.0 ± 9.7	17.7 ± 4.3	0.21 ± 0.08

[a] Average T of the low temperature flows for a particular nozzle and a particular gas. [b] Uncertainties in each value of T and density are $\pm 1\sigma$ (the standard deviation) of the measured T and density along the axis of the Laval expansion. [c] Uncertainties reported at the 1σ level for the linear least-squares fitting of the pseudo-first-order coefficients as a function of [CH_3CHO].

pressure limit of a reaction is the proxy method,[36,37] in which the loss of vibrationally excited NH_2 with CH_3CHO is measured. Here, the assumption is made that following the formation of the collision complex, the excess vibrational

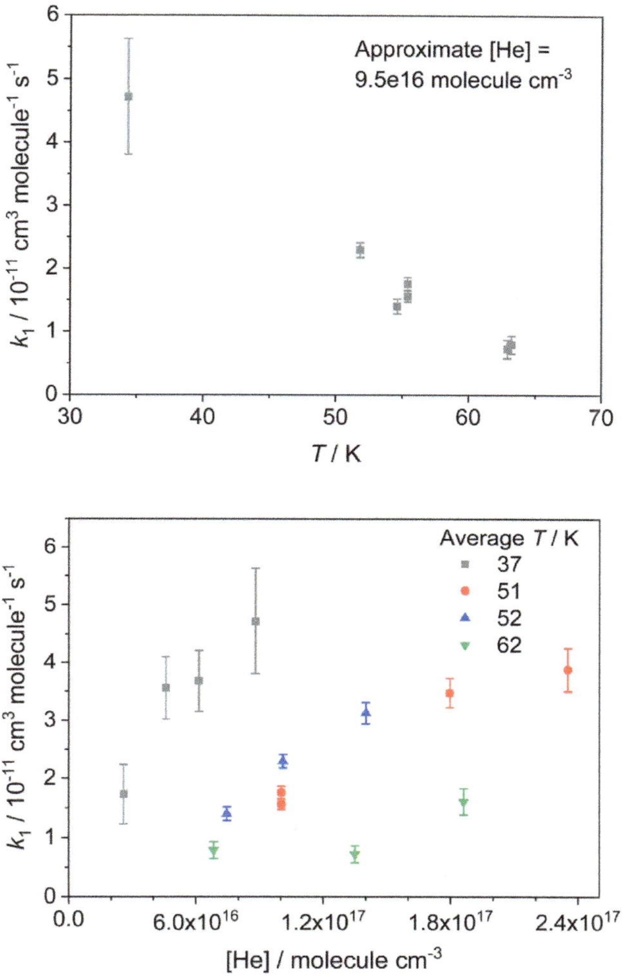

Fig. 3 Selected T and P dependent rate coefficients, collected in He bath gas, for the reaction NH_2 + CH_3CHO (R1). Top panel: T dependent rate coefficients collected at an approximate He density of 9.5×10^{16} molecules per cm^3; bottom panel: P dependent rate coefficients collected at approximate temperatures of 37 K (grey squares), 51 K (red circles), 52 K (blue upward triangles), and 62 K (green downward triangles). Each T corresponds to a different Laval nozzle (1–4 respectively).

energy in the NH_2 is rapidly redistributed among the many other modes in the complex, and any NH_2 formed following dissociation of the complex is in $\nu = 0$. Thus, the proxy method provides an estimate of the capture rate coefficient for complex formation, which can be used as the inverse Laplace transform (ILT) parameters for the initial association reaction of NH_2 with CH_3CHO in our MESMER calculations (Section 4).

Rate coefficients for the removal of NH_2 ($\nu = 0, x, 0$) with CH_3CHO, where $x =$ 0–4 and represents vibrational excitement of the ν_2 bending mode in NH_2, are presented in Fig. 4 and Table 2, together with the transitions and wavelengths

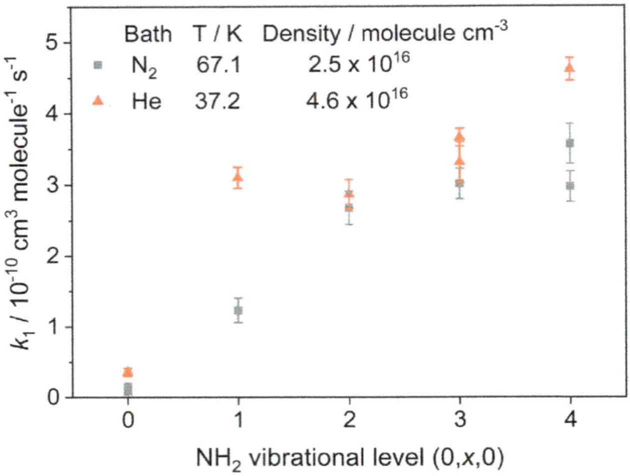

Fig. 4 Rate coefficients for the removal of NH_2 ($v = 0–4$) with CH_3CHO. Grey squares; $T = 67.1$ K, $[N_2] = 2.5 \times 10^{16}$ molecules per cm^3. Red triangles; $T = 37.2$ K, $[He] = 4.6 \times 10^{16}$ molecules per cm^3.

Table 2 Rate coefficients for the removal of NH_2 ($v_2 = 0–4$) by CH_3CHO

			$k^b/\times10^{-10}$ cm^3 per molecule per s	
v_2	Transition[a]	λ/nm	In N_2 at 67.1 [c] K	In He at 37.2 [d] K
0	0,9,0 ← 0,0,0	597.7	0.08 ± 0.04	0.54 ± 0.05
			0.15 ± 0.03	—
1	0,11,0 ← 0,1,0	591.1	1.23 ± 0.18	3.1 ± 0.15
2	0,12,0 ← 0,2,0	607.2	2.68 ± 0.23	—
		608.4	—	2.87 ± 0.2
3	0,13,0 ← 0,3,0	628.9	3.01 ± 0.21	3.31 ± 0.28
		627.4	—	3.66 ± 0.12
4	0,16,0 ← 0,4,0	611.0	3.57 ± 0.28	4.61 ± 0.16
		616.8	2.97 ± 0.21	—

[a] $A\ ^2A_1 \leftarrow X\ ^2B_1$ transition. [b] Uncertainties reported at the 1σ level for the linear least-squares fitting of the pseudo-first-order coefficients as a function of $[CH_3CHO]$. [c] Collected using nozzle 1 at $T = 67.1$ K and $[N_2] = 2.5 \times 10^{16}$ molecules per cm^3. [d] Collected using nozzle 1 at $T = 37.2$ K and $[He] = 4.6 \times 10^{16}$ molecules per cm^3.

employed.[21,38] In experiments monitoring removal of NH_2 ($\nu = 1–4$), no CH_4 was added to the low temperature flows, and a Perspex filter was used to collect the fluorescence and cut out the scattered laser light from the 213 nm photolysis pulse. As can be seen from Fig. 4 and Table 2, removal of vibrationally excited NH_2 by CH_3CHO is significantly faster than removal of ground state NH_2, with higher vibrational levels appearing to be removed faster than lower vibrational levels. However, the rate coefficient does appear to be reaching a limit (the high-pressure limit) at high ν; at 37.2 K in He this limit is practically reached by $\nu = 1$, and appears around 4×10^{-10} cm^3 per molecule per s (assessed visually), while at 67.1 K in N_2 the limit is not reached until around $\nu = 2$ or 3, and appears to be around

3.5×10^{-10} cm^3 per molecule per s. These results suggest that for the NH_2 + CH_3CHO system, only at the lowest temperatures are the initial collision complexes long-lived enough to ensure complete redistribution of the excess vibrational energy from the NH_2 species into the complex, and that at higher temperatures some complexes may dissociate before this has happened. Thus, proxy method estimations of the high-pressure limit for other systems carried out looking at the removal of $\nu = 1$ only, may underestimate the true limit. However, as the results at 67.1 K in N_2 show, the redistribution of energy from higher vibrational states appears to be faster, suggesting the high-pressure limit may still be estimated by looking at removal from successively higher vibrational levels. The slightly larger high-pressure limit at 37.2 K suggests a small negative temperature dependence in the initial association reaction between NH_2 and CH_3CHO.

3.3 OH/CH_3CO yields

OH LIF profiles produced following the photolysis of NH_3 in the presence of both CH_3CHO and O_2 are presented in Fig. 5. The OH observed is the result of CH_3CO, the product of reaction (R1a), reacting with O_2 to produce OH (R4):

$$CH_3CO + O_2 \rightarrow OH + \text{co-products} \tag{R4}$$

As the experiments were carried out under pseudo-first-order conditions (*i.e.* $[NH_2] \ll [CH_4]$, $[CH_3CHO]$, and $[O_2]$), the growth and loss of the OH signal can be described by a bi-exponential function of the form:

$$[OH]_t = f_{OH}\left(\frac{k'_{growth}}{k'_{loss} - k'_{growth}}\right)[NH_2]_0\left(\exp^{-k'_{growth}t} - \exp^{-k'_{loss}t}\right) \tag{E3}$$

where k'_{growth} and k'_{loss} are the pseudo-first-order rate coefficients for the reactions producing and removing OH, $[NH_2]_0$ is the initial amount of NH_2 produced following PLP of NH_3, and f_{OH} is the observed OH fluorescence signal, or OH yield (note, the reaction of CH_3CO + O_2 can also lead to other products). As OH itself reacts with CH_3CHO (R5), its primary loss is the sum of (R5) and diffusional losses. Although the OH profiles are not strictly bi-exponential in nature, as the growth of the OH signal is the result of series of subsequent reactions (vibrational relaxation of NH_2^* by CH_4 (R3), reaction of NH_2 with CH_3CHO to form CH_3CO (R1a), and reaction of CH_3CO with O_2 to form OH (R4)), they could be satisfactorily fit using eqn (E3), and the parameters k'_{growth}, k'_{loss}, and f_{OH} extracted.

The OH yields themselves do not give the yield for CH_3CO production from reaction (R1). To do this requires two additional pieces of knowledge; the first is an understanding of how many OH radicals have been produced relative to the initial number of NH_2 radicals produced, while the second is an understanding of the yield of OH from the reaction of CH_3CO with O_2 (ϕ_{OH}; R4). The second can be determined by measuring the rate coefficient for the removal of OH by CH_3CHO (R5) both with and without O_2 present.[39,40] In the experiments with O_2 present, CH_3CO formed by reaction (R5) will react with O_2, with a fraction regenerating OH (R4). If the concentration of O_2 is great enough such that the reaction between CH_3CO and O_2 is fast compared to the OH + CH_3CHO reaction (*i.e.* $k'_4 \gg k'_5$), then

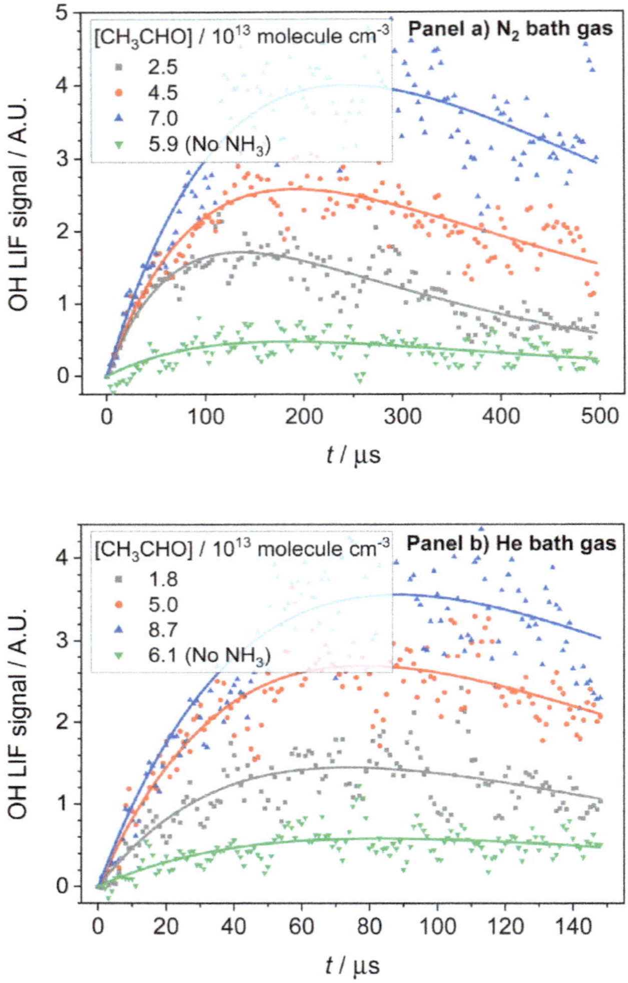

Fig. 5 OH traces collected following photolysis of NH_3 in the presence of varying amounts of CH_3CHO and O_2. Panel (a) traces collected at 67.1 K, a total N_2 density of 2.5×10^{16} molecules per cm^3, $[NH_3] = 1.4 \times 10^{13}$ molecules per cm^3, $[CH_4] = 8.3 \times 10^{14}$ molecules per cm^3, and $[O_2] = 8.2 \times 10^{14}$ molecules per cm^3. Panel (b) traces collected at 35.0 K, a total He density of 6.1×10^{16} molecules per cm^3, $[NH_3] = 1.2 \times 10^{13}$ molecules per cm^3, $[CH_4] = 8.4 \times 10^{14}$ molecules per cm^3, and $[O_2] = 8.3 \times 10^{14}$ molecules per cm^3. The green downward triangles in both plots are collected with no NH_3 present, indicating a small contribution of OH from the photolysis of CH_3CHO in the presence of O_2.

the OH removal will be determined by the fraction of (R5) that does not regenerate OH. As such, the reduction in the bimolecular rate coefficient for OH loss with CH_3CHO in the presence of O_2, k_{O_2}, as compared to without O_2, $k_{\text{no } O_2}$, gives the OH yield:

$$\phi_{OH} = 1 - k_{O_2} \Big/ k_{\text{no } O_2} \tag{E4}$$

$$OH + CH_3CHO \rightarrow CH_3CO, + \text{other products} \quad (R5)$$

Experiments monitoring the loss of OH by CH_3CHO, both with and without O_2 present, were carried out at $T = 67.1$ K in N_2 bath gas. OH LIF traces collected in these experiments (see Fig. S5†) could be fit with a single exponential loss in order to obtain the observed pseudo-first-order loss rate, k'_{obs}. Bimolecular plots obtained by plotting k'_{obs} *vs.* [CH_3CHO] can be seen in Fig. 6, in which the clear reduction in the bimolecular rate coefficient with O_2 can be seen. With no O_2 present, we obtain a value of $k_{5(\text{no } O_2)} = (9.81 \pm 0.57) \times 10^{-11}$ cm^3 per molecule per s. This rate coefficient is in good agreement with a recent literature study of the low temperature reaction of OH with C_3HCHO, in which they obtain a rate coefficient of $(1.25 \pm 0.13) \times 10^{-10}$ cm^3 per molecule per s at $T = 64$ K.[35] With O_2 present, we obtain a value of $k_{5(O_2)} = (5.4 \pm 4.2) \times 10^{-12}$ cm^3 per molecule per s, giving a value of $\phi_{OH} = 0.94 \pm 0.05$.

To address the first issue in determining the yield for CH_3CO production from (R1), that is understanding how many OH radicals have been produced relative to the initial number of NH_2 radicals produced, we used the following procedure. In these experiments, we are photolyzing a known amount of NH_3 to produce NH_2 radicals, which go on to form a number of OH radicals which we observe using LIF. Using both the concentration and absorption cross section of NH_3 we are able to estimate the number of NH_2 radicals produced in our system. This calculation also requires the energy of the photolysis laser, however as this is kept constant in our experiments, this will later cancel out in our calculations and can be ignored. To put the OH signal we observe on an absolute scale, we then calibrate our OH LIF signal in a separate experiment by photolyzing a known amount of *t*BuOOH under the same conditions and using the same photomultiplier settings. Again, as we know both the concentration and absorption cross section of *t*BuOOH, we

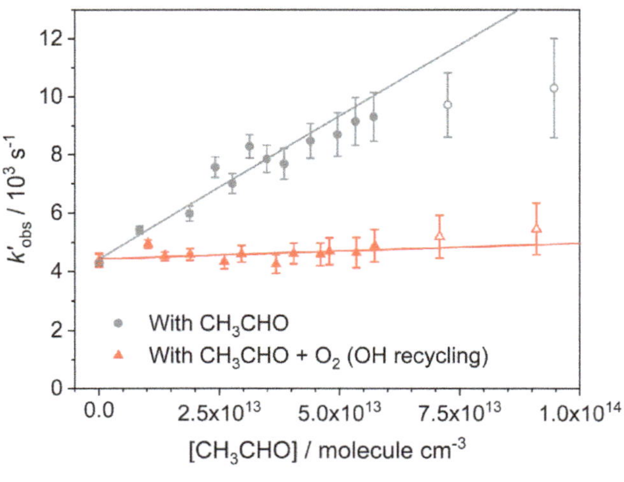

Fig. 6 Bimolecular plot of k'_{obs} *vs.* [CH_3CHO] for the reaction between OH + CH_3CHO (R5). Grey circles; OH loss with only CH_3CHO present. Red triangles; OH loss with both CH_3CHO and O_2 present. Curvature of the grey circles at [CH_3CHO] above 7×10^{13} molecules per cm^3 indicates significant dimer formation, with the linear fits obtained only using data points below this.

can determine the number of OH radicals present in our system giving us our OH LIF signal. In this manner we can determine a ratio of how many OH radicals we observe compared to the initial number of NH_2 radicals produced. In these calculations we account for the reduction in the initial NH_3 concentration in our system due to removal by CH_3CHO, and for the non-unity conversion of CH_3CO to OH by O_2. Using this method, we have determined CH_3CHO yields from reaction (R1) (*i.e.* the branching ratio for reaction (R1a)) to be 0.029 ± 0.014 at $T = 67.1$ K and $[N_2] = 2.50 \times 10^{16}$ molecules per cm^3, and 0.026 ± 0.008 at $T = 35.0$ K and $[He] = 6.14 \times 10^{16}$ molecules per cm^3. A more detailed account of how we determined these CH_3CO yields is given in the ESI in Section S6.†

4. Discussion

4.1 *Ab initio* calculations of the PES

A schematic of the full PES for the reaction between NH_2 and CH_3CHO can be seen in Fig. 7. The energies given are at the CCSD(T) level of theory and include ZPEs calculated at the M062x level. The optimised structures of the stationary points are given in Fig. S7 in the ESI,† together with the full molecular properties in Tables S8–S10.† As can be seen from Fig. 7, the reaction initially proceeds *via* the formation of one of two pre-reaction complexes (PRCs; labelled PRC1 and PRC2 in Fig. 7). PRC1, in which the NH_2 moiety is located over the acetyl H/C=O end of the acetaldehyde, is linked to the formation of the acetyl H-abstraction products NH_3 + CH_3CO *via* TS1 (+17.3 kJ mol^{-1}). PRC2, in which the NH_2 moiety is located above the methyl/C=O end of the acetaldehyde, is linked to two product channels; the methyl H-abstraction channel products NH_3 + CH_2CHO *via*

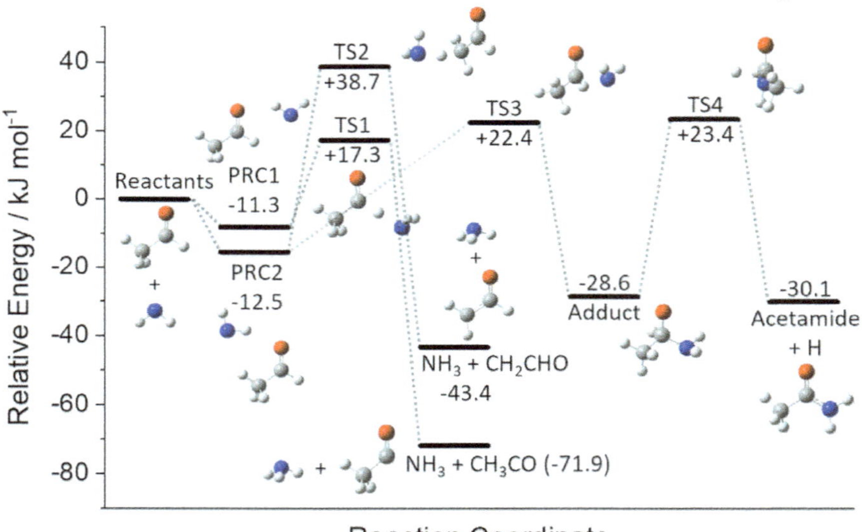

Fig. 7 Potential energy surface for NH_2 + CH_3CHO determined at the CCSD(T)/aug-cc-pVTZ//M062X/aug-cc-pVTZ level of theory. All energies are in kJ mol^{-1} and are corrected with scaled zero-point vibrational energy.

TS2 (+38.7 kJ mol^{-1}), and the addition–elimination channel products acetamide + H. The acetamide + H product channel requires the initial formation of a bound adduct from PRC2 *via* TS3 (+22.4 kJ mol^{-1}), followed by the elimination of an H atom to form products *via* TS4 (+23.4 kJ mol^{-1}). As both the H-abstraction channels involve the transfer of an H-atom, quantum mechanical tunnelling is likely to play a role at low temperatures.

There have been no previous theoretical investigations into the reaction between NH_2 and CH_3CHO; however, the reaction between NH_2 and formaldehyde (CH_2O) has been studied.[9,41–44] As expected, considering the similarity between the two systems, the PESs for the two reactions resemble each other remarkably well, in both the depth of the PRCs (where mentioned) and the heights of the barriers; the only significant difference is the additional methyl-H abstraction channel present in the case of NH_2 + CH_3CHO.

4.2 Rate theory calculations using MESMER

Using the calculated NH_2 + CH_3CHO PES surface given in Fig. 7, rate coefficients and branching ratios for the reaction over a wide range of temperatures and pressures have been calculated using the MESMER program.[31] The inverse Laplace transform (ILT) parameters for the association reaction of NH_2 with acetaldehyde, which take the form of a modified Arrhenius function $A(T/300)^n$ (with the activation energy being set to zero), were adjusted to give good agreement to the high pressure limits determined in Section 3.1, with the values $A = 2.5 \times 10^{-10}$ cm^3 per molecule per s and $n = -0.25$ used. Initial calculations carried out using the full PES have shown that the acetamide + H product channel (R1c) is predicted to be a very minor channel at all temperatures and pressures (over the range 20–400 K and 1×10^{11} to 1×10^{24} molecules per cm^3). Indeed, at a pressure of [Ar] = 1×10^{16} molecules per cm^3, the channel (R1c) only accounts for 0.1% of the product yield at 400 K, and decreases with decreasing temperature, becoming insignificant below ~150 K (see ESI Fig. S11†). This is due to the relatively high barrier to adduct formation (TS3, 22.4 kJ mol^{-1}) effectively turning off this channel at low temperatures as there is no coordinate for H-atom tunnelling under this barrier. As such, for the MESMER calculations discussed below, a simplified PES was used, in which the channel (R1c) was excluded (see ESI Fig. S12†).

Rate coefficients predicted by MESMER using the surface given in Fig. S12† and at a pressure of 1×10^{17} molecules per cm^3, are compared with our experimental data collected at a similar pressure in Fig. S13.† As can be seen from Fig. S13,† the experimental rate coefficients are significantly faster than those predicted by MESMER, by around 2 orders of magnitude. However, both the experimental and MESMER rate coefficients show a strong positive pressure dependence and a negative temperature dependence. This T and P dependence is the result of the formation of the weakly bound pre-reaction complexes on the PES. As the pressure is increased, there is more chance for the excess energy in the incoming reactants to be removed by collisions and for stabilization into the potential well to occur, while at the same time as the temperature is lowered, the rate of dissociation of the PRCs back to products will also decrease, increasing the effective loss rate of the reactants. Looking at species time profiles produced by MESMER, it can be seen that on the timescales of our experiments, as the reactants are lost we see significant population of the PRCs, which eventually go on to form products on longer timescales *via* quantum mechanical tunnelling. Thus, the fast removal rates of NH_2 we observe in our experiments correspond to the population of the PRCs. How MESMER calculates the stabilization of the reactants into the PRC wells is dependent on the description of the molecular parameters of the species involved in the reaction scheme, and in particular how these affect the density of states of the reactants and adducts. The significantly smaller rate coefficients predicted by MESMER compared to experiment suggest an underestimation in the calculation of the density of states of the PRCs as compared to the reactants.

The approximation of the harmonic oscillator was used for the vibrational frequencies in the calculation of the density of states. However, a few vibrational frequencies may be better represented by the hindered rotor (HR) approximation. Typically, treating low frequency vibrations as HRs increases the density of states. Fig. S14† shows all the rotations described with the HR approach. HR potentials were constructed through an *ab initio* calculation of a relaxed scan of the energy as the dihedral angle respective to the rotation is changed, with steps of 5°. A partial structure optimization was performed at the end of every step, until a 360° coverage was obtained. Both PRCs have 3 hindered rotors; rotation of the methyl group around the C–C bond, rotation of the terminal N–H around bond 7, and rotation of the N around bond 9 (see Fig. S14† for bond labels). For the hindered rotors around bond 9, we were only able to obtain a potential around ~180° of the dihedral when carrying out a relaxed scan, due to the O–H–N angle approaching linearity and causing difficulties in the representation of the dihedral in Gaussian. However, by carrying out a fixed scan an approximation of the potential around the full 360° was obtained. Surprisingly, the treatment of the rotations around bonds 4 and 7 as HRs, rather than harmonic vibrations, results in a marginal increase in the density of states in the PRCs. However, the treatment of the rotation around bond 9 as a HR does increase the density of states, especially for PRC2. This increase in the density of states results in around a 3 fold increase in the rate coefficients predicted by MESMER (between 20 and 100 K and at $[Ar] = 1 \times 10^{17}$ molecules per cm^3), bringing them closer to the experimental values (see Fig. S13†).

In order to obtain better agreement between the experimental and calculated rate coefficients, we have used MESMER to fit the experimental data by adjusting the well depths of the PRCs. In these calculations, we include the HR potentials and use the reduced $NH_2 + CH_3CHO$ PES (excluding channel (R1c), Fig. S12†), and the errors of the experimental rate coefficients were set to 20% (or kept as the experimental error if >20%). In order to reduce the computational expense when fitting the experimental data, rather than allowing the ZPEs of each PRC well to be varied independently, we set the ZPE of PRC2 to be defined from the ZPE of PRC1 minus 1.2 kJ mol^{-1}. In this manner we keep the absolute difference in energy between the two PRCs, while allowing the ZPEs of both to move up and down in tandem during the fitting.

The best fit to the experimental rate coefficients was obtained by moving the ZPEs of the PRCs down by 9.5 kJ mol^{-1}, to −20.8 and −22.0 kJ mol^{-1} for PRCs 1 and 2, respectively. The χ^2 parameter of the fit reduces from 856 (20.4 per point) using the original ZPEs down to 177 (4.2 per point) with the new values. Table S15† compares the experimental and calculated rate coefficients determined using these new values, and gives the % difference between the two. There is reasonable agreement between the rate coefficients, with an average difference of 20% for those using He as the bath, 30% using Ar, and 55% using N_2. The larger % difference using N_2 as a bath gas is likely the result of the larger errors in the experimental rate coefficients, which are on the lower edge of what we can reasonably measure in our experiments. This increase of 9.5 kJ mol^{-1} in the well depths of the PRCs is rather large. Typically, CCSD(T) calculations would be expected to be accurate to within 5 kJ mol^{-1}, so an adjustment of almost double this suggests either some sort of systematic error or interference in the experimental results, or some error in the description and treatment of the species in the calculation. Possible sources of error in the experimental results which have been discussed above, are in the acetaldehyde concentrations in our experiments, and in the possible formation of acetaldehyde dimers. Our acetaldehyde concentrations are confirmed using UV absorption spectroscopy, and furthermore we obtain a rate coefficient for OH + CH_3CHO in good agreement with another study.[35] Regarding dimers, as we are able to directly monitor the fluorescence from acetaldehyde we are able to determine at what [CH_3CHO] significant dimers begin to form, and ensure we only determine rate coefficients below this value where the second order plot is entirely linear. Another possible source of error is due to secondary chemistry in our experiments, such as from the removal of NH_2 with a photolysis product of CH_3CHO. However we deem this unlikely, as the cross-section for CH_3CHO at 213 nm is $\sim 5 \times 10^{-22}$ cm^{-2},[45] over 3 orders of magnitude smaller than for NH_3 at 213 nm, meaning that even at the highest CH_3CHO concentrations in our experiments we would expect a ratio of NH_2 to CH_3CHO photolysis products of 40 : 1. Possible sources of error in the MESMER calculations likely lie in the description of the HRs, in particular the HR relating to rotation around bond 9 in both PRCs. The magnitude and shape of the potential for this HR has been shown to have a strong effect on the calculated density of states of the PRCs, and as such the calculated rate coefficients are sensitive to this potential. Thus, this study highlights the importance of improved computational methods in calculating low frequency vibrational frequencies and hindered rotor potentials.

In addition to fitting the rate coefficients, we are also able to fit the experimentally determined yields for CH_3CO. As discussed above, at the temperatures and pressures in our experimental setup, the loss of NH_2 that we observe in our experiments is primarily due to the formation of the PRCs (and in particular PRC2), which on timescales significantly longer than our experiments will go on to form products (both channels (R1a) and (R1b)). However, as observed in our experiments, some CH_3CO is formed on the same timescale as NH_2 is lost, which for convenience we will call instant CH_3CO. This instant CH_3CO can be thought of as the CH_3CO produced from well-skipping, and is sensitive to the height of the acetyl H-abstraction transition state (TS1). Thus, we are able to fit the height of TS1 to the experimentally determined CH_3CO yields. This is crucially important for understanding the astrochemical implications of this reaction, as the height of TS1 greatly affects the low-pressure limiting rate coefficient, which is the relevant value for interstellar environments where there is no collisional stabilisation of the pre-reactive complex. As in the case of fitting the ZPEs of the PRCs, we have kept the absolute difference in the ZPEs of the TSs for the two H-abstraction channels constant, by defining the ZPE of TS2 from the ZPE of TS1 plus 21.4 kJ mol^{-1}, while allowing the ZPEs of both to move up and down in tandem during the fitting. When entering the CH_3CO yields to be fit into MESMER, the timescale for the yield must also be entered to ensure we are fitting to the instant CH_3CO, rather than the final CH_3CO yield. This timescale must be also be consistent with the excess reagent concentration entered in the model.

Table S16† compares the calculated instant CH_3CO yields to the experimental values following fitting of the ZPEs of the H-abstraction TSs (TS1 and TS2) to the experimental data. The best fit to the experimental yields is obtained by moving the ZPEs of the H-abstraction TSs down by 0.9 kJ mol^{-1}, to 16.4 and 37.8 kJ mol^{-1} for TSs 1 and 2, respectively. This small change in the heights of the H-abstraction TSs is well within the expected error of the calculated CCSD(T) energies.

Using the ZPEs of the PRCs and TSs obtained by fitting to the experimental data, we have calculated rate coefficients and branching ratios over a wide range of temperatures and pressures. Fig. 8 gives the low-pressure limiting rate coefficients which are applicable to the interstellar medium, while the rate coefficients over the full T and P range, which may be applicable to other high-pressure environments, can be seen in Fig. S17 in the ESI.† At the low-pressure limit, the only product channel open is the $CH_3CO + NH_3$ channel (R1a). Looking at Fig. 8, a turn around in the rate can be seen at around 200 K; above this temperature we see a small positive temperature dependence, whereas below this temperature we see a strong negative temperature dependence. This sharp increase in the rate coefficient at low temperatures is the result of the PRC wells on the PES, which are sufficiently long-lived at low temperatures to allow the H-atom to quantum-mechanically tunnel through the aldehydic H-abstraction barrier (TS1) to products. This mechanism has been reported previously for a range of low temperature H-abstraction reactions involving OH and oxygenated volatile organic compounds,[11–13,35] and for $NH_2 + CH_2O$.[9] Although the product channel (R1b), producing $CH_2CHO + NH_3$, also involves the transfer of an H atom

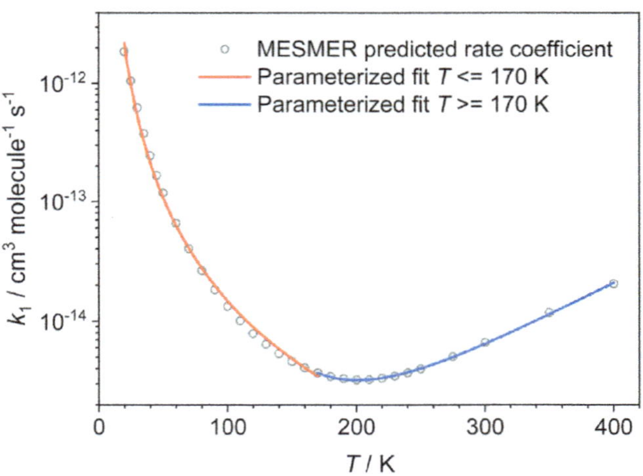

Fig. 8 Low-pressure limiting rate coefficients for the reaction between $NH_2 + CH_3CHO \rightarrow NH_3 + CH_3CO$ (R1a) calculated by MESMER, using the PRC and TS energies determined by fitting to the experimental data (see text for details).

and can occur by quantum mechanical tunnelling, due to the significantly larger barrier to methyl H-abstraction (TS2, 21.4 kJ mol^{-1} higher than TS1), and a similar imaginary frequency, the tunnelling rate through TS2 is insignificant when compared to the tunnelling rate through TS1. As such, the only products at the low-pressure limit are CH_3CO + NH_3 (R1a). The unique shape of the low-pressure limiting rate coefficient for (R1) does not allow for the parameterization of the data over the whole temperature range using a single modified Arrhenius equation. Instead, the data above and below 170 K were parameterized, giving (see red and blue solid lines in Fig. 8: units are cm^3 per molecule per s; errors are the 1σ level of a least squares fit to the data):

$$k_{(\mathrm{NH_2+CH_3CH\tilde{O}NH_3+CH_3CO})}\ (20 \leq T/\mathrm{K} \leq 170) = (7.23 \pm 0.64) \times 10^{-16} \times (T/300)^{(-2.56\pm0.15)} \times \exp^{[(22.0\pm7.8)/T]}$$

$$k_{(\mathrm{NH_2+CH_3CH\tilde{O}NH_3+CH_3CO})}\ (170 \leq T/K \leq 400) = (9.99 \pm 1.49) \times 10^{-18} \times (T/300)^{(9.70\pm0.17)} \times \exp^{[(1944\pm44)/T]}$$

These new rate coefficients will be submitted to the Kinetic Database for Astrochemistry (KIDA),[46] and the UMIST Database for Astrochemistry (UdFA).[47]

4.3 Astrochemical implications

To assess the astrophysical implications of the measured rate constant, we ran a single-point dark cloud astrochemical model with a temperature of 10 K, a gas number density of 10^4 cm^{-3}, and a cosmic-ray ionisation rate of 1.3×10^{-17} s^{-1}. We assume that the cloud has a high visual extinction and is fully shielded from any external sources of UV radiation. We ran four variants of the model with different assumptions for the chemistry: (i) a gas-phase only model using the network from Walsh *et al.*[48] with updates as described in Douglas *et al.*,[15] (ii) a gas–grain model with the same chemistry, (iii) a gas-phase only model including the new reaction and rate constant measured in this work, and (iv) a gas–grain model including the same update. We find that the abundance of the reactants, NH_2 and CH_3CHO, and that of one of the products, NH_3, are not significantly affected when the new reaction is included in the network. However, we do see a non-negligible effect on the gas-phase abundance of the radical, CH_3CO, as shown in Fig. 9. In scenario (i), there are no efficient gas-phase routes to form this radical in the Walsh *et al.*[48] network and its abundance is effectively zero. When gas–grain chemistry is included (scenario (ii); red solid line in Fig. 9), the abundance increases to $\sim10^{-26}$ to 10^{-24} with respect to H nuclei density; however, this range is still considered negligible. In this gas–grain model, the only source of gas-phase CH_3CO is *via* cosmic-ray-induced photodesorption from the ice phase. Radicals such as CH_3CO are efficiently hydrogenated in the ice even at 10 K, and so have a very short life-time on the surface and hence reach a negligible gas-phase abundance only. When the new reaction studied here is included in the gas-phase only model (scenario (iii); blue dotted line in Fig. 9), the abundance of CH_3CO increases from $\sim10^{-20}$ to 10^{-18} with production driven solely by the included reaction. Finally, when gas–grain chemistry is included together with the new reaction studied here (scenario (iv); blue solid line in Fig. 9), the abundance of CH_3CO increases further, to the arguably non-negligible range of 10^{-14} to 10^{-12} with respect to H nuclei density beyond a time of 10^6 years. This is driven

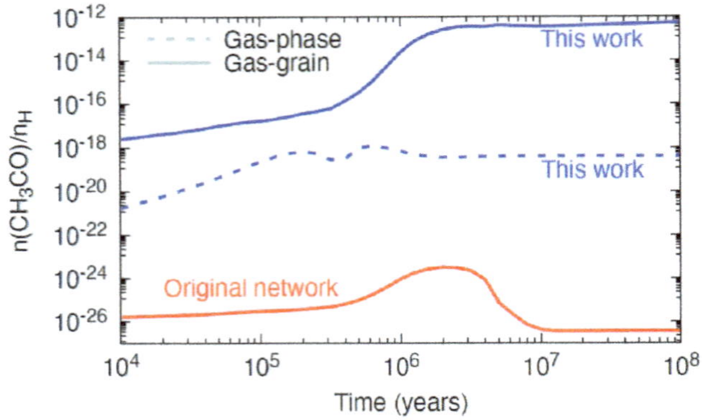

Fig. 9 The abundance of gas-phase CH_3CO with respect to hydrogen nuclei density as a function of time, a dark cloud model using three different variants of the chemical network from Walsh *et al.*[48] with: one using the gas–grain network from Walsh *et al.*[48] with the rate constants from Douglas *et al.*[15] (red solid line), one using a gas-phase only network including the gas-phase reaction studied in this work (NH_2 + CH_3CHO; dotted blue line), and one using a gas–grain network including the reaction studied in this work (solid blue line).

by the more efficient production of gas-phase CH_3CHO in the gas–grain model, for which the abundance is boosted by two orders of magnitude when grain-surface processes are included. The steady-state fractional abundance reached for gas-phase CH_3CO ($\sim 10^{-12}$) is approaching the threshold at which molecules become potentially observable in interstellar environments. For example, the methoxy radical, CH_3O, has been observed towards several dense cores, with an estimated fractional abundance of $\sim 10^{-12}$ to 10^{-11} with respect to H nuclei density.[49–51]

The model results for a dark cloud shown here, suggest that, whilst the NH_2 + CH_3CHO reaction is not a dominant loss mechanism for either reactant, nor a dominant production mechanism for NH_3, it is a potential production route for the gas-phase radical, CH_3CO, at least under dark cloud conditions. We leave exploration of the impact of this reaction on the chemistry in other astrophysical environments to future work. The gas-phase chemistry of organic radicals, such as CH_3CO, remains poorly constrained. A search of KIDA,[46] suggested only two gas-phase reactions forming this radical under astrophysical conditions (C_2H_3 + O and OH + H_2C_3O),[52,53] and no gas-phase destruction reactions.

5. Conclusions

The reaction between NH_2 and CH_3CHO has been investigated experimentally over the temperature and pressure range 29–107 K and 1.4–28.2 × 10^{16} molecules per cm^3, using a PLP-LIF technique. The low temperatures relevant to the ISM were achieved by a pulsed Laval nozzle expansion. Over the temperatures and pressures investigated, the reaction was shown to have a negative temperature dependence and a positive pressure dependence. Yields for the production of the CH_3CO product from the reaction were determined at 67.1 and 35.0 K, by monitoring the production of OH from the reaction of CH_3CO with O_2. The full

PES for the NH_2 + CH_3CHO system was determined using electronic structure calculations, and these combined with RRKM theory to obtain rate coefficients and branching ratios over a broad range of temperatures and pressures. The calculated rate coefficients have been shown to be sensitive to the inclusion of hindered rotor (HR) potentials in the rate theory calculations. In particular, the magnitude and shape of the HR potentials for the PRCs have been shown to have a strong effect on the calculated density of states of the PRCs, and as such the calculated rate coefficients are sensitive to this. Species time profiles produced by these calculations indicate that on the timescales of our experiments, the loss of NH_2 we observe is primarily due to the formation of the PRCs, and in particular PRC2, which on timescales significantly longer than our experiments will go on to form products (both channels (R1a) and (R1b)). However, as observed in our experiments, some CH_3CO is formed on the same timescale as NH_2 is lost; this instant CH_3CO can be thought of as the CH_3CO produced from well-skipping, and is sensitive to the height of the acetyl H-abstraction transition state (TS1). The experimentally determined rate coefficients and yields have been used to fit the calculated PES, from which low-pressure liming rate coefficients relevant to the ISM have been determined. These low-pressure limiting rate coefficients have been included in a single point dark cloud astronomical model, in which the reaction is shown to be a potential source of CH_3CO radicals under dark cloud conditions.

Conflicts of interest

There are no conflicts to declare.

Acknowledgements

We are grateful for the Science and Technology Facilities Council (STFC) for funding of this work under grant number ST/T000287/1. CW further acknowledges financial support from UKRI research and innovation grant number MR/T040726/1.

References

1 E. Herbst and E. F. van Dishoeck, Complex Organic Interstellar Molecules, in *Annual Review of Astronomy and Astrophysics*, ed. R. Blandford, J. Kormendy and E. van Dishoeck, 2009, vol. 47, pp. 427–480.
2 B. A. McGuire, *Astrophys. J. Suppl.*, 2022, **259**, 30.
3 A. I. Vasyunin and E. Herbst, *Astrophys. J.*, 2013, **769**, 34.
4 S. A. Sandford, M. Nuevo, P. P. Bera and T. J. Lee, *Chem. Rev.*, 2020, **120**, 4616–4659.
5 K. Altwegg, H. Balsiger, A. Bar-Nun, J. J. Berthelier, A. Bieler, P. Bochsler, C. Briois, U. Calmonte, M. R. Combi, H. Cottin, *et al.*, *Sci. Adv.*, 2016, **2**, e1600285.
6 R. Saladino, C. Crestini, S. Pino, G. Costanzo and E. Di Mauro, *Phys. Life Rev.*, 2012, **9**, 84–104.
7 R. H. Rubin, G. W. Swenson, R. C. Benson, H. L. Tigelaar and W. H. Flygare, *Astrophys. J.*, 1971, **169**, L39.

8 J. M. Hollis, F. J. Lovas, A. J. Remijan, P. R. Jewell, V. V. Ilyushin and I. Kleiner, *Astrophys. J.*, 2006, **643**, L25–L28.

9 K. M. Douglas, D. I. Lucas, C. Walsh, N. A. West, M. A. Blitz and D. E. Heard, *Astrophys. J.*, 2022, **937**, L16.

10 S. E. Taylor, A. Goddard, M. A. Blitz, P. A. Cleary and D. E. Heard, *Phys. Chem. Chem. Phys.*, 2008, **10**, 422–437.

11 R. L. Caravan, R. J. Shannon, T. Lewis, M. A. Blitz and D. E. Heard, *J. Phys. Chem. A*, 2015, **119**, 7130–7137.

12 J. C. Gomez Martin, R. L. Caravan, M. A. Blitz, D. E. Heard and J. M. C. Plane, *J. Phys. Chem. A*, 2014, **118**, 2693–2701.

13 R. J. Shannon, M. A. Blitz, A. Goddard and D. E. Heard, *Nat. Chem.*, 2013, **5**, 745–749.

14 R. J. Shannon, R. L. Caravan, M. A. Blitz and D. E. Heard, *Phys. Chem. Chem. Phys.*, 2014, **16**, 3466–3478.

15 K. Douglas, M. A. Blitz, W. H. Feng, D. E. Heard, J. M. C. Plane, E. Slater, K. Willacy and P. W. Seakins, *Icarus*, 2018, **303**, 10–21.

16 K. M. Douglas, M. A. Blitz, W. Feng, D. E. Heard, J. M. C. Plane, H. Rashid and P. W. Seakins, *Icarus*, 2019, **321**, 752–766.

17 N. A. West, T. J. Millar, M. Van de Sande, E. Rutter, M. A. Blitz, L. Decin and D. E. Heard, *Astrophys. J.*, 2019, **885**, 134.

18 S. Bull, I. Seregrennaja and P. Tsherbakora, *Khim. Prom.*, 1963, **7**, 507–509.

19 R. A. Copeland, D. R. Crosley and G. P. Smith, In laser-induced fluorescence spectroscopy of NCO and NH_2 in atmospheric pressure flames, in *Symposium (International) on Combustion*, Elsevier, 1985, pp. 1195–1203.

20 V. M. Donnelly, A. P. Baronavski and J. R. McDonald, *Chem. Phys.*, 1979, **43**, 271–281.

21 K. Yamasaki, A. Watanabe, T. Kakuda, A. Itakura, H. Fukushima, M. Endo, C. Maruyama and I. Tokue, *J. Phys. Chem. A*, 2002, **106**, 7728–7735.

22 P. Limão-Vieira, S. Eden, N. J. Mason and S. V. Hoffmann, *Chem. Phys. Lett.*, 2003, **376**, 737–747.

23 M. J. Frisch, G. W. Trucks, H. B. Schlegel, G. E. Scuseria, M. A. Robb, J. R. Cheeseman, G. Scalmani, V. Barone, B. Mennucci, G. A. Petersson, *et al.*, *Gaussian 09, Revision D.01*, Gaussian, Inc., Wallingford, CT, 2013.

24 T. H. Dunning Jr, *J. Chem. Phys.*, 1989, **90**, 1007–1023.

25 R. A. Kendall, T. H. Dunning Jr and R. J. Harrison, *J. Chem. Phys.*, 1992, **96**, 6796–6806.

26 D. E. Woon and T. H. Dunning Jr, *J. Chem. Phys.*, 1995, **103**, 4572–4585.

27 Y. Zhao and D. G. Truhlar, *Theor. Chem. Acc.*, 2008, **120**, 215–241.

28 NIST Computational Chemistry Comparison and Benchmark Database, http://cccbdb.nist.gov/.

29 R. J. Bartlett and G. D. Purvis, *Int. J. Quantum Chem.*, 1978, **14**, 561–581.

30 J. A. Pople, R. Krishnan, H. B. Schlegel and J. S. Binkley, *Int. J. Quantum Chem.*, 1978, **14**, 545–560.

31 D. R. Glowacki, C.-H. Liang, C. Morley, M. J. Pilling and S. H. Robertson, *J. Phys. Chem. A*, 2012, **116**, 9545–9560.

32 K. H. Gericke, L. M. Torres and W. A. Guillory, *J. Chem. Phys.*, 1984, **80**, 6134–6140.

33 T.-X. Xiang, K.-H. Gericke, L. M. Torres and W. A. Guillory, *Chem. Phys.*, 1986, **101**, 157–163.

34 W. Hack, H. Kurzke, P. Rouveirolles and H. G. Wagner, *Ber. Bunsen-Ges. Phys. Chem.*, 1986, **90**, 1210–1219.

35 S. Blazquez, D. Gonzalez, E. M. Neeman, B. Ballesteros, M. Agundez, A. Canosa, J. Albaladejo, J. Cernicharo and E. Jimenez, *Phys. Chem. Chem. Phys.*, 2020, **22**, 20562–20572.

36 D. H. Jaffer and I. W. M. Smith, *Faraday Discuss. Chem. Soc.*, 1979, **67**, 212–220.

37 M. Quack and J. Troe, *Ber. Bunsen-Ges. Phys. Chem.*, 1975, **79**, 170–183.

38 R. A. Copeland, D. R. Crosley and G. P. Smith, *Symp. (Int.) Combust., [Proc.]*, 1985, **20**, 1195–1203.

39 N. U. M. Howes, J. P. A. Lockhart, M. A. Blitz, S. A. Carr, M. T. Baeza-Romero, D. E. Heard, R. J. Shannon, P. W. Seakins and T. Varga, *Phys. Chem. Chem. Phys.*, 2016, **18**, 26423–26433.

40 L. Onel, L. Thonger, M. A. Blitz, P. W. Seakins, A. J. C. Bunkan, M. Solimannejad and C. J. Nielsen, *J. Phys. Chem. A*, 2013, **117**, 10736–10745.

41 V. Barone, C. Latouche, D. Skouteris, F. Vazart, N. Balucani, C. Ceccarelli and B. Lefloch, *Mon. Not. R. Astron. Soc.: Lett.*, 2015, **453**, L31–L35.

42 Q. S. Li and R. H. Lü, *J. Phys. Chem. A*, 2002, **106**, 9446–9450.

43 L. Song and J. Kästner, *Phys. Chem. Chem. Phys.*, 2016, **18**, 29278–29285.

44 F. Vazart, D. Calderini, C. Puzzarini, D. Skouteris and V. Barone, *J. Chem. Theory Comput.*, 2016, **12**, 5385–5397.

45 H. Keller-Rudek, G. K. Moortgat, R. Sander and R. Sörensen, *Earth System Science Data*, 2013, **5**, 365–373.

46 V. Wakelam, E. Herbst, J. C. Loison, I. W. M. Smith, V. Chandrasekaran, B. Pavone, N. G. Adams, M. C. Bacchus-Montabonel, A. Bergeat, K. Beroff, *et al.*, *Astrophys. J. Suppl.*, 2012, **199**, 21.

47 D. McElroy, C. Walsh, A. J. Markwick, M. A. Cordiner, K. Smith and T. J. Millar, *Astron. Astrophys.*, 2013, **550**, A36.

48 C. Walsh, H. Nomura and E. van Dishoeck, *Astron. Astrophys.*, 2015, **582**, A88.

49 J. Cernicharo, N. Marcelino, E. Roueff, M. Gerin, A. Jiménez-Escobar and G. M. Muñoz Caro, *Astrophys. J.*, 2012, **759**, L43.

50 I. Jiménez-Serra, A. I. Vasyunin, P. Caselli, N. Marcelino, N. Billot, S. Viti, L. Testi, C. Vastel, B. Lefloch and R. Bachiller, *Astrophys. J.*, 2016, **830**, L6.

51 A. Megías, I. Jiménez-Serra, J. Martín-Pintado, A. I. Vasyunin, S. Spezzano, P. Caselli, G. Cosentino and S. Viti, *Mon. Not. R. Astron. Soc.*, 2022, **519**, 1601–1617.

52 J.-C. Loison, M. Agúndez, N. Marcelino, V. Wakelam, K. M. Hickson, J. Cernicharo, M. Gerin, E. Roueff and M. Guélin, *Mon. Not. R. Astron. Soc.*, 2016, **456**, 4101–4110.

53 J.-C. Loison, M. Agúndez, V. Wakelam, E. Roueff, P. Gratier, N. Marcelino, D. N. Reyes, J. Cernicharo and M. Gerin, *Mon. Not. R. Astron. Soc.*, 2017, **470**, 4075–4088.

Faraday Discussions

ROYAL SOCIETY OF CHEMISTRY

PAPER

Searches for bridged bicyclic molecules in space—norbornadiene and its cyano derivatives†

Marie-Aline Martin-Drumel, [a] Jean-Thibaut Spaniol,[a] Helen Hölzel, [b] Marcelino Agúndez,[c] Jose Cernicharo,[c] Kasper Moth-Poulsen [bdef] and Ugo Jacovella [*a]

Received 26th January 2023, Accepted 14th February 2023

DOI: 10.1039/d3fd00016h

The norbornadiene (**NBD**) molecule, C_7H_8, owes its fame to its remarkable photoswitching properties that are promising for molecular solar-thermal energy storage systems. Besides this photochemical interest, **NBD** is a rather unreactive species within astrophysical conditions and it should exhibit high photostability, properties that might also position this molecule as an important constituent of the interstellar medium (ISM)—especially in environments that are well shielded from short-wavelength radiation, such as dense molecular clouds. It is thus conceivable that, once formed, **NBD** can survive in dense molecular clouds and act as a carbon sink. Following the recent interstellar detections of large hydrocarbons, including several cyano-containing ones, in the dense molecular cloud TMC-1, it is thus logical to consider searching for NBD—which presents a shallow but non-zero permanent electric dipole moment (0.06 D)—as well as for its mono- and dicyano-substituted compounds, referred to as **CN-NBD** and **DCN-NBD**, respectively. The pure rotational spectra of **NBD**, **CN-NBD**, and **DCN-NBD** have been measured at 300 K in the 75–110 GHz range using a chirped-pulse Fourier-transform millimetre-wave spectrometer. Of the three species, only **NBD** was previously studied at high resolution in the microwave domain. From the present measurements, the derived spectroscopic constants enable prediction of the spectra of all three species at various rotational temperatures (up to 300 K) in the spectral range mapped at high resolution by current radio observatories. Unsuccessful searches for these molecules were

[a]Université Paris-Saclay, CNRS, Institut des Sciences Moléculaires d'Orsay, 91405 Orsay, France. E-mail: ugo.jacovella@universite-paris-saclay.fr

[b]Department of Chemistry and Chemical Engineering, Chalmers University of Technology, 41296 Gothenburg, Sweden

[c]Instituto de Física Fundamental, CSIC, Department of Molecular Astrophysics, Serrano 121, E-28006 Madrid, Spain

[d]The Institute of Materials Science of Barcelona, ICMAB-CSIC, Bellaterra, 08193 Barcelona, Spain

[e]Catalan Institution for Research & Advanced Studies, ICREA, Pg. Lluís Companys 23, 08010 Barcelona, Spain

[f]Department of Chemical Engineering, Universitat Politècnica de Catalunya, EEBE, Eduard Maristany 10–14, 08019 Barcelona, Spain

† Electronic supplementary information (ESI) available. See https://doi.org/10.1039/d3fd00016h

conducted toward TMC-1 using the QUIJOTE survey, carried out at the Yebes telescope, allowing derivation of the upper limits to the column densities of 1.6×10^{14} cm^{-2}, 4.9×10^{10} cm^{-2}, and 2.9×10^{10} cm^{-2} for **NBD**, **CN-NBD**, and **DCN-NBD**, respectively. Using **CN-NBD** and cyano-indene as proxies for the corresponding bare hydrocarbons, this indicates that—if present in TMC-1—**NBD** would be at least four times less abundant than indene.

Introduction

The hunt for carbon-based molecules in space is a long-standing venture for astrochemists and astronomers.[1] Their identification is mandatory to understand the carbon life cycle at work in extraterrestrial environments and to determine the initial physicochemical conditions in star and planet formation that ultimately led to the conditions for life on Earth.[2] Nowadays, more than 270 gas phase molecular species are known to be present in the interstellar medium (ISM) with ~90% detected by radio astronomy *via* their pure rotational spectral signatures.[3] The detection of these molecules usually results from a collective endeavour within the fields of astronomy, laboratory spectroscopy, and quantum chemistry. The continuous advances in the sensitivity that can be achieved by radio telescopes have led to a large increase in the number of detected molecules within the last 2 years. Conceivably, the culmination of these recent successes is the discovery of numerous aromatics and polycyclic aromatics with an attached –CN function and even one "bare" polycyclic aromatic hydrocarbon (PAH)—indene—using the GOTHAM and QUIJOTE spectral surveys of TMC-1 carried out with the Green Bank and Yebes 40 m telescopes, respectively.[4–10]

The PAH hypothesis[11,12] is now unambiguously confirmed and together with the detection of large carbon cages in different space environments,[13–15] it is strongly tenable that other large carbonaceous systems must exist in space. In general, for molecules to build up large enough densities in space and be detected, they need to be formed efficiently and/or to survive (UV photostability, weak reactivity) long enough to accumulate. A family fulfilling at least the survival criterion is that of diamondoids, which are cage-like, ultra stable, saturated ringed hydrocarbons; the smallest unit being the adamantane ($C_{10}H_{16}$) molecule. Numerous hints allude to the presence of diamondoids in various objects in the universe. Indeed, they possess a strong aptitude for resisting harsh UV radiation[16] and were identified in meteorites[17–19] and in proto-planetary discs.[20] The detection of adamantane or any of its derivatives using radio astronomy, however, remains elusive.[21] From classical synthesis of diamantane ($C_{14}H_{20}$) and larger adamantane clusters ($C_{10+4n}H_{16+4n}$) in Earth laboratories, which typically involves the polymerisation of 2,5-norbornadiene (or 2,2,1-bicycloheptadiene, C_7H_8, hereafter **NBD**; see Zieliński *et al.*[22] and ref. therein), one can foresee **NBD** to be a piece of the complex puzzle generated by the chemical composition of the ISM.

The **NBD** molecule is well known by virtue of its remarkable photoswitching properties showing great promise for molecular solar-thermal energy storage systems.[23] Besides this photochemical interest, **NBD** is a rather nonreactive species by astrophysical standards. It should exhibit high-photostability properties that might also position this molecule as an important constituent of the ISM. **NBD** has its first electronic transition observed around 236 nm, which is symmetry forbidden and thus extremely weak. The absorption cross section only becomes significant below 200 nm.[24] Under irradiation at 235.7 nm, **NBD** exhibits a fragmentation yield—to form cyclopentadiene and acetylene—of 0.55. The remaining 0.45 yield is the isomerisation to quadricyclane or toluene.[25] Naturally, regions that are well shielded from short-wavelength radiation, such as dense molecular clouds, appear as the most suitable environments to find **NBD**. It is thus conceivable that, once formed, **NBD** can survive in dense molecular clouds and act as a carbon sink.

Because of its broad relevance to many fields, **NBD** has been the subject of several spectroscopic investigations. In the gas phase, electron diffraction and infrared spectroscopy have been employed to determine its structure (see, *e.g.*, ref. 26–28). The electronic spectrum of its cationic counterpart unveiling several interconversion routes on the $C_7H_8^+$ potential energy surface was reported.[29] One gas-phase study examined the photoisomerisation of a charged tagged **NBD** derivative as a function of the excitation wavelength.[30] The pure rotational spectrum of **NBD** (and its isotopologues) has been investigated by Fourier-transform microwave spectroscopy in the 7–17 GHz region.[31,32] These measurements were performed in waveguide cells, with the cells cooled down to −70 °C, hence relatively large quantum number values were probed (up to $J'' = 37$ and $K_a'' = 16$), allowing for the determination of quartic centrifugal distortion constants. Several rovibrational bands were subsequently observed using high-resolution Fourier-transform infrared spectroscopy, providing improved spectroscopic constants using ground-state combination differences owing to a significant increase in observed quantum number values (up to $J'' = 89$ and $K_a'' = 80$).[33] One can expect these constants to allow for reliable predictions of the pure rotational spectrum of **NBD** at higher frequencies than already observed, although no measurements have been reported to date to confirm this. Since many large carbon molecules have recently been detected by radio astronomy at millimetre wavelengths,[9] experimental measurements of **NBD** at these frequencies seem warranted to ensure the accuracy of the spectral predictions.

The shallow permanent electric dipole moment of **NBD** (0.06 D, measured accurately by Stark experiments[31]) implies that a considerable amount of the molecule must be present to be detected. Searching for its cyano derivatives—of much larger permanent dipole moment—may help to overcome this complication. The recent detection of several cyano-substituted molecules[4,6,7] renders the strategy of searching for cyano derivatives extremely tantalising and provides a two-fold piece of information: (i) the cyano derivatives are good trackers of aromatics that own no or weak permanent dipole moments; and (ii) the unexpected large abundance of the found interstellar cyano species informs on the potential chemical role of cyano-bearing molecules in space. Nevertheless, the hardship of reproducing the abundances of these cyano aromatics with current astrochemical models prevents their use as a direct proxy, *i.e.*, the abundance of the non-substituted counterparts cannot be retrieved reliably. Indene is currently the single polycyclic aromatic hydrocarbon for which both the pure hydrocarbon and a cyano-substituted counterpart have been detected in the same source, providing a direct determination of the ratio of their abundance.[34] Even though this observed ratio is reasonably close to the one predicted by computations, the detection of only one of the cyano-derivative isomers still challenges the theory. **NBD** thus appears as a potential highly strained hydrocarbon on which to further test the proxy hypothesis (*i.e.*, infer the abundance of a non-polar large hydrocarbon based on the interstellar abundance of its cyano derivatives). Literature on the cyano derivatives of **NBD** is rather scarce compared to their bare chromophore and mostly focused on their photoisomerisation properties[35,36]—of relevance to potential new energy storage solutions—and role in Diels–Alder cycloadditions.[37,38] Spectroscopic information on the gaseous compounds is, to the best of our knowledge, limited to a couple of electronic structure investigations.[39,40]

In this paper, we present the pure rotational spectra of **NBD** and two of its mono- and dicyano-substituted compounds, namely bicyclo[2.2.1]hepta-2,5-diene-2-carbonitrile and bicyclo[2.2.1]hepta-2,5-diene-2,3-dicarbonitrile, referred to as **CN-NBD** and **DCN-NBD**, respectively. Measurements were performed at 300 K in the 75–110 GHz range using a chirped-pulse (CP) Fourier-transform millimetre-wave (FTMM) spectrometer. From these measurements, the derived spectroscopic constants are used to predict the spectra of all three species at various rotational temperatures (up to 300 K) in the spectral range in which cold sources are currently mapped at high resolution by radio observatories. Searches for the three species have been undertaken in the 31–40 GHz region on the QUIJOTE line survey of TMC-1.[41] This survey has previously shown the presence of large cyclic molecules, such as indene,[9] which suggests that other large cyclic species, such as **NBD** or its cyano derivatives, could be present as well.

Methods

Sample synthesis

NBD was purchased from Sigma-Aldrich (≥97% purity) and used without further purification. **CN-NBD** and **DCN-NBD** were synthesised at Chalmers University.

CN-NBD was prepared following a procedure adapted from the literature;[42,43] the corresponding scheme is displayed is Fig. S1 in the ESI†. Briefly, propargyl alcohol (1.11 g; 19.8 mmol) was combined with cyclopentadiene (1.32 g, 20 mmol) and hydroquinone (a few crystals) in a 20 mL vial, suitable for microwave reactions, and sealed. The mixture was heated to 170 °C for 8 h. The crude product was purified *via* flash column chromatography in 100% dichloromethane. The resulting product, (±)-2-hydroxymethylbicyclo[2.2.1]-hepta-2,5-diene, is an oil and was obtained in a 7.8% yield (0.19 g, 1.55 mmol). ^{1}H NMR data (Fig. S2†) are in accordance with literature data.[42] (±)-2-Hydroxymethylbicyclo[2.2.1]-hepta-2,5-diene (0.19 g, 1.55 mmol) was then combined with CH_3CN (4 mL), H_2O (0.5 mL), TEMPO (12 mg, 77 mmol), NH_4OAc (0.475 g, 6.16 mmol), and $PhI(OAc)_2$ (1.06 g, 3.3 mmol), and stirred at room temperature for 1 h. Dichloromethane was added to the reaction mixture and the phases were separated. The aqueous phase was extracted with dichloromethane (repeated three times) and the solvent was evaporated. The crude product was purified using flash column chromatography in 100% petroleum spirit. The pure product, **CN-NBD**, was isolated in 11% yield (20 mg, 0.17 mmol). NMR data are in accordance with literature data.[44]

The synthesis for **DCN-NBD** was carried out according to literature procedures (Fig. S3†); ^{1}H and ^{13}C NMR characterisation data (Figs. S4 and S5†) are in accordance with the literature.[45,46]

Both samples were stored at −80 °C until their investigation at the University of Paris-Saclay.

Quantum chemical calculations

Calculations were performed using the Gaussian 16 suite of electronic structure programs.[47] Geometry optimisation and harmonic frequency analysis have been carried out at the ωB97X-D/cc-pVQZ level of theory,[48–50] a level that provides very reliable spectroscopic parameters for the spectral analysis of rigid molecules (dipole moment projections, rotational constants, and quartic centrifugal distortion constants).[51,52]

Millimetre-wave measurements

CP-FTMM measurements were performed in the 75–110 GHz spectral range using a commercial CP spectrometer from BrightSpec. Vapour pressure of each sample, typically of the order of 3 μbar, was introduced into a ~60 cm-long cell terminated by two PTFE windows. A slow flow was ensured by a turbomolecular pump (Pfeiffer, HiCube 80) to maintain stable pressure conditions during the acquisition and prevent cell outgassing. Spectra were recorded in the standard high dynamic range mode of the instrument consisting of a segmented acquisition approach with 30 MHz-wide segments. Excitation pulses of 0.5 μs (the longest possible in this mode of acquisition), 0.25 μs, and 0.1 μs length were found to maximise the signals of **NBD**, **CN-NBD**, and **DCN-NBD**, respectively. These decreasing values are consistent with the increasing permanent dipole moment value of the species—since molecules with lower dipole moments usually require a longer excitation pulse to achieve the highest molecular signal. The final CP spectra were obtained *via* the Fourier transform of 4 μs of the free induction decay, after applying a Blackmann–Harris filter. The spectra were averaged over 1 million, 640 thousand, and 500 thousand frames for **NBD**, **CN-NBD**, and **DCN-NBD**, respectively (2 to 1 h of acquisition time based on sample availability).

Results and discussion

Spectroscopic considerations

Cartesian coordinates of the optimised equilibrium structures can be found in Tables S1–S3 in the ESI.† The associated molecular representations of the three molecules are shown in Fig. 1. **NBD** has C_{2v} point group symmetry and possesses a small permanent dipole moment of 0.06 Debye along the *c*-axis of inertia (where the axis is going through the centre of the C_6H_6 ring and the carbon of the CH_2 moiety). **CN-NBD** has C_1 point group symmetry and its permanent dipole moment has projections along the three axes of inertia, with values of 4.67 Debye, 0.77 Debye, and 0.22 Debye along the *a*-, *b*-, and *c*-axis, respectively (the *a*-axis is parallel to the –CN bond). Finally, **DCN-NBD** has C_s point group symmetry and exhibits projections of its permanent dipole moment of 6.44 Debye and 0.86 Debye along the *a*- and *c*-axis of inertia (the *a* axis is perpendicular to the plane containing the two –CN bonds). Projections of the three molecular structures onto the three inertial planes, performed using the PMIFST software developed by Z. Kisiel,[53] are proposed in Fig. S6 in the ESI.† All three species are asymmetric top rotors with asymmetry parameters κ taking values of −0.2203,[31] −0.9426, and 0.0784, respectively. **NBD** possesses four pairs of exchangeable H nuclei with respect to the C_2-axis resulting in a statistical weight for the *ortho* : *para* levels of 17 : 15. This ratio, very close to unity, did not result in any tangible intensity

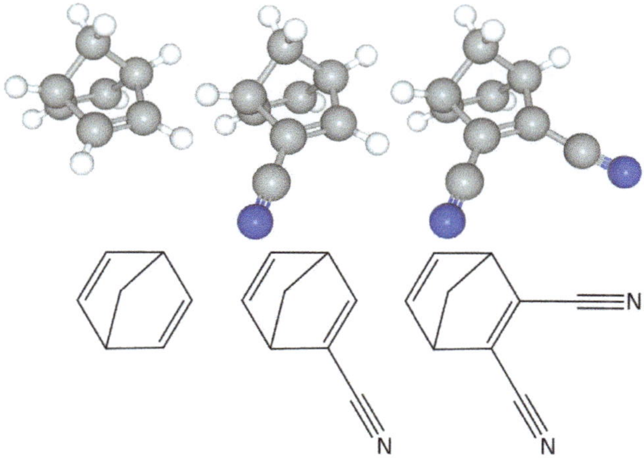

Fig. 1 Molecular representations of **NBD**, **CN-NBD**, and **DCN-NBD** (from left to right).

difference on the experimental spectrum but was nonetheless taken into account for the spectral predictions.

Rotational analysis

Experimental assignments have been performed using the PGOPHER software by exploiting Loomis–Wood diagrams.[54] PGOPHER was also used for spectral simulations. The SPFIT/SPCAT suite of programs[55] was used to adjust the rotational constants of all three species using a Watson A-reduced Hamiltonian in the I^r representation.

For **NBD**, literature pure rotation data[31] (25 transitions) and ground-state combination differences (GSCD) from rovibrational data[33] (369 GSCD/196 wavenumbers) were fitted together allowing for efficient searches at millimetre wavelengths. Both datasets were weighted according to their expected accuracy, *i.e.*, 1 kHz for data from Vogelsanger and Bauder[31] and 0.0001 cm^{-1} to 0.0003 cm^{-1} ($\sim$3–9 MHz) for GCSD from Sams and Blake.[33] GCSD are provided in MHz in ref. 33 but we favoured here wavenumber units in order to access individual root mean square values for the two datasets, as conventionally done using SPFIT. To reproduce these datasets at their experimental accuracy, all three rotational constants and five quartic centrifugal distortion parameters were adjusted. The resulting parameters are reported in Table 1 and are in excellent agreement with those reported by Sams and Blake.[33] The only difference is that we chose not to adjust the sextic centrifugal distortion parameter h_K in favour of a more constrained fit. All parameters are also in excellent agreement with those predicted from the calculations performed in this study. Using these reliable constants, spectroscopic assignments of the millimetre-wave spectrum of **NBD** were straightforward, with no notable divergence of the prediction. In total, 94 transitions (79 different frequencies) involving $J'' \leq 16$ and $K_a'' \leq 11$ were assigned on the spectrum. Frequency accuracy is estimated to be 30 kHz. These transitions were fitted together with the literature and GSCD data; the adjusted spectroscopic parameters are reported in Table 1. All transitions are reproduced at their

Table 1 Rotational constants and quartic centrifugal distortion constants (in MHz) of **NBD**, **CN-NBD**, and **DCN-NBD**. Predictions (Pred.) result from the ωB97X-D/cc-pVQZ calculations with Bayesian correction for the rotational constants;[51] harmonic equilibrium values are reported for the centrifugal distortion parameters. For **NBD**, values derived from the refit of the literature data (Lit.) are reported together with the combined fit from this work (TW). All adjusted parameters are reported with 1σ uncertainties expressed in the unit of the last digit in parentheses. The value in square brackets is fixed to the calculated value. Relevant fit parameters are reported at the bottom of the table

		NBD			CN-NBD		DCN-NBD	
Parameter		Pred.	Lit.^{a}	TW	Pred.	TW	Pred.	TW
A		4279.20	4273.62825(11)	4273.628151(89)	3829.02	3831.6473(15)	1519.59	1524.6688(45)
B		3590.28	3610.30044(10)	3610.300350(86)	1309.06	1316.14694(25)	1180.74	1186.9614(30)
C		3176.64	3186.43725(10)	3186.437158(87)	1235.56	1241.76251(23)	788.07	791.81703(12)
Δ_J	$\times 10^3$	0.3284	0.33376(21)	0.33379(21)	0.05950	0.063035(52)	0.3917	0.10090(83)
Δ_{JK}	$\times 10^3$	−0.07030	−0.07239(71)	−0.07226(70)	0.5917	0.62671(24)	−1.2814	−0.3979(45)
Δ_K	$\times 10^3$	0.2679	0.26745(61)	0.26725(59)	0.08843	0.0533(24)	0.9067	[0.9067]
δ_J	$\times 10^3$	0.004560	0.004777(82)	0.004790(80)	0.00308	0.003530(44)	0.1122	0.04124(42)
δ_K	$\times 10^3$	0.08956	0.09221(77)	0.09206(76)	0.06219	0.0805(53)	−1.0833	−0.08482(41)
N/n^b			394/221	438/300		1067/626		761/389
J''_{max}, $K''_{a\,max}$			89, 80	89, 80		74, 40		67, 21
RMS^c			0.00079	0.027		0.051		0.026
IR RMS^d			0.00014	0.00014				
σ^e			1.03	1.03		0.96		0.87

^{a} Refit of the literature data: pure rotation from Vogelsanger and Bauder[31] (in MHz) and ground-state combination differences from Sams and Blake[33] (in cm^{-1}). ^{b} Number of lines (N) and number of lines with different frequencies (n), unitless. ^{c} Root mean square of the microwave data, in MHz. ^{d} Root mean square of the infrared data, in cm^{-1}. ^{e} Weighted standard deviation, unitless.

expected accuracy with a final weighted standard deviation of 1.03. The addition of the millimetre-wave transitions allows for a slight, but not substantial, refinement of the parameters (Table 1). Nevertheless, accurate experimental frequencies of **NBD** are now available in the millimetre-wave domain, and spectral predictions in the range not covered yet in the laboratory, 17–75 GHz, should be extremely reliable. Frequency extrapolation always requires more caution, although for a molecule as rigid as **NBD**—only 8 parameters are required to reproduce 300 different frequencies with transitions involving J and K_a as high as 89 and 80, respectively—these are expected to be trustworthy as well.

For the two other species studied here, **CN-NBD** and **DCN-NBD**, since no literature data was available on the spectroscopic constants, spectral searches were initiated using the rotational constants derived from the calculated equilibrium structure. To more accurately predict the rotational constants in $v = 0$, the equilibrium A, B, and C values were scaled using the Bayesian correction factor recommended for the ωB97X-D/cc-pVQZ level of calculations (0.9866).[51] These predicted constants are reported in Table 1 (the equilibrium A_e, B_e, and C_e values are available in Table S4 in the ESI†). In both cases, the predictions allowed for a straightforward assignment of the pure rotational transitions.

For **CN-NBD**, a-type R-branch transitions were assigned first, owing to the large projection of the permanent dipole moment along that axis ($\mu_a = 4.67$). In total, 863 transitions (524 different frequencies) involving quantum numbers up to $J'' = 42$ and $K_a'' = 40$ were assigned with the frequency accuracy estimated to be 30 kHz. These transitions allowed for the determination of the rotational constants and quartic centrifugal distortion constants, with the exception of Δ_K (which is typical of a fit involving only a-type transitions). After this first iteration of assignments, b-type Q-branch transitions were identified on the spectrum. These transitions are about 30 times weaker than the a-type ones, in qualitative agreement with what is expected from the squared ratio of the dipole-moment projections $[(\mu_a/\mu_b)^2 = 36]$. 204 of these transitions (102 different frequencies) involving $J'' \leq 74$ and $K_a'' \leq 20$ were assigned. These lines exhibit a lower signal-to-noise ratio than the a-type ones and their frequency uncertainty is consequently estimated to be 100 kHz. No a-type Q-branch or b-type R-branch transitions were assigned, nor were any c-type ones (they are mostly overlapping with stronger b-type transitions). The addition of the b-type transitions to the fit enabled the determination of the Δ_K parameter (Table 1). No additional distortion parameter was required to reproduce the data at their experimental accuracy (the final weighted standard deviation is 0.96)—not even J-dependant ones, even though high J values are observed—which highlights the rigidity of the molecule. The adjusted set of parameters is in excellent agreement with the predicted one, with the relative error on the rotational constants on the order of 0.5% or better (against 1–1.25% for the pure equilibrium values), which is consistent with what has previously been observed using the Bayesian correction.[52] The experimentally-derived centrifugal distortion constants are also in good agreement with the calculations, both in sign and order of magnitude. Fig. 2 shows a comparison between portions of the experimental spectrum of **CN-NBD** and the simulation obtained using the optimised parameters.

Finally, for **DCN-NBD**, 761 a-type R-branch transitions (389 different frequencies) with $J'' \leq 67$ and $K_a'' \leq 21$ have been assigned. No b-type transitions were assigned because these weaker features are overlapping with the a-type

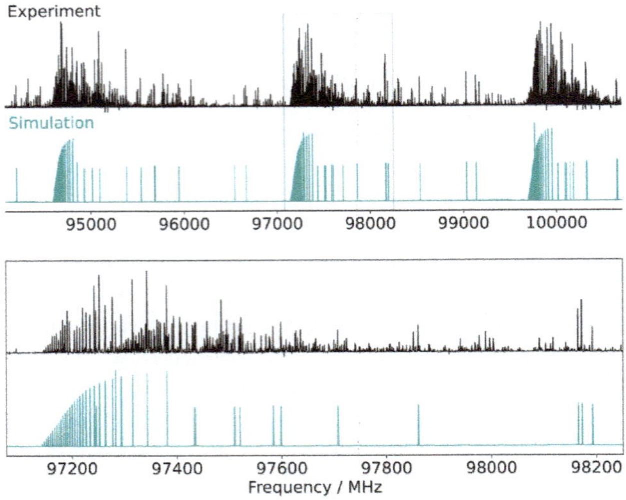

Fig. 2 Portions of the experimental spectrum (in black) of **CN-NBD** recorded using the CP-FTMM spectrometer and comparison with a 300 K simulation performed using the rotational constants derived from the spectral analysis (in green). The bottom panel shows the zoomed-in regions of the spectra around the $^{Q}R(J'' = 37)$ transition; the corresponding spectral window is highlighted by a grey area on the top panel. Transitions on the experimental spectrum not reproduced on the simulation arise from vibrational satellites (pure rotation in excited vibrational states) not investigated in this work.

transitions. The assigned transitions, of expected frequency accuracy of 30 kHz, have enabled the adjustment of the rotational and quartic centrifugal distortion constants of the molecule (Table 1). The Δ_K parameter could not be adjusted and was thus fixed to the calculated value. The derived rotational constants are again in excellent agreement with the prediction (with relative errors on the order of 0.5% or better). The centrifugal distortion parameters are further away from the calculated values than for the other species (by a factor of 3 to 10), while agreeing in sign, but fixing some distortion parameters (*e.g.*, δ_K, which is showing the worse agreement) to the calculated values does not allow a satisfactory fit to be obtained. Overall, all transitions are reproduced at their expected accuracy with a final root-mean-square of 26 kHz, corresponding to a weighted standard deviation close to unity.

All files relevant to the assignment and fit (linelists, parameters, and fit results) are available in the ESI† as ASCII files.

Astronomical implications and interstellar searches

Using the spectroscopic parameters determined for the three species in this study, reliable spectral predictions can be obtained for frequencies up to 110 GHz and temperatures up to 300 K (so as to simulate transitions involving the same range of quantum numbers as those observed experimentally). Frequency and/or temperature extrapolation may be feasible as well since the three species appear particularly rigid, although the accuracy of such predictions is hard to estimate. Catalogues of the frequencies at 10 K have been generated for the three molecules

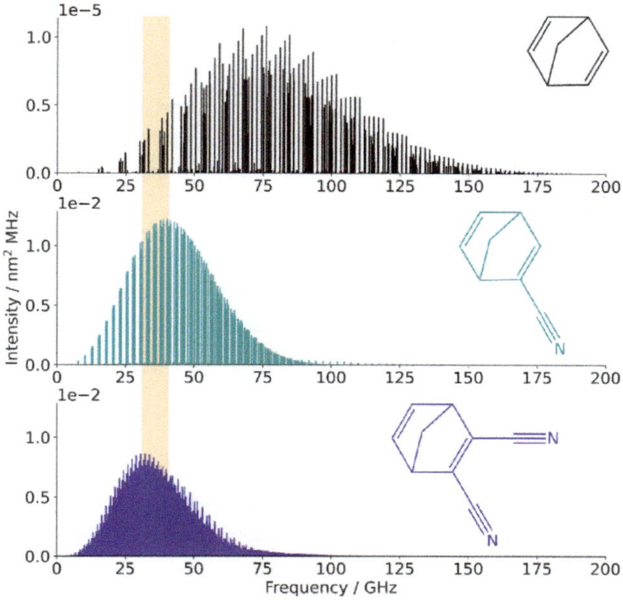

Fig. 3 Simulated spectra of **NBD**, **CN-NBD**, and **DCN-NBD** (from top to bottom) at 10 K. The orange area is the spectral regions where features of these species were searched for on the QUIJOTE survey.

using the SPCAT program and the partition functions reported in Table S5 in the ESI† (also generated using SPCAT); the corresponding traces are shown in Fig. 3. At this temperature, the strongest transitions of **NBD** lie around 75 GHz while the spectrum of the two –CN substituted species is shifted to lower frequencies (because of their heavier weight) with an optimum region for interstellar searches between ∼25 GHz and ∼50 GHz.

The frequency predictions of **NBD**, **CN-NBD**, and **DCN-NBD** were used to search for these three species toward the starless core TMC-1 using the QUIJOTE line survey of this source that is being carried out with the Yebes 40 m telescope.[41] Currently, the QUIJOTE data cover observing sessions from November 2019 to November 2022, and amount to a total on-source telescope time of 758 h. The

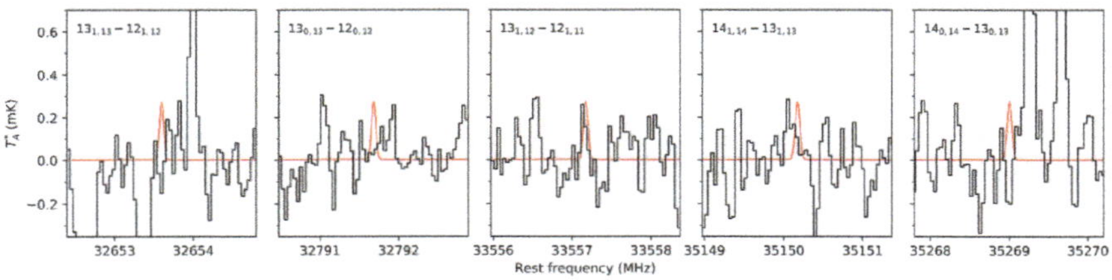

Fig. 4 QUIJOTE data at the position of some of the lines of **CN-NBD** predicted to be the most intense. Red traces are the synthetic line profiles calculated assuming a rotational temperature of 10 K, a line width of 0.6 km s^{-1}, and a column density equal to the 3σ upper limit derived, *i.e.*, 4.9 × 10^{10} cm^{-2}.

antenna temperature noise level is around 0.1 mK in the 31–40 GHz region and somewhat higher at frequencies above 40 GHz. We thus carried out the search for **NBD** and its cyano derivatives focusing on the low-frequency data of the QUIJOTE survey (region highlighted in orange in Fig. 3). At the current sensitivity level of the data, we do not detect the lines predicted to be the most intense in this frequency region for neither **NBD**, **CN-NBD**, nor **DCN-NBD**. Assuming a rotational temperature of 10 K and line width of 0.6 km s^{-1}, as found for indene,[9] we derive 3σ upper limits to the column densities of 1.6×10^{14} cm^{-2} for **NBD**, 4.9×10^{10} cm^{-2} for **CN-NBD**, and 2.9×10^{10} cm^{-2} for **DCN-NBD**. Fig. 4 illustrates the non-detection of the most intense lines of **CN-NBD** assuming the aforementioned conditions. The upper limit derived for **NBD** is quite high and thus not very informative due to the very low dipole moment of this molecule. For comparison, the column density derived for indene, a molecule that is also weakly polar, is 1.6×10^{13} cm^{-2}.[9] On the other hand, the upper limit imposed on the monocyano derivative, which is very polar, is much more informative. For comparison, the column density inferred for the monocyano derivative of indene is 2.1×10^{11} cm^{-2}.[34] If the abundance ratio between the monocyano derivative and the parent hydrocarbon behaves similarly for indene and **NBD**, then **NBD**—if present in TMC-1—is less abundant than indene by a factor of at least four.

Conclusions

The pure rotational spectra of **NBD**, **CN-NBD**, and **DCN-NBD** have been investigated at room temperature by CP-FTMM spectroscopy in the 75–110 GHz range allowing for a refinement of the spectroscopic constants of **NBD** and the first experimental determination of those of **CN-NBD** and **DCN-NBD**. Using these constants, transitions of the three molecules have been searched for, unsuccessfully, toward TMC-1 using the QUIJOTE survey. Upper limits to the column density of each species have been derived.

Conflicts of interest

There are no conflicts to declare.

Acknowledgements

The work at ISMO was performed using HPC resources from the "Mésocentre" computing center of CentraleSupélec and École Normale Supérieure Paris-Saclay supported by CNRS and Région Île-de-France (https://mesocentre.centralesupelec.fr/). It has also been supported by the Région Ile-de-France, through DIM-ACAV+, from the Agence Nationale de la Recherche (ANR-19-CE30-0017-01), and from the Programme National "Physique et Chimie du Milieu Interstellaire" (PCMI) of CNRS/INSU with INC/INP co-funded by CEA and CNES. The work at Chalmers was funded by the European Research Council (ERC) under grant agreement CoG, PHOTHERM – 101002131, the Catalan Institute of Advanced Studies (ICREA), and the European Union's Horizon 2020 Framework Programme under grant agreement no. 951801. We thank Anders Lennartsson for his help with the synthesis of **CN-NBD**.

Notes and references

1 C. S. Hansen, E. Peeters, J. Cami and T. W. Schmidt, *Commun. Chem.*, 2022, **5**, 94.

2 S. A. Sandford, M. Nuevo, P. P. Bera and T. J. Lee, *Chem. Rev.*, 2020, **120**, 4616–4659.

3 B. A. McGuire, *Astrophys. J., Suppl. Ser.*, 2022, **259**, 30.

4 B. A. McGuire, A. M. Burkhardt, S. Kalenskii, C. N. Shingledecker, A. J. Remijan, E. Herbst and M. C. McCarthy, *Science*, 2018, **359**, 202–205.

5 A. M. Burkhardt, R. A. Loomis, C. N. Shingledecker, K. L. K. Lee, A. J. Remijan, M. C. McCarthy and B. A. McGuire, *Nat. Astron.*, 2021, **5**, 181–187.

6 M. C. McCarthy, K. L. K. Lee, R. A. Loomis, A. M. Burkhardt, C. N. Shingledecker, S. B. Charnley, M. A. Cordiner, E. Herbst, S. Kalenskii, E. R. Willis, et al., *Nat. Astron.*, 2021, **5**, 176–180.

7 K. L. K. Lee, P. B. Changala, R. A. Loomis, A. M. Burkhardt, C. Xue, M. A. Cordiner, S. B. Charnley, M. C. McCarthy and B. A. McGuire, *Astrophys. J., Lett.*, 2021, **910**, L2.

8 B. A. McGuire, R. A. Loomis, A. M. Burkhardt, K. L. K. Lee, C. N. Shingledecker, S. B. Charnley, I. R. Cooke, M. A. Cordiner, E. Herbst, S. Kalenskii, et al., *Science*, 2021, **371**, 1265–1269.

9 J. Cernicharo, M. Agúndez, C. Cabezas, B. Tercero, N. Marcelino, J. R. Pardo and P. de Vicente, *Astron. Astrophys.*, 2021, **649**, L15.

10 A. M. Burkhardt, K. L. K. Lee, P. B. Changala, C. N. Shingledecker, I. R. Cooke, R. A. Loomis, H. Wei, S. B. Charnley, E. Herbst, M. C. McCarthy, et al., *Astrophys. J., Lett.*, 2021, **913**, L18.

11 L. Allamandola, A. Tielens and J. Barker, *Astrophys. J., Suppl. Ser.*, 1989, **71**, 733–775.

12 S. Schlemmer, D. Cook, J. Harrison, B. Wurfel, W. Chapman and R. Saykally, *Science*, 1994, **265**, 1686–1689.

13 B. Foing and P. Ehrenfreund, *Nature*, 1994, **369**, 296–298.

14 J. Cami, J. Bernard-Salas, E. Peeters and S. E. Malek, *Science*, 2010, **329**, 1180–1182.

15 E. K. Campbell, M. Holz, D. Gerlich and J. P. Maier, *Nature*, 2015, **523**, 322–323.

16 W. C. Saslaw and J. E. Gaustad, *Nature*, 1969, **221**, 160–162.

17 R. S. Lewis, T. Ming, J. F. Wacker, E. Anders and E. Steel, *Nature*, 1987, **326**, 160–162.

18 R. S. Lewis, E. Anders and B. T. Draine, *Nature*, 1989, **339**, 117–121.

19 E. Anders and E. Zinner, *Meteoritics*, 1993, **28**, 490–514.

20 A. P. Jones, *Astron. Astrophys.*, 2022, **665**, A21.

21 O. Chitarra, M.-A. Martin-Drumel, Z. Buchanan and O. Pirali, *J. Mol. Spectrosc.*, 2021, **378**, 111468.

22 A. Zieliński, X. Marset, C. Golz, L. M. Wolf and M. Alcarazo, *Angew. Chem., Int. Ed.*, 2020, **59**, 23299–23305.

23 M. Quant, A. Lennartson, A. Dreos, M. Kuisma, P. Erhart, K. Börjesson and K. Moth-Poulsen, *Chem. - Eur. J.*, 2016, **22**, 13265–13274.

24 B. O. Roos, M. Merchan, R. McDiarmid and X. Xing, *J. Am. Chem. Soc.*, 1994, **116**, 5927–5936.

25 B. Roquitte, *J. Phys. Chem.*, 1965, **69**, 2475–2477.

26 A. Yokozeki and K. Kuchitsu, *Bull. Chem. Soc. Jpn.*, 1971, **44**, 2356–2363.
27 I. W. Levin and W. Harris, *Spectrochim. Acta, Part A*, 1973, **29**, 1815–1834.
28 R. A. Shaw, C. Castro, R. Dutler, A. Rauk and H. Wieser, *J. Chem. Phys.*, 1988, **89**, 716–731.
29 U. Jacovella, G. da Silva and E. J. Bieske, *J. Phys. Chem. A*, 2019, **123**, 823–830.
30 U. Jacovella, E. Carrascosa, J. T. Buntine, N. Ree, K. V. Mikkelsen, M. Jevric, K. Moth-Poulsen and E. J. Bieske, *J. Phys. Chem. Lett.*, 2020, **11**, 6045–6050.
31 B. Vogelsanger and A. Bauder, *J. Mol. Spectrosc.*, 1988, **130**, 249–257.
32 G. Knuchel, G. Grassi, B. Vogelsanger and A. Bauder, *J. Am. Chem. Soc.*, 1993, **115**, 10845–10848.
33 R. L. Sams and T. A. Blake, *J. Mol. Spectrosc.*, 2020, **373**, 111354.
34 M. L. Sita, P. B. Changala, C. Xue, A. M. Burkhardt, C. N. Shingledecker, K. L. Kelvin Lee, R. A. Loomis, E. Momjian, M. A. Siebert, D. Gupta, E. Herbst, A. J. Remijan, M. C. McCarthy, I. R. Cooke and B. A. McGuire, *Astrophys. J., Lett.*, 2022, **938**, L12.
35 Y. Harel, A. W. Adamson, C. Kutal, P. A. Grutsch and K. Yasufuku, *J. Phys. Chem.*, 1987, **91**, 901.
36 F. Hemauer, U. Bauer, L. Fromm, C. Weiss, A. Leng, P. Bachmann, F. Duell, J. Steinhauer, V. Schwaab, R. Grzonka, A. Hirsch, A. Gorling, H.-P. Steinrueck and C. Papp, *ChemPhysChem*, 2022, **23**, e202200199.
37 E. Dalkilic and A. Dastan, *Tetrahedron*, 2015, **71**, 1966–1970.
38 H. Abdullah, R. Tia and E. Adei, *J. Phys. Org. Chem.*, 2021, **34**, e4259.
39 T. Kobayashi, Z. Yoshida, Y. Asako, S. Miki and S. Kato, *J. Am. Chem. Soc.*, 1987, **109**, 5103.
40 K. N. Houk, N. G. Rondan, M. N. Paddon-Row, C. W. Jefford, P. T. Huy, P. D. Burrow and K. D. Jordan, *J. Am. Chem. Soc.*, 1983, **105**, 5563.
41 J. Cernicharo, M. Agúndez, R. I. Kaiser, C. Cabezas, B. Tercero, N. Marcelino, J. R. Pardo and P. de Vicente, *Astron. Astrophys.*, 2021, **652**, L9.
42 C. Tanyeli, G. Çelikel and İ. M. Akhmedov, *Tetrahedron: Asymmetry*, 2001, **12**, 2305–2308.
43 J.-M. Vatèle, *Synlett*, 2014, **25**, 1275–1278.
44 W. Adam, L. Pasquato, B. Will and O. D. Lucchi, *Chem. Ber.*, 1987, **120**, 531–535.
45 O. Diels and K. Alder, *Justus Liebigs Ann. Chem.*, 1931, **490**, 236–242.
46 H. Taniguchi, T. Ikeda, Y. Yoshida and E. Imoto, *Bull. Chem. Soc. Jpn.*, 1977, **50**, 2694–2699.
47 M. J. Frisch, G. W. Trucks, H. B. Schlegel, G. E. Scuseria, M. A. Robb, J. R. Cheeseman, G. Scalmani, V. Barone, G. A. Petersson, H. Nakatsuji, X. Li, M. Caricato, A. V. Marenich, J. Bloino, B. G. Janesko, R. Gomperts, B. Mennucci, H. P. Hratchian, J. V. Ortiz, A. F. Izmaylov, J. L. Sonnenberg, D. Williams-Young, F. Ding, F. Lipparini, F. Egidi, J. Goings, B. Peng, A. Petrone, T. Henderson, D. Ranasinghe, V. G. Zakrzewski, J. Gao, N. Rega, G. Zheng, W. Liang, M. Hada, M. Ehara, K. Toyota, R. Fukuda, J. Hasegawa, M. Ishida, T. Nakajima, Y. Honda, O. Kitao, H. Nakai, T. Vreven, K. Throssell, J. A. Montgomery Jr, J. E. Peralta, F. Ogliaro, M. J. Bearpark, J. J. Heyd, E. N. Brothers, K. N. Kudin, V. N. Staroverov, T. A. Keith, R. Kobayashi, J. Normand, K. Raghavachari, A. P. Rendell, J. C. Burant, S. S. Iyengar, J. Tomasi, M. Cossi, J. M. Millam, M. Klene, C. Adamo,

R. Cammi, J. W. Ochterski, R. L. Martin, K. Morokuma, O. Farkas, J. B. Foresman and D. J. Fox, *Gaussian 16 Revision A.01*, 2016.
48 J.-D. Chai and M. Head-Gordon, *Phys. Chem. Chem. Phys.*, 2008, **10**, 6615.
49 T. H. Dunning, *J. Chem. Phys.*, 1989, **90**, 1007–1023.
50 D. E. Woon and T. H. Dunning, *J. Chem. Phys.*, 1993, **98**, 1358–1371.
51 K. L. K. Lee and M. McCarthy, *J. Phys. Chem. A*, 2020, **124**, 898–910.
52 Z. Buchanan, K. L. K. Lee, O. Chitarra, M. C. McCarthy, O. Pirali and M.-A. Martin-Drumel, *J. Mol. Spectrosc.*, 2021, **377**, 111425.
53 Z. Kisiel, *Spectroscopy from Space*, Springer, Dordrecht, Netherlands, 2001, pp. 91–106.
54 C. M. Western, *J. Quant. Spectrosc. Radiat. Transfer*, 2017, **186**, 221–242.
55 H. M. Pickett, *J. Mol. Spectrosc.*, 1991, **148**, 371–377.

Faraday Discussions

ROYAL SOCIETY OF CHEMISTRY

PAPER

High-resolution rovibrational and rotational spectroscopy of the singly deuterated cyclopropenyl cation, c-$C_3H_2D^+$†

Divita Gupta,[a] Weslley G. D. P. Silva,[a] José L. Doménech,[b] Eline Plaar,[a] Sven Thorwirth,[a] Stephan Schlemmer[a] and Oskar Asvany*[a]

Received 17th March 2023, Accepted 30th March 2023

DOI: 10.1039/d3fd00068k

Applying a novel action spectroscopic technique in a 4 K cryogenic ion-trap instrument, the molecule c-$C_3H_2D^+$ has been investigated by high-resolution rovibrational and pure rotational spectroscopy for the first time. In total, 126 rovibrational transitions within the fundamental band of the ν_1 symmetric C–H stretch were measured with a band origin centred at 3168.565 cm^{-1}, which were used to predict pure rotational transition frequencies in the ground vibrational state. Based on these predictions, 16 rotational transitions were observed between 90 and 230 GHz by using a double-resonance scheme. These new measurements will enable the first radio-astronomical search for c-$C_3H_2D^+$.

1 Introduction

In the late 1970s, a spectral line was detected at 85.3 GHz in a number of different interstellar environments.[1] Following its firm laboratory identification as the $2_{12} \rightarrow 1_{01}$ rotational transition of cyclopropenylidene (c-C_3H_2) by Thaddeus *et al.*,[2] it became clear that this molecule is ubiquitous in space. c-C_3H_2 is expected to be formed in the gas phase by the dissociative recombination of the cyclopropenyl cation with an electron: c-$C_3H_3^+$ + e^- → c-C_3H_2 + H (see *e.g.* Sipilä *et al.*[3] and references therein). Cyclopropenylidene is so abundant that even isotopically substituted versions were subsequently detected with a high signal-to-noise ratio, such as the two singly ^{13}C-substituted versions,[4] c-C_3HD,[5] and even c-C_3D_2.[6] The deuterated variants of c-C_3H_2 are thought to be formed exclusively *via* gas-phase reactions in cold environments, such as the cold cores TMC-1C and L1544, *via*

[a]*I. Physikalisches Institut, Universität zu Köln, Zülpicher Str. 77, 50937 Köln, Germany. E-mail: asvany@ph1.uni-koeln.de*

[b]*Instituto de Estructura de la Materia, IEM-CSIC, Serrano 123, 28006 Madrid, Spain*

† Electronic supplementary information (ESI) available. See DOI: https://doi.org/10.1039/d3fd00068k

deuterated versions of c-$C_3H_3^+$, followed again by dissociative recombination. Observational proof of these hypotheses could not be obtained to date, as c-$C_3H_3^+$ has no permanent electric dipole moment and can thus only be detected in the infrared region.[7] Its isotopically substituted versions, in particular c-$C_3H_2D^+$, have a small dipole moment, but high-resolution microwave laboratory data for these are still missing.

In this work, we provide the first rovibrational and pure rotational data for c-$C_3H_2D^+$, detected by a novel action spectroscopic method. The search for the spectral fingerprints was facilitated by the accurate *ab initio* predictions by Huang and Lee[8] reported previously.

2 Experimental methods

The rovibrational and rotational transitions of c-$C_3H_2D^+$ were measured in the 4 K 22-pole trap instrument COLTRAP, which has been previously described in detail by Asvany *et al.*[9,10] and will only be briefly explained here. Ions were generated in a storage ion source by electron impact ionization ($E_e \approx$ 30 eV) of a precursor 1 : 1 mixture of singly deuterated acetylene, HCCD (CDN isotopes Inc.; which over time had become a mixture of HCCH, HCCD and DCCD), and methane, CH_4. This mixture yielded excellent ion production conditions, possibly due to efficient reactions of the form CH_3^+ + HCCD → c-$C_3H_2D^+$ + H_2 (Ali *et al.*[11]). Every second, a pulse of up to a hundred thousand mass-selected c-$C_3H_2D^+$ ions (m = 40 u) was injected into the 4 K cold 22-pole ion trap. The trap was filled constantly with He ($n \approx 10^{13}$ cm^{-3}), and additionally a 1 : 3 Ne : He gas mixture was pulsed into the trap through a piezoelectrically actuated valve at the beginning of each trapping cycle.

Once trapped, the rovibrational transitions of c-$C_3H_2D^+$ were detected using a novel and very sensitive action spectroscopic method, called leak-out spectroscopy (LOS[12]). This method is based on the escape of a trapped ion after collision-induced transfer of vibrational into kinetic energy. After a cooling period of about 40 ms, the ions were irradiated for 500 ms by an IR beam traversing the trap. During this irradiation time, vibrationally excited c-$C_3H_2D^+$ ions will eventually be quenched by collisions with the Ne atoms present in the trap. Due to the neutral-to-ion mass ratio of 20 : 40, a substantial part of the vibrational energy is transferred into kinetic energy of the ion, namely a maximum of $20/(20 + 40) \times 3170$ cm^{-1} = 1057 cm^{-1} ≈ 0.131 eV. By keeping the potential difference between the exit electrode and the floating potential of the trap well below 131 mV, the ions may escape in that direction so that they fly towards the ion detector where they are counted. By repeating these trapping cycles at 1 Hz and counting the escaping c-$C_3H_2D^+$ ions as a function of the laser wavenumber, a rovibrational spectrum was recorded.

The narrow-band IR radiation was generated by a continuous-wave optical parametric oscillator (cw-OPO, Aculight Argos Model 2400, module C). The IR beam entered the vacuum environment of the ion-trap machine *via* a 0.6 mm thick diamond window (Diamond Materials GmbH), irradiated the trapped ions by crossing the 22-pole trap, and exited the trap instrument *via* a CaF_2 window, after which it was absorbed by a power meter. The detected power was on the order of 200 mW. The frequency of the IR radiation was measured by a Bristol model 621A wavemeter with an accuracy of about 0.001 cm^{-1} (in well-adjusted settings). We did additional calibration measurements with neutral C_2H_2 contained in a gas cell after the IR data reported here was measured. The exact frequencies for C_2H_2 are given in the HITRAN database[13] and, following this calibration, we shifted our data up by 0.007 cm^{-1}. With this, the accuracy of the given IR data should be on the order of 0.001 cm^{-1}.

For detecting pure rotational transitions, we used a novel double-resonance scheme[14] based on LOS, as recently described in Asvany *et al.*[15]. A rubidium-clock-referenced microwave synthesizer (Rohde & Schwarz SMF100A) driving an amplifier-multiplier chain (Virginia Diodes Inc. WR9.0M-AMC, WR4.3x2) was used to generate the mm-wave radiation. The radiation was focused by an ellipsoidal mirror ($f = 43.7$ mm (ref. 16)) before entering the vacuum environment through the diamond window. Both the IR and mm-wave radiation sources were used simultaneously, and their beams combined by a small hole in the ellipsoidal mirror through which the narrow IR beam could pass. The frequency of the IR photons is kept fixed on a rovibrational transition starting from a rotational level of the vibrational ground state, resulting in a detectable and constant LOS signal. The mm-wave photon then excites a rotational transition starting or ending on the rotational quantum state probed by the IR laser, thus decreasing or increasing the LOS signal, respectively. A rotational line can therefore be recorded by modulating this LOS signal, *i.e.*, by scanning the frequency of the mm-wave source, as will be shown in section 3.2.

3 Experimental measurements

3.1 Rovibrational measurements

c-$C_3H_2D^+$ is a near-oblate asymmetric top (C_{2v} symmetry) with 12 fundamental vibrational modes, none of which have been measured previously in the gas phase, to the best of our knowledge. A previous infrared experimental study for $C_3H_2D^+$ was done in 1986 by Craig *et al.*[17] in the solution phase, but its vibrational band centres are subject to shifts. Based on the spectroscopic predictions of Huang and Lee,[8] we detect the symmetric C–H stretch ν_1 (vibrational symmetry a_1, a-type transition) at high spectral resolution. The spectrum was measured in the range of our 3 μm OPO over multiple scans ranging over 1.5–2 cm^{-1} and is depicted in Fig. 1. In total, 126 lines were assigned in the ν_1 fundamental band of c-$C_3H_2D^+$, covering a range of 20 cm^{-1}. The observed and assigned line positions are given in the ESI† together with the fit residuals (see section 4 for the final fit).

Apart from the ν_1 lines of c-$C_3H_2D^+$, additional lines with seemingly characteristic patterns were observed in the spectrum. Based on our mass selectivity ($m = 40$ u) and by considering the possible isomers that could be formed in the source during the ionization of the precursor mixture, we believe that the remaining lines are presumably from singly deuterated l-$C_3H_3^+$, l-HDC_3H^+ or l-$H_2C_3D^+$. We tentatively assign the additional features to the l-HDC_3H^+ form (simulated as blue sticks in Fig. 1), based on theoretical predictions for both isomers of l-$C_3H_2D^+$ reported by Huang and Lee.[8] But due to perturbations in the measured spectrum, a definitive assignment needs further investigation, and we postpone the discussion of the l-$C_3H_2D^+$ species to a forthcoming publication. Additionally, due to the high sensitivity of LOS, some weak lines were seen that could not be assigned to either c-$C_3H_2D^+$ or l-HDC_3H^+. These might be perturbed

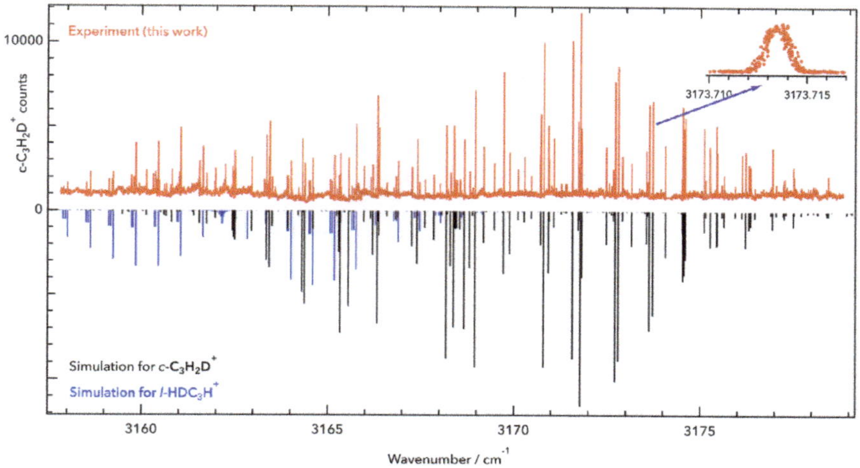

Fig. 1 Rovibrational spectrum of the symmetric C–H stretching vibration of c-$C_3H_2D^+$. Upper part: experimental data of this work (red). Lower part: PGOPHER simulation at a rotational temperature of $T = 10$ K (black sticks). The band origin was found to be at 3168.565 cm^{-1}. The zoomed-in inset shows the line $4_{13} \leftarrow 3_{12}$ of the $\nu_1 = 1 \leftarrow 0$ band at 3173.714 cm^{-1}. Some additional lines have been detected, which are tentatively assigned here to l-HDC_3H^+ (blue sticks). The red trace is provided as a separate data file in the ESI.†

lines of the l-HDC_3H^+ band, or some combination or overtone bands of lower vibrational states of c-$C_3H_2D^+$.

Because of the low temperature of the ion trap, the transitions exhibit narrow Doppler widths. One example is the line $4_{13} \leftarrow 3_{12}$ of the $\nu_1 = 1 \leftarrow 0$ band at 3173.714 cm^{-1}, which is depicted in the inset of Fig. 1. This line has a full width at half maximum (FWHM) of about 40 MHz, corresponding to a kinetic temperature of 12 K. This temperature is slightly higher than the nominal trap temperature due to known heating effects.[18]

3.2 Rotational measurements

c-$C_3H_2D^+$ is predicted to have a dipole moment component of 0.225 Debye along its *a*-axis[8] and should therefore feature weak a-type rotational transitions. Such rotational transitions can be measured in ion traps using double-resonance schemes,[14] as demonstrated in our group by several examples.[19,20] The double-resonance scheme using LOS, as applied in this work, has been demonstrated only recently.[15] Example transitions recorded by applying this double-resonance scheme are shown in Fig. 2.

Such spectra were recorded in multiple individual measurements in which the mm-wave frequency (blue arrows in Fig. 2) was stepped in an up-and-down manner several times in a given frequency window; the frequency steps were kept constant in individual experiments, and varied between 2 and 10 kHz. Selected rovibrational lines from the ν_1 band were used for the IR excitation (red arrows in Fig. 2). The spectroscopic data were normalized employing a frequency-switching procedure, *i.e.*, by dividing the c-$C_3H_2D^+$ counts monitored while scanning the spectral window of interest by the counts at an off-resonant mm-wave reference frequency. Therefore, the baselines in Fig. 2 are close to unity.

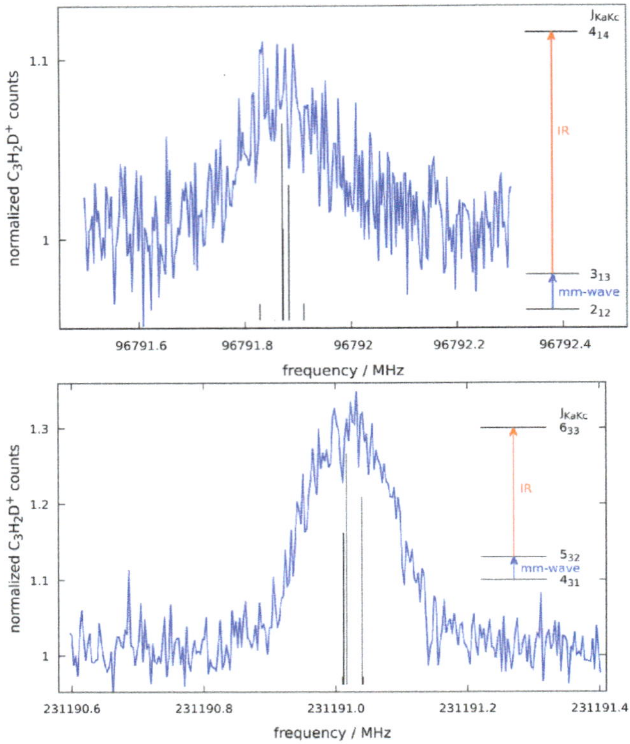

Fig. 2 Pure rotational transitions $J_{K_aK_c} = 3_{13} \leftarrow 2_{12}$ and $J_{K_aK_c} = 5_{32} \leftarrow 4_{31}$ of c-$C_3H_2D^+$ measured by a double-resonance scheme, in which the mm-wave excitation (blue arrows in insets) is followed by IR excitation (red arrows in insets) into the ν_1 vibrational band and subsequent leak-out from the trap. The step sizes are 3 kHz in both cases. The signal counts are normalized and therefore the baseline is unity. The calculated hyperfine structure of the lines are indicated by black sticks. The $J_{K_aK_c} = 3_{13} \leftarrow 2_{12}$ line is power-broadened.

Subsequently, the normalized counts of one measurement were averaged for each given frequency position. The obtained on-resonance signal enhancement is on the order of 8% to 30%, as seen in Fig. 2. Transition frequencies were determined by adjusting the parameters of an appropriate line-shape function (a Gaussian) to the experimental spectrum in a least-squares procedure. In total, 16 rotational lines have been detected, which are summarised in Table 1 and depicted in Fig. 3. The frequencies and their uncertainties in Table 1 result from several (about 3 to 10) independent measurements for each line.

4 Spectroscopic parameters

To obtain accurate spectroscopic parameters for the ground and first excited vibrational states of c-$C_3H_2D^+$, we carried out fits of the experimental transition frequencies using a Watson's S-reduced Hamiltonian in the I^r representation as implemented in the PGOPHER program.[21] Initially, the rovibrational lines were assigned using the assistance of *ab initio* calculations from Huang and Lee.[8] A first

Table 1 Measured low-lying rotational transitions of c-$C_3H_2D^+$ (in MHz). In total, 16 rotational lines have been measured directly by the double resonance technique. Uncertainties are given in parentheses. The obs-calc values are the difference between the observed transition frequency and the calculated frequency using the fitted rotational constants

$J_{K_aK_c} \leftarrow J_{K_aK_c}$	This work	obs-calc
$2_{11} \leftarrow 1_{10}$	90 293.296(15)	0.0283
$3_{13} \leftarrow 2_{12}$	96 791.884(15)	0.0235
$3_{03} \leftarrow 2_{02}$	97 972.838(15)	0.0190
$3_{22} \leftarrow 2_{21}$	118 087.007(10)	0.0090
$4_{14} \leftarrow 3_{13}$	125 054.348(10)	0.0071
$5_{24} \leftarrow 4_{23}$	180 356.9896(34)	−0.0024
$6_{16} \leftarrow 5_{15}$	180 733.6130(07)	−0.0005
$6_{06} \leftarrow 5_{05}$	180 737.9305(16)	0.0021
$4_{22} \leftarrow 3_{21}$	181 026.5963(24)	−0.0007
$5_{14} \leftarrow 4_{13}$	181 735.1075(07)	−0.0001
$4_{31} \leftarrow 3_{30}$	182 399.3395(19)	−0.0002
$3_{30} \leftarrow 2_{11}$	197 358.1392(25)	0.0001
$5_{33} \leftarrow 4_{32}$	202 928.5429(21)	0.0001
$7_{17} \leftarrow 6_{16}$	208 526.1574(26)	0.0003
$6_{15} \leftarrow 5_{14}$	208 882.1638(36)	0.0033
$5_{32} \leftarrow 4_{31}$	231 191.0231(09)	0.0001

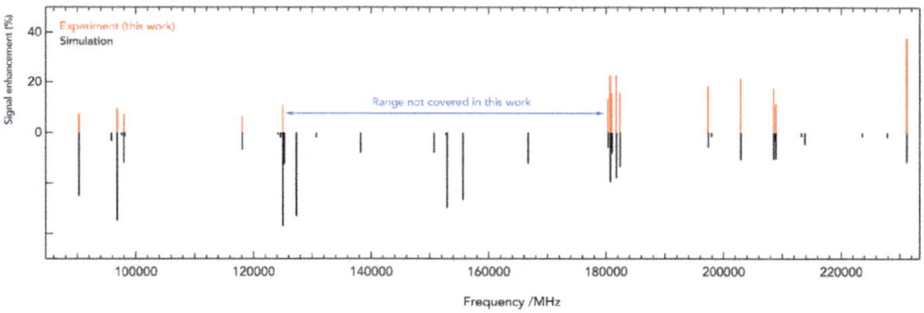

Fig. 3 Stick spectra of the pure rotational transitions of c-$C_3H_2D^+$ measured between 90 and 230 GHz. The experimental frequencies (in red, top) are compared to a PGOPHER simulation at 10 K (black, bottom). The experimental intensities are given as signal enhancement of the normalized ion counts (as in Fig. 2). The regions 90–125 GHz and 175–230 GHz were collected using two different multiplier chain settings and no experimental measurement was done for 125–170 GHz.

fit of these data was then used to predict the pure rotational transition frequencies of c-$C_3H_2D^+$ in the ground vibrational state. The best-fit ground-state molecular parameters were determined using the pure rotational transition frequencies given in Table 1 together with ground-state combination differences from the IR measurements. As the low-frequency measurements (up to 125 GHz) are affected by power broadening, transition frequencies are thought to be accurate only within 10–15 kHz, while the uncertainties for the measurements at higher frequencies are smaller. The overall rms of the fit is 0.0008 cm^{-1} for the IR combination differences and 11 kHz for the pure rotational data. The complete

Table 2 The best-fit spectroscopic parameters (in Watson's S-reduction) for the ground and first vibrationally excited state (C–H symmetric stretch) of c-$C_3H_2D^+$ obtained by fitting our data with the program PGOPHER.[21] All values are in MHz, except for the band origin of ν_1, which is given in cm^{-1}. The numbers in parentheses give the uncertainty of the last digits. Our values are compared to the accurate *ab initio* predictions of Huang and Lee[8]

	Experimental		Calculated[8]	
Parameter	Ground state	ν_1	Ground state	ν_1
ν_0	0	3168.56489(19)	0	3164.8/3173.1
A	30 747.25030(85)	30 586.58(69)	30 757	30 625
B	25 465.51589(40)	25 421.40(56)	25 478	25 442
C	13 897.86084(23)	13 859.82(24)	13 901	13 864
α^A		160.67(69)		131.92
α^B		44.12(56)		36.02
α^C		38.04(24)		37.57
D_J	0.014734(6)	[a]	0.0145	
D_{JK}	0.08525(5)	[a]	0.084	
D_K	−0.02599(7)	[a]	−0.026	
d_1	−0.010248(3)	[a]	−0.010018	
d_2	−0.005259(3)	[a]	−0.005106	
χ_{aa} (D)	0.187[b]			
χ_{bb} (D)	−0.100[b]			
χ_{cc} (D)	−0.087[b]			

[a] The distortion constants of the ν_1 state have been fixed to those determined for the ground state. [b] Nuclear quadrupole coupling constants of the deuterium nucleus calculated at the ae-CCSD(T)/cc-pwCVQZ level.

parameter set is given in Table 2. Hyperfine structure from the presence of the deuterium nucleus ($I_D = 1$) was not resolved in our measurements (*cf.* Fig. 2). For the sake of completeness, deuterium ($I_D = 1$) quadrupole coupling constants that were estimated at the CCSD(T)/cc-pwCVQZ level of theory (considering all electrons in the correlation treatment) using the CFOUR program[22] are also given in Table 2. The corresponding structural parameters at this level are $r_{C-C} = 1.36024$ Å and $r_{C-H} = 1.07828$ Å.

After obtaining the ground-state spectroscopic parameters, the band origin and rotational constants for the ν_1 vibrationally excited state were determined in a second fitting procedure. In this fit, the distortion constants in ν_1 have simply been fixed to those determined for the ground state. The derived spectroscopic parameters for the ground and ν_1 states are summarized in Table 2 along with the theoretical predictions from Huang and Lee.[8] The line lists of transition frequencies and residuals for both fits are provided in the ESI file.† Overall, the experimentally derived constants show excellent agreement with the calculated values, providing extra confirmation for the unequivocal detection of c-$C_3H_2D^+$ here.

5 Discussion and outlook

Leak-out spectroscopy (LOS) is a novel method for spectroscopy in ion traps, which features general applicability and high sensitivity. This is demonstrated here for the astronomically important c-$C_3H_2D^+$ molecule, which has evaded spectroscopic detection by other methods so far. The excellent signal quality (the background in Fig. 1 is on the order of 1000 c-$C_3H_2D^+$ counts per trapping cycle, while the signal is on the order of 11 000 counts for the strongest rovibrational lines) allowed the complete IR spectrum, shown in Fig. 1, to be recorded within a few days. Also, subsequent measurements of its 16 rotational lines in the ground state provide the first experimental rotational spectrum, which will be useful for the astronomical community.

As mentioned in the introduction, c-C_3H_2 is a molecule relevant for the chemistry of cold interstellar environments,[23,24] which is thought to be formed from c-$C_3H_3^+$ by dissociative recombination with an electron. But while the abundant c-C_3H_2 has a very large permanent dipole moment of 3.43 D (ref. 25) and was thus detected in the early days of radio astronomy, c-$C_3H_3^+$ (D_{3h} point-group symmetry) has no permanent dipole moment making its identification through rotational spectroscopy or radio astronomy impossible. The recently launched James Webb Space Telescope (JWST) offers higher sensitivity for the infrared region, which might aid the detection of c-$C_3H_3^+$ using its high-resolution rovibrational spectrum.[7]

Inclusion of one ^{13}C or one D atom in the structure of c-$C_3H_3^+$ imparts a non-zero dipole moment component in the isotopologues. In the case of c-$C_3H_2D^+$, this dipole moment is calculated to be 0.225 D,[8] which opens up the possibility for its search in various interstellar environments by its rotational fingerprints. This may be especially relevant for regions where other carbocations, such as l-C_3H^+ and the neutral c-C_3H_2 (and its isotopologues), have been detected.[6,26,27] These future detections could provide crucial inputs to improve low-temperature chemical models, and potentially highlight which reactions and their deuterated versions play a significant role in the hydrocarbon chemistry in these environments.

Among the different c-$C_3H_3^+$ isotopologues, given the dipole moment and elemental isotope abundances, it would seem that c-$C_3H_2D^+$ should be the easiest species to be identified in an astronomical spectrum. It is worth noting that its dipole moment is essentially twice that of carbon monoxide, which has a dipole moment of only 0.122 D.[24] For the warm carbon chain chemistry in the protostar L1527, a column density of $\sim 10^{12}$ cm^{-2} is determined for c-C_3HD by Sakai *et al.*,[26] implying a column density of $\sim 10^{10}$ cm^{-2} for c-$C_3H_2D^+$ (Aikawa, private communication, based on her model[28]). Similarly, for the colder pre-stellar core L1544, based on the abundance of c-C_3HD ($\sim$4–6 $\times$ 10^{12} cm^{-2}; Spezzano *et al.*,[6] and Giers *et al.*[29]), and available astrochemical models,[3,28] a column density in the order of 3×10^{10} cm^{-2} is estimated for c-$C_3H_2D^+$. This low column density in combination with the small dipole moment and the diluted spectral signature of this asymmetric top molecule results in integration times for strong selected lines in excess of several hundred hours for a single-dish telescope such as the IRAM 30 m facility (Schilke, Kim; private communication), rendering its current radio-astronomical detection very challenging. Indeed, this molecule has not been found in the deep surveys QUIJOTE[31] and GOTHAM[32] of TMC-1 (Cernicharo and McGuire, respectively, private communication).

Nevertheless, the first rotational spectra for this molecule obtained here lays the groundwork for future astronomical searches. The high sensitivity of the novel LOS method, with the possibility to be used also in an infrared-millimetre-wave double-resonance fashion, will also enable new investigations of challenging molecular ions relevant to astrochemistry. A recent example from our laboratory concerns another member of the $C_3H_3^+$ family, namely open-chain $H_2C_3H^+$, which is thought to be the precursor for l-C_3H_2 in space.[30] A publication about its rovibrational and rotational signatures is in preparation.

Author contributions

Divita Gupta: writing – original draft, methodology, investigation, formal analysis, visualization, validation, writing – review & editing. Weslley G. D. P. Silva: writing – original draft, formal analysis, investigation, visualization, validation, writing – review & editing. José L. Doménech: investigation, validation, writing – review & editing. Eline Plaar: investigation, writing – review & editing. Sven Thorwirth: validation, writing – review & editing. Stephan Schlemmer: conceptualization, funding acquisition, project administration, supervision, writing – review & editing. Oskar Asvany: conceptualization, funding acquisition, validation, methodology, supervision, writing – original draft, writing – review & editing.

Conflicts of interest

There are no conflicts to declare.

Acknowledgements

This work has been supported *via* Collaborative Research Centre 956, sub-project B2, funded by the Deutsche Forschungsgemeinschaft (DFG, project ID 184018867) and DFG SCHL 341/15-1 ("Cologne Center for Terahertz Spectroscopy"). The Cologne group acknowledges the ERC advanced grant (MissIons: 101020583) for funding. W. G. D. P. S. thanks the Alexander von Humboldt Foundation for support through a postdoctoral fellowship. J. L. D. acknowledges the support from the MCINN project PID2020-113084GB-I00 and the CSIC project ILINK+ LINKA20353. We thank Peter Schilke, Wonju Kim, Yuri Aikawa, José Cernicharo, Brett McGuire and Silvia Spezzano for discussions about the feasibility of detecting c-$C_3H_2D^+$ in space.

Notes and references

1 P. Thaddeus, M. Guelin and R. A. Linke, *Astrophys. J.*, 1981, **246**, L41–L45.
2 P. Thaddeus, J. M. Vrtilek and C. A. Gottlieb, *Astrophys. J.*, 1985, **299**, L63–L66.
3 O. Sipilä, S. Spezzano and P. Caselli, *Astron. Astrophys.*, 2016, **591**, L1.
4 J. Gómez-González, M. Guélin, J. Cernicharo, C. Kahane and M. Bogey, *Astron. Astrophys.*, 1986, **168**, L11.
5 M. B. Bell, P. A. Feldman, H. E. Matthews and L. W. Avery, *Astrophys. J.*, 1986, **311**, L89.
6 S. Spezzano, S. Brünken, P. Schilke, P. Caselli, K. M. Menten, M. C. McCarthy, L. Bizzocchi, S. P. Treviñõ-Morales, Y. Aikawa and S. Schlemmer, *Astrophys. J., Lett.*, 2013, **769**, L19.
7 D. Zhao, K. D. Doney and H. Linnartz, *Astrophys. J., Lett.*, 2014, **791**, L28.
8 X. Huang and T. J. Lee, *Astrophys. J.*, 2011, **736**, 33.
9 O. Asvany, F. Bielau, D. Moratschke, J. Krause and S. Schlemmer, *Rev. Sci. Instrum.*, 2010, **81**, 076102.

10 O. Asvany, S. Brünken, L. Kluge and S. Schlemmer, *Appl. Phys. B*, 2014, **114**, 203–211.
11 A. Ali, E. C. Sittler, D. Chornay, B. R. Rowe and C. Puzzarini, *Planet. Space Sci.*, 2013, **87**, 96–105.
12 P. C. Schmid, O. Asvany, T. Salomon, S. Thorwirth and S. Schlemmer, *J. Phys. Chem. A*, 2022, **126**, 8111–8117.
13 I. Gordon, L. Rothman, R. Hargreaves, R. Hashemi, E. Karlovets, F. Skinner, E. Conway, C. Hill, R. Kochanov, Y. Tan, P. Wcisło, A. Finenko, K. Nelson, P. Bernath, M. Birk, V. Boudon, A. Campargue, K. Chance, A. Coustenis, B. Drouin, J. Flaud, R. Gamache, J. Hodges, D. Jacquemart, E. Mlawer, A. Nikitin, V. Perevalov, M. Rotger, J. Tennyson, G. Toon, H. Tran, V. Tyuterev, E. Adkins, A. Baker, A. Barbe, E. Canè, A. Császár, A. Dudaryonok, O. Egorov, A. Fleisher, H. Fleurbaey, A. Foltynowicz, T. Furtenbacher, J. Harrison, J. Hartmann, V. Horneman, X. Huang, T. Karman, J. Karns, S. Kassi, I. Kleiner, V. Kofman, F. Kwabia-Tchana, N. Lavrentieva, T. Lee, D. Long, A. Lukashevskaya, O. Lyulin, V. Makhnev, W. Matt, S. Massie, M. Melosso, S. Mikhailenko, D. Mondelain, H. Müller, O. Naumenko, A. Perrin, O. Polyansky, E. Raddaoui, P. Raston, Z. Reed, M. Rey, C. Richard, R. Tóbiás, I. Sadiek, D. Schwenke, E. Starikova, K. Sung, F. Tamassia, S. Tashkun, J. Vander Auwera, I. Vasilenko, A. Vigasin, G. Villanueva, B. Vispoel, G. Wagner, A. Yachmenev and S. Yurchenko, *J. Quant. Spectrosc. Radiat. Transfer*, 2022, **277**, 107949.
14 O. Asvany and S. Schlemmer, *Phys. Chem. Chem. Phys.*, 2021, **23**, 26602–26622.
15 O. Asvany, S. Thorwirth, P. C. Schmid, T. Salomon and S. Schlemmer, 2023, submitted to PCCP.
16 P. Jusko, O. Asvany, A.-C. Wallerstein, S. Brünken and S. Schlemmer, *Phys. Rev. Lett.*, 2014, **112**, 253005.
17 N. C. Craig, J. Pranata, S. J. Reinganum, J. R. Sprague and P. S. Stevens, *J. Am. Chem. Soc.*, 1986, **108**, 4378–4386.
18 O. Asvany and S. Schlemmer, *Int. J. Mass Spectrom.*, 2009, **279**, 147–155.
19 C. Markus, S. Thorwirth, O. Asvany and S. Schlemmer, *Phys. Chem. Chem. Phys.*, 2019, **21**, 26406–26412.
20 P. Jusko, A. Stoffels, S. Thorwirth, S. Brünken, S. Schlemmer and O. Asvany, *J. Mol. Spectrosc.*, 2017, **332**, 59–66.
21 C. M. Western, *J. Quant. Spectrosc. Radiat. Transfer*, 2017, **186**, 221–242.
22 D. A. Matthews, L. Cheng, M. E. Harding, F. Lipparini, S. Stopkowicz, T.-C. Jagau, P. G. Szalay, J. Gauss and J. F. Stanton, *J. Chem. Phys.*, 2020, **152**, 214108.
23 J. Pety, P. Gratier, V. Guzmán, E. Roueff, M. Gerin, J. R. Goicoechea, S. Bardeau, A. Sievers, F. Le Petit, J. Le Bourlot, A. Belloche and D. Talbi, *Astron. Astrophys.*, 2012, **548**, A68.
24 J. Muenter, *J. Mol. Spectrosc.*, 1975, **55**, 490–491.
25 H. Kanata, S. Yamamoto and S. Saito, *Chem. Phys. Lett.*, 1987, **140**, 221–224.
26 N. Sakai, T. Sakai, T. Hirota and S. Yamamoto, *Astrophys. J.*, 2009, **702**, 1025.
27 S. Brünken, L. Kluge, A. Stoffels, O. Asvany and S. Schlemmer, *Astrophys. J., Lett.*, 2014, **783**, L4.
28 Y. Aikawa, E. Herbst, H. Roberts and P. Caselli, *Astrophys. J.*, 2005, **620**, 330.
29 K. Giers, S. Spezzano, F. Alves, P. Caselli, E. Redaelli, O. Sipilä, M. Ben Khalifa, L. Wiesenfeld, S. Brünken and L. Bizzocchi, *Astron. Astrophys.*, 2022, **664**, A119.

30 J. Cernicharo, C. A. Gottlieb, M. Guelin, T. C. Killian, P. Thaddeus and J. M. Vrtilek, *Astrophys. J., Lett.*, 1991, **368**, L43.
31 J. Cernicharo, M. Agúndez, C. Cabezas, N. Marcelino, B. Tercero, J. R. Pardo, R. Fuentetaja and P. de Vicente, *EPJ Web Conf.*, 2022, **265**, 00041.
32 B. A. McGuire, A. M. Burkhardt, R. A. Loomis, C. N. Shingledecker, K. L. K. Lee, S. B. Charnley, M. A. Cordiner, E. Herbst, S. Kalenskii, E. Momjian, E. R. Willis, C. Xue, A. J. Remijan and M. C. McCarthy, *Astrophys. J., Lett.*, 2020, **900**, L10.

Faraday Discussions

PAPER

Hunting for interstellar molecules: rotational spectra of reactive species†

Cristina Puzzarini, * Silvia Alessandrini, Luca Bizzocchi * and Mattia Melosso *

Received 25th February 2023, Accepted 13th March 2023
DOI: 10.1039/d3fd00052d

Interstellar molecules are often highly reactive species, which are unstable under terrestrial conditions, such as radicals, ions and unsaturated carbon chains. Their detection in space is usually based on the astronomical observation of their rotational fingerprints. However, laboratory investigations have to face the issue of efficiently producing these molecules and preserving them during rotational spectroscopy measurements. A general approach for producing and investigating unstable/reactive species is presented by means of selected case-study molecules. The overall strategy starts from quantum-chemical calculations that aim at obtaining accurate predictions of the missing spectroscopic information required to guide spectral analysis and assignment. Rotational spectra of these species are then recorded by exploiting the approach mentioned above, and their subsequent analysis leads to accurate spectroscopic parameters. These are then used for setting up accurate line catalogs for astronomical searches.

1 Introduction

Since the discovery of the first polyatomic molecules in the late 1960s and early 1970s, it soon became clear that the interstellar medium (ISM) is characterized by a rich chemistry. Because of the extreme conditions of the ISM, with temperatures ranging between 10 and 100 K, very low densities (from 10 to 10^8 molecules cm^{-3}), and ionizing radiation, the chemistry there occurring leads to the formation of exotic molecules.[1–4] Among the about 270 molecules detected in the ISM and circumstellar shells so far, a sizeable number of them are species that under terrestrial conditions are unstable and/or highly reactive.[5] The list includes radicals and ions, but also closed-shell unsaturated systems such as imines and carbon chains. Illustrative examples are the HC_5O radical,[6,7] the C_8H^- negative ion,[8,9] the HC_9N carbon chain,[10] the protonated HC_5NH^+ carbon chain,[11]

ROT&Comp Lab, Department of Chemistry "Giacomo Ciamician", University of Bologna, Via F. Selmi 2, I-40126 Bologna, Italy. E-mail: cristina.puzzarini@unibo.it; luca.bizzocchi@unibo.it; mattia.melosso2@unibo.it

† Electronic supplementary information (ESI) available. See DOI: https://doi.org/10.1039/d3fd00052d

substituted imines like allylimine,[12] and the unstable enol form of glycolaldehyde, *i.e.* (*Z*)-1,2-ethenediol enol.[13] These examples give an idea of the exotic chemistry taking place in the ISM and provide a clear evidence of the fact that unstable and/or reactive species are produced and survive long enough to be detected. In parallel, this chemistry also leads to small organic molecules (denoted as "interstellar" complex organic molecules, COMs[14]) such as *N*-methylformamide (CH_3NHCHO),[15] urea ($NH_2C(O)NH_2$),[16,17] ethanolamine ($HOCH_2CH_2NH_2$),[18] and *n*-propanol ($CH_3CH_2CH_2OH$).[19,20]

Molecular spectroscopy plays a central role in the discovery of interstellar molecules,[5,21] with the overwhelming majority of chemical species in the cosmos being discovered *via* their rotational signatures. Indeed, the astronomical observation of the rotational spectroscopic features of a given molecule provides unequivocal proof of its presence in the astronomical environment under consideration.[5,21–23] In principle, the systematic observation of a given astronomical source leads to a complete census of its molecular content. However, astronomical line surveys still reveal unknown features, the main reason being the lack of corresponding spectroscopic information.[5,21] As suggested above, interstellar species can be short-lived systems on Earth, this rendering their spectroscopic characterization a challenging task. Indeed, molecular species that are not yet studied in the laboratory show relevant difficulties in their production. This is the topic touched on in this work. Radicals and ions are transient species which are typically generated *in situ* using plasma techniques.[24–26] Neutral, closed-shell unsaturated species such as carbon chains are stable molecules that, however, tend to be highly reactive and thus need to be produced *in situ* as well.

In this contribution, a laboratory strategy pursued to obtain very accurate line catalogs for guiding the astronomical search of unstable and/or reactive species is presented. Illustrative systems that are produced either using flash vacuum pyrolysis[27–29] or DC discharge[24–26] are investigated. As examples of closed-shell reactive species, the DC_7N and ClC_3N carbon chains have been studied, while as examples of open-shell systems, the CCS and CH_2CN radicals are considered. The strategy relies on the interplay of experiment and theory, which is crucial because it often happens that, for this type of molecular species, no previous experimental data are available and the study starts from an accurate computational characterization.[12,28] In the following section (Methods), the strategy is introduced; then, the experimental and computational aspects are detailed. Subsequently, the results for the species mentioned above are presented and discussed. Finally, in the last section, conclusions are drawn.

2 Methods

As mentioned in the Introduction, an accurate spectroscopic characterization is required to guide astronomical observations. The interplay of experiment and theory in the field of rotational spectroscopy to be exploited in astronomical searches is graphically represented in Fig. 1. As shown there, the strategy starts with accurate quantum-chemical computations of the spectroscopic parameters (those of an effective Watson-type Hamiltonian) required for the prediction of the rotational spectrum. The simulated spectrum is used to plan the experiment and, once the experimental counterpart is available, is employed to guide spectral analysis and assignment. The subsequent step is the fitting of the assigned

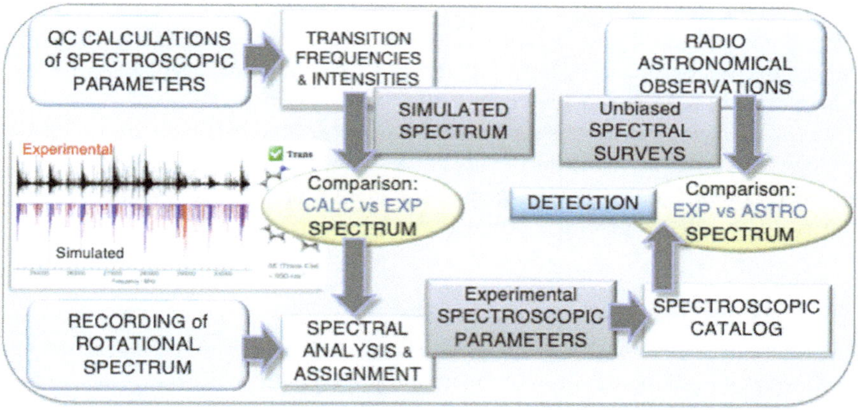

Fig. 1 Graphical representation of the interplay of experiment and theory exploited at the ROT&Comp laboratory in Bologna for hunting molecular species in the ISM.

transitions with a suitable effective Hamiltonian (the same used for simulation), with the computed parameters providing the starting point. The fitting procedure eventually leads to the set up of a spectroscopic line catalog, which is then used to search for the molecule in the astronomical line surveys.

An illustrative example of this strategy at work is provided by ref. 13 and 28: (*Z*)-1,2-ethenediol, a key prebiotic intermediate in the formose reaction, has been spectroscopically characterized in the laboratory,[28] thus providing the line list allowing its detection.[13] Fig. 2 provides a graphical representation of how the quantum-chemical calculations allowed an easy assignment of the experimental

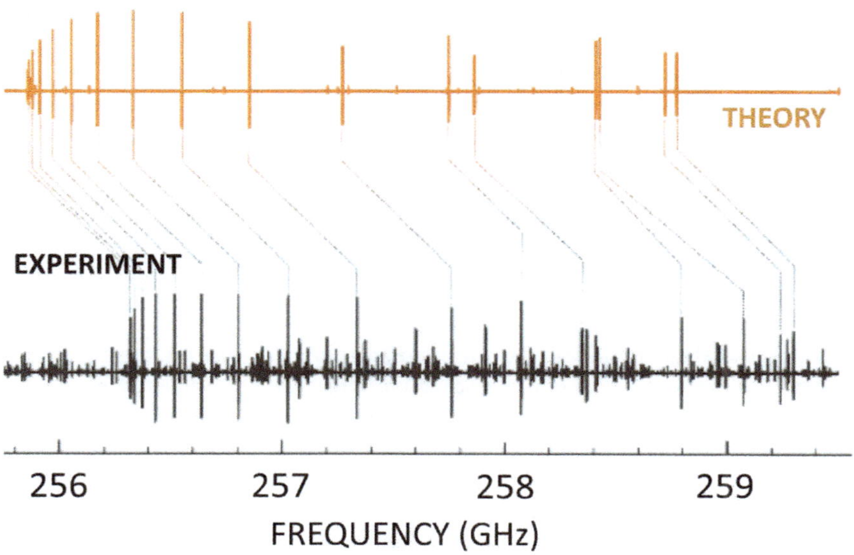

Fig. 2 Portion of the millimeter-wave spectrum of (*Z*)-1,2-ethenediol in the 256–259 GHz range: the experimental spectrum (in black) is compared with the computational simulation (in orange).

spectrum. In this respect, it has to be noted that no previous spectroscopic information was available for (*Z*)-1,2-ethenediol, which – being an unstable molecule – was produced by flash vacuum pyrolysis. Another recent example for this strategy at work is provided by ref. 12.

In the following section the experimental aspects of the strategy are presented, while in the subsequent one the details of the quantum-chemical calculations are provided. In this work, the focus is on unstable and/or reactive molecules. Therefore, the challenge of the production of these species is mainly addressed.

2.1 Experimental details

The rotational spectrometer employed in this work is a frequency-modulation millimeter/sub-millimeter-wave (mm/sub-mm) spectrometer working in the 75 GHz to 1.6 THz range, which is described in detail in ref. 30 and 31. In short, radiation is produced by a series of Gunn oscillators covering the 75–134 GHz range. Higher frequencies, up to the THz regime, are obtained by means of passive frequency multipliers. Frequency and phase are stabilized by a phase-lock loop. The detection system consists of either Schottky barrier diodes or a liquid-helium-cooled InSb hot electron bolometer (QMC Instr. Ltd type QFI/2). In all cases, the output signal is demodulated by a lock-in amplifier set at twice the modulation frequency, thus leading to second harmonic (2*f*) detection (nearly second derivative of the line profile). To produce unstable and/or reactive molecules, such as ions, radicals, imines and carbon chains, the mm/sub-mm spectrometer is combined with either a flash vacuum pyrolysis system[29] or a DC discharge apparatus.[25,26]

Concerning the flash vacuum pyrolysis (FVP) technique, two different apparatuses can be employed. The first system is constituted by a 30 cm long tubular oven which surrounds a quartz tube directly connected to one inlet of the spectrometer absorption cell. This latter is a glass tube 3.25 m long and 5 cm in diameter. The maximum temperature that can be reached by the oven is 1200 °C. In the second apparatus, a 1.5 m long quartz absorption cell is surrounded by a 90 cm tubular furnace, which is able to reach the same temperature as the first system. Since the cell and oven lengths are different, the heating is inhomogeneously distributed along the cell, with the pyrolysis reaction being confined to the 90 cm directly heated by the oven. To increase the length of the reactive area, a double-pass configuration is arranged,[32] with the effective absorption path of the pyrolysis products increasing to, at least, 1.8 m. For both apparatuses, in view of the unstable nature of the target molecules, measurements are performed in dynamic conditions, *i.e.* a tenuous flow of fresh pyrolysis products is continuously provided by the vacuum system. In all experiments, the production conditions are optimized by adjusting the temperature of the precursor, the temperature of the pyrolysis apparatus, and the pressure of the gaseous products inside the absorption cell.

Moving to the species investigated in this work, ClC_3N and DC_7N have been produced by means of FVP. Gaseous samples of ClC_3N have been produced by co-pyrolysis of carbon tetrachloride (CCl_4) and acetonitrile (CH_3CN) in a 3 : 1 abundance ratio. Typically, 60 mTorr (8 Pa) of CCl_4 and 20 mTorr (~3 Pa) of CH_3CN were flowed into the tubular quartz reactor heated at 1150 °C. The measurements were performed while continuously pumping the pyrolysis products through the absorption cell, which was kept at a pressure of ~5 mTorr (~1 Pa). ClC_3N is moderately stable in the gas phase. Trapping and conservation of the compound at liquid nitrogen temperature is also possible. Purification of the samples *via* low-temperature distillation has been attempted and proved it was possible to achieve a factor of 2 improvement in the intensity of the absorption signals. DC_7N was produced by co-pyrolysis of fully-deuterated toluene (toluene-d_8) and trichloroacetonitrile (Cl_3CCN) at 1200 °C. The mixture composition was adjusted to have a constant flow of 150 mTorr (20 Pa) of Cl_3CCN and 5–10 mTorr (~1 Pa) of toluene-d_8 in the pyrolysis reactor. This corresponds to a pressure of ~15 mTorr (~2 Pa) after the expansion of the reaction products in the absorption cell.

The mm/sub-mm spectrometer can also be equipped with a negative glow-discharge cell to produce either ionic or radical species. To improve the signal-to-noise ratio (S/N), measurements can be carried out at low temperature (*e.g.* by circulation of liquid nitrogen in an external plastic pipe tightly wound around the cell). Ions or radicals are prepared directly inside the absorption cell starting from a suitable mixture of gases and by applying a DC discharge of a few mA to some tens of mA. A buffer gas (usually, He or Ar) is often employed together with the precursor(s). Measurements are always performed in a continuous flow of gas, maintained by a diffusion pump, in order to constantly provide fresh precursor gases and remove degraded products. For ions, a longitudinal magnetic field can be applied to improve the S/N. In the case of radicals, this can be used to disentangle their signal from those of the closed-shell species whose transitions contaminate the spectral window. In this work, we considered a sulfur-containing radical, CCS, which has been produced using CS_2 as precursor and He as buffer gas in a 2 : 1 ratio. In most of the measurements, the combination of 30 mTorr (4 Pa) of CS_2 with 15 mTorr (2 Pa) of He has been employed, and a DC discharge of 70–75 mA applied.

2.2 Computational details

The pieces of information required for predicting rotational spectra are:

- rotational parameters,
- type of transitions observable (and their intensity),
- if relevant, hyperfine parameters.

In Table 1, the correlation between the spectroscopic and molecular parameters to be predicted and the quantities determinable by quantum chemistry is

Table 1 The relevant parameters in rotational spectroscopy and their computational requirements

Parameter	Symbol(s)	Computational task(s)
Rotational constants	A_e, B_e, C_e	Geometry optimization
Vibrational corrections to B_e values	ΔB_{vib}	Anharmonic force field
Quartic centrifugal-distortion constants	D_J, D_K, D_{JK}, …	Harmonic force field
Sextic centrifugal-distortion constants	H_J, H_K, H_{JK}, …	Anharmonic force field
Electric dipole moment	μ	First derivative of electronic energy with respect to electric-field components
Nuclear quadrupole coupling constants	χ_{ij}	Electric-field gradient (first derivative of electronic energy with respect to quadrupole moment)

summarized. While in the following the first and third point of the list above are discussed, when considering the type of transitions observable and their intensity, the crucial information is the electric dipole moment or better its components along the inertial axes.[33] As pointed out in Table 1, the dipole moment is a first order property which can be obtained from dedicated computations, but also as a byproduct in geometry optimizations.

Rotational parameters include rotational and centrifugal distortion constants. Rotational constants are by far the leading terms and can be seen as formed by two contributions: the dominant one (about 99%) which only depends on the equilibrium structure and the vibrational correction (*vide infra*) which depends on the coupling between the rotational and vibrational motion.[34,35] It is thus clear that accurate equilibrium structures are needed in order to get accurate equilibrium rotational constants.[34,36,37] To reach high accuracy, basis-set and electron-correlation effects must be accounted for simultaneously, and this is accomplished by exploiting quantum-chemical composite schemes.[38] In these approaches, the various contributions are evaluated separately at the highest possible level and then combined in order to obtain the best theoretical estimate, thereby exploiting the additivity approximation. In the literature, several composite schemes have been proposed in order to obtain the best possible compromise between accuracy and computational cost.[37,39–51]

In this work, we employed a so-called "gradient scheme" entirely based on coupled-cluster (CC) theory,[39,40] which is implemented in the CFOUR quantum-chemistry package.[52] In this approach, the additivity approximation is exploited at the energy-gradient level to be minimized for obtaining the equilibrium structure. In particular, we employed the CCSD(T)/CBS+CV scheme, where CCSD(T) stands for the CC singles and doubles (CCSD) approximation augmented by a perturbative treatment of triple excitations.[53] The CBS acronym implies the extrapolation to the complete basis set (CBS) limit, while CV refers to the incorporation of the core-valence correlation contribution. The energy gradient to be minimized contains three terms. The first two contributions are the gradients for the separate extrapolation of HF-SCF (Hartree-Fock Self Consistent Field) and CCSD(T) correlation energies to the CBS limit using an exponential extrapolation[54] for the former and the n^{-3} extrapolation scheme[55] for the latter. Extrapolation to the CBS limit requires a hierarchy of bases to be employed, with the correlation-consistent valence cc-pV*n*Z bases[56,57] being used in this work, with $n =$ T, Q, 5 for HF-SCF and $n =$ T, Q for CCSD(T). The third term incorporates the CV correlation effects because the CCSD(T) extrapolation to the CBS is performed within the frozen-core (fc) approximation. The CV correction is obtained as the difference of all-electron (ae) and fc CCSD(T) calculations using the same core-valence basis set, which is cc-p(w)CV*n*Z ($n =$ T, Q) in the present case.[58,59]

If required by the dimension of the molecule, to obtain accurate equilibrium structures while retaining a feasible computational cost, one can resort to the so-called "cheap-composite" scheme,[41,48] which – starting from fc-CCSD(T)/cc-pVTZ calculations – incorporates the extrapolation to the CBS limit and the effect of core correlation at the MP2 level (MP2 stands for second-order Møller–Plesset perturbation theory[60]). For larger systems for which even the "cheap-composite" scheme becomes too expensive, such as small polycyclic aromatic hydrocarbons (PAH),[61] a possible way-out is offered by the so-called "Lego brick" approach.[61,62] The idea at its basis is that the system can be seen as formed by different fragments (*i.e.*, the "Lego bricks"), whose accurate equilibrium geometries are available, with the template molecule (TM) approach[63] being employed to account for the modifications occurring when moving from the isolated fragment to the molecular system under consideration. To correct the linkage between different fragments, the linear regression (LR) model is then employed.[64,65] For the accurate determination of the equilibrium structure of the fragments, in this work, we have resorted to the so-called semi-experimental (SE) approach.[63,66,67] The SE equilibrium structural parameters are determined by a linear least-squares fit of the SE equilibrium rotational constants of different isotopologues. These are obtained by correcting the experimental ground-state rotational constants (B_0^{exp}) with computed vibrational ($\Delta B_{\mathrm{vib}}^{\mathrm{calc}}$) corrections.[66]

In the following, we refer to the "Lego brick" approach as TM+LR, or only TM whenever LR is not applied. While the reader is referred to ref. 61 and 62 for all details on this scheme, it has to be mentioned that the TM approach combines density functional theory (DFT) calculations with corrections from SE equilibrium structures of the fragments envisaged in the molecule under investigation. In this work, the DFT level considered is the double-hybrid rev-DSDPBEP86 (revDSD) functional[68] in conjunction with the jun-cc-pVTZ triple-zeta basis set.[69] DFT calculations also incorporate dispersion corrections by means of the Grimme's DFT-D3 scheme[70] together with the Becke–Johnson damping function.[71] In this work, the TM+LR approach has been employed for the accurate equilibrium structure determination of HC_7N, the envisaged fragments (HC≡C–, –C≡C–, and –C≡N) being taken from the HCCH and HCN molecules. Furthermore, in addition to the TM approach, it has been necessary to correct the bonds connecting the fragments. For HC_7N, all connectors are single C–C-bond types, the corrective factor being −0.00184 Å and taken from ref. 65.

Whenever we want to predict the experiment and, thus, to move from equilibrium to the vibrational ground state, computation of the vibrational corrections is required. According to vibrational perturbation theory to the second order (VPT2),[72] we can write:

$$B_0 = B_\mathrm{e} + \Delta B_\mathrm{vib} = B_\mathrm{e} - \frac{1}{2}\sum_r \alpha_r{}^i \tag{1}$$

where $\alpha_r{}^i$ denotes the vibration–rotation interaction constants, with i being the inertial axis (a, b or c) and the sum running over all the r vibrational modes. Since the largest contribution to $\alpha_r{}^i$ involves cubic force constants, anharmonic force field calculations are required for evaluating the $\Delta B_{\mathrm{vib}}^i$ term. In this work, we have employed both CCSD(T) and MP2 methods for its evaluation (details are provided alongside the discussion of the specific molecules).

As concerns centrifugal-distortion constants, their evaluation involves force field calculations: the harmonic one for the quartic terms and the anharmonic one for the sextic terms. This means that these parameters can be obtained as byproducts of the computations required for getting the vibrational corrections to the rotational constants.

Whenever a closed-shell molecule contains one or more atoms with non-null nuclear spin, electric and/or magnetic interactions that split rotational energy levels occur. The consequence is that rotational transitions are split as well, giving rise to the so-called hyperfine structure. In this study, the only interaction of interest is the nuclear quadrupole coupling because nitrogen, chlorine and deuterium are quadrupolar nuclei, *i.e.* their nuclear spin I is greater than 1/2. While interested readers are referred to ref. 33, 37 and 51, here we mention that, from a computational point of view, the prediction of nuclear quadrupole coupling constants requires the computation of the electric field gradient at the quadrupolar nucleus (see Table 1). Following the literature on this topic,[37,51,73] we have employed the CCSD(T) method in combination with core-valence basis sets and correlating all electrons. The situation is more involved when dealing with open-shell species because in such cases we need to account for the coupling of the rotational angular momentum with the electronic spin momentum and, if more than one unpaired electron is present, the coupling between the electronic spins.[26,33]

3 Results

In this section, the spectroscopic results are presented and discussed. First, the unsaturated carbon chains are addressed and, then, we move to the radical species considered in this work.

3.1 Unsaturated carbon chains: ClC_3N and DC_7N

The rotational spectrum of ClC_3N was previously investigated in the 13–25 GHz range,[74] but the frequencies of the recorded transitions were not provided. Therefore, the present results are exclusively based on our measurements. These have been performed using the spectrometer introduced in the Methods section, with ClC_3N being produced by FVP as explained above. Measurements have been carried out in the 82–111 GHz range, sampling transitions with rotational quantum number J values varying from 29 to 40. At such large values of J, the hyperfine structure (mainly due to chlorine) collapses and is no longer resolvable, as seen in Fig. 3. This figure provides an overview of the $J = 31 \leftarrow 30$ transition of $^{35}ClC_3N$ in the vibrational ground state, also pointing out the presence of rotational transitions in vibrational excited states. In view of the large natural abundance of the ^{37}Cl isotope, rotational spectra of both $^{35}ClC_3N$ and $^{37}ClC_3N$ have been recorded, with transition frequencies retrieved with an accuracy of 20 kHz for both isotopologues. The spectroscopic parameters obtained from their analysis are collected in Table 2, where they are compared to those from ref. 74 and our quantum-chemical results. Concerning the latter, the equilibrium rotational constant has been obtained at the CCSD(T)/CBS+CV level and augmented by the vibrational correction from fc-CCSD(T)/cc-pVTZ calculations, which also provided the sextic centrifugal distortion constant. The quartic centrifugal term has instead been computed at the CCSD(T)/cc-pCVTZ level of theory, with all electrons correlated. Finally, nitrogen and chlorine quadrupole coupling constants have been evaluated from ae-CCSD(T)/cc-pCVQZ calculations on top of the CCSD(T)/CBS+CV geometry.

The results collected in Table 2 point out the need of measurements at higher frequency in order to improve rotational constant determination, which is also related to the derivation of an improved value of the quartic centrifugal distortion constant. In this respect, it is interesting to note that, while our measurements are not able to determine the sextic centrifugal distortion constant with a meaningful

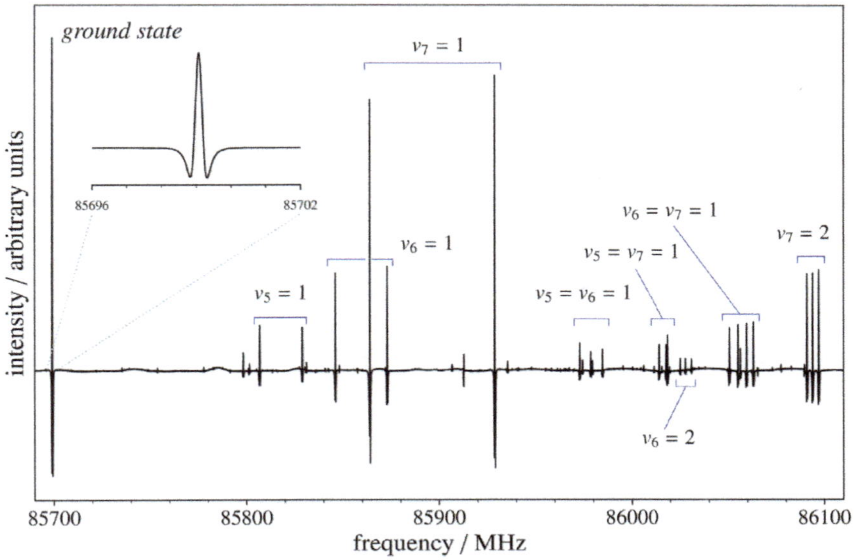

Fig. 3 Recording of the $J = 31 \leftarrow 30$ transition of $^{35}ClC_3N$. The strong leftmost feature belongs to the rotational spectrum in the vibrational ground state (also shown in the inset with a magnified *x*-axis scale). Several multiplets belonging to the rotational spectra in vibrational excited states due to the fundamentals, overtones, and combinations of the v_5, v_6, and v_7 bending modes are clearly visible towards higher frequencies.

Table 2 Vibrational ground-state spectroscopic parameters of $^{35}ClC_3N$ and $^{37}ClC_3N$

		This work		
Constant	Unit	Exp.[a]	Theo.[b]	Ref. 74
$^{35}ClC_3N$				
B_0	MHz	1382.32357(35)	1382.35	1382.328(2)
D_J	kHz	41.51(14)	39.6	0.054(5)
H_J	mHz	-0.3^{fix}	−0.3	—
χ(Cl)	MHz	—	−78.74	−75(4)
χ(N)	MHz	—	−4.28	—
$^{37}ClC_3N$				
B_0	MHz	1350.35289(25)	1350.38	1350.360(2)
D_J	kHz	39.835(92)	37.9	0.054(5)
H_J	mHz	-0.3^{fix}	−0.3	—
χ(Cl)	MHz	—	−62.06	−62(3)
χ(N)	MHz	—	−4.28	—

[a] Numbers in parentheses are one standard deviation in units of the last quoted digit. The "fix" superscript note indicates values held fixed at their computed value. [b] Rotational constants at the CCSD(T)/CBS+CV level augmented by fc-CCSD(T)/cc-pVTZ vibrational corrections. Quartic and sextic centrifugal distortion constants are at the ae-CCSD(T)/cc-pCVTZ and fc-CCSD(T)/cc-pVTZ levels, respectively. Nuclear quadrupole coupling constants at the ae-CCSD(T)/cc-pCVQZ level.

uncertainty, it is necessary to fix it at the computed value in order to improve the fit (and thus reproduce transition frequencies within their experimental error) and obtain the correct ratio for the *D* values of the two isotopic species. This simple example shows the important role played by accurate quantum-chemical computations also for guiding the fitting procedure for the derivation of the experimental spectroscopic parameters. Focusing on computational results, it is worth noting that, for both isotopologues, the rotational constant is predicted with a deviation as small as 0.002% (which means about 30 kHz in absolute terms). The error increases to 5% for the quartic centrifugal distortion constant, but still the accuracy is good enough to guide experiment and in particular, as mentioned above, to provide the reference value for the $D(^{35}ClC_3N)/D(^{37}ClC_3N)$ ratio. In Table 2, the nuclear quadrupole coupling constants are also reported. The only experimental values available are for $\chi(^{35}Cl)$ and $\chi(^{37}Cl)$, and they come from ref. 74. The experimental error is quite large (about 5%) and, according to the literature on this topic,[37,51,73,77] the computed values are probably more accurate.

The present work allowed us to set up an accurate line catalog to support astronomical observations, with transition frequencies predicted with the proper accuracy (*i.e.* uncertainties smaller than 100 kHz) up to 250 GHz. Even if only seven molecules containing chlorine have been detected in space,[5] the detection of CH_3Cl towards IRAS 16293 (ALMA observations) and in the coma of the 67P/Churyumov–Gerasimenko comet (ROSINA mass spectrometry measurements)[78] suggests that polyatomic molecules with chlorine can be discovered in the ISM.

In ref. 75, the rotational spectra of all single-substituted isotopologues of three linear cyanopolyynes, namely HC_7N, HC_9N, and $HC_{11}N$, have been investigated in the 6–18 GHz frequency range. These molecules were produced by high-voltage low-current discharge. Of interest to this work, the rotational transitions of DC_7N with *J* values varying from 6 to 12 were recorded, with the hyperfine structure due to N being resolved. The results obtained are collected in Table 3 together with the present experimental and computational outcomes.

In our experiment, DC_7N has been produced by FVP as explained in the Methods section. Measurements have been performed in the 80–90 GHz range (with transition frequencies retrieved with 15 kHz accuracy) and merged in

Table 3 Vibrational ground-state spectroscopic parameters of DC_7N

		This work		
Constant	Unit	Exp.[a]	Theo.[b]	Ref. 75
B_0	MHz	545.3151998(89)	545.47	545.31523(7)
D_J	Hz	3.7241(27)	2.98	3.85(28)
H_J	nHz	—	0.86	—
$\chi(N)$	MHz	−4.315(87)	−4.21	−4.33(22)
$\chi(D)$	MHz	—	−0.18	—

[a] Numbers in parentheses are one standard deviation in units of the last quoted digit. [b] Rotational constant from the TM+LR approach augmented by fc-MP2/cc-pVDZ vibrational correction. Quartic and sextic centrifugal distortion constants at the fc-MP2/cc-pVDZ level. Nuclear quadrupole coupling constants at the ae-CCSD(T)/cc-pCVTZ level on top of the TM+LR structure.

a global fit together with those (7.6–14.2 GHz) from ref. 75. Incorporation of millimeter-wave measurements in the fit has led to crucial improvement in the accuracy of the spectroscopic parameters (B_0 and D_J by one and two orders of magnitude, respectively) as well as in the line catalog. For example, transitions lying at about 100 GHz are predicted with an uncertainty of 0.9–1 MHz by the spectroscopic parameters of ref. 75 (see Cologne database catalog: https://cdms.astro.uni-koeln.de/cgi-bin/cdmssearch?file=c100502.cat), while the present work allows the prediction of the same lines with an uncertainty of about 10 kHz, with the modification in the transition frequencies (with respect to the Cologne database catalog[22]) of about 500 kHz. Such changes are relevant even for a long carbon chain like DC_7N because at temperatures of 100–200 K the most intense rotational transitions lies in the millimeter-wave region. Since our measurements involve high J values (ranging from 75 to 81), the hyperfine structure due to nitrogen could not be resolved. However, the improvement in the determination of the quartic centrifugal distortion constant also reduces the error affecting the nitrogen quadrupole coupling constant (because of a better constraint of the experimental data).

DC_7N provides a significant testing ground for the TM+LR approach described in the Methods section. The TM+LR structure, derived as explained in the methodology section, is reported in Fig. 4, where it is compared with the CCSD(T) geometry by Botschwina *et al.*[76] and the effective r_0 structure determined in ref. 75. In ref. 76, a fc-CCSD(T)/cc-pVTZ optimized geometry was empirically corrected by reducing the CH bond length by 0.0019 Å, triple CC bonds by 0.0064 Å, single CC bonds by 0.0060 Å, and the CN bond distance by 0.0069 Å. These corrections were meant to account for basis-set truncation and core-correlation effects. Despite the empiric nature of these corrections, the "corrected CCSD(T)/cc-pVTZ" structure shows only small deviations (about 2 to 5 mÅ) from our TM+LR equilibrium geometry. This has, however, a small but non-negligible impact on the rotational constant. As can be seen in Table 3, our theoretical prediction overestimates the experimental B_0 by only 158 kHz, which means less than 0.03% in relative terms. Using our computed vibrational correction (at the fc-MP2/cc-pVDZ level), the structure by Botschwina *et al.* leads to a B_0 underestimated by 0.05%. However, it has to be emphasized that our result has been obtained at the price of an extremely limited computational cost. The final comment concerns the experimental effective (r_0) structure obtained from a least-square fit of the rotational constants of all single-substituted isotopologues. In this case, larger deviations

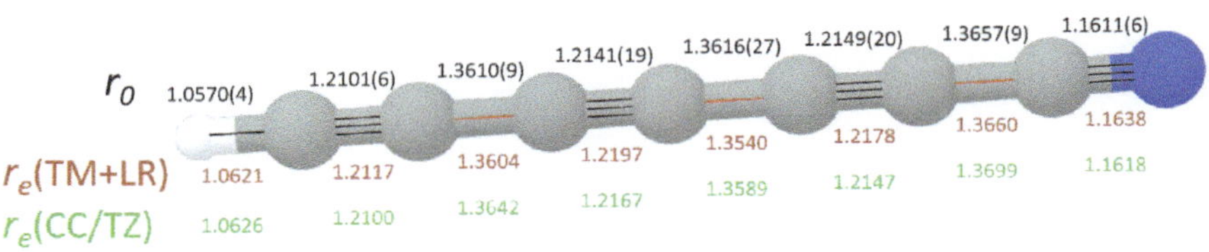

Fig. 4 Molecular structure of HC_7N. The effective structure (r_0) from ref. 75 and the corrected CCSD(T)/cc-pVTZ equilibrium geometry (in green; see text for details) from ref. 76 are compared to the TM+LR equilibrium structure (in red; this work. Bonds are depicted along the carbon chain: in black those corrected using the TM approach, in red those incorporating the LR correction).

from our TM+LR structure are noted, thus pointing out that vibrational effects need to be explicitly taken into account by means of the SE approach.[37]

3.2 Radical species: CCS and CH_2CN

For the first time in our lab, FVP has been employed to produce a radical species. The cyanomethyl radical, CH_2CN, has been produced using a 90 cm tubular furnace (see Methods for details) and starting from a small amount (about 3 mTorr, 0.4 Pa) of CH_3CN with the oven temperature set at 950 °C, with measurements performed as above in a constant flow of gas. An example of rotational spectrum recording is provided in Fig. 5, where it is noted that a good S/N could be obtained. The rotational spectrum of CH_2CN has already been well characterized in the centimeter-[79] and millimeter-wave[80] region. The first experimental work dates back to 1988 (ref. 81) and immediately led to the detection of CH_2CN in the ISM.[82] In ref. 80,81, the cyanomethyl radical was prepared using a DC discharge technique, as usually done in frequency-domain spectrometers.[26,83–86] In the microwave regime, where time-domain rotational spectroscopy is exploited, CH_2CN was produced using a pulsed-discharge nozzle.[79] While a flash pyrolysis microreactor exploiting a Chen nozzle combined with a broadband microwave spectrometer is a well established technique for the investigation of the rotational spectrum of radicals (see, for example, ref. 87 and 88), in millimeter-wave direct absorption spectrometers, employment of pyrolysis is very limited and only very few examples can be found in the literature.[89] However, our first application is very promising, with production conditions being easier to control and adjust with respect to DC discharge.

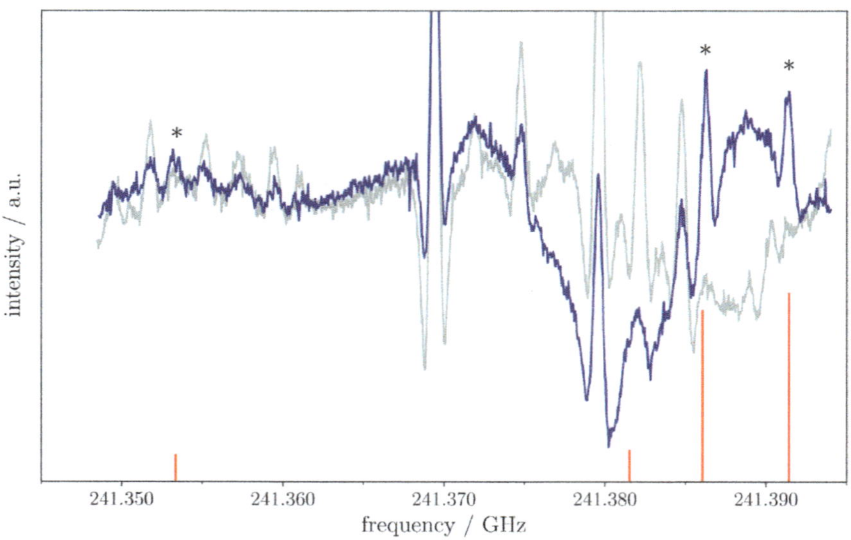

Fig. 5 Small portion of the CH_3CN spectrum around 241.37 GHz. The dark blue profile has been obtained with the oven temperature set at 950 °C: the lines belonging to the CH_2CN radical are made evident by asterisks. The grey profile is the recording with the FVP oven switched off.

First detections of the CCS radical date back to 1987 and refer to the TMC-1 molecular cloud[91] and the envelope of the C-rich star IRC+10 216.[92] Subsequently, the characterization of the rotational spectrum of CCS was extended to the millimeter-wave region up to 292 GHz.[90] However, extrapolations to frequencies much higher than 300 GHz should be viewed with caution. For this reason, with the aim of further improving its spectroscopic parameters and thus the line catalog to be used for astronomical observations, we have extended the analysis of the CCS rotational spectrum up to 686 GHz, this radical being produced by DC discharge as explained in the Methods section. The results of our spectroscopic analysis, which collects all data available on CCS, are reported in Table 4, where they are compared with those from ref. 90. The highly accurate computed parameters obtained in ref. 36 are also given. Concerning the latter, while interested readers are referred to the original paper, we recall that the equilibrium rotational constant was obtained at the CCSD(T)/CBS+CV+fT+fQ level (which incorporates into the CCSD(T)/CBS+CV approach, described in the Methods section, the effect of full treatment of triple and quadruple excitations) and corrected for vibrational effects at the ae-CCSD(T)/cc-pCVQZ level, which also provided the quartic and sextic centrifugal distortion constants. The comparison of Table 4 points out that the new measurements allowed us to improve all spectroscopic parameters and determine the sextic centrifugal distortion constant for the first time. It is also worth noting that the state-of-the-art computations of ref. 36 allowed the prediction of B_0 with an accuracy of 0.005%.

CCS is a $^3\Sigma^-$ radical in its electronic ground state. In such a case, the total angular momentum J is given by the coupling between the molecular angular momentum N and the electronic spin momentum S. The effective Hamiltonian operator thus contains three terms: in addition to the rotational Hamiltonian, with N being the associated quantum number (different from the J symbol used for closed-shell species), there are the Hamiltonians describing the electron spin-rotation and spin–spin interactions. The corresponding spectroscopic parameters are γ and λ, namely the electron spin-rotation and the spin–spin interaction constants, respectively, with γ_D and λ_D representing their centrifugal distortion correction. From Table 4, it is noted that γ and λ are determined with high accuracy and that their centrifugal distortion dependence needs to be incorporated in order to reproduce the recorded transitions within their experimental accuracy. Concerning the latter, for our measurements, uncertainty varies from 50

Table 4 Vibrational ground-state spectroscopic parameters of CCS

Constant/unit	This work Exp.[a]	Ref. 36 Theo.[b]	Ref. 90 Exp.[a]
B_0/MHz	6477.75023(15)	6477.403	6477.75036(71)
D_J/kHz	1.72791(18)	1.6	1.72796(95)
H_J/mHz	0.330(47)	0.02	—
λ/MHz	97 195.69(14)	—	97 196.07(77)
λ_D/kHz	26.97(17)	—	27.00(67)
γ/MHz	−14.7159(88)	−12.1	−14.737(49)
γ_D/mHz	43.4(46)	—	55(37)

[a] Numbers in parentheses are one standard deviation in units of the last quoted digit. [b] For details, see ref. 36.

kHz to 100 kHz depending on the intensity and S/N. The largest value applies to the fine component of the highest transition recorded at 685.5 GHz.

4 Conclusions

In this work, an overview of the strategy exploited at the ROT&Comp laboratory of the Department of Chemistry "Giacomo Ciamician", University of Bologna, for the spectroscopic characterization of unstable and reactive species has been provided. This strategy relies on the strong interplay of experiment and theory in the field of rotational spectroscopy. The examples addressed point out that our computational methodology is able to predict rotational constants with uncertainties well below 0.05% and the other spectroscopic parameters with discrepancies with respect to experiment of the order of a few percents. Consequently, computed spectroscopic parameters and predicted spectra are powerful tools for guiding the recording, the assignment and the analysis of the rotational spectra as well as for providing reliable data for missing information. The experimental analysis in terms of an effective Watson-type Hamiltonian then leads to the set up of spectroscopic line catalogs, where transition frequencies are provided with uncertainties ranging from a few kHz to a maximum of 100–150 kHz. Therefore, these line catalogs are highly suitable for guiding astronomical searches.

Author contributions

CP contributed to the conceptualization, project administration, methodology, data curation, validation, writing – original draft, writing – review & editing, and funding acquisition. SA, LB and MM contributed to the investigation, data curation, formal analysis, validation, and writing – review & editing.

Conflicts of interest

There are no conflicts to declare.

Acknowledgements

This work has been supported by MUR (PRIN Grant Number 202082CE3T) and by the University of Bologna (RFO funds). The COST Action CA21101 "COSY – Confined molecular systems: from a new generation of materials to the stars" is also acknowledged.

Notes and references

1 S. Yamamoto, *Introduction to Astrochemistry: Chemical Evolution from Interstellar Clouds to Star and Planet Formation*, 2017.
2 A. Negron-Mendoza and S. Ramos-Bernal, *First Steps in the Origin of Life in the Universe*, Dordrecht, 2001, pp. 145–150.
3 C. Puzzarini, *Front. Astron. Space Sci.*, 2020, **7**, 19.
4 E. Herbst, *Front. Astron. Space Sci.*, 2021, **8**, 776942.
5 B. A. McGuire, *Astrophys. J., Suppl. Ser.*, 2022, **259**, 30.

6 B. A. McGuire, A. M. Burkhardt, C. N. Shingledecker, S. V. Kalenskii, E. Herbst, A. J. Remijan and M. C. McCarthy, *Astrophys. J., Lett.*, 2017, **843**, L28.

7 J. Cernicharo, M. Agúndez, C. Cabezas, B. Tercero, N. Marcelino, R. Fuentetaja, J. R. Pardo and P. de Vicente, *Astron. Astrophys.*, 2021, **656**, L21.

8 S. Brünken, H. Gupta, C. A. Gottlieb, M. C. McCarthy and P. Thaddeus, *Astrophys. J.*, 2007, **664**, L43.

9 A. J. Remijan, J. M. Hollis, F. J. Lovas, M. A. Cordiner, T. J. Millar, A. J. Markwick-Kemper and P. R. Jewell, *Astrophys. J.*, 2007, **664**, L47.

10 N. W. Broten, T. Oka, L. W. Avery, J. M. MacLeod and H. W. Kroto, *Astrophys. J., Lett.*, 1978, **223**, L105–L107.

11 N. Marcelino, M. Agúndez, B. Tercero, C. Cabezas, C. Bermúdez, J. D. Gallego, P. deVicente and J. Cernicharo, *Astron. Astrophys.*, 2020, **643**, L6.

12 D. Alberton, L. Bizzocchi, N. Jiang, M. Melosso, V. M. Rivilla, A. P. Charmet, B. M. Giuliano, P. Caselli, C. Puzzarini, S. Alessandrini, L. Dore, I. Jiménez-Serra and J. Martín-Pintado, *Astron. Astrophys.*, 2023, **669**, A93.

13 V. M. Rivilla, L. Colzi, I. Jiménez-Serra, J. Martín-Pintado, A. Megías, M. Melosso, L. Bizzocchi, Á. López-Gallifa, A. Martínez-Henares, S. Massalkhi, B. Tercero, P. de Vicente, J.-C. Guillemin, J. G. de la Concepción, F. Rico-Villas, S. Zeng, S. Martín, M. A. Requena-Torres, F. Tonolo, S. Alessandrini, L. Dore, V. Barone and C. Puzzarini, *Astrophys. J., Lett.*, 2022, **929**, L11.

14 E. Herbst and E. F. van Dishoeck, *Annu. Rev. Astron. Astrophys.*, 2009, **47**, 427–480.

15 A. Belloche, A. A. Meshcheryakov, R. T. Garrod, V. V. Ilyushin, E. A. Alekseev, R. A. Motiyenko, L. Margulès, H. S. P. Müller and K. M. Menten, *Astron. Astrophys.*, 2017, **601**, A49.

16 A. Belloche, R. T. Garrod, H. S. P. Müller, K. M. Menten, I. Medvedev, J. Thomas and Z. Kisiel, *Astron. Astrophys.*, 2019, **628**, A10.

17 I. Jiménez-Serra, J. Martín-Pintado, V. M. Rivilla, L. Rodríguez-Almeida, E. R. Alonso Alonso, S. Zeng, E. J. Cocinero, S. Martín, M. Requena-Torres, R. Martín-Domenech and L. Testi, *Astrobiology*, 2020, **20**, 1048–1066.

18 V. M. Rivilla, I. Jiménez-Serra, J. Martín-Pintado, C. Briones, L. F. Rodríguez-Almeida, F. Rico-Villas, B. Tercero, S. Zeng, L. Colzi, P. de Vicente, S. Martín and M. A. Requena-Torres, *Proc. Natl. Acad. Sci. U. S. A.*, 2021, **118**, e2101314118.

19 A. Belloche, R. T. Garrod, O. Zingsheim, H. S. P. Müller and K. M. Menten, *Astron. Astrophys.*, 2022, **662**, A110.

20 I. Jiménez-Serra, L. F. Rodríguez-Almeida, J. Martín-Pintado, V. M. Rivilla, M. Melosso, S. Zeng, L. Colzi, Y. Kawashima, E. Hirota, C. Puzzarini, B. Tercero, P. de Vicente, F. Rico-Villas, M. A. Requena-Torres and S. Martín, *Astron. Astrophys.*, 2022, **663**, A181.

21 B. A. McGuire, *Astrophys. J., Suppl. Ser.*, 2018, **239**, 17.

22 H. S. P. Müller, F. Schlöder, J. Stutzki and G. Winnewisser, *J. Mol. Struct.*, 2005, **742**, 215–227.

23 J. Tennyson, *Astronomical Spectroscopy. An Introduction to the Atomic and Molecular Physics of Astronomical Spectroscopy*, 2019.

24 R. C. Woods, *J. Mol. Struct.*, 1983, **97**, 195–202.

25 G. Cazzoli, L. Cludi, G. Buffa and C. Puzzarini, *Astrophys. J., Suppl. Ser.*, 2012, **203**, 11.

26 G. Cazzoli, V. Lattanzi, T. Kirsch, J. Gauss, B. Tercero, J. Cernicharo and C. Puzzarini, *Astron. Astrophys.*, 2016, **591**, A126.

27 G. Winnewisser, F. Lewen, S. Thorwirth, M. Behnke, J. Hahn, J. Gauss and E. Herbst, *Chem.–Eur. J.*, 2003, **9**, 5501–5510.

28 M. Melosso, L. Bizzocchi, H. Gazzeh, F. Tonolo, J.-C. Guillemin, S. Alessandrini, V. M. Rivilla, L. Dore, V. Barone and C. Puzzarini, *Chem. Commun.*, 2022, **58**, 2750–2753.

29 A. Melli, M. Melosso, L. Bizzocchi, S. Alessandrini, N. Jiang, F. Tonolo, S. Boi, G. Castellan, C. Sapienza, J.-C. Guillemin, L. Dore and C. Puzzarini, *J. Phys. Chem. A*, 2022, **126**, 6210–6220.

30 M. Melosso, B. Conversazioni, C. Degli Esposti, L. Dore, E. Cané, F. Tamassia and L. Bizzocchi, *J. Quant. Spectrosc. Radiat. Transfer*, 2019, **222**, 186–189.

31 M. Melosso, L. Bizzocchi, F. Tamassia, C. Degli Esposti, E. Canè and L. Dore, *Phys. Chem. Chem. Phys.*, 2019, **21**, 3564–3573.

32 N. Jiang, M. Melosso, F. Tamassia, L. Bizzocchi, L. Dore, E. Canè, D. Fedele, J.-C. Guillemin and C. Puzzarini, *Front. Astron. Space Sci.*, 2021, **8**, 29.

33 W. Gordy and R. L. Cook, *Microwave Molecular Spectra*, Wiley, New York, 1984.

34 C. Puzzarini, M. Heckert and J. Gauss, *J. Chem. Phys.*, 2008, **128**, 194108.

35 C. Puzzarini and J. F. Stanton, *Phys. Chem. Chem. Phys.*, 2023, **25**, 1421–1429.

36 S. Alessandrini, J. Gauss and C. Puzzarini, *J. Chem. Theory Comput.*, 2018, **14**, 5360–5371.

37 C. Puzzarini, J. F. Stanton and J. Gauss, *Int. Rev. Phys. Chem.*, 2010, **29**, 273–367.

38 T. Helgaker, P. Jørgensen and J. Olsen, *Electronic-Structure Theory*, Wiley, Chichester, 2000.

39 M. Heckert, M. Kállay and J. Gauss, *Mol. Phys.*, 2005, **103**, 2109.

40 M. Heckert, M. Kállay, D. P. Tew, W. Klopper and J. Gauss, *J. Chem. Phys.*, 2006, **125**, 044108.

41 C. Puzzarini and V. Barone, *Phys. Chem. Chem. Phys.*, 2011, **13**, 7189–7197.

42 A. G. Császár, W. D. Allen and H. F. Schaefer III, *J. Chem. Phys.*, 1998, **108**, 9751–9764.

43 A. Karton, E. Rabinovich, J. M. L. Martin and B. Ruscic, *J. Chem. Phys.*, 2006, **125**, 144108.

44 A. Karton and J. M. L. Martin, *J. Chem. Phys.*, 2012, **136**, 124114.

45 A. Tajti, P. G. Szalay, A. G. Császár, M. Kállay, J. Gauss, E. F. Valeev, B. A. Flowers, J. Vázquez and J. F. Stanton, *J. Chem. Phys.*, 2004, **121**, 11599–11613.

46 K. A. Peterson, D. Feller and D. Dixon, *Theor. Chem. Acc.*, 2012, **113**, 1079.

47 W. J. Morgan, D. A. Matthews, M. Ringholm, J. Agarwal, J. Z. Gong, K. Ruud, W. D. Allen, J. F. Stanton and H. F. Schaefer, *J. Chem. Theory Comput.*, 2018, **14**, 1333–1350.

48 J. Lupi, S. Alessandrini, V. Barone and C. Puzzarini, *J. Chem. Theory Comput.*, 2021, **17**, 6974–6992.

49 X. Huang and T. J. Lee, *J. Chem. Phys.*, 2009, **131**, 104301.

50 Q. Cheng, R. C. Fortenberry and N. J. DeYonker, *J. Chem. Phys.*, 2017, **147**, 234303.

51 C. Puzzarini, J. Bloino, N. Tasinato and V. Barone, *Chem. Rev.*, 2019, **119**, 8131–8191.

52 D. A. Matthews, L. Cheng, M. E. Harding, F. Lipparini, S. Stopkowicz, T.-C. Jagau, P. G. Szalay, J. Gauss and J. F. Stanton, *J. Chem. Phys.*, 2020, **152**, 214108.

53 K. Raghavachari, G. W. Trucks, J. A. Pople and M. Head-Gordon, *Chem. Phys. Lett.*, 1989, **157**, 479–483.

54 D. Feller, *J. Chem. Phys.*, 1993, **98**, 7059–7071.

55 T. Helgaker, W. Klopper, H. Koch and J. Noga, *J. Chem. Phys.*, 1997, **106**, 9639–9646.

56 T. H. Dunning Jr, *J. Chem. Phys.*, 1989, **90**, 1007–1023.

57 A. K. Wilson, T. van Mourik and T. H. Dunning Jr, *J. Mol. Struct.: THEOCHEM*, 1996, **388**, 339–349.

58 D. E. Woon and T. H. Dunning Jr, *J. Chem. Phys.*, 1995, **103**, 4572–4585.

59 K. A. Peterson and T. H. Dunning Jr, *J. Chem. Phys.*, 2002, **117**, 10548–10560.

60 C. Møller and M. S. Plesset, *Phys. Rev.*, 1934, **46**, 618.

61 H. Ye, S. Alessandrini, M. Melosso and C. Puzzarini, *Phys. Chem. Chem. Phys.*, 2022, **24**, 23254–23264.

62 A. Melli, F. Tonolo, V. Barone and C. Puzzarini, *J. Phys. Chem. A*, 2021, **125**, 9904–9916.

63 M. Piccardo, E. Penocchio, C. Puzzarini, M. Biczysko and V. Barone, *J. Phys. Chem. A*, 2015, **119**, 2058–2082.

64 E. Penocchio, M. Piccardo and V. Barone, *J. Chem. Theory Comput.*, 2015, **11**, 4689–4707.

65 G. Ceselin, V. Barone and N. Tasinato, *J. Chem. Theory Comput.*, 2021, **17**, 7290–7311.

66 P. Pulay, W. Meyer and J. E. Boggs, *J. Chem. Phys.*, 1978, **68**, 5077–5085.

67 J. Demaison, J. E. Boggs and A. G. Császár, *Equilibrium Molecular Structures: from Spectroscopy to Quantum Chemistry*, CRC Press, 2016.

68 G. Santra, N. Sylvetsky and J. M. L. Martin, *J. Phys. Chem. A*, 2019, **123**, 5129–5143.

69 E. Papajak, J. Zheng, X. Xu, H. R. Leverentz and D. G. Truhlar, *J. Chem. Theory Comput.*, 2011, **7**, 3027–3034.

70 S. Grimme, J. Antony, S. Ehrlich and H. Krieg, *J. Chem. Phys.*, 2010, **132**, 154104.

71 S. Grimme, S. Ehrlich and L. Goerigk, *J. Comput. Chem.*, 2011, **32**, 1456–1465.

72 I. M. Mills, in *Molecular Spectroscopy: Modern Research*, ed. K. N. Rao and C. W. Matthews, Academic Press, 1972, vol. 1, pp. 115–140.

73 C. Puzzarini, G. Cazzoli, M. E. Harding, J. Vázquez and J. Gauss, *J. Chem. Phys.*, 2009, **131**, 234304.

74 T. Bjorvatten, *J. Mol. Struct.*, 1974, **20**, 75–82.

75 M. C. McCarthy, E. S. Levine, A. J. Apponi and P. Thaddeus, *J. Mol. Spectrosc.*, 2000, **203**, 75–81.

76 P. Botschwina, M. Horn, K. Markey and R. Oswald, *Mol. Phys.*, 1997, **92**, 381–392.

77 G. Cazzoli, C. Puzzarini and J. Gauss, *Mol. Phys.*, 2012, **110**, 2359–2369.

78 E. C. Fayolle, K. I. Öberg, J. K. Jørgensen, K. Altwegg, H. Calcutt, H. S. P. Müller, M. Rubin, M. H. D. van der Wiel, P. Bjerkeli, T. L. Bourke, A. Coutens, E. F. van Dishoeck, M. N. Drozdovskaya, R. T. Garrod, N. F. W. Ligterink, M. V. Persson, S. F. Wampfler and R. Team, *Nat. Astron.*, 2017, **1**, 703–708.

79 H. Ozeki, T. Hirao, S. Saito and S. Yamamoto, *Astrophys. J.*, 2004, **617**, 680.

80 S. Saito and S. Yamamoto, *J. Chem. Phys.*, 1997, **107**, 1732–1739.
81 S. Saito, S. Yamamoto, W. M. Irvine, L. M. Ziurys, H. Suzuki, M. Ohishi and N. Kaifu, *Astrophys. J.,Lett.*, 1988, **334**, L113–L116.
82 W. M. Irvine, P. Friberg, A. Hjalmarson, S. Ishikawa, N. Kaifu, K. Kawaguchi, S. C. Madden, H. E. Matthews, M. Ohishi, S. Saito, H. Suzuki, P. Thaddeus, B. E. Turner, S. Yamamoto and L. M. Ziurys, *Astrophys. J., Lett.*, 1988, **334**, L107–L111.
83 Y. Endo, S. Saito and E. Hirota, *J. Mol. Spectrosc.*, 1982, **92**, 443–450.
84 R. J. McMahon, M. C. McCarthy, C. A. Gottlieb, J. B. Dudek, J. F. Stanton and P. Thaddeus, *Astrophys. J.*, 2003, **590**, L61.
85 H. Ozeki, S. Bailleux and G. Wlodarczak, *Astron. Astrophys.*, 2011, **527**, A64.
86 L. Bizzocchi, M. Melosso, B. M. Giuliano, L. Dore, F. Tamassia, M.-A. Martin-Drumel, O. Pirali, L. Margulès and P. Caselli, *Astrophys. J., Suppl. Ser.*, 2020, **247**, 59.
87 N. M. Kidwell, V. Vaquero-Vara, T. K. Ormond, G. T. Buckingham, D. Zhang, D. N. Mehta-Hurt, L. McCaslin, M. R. Nimlos, J. W. Daily, B. C. Dian, J. F. Stanton, G. B. Ellison and T. S. Zwier, *J. Phys. Chem. Lett.*, 2014, **5**, 2201–2207.
88 C. Abeysekera, A. Hernandez-Castillo, J. F. Stanton and T. S. Zwier, *J. Phys. Chem. A*, 2018, **122**, 6879–6885.
89 L. Kolesniková, J. Varga, L. N. Stříteská, H. Beckers, H. Willner, F. Aubke and s. Urban, *J. Chem. Phys.*, 2009, **130**, 184309.
90 S. Yamamoto, S. Saito, K. Kawaguchi, Y. Chikada, H. Suzuki, N. Kaifu, S.-I. Ishikawa and M. Ohishi, *Astrophys. J.*, 1990, **361**, 318.
91 S. Saito, K. Kawaguchi, S. Yamamoto, M. Ohishi, H. Suzuki and N. Kaifu, *Astrophys. J., Lett.*, 1987, **317**, L115.
92 J. Cernicharo, M. Guelin, H. Hein and C. Kahane, *Astron. Astrophys.*, 1987, **181**, L9–L12.

Faraday Discussions

ROYAL SOCIETY OF CHEMISTRY

PAPER

An experimental and theoretical investigation of the $N(^2D) + C_6H_6$ (benzene) reaction with implications for the photochemical models of Titan†

Nadia Balucani, *[a] Adriana Caracciolo,‡[a] Gianmarco Vanuzzo, [a] Dimitrios Skouteris, [b] Marzio Rosi, [c] Leonardo Pacifici,[a] Piergiorgio Casavecchia, [a] Kevin M. Hickson, *[d] Jean-Christophe Loison [d] and Michel Dobrijevic[e]

Received 3rd March 2023, Accepted 11th April 2023

DOI: 10.1039/d3fd00057e

We report on a combined experimental and theoretical investigation of the $N(^2D) + C_6H_6$ (benzene) reaction, which is of relevance in the aromatic chemistry of the atmosphere of Titan. Experimentally, the reaction was studied (i) under single-collision conditions by the crossed molecular beams (CMB) scattering method with mass spectrometric detection and time-of-flight analysis at the collision energy (E_c) of 31.8 kJ mol^{-1} to determine the primary products, their branching fractions (BFs), and the reaction micromechanism, and (ii) in a continuous supersonic flow reactor to determine the rate constant as a function of temperature from 50 K to 296 K. Theoretically, electronic structure calculations of the doublet C_6H_6N potential energy surface (PES) were performed to assist the interpretation of the experimental results and characterize the overall reaction mechanism. The reaction is found to proceed *via* barrierless addition of $N(^2D)$ to the aromatic ring of C_6H_6, followed by formation of several cyclic (five-, six-, and seven-membered ring) and linear isomeric C_6H_6N intermediates that can undergo unimolecular decomposition to bimolecular products. Statistical estimates of product BFs on the theoretical PES were carried out under the conditions of the CMB experiments and at the temperatures relevant for Titan's atmosphere. In all conditions the ring-contraction channel leading to C_5H_5 (cyclopentadienyl) + HCN is dominant,

[a]*Dipartimento di Chimica, Biologia e Biotecnologie, Università degli Studi di Perugia, 06123 Perugia, Italy. E-mail: nadia.balucani@unipg.it*

[b]*Master-Tec SrL, Via Sicilia, 41, 06128 Perugia, Italy*

[c]*Dipartimento di Ingegneria Civile e Ambientale, Università degli Studi di Perugia, 06100, Perugia, Italy*

[d]*Univ. Bordeaux, CNRS, Bordeaux INP, ISM, UMR 5255, F-33400 Talence, France*

[e]*Univ. Bordeaux, CNRS, LAB, UMR 5804, F-33600 Pessac, France*

† Electronic supplementary information (ESI) available: Fig. S1–S9; Tables S1 and S2. See DOI: https://doi.org/10.1039/d3fd00057e

‡ Present address: Institute of Chemistry and Chemical Engineering, École Polytechnique Fédérale de Lausanne, 1015 Lausanne, Switzerland.

while minor contributions come from the channels leading to o-C_6H_5N (o-N-cycloheptatriene radical) + H, C_4H_4N (pyrrolyl) + C_2H_2 (acetylene), C_5H_5CN (cyano-cyclopentadiene) + H, and p-C_6H_5N + H. Rate constants (which are close to the gas kinetic limit at all temperatures, with the recommended value of $2.19 \pm 0.30 \times 10^{-10}$ $cm^3\ s^{-1}$ over the 50–296 K range) and BFs have been used in a photochemical model of Titan's atmosphere to simulate the effect of the title reaction on the species abundances as a function of the altitude.

1. Introduction

The study of the atmospheric chemistry of the planets of the Solar System can help us to understand the evolution of our own atmosphere and its primordial composition before the emergence of life drastically altered it.[1] Among the others, the case of Titan (a massive moon of Saturn with a dense N_2-dominated atmosphere) is particularly interesting because of the rich organic chemistry occurring in its upper atmosphere.[1–4] After several space missions and ground-based observations, involving also ALMA in recent years,[5] our knowledge of the atmosphere of Titan has reached an unprecedented level[2,3] and new measurements from the James Webb Space Telescope promise to enrich further the list of organic molecules that have been identified so far.[6] In addition to nitrogen (accounting for *ca.* 95–98% of the composition depending on the altitude considered), the second most abundant component is methane (varying from *ca.* 2% in the upper atmosphere up to *ca.* 5% close to the surface). The chemistry initiated by the activation of N_2 and CH_4 by UV photons or energetic particles leads to the formation of trace components including higher hydrocarbons, nitriles and other N-containing organic molecules.[7–9] Since the first missions, we learnt that Titan is covered by organic haze made up of aerosols. After early laboratory experiments trying to simulate the conditions of the atmosphere of Titan,[10] polycyclic aromatic hydrocarbons (PAHs) were proposed as the main component of that aerosol. The presence of PAHs would imply that aromatic rings are formed starting from the main components of Titan and the subsequent detection of benzene, the simplest aromatic hydrocarbon (with a first tentative detection by the Infrared Space Observatory,[11] later confirmed by the Composite Infrared Spectrometer (CIRS) onboard Cassini[12] in the stratosphere and by the analysis of the data recorded by the Ion Neutral Mass Spectrometer onboard Cassini[13] in the thermosphere), supported this interpretation. We now know that benzene is quite diffuse in Titan as it has been identified in the upper atmosphere (peak around 1000 km),[13] in the mesosphere,[14] and in the stratosphere of Titan,[12] and even on the surface as seen upon the impact of the Huygens probe.[15] Furthermore, benzene ice clouds have been detected in the stratosphere.[16] Various formation routes of benzene have been considered. In the photochemical model we will use in this work, benzene is formed by both ionic production pathways in the ionosphere (mainly by the reaction $C_4H_3^+ + C_2H_4$ and to a lesser extent by $C_3H_5^+ + C_2H_2$ and $C_4H_2^+ + C_2H$) and neutral pathways ($C_2H_3 + C_4H_3$ and $C_3H_3 + C_3H_3$) with a similar global efficiency.[17]

The detection of N-containing aromatic compounds is also of interest to the present work. The presence of the N-heterocyclic molecules pyridine and pyrimidine was considered after the analysis of Cassini mass spectra where ions at m/z 80 and 81 (associated with $C_5H_5NH^+$ and $C_4H_4N_2H^+$, and possibly corresponding to protonated pyridine and pyrimidine) were detected,[18] but remote spectroscopy could not confirm this suggestion and only upper limits were derived.[19] However, the analysis of the strong unidentified emission near 3.28 μm in the upper daytime atmosphere reported by Dinelli *et al.*[20] (extending from 600 km up to 1250 km) led to the suggestion that, in addition to PAHs, several N-containing PAHs (such as N-heterocyclic compounds like phenanthridine or aromatic amines like 2-naphthylamine) are the carrier of this band which cannot be explained with non-aromatic species.[21] Likewise, the analysis of the data recorded by the Cassini Plasma Spectrometer Ion Beam Spectrometer strongly suggests that PAHs and N-containing PAHs are responsible for the heavy positive ions signal (170–310 Da).[22]

Therefore, the incorporation of nitrogen in aromatic compounds needs to be explained. Molecular nitrogen is not a reactive species, but it can be activated either by dissociation or ionization induced by EUV photons, cosmic rays and solar winds as well as other energetic particles like electrons from the magnetosphere of Saturn.[7–9] Interestingly, photodissociation, dissociative photoionization, cosmic ray induced dissociation, electron impact as well as N_2^+ dissociative recombination can produce atomic nitrogen in the first electronically excited ^{2}D state in similar amounts to the ground ^{4}S state.[7–9,23] N(^{4}S) exhibits very low reactivity with closed shell molecules,[23–25] while N(^{2}D) is reactive with the typical molecules identified in the atmosphere of Titan, including CH_4 and H_2.[26–28] N(^{2}D) is metastable but has a radiative lifetime long enough (6.1×10^4 s and 1.4×10^5 s for the $^2D_{3/2}$ and $^2D_{5/2}$ fine structure states, respectively)[29] to react in binary collisions with other constituents of the upper atmosphere of Titan.[7–9]

In the Perugia laboratory, featuring a crossed molecular beam (CMB) apparatus with mass spectrometric detection, some of the present authors have investigated numerous reactions between N(^{2}D) and molecular species of interest in the atmospheric chemistry of Titan.[26–28,30–40] Interestingly, in all the investigated N(^{2}D) reactions with C-containing molecules the formation of products containing a novel C–N bond has been observed.[26,31–40] Dedicated electronic structure calculations of the reactive potential energy surface (PES) have also been performed for those reactions.

In the Bordeaux laboratory, featuring a continuous supersonic flow apparatus, the kinetics of several reactions involving N(^{2}D) and hydrocarbons[36,41–43] have been determined in a range of temperatures, *T*, from 50 K to 296 K, encompassing those of relevance for Titan and the results have been incorporated in a photochemical model developed in-house.[8] These studies have shown that most reactions of N(^{2}D) are faster than what was believed, approaching the gas kinetic limit down to very low *T*. In some cases, a significant effect in the photochemical model outcomes was observed.[36,41–43]

In the case of the reaction N(^{2}D) + allene, we have already combined the Perugia and Bordeaux expertise to derive the rate coefficients and the product branching fractions (BFs) as a function of the temperature. The results have been included in the photochemical model and the effects analysed.[36] This approach is pivotal to improve current photochemical models of Titan's atmosphere because many relevant elementary reactions have never been investigated in laboratory experiments and are included in chemical networks of the models with estimated values of rate constants and product BFs.

In the present work, we use the same combined approach to study the reaction N(^{2}D) + C_6H_6 with the aim to understand the chemistry of N(^{2}D) with aromatic species after that with aliphatic molecules. In particular, we wish to explore whether (1) the reaction of N(^{2}D) with the simplest aromatic is fast as in the case of aliphatic unsaturated hydrocarbons, also at the very low *T* typical of Titan (94–200 K); (2) the aromatic ring is preserved upon the chemical attack by the very reactive species N(^{2}D) or the reaction causes a ring contraction and/or the loss of aromaticity (both cases have been previously observed in reactions involving benzene and other atomic/radical species[44–49]); (3) the N atom can be incorporated in the ring possibly forming pyridine in a N/CH exchange channel (preliminary calculations performed *ad hoc* have indicated that this reaction path is exothermic and correlates with the reactants[50,51]); (4) other N-containing cyclic organic molecules can be formed (the same preliminary calculations have already indicated that there are other open reactive channels;[50,51] the PES derived by another research group that also performed statistical estimates of the product branching fractions confirmed that[52]). If confirmed that N is incorporated into the aromatic ring or in other cyclic non-aromatic compounds, this reaction could be added to the list of processes generating the precursors of the N-containing PAHs observed in Titan, together with other routes characterized by the CMB method in the group of Kaiser[53–57] or inferred by plasma studies in air.[58,59] Finally, the rate constant and product BFs as a function of temperature derived in this study are used in a photochemical model of Titan's atmosphere to simulate the effect of the title reaction on the species abundances (including new products formed) as a function of the altitude.

The Perugia group has already investigated the related reactions between benzene and atomic oxygen[48,49] and the reaction between N(^{2}D) and pyridine.[40] In both cases, ring-contraction mechanisms have been found to be important. Similarities and differences will be noted in the paper.

According to the present theoretical calculations the N(^{2}D) + benzene reaction exhibits numerous energetically open channels, with the most relevant ones being the following (the reported enthalpies of reaction are those calculated in the present work at the CCSD(T)-CBS level, see Section 3.1):

$$\mathrm{N(^2D) + C_6H_6 \rightarrow C_5H_5 + HCN} \qquad \Delta H_0^0 = -397\ \mathrm{kJ\ mol^{-1}} \tag{1}$$

$$\mathrm{N(^2D) + C_6H_6 \rightarrow C_5H_5 + HNC} \qquad \Delta H_0^0 = -337\ \mathrm{kJ\ mol^{-1}} \tag{2}$$

$$\mathrm{N(^2D) + C_6H_6 \rightarrow C_5H_5CN + H} \qquad \Delta H_0^0 = -311\ \mathrm{kJ\ mol^{-1}} \tag{3}$$

$$\mathrm{N(^2D) + C_6H_6 \rightarrow C_4H_4N + C_2H_2} \qquad \Delta H_0^0 = -269\ \mathrm{kJ\ mol^{-1}} \tag{4}$$

$$\mathrm{N(^2D) + C_6H_6 \rightarrow C_5H_6 + CN} \qquad \Delta H_0^0 = -198\ \mathrm{kJ\ mol^{-1}} \tag{5}$$

$$\mathrm{N(^2D) + C_6H_6 \rightarrow C_4H_4NCCH + H} \qquad \Delta H_0^0 = -172\ \mathrm{kJ\ mol^{-1}} \tag{6}$$

$$\mathrm{N(^2D) + C_6H_6 \rightarrow C_6H_5N + H} \qquad \Delta H_0^0 = -157\ \mathrm{kJ\ mol^{-1}} \tag{7}$$

$$\mathrm{N(^2D) + C_6H_6 \rightarrow \mathit{o}\text{-}C_6H_5N + H} \qquad \Delta H_0^0 = -132\ \mathrm{kJ\ mol^{-1}} \tag{8}$$

$$\mathrm{N(^2D) + C_6H_6 \rightarrow \mathit{p}\text{-}C_6H_5N + H} \qquad \Delta H_0^0 = -110\ \mathrm{kJ\ mol^{-1}} \tag{9}$$

$$\mathrm{N(^2D) + C_6H_6 \rightarrow C_5H_5N + CH} \qquad \Delta H_0^0 = -45\ \mathrm{kJ\ mol^{-1}} \tag{10}$$

The paper is organized as follows. In Sections 2 and 3 we describe the experimental techniques and the theoretical methods, respectively, adopted to investigate the title reaction. In Section 4 we report the experimental results. In Section 5 we describe the PES used to interpret the experimental results and the predictions of statistical simulations of product BFs. A general discussion of the results is given in Section 6, and the effects of the title reaction in a photochemical model of Titan's atmosphere is then examined in Section 7, before the conclusions.

2. Experimental methods

2.1 The crossed molecular beam experiments

The scattering experiments were carried out by using a CMB apparatus that has been described in detail elsewhere.[60,61] Briefly, two well-collimated, in angle and velocity, continuous supersonic beams of the reactants are crossed at 90° in a large scattering chamber with a background pressure of 2×10^{-6} mbar in operating conditions, which assures single-collision conditions. The detection system consists of an electron impact ionizer, a quadrupole mass filter and a Daly-type detector. The ionizer is located in the innermost region of a triply differentially pumped ultra-high-vacuum chamber, which is maintained in the 10^{-11} mbar pressure range in operating conditions by extensive turbo- and cryo-pumping. The whole detector unit can be rotated in the collision plane around an axis passing through the collision center and product/reactant velocity distributions can be derived from time-of-flight (TOF) measurements. Beam conditions and other details of the experimental conditions are given in the ESI.† The resulting collision energy, E_c, was 31.8 kJ mol^{-1}.

The laboratory (LAB) angular distribution, $N(\Theta)$, was measured with respect to the nitrogen beam direction. The secondary target beam (benzene beam) was modulated at 160 Hz with a tuning fork chopper for background subtraction. Velocity distributions of the products were obtained at selected LAB angles by measuring time-of-flight, $N(\Theta, t)$ distributions using the cross-correlation TOF technique with four 127 bit pseudorandom sequences.[60] High-time resolution was achieved by spinning the TOF disk, located at the entrance of the detector, at 328 Hz corresponding to a dwell time of 6 μs per channel. TOF counting times varied from 60 to 180 min depending upon signal intensity. The analysis of raw data has been performed by the usual forward convolution approach where trial CM product angular, $T(\theta)$, and translational energy, $P(E'_T)$, distributions are assumed, averaged, and transformed to the LAB frame for comparison with the experimental data until the best-fit of the LAB distributions is achieved.[60,61]

2.2 Kinetic experiments

The kinetic experiments described here were performed using an existing continuous supersonic flow or Laval nozzle apparatus. The principal attributes of this system have been presented in earlier work,[62] while the main modifications that have been made to allow us to follow the kinetics of atomic radical reactions are described in later work.[63–66] Three different Laval nozzles based on the carrier gases Ar or N_2 were employed during this investigation to access four different low temperature flows between 50 and 177 K (one nozzle was used with both Ar and N_2 to generate flows with characteristic temperatures of 127 and 177 K, respectively).

The nozzle properties are summarized in Table 1 of Nuñez-Reyes *et al.*[43] Room temperature measurements (296 K) were conducted with Ar as the carrier gas in the absence of a Laval nozzle and by reducing the flow velocity to avoid pressure gradients in the reactor. $N(^2D)$ atoms are only slowly quenched by N_2 and Ar[67] so the use of these carrier gases did not adversely affect the excited state populations. Details of the methods used for the generation and detection of $N(^2D)$ atoms, as well as for the introduction of the benzene co-reagent are provided in the ESI.†

3. Theoretical methods

3.1 Electronic structure calculations of potential energy surface (PES)

The theoretical investigation of the reaction between $N(^2D)$ and C_6H_6 was performed adopting a well-established computational strategy previously used for the investigation of many other bimolecular reactions.[26,31,34–39,68–70] The PES of the $N(^2D)$ + C_6H_6 system was first characterized by locating the lowest stationary points (minima and transition states) at the B3LYP level of theory in conjunction with the 6-311+G** basis set.[71,72] The optimization of all the stationary points was then refined by the use of the same hybrid density functional with a larger basis set, *i.e.* the Dunning correlation consistent valence polarized aug-cc-pVTZ basis set.[73–75] At the same level of theory we have computed the harmonic vibrational frequencies in order to check the nature of the stationary points, *i.e.* minimum if all the frequencies are real, saddle point if there is one, and only one, imaginary frequency. The assignment of the saddle points was performed using intrinsic reaction coordinate (IRC) calculations.[76,77] The energy of all the stationary points was computed then at the CCSD(T) level[78–80] using the same basis set aug-cc-pVTZ.[73–75] Both the B3LYP and the CCSD(T) energies were corrected to 0 K by adding the zero point energy (ZPE) correction computed using the scaled harmonic vibrational frequencies evaluated at the B3LYP level. The energy of $N(^2D)$ was estimated by adding the experimental[81] separation $N(^4S)$–$N(^2D)$ of 230 kJ mol^{-1} to the energy of $N(^4S)$ at all levels of calculation. Preliminary reports on a portion of the PES have been already given,[50,51] but the full PES and statistical calculations based on the Rice–Ramsperger–Kassel–Marcus (RRKM) method are presented here. Furthermore, for selected stationary points, which were demonstrated to be relevant for the main reactive channels from RRKM calculations using the CCSD(T) PES, more accurate calculations were performed at the CCSD(T) level corrected with a Density Fitted (DF) MP2 extrapolation to the complete basis set (CBS) and with corrections for core electrons excitations. In particular, we compute the energies as:

$$E_{\mathrm{CBS}} = E(\mathrm{CCSD(T)/aug\text{-}cc\text{-}pVTZ}) + [E(\mathrm{CCSD(T,core)/cc\text{-}pVTZ}) - E(\mathrm{CCSD(T)/cc\text{-}pVTZ})] + [E(\mathrm{DF\text{-}MP2/CBS}) - E(\mathrm{DF\text{-}MP2/aug\text{-}cc\text{-}pVTZ})] \quad (11)$$

where E(DF-MP2/CBS) is defined as:

$$E(\mathrm{DF\text{-}MP2/CBS}) = E[(\mathrm{DF\text{-}MP2})/\mathrm{aug\text{-}cc\text{-}pVQZ}] + 0.5772 \times [E(\mathrm{DF\text{-}MP2/aug\text{-}cc\text{-}pVQZ}) - E(\mathrm{DF\text{-}MP2/aug\text{-}cc\text{-}pVTZ})] \quad (12)$$

The E(DF-MP2/CBS) extrapolation was performed using Martin's two parameter scheme[82] and the related energies were used for the kinetic (statistical) investigation described below. These energies will be denoted as CBS although they include also the core–valence correlation correction. The accuracy of these very accurate calculations should be within ±5 kJ mol^{-1}.

The DFT and CCSD(T) calculations were done using Gaussian 09 (ref. 83) while the CBS calculations were done with MOLPRO.[84,85] The analysis of the vibrational frequencies was performed using Molekel.[86]

3.2 Statistical estimates of the product branching fractions

RRKM calculations for the $N(^2D)$ + C_6H_6 reaction were performed using a code implemented in our group for this purpose.[26,31,68–70] The microcanonical rate constant $k(E)$ for a specific reaction at a specific total energy is given by the expression $k(E) = \frac{N_{TS}(E)}{h\rho_T(E)}$ where $N(E)$ represents the sum of states at the transition state at energy E, $\rho(E)$ is the density of states of the reactant, and h is Planck's constant. $N(E)$ is obtained by integrating the relevant density of states up to energy E, and the rigid rotor/harmonic oscillator model is assumed. Since many reaction channels are H-displacement processes, tunnelling (as well as quantum reflection) was included in the RRKM calculations by using the corresponding imaginary frequency of the transition state and calculating the tunneling probability for the corresponding Eckart barrier.

For the cases in which we were not able to locate a clear transition state in the exit channel, the corresponding microcanonical rate constant was obtained through a variational approach: $k(E)$ was evaluated at various points along the reaction coordinate and the point which minimizes the rate constant was chosen, in accordance with the variational theory.[87] For dissociation steps in which, due to difficulties in the electronic structure calculations, no intermediates points are available, the products at infinite separation were taken into account as a possible "transition state". After the calculation of all microcanonical rate constants, a Markov (stochastic) matrix was set up for all intermediates and final channels to derive the product branching fractions for the overall reaction. $k(E)$ is subsequently Boltzmann averaged for each temperature of interest to yield $k(T)$.

4. Experimental results

4.1 CMB experiments

Reactive scattering signals were observed at m/z = 91 ($C_6H_5N^+$), m/z = 90 ($C_6H_4N^+$), m/z = 66 ($C_5H_6^+$), and m/z = 65 ($C_5H_5^+$) (with relative intensities of 1.00, 0.17, 0.09, and 0.78 at $\Theta = 48°$ using an energy of the ionizing electrons of 70 eV). Full sets of angular and TOF distributions were determined only at m/z = 91, 90, and 65. The signal at m/z = 66 was too weak to perform scattering measurements while angular and TOF distributions at m/z = 79 (corresponding to pyridine mass from channel (10)) could not be achieved because of the strong interfering signal due to the elastic scattering of benzene with one ^{13}C (the m/z of $^{13}CC_5H_6^+$ is 79).

Fig. 1a depicts the velocity vector (so called "Newton") diagram of the experiment, while Fig. 1b and c show the LAB angular distributions measured at m/z = 91 and m/z = 65, respectively. The circles in the Newton diagram delimit the maximum speed of the indicated heavy products with gross formula C_6H_5N formed in the H-displacement channels (3) and (6–9) (with five possible

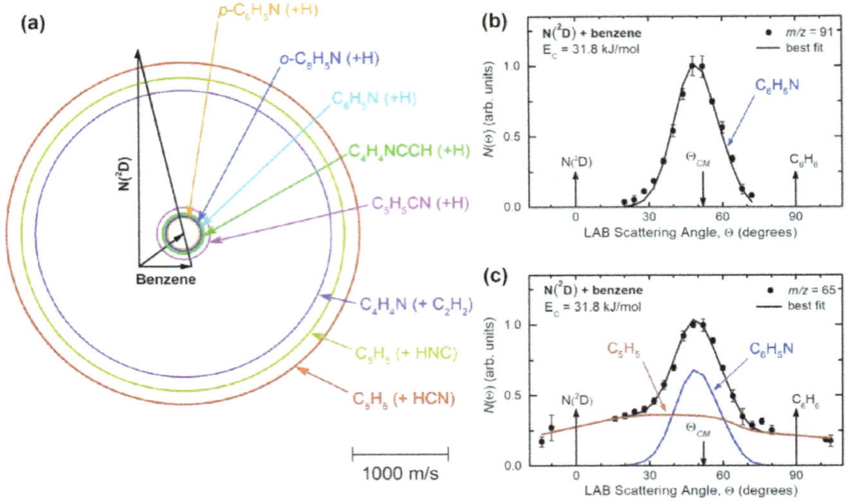

Fig. 1 (a) Velocity vector diagram ("Newton" diagram) for the $N(^2D) + C_6H_6$ (benzene) reaction at $E_c = 31.8$ kJ mol^{-1} with superimposed color-coded circles that delimit the speed in the CM frame of the indicated primary products. Product LAB angular distributions at (b) $m/z = 91$ and (c) $m/z = 65$. The black solid curves superimposed on the experimental data (dots) are LAB angular distributions calculated by using the best-fit CM functions shown in Fig. 3. Partial contributions, as resulted from the best-fit procedure and originating from H-displacement (blue) and ring-contraction (red) channels, are also shown in the $m/z = 65$ angular distribution (see text).

structures: 2-cyano-cyclopentadiene, 1-ethynyl-pyrrole, phenylnitrene, and *ortho-/para*-N-cycloheptatriene radicals), C_5H_5 formed in the HCN/HNC elimination channels (1) and (2), and C_4H_4N (pyrrolyl) formed in the C_2H_2 elimination channel (4). The maximum CM speed was obtained by assuming that all the available energy, E_{TOT} $(= E_c - \Delta H_0^0)$, for the corresponding channel is released into product translational energy.

The solid dots in Fig. 1b and c represent the average of five angular scans with counting times of 100 s at each angle per scan (the error bars represent the standard deviation). On the basis of energy and linear momentum conservation laws, the reactive signal recorded at $m/z = 91$ corresponds unambiguously to the heavy co-product(s) of the H-displacement channel(s). Because of the limited experimental resolution and the similarity in the energetics and dynamics of these channels, we were not able to disentangle them based only on the analysis of the experimental data and the data could be fit using a single set of best-fit $T(\theta)$ and $P(E_T^{'})$.

The $m/z = 90$ angular distribution (not shown) was identical to that at $m/z = 91$ indicating that the reactive signal at $m/z = 90$ originates from dissociative ionization of the C_6H_5N product(s) from the H-displacement channels. Therefore, no H_2 elimination was seen to occur.

During the data analysis of the $m/z = 65$ LAB distributions, it was necessary to consider two contributions. According to that analysis, the angular distribution measured at $m/z = 65$ is the result of one component peaking around Θ_{CM} (similar in all respects to the angular distribution recorded at m/z = 91 and, therefore, originating from the products of the H-displacement channels *via* dissociative ionization) while two wings of comparable intensity originate from other product channels. The presence of two different contributions is more clearly visible in the TOF spectra (see Fig. 2) collected at six different LAB angles (24°, 32°, 40°, 48°, 54°, and 64°). A fast and a slow peak are easily visible from 40° to 54° with the relative importance varying with the angle. Considering the kinematics of the various possible channels, this additional contribution can be ascribed to the occurrence of either channel (1) and/or channel (2). Possibly, also channels (4), associated with C_2H_2 formation, and (5), associated with CN elimination, could contribute to the m/z = 65 distributions *via* the −1 daughter ions of C_4H_4N (pyrrolyl) and C_5H_6 (cyclopentadiene) products, respectively. In the TOF distributions, the fast peak is associated to channels (1), (2), (4) and (5) while the slower peak is associated to the H-displacement channels and has exactly the same characteristics of the analogous peak measured at m/z = 91 (see Fig. S1 in the ESI†). In the TOF distributions at angles outside the scattering angular range of C_6H_5N (Θ = 24° and, to some extent, 32° and 64°), only the contribution originating from channel (1) – or (2), (4) and (5) – is visible. During the analysis, the two contributions were disentangled and the CM total flux $I_{CM}(\theta, E'_T)$ was derived by considering the relation $I_{CM}(\theta, E'_T) = [T(\theta)P(E'_T)]_{C_5H_5} + \alpha[T(\theta)P(E'_T)]_{C_6H_5N}$, where α weighs the contribution of the H-displacement channel(s) to the signal recorded

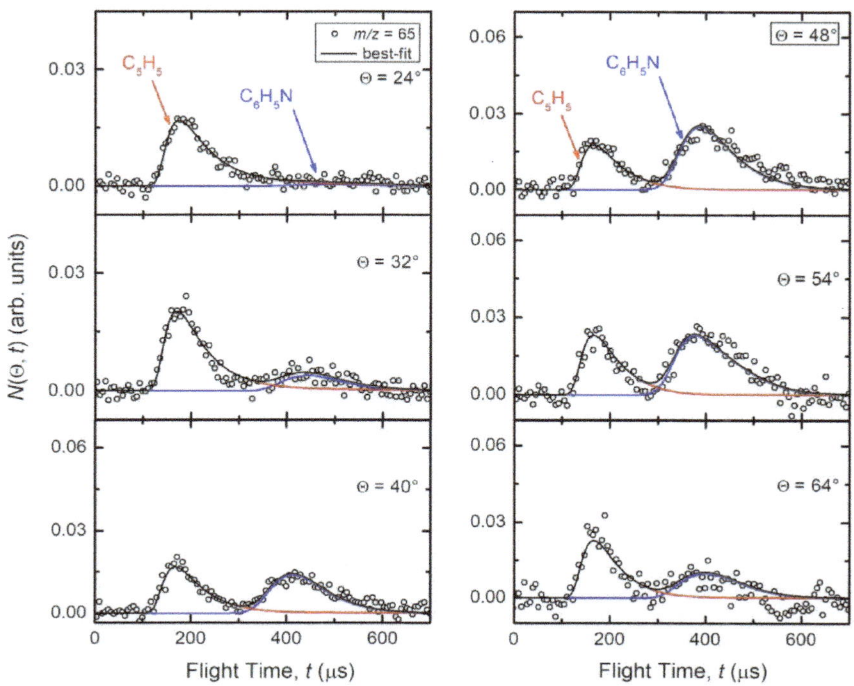

Fig. 2 TOF distributions measured at m/z = 65 at the six indicated LAB angles. The best-fit is obtained by considering two different sets of CM functions (blue lines: H-displacement channels; red lines: ring-contraction channels).

at $m/z = 65$ and is a best-fit parameter.[60,61] The best-fit value of $\alpha(C_6H_5N)$ was minor being only 0.032. However, as can be expected on the basis of energy and momentum conservation and of the relation between the CM and LAB reactive fluxes,[60,61] the signal intensity of the C_6H_5N co-products of the indicated five H-displacement channels appears to be strongly enhanced in the LAB system because of the small Newton circles compared to those associated to the C_5H_5 scattered by the relatively heavy HCN (or HNC) co-product. If we refer to the total available energy for the *o*-C_6H_6CN + H channel, the resulting average fraction of the total available energy channeled into product translation, $\langle f_T \rangle$, is 0.33.

4.1.1 Product angular and translational energy distributions in the CM frame. In Fig. 3a the best-fit CM product angular distributions, $T(\theta)$, for the C_6H_5N (top panel) and C_5H_5 (bottom panel) products are depicted, while in Fig. 3b the product translational energy distributions, $P(E'_T)$, are displayed for the H-displacement (top panel) and C_5H_5-forming (bottom panel) channels. The best-fit $T(\theta)$ function for the C_6H_5N product(s) is strongly forward peaked. It indicates that the H-displacement channels proceed *via* an osculating complex mechanism.[88] The $P(E'_T)$ function for the H-displacement channel(s) peaks at about 35 kJ mol^{-1} and dies off at about 135 kJ mol^{-1}. Within the error bars, this cut-off energy is consistent with the total available energy for the formation of the two isomers *o*-C_6H_5N and *p*-C_6H_5N, while it is substantially lower than the maximum total available energy for the C_5H_5CN + H channel and slightly lower than those for the C_4H_4NCCH + H channel and C_6H_5N + H channel.

4.1.2 Experimental branching fractions. The product BFs were derived following the procedure developed by Schmoltner *et al.*[89] and widely applied by us to a series of multichannel bimolecular reactions.[48,49,90,94] The overall H-

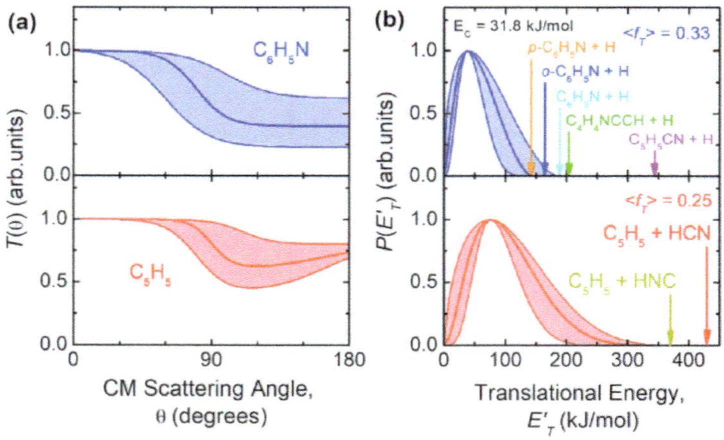

Fig. 3 (a) Best-fit CM angular distribution of the C_6H_5N (top panel) and C_5H_5 (bottom panel) products for the N(^{2}D) + benzene reaction at $E_c = 31.8$ kJ mol^{-1}. (b) Best-fit CM product translational energy distributions for the H-displacement channels (top panel) and for the C_5H_5 (cyclopentadienyl) ring-contraction channel (bottom panel). The arrows mark the total available energy for the indicated product channels. The average fraction released into product translational energy, $\langle f_T \rangle$, is also indicated considering the case of the channel leading to the formation of *o*-C_6H_5N. The shaded areas in (a) and (b) represent the error bars of the CM functions.

displacement channels were found to be minor accounting for BF = 0.08 ± 0.04, while the C_5H_5 (cyclopentadienyl) forming channel(s) (and/or channels (4) and (5)) was determined to be dominant with a BF of 0.92 ± 0.04. Based on only our experimental results, we could not quantify the relative importance of either the various possible H-displacement channels (3, 6–9) or the relative importance of channels (1, 2, 4, and 5).

4.2 Kinetic results

Excess concentrations of both co-reagents NO and C_6H_6 (with respect to the minor reagents $C(^3P)$ and $N(^2D)$) were used during these experiments, allowing the pseudo-first-order approximation to be applied. As a result, the $N(^2D)$ fluorescence emission signal, $I_{N(^2D)}$ (assumed to be proportional to $[N(^2D)]$), obeys a biexponential rate law with the following form

$$I_{N(^2D)} = A\left(\exp\left(-k_a^{'}t\right) - \exp\left(-k_b^{'}t\right)\right) \tag{13}$$

with A equal to the maximum $N(^2D)$ signal amplitude, $k_a^{'}$ and $k_b^{'}$ are the pseudo-first-order rate constants for $N(^2D)$ loss and formation, respectively, and t is the reaction time corresponding to the delay between photolysis and probe lasers. Nevertheless, as the first few microseconds of $N(^2D)$ formation could not be exploited due to PMT saturation issues, a single exponential rate law of the type

$$I_{N(^2D)} = A\exp\left(-k_a^{'}t\right) \tag{14}$$

was used instead to fit to the $N(^2D)$ fluorescence curves. In this instance, the starting point of the fit was chosen carefully to avoid fitting to the rising part of the $N(^2D)$ temporal profiles. Some typical examples of the $N(^2D)$ fluorescence

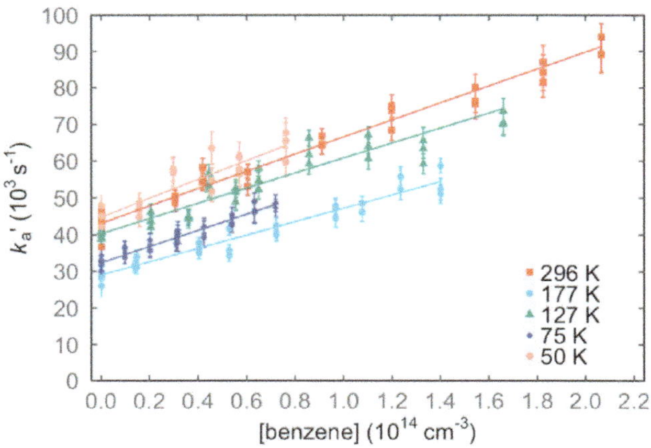

Fig. 4 Plots of the pseudo-first-order rate constant, k'_a, *versus* $[C_6H_6]$ for a range of temperatures: (red squares) 296 K; (light blue circles) 177 K; (green triangles) 127 K; (dark blue diamonds) 75 K; (pink circles) 50 K. The solid lines are weighted linear least-squares fits to the individual datasets. The error bars on individual data points are derived from exponential fits to traces such as those displayed in Fig. S2† and are shown at the level of a single standard deviation.

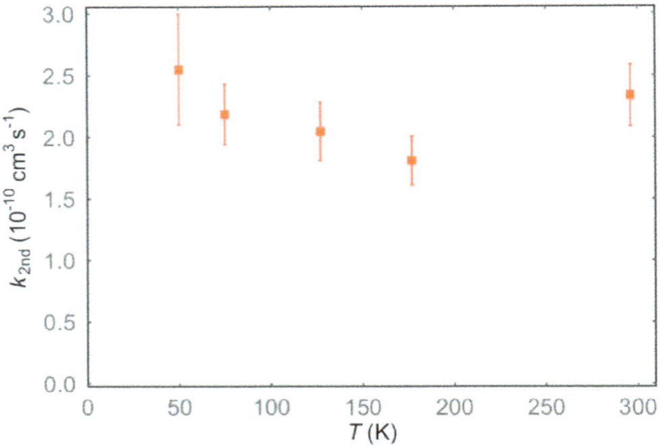

Fig. 5 Second-order rate constants for the N(^{2}D) + C_6H_6 reaction as a function of temperature. Error bars represent the combined statistical uncertainty at the level of a single standard deviation in addition to an estimated systematic error of 10%.

signal as a function of time for different C_6H_6 concentrations recorded at 127 K are displayed in Fig. S2.†

k_a' is actually made up of several different contributions, namely N(^{2}D) loss by reaction with C_6H_6, $k_{N(^2D)+C_6H_6}[C_6H_6]$, N(^{2}D) loss by reaction with NO, $k_{N(^2D)+NO}$[NO] and N(^{2}D) loss by other processes including reactions with impurities and loss by diffusion. The first two terms dominate here, with the third term negligible in comparison. As [NO] is fixed for any single series of experiments, the difference in decay time between the traces in Fig. S2† is entirely due to the change in C_6H_6 concentration.

Values of the second-order rate constant for the N(^{2}D) + C_6H_6 reaction at a given temperature $k_{N(^2D)+C_6H_6}(T)$ were obtained by recording several decay curves similar to those shown in Fig. S2† at various C_6H_6 concentrations while keeping the NO concentration constant. The results of these experiments are displayed in Fig. 4, where k_a' is plotted as a function of [benzene] for five temperatures from 50

Table 1 Second-order rate constants for the N(^{2}D) + C_6H_6 reaction alongside other relevant information

T/K	N^b	[C_6H_6]/10^{14} cm^{-3}	[NO]/10^{14} cm^{-3}	$k_{N(^2D)+C_6H_6}$/10^{-11} cm^3 s^{-1}
296	30	0–2.1	6.4	(23.4 ± 2.5)[c]
177 ± 2[a]	33	0–1.4	4.1	(18.1 ± 2.0)
127 ± 2	33	0–1.7	4.6	(20.5 ± 2.4)
75 ± 2	30	0–0.7	3.0	(21.9 ± 2.4)
50 ± 1	18	0–0.8	4.2	(25.5 ± 4.5)

[a] Uncertainties on the temperatures represent the statistical (1σ) errors obtained from Pitot tube measurements of the impact pressure. [b] Number of individual measurements. [c] Uncertainties on the measured rate constants represent the combined statistical (1σ) and estimated systematic (10%) errors.

K to 296 K. The solid lines represent weighted fits to the individual datasets. The gradients of these lines yield the values for $k_{N(^2D)+C_6H_6}$ with the intercept corresponding essentially to $k_{N(^2D)+NO}$[NO]. The derived second-order rate constants are plotted as a function of temperature in Fig. 5 and are summarized in Table 1 along with other relevant information. To the best of our knowledge, there are no earlier experimental studies of the kinetics of the $N(^2D) + C_6H_6$ reaction. As the reaction rate does not display a large temperature dependence, we recommend the use of a temperature independent rate constant in photochemical models with a value of $(21.9 \pm 3.0) \times 10^{-11}$ cm^3 s^{-1} over the 50–296 K range.

5. Theoretical results

5.1 The potential energy surface

The theoretical characterization of the PES of the system $N(^2D)$ + benzene leads to the identification of 18 minima, 25 transition states and 10 exit channels. A simplified scheme of the PES is shown in Fig. 6 while the complete PES is reported in Fig. S3–S6 of the ESI.† The structures of the most relevant minima and transition states, as well as those of several unusual products, are also shown in the ESI (Fig. S7–S9).† The numbering of the minima, the transition states and the products channels is the same as that adopted in the previous work of Chin *et al.*[52] to make the comparison easier. All the pathways considered in the work of Chin *et al.* are present in our calculations, but additional paths have also been identified.

We performed preliminary statistical calculations, including all the computed reactive channels at the CC level, to evaluate the BFs of all the channels. In this way, we selected only the exit channels for which a non-negligible BF was obtained. Then the energy of these channels was recalculated at the very accurate CBS level (these are the values reported in Fig. 6).

In the calculations at the B3LYP level, the interaction of nitrogen in its excited ^2D state with C_6H_6 leads to a van der Waals complex with a very long (longer than 3 Å) N–C distance. The stability of this complex is strongly overestimated because of the very poor description of nitrogen in its excited ^2D state. Since the presence of this initial complex has no influence on the reactive channels, we have not considered it in the higher-level calculations. As shown in Fig. 6, the interaction of $N(^2D)$ with benzene leads to the formation of the i1 intermediate (more stable than the reactants by 218 kJ mol^{-1}) in which one of the carbon atoms has changed hybridization from sp^2 to sp^3 making a new bond with the nitrogen atom. The i1 intermediate, overcoming a small barrier of only 12 kJ mol^{-1}, can isomerize to i2 which, following the path TS13 → i12 → TS18 → i15 → TS19 → i16 → TS20, finally leads to the products cyclopentadienyl radical and hydrogen cyanide (P9) (channel (1)). RRKM calculations suggest it to be the main reaction channel (see below). Once intermediate i12 is formed, however, it can also lose an H-atom without any barrier from C1, C2 or C3 giving rise to three different products, the most stable of which is the one associated with the H-loss from C1 (P15), the second from C3 (P16) while the loss of H from C2 is the least favored.

The two last exit channels P15 and P16, which are originating in a barrier-less fashion from the 7-member ring intermediate i12, were not considered in the work of Chin *et al.*[52] i12 can also isomerize to intermediate i13, overcoming a barrier of 205 kJ mol^{-1}, which, following the path TS15 → i14 → TS16, gives rise

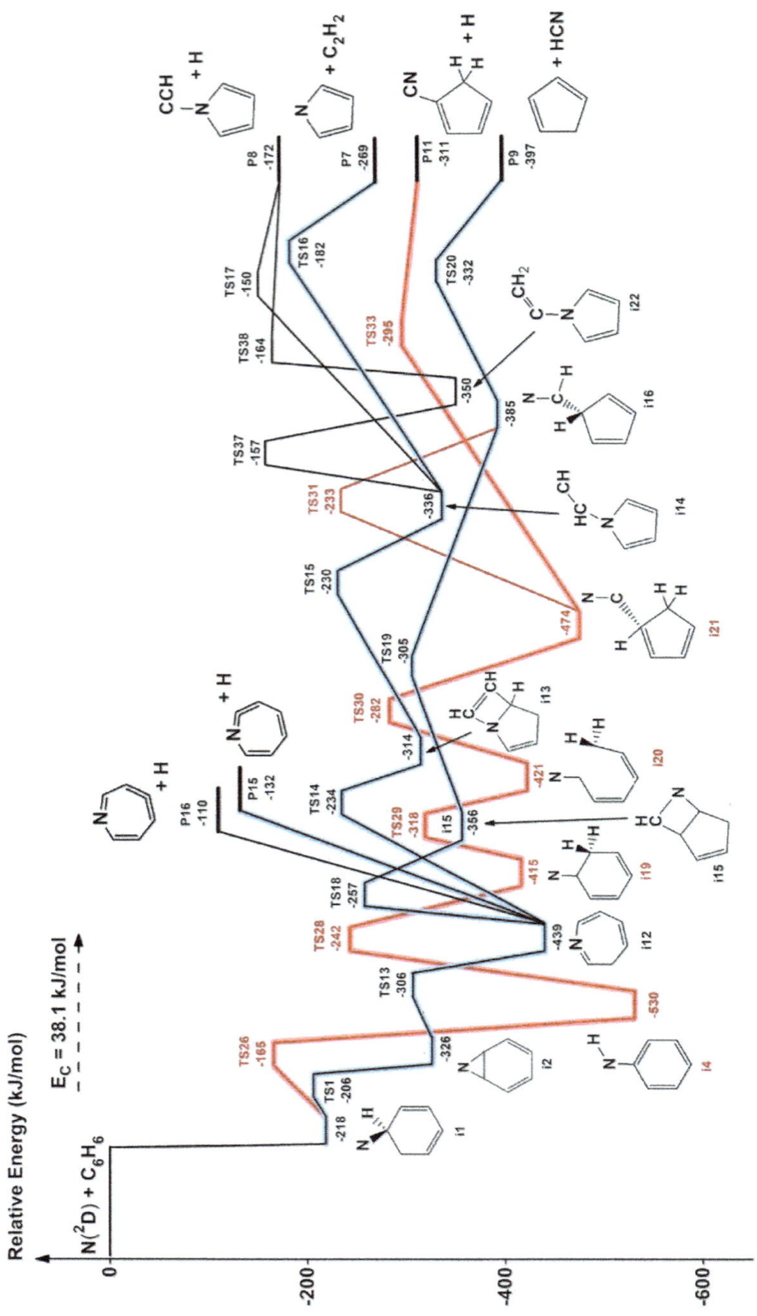

Fig. 6 Potential energy surface for the reaction N(^{2}D) + C_6H_6. Relative energies computed at CBS level with inclusion of core–valence correlation and ZPE correction. The product channels P11, P15, and P16 were not present in the work of Chin *et al.*[52] due to the overlooking of some pathways (see text). For the sake of simplicity, a few pathways have been neglected; only the channels predicted statistically to be relevant at E_c = 31.8 kJ mol^{-1} are shown (for more details, see ESI†).

to the products C_4H_4N (pyrrolyl) and acetylene (P7) (channel 4). i14 can also give rise to the products $C_4H_4NCCH + H$ (P8) (channel 6) through TS17 or following the path TS37 → i22 → TS38. This last pathway was not considered by Chin *et al.*[52] Finally, i1, by overcoming a barrier of 53 kJ mol^{-1}, can isomerize also to the very stable intermediate i4 which, following the path TS28 → i19 → TS29 → i20 → TS30 → i21 → TS33, gives rise to the products $C_5H_5CN + H$ (P11) (channel 3). i21 can also give rise to the products C_5H_5 + HCN (P9) (channel 1) through the pathway TS31 → i16 → TS20. These last two reactive channels, originating from i4, were not considered in the work of Chin *et al.*[52]

Comparing the energies reported in Fig. 6 with those reported in Fig. 1 and 2 of the article by Chin *et al.*[52] we can notice an almost constant difference of about 20 kJ mol^{-1}, the energies of Chin *et al.* being systematically lower than ours. However, this difference is due almost entirely to the error that Chin *et al.*[52] made in computing the energy of N(^{2}D). They computed the energy of N(^{2}D) at the same level of the other points reported in the PES, while we estimated this energy by adding the experimental excitation energy to the computed energy of N(^{4}S). We evaluated an energy difference of 25 kJ mol^{-1} at the CBS level between the experimental excitation energy of N(^{2}D) and that calculated. This is the most important contribution to the difference between the energies reported by Chin *et al.*[52] and those calculated in this work. The rest of the difference is due to the fact that we included in our calculations also the core–valence correlation contribution. We believe therefore that our results are more accurate than those reported by Chin *et al.*[52] and the error is within ±5 kJ mol^{-1}.

5.2 RRKM estimates of the product branching fractions

RRKM estimates of product branching fractions were performed considering the collision energy of the CMB experiment (31.8 kJ mol^{-1}) and for three different temperatures corresponding to the surface temperature of Titan (94 K), its stratospheric temperature (175 K) and the temperature at 1000 km of altitude where benzene has its abundance peak (200 K). The calculated BFs are referring to the limit of zero pressure which is a good approximation for CMB experiments and the stratosphere and thermosphere of Titan. All the possible elementary processes, including back-dissociation for the first intermediate i1 and back-isomerization of all intermediates, have been considered in the RRKM calculations to obtain the branching fractions, reported in Table 2. The results reported here have been obtained by using the CBS values of energy. Preparatory RRKM calculations on the CC PES, however, provided very similar results. Under all the considered conditions, the dominant channel is the one leading to C_5H_5 + HCN (1) with a BF varying from 0.860 at 94 K to 0.745 under the conditions of the CMB experiment. The other most important channels are channel (4) leading to $C_4H_4N + C_2H_2$ (BF varying from 0.051 at 94 K to 0.056 at the experimental E_c) and channel (8) leading to *o*-C_6H_5N + H, which has the strongest energy/temperature dependence (BF varying from 0.048 at 94 K to 0.135 at the experimental E_c) as already seen in the case of other H-displacement channels.[37,38,48,49] This can be rationalized by considering that this channel originates from the dissociation of the i12 intermediate which is readily formed by a two-step isomerization of the initial intermediate i1. The competition affects almost only channel (1) which is, indeed, the lowest energy pathway originating from the same intermediate i12.

Table 2 Product branching fractions (BFs) of the N(^{2}D) + benzene reaction at $E_c = 31.8$ kJ mol^{-1}, compared to statistical RRKM/ME results in CMB conditions and at the temperatures of 200, 175 and 94 K, corresponding to the temperature of Titan at 1000 km, in the stratosphere and at the surface, respectively. The statistical results of ref. 52 in the range of energies from 41.8 to 0 kJ mol^{-1} on a theoretical PES are also reported

Channel number and products	CMB Expt, $E_c = 31.8$ kJ mol^{-1}	RRKM, $E_c = 31.8$ kJ mol^{-1}	RRKM[a], $E_c = 41.8$–0 kJ mol^{-1}	RRKM, 94 K	RRKM, 175 K	RRKM, 200 K
(3) $C_5H_5CN + H$	**0.08 ± 0.04**	**0.053**	0.030–0.023	0.039	0.040	**0.040**
(6) $C_4H_4NCCH + H$		*Negligible*	0.006–0.004	*Negligible*	*Negligible*	*Negligible*
(7) $C_6H_5N + H$		*Negligible*	*Negligible*	*Negligible*	*Negligible*	*Negligible*
(8) *o*-$C_6H_5N + H$		**0.135**	nd	0.048	0.053	**0.058**
(9) *p*-$C_6H_5N + H$		**0.011**	nd	0.003	0.003	**0.003**
(1) $C_5H_5 + HCN$	**0.92 ± 0.04**	**0.745**	0.889–0.915	0.860	0.852	**0.849**
(2) $C_5H_5 + HNC$		*Negligible*	*Negligible*	*Negligible*	*Negligible*	*Negligible*
(4) $C_4H_4N + C_2H_2$		**0.056**	0.075–0.058	0.051	0.051	**0.052**
(5) $C_5H_6 + CN$		*Negligible*	*Negligible*	*Negligible*	*Negligible*	*Negligible*
(10) $C_5H_5N + CH$	nd	*Negligible*	*Negligible*	*Negligible*	*Negligible*	*Negligible*

[a] Ref. 52.

The yield of channel (9), also originating from i12, increases with the available energy, even though it remains marginal under all the conditions considered. All the other contributions show little dependence on the energy available to the system.

The BFs derived from the present RRKM calculations are reported for all ten exothermic reactive channels in Table 2, where they are also compared with both the experimental BFs and theoretical BFs from a previous theoretical study.[52] There is a substantial agreement with those results,[52] if one considers that channels (8) and (9) were not considered in that study. As already commented, it is the presence of these alternative dissociation channels (in particular channel (8)) that reduces the yield of C_5H_5 + HCN and this nicely explains the difference in the BF for channel (1) in the two cases.

6. Discussion

The rate coefficients determined in this work vary between 1.8 and 2.6 $\times$ 10^{-10} cm^3 s^{-1} with no obvious trend with the temperature. Therefore, the title reaction is fast in the entire range of temperature investigated and close to the gas kinetic limit. According to the PES derived in this work, as well as that derived by Chin *et al.*[52] the title reaction is characterized by the absence of an energy barrier in the entrance channel with no other major barriers leading to exothermically accessible products. This is perfectly in line with the experimental results. If we compare the rate constant of this reaction with those already investigated by us with the same experimental method and concerning the reactions of N(^{2}D) with hydrocarbons relevant for Titan, we note that the rate constants of the title reaction have values comparable to those of the N(^{2}D) reactions with unsaturated hydrocarbons (C_2H_2, C_2H_4 and allene)[36,42,43] and the same behavior with the temperature, while they are significantly larger than those determined for the N(^{2}D) reactions with saturated hydrocarbons (CH_4, C_2H_6, and C_3H_8)[41] which are smaller and characterized by a significant decrease as the temperature falls. This comparison sustains again what is seen by the theoretical calculations, that is, the N(^{2}D) reactions are barrierless when they involve unsaturated hydrocarbons and proceed with an addition mechanism guided by the electron density of multiple C–C bonds, while they are characterized by a small entrance barrier when they involve hydrocarbons with only sigma bonds and the insertion mechanism is the only option.

The analysis of the CMB results points to the occurrence of two groups of mechanisms: (1) the H-displacement channels, which can be accompanied by the formation of five different isomeric molecular products, and (2) the ring contraction channels with the elimination of a molecular moiety, which actually correspond to four possible routes leading to the cyclopentadienyl radical and HCN (1) or HNC (2), pyrrolyl radical and C_2H_2, (4) and cyclopentadiene and CN (5). The yield of the H-displacement channels is minor (overall BF = 0.08 ± 0.04) (see Table 2), while the dominant reaction channel is the one leading to the destruction of the aromatic ring *via* a ring-contraction process (overall BF = 0.92 ± 0.04). According to the theoretical calculations, the BF of HNC is negligible, while it is 0.745 for HCN production under the conditions of the CM experiments (E_c = 31.8 kJ mol^{-1}) (see Table 2). The much more favorable formation of HCN with respect to HNC mainly originates from the very unfavorable competition of the isomerization of the initial addition intermediate i1 (located at −218 kJ mol^{-1}) to i4 (the precursor leading to HNC formation *via* TS4 → i5 → TS5 → i6 → TS11 (see Fig. S4 in the ESI†)) with respect to the isomerization of i1 to i2 (the precursor leading to HCN formation) *via* TS1. TS1 is located much lower in energy (at −206 kJ mol^{-1}, see Fig. 6) with respect to TS4 (see Fig. 6) (−182 kJ mol^{-1} at the CC level, see Fig. S5 in ESI†). Theory predicts that after the facile isomerization of the initial benzazirine intermediate (i2) to the seven-member ring N-cycloheptatriene intermediate (i12), the latter can evolve to bimolecular products *via* three main competitive pathways: (i) the main one leading to C_5H_5 + HCN (channel (1)) (BF = 0.745) *via* a series of four isomerizations (see Fig. 6); (ii) the second one leading, in a barrier-less fashion, dominantly (BF = 0.135) to *o*-C_6H_5N (*ortho*-N-cycloheptatriene radical) + H, and in minor part (BF = 0.011) to *p*-C_6H_5N (*para*-N-cycloheptatriene radical) + H; (iii) the third one leading to C_4H_4N (pyrrolyl radical) + C_2H_2 (acetylene) (BF = 0.056). Therefore, the theory predicts that a small contribution to the reactive scattering comes from channel (4). Indeed, we have some experimental evidence that the pyrrolyl radical (m/z = 66) is formed, because we observed at the CM angle a small signal intensity at m/z = 66, larger than that due to the ^{13}C natural isotopic abundance of C_5H_5 (cyclopentadienyl) (m/z = 65). Considering the theoretical BFs, the yield of HCN channel (1) is more than 13 times larger than that of the C_2H_2 channel (4) (see Table 2) and, therefore, it is not surprising that the signal at m/z = 66 is about nine times smaller than the one at m/z = 65. The small signal at m/z = 66 could also be associated to the C_5H_6 (cyclopentadiene) + CN formation channel (5), but for this channel the theory predicts a negligible BF.

Regarding the H-displacement channels, according to our electronic structure calculations five isomers with gross formula C_6H_5N can be formed (see Section 5 and Fig. 6). A satisfactory fit of the LAB angular and TOF distributions was achieved by using a single set of CM functions for the H-displacement channel(s), which implies that our data are not sensitive enough to allow disentangling of the possible different contributions to the signal at the same m/z, because the enthalpies of reaction of these channels are not very different and the reaction mechanism is the same. Among the possible H-displacement channels (3) and (6)–(9), those predicted by RRKM as more significant are *o*-C_6H_5N + H (8) (BF = 0.135) and C_5H_5CN (2-cyano-cyclopentadiene) + H (3) (BF = 0.053), with *p*-C_6H_5N + H (9) being minor (BF = 0.011). The other two H-displacement channels, (6) and (7), have a negligible BF (this is due to the high isomerization barriers along the corresponding reaction pathways). The reason why channel (8) is predicted to be more abundant than channel (3) is because the former arises from the initial intermediate i2 without an exit barrier and the latter from the initial intermediate i4, which is formed from i1 with much less probability than i2, as already discussed.

It is interesting to examine the variation of the product BFs with energy (temperature). As Table 2 shows, the BF of the dominant C_5H_5 + HCN channel (1) is predicted to increase substantially with the lowering of the temperature (energy), while the BFs for the H-forming channels, especially the main channel (8), decrease substantially. Our predicted trend is in agreement with that predicted by Chin *et al.*[52] (see Table 2). However, we emphasize that the main H-displacement channel (8) was not revealed by the study of Chin *et al.*, who overall underestimate substantially (by a factor larger than five at E_c = 31.8 kJ mol^{-1})

the H-forming channels with respect to our findings, and therefore overestimate somewhat the C_5H_5 + HCN channel. Notably, however, the present and Chin *et al.* studies predict a similar BF for the pyrrolyl + acetylene channel (4) (see Table 2).

A final comment is in order concerning the channel leading to C_5H_5N (pyridine) + CH. We could not investigate it experimentally because of elastic interference from the secondary beam. However, the theoretical BF resulted to be negligible because of the presence of very high isomerization barriers (especially TS9) to its precursor i9 (see Fig. S5 in the ESI†). Therefore, the title reaction cannot be considered a formation route of this species.

In the Perugia laboratory we have recently investigated two related reactions, that is the N(^{2}D) reaction with another small aromatic (pyridine)[40] and the reaction of both ground and electronically excited atomic oxygen with benzene.[48,49] The experimental BFs for the reaction with pyridine are quite different with respect to the BFs of the title reaction. In that case, indeed, the H-displacement channels (that could be associated with four distinct isomeric co-products) are dominant with a BF of *ca.* 0.65 at a comparable collision energy (33.5 kJ mol^{-1}) while the BF of the analogous ring-contraction channel (associated with the formation of C_4H_4N (pyrrolyl radical) and HCN) is only 0.35. The PES of N(^{2}D) + pyridine has not been characterized to date and, therefore, we cannot provide a reason for such a different behavior. Interestingly, the best fit CM angular distributions were found to be symmetric for both the H-displacement channels and the ring-contraction channels, while in the present case they both show some forward bias. Despite the expected similarities, the detailed reaction mechanism is different in the two cases. The comparison with the O(^{3}P,^{1}D) + C_6H_6 reactions (investigated at $E_c = 34.3$ kJ mol^{-1}) is also interesting. In that case, for the reaction of O(^{3}P) the experimental BFs are 0.66 for the H-displacement channel and 0.32 for the ring-contraction channel (leading to C_6H_5 + CO after intersystem crossing to the underlying singlet PES). But if we focus on the reaction of O(^{1}D), then the BFs are very similar to those obtained here, with a BF for the H-displacement channel of only 0.04 and a BF for the ring-contraction channel (leading to C_5H_6 + CO and to C_5H_5 + CO + H) of 0.96. In conclusion, the detailed shape of the underlying PESs and the reaction mechanism with its variation with the total available energy are essential in determining the product BFs and also the extent of the aromatic ring preservation mechanism.

7. Implications for the modeling of the atmosphere of Titan

To examine the effects of the present experimental determinations on the chemistry of Titan's atmosphere, we included the N(^{2}D) + benzene reaction in the 1D-photochemical model described in Dobrijevic *et al.*,[92] which treats the chemistry of neutrals and cations (anions are not considered in this study as they have only a minor influence on the overall chemistry), and the coupling between them from the lower atmosphere to the ionosphere. Details of the recent updates of the chemical scheme are presented in Benne *et al.*,[93] with a subsequent update in Vanuzzo *et al.*[36] Two different simulations were performed during this investigation. The first simulation represents the nominal model where the N(^{2}D) + benzene reaction is neglected. For the second one, we included the N(^{2}D) +

benzene reaction using the rate constants and simplified BFs determined in this study, as we have assumed that the reaction leads to the production of o-C_6H_5N + H with a yield of 9% and of C_5H_5 + HCN with a yield of 91%. In addition, we completed the chemical network for o-C_6H_5N, the main fate of which is the reaction with H atoms giving mainly C_5H_5 + HCN considering the PES of the C_6H_6N system presented in Fig. 6.

Overall, we verify that the inclusion of the $N(^2D)$ + benzene reaction has a minor effect on the chemistry of Titan's atmosphere. Firstly, even if the flux of the $N(^2D)$ + benzene reaction is not negligible, it corresponds to only 5% of the destruction pathways of benzene and therefore has little effect on its abundance. Secondly, in the thermosphere there are other sources of C_5H_5, which is mainly produced by the photodissociation of C_5H_6 which is, in turn, mostly produced by the ion–molecule reaction $C_3H_5^+ + C_2H_4 \rightarrow$ c-$C_5H_7^+ + H_2$ followed by the electronic dissociative recombination of $C_5H_7^+$ (1-cyclopentene-2-ylium). The $N(^2D)$ + benzene reaction produces new species however, here identified with o-C_6H_5N for simplicity, that reach an abundance relative to N_2 of 4×10^{-10} around 1150 km. At that altitude, the main fate of o-C_6H_5N will be to react with H to produce C_5H_5 + HCN and, therefore, is not expected to form new potentially detectable molecules or to be a significant source of the N-containing aromatic species identified in the thermosphere.[21,22] However, even though the main sink of o-C_6H_5N is the reaction with H, the reaction of o-C_6H_5N with CH_3 can give an important contribution. In the current model, the products of this reaction are not explicitly described. Instead, these are included in a generic aromatic compound encompassing all the unknown aromatics of the model (a minor generic compound compared to benzene). It will be important in the future to study this reaction which could potentially lead to aromatic nitrogen heterocycles and, in particular, to the ethenyl-pyridine molecule.

Furthermore, roughly 4% of the reactive flux is actually leading to 1-cyano-1,3-cyclopentadiene, even though we have not explicitly included channel (3) in our model for simplicity (all the H-displacement flux was associated to the formation of o-C_6H_5N). 1-Cyano-1,3-cyclopentadiene has just been detected (together with its less stable 1-cyano-2,4-cyclopentadiene isomer) in the interstellar medium towards the Taurus Molecular Cloud.[94,95] Having a very strong dipole moment and considering that its reaction with H should not be efficient under the conditions of Titan,[96] this species might be a target for remote detection. In this respect, we note that the James Webb Space Telescope has been recently directed towards the atmosphere of Titan searching for new species.[6]

8. Conclusions

The reaction $N(^2D) + C_6H_6$ was investigated in a combined experimental and theoretical approach. Experimental rate constants and product BFs are in line with the prediction based on a novel PES. The kinetic experiments provided us with the rate constant as a function of temperature from 50 K to 296 K, a range that includes the values relevant to the atmosphere of Titan (the recommended value in the 50–296 K range is $2.19 \pm 0.30 \times 10^{-10}$ cm^3 s^{-1}). These values, in the gas-kinetic limit, confirm the barrierless nature of the reaction, as indicated by the electronic structure calculations of the PES. The CMB experiments, complemented by statistical (RRKM/ME) calculations of the product BFs on the high-level *ab initio* PES, have identified the main reaction channels. The ring-contraction mechanism was found to be dominant in the CMB experiments (with a BF of 0.92) in good agreement with RRKM predictions according to which the channel leading to cyclopentadienyl + HCN (BF = 0.745) dominates and some contribution from the channel leading to pyrrolyl + C_2H_2 (BF = 0.056) is also possible. H-Displacement channels give a small contribution in the CMB experiments as well as in the RRKM estimates that predict *o*-N-cycloheptatriene as the main molecular product. Pyridine is not formed through an N/CH exchange channel. The effect of the energy available to the system on BFs is such that can only be seen in the competition between the cyclopentadienyl + HCN channel and the H-displacement channel leading to *o*-C_6H_5N. Under the conditions relevant for Titan, the yield of C_5H_5 increases up to *ca.* 0.85 while that of *o*-C_6H_5N becomes *ca.* 0.05–0.06. The BFs of the other channels are not so sensitive to the temperature.

The recommended value of the rate constant and the derived product BFs have been used in a 1-D photochemical model of Titan's atmosphere to simulate the effect of the title reaction on the species abundances (including any new products formed) as a function of the altitude. It is found that the inclusion of the N(^{2}D) + benzene reaction has only a minor effect on the amount of benzene but new N-bearing species could be formed and accumulate. Remote detection of 1-cyano-1,3-cyclopentadiene may be feasible due to its large dipole moment.

A more general conclusion is that aromatic rings are not so resistant to the chemical attack as commonly believed, at least not in the case of very energetic reactants like electronically excited species. This is the fourth case we have investigated where ring contraction of small aromatics (benzene or pyridine) is a significant (if not the dominant) reaction pathway, after O(^{3}P,^{1}D) + benzene (significant in the case of the ^{3}P reaction, dominant in the case of the ^{1}D reaction),[48,49] O(^{3}P,^{1}D) + pyridine (largely dominant in both cases)[91] and N(^{2}D) + pyridine (significant).[40] It will be interesting to verify whether a similar mechanism is present also in the case of polycyclic aromatic hydrocarbons which are supposed to accumulate in various environments because of their chemical stability.

Conflicts of interest

There are no conflicts to declare.

Acknowledgements

This work was supported by the Italian Space Agency (ASI, DC-VUM-2017-034, Grant no. 2019-3 U.0 Life in Space). The authors acknowledge support from also the Italian MUR, University of Perugia within the program "Department of Excellence-2018-2022-Project AMIS" and the "Dipartimento di Ingegneria Civile e Ambientale" of the University of Perugia within the project "Dipartimenti di Eccellenza 2018–2022" and the Herla Project http://www.hpc.unipg.it/hosting/vherla/vherla.html Università degli Studi di Perugia for allocated computing time. K. M. H. acknowledges support from the French program "Physique et Chimie du Milieu Interstellaire" (PCMI) of the CNRS/INSU with the INC/INP co-funded by the CEA and CNES as well as funding from the "Program National de Planétologie" (PNP) of the CNRS/INSU.

References

1 N. Balucani and D. Skouteris, Gas-phase Prebiotic Chemistry Driven by Ultraviolet Photolysis of Simple Molecules, in *Prebiotic Photochemistry: from Urey–Miller-like Experiments to Recent Findings*, ed. F. Saija and G. Cassone, Comprehensive Series in Photochemical and Photobiological Science No. 20, Royal Society of Chemistry, Cambridge, ch. 3, 2021, pp. 37–59.
2 *Titan: Surface, Atmosphere and Magnetosphere*, ed. I. Mueller-Wodarg, C. Griffith, E. Lellouch and T. Cravens, Cambridge University Press, 2014.
3 *Titan from Cassini-Huygens*, ed. R. Brown, J. P. Lebreton and J. Waite, Springer, Heidelberg, 2010.
4 N. Balucani, *Chem. Soc. Rev.*, 2012, **41**, 5473–5483.
5 A. E. Thelen, C. A. Nixon, R. G. Cosentino, M. A. Cordiner, N. A. Teanby, C. E. Newman, P. G. J. Irwin and S. B. Charnley, *Planet. Sci. J.*, 2022, **3**, 146.
6 C. A. Nixon, *et al.*, *Publ. Astron. Soc. Pac.*, 2016, **128**, 018007.
7 V. Vuitton, R. V. Yelle, S. J. Klippenstein, S. M. Hörst and P. Lavvas, *Icarus*, 2019, **324**, 120–197.
8 J. C. Loison, E. Hébrard, M. Dobrijevic, K. M. Hickson, F. Caralp, V. Hue, G. Gronoff, O. Venot and Y. Bénilan, *Icarus*, 2015, **247**, 218–247.
9 K. Willacy, S. Chen, D. J. Adams and Y. L. Yung, *Astrophys. J.*, 2022, **933**, 230.
10 C. Sagan, B. N. Khare, W. R. Thompson, G. D. McDonald, M. R. Wing, J. L. Bada, T. Vo-Dinh and E. T. Arakawa, *Astrophys. J.*, 1993, **414**, 399–405.
11 A. Coustenis, A. Salama, B. Schulz, S. Ott, E. Lellouch, Th. Encrenaz, D. Gautier and H. Feuchtgruberc, *Icarus*, 2003, **161**, 383–403.
12 A. Coustenis, *et al.*, *Icarus*, 2007, **189**, 35–62.
13 J. H. Waite Jr, *et al.*, *Science*, 2007, **316**, 870–875.
14 T. T. Koskinen, R. V. Yelle, D. S. Snowden, P. Lavvas, B. R. Sandel, F. J. Capalbo, Y. Benilan and R. A. West, *Icarus*, 2011, **216**, 507–534.
15 H. B. Niemann, *et al.*, *Nature*, 2005, **438**, 779–784.
16 S. Vinatier, B. Schmitt, B. Bézard, P. Rannou, C. Dauphin, R. de Kok, D. E. Jennings and F. M. Flasar, *Icarus*, 2018, **310**, 89–104.
17 J. C. Loison, M. Dobrijevic and K. M. Hickson, *Icarus*, 2019, **329**, 55–71.
18 V. Vuitton, R. V. Yelle and M. J. McEwan, *Icarus*, 2007, **191**, 722–742.
19 C. A. Nixon, A. E. Thelen, M. A. Cordiner, Z. Kisiel, S. B. Charnley, E. M. Molter, J. Serigano, P. G. J. Irwin, N. A. Teanby and Y.-J. Kuan, *Astrophys. J.*, 2020, **160**, 205.
20 B. M. Dinelli, M. López-Puertas, A. Adriani, M. L. Moriconi, B. Funke, M. García-Comas and E. D'Aversa, *Geophys. Res. Lett.*, 2013, **40**, 1489–1493.
21 M. López-Puertas, B. M. Dinelli, A. Adriani, B. Funke, M. García-Comas, M. L. Moriconi, E. D'Aversa, C. Boersma and L. J. Allamandola, *Astrophys. J.*, 2013, **770**, 132.
22 R. P. Haythornthwaite, A. J. Coates, G. H. Jones, A. Wellbrock, J. H. Waite, V. Vuitton and P. Lavvas, *Planet. Sci. J.*, 2021, **2**, 26.
23 O. Dutuit, *et al.*, *Astrophys. J. Suppl. Ser.*, 2013, **204**, 20.
24 K. Sato, K. Misawa, Y. Kobayashi, M. Matsui, S. Tsunashima, Y. Kurosaki and T. Takayanagi, *J. Phys. Chem. A*, 1999, **103**, 8650–8656.

25 T. Suzuki, Y. Shihira, T. Sato, H. Umemoto and S. Tsunashima, *J. Chem. Soc., Faraday Trans.*, 1993, **89**, 995–999.
26 N. Balucani, A. Bergeat, L. Cartechini, G. G. Volpi, P. Casavecchia, D. Skouteris and M. Rosi, *J. Phys. Chem. A*, 2009, **113**, 11138–11152.
27 N. Balucani, P. Casavecchia, L. Bañares, F. J. Aoiz, T. Gonzales-Lezana, P. Honvault and J. M. Launay, *J. Phys. Chem. A*, 2006, **110**, 817–829.
28 N. Balucani, M. Alagia, L. Cartechini, P. Casavecchia, G. G. Volpi, L. A. Pederson and G. C. Schatz, *J. Phys. Chem. A*, 2001, **105**, 2414–2422.
29 Y. Ralchenko, A. E. Kramida and J. Reader, *NIST Atomic Spectra Database (Ver. 4.0.1)*, National Institute of Standards and Technology, Gaithersburg, MD, 2006.
30 Z. Homayoon, J. M. Bowman, N. Balucani and P. Casavecchia, *J. Phys. Chem. Lett.*, 2014, **5**, 3508–3513.
31 N. Balucani, F. Leonori, R. Petrucci, M. Stazi, D. Skouteris, M. Rosi and P. Casavecchia, *Faraday Discuss.*, 2010, **147**, 189–216.
32 N. Balucani, M. Alagia, L. Cartechini, P. Casavecchia, G. G. Volpi, K. Sato, T. Takayanagi and Y. Kurosaki, *J. Am. Chem. Soc.*, 2000, **122**, 4443–4450.
33 N. Balucani, L. Cartechini, M. Alagia, P. Casavecchia and G. G. Volpi, *J. Phys. Chem. A*, 2000, **104**, 5655–5659.
34 N. Balucani, D. Skouteris, F. Leonori, R. Petrucci, M. Hamberg, W. D. Geppert, P. Casavecchia and M. Rosi, *J. Phys. Chem. A*, 2012, **116**, 10467–10479.
35 L. Mancini, G. Vanuzzo, D. Marchione, G. Pannacci, P. Liang, P. Recio, M. Rosi, D. Skouteris, P. Casavecchia and N. Balucani, *J. Phys. Chem. A*, 2021, **125**, 8846–8859.
36 G. Vanuzzo, L. Mancini, G. Pannacci, P. Liang, D. Marchione, P. Recio, Y. Tan, M. Rosi, D. Skouteris, P. Casavecchia, N. Balucani, K. M. Hickson, J.-C. Loison and M. Dobrijevic, *ACS Earth Space Chem.*, 2022, **6**, 2305–2321.
37 P. Liang, L. Mancini, D. Marchione, G. Vanuzzo, F. Ferlin, P. Recio, Y. Tan, G. Pannacci, L. Vaccaro, M. Rosi, P. Casavecchia and N. Balucani, *Mol. Phys.*, 2022, **120**, e1948126.
38 G. Vanuzzo, D. Marchione, L. Mancini, P. Liang, G. Pannacci, P. Recio, Y. Tan, M. Rosi, D. Skouteris, P. Casavecchia and N. Balucani, *J. Phys. Chem. A*, 2022, **126**, 6110–6123.
39 M. Rosi, S. Falcinelli, P. Casavecchia, N. Balucani, P. Recio, A. Caracciolo, G. Vanuzzo, D. Skouteris and C. Cavallotti, in *Computational Science and its Applications – ICCSA 2021*, ed. O. Gervasi, B. Murgante, S. Misra, C. Garau, I. Blečić, D. Taniar, B. O. Apduhan, A. M. A. C. Rocha, E. Tarantino, C. M. Torre, Springer International Publishing, Cham, 2021, pp. 620–631.
40 P. Recio, D. Marchione, A. Caracciolo, V. J. Murray, L. Mancini, M. Rosi, P. Casavecchia and N. Balucani, *Chem. Phys. Lett.*, 2021, **779**, 138852.
41 D. Nuñez-Reyes, J.-C. Loison, K. M. Hickson and M. Dobrijevic, *Phys. Chem. Chem. Phys.*, 2019, **21**, 6574–6581.
42 K. M. Hickson, C. Bray, J.-C. Loison and M. Dobrijevic, *Phys. Chem. Chem. Phys.*, 2020, **22**, 14026–14035.
43 D. Nuñez-Reyes, J.-C. Loison, K. M. Hickson and M. Dobrijevic, *Phys. Chem. Chem. Phys.*, 2019, **21**, 22230–22237.
44 N. Balucani, O. Asvany, A. H. H. Chang, S. H. Lin, Y. T. Lee, R. I. Kaiser, H. F. Bettinger, P. V. R. Schleyer and H. F. Schaefer III, *J. Chem. Phys.*, 1999, **111**, 7457–7471.

45 F. Zhang, Y. Guo, X. Gu and R. I. Kaiser, *Chem. Phys. Lett.*, 2007, **440**, 56–63.
46 D. S. N. Parker, B. B. Dangi, N. Balucani, D. Stranges, A. M. Mebel and R. I. Kaiser, *J. Org. Chem.*, 2013, **78**, 11896–11900.
47 H. F. Bettinger, P. V. R. Schleyer, H. F. Schaefer III, P. R. Schreiner, R. I. Kaiser and Y. T. Lee, *J. Chem. Phys.*, 2000, **113**, 4250–4264.
48 C. Cavallotti, C. De Falco, L. Pratali Maffei, A. Caracciolo, G. Vanuzzo, N. Balucani and P. Casavecchia, *J. Phys. Chem. Lett.*, 2020, **11**, 9621–9628.
49 G. Vanuzzo, A. Caracciolo, T. K. Minton, N. Balucani, P. Casavecchia, C. de Falco, A. Baggioli and C. Cavallotti, *J. Phys. Chem. A*, 2021, **125**, 8434–8453.
50 N. Balucani, L. Pacifici, D. Skouteris, A. Caracciolo, P. Casavecchia and M. Rosi, in *Computational Science and its Applications – ICCSA 2018*, ed. O. Gervasi, B. Murgante, S. Misra, E. Stankova, C. M. Torre, A. M. A. C. Rocha, D. Taniar, B. O. Apduhan, E. Tarantino and Y. Ryu, Springer International Publishing, Cham, , pp. 763–772.
51 N. Balucani, L. Pacifici, D. Skouteris, A. Caracciolo, P. Casavecchia, S. Falcinelli and M. Rosi, in *Computational Science and its Applications – ICCSA 2019*, ed. S. Misra, O. Gervasi, B. Murgante, E.Stankova, V. Korkhov, C. Torre, A. M. A. C. Rocha, D. Taniar, B. O. Apduhan and E. Tarantino, Springer International Publishing, Cham, 2019, pp. 316–324.
52 C.-H. Chin, T. Zhu and J. Z. H. Zhang, *Phys. Chem. Chem. Phys.*, 2021, **23**, 12408–12420.
53 D. S. N. Parker, R. I. Kaiser, O. Kostko, T. P. Troy, M. Ahmed, A. M. Mebel and A. G. G. M. Tielens, *Astrophys. J.*, 2015, **803**, 53.
54 L. Zhao, *et al.*, *Phys. Chem. Chem. Phys.*, 2021, **23**, 18495–18505.
55 D. S. N. Parker, T. Yang, B. B. Dangi, R. I. Kaiser, P. P. Bera and T. J. Lee, *Astrophys. J.*, 2015, **815**, 115.
56 B. J. Sun, C. H. Huang, S. Y. Chen, S. H. Chen, R. I. Kaiser and A. H. H. Chang, *J. Phys. Chem. A*, 2014, **118**, 7715–7724.
57 D. S. N. Parker, R. I. Kaiser, O. Kostko, T. P. Troy, M. Ahmed, B.-J. Sun, S.-H. Chen and A. H. H. Chang, *Phys. Chem. Chem. Phys.*, 2015, **17**, 32000–32008.
58 A. I. Begley, N. S. Shuman, B. A. Long, R. Kämpf, L. Gyr, A. A. Viggiano and R. Zenobi, *J. Phys. Chem. A*, 2022, **126**, 1743–1754.
59 Z. Zhang, X. Gong, S. Zhang, H. Yang, Y. Shi, C. Yang, X. Zhang, X. Xiong, X. Fang and Z. Ouyang, *Sci. Rep.*, 2013, **3**, 3481.
60 P. Casavecchia, F. Leonori, N. Balucani, R. Petrucci, G. Capozza and E. Segoloni, *Phys. Chem. Chem. Phys.*, 2009, **11**, 46–65.
61 P. Casavecchia, F. Leonori and N. Balucani, *Int. Rev. Phys. Chem.*, 2015, **34**, 161–204.
62 N. Daugey, P. Caubet, B. Retail, M. Costes, A. Bergeat and G. Dorthe, *Phys. Chem. Chem. Phys.*, 2005, **7**, 2921–2927.
63 K. M. Hickson, J.-C. Loison, P. Larregaray, L. Bonnet and V. Wakelam, *J. Phys. Chem. A*, 2022, **126**, 940–950.
64 Y. Wu, J. Cao, H. Ma, C. Zhang, W. Bian, D. Nuñez-Reyes and K. M. Hickson, *Sci. Adv.*, 2019, **5**, eaaw0446.
65 D. Nuñez-Reyes and K. M. Hickson, *Phys. Chem. Chem. Phys.*, 2018, **20**, 17442–17447.
66 K. M. Hickson and J.-C. Loison, *J. Phys. Chem. A*, 2022, **126**, 3903–3913.
67 C. L. Lin and F. Kaufman, *J. Chem. Phys.*, 1971, **55**, 3760–3770.

68 D. Marchione, L. Mancini, P. Liang, G. Vanuzzo, F. Pirani, D. Skouteris, M. Rosi, P. Casavecchia and N. Balucani, *J. Phys. Chem. A*, 2022, **126**, 3569–3582.

69 D. Skouteris, N. Balucani, C. Ceccarelli, N. Faginas Lago, C. Codella, S. Falcinelli and M. Rosi, *Mon. Not. R. Astron. Soc.*, 2019, **482**, 3567–3575.

70 D. Skouteris, N. Balucani, N. Faginas-Lago, S. Falcinelli and M. Rosi, *Astron. Astrophys.*, 2015, **584**, A76.

71 A. D. Becke, *J. Chem. Phys.*, 1993, **98**, 1372–1377.

72 P. J. Stephens, F. J. Devlin, C. F. Chabalowski and M. J. Frisch, *J. Phys. Chem.*, 1994, **98**, 11623–11627.

73 T. H. Dunning Jr, *J. Chem. Phys.*, 1989, **90**, 1007–1023.

74 D. E. Woon and T. H. Dunning Jr, *J. Chem. Phys.*, 1993, **98**, 1358–1371.

75 R. A. Kendall, T. H. Dunning Jr and R. J. Harrison, *J. Chem. Phys.*, 1992, **96**, 6796–6806.

76 C. Gonzalez and H. B. Schlegel, *J. Chem. Phys.*, 1989, **90**, 2154–2161.

77 C. Gonzalez and H. B. Schlegel, *J. Phys. Chem.*, 1990, **94**, 5523–5527.

78 R. J. Bartlett, *Annu. Rev. Phys. Chem.*, 1981, **32**, 359–401.

79 K. Raghavachari, G. W. Trucks, J. A. Pople and M. Head-Gordon, *Chem. Phys. Lett.*, 1989, **157**, 479–483.

80 J. Olsen, P. Jørgensen, H. Koch, A. Balkova and R. J. Bartlett, *J. Chem. Phys.*, 1996, **104**, 8007–8015.

81 C. E. Moore, *Atomic Energy Levels, Natl. Bur. Stand. (U.S.) Circ. N. 467*, U.S., GPO, Washington, DC, 1949.

82 J. M. L. Martin, *Chem. Phys. Lett.*, 1996, **259**, 669–678.

83 M. Frisch, *et al.*, *Gaussian 09, Revision D.01*, 2009.

84 H.-J. Werner, *et al.*, *MOLPRO, Version 2021.2, a Package of Ab Initio Programs*, 2010, see: https://www.molpro.net.

85 H.-J. Werner, *et al.*, *J. Chem. Phys.*, 2020, **152**, 144107.

86 P. Flükiger, H. P. Lüthi, S. Portmann and J. Weber, *MOLEKEL 4.3*, Swiss Center for Scientific Computing, Manno, Switzerland, 2000–2002.

87 S. J. Klippenstein, *J. Chem. Phys.*, 1992, **96**, 367–371.

88 G. A. Fisk, J. D. McDonald and D. R. Herschbach, General discussion, *Discuss. Faraday Soc.*, 1967, **44**, 228–230.

89 A. M. Schmoltner, P. M. Chu and Y. T. Lee, *J. Chem. Phys.*, 1989, **91**, 5365–5373.

90 A. Caracciolo, G. Vanuzzo, N. Balucani, D. Stranges, S. Tanteri, C. Cavallotti and P. Casavecchia, *Chin. J. Chem. Phys.*, 2019, **32**, 113–122.

91 P. Recio, S. Alessandrini, G. Vanuzzo, G. Pannacci, A. Baggioli, D. Marchione, A. Caracciolo, V. J. Murray, P. Casavecchia, N. Balucani, C. Cavallotti, C. Puzzarini and V. Barone, *Nat. Chem.*, 2022, **14**, 1405.

92 M. Dobrijevic, J. Loison, K. Hickson and G. Gronoff, *Icarus*, 2016, **268**, 313–339.

93 B. Benne, M. Dobrijevic, T. Cavalié, J.-C. Loison and K. M. Hickson, *Astron. Astrophys.*, 2022, **667**, A169.

94 M. C. McCarthy, *et al.*, *Nat. Astron.*, 2021, **5**, 176–180.

95 K. L. K. Lee, P. B. Changala, R. A. Loomis, A. M. Burkhardt, C. Xue, M. A. Cordiner, S. B. Charnley, M. C. McCarthy and B. A. McGuire, *Astrophys. J. Lett.*, 2021, **910**, L2.

96 Q. Mao, L. Cai and H. Pitsch, *Combust. Flame*, 2020, **222**, 423–433.

Faraday Discussions

PAPER

Experimental radiative cooling rates of a polycyclic aromatic hydrocarbon cation

José E. Navarro Navarrete,[a] James N. Bull, [b] Henrik Cederquist,[a] Suvasthika Indrajith,[a] MingChao Ji, [a] Henning T. Schmidt, [a] Henning Zettergren, [a] Boxing Zhu[a] and Mark H. Stockett [*a]

Received 10th January 2023, Accepted 6th February 2023
DOI: 10.1039/d3fd00005b

Several small Polycyclic Aromatic Hydrocarbons (PAHs) have been identified recently in the Taurus Molecular Cloud (TMC-1) using radio telescope observations. Reproducing the observed abundances of these molecules has been a challenge for astrochemical models. Rapid radiative cooling of PAHs by Recurrent Fluorescence (RF), the emission of optical photons from thermally populated electronically excited states, has been shown to efficiently stabilize small PAHs following ionization, augmenting their resilience in astronomical environments and helping to rationalize their observed high abundances. Here, we use a novel method to experimentally determine the radiative cooling rate of the cation of 1-cyanonaphthalene ($C_{10}H_7CN$, 1-CNN), the neutral species of which has been identified in TMC-1. Laser-induced dissociation rates and kinetic energy release distributions of 1-CNN cations isolated in a cryogenic electrostatic ion-beam storage ring are analysed to track the time evolution of the vibrational energy distribution of the initially hot ion ensemble as it cools. The measured cooling rate is in good agreement with the previously calculated RF rate coefficient. Improved measurements and models of the RF mechanism are needed to interpret astronomical observations and refine predictions of the stabilities of interstellar PAHs.

1 Introduction

In 2021, after decades of inconclusive searches, astronomers identified the first specific Polycyclic Aromatic Hydrocarbon (PAH) molecules in space.[1] Using the Greenbank Radio Telescope, McGuire *et al.* identified two isomers of the small PAH cyanonaphthalene (CNN, $C_{10}H_7CN$) in the Taurus Molecular Cloud (TMC-1). Several other aromatics have been identified using similar methods[2] and many more likely await discovery. Meanwhile, JWST promises to provide new insight into the properties and evolution of interstellar PAHs through observations of their infrared emission bands.[3,4]

[a]*Department of Physics, Stockholm University, Stockholm, Sweden. E-mail: Mark.Stockett@fysik.su.se*
[b]*School of Chemistry, University of East Anglia, Norwich, UK*

At the dawn of this new era, we find challenges to the established understanding of the interstellar organic inventory. Indeed, McGuire *et al.* themselves could not explain the observed abundance of CNN through their state-of-the-art astrochemical modeling.[1] Their model, which performs well for smaller nitriles,[5] underestimates the observed abundance of CNN by six orders of magnitude.[1]

In a recent report,[6] we addressed this gap by elucidating the main destruction pathway for one of the two CNN isomers, which are expected to behave similarly and indeed have comparable abundances in TMC-1.[1] We found, contrary to the explicit assumption of McGuire *et al.*, that 1-cyanonaphthalene (1-CNN) is efficiently stabilized by rapid radiative cooling following ionization, closing off some of the reaction channels assumed to deplete CNN from TMC-1. Specifically, the model of McGuire *et al.* includes several charge transfer reactions:

$$A^+ + C_{10}H_7CN \rightarrow A + C_{10}H_7CN^+ \tag{1}$$

where $A = C$, H, or He, and where the resulting $C_{10}H_7CN^+$ is assumed to disintegrate into linear fragments. We show that the excess energies of these reactions (the difference between the ionization energies of the reactants) is insufficient to induce such thorough fragmentation and will mainly activate the lowest-energy unimolecular dissociation channel:

$$C_{10}H_7CN^+ \rightarrow C_{10}H_6^+ + HCN + \varepsilon, \tag{2}$$

where ε is the kinetic energy released in the reaction. Owing to efficient radiative stabilization by Recurrent Fluorescence (RF), the emission of optical photons from thermally populated electronically excited states,[7,8] even this low-energy channel is closed for collision with C^+ and at least partially closed for H^+.[6]

Crucially, vibronic coupling greatly increases the RF rate. RF, which is common for small PAHs,[9–16] has important implications for astrochemistry beyond the case of CNN in TMC-1. Stabilized by RF, small PAHs may be much more abundant in space than hitherto thought, including in more diffuse regions where it has long been held that PAHs must include at least ≈ 50 carbon atoms to survive in the UV radiation field there.[17] Better understanding of the RF mechanism and its competition with destructive reaction pathways can serve to refine candidate lists for astronomical searches for interstellar PAHs.

Recurrent Fluorescence of PAHs excited by UV photons has been suggested as the source of the Extended Red Emission (ERE)[18,19] observed in the red rectangle and other reflection nebulae. The ERE is a broad feature whose peak wavelength shifts in different astronomical environments.[18] Improved models of RF in PAHs, including vibronic couplings and anharmonic effects, and predicting excitation energy dependent emission spectra, are needed for quantitative comparison to astronomical observations.

In the present contribution, we extend our previous study by measuring the laser-induced dissociation rate and Kinetic Energy Release (KER) distributions of 1-CNN cations isolated in a cryogenic electrostatic ion-beam storage ring. This novel combination of techniques allows us to track the evolution of the internal energy distribution of the ions over four orders of magnitude in time after ionization. The measured cooling rate allows us to experimentally confirm our previously calculated Recurrent Fluorescence rate coefficient. Such laboratory studies are essential to understanding observations in the era of high-resolution astrochemistry.

2 Experiments

All experiments were conducted at the Double ElectroStatic Ion Ring ExpEriment (DESIREE) infrastructure at Stockholm University.[20,21] Cryogenic cooling of the storage ring, which is schematically shown in Fig. 1, to a temperature of approximately 13 K results in a residual gas density on the order of $\sim 10^4$ cm^{-3}, consisting mostly of H_2.[21] These conditions enable isolation of highly excited ions or reactive ions in a collision free environment for hours.[22,23] Electrostatic ion storage devices feature sampling times exceeding 1 s, orders of magnitude longer than conventional mass spectrometers, enabling time-resolved observations of delayed processes such as unimolecular dissociation[16] and thermionic electron emission.[24] DESIREE has also been used for vibrational energy dependent action spectroscopy of astrophysically relevant ions including PAHs,[25,26] substituted PAHs thought to form in interstellar ices,[27,28] and carbon cluster anions.[29,30]

The methods used here have largely been described previously.[6,16] Briefly, 1-CNN (Sigma-Aldrich, >96%) was sublimed from powder in a resistively heated oven coupled to an electron cyclotron resonance (ECR) ion source (Pantechnik Monogan) using helium as a support gas. Cations extracted from the source were accelerated to 34 keV kinetic energy. Mass-selected beams of 1-CNN$^+$ ($m/z = 153$) were stored in the DESIREE ion storage ring illustrated in Fig. 1.

The ion production method is known to produce ensembles of ions with broad vibrational energy distributions similar to Boltzmann distributions with temperatures of a few thousand Kelvin.[16] A small fraction of ions are produced with vibrational energies within a window such that their dissociation rate coefficients are low enough that they reach the ring, but high enough so that they may dissociate before they are stabilized by radiative cooling. These ions contribute a rapidly decreasing yield of neutral fragments referred to as spontaneous decay in the first tens of milliseconds of ion storage. The majority of ions remain stored after the disappearance of this spontaneous signal.

Stored ions were overlapped collinearly with light from an Optical Parametric Oscillator (OPO) laser system in the interaction arm of the storage ring (upper straight section in Fig. 1). The OPO was operated at 250 Hz repetition rate and produced pulses of $\sim$5 ns duration and $\sim$5 μJ per pulse at 420 nm.

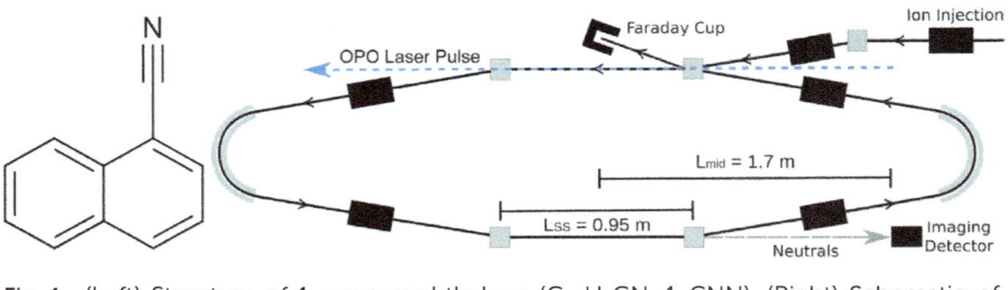

Fig. 1 (Left) Structure of 1-cyanonaphthalene ($C_{10}H_7CN$, 1-CNN). (Right) Schematic of DESIREE storage ring.

Neutral fragments emitted in the observation arm of the storage ring (lower straight section in Fig. 1), which may occur up to several hundred microseconds after excitation, are unaffected by the electrostatic fields and continue on straight trajectories until they reach an imaging detector system with a triple stack of custom-made microchannel plates with ultra-high dynamic range (Photonis), a phosphor screen anode, and optical lenses projecting the phosphorescence through the DESIREE vacuum chamber windows and onto a CMOS camera (Photon Focus). The $\delta t = 2.0$ ms camera exposures were synchronized with the laser pulses, such that the laser-induced neutral yield resulting from each laser shot was collected in a single frame. Frames for a given laser firing time were summed over more than 50 000 repeat injections of 1-CNN$^+$, each of 400 ms storage duration.

For each summed frame, three-dimensional Newton spheres were reconstructed by applying an inverse Abel transform, using the 'three-point' algorithm implemented in the PyAbel package[31] written in Python. The density distribution is related to the KER distribution by:

$$\varepsilon(r_{3D}) = \frac{m_{\text{neut}}}{m_{\text{cat}}} E_{\text{Acc}} \left(\frac{r_{3D}}{L}\right)^2 \tag{3}$$

where $\varepsilon(r_{3D})$ is the KER associated with a radial slice of the Newton sphere of radius r_{3D}, m_{neut} and m_{cat} are the masses of the neutral and cationic reaction products, $E_{\text{Acc}} = 34$ keV is the beam energy, and L is the distance traveled by the products from the point of reaction to the detector. For clarity of presentation, the KER distributions in Fig. 7 are plotted against an ε scale calculated according to eqn (3) with $L = L_{\text{mid}}$, where $L_{\text{mid}} = 1.7$ m is the distance from the detector to the mid-point of the observation arm. Our analysis accounts for dissociation occurring along the full length of the observation arm by summing contributions to the Newton sphere density distribution in the detector plane from points at distances in the range $L_{\text{mid}} \pm L_{\text{SS}}/2$, where $L_{\text{SS}} = 0.95$ m is the length of the straight section seen by the detector (see Fig. 1). In the present case, the procedure gives a nearly insignificant correction relative to assuming all decays occur at L_{mid}.

3 Results

3.1 Dissociation rates

The spontaneous dissociation rate of the source-heated ions $R(t)$ is shown in the upper panel of Fig. 2. A constant background rate due to detector dark noise and collisions with residual gas has been subtracted. Due to the broad distribution $g(E, t)$ of vibrational energies E and the rapid variation of the dissociation rate coefficient $k_{\text{diss}}(E)$ with energy, the dissociation rate $R(t) \propto \int k_{\text{diss}}(E) g(E, t) \text{d}E$ does not follow simple exponential decay[32] but rather has the approximate time dependence[33]

$$R(t) = r_0 t^{-1} \text{e}^{-k_c t} \tag{4}$$

where t is the time after ion formation and $k_c = 302.9(4)$ s^{-1} is the critical rate coefficient at which dissociation and radiative cooling are competitive. The dashed line in the upper panel of Fig. 2 is a fit of eqn (4) to the data. Note that at longer times significant deviation from the exponential quenching behaviour is

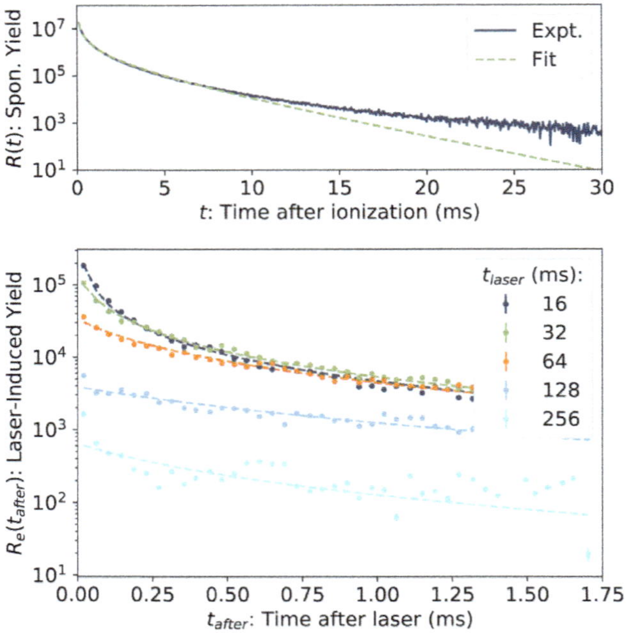

Fig. 2 (Top) Spontaneous dissociation rate $R(t)$, fit of eqn (4). (Bottom) Examples of laser-induced dissociation rates $R_e(t_{after})$ with fits of eqn (8).

observed. This effect is discussed in Section 4. The constant r_0 in eqn (4) contains the instrumental parameters and is given by:

$$r_0 = \eta_{det} N(t) \gamma_0(t) \frac{L_{SS}}{C} \tag{5}$$

where $\eta_{det} = 0.34(3)$ is the efficiency for detection of HCN (eqn (2)), $C = 8.7$ m is the circumference of the storage ring, $L_{SS} = 0.95$ m is the length of the stored beam viewed by the detector, and $N(t)$ is the average number of stored ions remaining in the ring at time t. The latter is determined from the count rate $R(t)$, measured during ion storage, and the terminal ion beam current, measured at the end of each injection–storage cycle using the Faraday cup shown in Fig. 1. The number $N(t)$ of ions stored during in the spontaneous decay measurement, relative to the initial number $N(120\ \mu s)$, is plotted in Fig. 3. The factor $\gamma_0(t)$ is the ensemble average destruction probability:[34]

$$\gamma_0(t) = \int g(E, t)(1 - e^{-k_{diss}(E)t})dE / \int g(E, t)dE. \tag{6}$$

Put another way, it is the fraction of stored ions with vibrational energies $E \approx E_m$ consistent with lifetimes equal to the observation time, *i.e.* $k_{diss}(E_m) = t^{-1}$.[32] Simulated vibrational energy distributions $g(E, t)$ and $g_m(E, t) = g(E, t)(1 - e^{-k_{diss}(E)t})$ are plotted in the upper panel of Fig. 4 for a time $t = 120\ \mu s$ corresponding to the first pass of the ions through the storage ring. Also indicated are the vibrational energies E_{avg} and E_m averaged over the distributions $g(E, t)$ and $g_m(E, t)$, respectively.

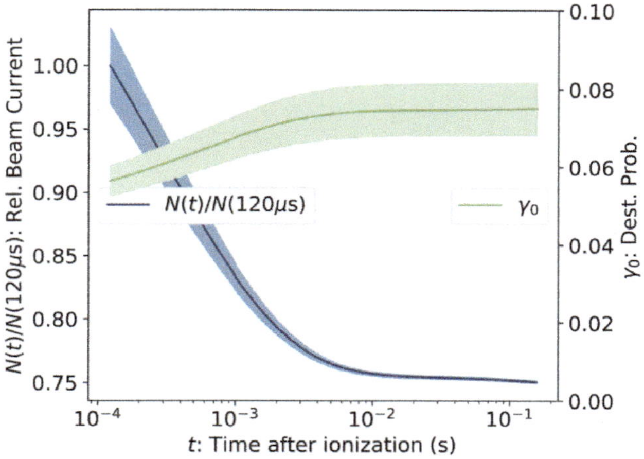

Fig. 3 (Left axis) Number of ions stored in the measurement of $R(t)$, relative to the value at the first time point at $t = 120$ μs. (Right axis) Ensemble averaged destruction probability γ_0, determined from eqn (5). The shaded areas give the uncertainties.

In the lower panel of Fig. 2, several examples of laser-induced decay rates R_e following single-photon absorption are plotted as functions of $t_{after} = t - t_{laser}$, the time after the laser was fired. At early laser firing times t_{laser} the laser-induced rate $R_e(t_{after})$ resembles the spontaneous decay rate $R(t)$, while at later times it converges on an exponential decay rate. We assume[35] that the portion of the ions which absorb a single photon of energy $h\nu$ is re-heated such that its internal energy distribution resembles that the full ensemble had at some earlier time $t_{laser} - \Delta t$, *i.e.*

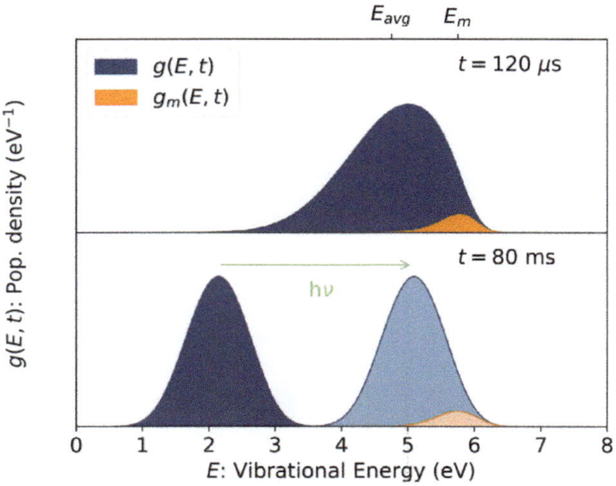

Fig. 4 (Top) Vibrational energy distribution of the stored ion ensemble, $g(E, t)$, and of the portion of the ensemble with energies $E \approx E_m$, $g_m(E, t)$, at $t = 120$ μs corresponding to the first pass of the ions through the ring. (Bottom) Vibrational energy distribution at a later time. Following the absorption of a single 420 nm photon, the ions in this example are reheated to the lighter shaded distribution and have the same value of E_m as in the upper panel.

$$g(E + h\nu, t_{\text{laser}}) \propto g(E, t_{\text{laser}} - \Delta t) \tag{7}$$

and that the laser-induced decay rate is given by

$$R_e(t_{\text{after}}) = \frac{p_1 L_{SS}}{C} R(t + t_0) = \frac{p_1 L_{SS} r_0}{C} (t + t_0)^{-1} e^{-k_c(t+t_0)} \tag{8}$$

where $t_0 = t_{\text{laser}} - \Delta t$ is the time after formation to which the distribution is back-shifted, and p_1 is the probability of absorbing a single photon. The factor L_{SS}/C enters eqn (8) as the laser and ion beams are only overlapped in the outer straight section of the storage ring (see Fig. 1). The solid lines in the lower panel of Fig. 2 are fits to eqn (8) with the parameter k_c fixed to the value extracted from the fit of eqn (4) to the spontaneous decay rate. However, as eqn (4) gives a poor fit to $R(t)$ at later times, the values of t_0 were determined instead by fitting $R_e(t_{\text{after}})$ directly to $R(t)$, as illustrated in Fig. 5. A simultaneous fit to 100 laser-induced decay curves using a single constant value of $p_1 = 3.04(3) \times 10^{-3}$ in eqn (8) yielded the results shown in Fig. 6.

3.2 Kinetic energy release distributions

Examples of the kinetic energy release distributions $P_e(\varepsilon)$ resulting from dissociation of laser-excited 1-CNN$^+$ are shown in Fig. 7. In our previous study of spontaneous dissociation of source-heated 1-CNN$^+$, we found that the KER distributions were well represented by the transition state model of Hansen,[36] which includes both tunneling through and reflection from a parabolic barrier:

$$P(\varepsilon) \propto \frac{e^{\beta'}}{e^{\beta'} + 1} e^{-(\varepsilon - \Delta E)/k_B T^{\ddagger}}, \quad \text{where } \beta' = 4\pi \frac{\Delta E}{\hbar\omega}\left(\sqrt{\frac{\varepsilon}{\Delta E}} - 1\right). \tag{9}$$

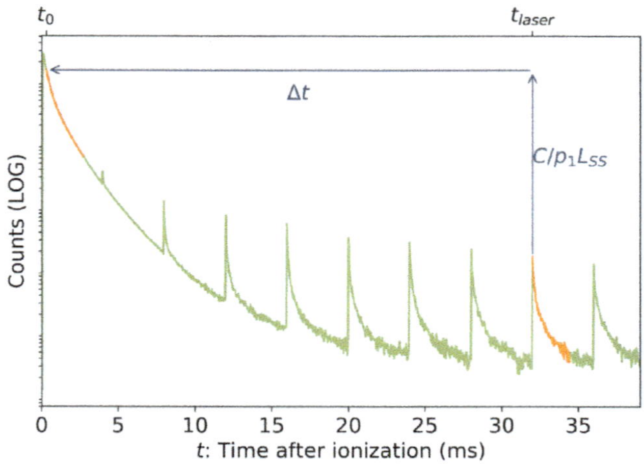

Fig. 5 Graphical example of method for determining back-shifted times t_0 by fitting the laser-induced dissociation rate $R_e(t_{\text{after}})$ to the spontaneous dissociation rate of source-heated ions $R(t)$. In this example, $t_{\text{laser}} = 32$ ms and $t_0 = 320$ μs.

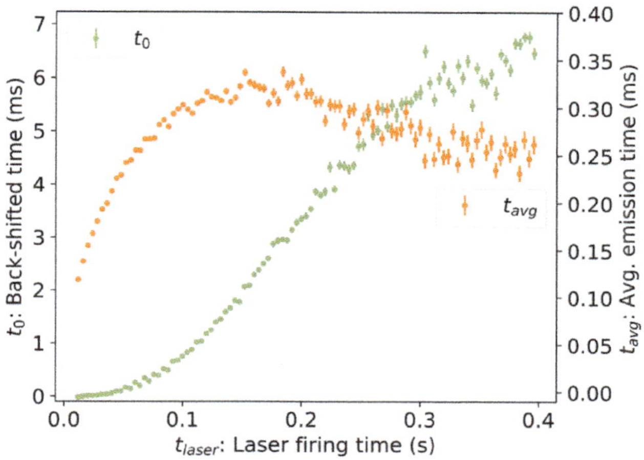

Fig. 6 (Left axis) Back-shifted times t_0 to which the vibrational energy distribution is reheated following photon absorption at time t_{laser}, determined from laser-induced decay curves $R_e(t_{after})$. (Right axis) Average emission times t_{avg}.

We adopt the parameters from our earlier study,[6] which found a rather small reverse barrier height $\Delta E = 7.6(4)$ meV and a transition state frequency $\hbar\omega = 350(20)\ cm^{-1}$ from simultaneous fits to the KER distributions for spontaneously decaying 1-CNN^+ up to 20 ms after ionization. That study also found the dissociation rate coefficient to follow a modified Arrhenius expression:

$$k_{diss}(T^{\ddagger}) = \frac{k_B T^{\ddagger}}{h}\ e^{-E_a/k_B T^{\ddagger}} \tag{10}$$

with an activation energy for HCN-loss of $E_a = 3.16(4)$ eV. In the present contribution, all 100 laser-induced KER distributions are simultaneously fit subject to

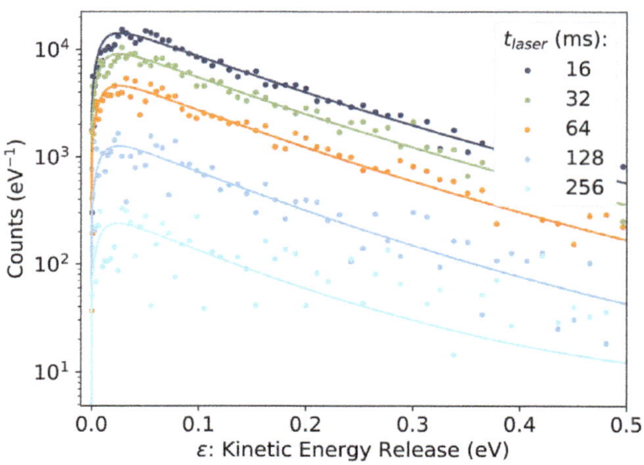

Fig. 7 Examples of KER distributions $P_e(\varepsilon)$ resulting from dissociation of laser-excited 1-CNN^+. Solid lines are fits to eqn (9).

the physical constraint that the integrated distribution be equal to the integral of the count rate:

$$\int P_e(\varepsilon, t)\mathrm{d}\varepsilon = \int R_e(t)\mathrm{d}t = \frac{p_1 L_{SS} r_0}{C} k_{\mathrm{diss}}(T^{\ddagger})\delta t \tag{11}$$

where $\delta t = 2$ ms is the integration time of the camera and the value of $p_1 = 3.04(3) \times 10^{-3}$ determined from the fit of the laser-induced decay rates was held fixed. The values of $T^{\ddagger}$ resulting from the fit are plotted in Fig. 8.

To convert the transition state temperatures $T^{\ddagger}$ to the typical energies E_m of dissociation ions, the caloric curve $E(T)$ is first computed assuming Boltzmann statistics:

$$E(T) = \frac{\int E'\rho(E')\mathrm{e}^{-E'/k_B T}\,\mathrm{d}E'}{\int \rho(E')\mathrm{e}^{-E'/k_B T}\,\mathrm{d}E'} \tag{12}$$

where $\rho(E)$ is the vibrational level density calculated using the Beyer–Swinehart algorithm[37] and vibrational frequencies calculated at the B3LYP/6-31G(d,p) level of Density Functional Theory (DFT) as implemented in Gaussian 16.[38] Finally, the energy E_m of the precursor ion includes a finite heat bath correction due to the energy required to reach the transition state:[39]

$$E_m = E(T^{\ddagger}) + \frac{E_a}{2} + \frac{{E_a}^2}{12(E(T^{\ddagger}) + E_a/2)}. \tag{13}$$

In this energy range, the values of E_m depend linearly on $T^{\ddagger}$, as indicated by the second vertical axis in Fig. 8.

In Fig. 9, the experimentally-determined dissociation rate coefficient k_{diss} extracted using eqn (11) is plotted against the vibrational energy E_m from eqn (13), labeled E to emphasize that this is an intrinsic property not dependent on the

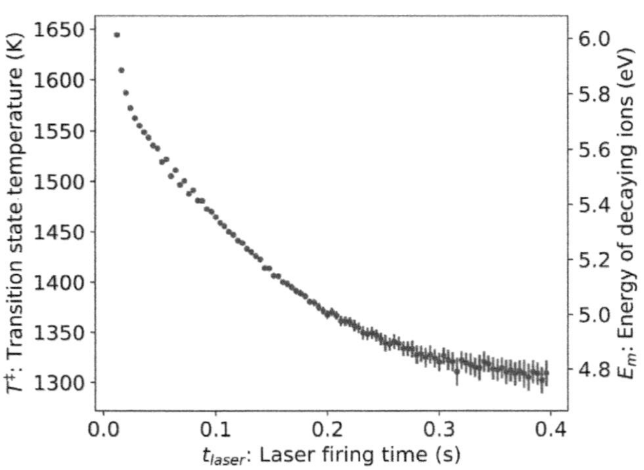

Fig. 8 Transition state temperatures $T^{\ddagger}$ (left axis) and corresponding vibrational energies E_m (right axis) of ions decaying following photon absorption at time t_{laser}, determined from KER distributions $P_e(\varepsilon)$.

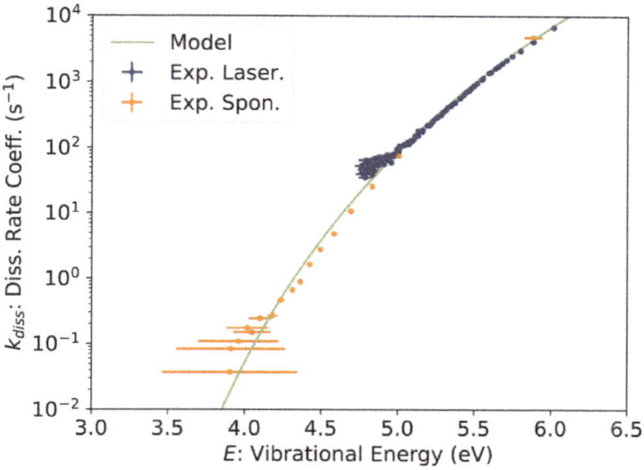

Fig. 9 Unimolecular dissociation rate coefficient $k_{diss}(E)$ from KER measurements.

internal energy distribution. The values from our previous study of spontaneous dissociation of 1-CNN$^+$ are also reproduced, along with our model dissociation rate coefficient in microcanonical form:

$$k_{\mathrm{diss}}(E) = A_{1000\mathrm{K}}^{\mathrm{diss}} \frac{\rho(E - E_{\mathrm{a}})}{\rho(E)}, \tag{14}$$

where a nominal value of the pre-exponential factor $A_{1000\mathrm{K}}^{\mathrm{diss}} = k_{\mathrm{B}}[1000\mathrm{K}]/h = 2 \times 10^{13}\ \mathrm{s}^{-1}$ was adopted.[6]

3.3 Cooling rate

To determine the cooling rate, we combine the back-shifted times t_0 from the decay rates with the corresponding energies E_{m} from the KER distributions. As the images for the KER measurement are integrated over the full decay curve following each laser shot, over which time the energy is changing rapidly, we compute an average emission time $t_{\mathrm{avg}} = \int tR_{\mathrm{e}}(t)\mathrm{d}t/\int R_{\mathrm{e}}(t)\mathrm{d}t$, plotted in Fig. 6. The time associated with each value of E_{m} is thus $t_0 + t_{\mathrm{avg}}$. These values, labeled $E_{\mathrm{m}}(t_0)$ for simplicity, are plotted in Fig. 10.

Also plotted in Fig. 10 are values from our previous study of spontaneously dissociating 1-CNN$^+$. The values $E_{\mathrm{m}}(t)$ from 2.5–50 ms were recorded during the measurement of the spontaneous decay rate $R(t)$ reproduced in Fig. 2. The same correction for the average emission time during the camera exposure time has been applied. The point at 120 μs is from a separate single-pass measurement.

Finally, the energies of the ions at the times the laser is fired, $E_{\mathrm{m}}(t_{\mathrm{laser}})$, are plotted in Fig. 10. Naively, one might assume $E_{\mathrm{m}}(t_{\mathrm{laser}}) = E_{\mathrm{m}}(t_0) - h\nu$. However, as illustrated in Fig. 4, the vibrational energy distribution of the laser-reheated ions is significantly narrower than the original distribution of the hot ions as they cool. The high-energy tail of the reheated distribution, $g_{\mathrm{m}}(E + h\nu, t_{\mathrm{laser}})$, is thus shifted up by less than if the full original distribution $g(E, t_0)$ had simply shifted down by $h\nu$. We note that the re-heated distribution $g(E + h\nu, t_{\mathrm{laser}})$, unlike the original

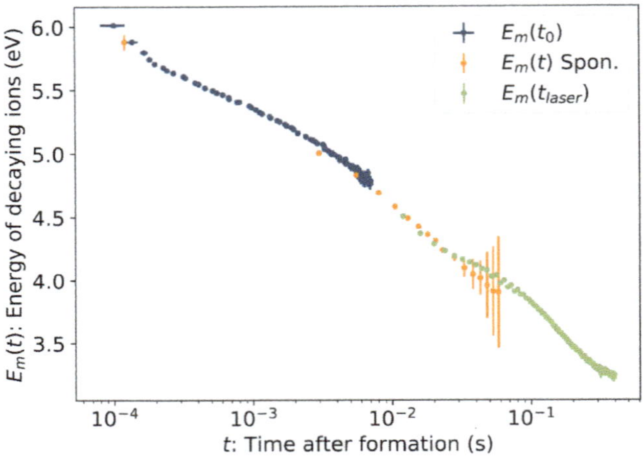

Fig. 10 Energies E_m of ions decaying at time t. Blue symbols $E_m t_0$ are the energies from Fig. 8 plotted against the back-shifted time $t_0 + t_{avg}$. The green symbols $E_m(t_{laser})$ give the energy in eqn (15) as a function of laser firing time. The orange symbols $E_m(t)$ are the energies of spontaneously decaying 1-CNN$^+$ determined previously.[6]

distribution $g(E, t_0)$, is approximately Gaussian and for such a distribution $E_m = E_{avg} + E_a$. Thus,

$$E_m(t_{laser}) = E(T^{\ddagger}) + E_a - h\nu. \tag{15}$$

These values are plotted in Fig. 10.

The time derivative of $E_m(t)$ is the rate at which the high-energy tail of the vibrational energy distribution $g_m(E, t)$ moves towards lower energy, be it by depletion of hot ions by dissociation or transposition of population by radiative cooling, and is called the energy shift rate. At any given time, only a small fraction γ_0 of the stored ion ensemble is part of the tail and subject to shift. Thus the total energy loss rate, which is directly comparable to the absolute dissociation and radiative cooling rate coefficients, is given by $\gamma_0^{-1}\, dE_m/dt$ and is plotted in Fig. 11. For clarity, points derived from laser firing times $t_{laser} > 300$ ms, which have large uncertainties due to the small changes in t_0 (see Fig. 6), are excluded from the plot but are included in the further analysis below.

Included in Fig. 11 are the calculated radiated powers P_{IR} and P_{RF} based on our previous study. Briefly, the IR (vibrational) radiative cooling rate coefficient is computed under the Simple Harmonic Cascade approximation:

$$k_{IR}(E) = \sum_s k_s = \sum_s A_s^{IR} \sum_{v=1}^{v \le E/h\nu_s} \frac{\rho(E - vh\nu_s)}{\rho(E)}, \tag{16}$$

where v is the vibrational quantum number, and $h\nu_s$ and A_s are the transition energy and Einstein coefficient of vibrational mode s, respectively, calculated at the B3LYP/6-31G(d,p) level of Density Functional Theory (DFT) as implemented in Gaussian 16.[38] The radiated power $P_{IR} = \sum_s k_s h\nu_s$. The RF (electronic) cooling rate coefficient is

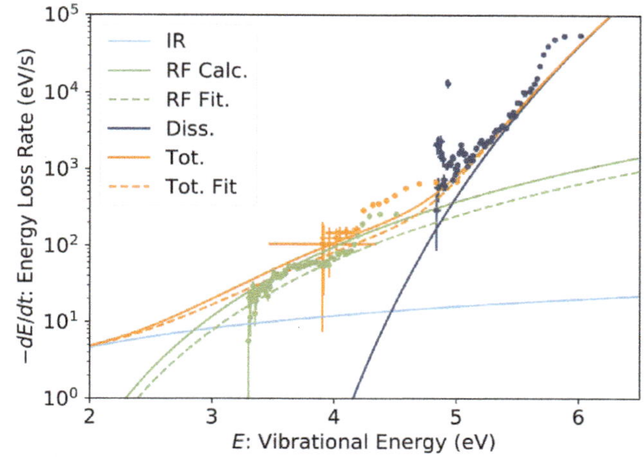

Fig. 11 Energy loss rate for 1-CNN^+. Symbol colours for experimental data correspond to those in Fig. 10. Solid lines are calculated energy loss rates due to dissociation, RF and IR radiative cooling. The dashed lines are a fit to the experimental data taking the oscillator strength f of the RF transition as a free parameter.

$$k_{\mathrm{RF}}(E) = A^{\mathrm{RF}}\frac{\rho(E - h\nu_{\mathrm{el}})}{\rho(E)}, \tag{17}$$

where the electronic transition energy $h\nu_{\mathrm{el}} = 1.10$ eV was computed at the equilibrium geometry of the lowest-lying L_α excited state using EOM-CCSD/cc-pVDZ calculations performed in CFOUR.[40] The Einstein coefficient is given by

$$A^{\mathrm{RF}} = \frac{2\pi\nu_{\mathrm{el}}{}^2 e^2}{\varepsilon_0 m_e c^3} f, \tag{18}$$

where the oscillator strength $f = 0.011$ was calculated using a Franck–Condon–Herzberg–Teller simulation[41] at the ωB97X-D/cc-pVDZ level of DFT. The radiated power is $P_{\mathrm{RF}} = k_{\mathrm{RF}} h\nu_{\mathrm{el}}$. The energy shift rate due to dissociation is $P_{\mathrm{diss}} = k_{\mathrm{diss}}E$ where the dissociation rate coefficient is given in eqn (14).

The calculated energy loss rates agree well with the experimental values. In the range from 3 to 5 eV, the experimental points are up to two orders of magnitude greater than can be attributed to IR radiative cooling, and agrees well with our modeled P_{RF}. In our previous report, we found that Herzberg–Teller vibronic coupling increases the oscillator strength of the RF transition from 1×10^{-4} to the value $f = 0.011$ used in the calculated P_{RF}. If the oscillator strength is taken as a fitting parameter, holding the RF transition energy $h\nu_{\mathrm{el}}$, P_{IR} and P_{diss} constant, we find $f = 0.0074(3)$ to agree best with the experimental data, close to our calculated value. Energy loss rates with this lower oscillator strength are plotted with dashed lines in Fig. 11. Both modeled and fitted curves underpredict the measured loss rate in the range above 4 eV. Excluding the points derived from $E_{\mathrm{m}}(t_{\mathrm{laser}})$ (green symbols) covering the lower energy range from the fit (not shown) gives a somewhat higher $f = 0.0161(12)$. This range of values may suggest a dependence of the vibronic coupling strength, and hence f, the on vibrational energy, which could explain the non-exponential quenching of $R(t)$ (Fig. 2).

4 Conclusions

By direct measurement of the energy loss rates of 1-CNN cations isolated in a cryogenic electrostatic ion-beam storage ring, we have confirmed that 1-CNN$^+$ is rapidly stabilized by recurrent fluorescence in the crucial energy range from 3–5 eV. Given the 8.6 eV ionization energy of 1-CNN, RF closes off some of the collisional destruction channels included in the model of McGuire *et al.*,[1] and completely forestalls photodissociation at photon energies found in molecular clouds.[42]

The non-exponential quenching of the spontaneous dissociation rate $R(t)$ (Fig. 2) has been observed previously for other PAHs, and has been attributed to sequential fragmentation of ionic products close enough in mass to the precursor to remain stored in the ring.[27] An alternative hypothesis is that the simple statistical model of RF (eqn (17)), which was first proposed by Boissel *et al.*,[8] does not fully capture the photophysics of RF in PAHs. Energy dependent oscillator strengths and/or emission wavelengths could lead to competition between dissociation and RF over a broader range of energies and thus non-exponential quenching. In the present results, the measured power radiated through RF is in good agreement with the simple model of Boissel *et al.*, with an oscillator strength lowest optical transition in 1-CNN$^+$ of $f = 0.0074(3)$. This is consistent with our previously calculated value of $f = 0.011$, nearly two orders of magnitude greater than if Herzberg–Teller vibronic coupling is neglected.[6] Improvements to the RF model, including energy-dependent f-values and emission wavelengths, can be expected to improve agreement with laboratory data. Consideration of Herzberg–Teller coupling is essential to modeling the radiative stabilization of isolated PAHs, as well as their optical spectra.

While the radiated power is the key quantity for predicting stability of isolated PAHs in astronomical environments, it does not completely constrain the emission mechanism. For example, more frequent emission of somewhat lower-energy photons would give the same power. Additional experiments of the present type on other PAHs, as well as, crucially, direct-detection measurements of dispersed RF spectra, will enable more quantitative comparison to astronomical observations and models of *e.g.* the Extended Red Emission.

Author contributions

JENN: investigation, formal analysis, software, writing – review and editing. JNB: conceptualization, data curation, formal analysis, funding acquisition, investigation, project administration, writing – review and editing. HC: funding acquisition, resources, writing – review and editing. SI: investigation, writing – review and editing. MCJ: investigation, writing – review and editing. HTS: funding acquisition, resources, supervision, writing – review and editing. HZ: funding acquisition, resources, supervision, writing – review and editing. BZ: investigation, writing – review and editing. MHS: conceptualization, data curation, formal analysis, funding acquisition, investigation, methodology, project administration, software, supervision, visualization, writing – original draft.

Conflicts of interest

There are no conflicts to declare.

Acknowledgements

This work was supported by Swedish Research Council grant numbers 2016-03675 (MHS), 2018-04092 (HTS), 2019-04379 (HC), 2020-03437 (HZ), Knut and Alice Wallenberg Foundation grant number 2018.0028 (HC, HTS, and HZ), Olle Engkvist Foundation grant number 200-575 (MHS), and Swedish Foundation for International Collaboration in Research and Higher Education (STINT) grant number PT2017-7328 (JNB and MHS). We acknowledge the DESIREE infrastructure for provisioning of facilities and experimental support, and thank the operators and technical staff for their invaluable assistance. The DESIREE infrastructure receives funding from the Swedish Research Council under the grant numbers 2017-00621 and 2021-00155. This article is based upon work from COST Action CA18212 – Molecular Dynamics in the GAS phase (MD-GAS), supported by COST (European Cooperation in Science and Technology).

Notes and references

1 B. A. McGuire, R. A. Loomis, A. M. Burkhardt, K. L. K. Lee, C. N. Shingledecker, S. B. Charnley, I. R. Cooke, M. A. Cordiner, E. Herbst, S. Kalenskii, M. A. Siebert, E. R. Willis, C. Xue, A. J. Remijan and M. C. McCarthy, *Science*, 2021, **371**, 1265–1269.

2 J. Cernicharo, M. Agúndez, C. Cabezas, B. Tercero, N. Marcelino, J. R. Pardo and P. de Vicente, *Astron. Astrophys.*, 2021, **649**, L15.

3 T. S.-Y. Lai, L. Armus, V. U, T. Díaz-Santos, K. L. Larson, A. Evans, M. A. Malkan, P. Appleton, J. Rich, F. Müller-Sánchez, H. Inami, T. Bohn, J. McKinney, L. Finnerty, D. R. Law, S. T. Linden, A. M. Medling, G. C. Privon, Y. Song, S. Stierwalt, P. P. van der Werf, L. Barcos-Muñoz, J. D. T. Smith, A. Togi, S. Aalto, T. Böker, V. Charmandaris, J. Howell, K. Iwasawa, F. Kemper, J. M. Mazzarella, E. J. Murphy, M. J. I. Brown, C. C. Hayward, J. Marshall, D. Sanders and J. Surace, *Astrophys. J., Lett.*, 2022, **941**, L36.

4 V. U, T. Lai, M. Bianchin, R. P. Remigio, L. Armus, K. L. Larson, T. Díaz-Santos, A. Evans, S. Stierwalt, D. R. Law, M. A. Malkan, S. Linden, Y. Song, P. P. van der Werf, T. Gao, G. C. Privon, A. M. Medling, L. Barcos-Muñoz, C. C. Hayward, H. Inami, J. Rich, S. Aalto, P. Appleton, T. Bohn, T. Böker, M. J. I. Brown, V. Charmandaris, L. Finnerty, J. Howell, K. Iwasawa, F. Kemper, J. Marshall, J. M. Mazzarella, J. McKinney, F. Muller-Sanchez, E. J. Murphy, D. Sanders and J. Surace, *Astrophys. J., Lett.*, 2022, **940**, L5.

5 B. A. McGuire, A. M. Burkhardt, S. Kalenskii, C. N. Shingledecker, A. J. Remijan, E. Herbst and M. C. McCarthy, *Science*, 2018, **359**, 202–205.

6 M. H. Stockett, J. N. Bull, H. Cederquist, S. Indrajith, M. Ji, J. E. Navarro Navarrete, H. T. Schmidt, H. Zettergren and B. Zhu, *Nat. Commun.*, 2023, **14**, 395.

7 A. Léger, P. Boissel and L. d'Hendecourt, *Phys. Rev. Lett.*, 1988, **60**, 921–924.

8 P. Boissel, P. de Parseval, P. Marty and G. Lefèvre, *J. Chem. Phys.*, 1997, **106**, 4973–4984.

9 S. Martin, J. Bernard, R. Brédy, B. Concina, C. Joblin, M. Ji, C. Ortega and L. Chen, *Phys. Rev. Lett.*, 2013, **110**, 063003.

10 M. Ji, G. Montagne, R. Brédy, J. Bernard, L. Chen and S. Martin, *Phys. Scr.*, 2013, **T156**, 014092.

11 S. Martin, M. Ji, J. Bernard, R. Brédy, B. Concina, A. R. Allouche, C. Joblin, C. Ortega, G. Montagne, A. Cassimi, Y. Ngono-Ravache and L. Chen, *Phys. Rev. A*, 2015, **92**, 053425.
12 R. Brédy, C. Ortéga, M. Ji, L. Chen, J. Bernard, A. R. Allouche, C. Joblin, A. Cassimi and S. Martin, *J. Phys.: Conf. Ser.*, 2015, **583**, 012042.
13 M. Ji, J. Bernard, L. Chen, R. Brédy, C. Ortéga, C. Joblin, A. Cassimi and S. Martin, *J. Chem. Phys.*, 2017, **146**, 044301.
14 J. Bernard, L. Chen, R. Brédy, M. Ji, C. Ortéga, J. Matsumoto and S. Martin, *Nucl. Instrum. Methods Phys. Res., Sect. B*, 2017, **408**, 21–26.
15 M. Saito, H. Kubota, K. Yamasa, K. Suzuki, T. Majima and H. Tsuchida, *Phys. Rev. A*, 2020, **102**, 012820.
16 M. H. Stockett, J. N. Bull, J. T. Buntine, E. Carrascosa, M. Ji, N. Kono, H. T. Schmidt and H. Zettergren, *J. Chem. Phys.*, 2020, **153**, 154303.
17 A. Li, *Nat. Astron.*, 2020, **4**, 339–351.
18 A. N. Witt and T. S.-Y. Lai, *Astrophys. Space Sci.*, 2020, **365**, 58.
19 T. S.-Y. Lai, A. N. Witt, C. Alvarez and J. Cami, *Mon. Not. R. Astron. Soc.*, 2020, **492**, 5853–5864.
20 R. D. Thomas, H. T. Schmidt, G. Andler, M. Björkhage, M. Blom, L. Brännholm, E. Bäckström, H. Danared, S. Das, N. Haag, P. Halldén, F. Hellberg, A. I. S. Holm, H. A. B. Johansson, A. Källberg, G. Källersjö, M. Larsson, S. Leontein, L. Liljeby, P. Löfgren, B. Malm, S. Mannervik, M. Masuda, D. Misra, A. Orbán, A. Paál, P. Reinhed, K.-G. Rensfelt, S. Rosén, K. Schmidt, F. Seitz, A. Simonsson, J. Weimer, H. Zettergren and H. Cederquist, *Rev. Sci. Instrum.*, 2011, **82**, 065112.
21 H. T. Schmidt, R. D. Thomas, M. Gatchell, S. Rosén, P. Reinhed, P. Löfgren, L. Brännholm, M. Blom, M. Björkhage, E. Bäckström, J. D. Alexander, S. Leontein, D. Hanstorp, H. Zettergren, L. Liljeby, A. Källberg, A. Simonsson, F. Hellberg, S. Mannervik, M. Larsson, W. D. Geppert, K. G. Rensfelt, H. Danared, A. Paál, M. Masuda, P. Halldén, G. Andler, M. H. Stockett, T. Chen, G. Källersjö, J. Weimer, K. Hansen, H. Hartman and H. Cederquist, *Rev. Sci. Instrum.*, 2013, **84**, 055115.
22 E. Bäckström, D. Hanstorp, O. M. Hole, M. Kaminska, R. F. Nascimento, M. Blom, M. Björkhage, A. Källberg, P. Löfgren, P. Reinhed, S. Rosén, A. Simonsson, R. D. Thomas, S. Mannervik, H. T. Schmidt and H. Cederquist, *Phys. Rev. Lett.*, 2015, **114**, 143003.
23 M. Gatchell, J. Ameixa, M. Ji, M. H. Stockett, A. Simonsson, S. Denifl, H. Cederquist, H. T. Schmidt and H. Zettergren, *Nat. Commun.*, 2021, **12**, 6646.
24 M. H. Stockett, J. N. Bull, J. T. Buntine, E. Carrascosa, E. K. Anderson, M. Gatchell, M. Kaminska, R. F. Nascimento, H. Cederquist, H. T. Schmidt and H. Zettergren, *Eur. Phys. J. D*, 2020, **74**, 150.
25 M. H. Stockett, M. Björkhage, H. Cederquist, H. Schmidt and H. Zettergren, *Faraday Discuss.*, 2019, **217**, 126–137.
26 M. H. Stockett, M. Björkhage, H. Cederquist, H. T. Schmidt and Z. Henning, *Proc. Int. Astron. Union*, 2019, **15**, 127–131.
27 M. H. Stockett, J. N. Bull, H. Schmidt and H. Zettergren, *Phys. Chem. Chem. Phys.*, 2022, **24**, 12002–12010.
28 B. Zhu, J. N. Bull, M. Ji, H. Zettergren and M. H. Stockett, *J. Chem. Phys.*, 2022, **157**, 044303.

29 J. N. Bull, M. S. Scholz, E. Carrascosa, M. K. Kristiansson, G. Eklund, N. Punnakayathil, N. de Ruette, H. Zettergren, H. T. Schmidt, H. Cederquist and M. H. Stockett, *J. Chem. Phys.*, 2019, **151**, 114304.
30 B. Zhu, J. N. Bull, J. E. Navarro Navarrete, A. F. Schmidt-May, H. Cederquist, H. T. Schmidt, H. Zettergren and M. H. Stockett, *J. Chem. Phys.*, 2022, **157**, 174308.
31 S. Gibson, D. D. Hickstein, R. Yurchak, M. Ryazanov, D. Das and G. Shih, *PyAbel: v0.8.4*, 2021, DOI: 10.5281/zenodo.7438595.
32 K. Hansen, J. U. Andersen, P. Hvelplund, S. P. Møller, U. V. Pedersen and V. V. Petrunin, *Phys. Rev. Lett.*, 2001, **87**, 123401.
33 J. Andersen, H. Cederquist, J. Forster, B. Huber, P. Hvelplund, J. Jensen, B. Liu, B. Manil, L. Maunoury, S. Brøndsted Nielsen, U. Pedersen, H. Schmidt, S. Tomita and H. Zettergren, *Eur. Phys. J. D*, 2003, **25**, 139–148.
34 K. Hansen, *Phys. Rev. A*, 2020, **102**, 052823.
35 A. E. K. Sundén, M. Goto, J. Matsumoto, H. Shiromaru, H. Tanuma, T. Azuma, J. U. Andersen, S. E. Canton and K. Hansen, *Phys. Rev. Lett.*, 2009, **103**, 143001.
36 K. Hansen, *Chem. Phys. Lett.*, 2018, **693**, 66–71.
37 T. Beyer and D. F. Swinehart, *Commun. ACM*, 1973, **16**, 379.
38 M. J. Frisch, G. W. Trucks, H. B. Schlegel, G. E. Scuseria, M. A. Robb, J. R. Cheeseman, G. Scalmani, V. Barone, B. Mennucci, G. A. Petersson, H. Nakatsuji, M. Caricato, X. Li, H. P. Hratchian, A. F. Izmaylov, J. Bloino, G. Zheng, J. L. Sonnenberg, M. Hada, M. Ehara, K. Toyota, R. Fukuda, J. Hasegawa, M. Ishida, T. Nakajima, Y. Honda, O. Kitao, H. Nakai, T. Vreven, J. A. Montgomery Jr, J. E. Peralta, F. Ogliaro, M. Bearpark, J. J. Heyd, E. Brothers, K. N. Kudin, V. N. Staroverov, R. Kobayashi, J. Normand, K. Raghavachari, A. Rendell, J. C. Burant, S. S. Iyengar, J. Tomasi, M. Cossi, N. Rega, J. M. Millam, M. Klene, J. E. Knox, J. B. Cross, V. Bakken, C. Adamo, J. Jaramillo, R. Gomperts, R. E. Stratmann, O. Yazyev, A. J. Austin, R. Cammi, C. Pomelli, J. W. Ochterski, R. L. Martin, K. Morokuma, V. G. Zakrzewski, G. A. Voth, P. Salvador, J. J. Dannenberg, S. Dapprich, A. D. Daniels, Ã. Farkas, J. B. Foresman, J. V. Ortiz, J. Cioslowski and D. J. Fox, *Gaussian 16 Revision B.01*, Gaussian Inc., Wallingford CT, 2016.
39 J. Andersen, E. Bonderup and K. Hansen, *J. Chem. Phys.*, 2001, **114**, 6518–6525.
40 J. F. Stanton, J. Gauss, L. Cheng, M. E. Harding, D. A. Matthews and P. G. Szalay, *CFOUR, Coupled-Cluster Techniques for Computational Chemistry, a Quantum-Chemical Program Package.*
41 F. Santoro, A. Lami, R. Improta, J. Bloino and V. Barone, *J. Chem. Phys.*, 2008, **128**, 224311.
42 K. I. Öberg, *Chem. Rev.*, 2016, **116**, 9631–9663.

Faraday Discussions

ROYAL SOCIETY OF CHEMISTRY

PAPER

Direct frequency comb spectroscopy of HCN to evaluate line lists

D. Michelle Bailey, [a] Eric M. Crump, [ab] Joseph T. Hodges [a] and Adam J. Fleisher *[a]

Received 30th January 2023, Accepted 14th February 2023

DOI: 10.1039/d3fd00019b

We report direct frequency comb spectroscopy of the $2\nu_1$ band of $H^{13}CN$ in the short-wave infrared ($\lambda = 1.56$ μm) towards experimental validation of molecular line lists that support observatories like JWST. The laboratory measurements aim to test spectral reference data generated from an experimentally accurate potential energy surface (PES) and an *ab initio* dipole moment surface (DMS) calculated from quantum chemistry theory. Benchmarking theory with experiment will improve confidence in new astrophysics and astrochemistry inferred from spectroscopic observations of HCN and HNC. Here we describe our instrumentation and initial results using a cross-dispersed spectrometer with a virtually imaged phased array (VIPA).

Introduction

Hydrogen cyanide (HCN) has been observed in dense interstellar clouds,[1] star-forming regions,[2] comets,[3] the atmosphere of Saturn's moon Titan,[4] the atmosphere of Pluto[5] and exoplanetary atmospheres.[6,7] When detected with the hydrogen isocyanide (HNC) isomer, the relative abundances of HCN and HNC and their respective isotopic compositions are used as indicators of local temperature and chemistry.[8] For example, HNC is suggested as a tracer to identify proto-brown dwarfs[9] and as a useful chemical clock for star formation.[2] In thermal environments, the HCN/HNC equilibrium constant becomes an important parameter.[10]

In general, observational research into the chemistry or physics of many of these objects and associated phenomena depends upon models that use spectral reference data. Therefore, complete molecular line lists to model extreme environments, generated from theory and benchmarked by low-uncertainty experiments, are broadly required.[11]

For HCN and HNC, there are several sources of spectral reference data for modeling astrophysical observations. They include HITRAN,[12] GIESA[13] and ExoMol.[14] In 2018, a calculated line-list for HCN and HNC showed improved agreement

[a]*Material Measurement Laboratory, National Institute of Standards and Technology, Gaithersburg, MD 20899, USA. E-mail: adam.fleisher@nist.gov*

[b]*Department of Chemistry and Biochemistry, University of Maryland, College Park, MD 20742, USA*

with experiments[15] and was merged into HITRAN2020.[12] There, calculated transition intensities for $H^{13}CN$ were in good agreement with the transition intensities measured by dual electro-optic frequency comb spectroscopy.[16] A subsequently larger and more comprehensive update to the experimentally accurate potential energy surface (PES) and *ab initio* dipole moment surface (DMS) for HCN was recently completed and presented as the MOMeNT-90 line list[17] – intended for integration with a future HITRAN update. Importantly, the best theory for the molecular PES and DMS can include both HCN and HNC isomers. Therefore, benchmarking spectral reference data for either species can improve our ability to predict and model spectroscopic observations of both isomers.

Here we report direct frequency comb spectroscopy to benchmark HCN line lists calculated from theory. Specifically, we use a free-running mode-locked laser and a cross-dispersed spectrometer to resolve molecular absorption by rovibrational transitions. Our comb-mode-resolved spectrometer enables precise molecular spectroscopy of the $2\nu_1$ band of $H^{13}CN$ in the short-wave infrared near a wavelength of $\lambda = 1.56$ μm. We report on spectrometer performance and compare the HITRAN2020 line list with our experimental results. Progress towards the acquisition of metrology-grade HCN and HNC spectral reference data using direct frequency comb spectroscopy in both the short-wave and mid-wave infrared is discussed.

Experiment

Optical frequency comb laser source

Here we use an optical frequency comb[18,19] to perform direct molecular spectroscopy.[20] Illustrated in Fig. 1 are both the time-domain and frequency-domain

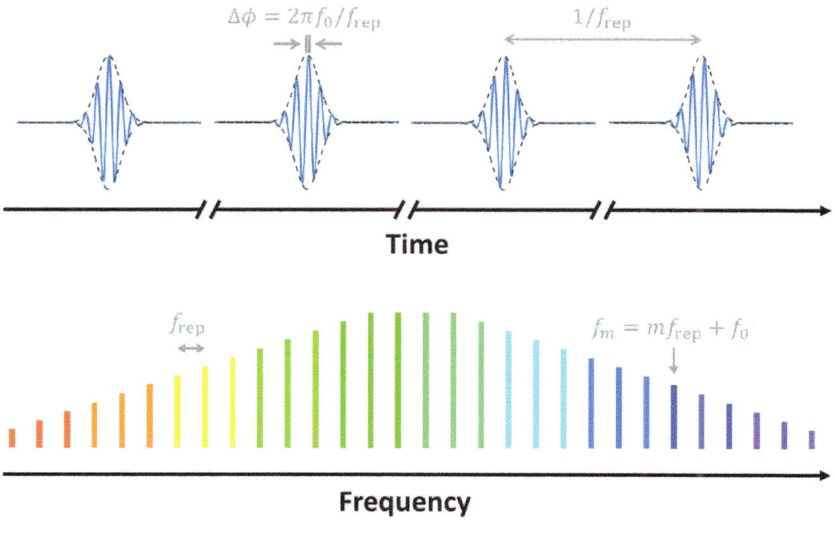

Fig. 1 Illustration of an optical frequency comb. The top panel shows a time-domain representation of a train of laser pulses with an optical carrier frequency (blue line) and pulse envelope (black dashed line). The phase shift between the carrier and envelope is $\Delta\phi = 2\pi f_0/f_{rep}$. The bottom panel shows the corresponding frequency domain spectrum of the train of pulses, revealing individual comb teeth separated by the laser repetition rate f_{rep} and each with unique optical frequencies $f_m = mf_{rep} + f_0$, where m is the tooth index and f_0 is the carrier envelope offset frequency.

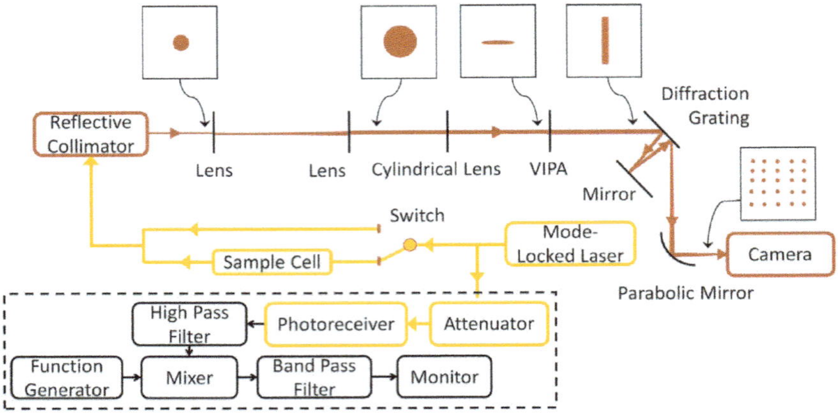

Fig. 2 Direct frequency comb spectrometer. Shown in the block diagram are the mode-locked laser source, gas sample and reference beam paths, sample cell, beam shaping optics and the cross-dispersed spectrometer. Fiber-coupled and free-space laser paths are illustrated with yellow and red arrows, respectively, and electronic signals for comb repetition rate monitoring are illustrated with black arrows. Image insets illustrate how the free-space laser beam profile transforms along its optical path, resulting in comb-mode-resolved images being recorded by the infrared camera. Optical dispersion within the beam profiles is not illustrated.

representations of the high resolution and broadband laser source. In this work, we used mode-locked lasers as the frequency comb sources. In the time domain, the optical frequency comb emitted from a laser cavity comprises a series of pulses with a fixed period of $1/f_{rep}$, where f_{rep} is the comb repetition rate. Illustrated in Fig. 1, the optical carrier frequency of the light pulses (blue line) and the laser pulse envelope (black dashed line) has a pulse-to-pulse phase shift defined as $\Delta\phi = 2\pi f_0/f_{rep}$, where f_0 is the carrier-envelope-offset frequency.

The Fourier transform of the train of stable laser pulses yields a series of narrow-linewidth lasers in the frequency domain, each with a unique mode index m. Each tooth has a fixed frequency spacing of f_{rep} and an absolute optical frequency of $f_m = mf_{rep} + f_0$. More details regarding frequency combs and their applications can be found in recent review articles (*e.g.*, ref. 21 and 22).

Direct frequency comb spectroscopy

The direct frequency comb spectrometer reported here comprised a mode-locked laser source, repetition-rate-monitoring electronics, fiber-coupled gas sample cell and cross-dispersed spectrometer. An illustration is shown in Fig. 2.

The frequency comb was a solid-state mode-locked laser (Menhir Photonics; MENHIR-1550)† with repetition rate of $f_{rep} = 1.00$ GHz. The value of f_{rep} was monitored by measuring the beat frequency $f_{beat} = 3 \times (f_{rep} - f_{LO})$, where $f_{LO} = 0.986\,580\,7$ GHz. Therefore, $f_{beat} \approx 40$ MHz. The carrier envelope offset frequency (f_0) of the mode-locked laser was not stabilized and its value not monitored during

† Certain commercial equipment is identified in this paper to foster understanding. Such identification does not imply recommendation or endorsement by the National Institute of Standards and Technology (NIST), nor does it imply that the equipment identified is necessarily the best available for the purpose.

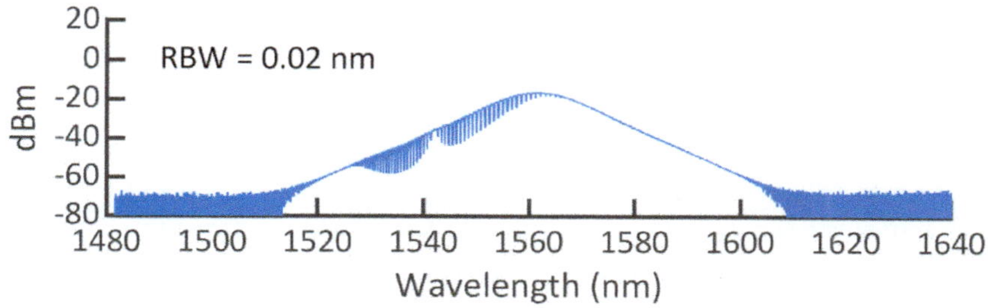

Fig. 3 Optical power spectrum of the mode-locked laser frequency comb after passing through the $H^{13}CN$ multipass gas-sample cell. RBW, resolution bandwidth of the optical spectrum analyzer.

measurements. As a result, each spectral acquisition comprises a highly precise grid of relative optical frequencies emitted from the mode-locked laser and defined by f_{rep}, but with an uncertain absolute offset frequency, f_0. We note that the mode-locked laser used here can be self-referenced using an f–$2f$ interferometer, thus creating an absolute frequency comb.[23]

The fiber-coupled output of the frequency comb was connected to a 1×2 fiber optic switch. One arm of the switch output was fiber-coupled to a commercial multipass gas sample cell (Wavelength References; HCN-13-100), whereas the second arm bypassed the gas sample cell and was used as a spectral reference channel. The gas sample cell was also discussed in a prior publication[16] and contained $H^{13}CN$ of molecular purity of >95% and isotopic purity of 99% at a pressure of 13.2 kPa $\pm$ 0.3 kPa. The absorption path length was $L = 79$ cm $\pm$ 1 cm. The optical power spectrum of the comb after passing through the $H^{13}CN$ sample is shown in Fig. 3.

The sample and reference arms were recombined in fiber and the collimated output was sent into free space towards a pair of lenses for beam expansion to a 20 mm collimated beam diameter. The larger-diameter beam was then sent to the spectrometer which ultimately resolved individual frequency comb modes on a two-dimensional detector array.

Cross-dispersed spectrometer and image analysis

The cross-dispersed spectrometer comprised a cylindrical lens, virtually imaged phased array (VIPA) etalon, double-passed ruled diffraction grating, imaging lens and infrared camera. The experimental set-up shares some similarities with several other previously reported comb-mode-resolved spectrometers.[24–29]

The cylindrical lens creates a line-focused beam profile at the entrance of the VIPA etalon, as illustrated in the top of Fig. 2. The VIPA etalon (Light Machinery; OP-6721-6743-8) is oriented such that the incident beam enters $\approx 1°$ off-normal. Transmission of the high-dispersion VIPA is governed by its dispersion law[30] and its characteristics determine several important spectrometer parameters. These include the VIPA etalon linewidth (full-width at half maximum), $\delta\nu_{VIPA} = 0.107$ GHz; free-spectral-range (FSR), $\nu_{FSR} = 15$ GHz; and finesse, $\mathcal{F} = 140$.

Dispersion from the VIPA is further separated – or cross-dispersed – using a ruled diffraction grating to create the two-dimensional image. Rotation of the diffraction grating results in a different portion of the broadband laser source being incident on the camera. Therefore, broadband optical tuning of the spectrometer was achieved by rotating the diffraction grating to a new position.

An example image from the cross-dispersed spectrometer is show in Fig. 4. Each gray dot is signal from a unique optical frequency comb tooth convolved with the instrument line shape (ILS) function. The ILS is a combination of the VIPA etalon linewidth and any additional broadening components such as standard single slit transmission broadening found in grating spectrometers. The composite ILS is modeled here by a simple Voigt function fitted for each imaged comb tooth along the vertical VIPA-dispersion direction. An example ILS fit is shown in Fig. 4 for a representative column of pixels from a typical reference image without molecular absorption.

With $\nu_{\mathrm{FSR}} = 15$ GHz and $f_{\mathrm{rep}} = 1.00$ GHz, there are 15 unique optical frequency comb teeth that comprise each column of unique comb teeth in Fig. 4. The VIPA orders are spatially separated in the horizontal dimension by the grating, and each image comprises up to 30 VIPA orders. Therefore, our spectrometer covers a bandwidth of 450 GHz at a single grating position, or 15 cm^{-1} in the short-wave infrared near 6410 cm^{-1}. The fitted full-width at half maximum of the composite ILS is nominally 0.44 GHz across the majority of the image, yielding a spectrometer resolving power of $R = \nu/\delta\nu = 4.4 \times 10^5$.

At an image integration time of 100 μs, we observe several thousand camera counts at the peak of each comb tooth. Here, the spectrometer bandwidth and resolving power are limited by our initial optical alignment and the choice of available imaging and beam shaping optics and can be improved by the inclusion of achromatic lenses and the subsequent reduction of optical aberrations near the edges of the image at the camera focal plane array.

Spectral modeling

To model the experimental transmission spectrum, the cross-dispersed spectrometer ILS function, $G(\tilde{\nu})$, can be convolved with the molecular transmission spectrum simulation as sampled at each comb tooth.[31] This is written in eqn (1):

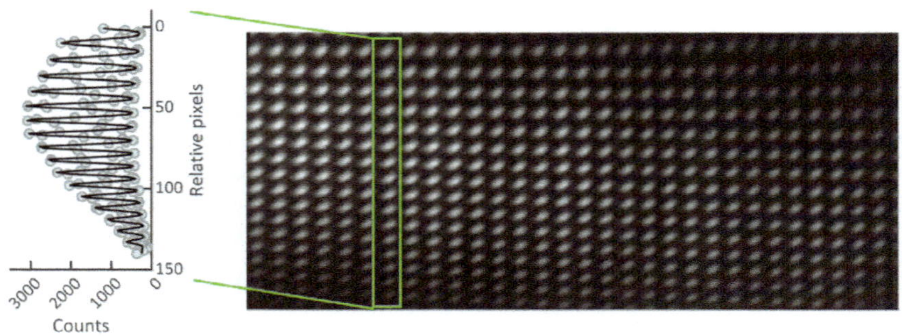

Fig. 4 Cross-dispersed spectrometer reference image with comb-mode resolution. The light gray dots on a field of black background are the resolved comb teeth, each with a unique and precisely known optical frequency. The vertical axis in the image is the VIPA dispersion direction and the horizontal axis is the grating dispersion direction. Counts from a column of pixels comprising one VIPA etalon free-spectral range are highlighted by the green box. On the far-left hand side is a rotated plot of camera counts *vs.* relative pixel position (gray dots) and the sum of the fitted instrument line shape (ILS) functions (black line).

$$T(\tilde{\nu}) = \exp\{-\alpha(\tilde{\nu})L\} \otimes G(\tilde{\nu}). \tag{1}$$

Above, $\alpha(\tilde{\nu})$ is the absorption coefficient as a function of wavenumber, $\tilde{\nu}$, and L is the sample path length.

Alternatively, the ILS can be effectively deconvolved from the columns in each experimental image (*e.g.*, Fig. 4) by the sifting property of a delta function. By assuming that the comb tooth linewidth is negligible compared to the ILS width, the modeled molecular transmission spectrum can be directly compared with the deconvolved comb tooth intensities.

The choice of modeling approach may depend upon the resolution of the cross-dispersed spectrometer compared to the comb mode spacing. Here in the short-wave infrared, with a nominal resolution of 0.4 GHz and a comb mode spacing of $f_{\text{rep}} = 1.00$ GHz, the comb modes may be considered resolved and effective deconvolution pursued.

Analysis of spectral noise limits

Here we present a brief analysis of the anticipated noise limits based on infrared camera manufacturer specifications and observations, optical frequency comb power measurements and estimated spectrometer throughput. We consider the following sources: detector noise, photon shot noise and quantization noise due to finite camera bit depth.

To estimate the influence of detector noise, we converted the camera noise-equivalent temperature difference of NETD = 60 mK to an irradiance noise from a blackbody source at a laboratory temperature of T = 297 K. Here we assume that only photons within the camera wavelength responsivity range of 1.5 μm to 5.1 μm contribute to the noise. To convert the calculated blackbody radiance[32] to irradiance at the planar detector array, we multiplied by the conical solid angle $\Omega = 0.15\pi$ sr – accounting for the aperture-limited detector field-of-view of $\theta = 45°$. Then we estimated the change in irradiance with respect to temperature ($\mathrm{d}E_{\mathrm{e,bb}}/\mathrm{d}T$) near T = 297 K. The detector irradiance noise was compared to the irradiance from a single comb tooth of index m, $E_{\mathrm{e},m} = P_m/A_m$, where $P_m = 10$ nW is the estimated optical power per comb tooth and $A_m = 2300\ \mu\mathrm{m}^2$ is the estimated area covered by a resolved comb tooth of imaged radius $r = 27$ μm (equivalent to 1.8 pixels):

$$(\mathrm{SNR})_{\mathrm{d}} = E_{\mathrm{e},m} \Big/ \left(\frac{\mathrm{d}E_{\mathrm{e,bb}}}{\mathrm{d}T}\mathrm{NETD}\right) \approx 2100. \tag{2}$$

The photon shot-noise-limited SNR is estimated from the square-root of the maximum time-integrated number of photoelectrons per pixel. First, we estimate the maximum number of time-integrated photons per pixel as

$$N_{\text{photons}} = P_m \Delta t G^{\mathrm{p}}(0)/(h\nu), \tag{3}$$

where Δt is the camera integration time, $G^{\mathrm{p}}(0)$ is the maximum of the area-normalized ILS function in units of pixel^{-1}, h is the Planck constant, $\nu = c/\lambda$ is the photon frequency, c is the speed of light. Then, for Δt = 100 μs, $G^{\mathrm{p}}(0) = 0.23$ pixel^{-1}, $\lambda = 1.56$ μm and an InSb detector quantum efficiency of $\eta_{\mathrm{QE}} = 0.95$:

$$(\mathrm{SNR})_s = \sqrt{\eta_{\mathrm{QE}} N_{\mathrm{photons}}} \approx 1300. \tag{4}$$

Finally, the quantization-noise-limited SNR due to finite camera bit depth of $n = 14$ bits is estimated from the quantity $\eta_{\mathrm{QE}} N_{\mathrm{photons}}$ divided by the charge capacity of each pixel, $N_{e^-} = 6.5 \times 10^6$, multiplied by the maximum number of digitizer counts, $C_{\max} = 2^n = 16\,384$:

$$(\mathrm{SNR})_q = \frac{\eta_{\mathrm{QE}} N_{\mathrm{photons}}}{N_{e^-}} C_{\max} \approx 4300. \tag{5}$$

From the quadrature sum of noise levels, we anticipate an achievable single-image SNR $\approx$ 1100 for P_m = 10 nW. The SNR on a transmission spectrum calculated from the ratio of two such images would then be SNR = 1100/2 $\approx$ 780, resulting in a transmission spectrum baseline noise of 1/SNR = 0.0013.

Results and discussion

The $2\nu_1$ band of $H^{13}CN$ near λ = 1.56 μm

The direct frequency comb spectrum of the $H^{13}CN$ recorded over a wavelength (wavenumber) range of λ = 1.548 μm to λ = 1.565 μm ($\tilde{\nu}$ = 6388 cm^{-1} to $\tilde{\nu}$ = 6458 cm^{-1}) is plotted in the top panel of Fig. 5. The transmission spectrum – recorded in 200 μs per grating position, over a total of six positions – is plotted as blue dots. The HITRAN2020 simulation is plotted as the red line and inverted for clarity.

At each grating position, the experimental transmission spectrum was calculated from the ratio of two images, obtained by alternating the 1 × 2 fiber optic switch between sample and reference paths. Each image was background-subtracted using a blank camera image (no comb light) recorded before or after the probe and reference image series, and normalized counts from each column were fitted to a 3rd-order polynomial baseline function along with the HITRAN2020 model simulated at experimental conditions using Voigt line profiles.

The observed SNR for the data in Fig. 5 is below the projections reported earlier in the paper. One source of reduced SNR is the optical transmission through the sample arm of the fiber-optic switch illustrated in Fig. 2. The comb-tooth counts in the sample probe images are a factor 30 lower than in the reference image – a result of broadband optical attenuation likely due to a malfunctioning or damaged fiber component. Reducing the power per comb mode from P_m = 10 nW to P_m = 0.33 nW reduces the expect detector-noise-limit to SNR $\approx$ 71 – or a baseline noise level of 1/SNR = 0.014.

This lower SNR predicted at the observed lower optical power is commensurate with the standard deviation of the fitted residuals found for the P 25e transition of the $2\nu_1$ band of $H^{13}CN$. The observed-minus-calculated residuals are plotted in the bottom panel of Fig. 5 as a black line, offset for clarity by +1.1. The standard deviation of the residuals from a Voigt line profile is 0.017, for a maximum spectral SNR $\approx$ 58. Progress towards closing this SNR gap for the two-image transmission spectrum is ongoing. The results will improve our ability to test various HCN line lists against precision direct frequency comb spectroscopy.

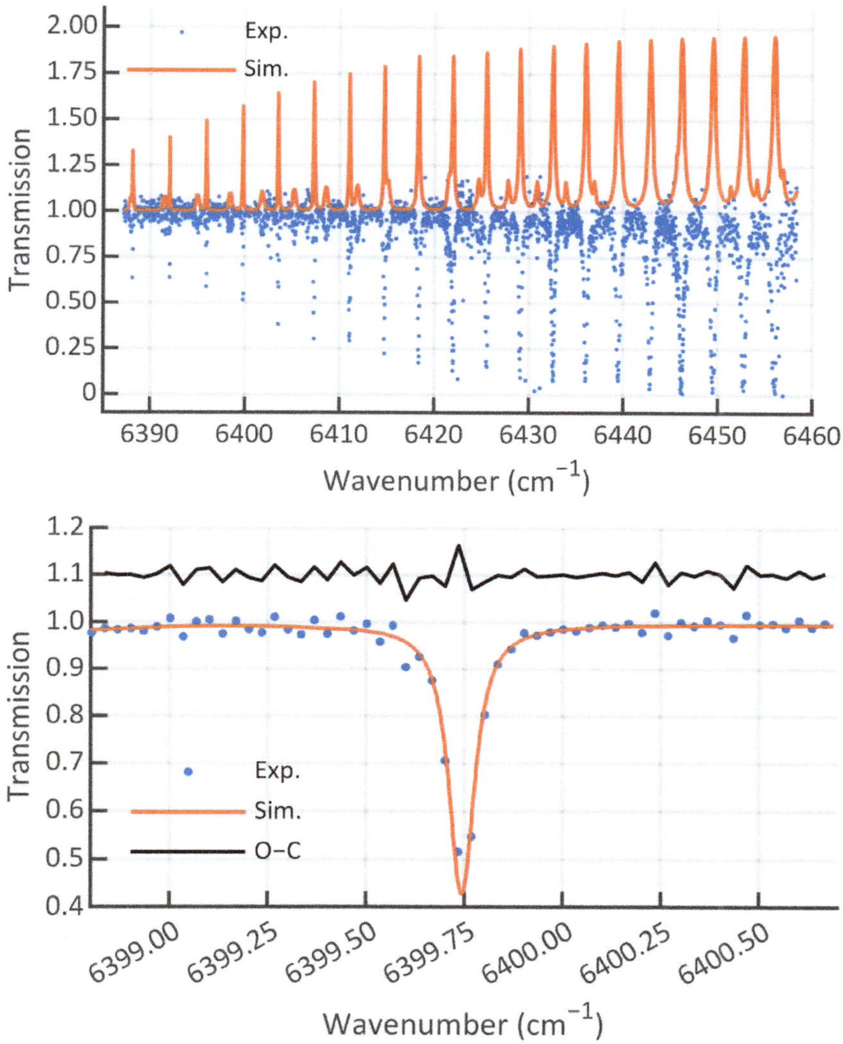

Fig. 5 Direct frequency comb spectroscopy of the $2\nu_1$ band of $H^{13}CN$ near $\lambda = 1.56$ µm. Top panel, comb-tooth-resolved transmission spectrum of $H^{13}CN$ (experiment, blue dots) and HITRAN2020 simulation (inverted; simulation, red line). Bottom panel, a zoom-in near the P 25e transition at 6399.744 cm^{-1}. There, the standard deviation of the observed-minus-calculated residuals (offset by +1.1; O–C, black line) is 0.017.

Outlook

Mid-infrared direct frequency comb spectroscopy

In addition to the short-wave infrared direct frequency comb spectroscopy of $H^{13}CN$ reported here, we also introduce our instrumentation for precision mid-wave infrared spectroscopy in the vicinity of the $3\nu_2$ band of $H^{12}CN$ near $\lambda = 4.73$ µm (2115 cm^{-1}).

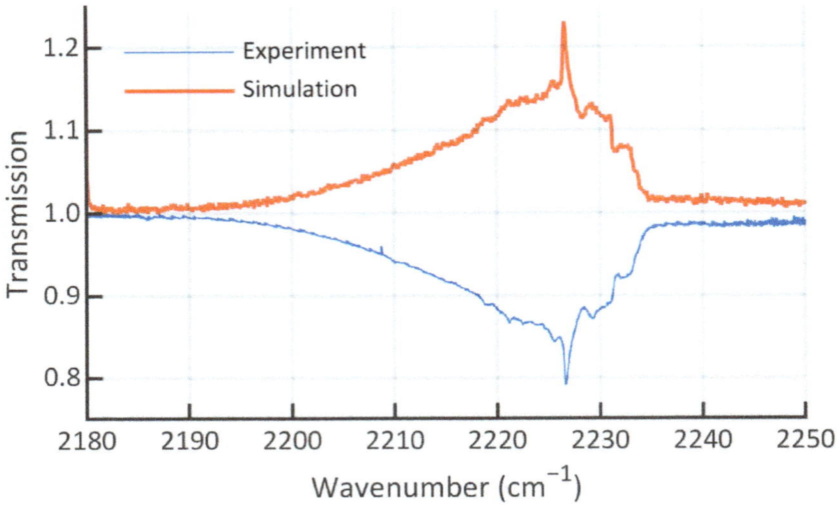

Fig. 6 Direct frequency comb spectrum of SF_6 (experiment, blue line) recorded in the mid-wave infrared at a wavelength near 4.49 μm. Also plotted is the spectrum from the PNNL database (inverted; simulation, red line),[34] convolved with our estimated instrument line shape (ILS) function.[31]

The laser source for our mid-wave infrared system is a commercial difference frequency generation comb (Menlo Systems; Mid-IR Comb) with tuneable output from 4.45 μm to 4.75 μm and instantaneous bandwidth of >90 cm^{-1}. The comb has a repetition rate of $f_{rep} = 0.250$ GHz and a carrier envelope offset frequency equal to zero by design.[33] The cross-dispersed spectrometer comprises a VIPA etalon (Light Machinery; OP-7553-3000-1), ruled diffraction grating and camera, and performs over a bandwidth of 30 cm^{-1} per grating position as previously demonstrated on N_2O.[31]

Beyond quantitative line-by-line analysis of rotationally resolved vibrational spectra, cross-dispersed spectrometers can also be leveraged to study broadly absorbing species. Shown in Fig. 6 is an experimental transmission spectrum for a pure sample of sulfur hexafluoride (SF_6) recorded at a pressure of 99.8 kPa and a temperature of 297 K (blue line) compared to reference data from the Pacific Northwest National Laboratory (PNNL) gas-phase spectral database (red line, inverted).[34] PNNL absorbance data was converted to spectral transmittance[35,36] and then convolved with the median value for the empirically derived ILS function from contemporary experiments on N_2O reported in Bailey *et al.*[31] The sample path length was 7 cm.

The spectrum of SF_6 is reported here as a test case for our mid-infrared system. The results illustrate potential for quantitative evaluation of a range of molecular targets including complex organic molecules like polycyclic aromatic hydrocarbons (PAHs) and simple alcohols (*e.g.*, methanol) – many of which are useful to probe astrochemical and astrophysical phenomena.[37]

Conclusions

Presented here is an experimental setup to perform direct frequency comb spectroscopy of $H^{13}CN$ in the short-wave infrared near a wavelength of 1.56 μm.

We used a mode-locked laser with repetition rate of 1.00 GHz as the laser source and a cross-dispersed design with a virtually imaged phased array as the spectrometer. The studied ^{13}C-enriched HCN gas sample was contained within a commercial glass-sealed and fiber-coupled optical multipass cell. We report an initial spectrum of $H^{13}CN$ in the short-wave infrared and compare with a model from HITRAN2020. The initial results show good agreement with the database, but at reduced experimental signal-to-noise ratio due to optical power losses upstream of the spectrometer.

Future work will focus on applying this experimental setup to measure high-precision and low-uncertainty spectral reference data including parameters like intensities, positions and line-shapes. We also aim to compare our results with those obtained from dual electro-optic frequency comb spectroscopy (*e.g.*, ref. 16), and to apply our cross-dispersed spectrometer approach to the direct frequency comb spectroscopy of $H^{12}CN$ in the mid-wave infrared near a wavelength of 4.73 μm.

Author contributions

D. Michelle Bailey: investigation, formal analysis, methodology, software, visualization, writing – reviewing and editing. Eric M. Crump: validation, visualization, writing – reviewing & editing. Joseph T. Hodges: formal analysis, methodology, supervision, writing – reviewing & editing. Adam J. Fleisher: conceptualization, investigation, formal analysis, data curation, funding acquisition, supervision, visualization, writing – original draft, writing – reviewing & editing.

Conflicts of interest

There are no conflicts to declare.

Acknowledgements

We acknowledge Eduardo Oteiza (Vescent Photonics) and Florian Emaury (Menhir Photonics) for loaning the mode-locked laser (MENHIR-1550), Ibrahim Sadiek (Leibniz Institute for Plasma Science and Technology, INP Greifswald, Germany) for discussing comb-mode-resolved spectrometers and Eisen Gross (National Institute of Standards and Technology, NIST) and Jake Connors (NIST) for reading the manuscript. Funding was provided by NIST and we acknowledge an award from the National Aeronautics and Space Administration (NASA) Astrophysics Research and Analysis (APRA) program NH21ZDA001N-APRA.

Notes and references

1 T. Horota, S. Yamamoto, H. Mikami and M. Ohishi, *Astrophys. J.*, 1998, **503**, 717.

2 S. Nickerson, N. Rangwala, S. W. J. Colgan, C. DeWitt, X. Huang, K. Acharyya, M. Drozkovskaya, R. C. Fortenberry, E. Herbst and T. J. Lee, *Astrophys. J.*, 2021, **907**, 51.

3 M. A. Cordiner, M. Y. Palmer, M. de Val-Borro, S. B. Charnley, L. Paganini, G. Villanueva, D. Bockelée-Morvan, N. Biver, A. J. Remijan, Y.-J. Kuan, *et al.*, *Astrophys. J., Lett.*, 2019, **870**, L26.

4 M. Rengel, D. Shulyak, P. Hartogh, H. Sagawa, R. Moreno, C. Jarchow and D. Breitschwerdt, *Astron. Astrophys.*, 2022, **658**, A88.

5 E. Lellouch, M. Gurwell, B. Butler, T. Fouchet, P. Lavvas, D. F. Strobel, B. Sicardy, A. Moulet, R. Moreno, D. Bockelée-Morvan, *et al.*, *Icarus*, 2017, **286**, 289.

6 A. Tsiaras, M. Rocchetto, I. P. Waldmann, O. Venot, R. Varley, G. Morello, M. Damiano, G. Tinetti, E. J. Barton and S. N. Yurchenko, *Astrophys. J.*, 2016, **820**, 99.

7 P. Giacobbe, M. Brogi, S. Gandhi, P. E. Cubillos, A. S. Bonomo, A. Sozzetti, L. Fossati, G. Guilluy, I. Carleo, M. Rainer, *et al.*, *Nature*, 2021, **592**, 205.

8 V. S. Magalhães, P. Hily-Blant, A. Faure, M. Hernandez-Vera and F. Lique, *Astron. Astrophys.*, 2018, **615**, A52.

9 B. Riaz, W.-F. Thi and P. Caselli, *Mon. Not. R. Astron. Soc.*, 2018, **481**, 4662.

10 R. J. Barber, J. K. Strange, C. Hill, O. L. Polyansky, G. Ch. Mellau, S. N. Yurchenko and J. Tennyson, *Mon. Not. R. Astron. Soc.*, 2014, **437**, 1828.

11 J. J. Fortney, T. D. Robinson, S. Domagal-Goldman, A. D. Del Genio, I. E. Gordon, E. Gharib-Nezhad, N. Lewis, C. Sousa-Silva, V. Airapetian, B. Drouin, *et al.*, *arXiv*, 2019, preprint, arXiv:1905.07064 [astro-ph.EP], DOI: 10.48550/arXiv.1905.07064.

12 I. E. Gordon, L. S. Rothman, R. J. Hargreaves, R. Hashemi, E. V. Karlovets, F. M. Skinner, E. K. Conway, C. Hill, R. V. Kochanov, Y. Tan, *et al.*, *J. Quant. Spectrosc. Radiat. Transfer*, 2022, **277**, 107949.

13 T. Delahaye, R. Armante, N. A. Scott, N. Jacquinet-Husson, A. Chédin, L. Crépeau, C. Crevoisier, V. Douet, A. Perrin, A. Barbe, *et al.*, *J. Mol. Spectrosc.*, 2021, **380**, 111510.

14 J. Tennyson, S. N. Yurchenko, A. F. Al-Refaie, V. H. J. Clark, K. L. Chubb, E. K. Conway, A. Dewan, M. N. Gorman, C. Hill, A. E. Lynas-Gray, *et al.*, *J. Quant. Spectrosc. Radiat. Transfer*, 2020, **255**, 107228.

15 V. Yu. Makhnev, A. A. Kyuberis, O. L. Polyansky, I. I. Mizus, J. Tennyson and N. F. Zobov, *J. Mol. Spectrosc.*, 2018, **353**, 40.

16 P. Guay, J. Genest and A. J. Fleisher, *Opt. Lett.*, 2018, **43**, 1407.

17 G. Ch. Mellau, V. Yu. Makhnev, I. E. Gordon, N. F. Zobov, J. Tennyson and O. L. Polyansky, *J. Quant. Spectrosc. Radiat. Transfer*, 2021, **270**, 107666.

18 J. L. Hall, *Rev. Mod. Phys.*, 2006, **78**, 1279.

19 T. W. Hänsch, *Rev. Mod. Phys.*, 2006, **78**, 1297.

20 A. Foltynowicz, P. Masłowski, T. Ban, F. Adler, K. C. Cossel, T. C. Briles and J. Ye, *Faraday Discuss.*, 2011, **150**, 23.

21 N. Picqué and T. W. Hänsch, *Nat. Photon.*, 2019, **13**, 146.

22 T. Fortier and E. Baumann, *Commun. Phys.*, 2019, **2**, 153.

23 D. M. B. Lesko, A. J. Lind, N. Hoghooghi, A. Kowligy, H. Timmers, P. Sekhar, B. Rudin, F. Emaury, G. B. Rieker and S. Diddams, *OSA Continuum*, 2020, **3**, 2070.

24 S. A. Diddams, L. Hollberg and V. Mbele, *Nature*, 2007, **445**, 627.

25 L. Nugent-Glandorf, T. Neely, F. Adler, A. J. Fleisher, K. C. Cossel, B. Bjork, T. Dinneen, J. Ye and S. A. Diddams, *Opt. Lett.*, 2012, **37**, 3285.

26 G. Kowzan, K. F. Lee, M. Paradowska, M. Borkowski, P. Ablewski, S. Wójtewicz, K. Stec, D. Lisak, M. E. Fermann, R. S. Trawiński and P. Masłowski, *Opt. Lett.*, 2016, **41**, 974.

27 F. Karim, S. K. Scholten, C. Perrella and A. N. Luiten, *Phys. Rev. Appl.*, 2020, **14**, 024087.

28 F. C. Roberts, H. J. Lewandowski, B. F. Hobson and J. H. Lehman, *Mol. Phys.*, 2020, **118**, e1733116.

29 I. Sadiek, N. Lang and J. H. van Helden, An Air-spaced Virtually Imaged Phased Array Spectrometer Based on a Mid-Infrared Optical Frequency Comb, in *Optical Sensors and Sensing Congress 2022*, AIS, LACSEA, Sensors, ES, Technical Digest Series (Optica Publishing Group), 2022, paper LM4B.5.

30 S. Xiao, A. M. Weiner and C. Lin, *IEEE J. Quantum Electron.*, 2004, **40**, 420.

31 D. M. Bailey, G. Zhao and A. J. Fleisher, *Anal. Chem.*, 2020, **92**, 13759.

32 https://www.spectralcalc.com/blackbody_calculator, accessed 24 January 2023.

33 M. Zimmermann, C. Gohle, R. Holzwarth, T. Udem and T. W. Hänsch, *Opt. Lett.*, 2004, **29**, 310.

34 S. W. Sharpe, T. J. Johnson, R. L. Sams, P. M. Chu, G. C. Rhoderick and P. A. Johnson, *Appl. Spectrosc.*, 2004, **58**, 1452.

35 J. J. Harrison and P. F. Bernath, *J. Quant. Spectrosc. Radiat. Transfer*, 2010, **111**, 1282.

36 H. Yi and A. J. Fleisher, *J. Chem. Phys.*, 2019, **151**, 234202.

37 K. I. Öberg, *Chem. Rev.*, 2016, **116**, 9631.

Faraday Discussions

PAPER

Size distribution of polycyclic aromatic hydrocarbons in space: an old new light on the 11.2/3.3 μm intensity ratio†

Alexander K. Lemmens, [ab] Cameron J. Mackie, [cd] Alessandra Candian, [ae] Timothy M. J. Lee, [f] Alexander G. G. M. Tielens, [g] Anouk M. Rijs [h] and Wybren Jan Buma *[ab]

Received 31st December 2022, Accepted 2nd February 2023
DOI: 10.1039/d2fd00180b

The intensity ratio of the 11.2/3.3 μm emission bands is considered to be a reliable tracer of the size distribution of polycyclic aromatic hydrocarbons (PAHs) in the interstellar medium (ISM). This paper describes the validation of the calculated intrinsic infrared (IR) spectra of PAHs that underlie the interpretation of the observed ratio. The comparison of harmonic calculations from the NASA Ames PAH IR spectroscopic database to gas-phase experimental absorption IR spectra reveals a consistent underestimation of the 11.2/3.3 μm intensity ratio by 34%. IR spectra based on higher level anharmonic calculations, on the other hand, are in very good agreement with the experiments. While there are indications that the 11.2/3.3 μm ratio increases systematically for PAHs in the relevant size range when using a larger basis set, it is unfortunately not yet possible to reliably calculate anharmonic spectra for large PAHs. Based on these considerations, we have adjusted the intrinsic ratio of these modes and incorporated this in an interstellar PAH emission model. This corrected model implies that typical PAH sizes in reflection nebulae such as NGC 7023 – previously inferred to be in the range of 50 to 70 carbon atoms per PAH are actually in the range of 40 to 55 carbon atoms. The higher limit of this range is close to the size of the C_{60} fullerene (also

[a] *Van 't Hoff Institute for Molecular Sciences, University of Amsterdam, 1098 XH Amsterdam, The Netherlands. E-mail: w.j.buma@uva.nl*

[b] *Radboud University, Institute for Molecules and Materials, FELIX Laboratory, 6525 ED Nijmegen, The Netherlands*

[c] *Kenneth S. Pitzer Center for Theoretical Chemistry, Department of Chemistry, University of California, Berkeley, California 94720, USA*

[d] *Chemical Sciences Division, Lawrence Berkeley National Laboratory, Berkeley, California 94720, USA*

[e] *Anton Pannekoek Institute for Astronomy, University of Amsterdam, 1098 XH Amsterdam, The Netherlands*

[f] *NASA Ames Research Center, Moffett Field, California 94035-1000, USA*

[g] *Leiden Observatory, Leiden University, 2333 CA Leiden, The Netherlands*

[h] *Division of BioAnalytical Chemistry, AIMMS Amsterdam Institute of Molecular and Life Sciences, Faculty of Science, Vrije Universiteit Amsterdam, De Boelelaan 1081, 1081 HV Amsterdam, The Netherlands*

† Electronic supplementary information (ESI) available. See DOI: https://doi.org/10.1039/d2fd00180b

detected in reflection nebulae), which would be in line with the hypothesis that, under appropriate conditions, large PAHs are converted into the more stable fullerenes in the ISM.

1. Introduction

The most widely accepted explanation for the emission of distinct infrared (IR) features, observed towards many different phases of the interstellar medium (ISM), is the PAH hypothesis.[1–3] From different regions of the ISM a small, but clear variation of spectral profiles is observed, which is attributed to a variation in the population of PAHs in terms of size, structure and charge state.[4,5] This variation is thought to be controlled by the local physical conditions.[6,7] Since the introduction of the PAH hypothesis, much effort has gone into modeling the mechanisms of molecular IR emission aiming to place better constraints on the molecular inventory of the ISM and to learn how they are affected by their environment. Apart from the charge state, abundance and structure, one of the constraints that receives a lot of attention is the size of PAHs, mostly in terms of the number of carbon atoms.

The IR emission by isolated PAHs starts with the absorption of a UV photon. Because under the dilute conditions of the ISM the collisional cooling rate is much smaller than the spontaneous emission rate, most energy is lost by vibrational relaxation which is accompanied by the fluorescence of IR photons. The observed IR spectrum is an average of various PAH species relaxing and emitting IR photons *via* different relaxation cascades.[8] Interstellar PAHs are exposed to a spectrum of UV photon energies, but in first approximation and because the UV photon energy does not vary a lot within objects the UV photon energy is often taken as a single value.[6]

Currently, the commonly accepted approach for deriving the size distribution of PAHs involves the intensity ratio of the 3.3 μm band to that of a band at a longer wavelength, typically the 11.2 μm band.[2,9–16] The 3.3 μm peak corresponds to CH stretch vibrations and the relative intensity of this band was found to have the strongest (inverse) dependence on the number of carbon atoms within a PAH.[9] The 11.2 μm peak is due to the CH out-of-plane (OOP) mode from solo Hs. The complete CH OOP region, which includes also the duo, trio and quartet H OOP modes, has been nicely decomposed and analyzed by Bauschlicher *et al.* and ranges from about 650–950 cm^{-1} (15.4 to 10.5 μm).[17]

The main reason behind the sensitivity of the 11.2/3.3 μm intensity ratio to molecular size is that smaller PAHs reach a higher temperature upon absorption of the same UV photon due to their smaller heat capacity, and therefore emit more photons at higher frequencies than larger PAHs.[4] The heat capacity, and thereby the cascade starting temperature that results from the absorption of a particular UV photon, can be described in terms of isolated harmonic oscillators or alternatively be calculated using more general PAH properties (approximation by Stein and Dwek *et al.*).[18–20] Additional advantages of using the 11.2/3.3 μm intensity ratio as a size tracer are that both bands originate mainly from CH bonds in neutral species and that they are spectrally well separated.[6] Moreover, the influence of the number of hydrogen atoms or of the PAH molecular structure on the ratio is minimal.[12]

The NASA Ames PAH IR spectroscopic database (PAHdb) supports a number of analyses of the observed 11.2/3.3 μm ratio.[21] It has been used to determine the PAH sizes in several astrophysical objects[12,22–25] and quite recently Maragkoudakis *et al.* explored extensively how the 11.2/3.3 μm ratio relates to the PAH size distribution.[9] The calculated 0 K absorption intensities (hereafter called intrinsic strengths) that underlie the emission models not only affect the emission ratio, but also the band profile. By artificially adjusting the intrinsic ratio it was found that a large 11.2/3.3 μm ratio results in a red-wing at the 11.2 μm feature which fades away at smaller ratios.[26]

The IR intensities in the PAHdb that are employed in the emission models are calculated at a relatively low level of theory (B3LYP/4-31G) and are based on the harmonic approximation. Because of the significant effect of intrinsic strengths on the outcome of the emission models in terms of ratio and peak profile, it is of key importance to validate these calculations with suitable experiments. Even more so because it was noticed already in the 90s by Langhoff and later by others that the intensity ratios derived at different levels of theory can vary significantly.[27–30] First, the use of a small basis set results in an overprediction of CH stretch intensities by about 20% when compared to matrix-isolation absorption spectra.[27] Secondly, it was recently shown that anharmonicity leading to phenomena such as combination bands and resonances has a dominant effect on the CH stretch region of the IR spectrum. These anharmonic contributions can carry significant intensity.[28,31]

Moreover, also from an experimental perspective it has been concluded that one needs to be cautious. Compared to the gas phase, the CH stretch intensity in IR absorption measurements in the solid phase is about a factor 3 lower.[32] In matrix-isolation IR absorption measurements, on the other hand, the CH OOP intensity is about a factor 5 lower. Both experimental conditions will thus have a large effect on the 11.2/3.3 μm ratio.[32]

The aim of this study is to compare theory that is commonly used to interpret astronomical observations to experimental data in terms of intensity, utilizing direct absorption techniques for reliable intensities over a wide IR wavelength range. The theory at a (relatively) low vibrational temperature should be validated before extrapolating the models to describe the emission of highly excited PAHs and using these models to put constraints on the size distribution of PAHs. Gas-phase FTIR spectroscopy provides a good benchmark for oscillator strengths, because at the relatively low experimental temperatures there is no significant occupation of higher vibrational levels. In first approximation, a higher temperature will thus only lead to band broadening.[32,33] Moreover, with FTIR spectroscopy the data are collected simultaneously over a wide spectral range. We will show that the present analysis leads to the conclusion that previously employed emission models need to be reconsidered, and, importantly, that the application of modified models predicts notably smaller PAH sizes than before.

2. Methods

We compared (relative) intensities from the IR spectra of PAHs from four different experimental and theoretical sources, namely the NASA Ames PAH IR spectroscopic database (PAHdb, version 3.20),[21] anharmonic calculations extracted from the work by Mackie *et al.*[34] or performed here using the Gaussian 16 software package, experimental gas-phase FT-IR absorption spectra from the NIST database[35] and experimental matrix isolation IR spectra extracted from the PAHdb.[36] To study the relation between molecular size and the intrinsic 3.3/11.2 μm ratio, we extracted theoretical PAH spectra from the PAHdb of neutral PAHs that do not contain any oxygen or nitrogen atoms (Fig. 1). The intensity is integrated between 650–950 cm^{-1} (15.4 and 10.5 μm) and will be indicated in the following as 'oop'. The integrated range for the CH stretch region, indicated in the following as 'stretch', spans from 2750–3250 cm^{-1} (3.6–3.1 μm). For the ratio in emission (Fig. 3) we selected neutral PAHs that contain more than one hydrogen atom, no aliphatic groups and did not include Fe, Si, Mg, N, or O substituted species. Finally, we only considered PAHs with solo (more than 4) and duo hydrogens, as they are generally more compact and thus more stable. One can reasonably expect that this restriction will only affect to a minor extent the conclusions as the structure of the PAH should not have a large influence on the oop/stretch emission ratio.[27,32] Calculations referred to as anharmonic are either described previously and extracted from ref. 34, or they have been performed at the anharmonic[37,38] B3LYP[39]/N07D[40,41] level of theory using Gaussian 16 (ref. 42) as indicated.

3. Results and discussion

3.1 Theory and gas-phase experiments

In order to investigate the intrinsic intensity ratio as a function of PAH size, Fig. 1a shows the band strength ratio extracted from the (harmonically calculated) PAH database for the complete available size range. Fig. 1a shows that this ratio varies by a factor of about 3 between the smallest and largest PAHs considered here. Interestingly, a large variation of the band strength ratio is associated with small PAHs containing up to about 35 carbon atoms, which results in a larger uncertainty in the calculated emission ratio over this size range. The color scale of Fig. 1a corresponds to the number of trio hydrogens in the PAH. A clear separation of PAHs that contain more than 4 trio hydrogens is visible from PAHs with 4 or less trio hydrogens. The presence of trio hydrogens increases the oop/stretch ratio by suppressing the CH stretch band intensity.

In the most stable PAH species, one does not expect many trio hydrogens as they are in general associated with an irregular shape rather than a compact one.[43] To further investigate the relation between the intrinsic oop/stretch ratio and PAH size, PAHs with more than 4 trio hydrogens have therefore been omitted in Fig. 1b. Moreover, to remove the bias towards the more numerous smaller PAHs an average has been taken for fixed sizes (tolerance 5 carbon atoms). Quite interestingly, we then find that the oop/stretch ratio and the number of carbon atoms in a PAH exhibit a clear drop with size that converges to an intrinsic oop/stretch ratio of 0.35 for large PAHs. This drop can be well represented by an exponential function given by

$$R_{\text{oop/stretch}} = 0.63\mathrm{e}^{-\frac{N_\mathrm{C}}{58.1}} + 0.35$$

With $R_{\text{oop/stretch}}$ the intrinsic oop/stretch intensity ratio and N_C the number of carbon atoms.

The large number of theoretical IR spectra in the PAHdb have provided extensive insight into the interstellar PAH population and spectroscopic properties of PAHs.

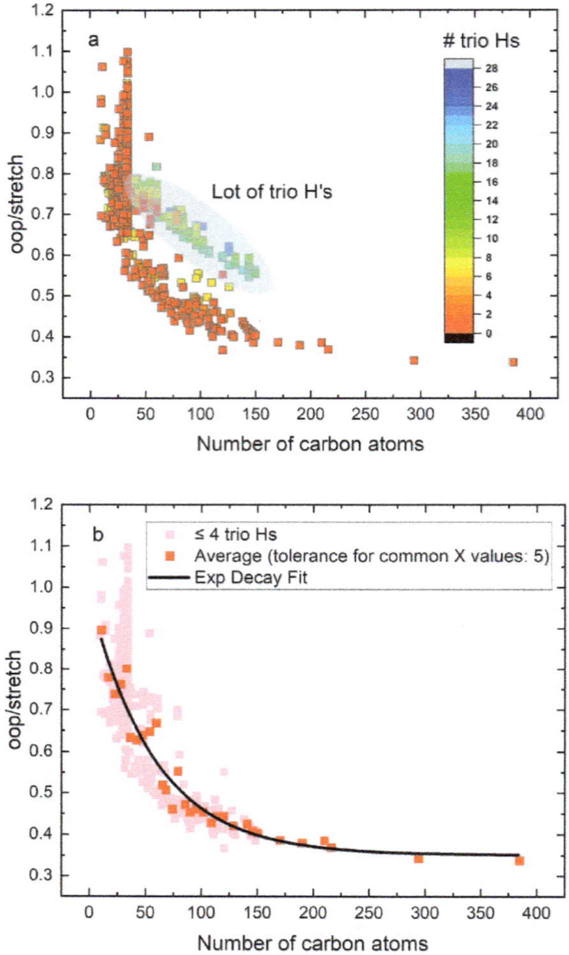

Fig. 1 (a) The intrinsic oop/stretch intensity ratio of neutral PAHs in the PAHdb as a function of the number of carbon atoms. The color scale corresponds to the number of trio hydrogens in the PAH. (b) Same plot as (a), but leaving out PAHs that contain more than 4 trio hydrogens (light red). The dark red symbols represent the average for fixed sizes (size tolerance is 5 carbon atoms). The black line is an exponential decay fit to the latter.

However, considering the intensities, the harmonic, relatively low-level calculations are known to consistently overestimate the CH stretch band intensity.[29] To corroborate these findings, in Fig. 2 we compare the calculated ratio from the PAHdb with the ratio determined from gas-phase direct absorption measurements from the NIST database.[35] As discussed in the introduction, anharmonic calculations lead to a significant improvement in predicting IR spectra. The improvement mainly lies in the addition of combination bands and resonances that carry a significant amount of intensity, especially in the CH stretch region. Fig. 2 thus reports on a comparison in terms of intensities resulting from such calculations. Finally, a comparison is made between the gas-phase absorption measurements to cold, matrix isolation absorption measurements.

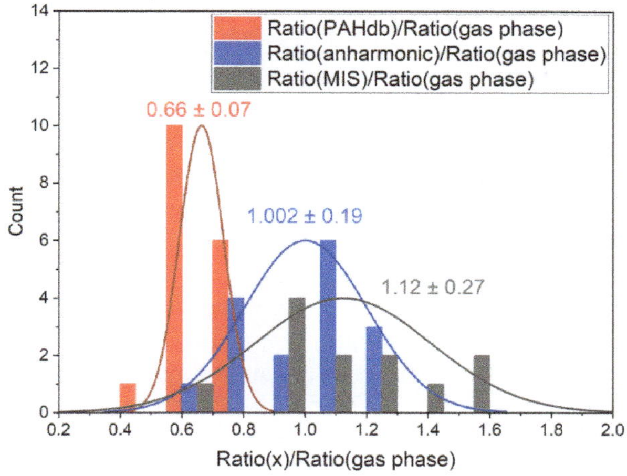

Fig. 2 Histogram of the oop/stretch intensity ratio determined using different theoretical and experimental methods divided by the gas-phase direct absorption measurements. The bin size is 0.15. The intensities were integrated between 650–950 cm^{-1} (15.4–10.5 μm) and 2750–3250 cm^{-1} (3.6–3.1 μm). The PAH species that are taken into consideration are listed in the ESI.† The mean and deviation of the distributions are indicated in the plot.

The oop/stretch intensity ratio determined using the theoretical spectra of the PAHdb is smaller than the experimental gas-phase ratio by a factor of 0.66. This corroborates the previous findings that the CH stretch intensities are overestimated, although this overestimation appears to be even slightly larger than reported previously.[27] Interestingly, the ratio is quite consistently off, which supports the approach to scale the band strengths computed at the B3LYP/4-31G level to obtain more accurate ratios. This approach was used before to obtain band strengths for methylated PAHs that were as accurate as those calculated at far more expensive levels (MP2/6-311+G(3df,3pd)).[44] We compare the DFT calculated ratios for the coronene family for different basis sets and find that upon increasing the basis set the oop/stretch intensity ratio increases (see Table S2†).

The ratios that result from anharmonic calculations show on average a remarkably good agreement with the gas-phase absorption experiments in terms of accuracy. It should be noted that a larger basis set (N07D, which is a modification of 6-31G*) is used in these anharmonic calculations than has been used for the PAHdb and such a larger basis set is expected to lead to a more accurate ratio, mainly resulting from a smaller stretch intensity.[29] Nevertheless, examination of the effects of the inclusion of anharmonicity on the ratio is of great interest as a significant amount of intensity is involved in the combination bands and resonances. The ratio determined from matrix isolation spectroscopy measurements shows the largest spread with respect to gas-phase absorption experiments. The comparison of the ratios of individual PAHs can be found in Fig. S1 in the ESI.†

3.2 Correcting for systematic deviation

In view of the above observations, it becomes quite interesting to determine how the underestimation of the oop/stretch intensity ratio affects conclusions on the

PAH size distributions. To this purpose, we used the harmonically calculated intrinsic band strengths and applied a full temperature cascade emission model from the PAHdb at 6.5 eV on a large selection of PAHs (for details, see the Methods section). The results are shown in Fig. 3 (grey squares). In order to correct for the underestimation of the intrinsic oop/stretch intensity ratio by low-level calculations, a coarse modification has been made by scaling the emission ratio with the factor by which the low-level theory calculations underestimate the intrinsic ratio (1/0.66, light-green squares in Fig. 3). Such an approach was used before for methylated PAHs, where the band strengths of inexpensive calculations were scaled to match the more expensive predictions.[44] This figure also displays the results of a more accurate approach in which the intrinsic ratio is scaled and afterwards the full temperature cascade emission model is applied (dark-green squares). Aiming to facilitate a readout of these plots and serving as a guide to the eye, the data sets have been fitted with an exponential function.

Since the average absorbed UV photon energy varies between astronomical objects, we have calculated the emission ratio as a function of molecular size for cascade starting energies of 6.5 and 8 eV, these energies being representative for the average photon energy of a reflection nebula such as NGC 7023 (ref. 12) and for a Photon Dominated Region such as the Orion Bar,[24] respectively. We assume one-photon absorption processes as cooling rates are high compared to absorption rates in the objects studied here.[6] Multiphoton events would result in a decrease of the oop/stretch ratio. The results of these calculations are shown in Fig. 4 with the ratio of correcting both before and after applying the emission model plotted as transparent and opaque color, respectively. While correcting the

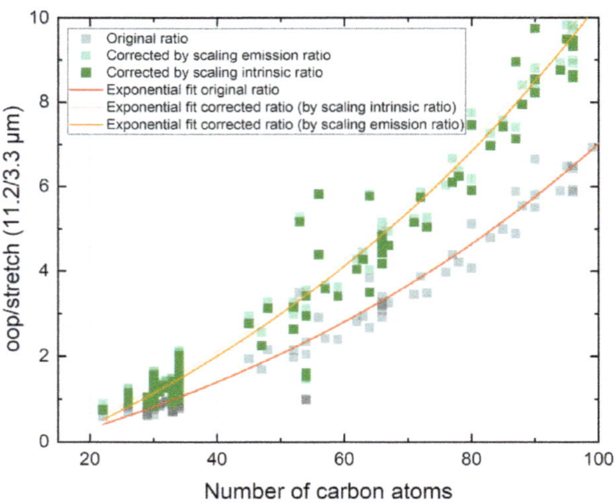

Fig. 3 Grey squares indicate the oop/stretch intensity ratio as a function of the number of carbon atoms determined using the calculated intrinsic band strengths and a full temperature cascade emission model from the PAHdb on a large selection of PAHs (for details, see the Methods section). The green squares are scaled by a factor of 1/0.66 (see text) to correct for the systematic underestimation of the intrinsic oop/stretch intensity ratio by low-level calculations. Intensities were integrated between 830–985 cm^{-1} (12.05–10.15 μm) and 2950–3115 cm^{-1} (3.4–3.2 μm).

intensity ratio before applying the emission model is most appropriate, we also include the results derived by applying the correction factor after applying the emission models as a convenient way to interpret the results of the current PAHdb. An exponential function to facilitate readout has been fitted to all plots and is provided in the ESI.†

With both the 'original' model and the corrected model, the 11.2/3.3 μm intensity ratio can now be translated into a PAH size distribution. The two horizontal solid black lines in Fig. 4 are typical lower and higher limits of observed intensity ratios in NGC 7023.[12] The two horizontal dotted black lines indicate lower and higher limits of the intensity ratios in the Orion Bar.[24] For NGC 7023, using the 'original' model and taking the indicated upper and lower observed ratio limits, one would conclude that typical PAH sizes are in the 50–70 carbon atoms range. Importantly, however, we find that when using the corrected model which takes the systematic underestimation of the intensity ratio into account, this range is reduced to 40–55 carbon atoms. For the Orion Bar region, the PAH size range is determined to be between 37–53 carbon atoms.

The results of the present model are quite remarkable but need to be substantiated further with more extensive analyses that incorporate accurate high-level anharmonic calculations on PAHs with a wide range of sizes in the emission model. Such analyses are now indeed high on our priority list. However, as for larger PAHs these calculations are not available yet, this model awaits completion. Nevertheless, our results do indicate that the underestimation of the

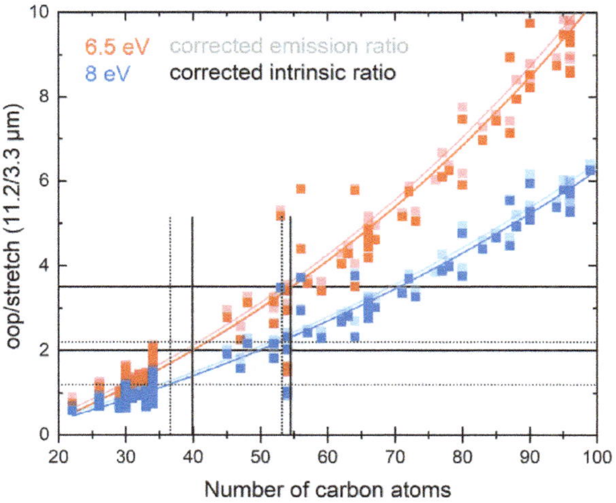

Fig. 4 The oop/stretch intensity ratio as a function of the number of carbon atoms determined using the calculated band strengths and a full temperature cascade emission model from the PAHdb on a large selection of PAHs. The transparent squares correspond to the ratio scaled after applying the emission model, the opaque squares to scaling the intrinsic strengths before applying the emission model, both cases aiming to correct for the systematic underestimation of the intrinsic oop/stretch intensity ratio by low-level calculations. The solid and dotted black horizontal lines are examples of lower and higher limits of the observed intensity ratio in NGC 7023 and the Orion Bar, respectively.[12,24] The exponential fits indicated with the colored solid lines are phenomenological and meant to facilitate readout (see Table S1† for fit parameters).

oop/stretch intensity ratio by low-level calculations translates into a significant overestimation of the PAH size distribution. Equally important to notice is that our results are still in line with the hypothesis that an interconversion between large PAHs and the even more stable C_{60} fullerene takes place at a certain level of the UV radiation field.[7] From the higher limit of about 55 carbon atoms that our results imply, it appears that PAHs containing more than 60 carbon atoms should indeed be readily converted into fullerene species.

4. Conclusions

By comparing calculations from the NASA Ames PAH IR spectroscopic database (PAHdb) and state-of-the-art anharmonic calculations with gas-phase direct absorption IR spectra from the NIST chemistry webbook, this study validates the predicted intrinsic oop/stretch intensity ratio of PAHs. Currently, this ratio is recognized as the most reliable tracer for determining the size distribution of PAHs in the interstellar medium. Our studies show that the relatively low-level calculations that have been employed so far to interpret astronomical observations consistently underestimate the intrinsic intensity ratio by about 34%. Predictions made with a larger basis-set and taking anharmonic effects such as combination bands and resonances into account are, in contrast, in good agreement with the experiment. Unfortunately, the high-level anharmonic calculations are not available yet for large PAHs. Nevertheless, the consistency of the underestimation of the intensity ratio allows for making an *ad hoc*, coarse correction to the emission model based on the PAHdb. Such a corrected ratio–PAH size relation suggests that the size distribution, which up till now has been taken to be in the 50–70 carbon atoms range, could very well need to be adjusted to the 40–55 range. It is quite satisfying to notice that the higher limit of the PAH size conforms to the hypothesis that an interconversion takes place between larger sized PAHs and fullerenes in the interstellar medium.

Conflicts of interest

There are no conflicts to declare.

Acknowledgements

We dedicate this article to Tim Lee who passed away November 3, 2022. His contributions have been key to advancing quantum chemistry in general and astrochemistry in particular. He will be sorely missed. We would like to thank Dr Christof Iserlohe for his computational support. This work is part of the Dutch Astrochemistry Network (DAN) II (project no. 648.000.029).

References

1 A. G. G. M. Tielens, *Annu. Rev. Astron. Astrophys.*, 2008, **46**, 289–337.
2 L. J. Allamandola, A. G. G. M. Tielens and J. R. Barker, *Astrophys. J., Suppl. Ser.*, 1989, **71**, 733–775.
3 A. Leger and J. L. Puget, *Astron. Astrophys.*, 1984, **137**, L5–L8.

4 E. Peeters, C. Mackie, A. Candian and A. G. G. M. Tielens, *Acc. Chem. Res.*, 2021, **54**, 1921–1933.
5 F. Galliano, S. C. Madden, A. G. G. M. Tielens, E. Peeters and A. Jones, *Astrophys. J.*, 2008, **679**, 310–345.
6 A. G. G. M. Tielens, *Molecular Astrophysics*, Cambridge University Press, Cambridge, 2021.
7 O. Berné and A. G. G. M. Tielens, *Proc. Natl. Acad. Sci. U. S. A.*, 2012, **109**, 401–406.
8 C. J. Mackie, T. Chen, A. Candian, T. J. Lee and A. G. G. M. Tielens, *J. Chem. Phys.*, 2018, **149**, 134302.
9 A. Maragkoudakis, E. Peeters and A. Ricca, *Mon. Not. R. Astron. Soc.*, 2020, **494**, 642–664.
10 W. A. Schutte, A. G. G. M. Tielens and L. J. Allamandola, *Astrophys. J.*, 1993, **415**, 397.
11 A. Ricca, C. W. Bauschlicher, C. Boersma, A. G. G. M. Tielens and L. J. Allamandola, *Astrophys. J.*, 2012, **754**, 75.
12 B. A. Croiset, A. Candian, O. Berné and A. G. G. M. Tielens, *Astron. Astrophys.*, 2016, **590**, A26.
13 M. Jourdain de Muizon, P. Cox and J. Lequeux, *Astron. Astrophys., Suppl. Ser.*, 1990, **83**, 337–355.
14 T. I. Mori, I. Sakon, T. Onaka, H. Kaneda, H. Umehata and R. Ohsawa, *Astrophys. J.*, 2012, **744**, 68.
15 C. Pech, C. Joblin and P. Boissel, *Astron. Astrophys.*, 2002, **388**, 639–651.
16 H. Andrews, C. Boersma, M. W. Werner, J. Livingston, L. J. Allamandola and A. G. G. M. Tielens, *Astrophys. J.*, 2015, **807**, 24.
17 C. W. Bauschlicher, E. Peeters and L. J. Allamandola, *Astrophys. J.*, 2009, **697**, 311–327.
18 S. E. Stein, J. N. Spencer, J. R. Seigart, E. Brown, R. L. Sensing and S. E. Stein, *J. Phys. Chem.*, 1978, **82**, 566–571.
19 E. Dwek, R. G. Arendt, D. J. Fixsen, T. J. Sodroski, N. Odegard, J. L. Weiland, W. T. Reach, M. G. Hauser, T. Kelsall, S. H. Moseley, R. F. Silverberg, R. A. Shafer, J. Ballester, D. Bazell and R. Isaacman, *Astrophys. J.*, 1997, **475**, 565–579.
20 C. W. Bauschlicher, C. Boersma, A. Ricca, A. L. Mattioda, J. Cami, E. Peeters, D. Armas, G. P. Saborido, D. M. Hudgins and L. J. Allamandola, *Astrophys. J., Suppl. Ser.*, 2011, **189**, 341–351.
21 C. W. Bauschlicher, A. Ricca, C. Boersma and L. J. Allamandola, *Astrophys. J., Suppl. Ser.*, 2018, **234**, 32.
22 C. Boersma, J. Bregman and L. J. Allamandola, *Astrophys. J.*, 2015, **806**, 121.
23 B. T. Draine, A. Li, B. S. Hensley, L. K. Hunt, K. Sandstrom and J.-D. T. Smith, *Astrophys. J.*, 2021, **917**, 3.
24 C. Knight, E. Peeters and A. G. G. M. Tielens, *Mon. Not. R. Astron. Soc.*, 2022, **509**, 3523–3546.
25 C. Knight, E. Peeters, D. J. Stock, W. D. Vacca and A. G. G. M. Tielens, *Astrophys. J.*, 2021, **918**, 8.
26 C. J. Mackie, A. Candian, T. J. Lee and A. G. G. M. Tielens, *Theor. Chem. Acc.*, 2021, **140**, 124.
27 S. R. Langhoff, *J. Phys. Chem.*, 1996, **100**, 2819–2841.

28 C. J. Mackie, A. Candian, X. Huang, E. Maltseva, A. Petrignani, J. Oomens, W. J. Buma, T. J. Lee and A. G. G. M. Tielens, *J. Chem. Phys.*, 2015, **143**, 224314.

29 H. Geindre, A. R. Allouche and D. Peláez, *J. Comput. Chem.*, 2021, **42**, 1018–1027.

30 C. J. Mackie, A. Candian, T. J. Lee and A. G. G. M. Tielens, *J. Phys. Chem. A*, 2022, **126**, 3198–3209.

31 A. K. Lemmens, D. B. Rap, J. M. M. Thunnissen, C. J. Mackie, A. Candian, A. G. G. M. Tielens, A. M. Rijs and W. J. Buma, *Astron. Astrophys.*, 2019, **628**, A130.

32 C. Joblin, L. D'Hendecourt, A. Leger and D. Defourneau, *Astron. Astrophys.*, 1994, **281**, 923–936.

33 C. Joblin, P. Boissel, A. Léger, L. D'Hendecourt and D. Défourneau, *Astron. Astrophys.*, 1995, **299**, 835–846.

34 C. J. Mackie, *The Anharmonic Infrared Spectra of Polycyclic Aromatic Hydrocarbons*, 2018.

35 NIST Mass Spectrometry Data Center, W. E. Wallace, director, in *NIST Chemistry WebBook, NIST Standard Reference Database Number 69*, ed. P. J. Linstrom and W. G. Mallard, National Institute of Standards and Technology, Gaithersburg, MD, Infrared Spectra, accessed Jan 2022.

36 A. L. Mattioda, D. M. Hudgins, C. Boersma, A. Ricca, E. Peeters, J. Cami, F. Sánchez de Armas, G. Puerta Saborido, C. W. Bauschlicher and L. J. Allamandola, *Astrophys. J.*, 2020, **251**, 22.

37 V. Barone, *J. Chem. Phys.*, 2005, **122**, 014108.

38 V. Barone, M. Biczysko and J. Bloino, *Phys. Chem. Chem. Phys.*, 2014, **16**, 1759–1787.

39 A. D. Becke, *J. Chem. Phys.*, 1993, **98**, 5648–5652.

40 V. Barone, P. Cimino and E. Stendardo, *J. Chem. Theory Comput.*, 2008, **4**, 751–764.

41 C. J. Mackie, A. Candian, X. Huang, E. Maltseva, A. Petrignani, J. Oomens, W. J. Buma, T. J. Lee and A. G. G. M. Tielens, *Phys. Chem. Chem. Phys.*, 2018, **20**, 1189–1197.

42 M. J. Frisch, G. W. Trucks, H. B. Schlegel, G. E. Scuseria, M. A. Robb, J. R. Cheeseman, G. Scalmani, V. Barone, G. A. Petersson, H. Nakatsuji, X. Li, M. Caricato, A. V. Marenich, J. Bloino, B. G. Janesko, R. Gomperts, B. Mennucci, H. P. Hratchian, J. V. Ortiz, A. F. Izmaylov, J. L. Sonnenberg, D. Williams-Young, F. Ding, F. Lipparini, F. Egidi, J. Goings, B. Peng, A. Petrone, T. Henderson, D. Ranasinghe, V. G. Zakrzewski, J. Gao, N. Rega, G. Zheng, W. Liang, M. Hada, M. Ehara, K. Toyota, R. Fukuda, J. Hasegawa, M. Ishida, T. Nakajima, Y. Honda, O. Kitao, H. Nakai, T. Vreven, K. Throssell, J. A. Montgomery Jr, J. E. Peralta, F. Ogliaro, M. J. Bearpark, J. J. Heyd, E. N. Brothers, K. N. Kudin, V. N. Staroverov, T. A. Keith, R. Kobayashi, J. Normand, K. Raghavachari, A. P. Rendell, J. C. Burant, S. S. Iyengar, J. Tomasi, M. Cossi, J. M. Millam, M. Klene, C. Adamo, R. Cammi, J. W. Ochterski, R. L. Martin, K. Morokuma, O. Farkas, J. B. Foresman and D. J. Fox, *Gaussian 16, Revision A.03*, Gaussian, Inc., Wallingford CT, 2016.

43 J. Bouwman, P. Castellanos, M. Bulak, J. Terwisscha van Scheltinga, J. Cami, H. Linnartz and A. G. G. M. Tielens, *Astron. Astrophys.*, 2019, **621**, A80.

44 X. J. Yang, A. Li, R. Glaser and J. X. Zhong, *Astrophys. J.*, 2017, **837**, 171.

Faraday Discussions

DISCUSSIONS

Laboratory astrochemistry of the gas phase: general discussion

Nadia Balucani, Michelle Brann, Sandra Brünken, Cecilia Ceccarelli, Martin Cordiner, Eric M. Crump, Kevin M. Douglas, Adam J. Fleisher, Athena Flint, Jack Fulker, Robin T. Garrod, Murthy S. Gudipati, Divita Gupta, Joshua Halpern, Dwayne E. Heard, Eric Herbst, Emily K. Hockey, Ko-Yun Huang, Ugo Jacovella, Inga Kamp, Alexander K. Lemmens, Nikku Madhusudhan, Martin R. S. McCoustra, Brett McGuire, Anthony Meijer, Cristina Puzzarini, Daniël B. Rap, Ian R. Sims, Mark H. Stockett, Ardjan Sturm, Arthur G. Suits, Ewine F. van Dishoeck, Serena Viti, Nicholas Walker, Susanna Widicus Weaver, Laurent Wiesenfeld and Olivia H. Wilkins

DOI: 10.1039/d3fd90025h

Emily K. Hockey opened discussion of the paper by Kevin M. Douglas: I was wondering about the interaction the bath gas (argon in this case) is having with your reactants before or after forming your pre-reactive complex. Has this been considered? I think reassessing this interaction in your beam would adjust the depth of the pre-reactive complex wells in the manner you arbitrarily assume in the paper (https://doi.org/10.1039/d3fd00046j). Using an implicit solvation model might also be helpful.

Kevin M. Douglas answered: In our low temperature gas expansions, the bath gas only interacts with the reactants and the pre-reaction complexes *via* collisions, that is it is only acting as a bath gas. Only at significantly lower temperatures or higher pressures than those in our gas expansions would we expect there to be any other interactions, such as solvation effects or the formation of bath gas–reagent complexes, as preliminary calculations have shown that Ar is only bound to CH_3CHO by less than 2 kJ mol^{-1}.

Ko-Yun Huang opened discussion of the paper by Sandra Brünken: There is already some discussion about this in the article (https://doi.org/10.1039/d3fd00015j). I am curious about the potential of using differences in fragmentation products – such as HCN/HNC ratios – in broader applications, such as differentiating extreme environments such as near accreting supermassive blackholes where high amounts of X-rays, high gas temperatures, or high cosmic-ray ionization rates may generally be expected. Are there any specific fragmentation products of pyridine and benzonitrile that would be informative in these physical conditions?

Sandra Brünken replied: This is an interesting question. There can be similarities in the product structures upon fragmentation induced by various events. In any case, experimental studies are required to verify the structures of the fragments. As an example, in the case of soft-X-ray ionization, see ref. 1, they observed that electronically excited ions undergo internal conversion into vibrationally excited ions in the electronic ground state. These would then cool through loss of H atoms similarly to vibrationally excited molecules excited by UV photons, collisions with electrons or in collision-induced dissociation (CID).

1 G. Reitsma, *et al.*, *Phys. Rev. Lett.*, 2014, **113**, 053002, DOI: 10.1103/PhysRevLett.113.053002.

Inga Kamp enquired: When you study the fragmentation of pyridine and benzonitrile as a function of electron impact energy, do you also quantify the amount of products you get as a function of energy of incoming electron? Do you see differences in some cases, *e.g.* different HCN/HNC losses? Or changes in C_2H_2 *versus* H loss?

Sandra Brünken responded: Our measured breakdown curves show that different fragmentation channels and fragment ratios are clearly seen as a function of electron impact energy, in agreement with earlier studies (see the supplementary information of the paper: https://doi.org/10.1039/d3fd00015j). We can see the changes in C_2H_2 *vs.* H-loss, as they differ by mass. This is not the case for HCN *vs.* HNC loss which lead to the same m/z of the ionic fragment, and which we have investigated using calculations. Theoretically we performed a benchmark for benzonitrile (see Supplementary Table 1 of the paper) to determine at which energy/time we could work. We did not perform a systematic study as a function of energy (same duration for a given set of parameters) but that could be interesting indeed. However, we should increase the energy in order to obtain more fragments. Indeed, in the conditions of the simulations reported in the article (minimum internal energy to see some dissociation), the number of fragments is too small to propose a trend.

Nicholas Walker asked: The calculated and experimental IR spectra for the linear fragment are in excellent agreement. On the other hand, there are a small number of bands in the experimental IR spectra for the m/z 77 amu and m/z 102 amu fragments (cyclic structures) from benzonitrile that aren't well-matched by the calculated spectra. Do you know the reason for this? Is it possible there are multiple different structures contributing to the product yield at that m/z?

Sandra Brünken answered: Yes, as discussed in the paper (https://doi.org/10.1039/d3fd00015j), there are multiple species that are responsible for the features of m/z 77 and m/z 102. In Fig. 4 of the paper we only present the calculated spectra of the dominant isomers. The remaining bands could come from an unidentified isomer, but more likely, given the amount of features, from anharmonic contributions, *i.e.* combination and overtone bands, which were not treated explicitly for those two *m*/*z* channels.

Dwayne E. Heard opened discussion of the paper by Arthur G. Suits: Following the laser photolysis of your precursor BrCN – what vibration levels are formed, and do you need to worry about any vibrational relaxation of CN (ν) in your kinetics analysis using the simultaneous kinetics and ring-down (SKaR) method?

Arthur G. Suits responded: The photolysis at 248 nm produces chiefly $\nu = 0$ and 1, 72 : 24 with only 3% $\nu = 2$.[1] Given the low concentration of higher vibrational levels and the long timescale for collisional relaxation of the high-frequency CN vibration, we believe relaxation from higher vibrational levels does not play a role in our experiment. In other experiments where fast reactive vibrational relaxation can contribute we may need to take account of this.

1 W. H. Fisher, R. Eng, T. Carrington, C. H. Dugan, S. V. Filet and C. M. Sadowski, Photodissociation of BrCN and ICN in the a continuum: vibrational and rotational distributions of CN (X $^2\Sigma^+$), *Chem. Phys.*, 1984, **89**, 457–471.

Dwayne E. Heard questioned: For the collision between CN ($\nu = 1$) and your reactants, there is presumably no role for vibrational relaxation back to CN ($\nu = 0$) rather than a reactive collision? Is vibrational relaxation much slower than the reactive channel and can be discounted completely?

Arthur G. Suits answered: Yes, given the 75 kcal mol^{-1} entrance channel well, we believe vibrational quenching *via* the adduct, *i.e.*, redissociation to reactants accompanied by CN vibrational relaxation, may safely be neglected. We have not examined that specifically as part of the theoretical investigation.

Ian R. Sims asked: In principle the absolute nature of the cavity ringdown technique means that you could use this new apparatus for measuring reaction product branching fractions. What are the prospects for this?

Arthur G. Suits replied: To be honest I am not sanguine about the prospects of UF-CRDS for product branching measurements except in limited cases where some products happen to have transitions spanned by the same laser and mirror set. The strength of the method is the ability to do SKaR, which allows for measuring the complete bimolecular kinetics in a couple of hours and with very little gas consumption. If we can implement that in the mid-IR then many species will be available for detection.

Ian R. Sims remarked: In general, vibrationally state-specific studies of CN reactions have shown that the kinetics follows the principle of vibrational adiabaticity, at least in the entrance channel part of the potential which determines the rate of reaction.[1]

1 I. R. Sims and I. W. M. Smith, *J. Chem. Soc., Faraday Trans. 2*, 1988, **84**, 527–539.

Arthur G. Suits responded: The question of vibrational adiabaticity is an interesting one. The kinetics of CN ($\nu = 1$) reactions have, to our knowledge, all been measured at room temperature and above. It is now well established, as we say in the paper (https://doi.org/10.1039/d3fd00029j), that the rates for CN complex-forming reactions are controlled by two transition states (TSs): an outer one at low temperature and an inner one at high temperature. The arguments you cite about the influence of vibrational adiabaticity enhancing the rate of the reaction for $\nu = 1$ (referring to reaction with O_2) appear to be focused on the inner TS and assume a softening of the CN vibrational frequency at that TS enhances the rate. Although vibrational adiabaticity at the outer TS seems a very good picture for low temperature, its impact on the rate is likely to be negligible as the vibrational frequency will be largely unaffected there.

Martin Cordiner continued discussion of the paper by Sandra Brünken: I'm interested in your result that HNC is produced as the dominant fragmentation pathway for benzonitrile. In comets, the origin of HNC is a mystery, since ALMA observations show it does not come from nucleus sublimation but is produced in the coma from an unknown process – presumably thermal or photo-fragmentation of a larger precursor species. Are you aware of any other complex nitriles (*e.g.* nitrogen-bearing aromatics) for which HNC is a significant dissociation pathway?

Sandra Brünken replied: It is very likely that other nitrogen-containing PAHs show a similar preference for HNC over HCN loss. There could be a difference in fragmentation behaviour between molecules with a cyano-group (or amino-group) *vs.* incorporated nitrogen atoms. For the former, only hydrogen migration processes are required for the loss of HNC, whereas for the latter, ring-opening and thereby loss of aromaticity can play a role, see a discussion for the case of quinoline where dominantly HCN loss was observed in Ramanathan *et al.*[1] The same experimental approach as shown in our paper (https://doi.org/10.1039/d3fd00015j) has been applied for aniline and points towards the formation of HNC as the neutral fragment,[2] as also discussed in ref. 3. Preliminary calculations on cyanonaphthalene in our work showed that the HNC loss channel is dominant (see the Astrophysical implications and conclusions section).

1 K. Ramanathan, *et al.*, *J. Chem. Phys.*, 2022, **157**, 064303, DOI: 10.1063/5.0092161.
2 D. Zeh, *et al.*, *J. Mol. Spectrosc.*, 2021, **378**, 111453, DOI: 10.1016/j.jms.2021.111453.
3 P. C. Burgers, *et al.*, *Chem. Phys. Lett.*, 1983, **102**, 1–3, DOI: 10.1016/0009-2614(83)80645-9.

Nadia Balucani continued discussion of the paper by Kevin M. Douglas: Looking at the scheme of the potential energy surface in your manuscript (https://doi.org/10.1039/d3fd00046j), it seems like there is a missing channel associated to the C–C bond breaking of the adduct with the formation of CH_3 + NH_2CHO. Since C–C sigma bonds are weaker than C–H sigma bonds I would expect that channel to be open.

Kevin M. Douglas answered: Yes, you are correct there are two possible product channels coming from the adduct, one in which the C–H bond breaks forming H + CH_3CONH_2 (H + acetamide), and a second in which the C–C bond breaks forming CH_3 + NH_2CHO (methyl radical + formamide). We have carried out some

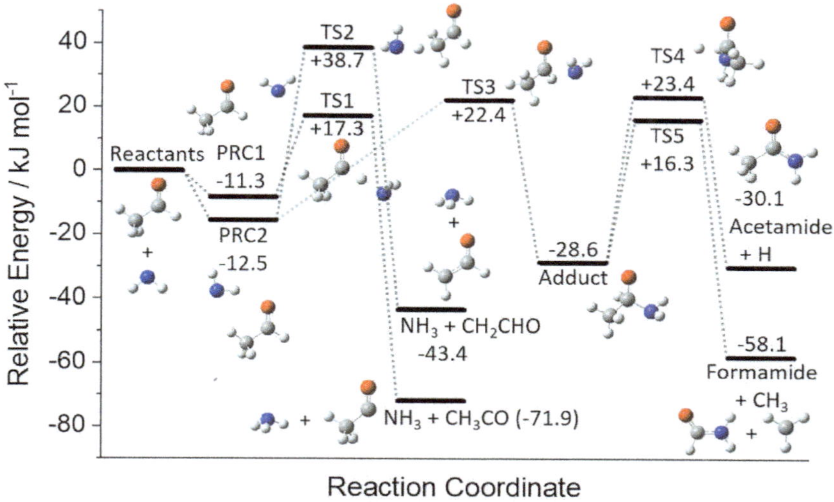

Fig. 1 Potential energy surface (PES) including the formamide + CH_3 product channel.

additional calculations and yes, the TS in which the C–C bond breaks is slightly lower than the TS in which the C–H bond breaks (TS5 = 16.3 kJ mol^{-1} *vs.* TS4 = 23.4 kJ mol^{-1}, see Fig. 1 here). However, as with the acetamide + H product channel, the formamide + CH_3 product channel first requires the formation of the adduct. As is discussed in the main paper (https://doi.org/10.1039/d3fd00046j), there is a relatively high barrier to adduct formation (TS3, 22.4 kJ mol^{-1}), effectively turning off this channel at low temperatures as there is no coordinate for H-atom tunnelling under this barrier. Indeed, using the surface given in Fig. 1 here, MESMER calculations indicate that the formamide + CH_3

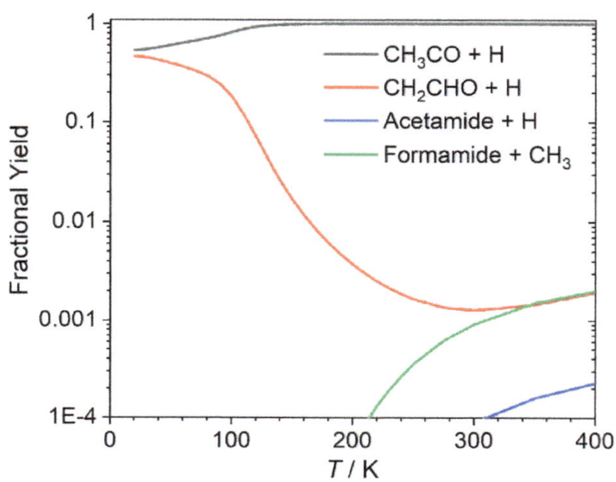

Fig. 2 Fractional yield of product channels *vs.* temperature at $P = 1 \times 10^{16}$ molecule cm^{-3} and with Ar bath gas.

channel only accounts for 0.2% of the total product yield at 400 K, while the acetamide + H channel only accounts for 0.02%, with both decreasing with temperature and becoming insignificant below ~150 K (see Fig. 2 here). As such, this additional product channel does not affect the results and conclusions of the main paper, and will only be important at higher temperatures.

Nadia Balucani said: My question concerns the comparison with other results from your group on the related reaction NH_2 + H_2CO.[1] I understand that the experiment on NH_2 + H_2CO was much more difficult because of the tendency of H_2CO to form dimers. Indeed, you were only able to provide an upper limit of 6×10^{-12} cm^3 s^{-1} for one value of the temperature (around 30 K). In this case, the experimental data are much clearer. If we compare your upper limit for NH_2 + H_2CO and the value of k you have derived for the reaction NH_2 + CH_3CHO around 30 K there is at least one order of magnitude of difference. Did you expect such a difference? There are other cases in the literature[2,3] (for instance the reactions with the isoelectronic OH radicals) in which the rate coefficients for H_2CO and CH_3CHO are not so different. In addition to that, the comparison between experimental and theoretical results that you have presented here required some adjustment of the calculated energy for the pre-reactive complexes of this reaction because the experimental rate coefficients are much faster than the calculated ones. This is probably caused by the limited accuracy of the method when applied to the case of weakly bound species or also by the choice of using MESMER[4] to evaluate the reaction rate coefficients when the energy randomization is not necessarily achieved (once again because we are considering species which are not covalently bound). Presumably, similar problems exist also for the calculations related to NH_2 + H_2CO. In that case, you used the theoretical values in the astrochemical models to peremptorily exclude that the NH_2 + H_2CO reaction could be a formation route of interstellar formamide.[1] Wasn't it too bold considering that you do not have experimental support? I remind you that the rate coefficients calculated by Skouteris *et al.*[5] were able to reproduce the spatial segregation of formamide in the L1157-B1 shocked region[6] as well as the relation between the deuteration degree of formamide and formaldehyde.[5,7] Both are stiff constraints.[7]

1 K. M. Douglas, D. I. Lucas, C. Walsh, N. A. West, M. A. Blitz and D. E. Heard, *Astrophys. J.*, 2022, **937**, L16.
2 A. J. Ocaña, E. Jiménez, B. Ballesteros, A. Canosa, M. Antiñolo, J. Albaladejo, M. Agúndez, J. Cernicharo, A. Zanchet, P. del Mazo, O. Roncero and A. Aguado, *Astrophys. J.*, 2017, **850**, 28.
3 S. Blázquez, D. González, E. M. Neeman, B. Ballesteros, M. Agúndez, A. Canosa, J. Albaladejo, J. Cernicharo and E. Jiménez, *Phys. Chem. Chem. Phys.*, 2020, **22**, 20562–20572.
4 D. R. Glowacki, C.-H. Liang, C. Morley, M. J. Pilling and S. H. Robertson, *J. Phys. Chem. A*, 2012, **116**, 9545–9560.
5 D. Skouteris, F. Vazart, C. Ceccarelli, N. Balucani, C. Puzzarini and V. Barone, *Mon. Not. R. Astron. Soc.: Lett.*, 2017, **468**, L1.
6 C. Codella, C. Ceccarelli, P. Caselli, N. Balucani, V. Barone, *et al.*, *Astron. Astrophys.*, 2017, **605**, L3.
7 A. López-Sepulcre, N. Balucani, C. Ceccarelli, C. Codella, F. Dulieu and P. Theulé, *ACS Earth Space Chem.*, 2019, **3**, 2122–2137.

Kevin M. Douglas responded: In both our current study on NH_2 + CH_3CHO (https://doi.org/10.1039/d3fd00046j) and our previous study on NH_2 + CH_2O,[1] we are very careful to operate under experimental conditions where dimers play no role. The observation of no removal of NH_2 in the presence of CH_2O is clear-cut evidence of low reactivity as it has been established that only CH_2O is present, there is no $(CH_2O)_2$.[1]

To answer your questions regarding comparing the rate coefficients for NH_2 with CH_3CHO and CH_2O: the upper limit we determine for the rate coefficient for the reaction between $NH_2 + CH_2O$ is 6×10^{-12} cm^3 molecule^{-1} s^{-1}, determined using He as a bath gas at total density of $\sim 5 \times 10^{16}$ molecules per cm^3 and a temperature of ~34 K.[1] The rate coefficient we determine for the reaction between $NH_2 + CH_3CHO$ under similar conditions, that is using He as a bath gas at a total density of $\sim 5 \times 10^{16}$ molecules per cm^3 and a temperature of ~36 K, is 3.6×10^{-11} cm^3 molecule^{-1} s^{-1}; this is around 6 times faster than the upper limit we determine for $NH_2 + CH_2O$ (although it should be noted this is an upper limit, and the true rate coefficient is likely to be slower). We see a similar difference in the zero-pressure rate coefficients that we calculate for each reaction, with the rate coefficient for $NH_2 + CH_3CHO$ being around 4–6 times faster than that for NH_2 + CH_2O at low temperatures (*e.g.* at 20 K the zero-pressure rate coefficient for NH_2 + CH_3CHO is 1.9×10^{-12} cm^3 molecule^{-1} s^{-1}, while for $NH_2 + CH_2O$ it is 4.5×10^{-13} cm^3 molecule^{-1} s^{-1}). This is not so different to comparing the rate coefficients for the isoelectronic OH with CH_3CHO and CH_2O; at 20 K the rate coefficient for OH + CH_3CHO (ref. 2) is around 3.5 times faster than for OH + CH_2O (ref. 3) ($\sim 3.5 \times 10^{-10}$ cm^3 molecule^{-1} s^{-1} *vs.* $\sim 1.0 \times 10^{-10}$ cm^3 molecule^{-1} s^{-1} respectively). Our results indicate that the reactions of NH_2 with CH_3CHO and CH_2O have a strong pressure dependence over the pressure range accessible in our experiments, while the reactions of OH are effectively pressure independent.[2–4] This strong pressure dependence indicates that loss of NH_2 in our experiments is primarily due to stabilisation into the pre-reaction complex (PRC) wells. Our work on the $NH_2 + CH_3CHO$ system has shown that this stabilization is sensitive to the density of states of the PRCs, which in turn is sensitive to the low frequency vibrations in these species. The larger CH_3CHO molecule will form PRCs with NH_2 that have more vibrational frequencies than the smaller $CH_2O\cdot NH_2$ PRCs. It is therefore not surprising that the larger $CH_3CHO\cdot NH_2$ PRCs would have smaller low frequency vibrations than the $CH_2O\cdot NH_2$ PRCs, and as such a larger density of states, making stabilization into the $CH_3CHO\cdot NH_2$ PRCs faster than into the $CH_2O\cdot NH_2$ PRCs. As such, a difference in rate coefficient of around 5 or even 10 times is not too surprising.

To answer your question regarding the $NH_2 + CH_2O$ system and whether it is a possible source of formamide under interstellar conditions: The reaction between $NH_2 + CH_2O$ has two primary exothermic channels: a hydrogen-abstraction channel leading to NH_3 + CHO, and an addition–elimination channel, in which the NH_2 first attacks the C on the CH_2O to form a bound adduct, which then goes on to eliminate an H atom and produce formamide + H.[1] Both these channels have significant barriers: a 23 kJ mol^{-1} barrier to H-abstraction products, and an 18 kJ mol^{-1} barrier to adduct formation (as calculated by our group in our previous paper,[1] by Vazart *et al.*,[5] and by Song *et al.*[6]). It is the presence of this significant barrier to adduct formation on the PES that leads from $NH_2 + CH_2O$ to formamide + H why we exclude that this reaction could be a formation route of interstellar formamide. Even adjusting this barrier down by the 9.5 kJ mol^{-1} that we were required to move our PRC wells down by on the

NH_2 + CH_3CHO surface (see main paper: https://doi.org/10.1039/d3fd00046j) would still result in an 8.5 kJ mol^{-1} barrier on the pathway to formamide + H from NH_2 + CH_2O. As there is no coordinate for H-atom tunnelling under this barrier, this channel simply cannot go at the low temperatures of the interstellar medium. We should note that with only a small change in the size and shape of some of the hindered rotors used in the rate theory calculations for NH_2 + CH_3CHO, we were only required to increase the well depths of the PRCs by less than 5 kJ mol^{-1} in order to get good agreement between the calculated and experimental rate coefficients. This suggests that we are not fully capturing some of these low frequency modes in our calculations, and that we have had to change the depth of the PRCs by more than would be expected to account for this. Indeed, only a very minor adjustment of the heights of the H-abstraction transition states was required (<1 kJ mol^{-1}) in order to get good agreement between the calculated and experimental branching ratios. It should also be noted that changing the depths of the PRCs for the NH_2 + CH_2O system would also not alter our result that this reaction is not a source of interstellar formamide, this would simply increase the rate of stabilization into the PRC wells. Some preliminary MESMER calculations on the NH_2 + CH_2O system indicated that even moving the height of the TS leading to the adduct down to 2.5 kJ mol^{-1} (a reduction of 15.5 kJ mol^{-1}!) still did not lead to the formation of formamide at low temperatures. Only the H-abstraction channel producing NH_3 + CHO, in which the transfer of an H atom can occur by quantum mechanical tunnelling through a barrier, is open at low temperatures. We see the same result when looking at the NH_2 + CH_3CHO system: under interstellar conditions the reaction is not a route to acetamide or formamide, as both these channels require the initial formation of an adduct, for which there is a significant barrier on the PES (22 kJ mol^{-1}). This barrier is insurmountable at the low temperatures of the interstellar medium (ISM), and as there is no coordinate for H-atom tunnelling under this barrier, this channel simply cannot go under the conditions of the ISM. Only the H-abstraction channel forming NH_3 + CH_3CO, in which the transfer of an H atom can occur by quantum mechanical tunnelling through a barrier, is open at these low temperatures.

We would like to point out that in the calculation of rate coefficients for the reaction of NH_2 with CH_2O by Skouteris *et al.*,[7] the barrier to adduct formation is completely omitted. In this paper, it is argued that when zero-point energy is excluded, the barrier to adduct formation is likely submerged. However, rather than including a submerged barrier in the calculations, the barrier to adduct formation is completely omitted, and as such they calculate fast rate coefficients for the formation of formamide from this reaction. As we discuss in our previous paper,[1] we argue against the exclusion of this barrier to adduct formation in the determination of rate coefficients for this reaction, and against the exclusion of zero-point energy in the determination of the barrier heights on the PES.

1 K. M. Douglas, D. I. Lucas, C. Walsh, N. A. West, M. A. Blitz and D. E. Heard, The Gas-phase Reaction of NH_2 with Formaldehyde (CH_2O) is not a Source of Formamide (NH_2CHO) in Interstellar Environments, *Astrophys. J., Lett.*, 2022, **937**, L16, DOI: 10.3847/2041-8213/ac8cef.

2 S. Blazquez, D. Gonzalez, E. M. Neeman, B. Ballesteros, M. Agundez, A. Canosa, J. Albaladejo, J. Cernicharo and E. Jimenez, Gas-phase kinetics of CH_3CHO with OH radicals between 11.7 and 177.5 K, *Phys. Chem. Chem. Phys.*, 2020, **22**, 20562–20572, DOI: 10.1039/d0cp03203d.

3 A. J. Ocaña, E. Jiménez, B. Ballesteros, A. Canosa, M. Antiñolo, J. Albaladejo, M. Agúndez, J. Cernicharo, A. Zanchet, P. del Mazo, O. Roncero and A. Aguado, Is the Gas-phase OH + H_2CO Reaction a Source of HCO in Interstellar Cold Dark Clouds? A Kinetic, Dynamic, and Modeling Study, *Astrophys. J.*, 2017, **850**, 28, DOI: 10.3847/1538-4357/aa93d9.
4 E. Vöhringer-Martinez, B. Hansmann, H. Hernandez, J. S. Francisco, J. Troe and B. Abel, Water Catalysis of a Radical-Molecule Gas-Phase Reaction, *Science*, 2007, **315**, 497–501, DOI: 10.1126/science.1134494.
5 F. Vazart, D. Calderini, C. Puzzarini, D. Skouteris and V. Barone, State-of-the-Art Thermochemical and Kinetic Computations for Astrochemical Complex Organic Molecules: Formamide Formation in Cold Interstellar Clouds as a Case Study, *J. Chem. Theory Comput.*, 2016, **12**, 5385, DOI: 10.1021/acs.jctc.6b00379.
6 L. Song and J. Kästner, Formation of the prebiotic molecule NH_2CHO on astronomical amorphous solid water surfaces: accurate tunneling rate calculations, *Phys. Chem. Chem. Phys.*, 2016, **18**, 29278, DOI: 10.1039/c6cp05727f.
7 D. Skouteris, F. Vazart, C. Ceccarelli, N. Balucani, C. Puzzarini and V. Barone, New quantum chemical computations of formamide deuteration support gas-phase formation of this prebiotic molecule, *Mon. Not. R. Astron. Soc.: Lett.*, 2017, **468**, L1, DOI: 10.1093/mnrasl/slx012.

Nadia Balucani enquired: I have again a question on the reliability of MESMER[1] or any other statistical code to treat the cases of weakly, non-covalently bound intermediates. How do we know that intramolecular vibrational redistribution is complete in these cases? Indeed, there are plenty of cases where a non-statistical behaviour has been seen even when strongly bound intermediates are formed during a reaction (see, for instance, ref. 2–4). Also, in a recent investigation, we had experimental evidence that the energy is not randomized in the pre-reactive complex for the reaction $O(^3P)$ + pyridine.[5]

1 D. R. Glowacki, C.-H. Liang, C. Morley, M. J. Pilling and S. H. Robertson, *J. Phys. Chem. A*, 2012, **116**, 9545–9560.
2 F. Leonori, D. Skouteris, R. Petrucci, P. Casavecchia, M. Rosi and N. Balucani, *J. Chem. Phys.*, 2013, **138**, 024311.
3 N. Balucani, A. Bergeat, L. Cartechini, G. G. Volpi, P. Casavecchia, D. Skouteris and M. Rosi, *J. Phys. Chem. A*, 2009, **113**, 11138–11152.
4 Z. Homayoon, J. M. Bowman, N. Balucani and P. Casavecchia, *J. Phys. Chem. Lett.*, 2014, **5**, 3508–3513.
5 P. Recio, S. Alessandrini, G. Vanuzzo, G. Pannacci, A. Baggioli, D. Marchione, A. Caracciolo, V. J. Murray, P. Casavecchia, N. Balucani, C. Cavallotti, C. Puzzarini and V. Barone, *Nat. Chem.*, 2022, **14**, 1405.

Kevin M. Douglas replied: Regarding the reliability of Rice–Rampsberger–Kessel–Marcus (RRKM) theory for predicting rate coefficients there has been much speculation on when the underlying assumptions may break down if there is non-statistical behaviour. Remarkably, there are very few examples of non-statistical behaviour. One example that stands out is the reaction between H atoms and vibrationally excited HOD.[1] The main thrust of non-RRKM behaviour is the breaking of the ergodicity assumption, *i.e.* the complete re-distribution of vibrational energy within the molecule. The above example arises from ν(O–H) and ν(O–D) within H–O–D not mixing as the vibrational frequencies are so different. But this rare example is a "small molecule" effect. With larger systems anharmonicities ensure that mixing between the vibrational modes should be efficient. The cited examples of non-RRKM behaviour are from crossed-molecular beam (CMB) experiments, where rate coefficients are not measured, and the distribution of energy within the reagents is far from Boltzmann. In fact, the translational energy in CMB experiments is more than 1000 K, so in these experiments it is difficult to say much about the role of van der Waals (vdW) complexes. Laval experiments measure reaction rate coefficients at low temperatures, and it is these results that have highlighted the important role of vdW complexes. The most interesting of these low-temperature reactions are those characterized by initial formation of a vdW complex, in which there is a molecular coordinate which facilitates subsequent quantum mechanical tunnelling to overcome the overall barrier to products. OH + CH_3OH is the best-known example of this type of reaction[2] and regarding modelling the rate coefficient data with statistical rate theory,[3] it captures the low temperature increase but does not fully reproduce the size of this increase. Potential reasons for this failure are an incomplete description of the important low-frequency vibrations, and of the tunnelling mechanism. Non-RRKM *via* incomplete internal vibrational relaxation of the complex is another possibility but it is noted that the Ring Polymer Dynamics[4] study on this reaction did not indicate non-RRKM behaviour. Another possibility that has been raised[5] is that the reaction has a contribution from OH + $(CH_3OH)_2$, methanol dimer, which may form in the Laval experiment, but we are very careful to operate under experimental conditions where the dimer plays no role. So, both experiments and theory need to be aware of the various pitfalls. In our previous study[6] on the NH_2 + CH_2O system, the observation of no removal of NH_2 in the presence of CH_2O is clear-cut evidence of low reactivity as it has been established that only CH_2O is present, there is no $(CH_2O)_2$. This experimental observation is reproduced in the MESMER calculations.

1 A. Sinha, M. C. Hsiao and F. F. Crim, Bond-selected bimolecular chemistry: H + HOD $(4\nu_{OH}) \rightarrow$ OD + H_2, *J. Chem. Phys.*, 1990, **92**, 6333–6335, DOI: 10.1063/1.458312.
2 R. J. Shannon, M. A. Blitz, A. Goddard and D. E. Heard, Accelerated chemistry in the reaction between the hydroxyl radical and methanol at interstellar temperatures facilitated by tunnelling, *Nat. Chem.*, 2013, **5**, 745–749, DOI: 10.1038/nchem.1692.
3 A. J. Ocana, S. Blazquez, A. Potapov, B. Ballesteros, A. Canosa, M. Antinolo, L. Vereecken, J. Albaladejo and E. Jimenez, Gas-phase reactivity of CH_3OH toward OH at interstellar temperatures (11.7–177.5 K): experimental and theoretical study, *Phys. Chem. Chem. Phys.*, 2019, **21**, 6942–6957, DOI: 10.1039/c9cp00439d.
4 O. Roncero, A. Zanchet and A. Aguado, Low temperature reaction dynamics for CH_3OH + OH collisions on a new full dimensional potential energy surface, *Phys. Chem. Chem. Phys.*, 2018, **20**, 25951–25958, DOI: 10.1039/c8cp04970j.
5 W. Siebrand, Z. Smedarchina, E. Martinez-Nunez and A. Fernandez-Ramos, Methanol dimer formation drastically enhances hydrogen abstraction from methanol by OH at low temperature, *Phys. Chem. Chem. Phys.*, 2016, **18**, 22712–22718, DOI: 10.1039/c6cp04173f.
6 K. M. Douglas, D. I. Lucas, C. Walsh, N. A. West, M. A. Blitz and D. E. Heard, The Gas-phase Reaction of NH_2 with Formaldehyde (CH_2O) is not a Source of Formamide (NH_2CHO) in Interstellar Environments, *Astrophys. J., Lett.*, 2022, **937**, L16, DOI: 10.3847/2041-8213/ac8cef.

Cristina Puzzarini asked: As a follow-up to previous comments/questions, I am wondering whether you have checked the structure and energetics of the pre-reactive complex at a higher level of theory. Based on my experience, global hybrid density functionals are not suitable for the structural determination of pre-reactive complexes where non-covalent interactions play such a crucial role. Global hybrid density functionals might lead to incorrect intermolecular distances which in turn might cause problems in the subsequent energy calculations at higher level. Hybrid functionals can be so wrong that sometimes minima at the DFT level are actually transition states at a higher level of theory. If the pre-reactive complex is so important to explaining deviations from experimental results, I would improve the level of theory for it. Very likely, all stationary points of the reactive potential energy surface deserve to be better characterized. Furthermore, I would also perform improved calculations for energetics because it is well known that CCSD(T) in conjunction with a triple-zeta basis set does not provide accurate results. In particular, the accuracy required to explain kinetic experiments is not met.

Kevin M. Douglas responded: We agree that density functional theory (DFT) can result in inaccuracies in the structures, and thus the vibrational frequencies, of our pre-reaction complexes (PRCs), and that CCSD(T) in conjunction with a triple-zeta basis set may not always provide accurate electronic energies. However, the experimental results indicate the reaction exhibits a negative temperature dependence and positive pressure dependence, which can only be explained by the presence of both a PRC and a positive energy barrier (or more than one of each) on the potential energy surface (PES), which our calculations confirm. Although there may be inaccuracies in our calculated energies, as we are fitting the PES to experimental rate coefficients and branching ratios, this should not affect the temperature and pressure dependent rate coefficients that we predict. Furthermore, any inaccuracies in our calculated vibrational frequencies (and thus the calculated density of states) are likely to be minor when compared to the inaccuracies in determining accurate hindered rotor potentials, which we have shown our calculated rate coefficients are sensitive to. We have also carried out some additional calculations in which we have optimized the structures of the reactants and the two PRCs at the B2PLYP/aug-cc-pVTZ and the MP2/aug-cc-pVTZ levels of theory, and both give zero-point corrected energies of the PRCs that are within 2 kJ mol^{-1} of those we calculate in the main paper (https://doi.org/10.1039/d3fd00046j).

Cristina Puzzarini opened discussion of the paper by Ugo Jacovella: I have worked on cyanoethynylbenzene (HCC–C_6H_4–CN) isomers[1] with Marie-Aline Martin-Drumel and noticed problems, at the DFT level using the B3LYP functional, on quartic centrifugal distortion constants for the *ortho* species (only for the *ortho* species, and not for *meta* and *para* ones). I do not know whether the problem is related to B3LYP or the Gaussian code used for computations. However, like in your work (https://doi.org/10.1039/d3fd00016h), for the *ortho* species there are large deviations between experiment and theory (up to one order of magnitude) for the quartic centrifugal distortion constants. At the moment, I have only some evidence, but it seems a problem related to molecules with substituents in the *ortho* position. MP2 calculations, already with a double-zeta basis set, show instead a very good agreement with experiment. Do you have any further evidence on the discrepancy between experiment and theory for centrifugal distortion constants?

1 J.-T. Spaniol, K. L. K. Lee, O. Pirali, C. Puzzarini and M.-A. Martin-Drumel, *Phys. Chem. Chem. Phys.*, 2023, **25**, 6397, DOI: 10.1039/d2cp04825f.

Ugo Jacovella answered: We have seen other species for which the experimental centrifugal distortion constants are in disagreement with the calculated ones using ωB97X-D. However, we have not performed systematic tests to understand the issue, which appears Gaussian-related.

Cecilia Ceccarelli continued discussion of the paper by Kevin M. Douglas: Can you please explain what reactions have been added on the grain surface network in the predictions shown by the blue solid line in Fig. 9 of your paper (https://doi.org/10.1039/d3fd00046j)?

Kevin M. Douglas replied: Fig. 9 in the paper (https://doi.org/10.1039/d3fd00046j) presents the abundance of gas-phase CH_3CO from a dark cloud model. The model uses the network from Walsh *et al.* (2015)[1] with updates as described in Douglas *et al.* (2022).[2] The variants of the model we ran were two with gas-phase only chemistry, one with and one without the gas-phase reaction between NH_2 + CH_3CHO, and two gas–grain models, again one with and one without the NH_2 + CH_3CHO gas-phase reaction. No other changes were made to either the gas-phase or the grain-surface reactions in the model. In the gas–grain models, when the NH_2 + CH_3CHO reaction was not included (red solid line in Fig. 9) we see a CH_3CO radical abundance of $\sim 10^{-26}$ to 10^{-24}, a range which is still considered negligible. In this run, the only source of gas-phase CH_3CO is *via* cosmic-ray-induced photodesorption from the ice phase. When we include the gas-phase reaction between NH_2 + CH_3CHO in the gas–grain model (blue solid line in Fig. 9), we now see an arguably non-negligible abundance of CH_3CO in the range of 10^{-14} to 10^{-12}. This is driven by the more efficient production of gas-phase CH_3CHO in the gas–grain model (which increases by two orders of magnitude when compared to the gas-phase only model), which in turn results in an increase in the CH_3CO abundance *via* its gas-phase reaction with NH_2.

1 C. Walsh, H. Nomura and E. van Dishoeck, *Astron. Astrophys.*, 2015, **582**, A88.
2 K. M. Douglas, D. I. Lucas, C. Walsh, N. A. West, M. A. Blitz and D. E. Heard, *Astrophys. J., Lett.*, 2022, **937**, L16.

Daniël B. Rap continued discussion of the paper by Ugo Jacovella: What would be the relevant formation pathway towards these bridged bicyclic molecules in cold interstellar environments? Could top-down fragmentation processes occur or bottom-up approaches including reactions with building blocks such as benzene/cyclopentadiene and acetylene. In the latter case, the induced bond strain in these molecules could be energetically unfavorable?

Ugo Jacovella responded: Indeed, the induced bond strain resulting from a bottom-up approach seems to disfavour this pathway. A top-down approach seems much more attractive, in particular because norbornadiene (NBD) is often formed by isomerisation of C_7H_8 (toluene, cycloheptatriene, *o*-isotoluene, *etc.*) when they are energised. Furthermore, it is speculated that C_7H_8 species could be formed from destruction of larger aromatics. The NBD formation route could also begin on the $C_7H_8^+$ potential energy surface, and then be neutralised by electron capture.

Ian R. Sims asked: You only studied one of the dicyano isomers of NBD – was the chemical synthesis specific for this isomer, and would it be possible/interesting to study other isomers?

Ugo Jacovella replied: The reason we have only studied one dicyano (DCN) isomer is that it is the easiest to synthesize. To the best of our knowledge, no synthetic route is known for the others. Since this specific DCN isomer was not detected, neither was CN-NBD. We doubt that the study of other DCN isomers is relevant in an astrochemical context.

Adam J. Fleisher continued discussion of the paper by Arthur G. Suits: General questions on the experimental set-up. Is the simultaneous kinetics and ring-down (SKaR) performed at a single frequency? Because the time-dependent concentration of the absorber $[A](t)$ – as defined in eqn (3) of the paper (https://doi.org/10.1039/d3fd00029j) – is frequency dependent *via* the cross-section, σ. What experimental steps are taken to ensure that you always probe the same cavity resonance?

Arthur G. Suits answered: In these experiments the distributed feedback (DFB) probe laser is stabilized with a free-running control loop that adjusts its temperature to match a set frequency when monitored by a wave meter. A small frequency error would have an effect equivalent to a variation in the CN density such as induced by fluctuations in excimer power. Averaging an ensemble of cavity ring-down spectroscopy (CRDS) signals differing only in the amplitude of the time-dependent absorption, either due to excimer power fluctuation or frequency noise in the probe laser, has a similar effect on the kinetics deduced from the average. We have simulated this and find when the amplitude is changed by 20% for a 100 shot average, the inferred rate differs by less than 1%. We find the average of fluctuating ringdown signals referenced to an empty cavity ringdown can still be well approximated by the rise–fall model function with a single effective rise and fall rate that are good approximations to the "actual" rise and fall rates, and the fit amplitude that converges appropriately to the actual average intensity of the fluctuating signal.

Adam J. Fleisher asked: Follow-up question. You mentioned work towards the mid-infrared. What are the prospects of performing SKaR measurements in the mid-infrared? Can empty-cavity ring-down time constants of sufficiently long decay times be achieved?

Arthur G. Suits replied: We have not yet tested our mid-IR mirrors, but we do not expect them to perform as well as our current near-IR set. It may be difficult to perform SKaR if the ringdown times are very short. However, they need not approach 100 μs as we now achieve. Much shorter ringdown times might still permit SKaR if the variation in those times can be minimized and the sample densities adjusted accordingly. In any case, we could perform point-wise measurements as a function of delay if needed. Alternatively, I believe you will be gaining experience with very high performance mid-IR mirrors coming available. We look forward to learning from your experience with them and perhaps consider implementing them ourselves if costs permit.

Dwayne E. Heard enquired: In your paper (https://doi.org/10.1039/d3fd00029j) you state that the total CN concentration generated by photolysis of BrCN is around 2–3 × 10^{11} molecules per cm^3. What is the ultimate limit of detection for your CRDS method, and do you think your method offers promise for detection of products of chemical reactions at very low temperatures?

Arthur G. Suits responded: We have previously discussed sensitivity issues for UF-CRDS (ref. 1) and estimated a minimum absorption coefficient of 2×10^{-9} cm^{-1}, or 0.2 ppm per pass. We note our CN ($\nu = 1$) probed transition is a weak electronic transition rather than an overtone. In a sense these are proof-of-principle experiments. As I have noted, our interest in expanding the range of targets here is to go to the mid-IR. There we hope to have similar sensitivity and a broader range of molecules we can detect, although a limitation will still be the limited range of suitability of a given mirror set.

1 N. Suas-David, S. Thawoos and A. G. Suits, *J. Chem. Phys.*, 2019, **151**, 244202 (ref. 24 in the paper).

Anthony Meijer continued discussion of the paper by Sandra Brünken: Your paper (https://doi.org/10.1039/d3fd00015j) notes at the start that the fact that pyridine has not been detected yet is an unresolved issue. It shows the different dissociation pathways for benzonitrile and pyridine, which would allow for the dissociation products to be detected. However, it is not clear to me whether pyridine is more likely to dissociate than benzonitrile from your experiments and calculations, so the conclusion must be drawn that differential efficiency of dissociation of benzonitrile and pyridine is not the cause of the non-detection of pyridine. Would you like to comment on that and whether any aspects of your experiments/calculations still would allow a potential differential efficiency to be operational?

Sandra Brünken replied: The previously determined (partial) ionization cross-sections of both pyridine and benzonitrile are similar. However, as we discuss in our study, the HCN fragmentation trajectory from pyridine$^{\bullet+}$ includes a reverse barrier of formation, which is not present for benzonitrile^{+}. This could lead to a less likely backwards formation of pyridine *vs.* benzonitrile in reactions of the respective fragments with HCN.

Anthony Meijer asked: Is the basis set used in the calculation (which is of double-zeta quality) good enough to get good transition state energies? Have you done any benchmarking at a higher level of theory?

Sandra Brünken answered: The used level of theory provides a general understanding of the reaction pathways and gives sufficiently accurate energies for the transition states. The DFT method may underestimate the transition states and single-point energies compared to higher levels of theory which can give more accurate values, but this will not significantly change the conclusions drawn here. Here we aimed for a comparative study between two molecules and used the same level of theory for both cases and for all points.

So, no, we have not done any benchmarking in this specific case. We have performed such a benchmarking on different higher-levels of theory in another study where the reaction between pyridine$^{\bullet+}$ and acetylene showed a significant endothermic transition state, which was worth investigating in more detail.[1]

1 D. B. Rap, *et al.*, *Nat. Astron.*, 2022, **6**, 1059–1067, DOI: 10.1038/s41550-022-01713-z.

Daniël B. Rap commented: The variety of molecules that are formed in these fragmentation studies do not always allow the prediction of infrared frequencies for all species using a single theory. However, in the case of electronic energies, other methods than those applied here could help getting more accurate values, but with a computational cost. In the current paper (https://doi.org/10.1039/d3fd00015j), we have provided a general energetic overview of the pathways. In other studies, we have used more expensive theories in order to more accurately evaluate crucial reaction barriers.

Anthony Meijer added: Perhaps your earlier work would allow you to give an indication of the quality of the current calculations and give some insights into how the estimate of the error affects the conclusions of the current work.

Mark H. Stockett enquired: At the end of your paper (https://doi.org/10.1039/d3fd00015j) you give a preview of future studies of cyanonaphthalene using the same combination of methods used for the smaller molecules pyridine and benzonitrile. The results of the DFTB simulations show nearly 100% HNC-loss, which is intriguing for its connection to the relative abundances of HNC/HCN, a classic question in astrochemistry. However, breakdown curves measured using collision induced dissociation mass spectrometry (see Fig. 3 here) show 30% H-loss.[1] Is this a general problem with DFTB (*i.e.* not reproducing lab branching fractions) or is it something specific to H-loss, which is also difficult to

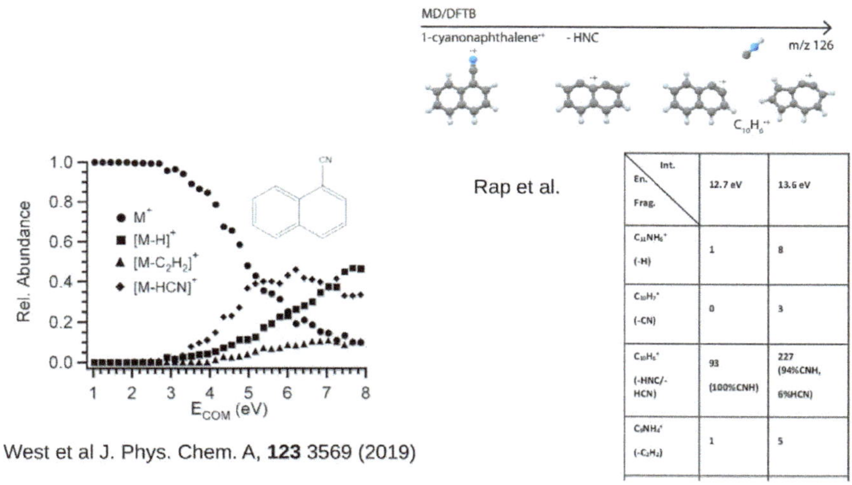

En. \ Int. / Frag.	12.7 eV	13.6 eV
$C_{11}NH_6^+$ (-H)	1	8
$C_{10}H_7^+$ (-CN)	0	3
$C_{10}H_6^+$ (-HNC/-HCN)	93 (100%CNH)	227 (94%CNH, 6%HCN)
$C_9NH_4^+$ (-C_2H_2)	1	5

Fig. 3 (Left) Collision-induced dissociation breakdown curves for ionized 1-cyanonaphthalene. Reprinted with permission from West *et al.*, *J. Phys. Chem. A*, 2019, **123**, 3569.[1] Copyright 2019 American Chemical Society. (Right, top) Cationic fragment structures upon the loss of HNC from canonical 1-cyanonaphthalene$^{·+}$ as observed in MD/DFTB simulations at 12.7 eV internal energy of the respective cation using Set1 parameters; (right, bottom) Stoichiometry of the fragments resulting from the dissociation of 1-cyanonaphthalene$^{·+}$ (pools of 720 simulations of 1 ns) obtained from MD/DFTB simulations at two internal energies of the respective cation. Reproduced from https://doi.org/10.1039/d3fd00015j.

model in *e.g.* RRKM? Returning to the main results of your paper, the H-loss fragments seemed to me to be the ones with the worst agreement between calculations and experiment. Is there something about H-loss that is not captured by our theories?

1 B. J. West, L. Lesniak and P. M. Mayer, *J. Phys. Chem. A*, 2019, **123**, 3569–3574, DOI: 10.1021/acs.jpca.9b01879.

Sandra Brünken responded: Actually, the loss of H is a bottleneck for closed shell-DFT methods – hence for closed-shell SCC-DFTB that has the drawbacks of DFT – as it is a radical fragment. Indeed, such mono-determinantal methods lead to wrong dissociation behavior and in particular to an overestimation of the dissociation energy. That can be counteracted using unrestricted DFT(B) methodology but that leads to a severe convergence issue during dissociation. In our simulations, we use a Fermi temperature to avoid the convergence issue and using such an approach can be viewed as a kind of electronic excitation. In order to correctly describe dissociation energies, we should use a multireference *ab initio* method, with which nanoscale BOMD simulations are totally out of reach given the number of degrees of freedom.

Recently, we have measured a mass spectrum of the cyanonaphthalene dissociative ionization upon electron impact ionization and indeed, besides the HCN/HNC loss channel, also a large contribution of H-loss is present.

Ugo Jacovella addressed Mark H. Stockett and Sandra Brünken: In the West *et al.* paper,[1] dissociative branching ratios are derived from collision-induced dissociation decay curves. This is therefore a process that takes place on the potential energy surface of the electronic ground state (probing the high vibrational excited states of the neutral molecule), whereas in the present paper (https://doi.org/10.1039/d3fd00015j) dissociation is induced by electron impact, so the potential energy surfaces of the excited electronic states are explored and may therefore lead to different dissociative branching ratios. No direct comparison can be made between the two experiments.

1 B. J. West, L. Lesniak and P. M. Mayer, *J. Phys. Chem. A*, 2019, **123**, 3569–3574, DOI: 10.1021/acs.jpca.9b01879.

Mark H. Stockett replied: This is true, however I was referring to the DFTB simulations which as I understand it run on ground state potentials.

Sandra Brünken added: Ugo's comment on Mark Stockett's question is an interesting one. Indeed, upon electron impact ionization the ions might be electronically excited, and only if fast internal conversion is present, transferring the electronic energy to vibrational energy in the electronic ground state, we can expect a similarity to the CID measurements. It might be interesting to perform calculations, using time-dependent DFT or kinetic Monte-Carlo models, to study the electronic de-excitation. In view of the paper of Stockett *et al.* in this issue (https://doi.org/10.1039/d3fd00005b), one might also have to include radiative de-excitation processes as a competing process to dissociation to obtain a full picture.

Mark H. Stockett asked: Have you carried on with calculations of the potential energy surfaces for cyanonaphthalene? What are the prospects?

Sandra Brünken replied: We did not perform PES calculations yet, but this would be interesting. Experimental fragmentation studies are a logical and necessary next step in order to elucidate the structure of the fragment, and we have started with those.

Jack Fulker communicated: I was wondering if these experiments and calculations could be run on pyridine and benzonitrile clusters/aggregates with water. I know that species like benzene, xylene and phenol can form these clusters in the gas phase and that the addition of water can tune the electronic properties of the species, so I'm curious if water could influence the fragmentation channels of nitrogen containing aromatics.

Sandra Brünken communicated in reply: Similar simulations can indeed be performed on such molecular clusters. One would likely first see the evaporation of water molecules at low internal excitation energy so it is unclear whether we will be able to see an influence of "solvation" on the dissociation paths. The binding energies of H_2O on PAH cations should be low, see, *e.g.* ref. 1 and 2, where the binding energy to the naphthalene cation was found to be only around 2800 cm^{-1}. The solvation of a molecule with water may thus act as a "protection mechanism", with the water molecules as favoured leaving groups upon excitation. Probably a different approach where you selectively populate some excited states of the systems could be more suitable.

1 K. Chatterjee and O. Dopfer, *Phys. Chem. Chem. Phys.*, 2017, **19**, 32262–32271, DOI: 10.1039/c7cp06893j.
2 K. Chatterjee and O. Dopfer, *Chem. Sci.*, 2018, **9**, 2301–2318, DOI: 10.1039/c7sc05124g.

Ugo Jacovella enquired: It appears that the times used in the MD simulations are several orders of magnitude shorter than the times between the birth of the ions and the quadrupole mass filter (the time for the ions to dissociate) and the 22-pole trap (still the time for the fragment ions to explore their potential energy surface). Furthermore, in the MD simulation, a single initial kinetic energy is used to reproduce the experimental branching ratios, but in the experiment, ions are generated with a wide range of kinetic energies. Therefore, how confident are you in the support provided by MD simulations for your experiments and what were the motivations for choosing this particular approach?

Sandra Brünken replied: The duration of the MD/SCC-DFTB simulation cannot exceed the nanosecond timescale which is different than the microsecond timescale of the experiment. This is an issue for this approach which is already discussed in the article (https://doi.org/10.1039/d3fd00015j). More detail is discussed by Banhatti *et al.*[1] If we can have an initial distribution of internal energy once the ions are formed (*i.e.* amount of internal energy transferred to the ions) then yes the best approach is to use an energy distribution. But such an energy distribution is not easy to estimate (see for instance the paper by Simon *et al.*[2] where they used an energy distribution estimated after a modelling by the experimentalists). The goal in the present article was to get insights into mechanisms and possibly branching ratios of the different pathways at the lowest possible energy.

1 S. Banhatti, *et al.*, *Phys. Chem. Chem. Phys.*, 2022, **24**, 27343–27354, DOI: 10.1039/d2cp03835h.
2 A. Simon, *et al.*, *Theor. Chem. Acc.*, 2018, **137**, 106, DOI: 10.1007/s00214-018-2287-z.

Mark H. Stockett remarked: We do sometimes DFT-B then classical MD simulations which run longer.[1] The classical simulations wouldn't capture excited state processes, but slow isomerizations would show up.

1 M. Gatchell, *et al.*, *Phys. Rev. A: At., Mol., Opt. Phys.*, 2015, **92**, 050702.

Sandra Brünken responded: We have actually used the same level of theory as in the given reference. We agree that SCC-DFT-B cannot tackle electronic excited states. Also, classical MD can hardly describe monomolecular reactivity, which is quite complex when many different atoms are present (here C, N and H). To capture the electronic excitation effects, one should perform time-dependent DFT (TDDFT) simulations, which are, however, very expensive. TDDFTB MD has recently been used to study the energy relaxation of photo-excited PAHs.[1,2] But the timescale is short and no dissociation was observed. Besides, the description of TD-DFTB electronic excited states is challenging. It is true that the fragmentation paths could depend on how the electronic states are populated (see ref. 3 as an example of a dication, and ref. 4) and this has not been investigated in this work. The presented MD simulations do not completely cover all the processes occurring in electron impact ionization. Besides not describing possible ionization, we neglect the excited states' contribution to the process. For PAH radical cations such as cationic cyanonaphthalene, our approach is probably reasonable, assuming the general statement that internal conversion is very effective and that the excited electronic states reached by excitation rapidly decay non-adiabatically to the ground state. The non-radiative lifetime for PAH radical cations in excited electronic states is indeed expected to be very short (less than several tens of fs)[5,6] because of numerous conical intersections, where non-adiabatic decay prevails.[7,8] However, the presence of conical intersections may lead to overexcitation in specific modes, which could orient further fragmentation. In the case of smaller systems such as pyridine and cyanobenzene it would be interesting to probe the role of electronic excited states on the dissociation paths.

1 E. Posenitskiy, *et al.*, *Phys. Chem. Chem. Phys.*, 2019, **21**, 12139–12149, DOI: 10.1039/c9cp00603f.
2 E. Posenitskiy, *et al.*, *J. Chem. Phys.*, 2020, **152**, 074306, DOI: 10.1063/1.5135369.
3 P. López-Tarifa, *et al.*, *Cent. Eur. J. Phys.*, 2014, **12**, 97–102, DOI: 10.2478/s11534-014-0428-0.
4 J. Nirasok, *et al.*, *J. Photochem. Photobiol., A*, 2023, **436**, 114354, DOI: 10.1016/j.jphotochem.2022.114354.
5 S. N. Reddy and S. Mahapatra, *J. Phys. Chem. A*, 2013, **117**, 8737–8749, DOI: 10.1021/jp4033645.
6 S. N. Reddy and S. Mahapatra, *J. Phys. Chem. B*, 2015, **119**, 11391–11402, DOI: 10.1021/acs.jpcb.5b03614.
7 K. F. Hall, *et al.*, *J. Phys. Chem. A*, 2006, **110**, 13591–13599, DOI: 10.1021/jp064711g.
8 A. M. Tokmachev, *et al.*, *J. Phys. Chem. A*, 2008, **112**, 10881–10886, DOI: 10.1021/jp8044109.

Divita Gupta continued discussion of the paper by Ugo Jacovella: As Ian R. Sims already commented, I don't doubt that the cyano species will be formed under low-temperature conditions from the reaction of CN with the parent cyclic molecule given other similar studies.[1–3] However, I wanted to ask if using the cyano species is a good proxy for this molecule. As you provide upper limits for the parent molecule but there is no proof yet that this will hold true under TMC-1 conditions where models have had difficulties predicting the correct abundances.

1 I. R. Cooke, *et al.*, *Astrophys. J., Lett.*, 2020, **891**, L41.
2 J. P. Messinger, *et al.*, *J. Phys. Chem. A*, 2020, **124**(39), 7950–7958.
3 A. J. Trevitt, *et al.*, *J. Phys. Chem. A*, 2010, **114**(4), 1749–1755.

Ugo Jacovella replied: I recognize that if we had only detected CN-NBDs, the link with the abundance of bare NBDs would have been difficult to establish. However, the idea behind the proxy is that the CN substitution greatly enhances the dipole moment and would greatly facilitate detection.

Divita Gupta asked: Is your sample liquid or solid? This will be useful to know as doing (neutral–neutral) reaction studies with liquid/solid samples is generally more challenging.

Ugo Jacovella responded: NBD and CN-NBD are liquid at room temperature. The DCN sample is solid at room temperature.

Alexander K. Lemmens continued discussion of the paper by Arthur G. Suits: You use an intense laser source to incite the reaction by photolysis of one of the precursors to form CN. Does that laser affect the butadiene, either fragmenting it or bringing it to a excited state?

Arthur G. Suits answered: To our knowledge, the photo absorption cross sections for 1,3- and 1,2-butadiene have not been reported at 248 nm, and the reason appears to be that the absorptions there are negligible. The temperature-dependent absorption cross section for 1,3-butadiene has been reported by Fahr and Nayak down to 218 K and 240 nm.[1] There, the cross section is 2.8×10^{-20} cm^2. Even assuming this cross section, the impact on our measured rate would be within our stated error. However, between 235 nm and 240 nm the cross section has dropped by a factor of 3, and from 273 K to 218 K another factor of three. Extrapolating these trends to 248 nm and 70 K strongly suggests there is negligible absorption of the excimer beam by the 1,3-butadiene. Although we can find no absorption data for 1,2-butadiene at 248 nm, where they have both been measured, the 1,2 butadiene absorption is much lower than that of the conjugated isomer.[2] This argues that there will be no impact on the measured rate in that case either.

1 A. Fahr and A. K. Nayak, *Chem. Phys.*, 1994, **189**, 725–731.
2 L. C. Jones and L. W. Taylor, *Anal. Chem.*, 1955, **27**, 228–237.

Joshua Halpern commented: Prof. Heard raised the point that CN produced in the photolysis of BrCN at 248 nm is rotationally hot and this might present some problems for interpreting the results of Prof. Suits' experiment.[1] Parenthetically it is also translationally hot. NCCN at 193 nm is a much cleaner photolytic source of CN radicals ($\nu = 0, 1$) with relatively little vibrational and rotational excitation. If the parent NCCN is cooled, either in a molecular expansion or using a Laval nozzle, the proportion of vibrationally excited CN radicals is smaller and some rotational cooling is observed in the nascent photofragment.[2] A quick search shows that NCCN may be available[3] (probably at high cost) although lecture bottles may still be lurking in old gas closets.

1 J. A. Russell, I. A. McLaren, W. M. Jackson and J. B. Halpern, Photolysis of cyanogen bromide between 193 and 266 nm, *J. Phys. Chem.*, 1987, **91**(12), 3248–3253.
2 R. Lu, J. B. Halpern and W. M. Jackson, Photodissociation of C_2N_2, chlorine cyanide (ClCN), and bromine cyanide (BrCN) in a pulsed molecular beam, *J. Phys. Chem.*, 1984, **88**(16), 3419–3425, DOI: 10.1021/j150660a011.
3 https://pubchem.ncbi.nlm.nih.gov/compound/Cyanogen#section+=+Associated-Chemicals.

Arthur G. Suits responded: In my response to Prof. Heard I discussed the role of possible higher vibrational excitation in CN which we feel we may neglect. As you describe, the CN from BrCN photolysis is both translationally and rotationally excited. In the high density of the flow, the translation and rotation are rapidly relaxed. Nevertheless, we do see, and model, the effects of the rotational relaxation and that is part of our analysis. Although NCCN has some virtues as you describe, application to the study would not be feasible owing to the large absorption cross section of butadienes at 193 nm as discussed in response to Dr Lemmens' question.

Ian R. Sims asked: In your article (https://doi.org/10.1039/d3fd00029j) you state that you mildly focused your excimer photolysis laser. Did you check for multiphoton effects?

Arthur G. Suits replied: We have not specifically checked for multiphoton effects in this case although we have examined other reactions at varying laser power without evidence of any effects. The cross section for BrCN is quite low at 248 nm, hence the use of high laser fluence, but for other reasons discussed above it is the best choice for us now.

Nadia Balucani enquired: Some years ago, in our laboratory we investigated several CN reactions by using a CN beam containing rovibrationally hot CN radicals.[1–4] Differently from your case, we could not see any effect of the vibrational excitation of CN on the reaction mechanism while we could see the effect of rotational excitation.[3,4] More precisely, we saw that isocyanoethylene could be formed with a small yield in the reaction CN + C_2H_4, possibly because excited rotational motion was making the attack on the N-side of CN less unfavorable. Have you had any evidence of isocyano isomer formation in you experiments?

1 P. Casavecchia, N. Balucani, L. Cartechini, G. Capozza, A. Bergeat and G. G. Volpi, *Faraday Discuss.*, 2001, **119**, 27.
2 F. Leonori, K. H. Hickson, S. Le Picard, X. Wang, R. Petrucci, P. Foggi, N. Balucani and P. Casavecchia, *Mol. Phys.*, 2010, **108**, 1097–1113.
3 F. Leonori, R. Petrucci, X. Wang, P. Casavecchia and N. Balucani, *Chem. Phys. Lett.*, 2012, **553**, 1–5.
4 N. Balucani, F. Leonori, R. Petrucci, X. Wang, P. Casavecchia, D. Skouteris, A. F. Albernaz and R. Gargano, *Chem. Phys.*, 2015, **449**, 34–42.

Arthur G. Suits responded: We have not considered the isocyano species, and as we are monitoring the reactant we have no evidence regarding product formation. However, given the energy differences we believe it is unlikely to play a role at low temperature.

Nadia Balucani asked: Does the reaction occur *via* intersystem crossing?

Arthur G. Suits answered: The quartet surface is very high in energy and does not correlate to products, so intersystem crossing is not likely to play any role in this reaction.

Nicholas Walker continued discussion of the paper by Ugo Jacovella: I wonder how easy it is (with a flow cell method) to perform measurements when only a small amount of sample is available, or where you wish to study (for example) a ^{13}C isotopologue in natural abundance. Do you think you have the sensitivity to study ^{13}C isotopologues of dicyano-norbornadiene in natural abundance – or is this out of reach?

Ugo Jacovella replied: Isotopic measurements of non-enriched samples are often challenging when a limited amount of sample is available. In the case of the cyano norbornadiene derivatives, we have reached signal-to-noise ratios (SNRs) of the order of 100 for the strongest lines in about 2 hours of acquisition. It is thus close to what is needed for detecting ^{13}C species in natural abundance, although a proper spectroscopic assignment would require more averaging thus more sample. For other molecules that we have been studying using chirped-pulse spectroscopy, we have observed that ^{13}C species can be detected after a few hours (4–10) of acquisition. Other groups have specialized in such measurements.

Susanna Widicus Weaver remarked: The JPL molecular spectroscopy group has built fully-sealed glass cells so that they can take rotational spectra of rare isotopologues or molecules with very small amounts of available sample. Perhaps this could be implemented for infrared measurements? Even with that setup, isotopologues in natural abundance are very hard to study without enhancing the signal *via* isotopic enrichment.

Ugo Jacovella responded: We agree that spectroscopy of isotopic species in natural abundance is very challenging. The sensitivity offered by chirped-pulse spectroscopy now allows such measurements assuming long acquisitions can be performed (so as to maximize the SNR) and that the line density is low enough that isotopic lines are not below the confusion limit. In the case of NBD and its cyano derivatives, the sample availability prevented deep acquisitions and no search for isotopologues was thus undertaken.

Ewine F. van Dishoeck addressed Brett McGuire: I would like to broaden the discussion for a moment. The papers presented in this session describe beautiful and sophisticated experiments, addressing molecular physics questions in exquisite detail. But the motivation in the introductions of these papers often cites the detection of benzonitrile in the cold dark cloud TMC-1 by McGuire *et al.*[1]

Could Brett comment on whether these experiments help to explain the presence of benzonitrile in TMC-1 and what further experiments would be needed?

1 B. A. McGuire, *et al.*, *Science*, 2018, **359**, 202.

Brett McGuire replied: Certainly the experiments and studies discussed here are quite useful in our quest to understand the chemistry leading to aromatic and polyaromatic chemistry in TMC-1. In my opinion, the biggest current issue is one of scale. These new species are, in terms of complexity, so incredibly far beyond what has been previously considered in detailed chemical models that we simply lack the foundation to understand what we're missing. We need laboratory, modeling, and quantum chemical studies looking at the formation of the first few aromatic rings. These are likely to identify key intermediates, such as the propargyl radical, for which similar laboratory, modeling, and quantum chemical studies need to be performed. There are a few groups in the world valiantly working in these areas (see, *e.g.*, ref. 1 among many others), but the amount of space that needs to be explored is just enormous. We are most likely looking at a slow but steady build-up of the chemical complexity picture over the coming years as more and more species are identified and the critical intermediate species become clear. Observationally, we simply need more information on reactive intermediates and the limits of chemical complexity. Can we make three-ring species in detectable quantities, for example?

1 W. Li, *et al.*, *Phys. Chem. Chem. Phys.*, 2023, **25**, 4141.

Ugo Jacovella asked: Most of the known bottom-up pathways to form benzonitrile have been implemented in the astrochemical model in order to reproduce the abundance of benzonitrile in TMC-1. The model still fails by several orders of magnitude. It seems that top-down routes should be added, but it is true that this requires an enormous amount of work. However, most of the ions observed in TMC-1 are in fact protonated species (perhaps a bias from the available laboratory data?). If protonated benzonitrile is formed by reaction with a proton donor (H_3^+ or/and HCO^+) for instance. Protonated benzonitrile has a facile dissociation path to form phenylium ($C_6H_5^+$),[1] an intermediate known to form benzene. As the reaction of benzene with CN is considered to be the main source of benzonitrile formation, one of the routes of destruction will also be one of the first steps in its reformation. Is this pathway included in the model?

1 U. Jacovella, J. A. Noble, A. Guliani, C. S. Hansen, A. J. Trevitt, J. Mouzay, I. Couturier-Tamburelli, N. Pietri and L. Nahon, *Astron. Astrophys.*, 2022, **657**, A85, DOI: 10.1051/0004-6361/202142206.

Brett McGuire answered: The models used by GOTHAM right now include the first-pass attempts we made for the initial detection for some low-hanging fruit reactions of benzonitrile, as well as one additional update some papers later. Consulting with C. Xue, I. Cooke, and A. Byrne, our models do currently have the destruction of benzonitrile by ions (H_3^+, He^+, HCO^+, and so on), but these all form $C_6H_5^+$ in our network, not protonated benzonitrile. We entirely agree that protonated benzonitrile reactions may be a major missing set of routes and intend to add these as part of a much larger, more dedicated effort to revisit benzonitrile chemistry in our network alongside related species (benzene, benzyne, *etc.*) that are also likely not completely represented yet.

Regarding the inclusion of top-down chemistry, we certainly agree that this would be a valuable route for exploration as well. Without advances in automated network generation tools, however, anything more than a token effort (or perhaps something that includes substantial over-generalization) is likely a PhD-thesis level project (or more), given the complexity of the problem. In particular, it is extremely challenging to decide which larger PAHs to include (as we have little hard evidence to suggest specific species), as well as which breakdown routes would be operable in TMC-1 and with what efficiencies.

Sandra Brünken requested: Could you comment on the prospects to use astronomical observations and obtained molecular abundances, for example in TMC-1, to pin-point specific formation pathways, for example towards benzonitrile? In particular, what are the chances to use the observations to distinguish between solid state and gas phase, neutral–radical and neutral–ion pathways?

Brett McGuire replied: This is tricky. When a chemical mechanism is proposed and implemented in a model, an agreement between the modeled abundances and those derived from observations can certainly suggest that the proposed pathway is reasonable, although I would not say it is proof that mechanism is correct. As well, I would not say that observations can realistically be used to pin-point specific pathways. Rather, when our observations are in disagreement with models and mechanisms, this is often an indication that we are missing a pathway (formation or destruction, gas or grain) or that the kinetics (rate, branching ratios), or thermodynamics of included reactions is not correct. The specific molecules which are over/under predicted might provide clues as to where further efforts should be focused.

Ian R. Sims asked: In a study motivated by your ground-breaking detection of cyanobenzene in TMC-1,[1] we measured the rate constant of the principal reaction responsible for its formation, CN + C_6H_6, down to 15 K, finding it to be very fast and essentially temperature independent over our temperature range (15–295 K).[2] Each study of this kind requires considerable resources and time, and we would appreciate your input to identify the most important targets that we could address in the future, in our case for neutral–neutral reactions.

1 B. A. McGuire, A. M. Burkhardt, S. Kalenskii, C. N. Shingledecker, A. J. Remijan, E. Herbst and M. C. McCarthy, *Science*, 2018, **359**, 202–205, DOI: 10.1126/science.aao4890.
2 I. R. Cooke, D. Gupta, J. P. Messinger and I. R. Sims, *Astrophys. J., Lett.*, 2020, **891**, L41, DOI: 10.3847/2041-8213/ab7a9c.

Brett McGuire responded: At least in GOTHAM, most of our detections are made on –CN substituted species, and thus in practice the rates of formation of these species (including precursors) are going to be the most impactful. That said, it's not clear to me that there should be a great variation in these rates from species to species. If great effort is expended and all the rates are essentially the same as C_6H_6 + CN to form benzonitrile, then in some sense that's not had a huge impact on the models. Perhaps from this perspective, a critical examination of whether these rates are expected to vary significantly from predictions/extrapolations from others would likely be quite worthwhile.

Beyond that, we are currently developing frameworks for detailed sensitivity analyses of our reaction networks to identify those pathways that have the greatest impact on downstream (P)AH abundances. Once identified, we'll certainly highlight these as prime targets for laboratory study.

Cecilia Ceccarelli enquired: TMC-1 is a peculiar region, based on the previous studies which suggest that it is very young, in a stage when gaseous carbon is not yet locked into CO and, hence, it is available for gas-phase reactions to occur. Is this the reason why so many large molecules are found in the gas phase?

Having more sources, evidently, will help to have more constraints and even suggest possible routes of molecule formation. For example, the scheme of the formation of methyl formate from dimethyl ether proposed by Balucani *et al.*[1] was triggered by the tight correlation found between their measured abundances by Jaber *et al.*[2] Of course, I am not saying that a correlation between two species necessarily implies a direct chemical link (mother–daughter or sisters), but it is still important information.

1 N. Balucani, *et al.*, *Mon. Not. R. Astron. Soc.: Lett.*, 2015, **449**, L16.
2 A. A. Jaber, *et al.*, *Astrophys. J.*, 2014, **791**, 29.

Brett McGuire replied: We've shown previously[1] that at least benzonitrile appears to be widespread and abundant in dark molecular clouds, and perhaps even in some protostellar sources (although it is not yet clear whether the benzonitrile is only detected in the remaining dark cloud surrounding the protostar). Additional observations since that time have only reinforced the widespread nature in dark clouds (A. Burkhardt, private communication). Thus, I think it is safe to say there is nothing particularly unique about TMC-1 to indicate that this chemistry is only operative there. Most of our models (and others) put the age of the source somewhere around a few 10^5 years (see, *e.g.*, ref. 2).

1 A. M. Burkhardt, *et al.*, *Nat. Astron.*, 2021, **5**, 181.
2 C. N. Shingledecker, *et al.*, *Astron. Astrophys.*, 2021, **652**, L12.

Martin R. S. McCoustra commented: The polymerisation of acetylene (C_2H_2) *via* Diels–Alder chemistry can occur in the gas phase but is also known to be catalysed on surfaces. Sarre and co-workers[1–3] have reported studies of the formation of benzene (C_6H_6) and polycyclic aromatic hydrocarbons catalysed by the surfaces of likely grain minerals. Plane and co-workers report kinetic measurements on C_6H_6 formation on likely grain surfaces.[4] This may be the most efficient route to these species in some warmer and denser environments.

1 M. Tian, B. S. Liu, M. Hammonds, N. Wang, P. J. Sarre and A. S.-C. Cheung, *Phys. Chem. Chem. Phys.*, 2012, **14**, 6603–6610, DOI: 10.1039/c2cp23309f.
2 M. Tian, B. S. Liu, M. Hammonds, N. Wang, P. J. Sarre and A. S.-C. Cheung, *Philos. Trans. R. Soc., A*, 2013, **371**, 20110590.
3 T. Q. Zhao, Q. Li, B. S. Liu, R. K. E. Gover, P. J. Sarre and A. S.-C. Cheung, *Phys. Chem. Chem. Phys.*, 2016, **18**, 3489–3496, DOI: 10.1039/c5cp06425b.
4 V. L. Frankland, A. D. James, J. D. Carrillo Sánchez, T. P. Mangan, K. Willacy, A. R. Poppe and J. M. C. Plane, *Icarus*, 2016, **278**, 88–99.

Ewine F. van Dishoeck responded: Many thanks for this comment and for the references.

Inga Kamp added: This is something we can also investigate in our thermochemical disk models.

Eric Herbst addressed Brett McGuire: Is it difficult to make atomic and molecular ions in the interstellar medium? It is certainly true that ions are not as abundant as neutral species. In diffuse clouds, the fractional ionization is close to 10^{-4} whereas in dense clouds the number is 10^{-7} or less. But these numbers are low because ions are reactive, not because it is difficult to produce them. Ions are produced copiously by radiation involving UV photons, X-rays, and cosmic rays. Moreover, photon dominated regions (PDRs) occupy large regions of clouds where C^+ is quite abundant.

Brett McGuire answered: I agree it is not difficult to make atomic and molecular ions in the ISM. I think this question may have been in response to my statement that it's quite difficult to make them in easily detectable/study-able quantities in the laboratory, at least for traditional Fourier-Transform Microwave Spectroscopy experiments. That then in turn limits our ability to study them in space, due to the lack of available spectra.

Dwayne E. Heard addressed Cecilia Ceccarelli, Eric Herbst, Robin T. Garrod and Ewine F. van Dishoeck: I would like to draw a parallel with observations in Earth's atmosphere, where the quantitative observation of the abundance of key intermediates, for example the OH radical (which we have measured for over 25 years, see for example ref. 1), have been made in an attempt to understand the chemistry better. Certainly there have been large differences between measurements and models for the composition of Earth's atmosphere, as has also been shown in some of the papers in this session for astrochemical environments. The chemical mechanisms for atmospheric models are now very complex. A strategy that has met with some success has been first to take your equipment to an environment where the chemistry is felt to be relatively simple, for example the remote marine environment, make measurements and compare with models, in order to establish a baseline of understanding for a smaller number of processes. Then after this, move on to environments which are more complex, but perhaps where certain types of chemistry dominate, for example in forested or polar regions, to understand those areas. Finally, then moving onto urban environments where both the emissions and the chemistry are very complex, and use our baseline understanding as a starting point to try to unravel the ever more complex chemistry. Although this is a simplification of course, I wonder if this type of strategy has also been used in an astrochemical context? That is, to try to understand a smaller subset of unknowns first in an environment which is less complex and better characterised, and then build on this understanding for more complex environments?

1 D. Stone, L. K. Whalley and D. E. Heard, *Chem. Soc. Rev.*, 2012, **41**, 6348–6404, DOI: 10.1039/c2cs35140d.

Cecilia Ceccarelli replied: I fully agree with your comment and the approach of proceeding in small steps. This is the approach that our group took in these past years to understand the chemistry of the interstellar complex organic molecules. Instead of using complex models that include a multitude of parameters (relatively poorly constrained) and processes (again, poorly constrained), we consider the smallest problems, for example, a few specific molecules at time, and couple observations with basic modeling, to constrain first the processes. The examples of ethanol, glycolaldehyde and acetaldehyde, discussed in Section 5.1 of my lecture (https://doi.org/10.1039/d3fd00106g), or of formamide, in Section 5.2, aim to show this approach. In my view, more sophisticated models are important to have a general look at the major processes that enter in the final model predictions, like for example the process of molecular cloud formation (the very recent work, for example, by Taillard *et al.*[1]).

1 A. Clément, A. Taillard, V. Wakelam, P. Gratier, J.-C. Loison, E. Dartois, F. Dulieu, J. A. Noble and M. Chabot, *Astron. Astrophys.*, 2023, **675**, A165, DOI: 10.1051/0004-6361/202346188.

Eric Herbst responded: To the best of my knowledge, this approach has been only partially successful for interstellar cloud chemistry. Work on reducing networks, as discussed by Dr Heard, has been studied by Tom Hartquist and collaborators. Look at the work by Ruffle *et al.*[1] in which a reduced network is used for CO production and removal.

1 D. P. Ruffle, *et al.*, *Astron. Astrophys*, 2002, **381**, L13–L16.

Robin T. Garrod said: I would add to Eric's response that dark interstellar clouds are typically taken as the simplest baseline environment for interstellar chemistry in moderately dense conditions, as these objects are typically quiescent (*i.e.* non-star forming), long lived (>10^6 year), and sufficiently well shielded to avoid external UV photons having a significant influence on the chemistry. This has allowed the models to be treated with simple physical/dynamical setups. The chemical networks have tended to do reasonably well over the years in broadly reproducing the abundances of many observed species, typically through gas-phase chemistry. However, the chemistry is still time dependent, with some sources/species showing so called "early-time" or "late-time" behaviour. Our more recent concerns over the feedback of grain-surface chemistry into the gas phase (*e.g.* the methanol problem; see ref. 1) has increased the complexity of the models and demonstrated new areas of uncertainty. But dark clouds, along with diffuse clouds, have tended to be the simplest environments for testing the networks, and will likely remain so, in spite of being perhaps less exciting than star-forming cores or disks.

1 R. Garrod, *et al.*, *Faraday Discuss.*, 2006, **133**, 51.

Ewine F. van Dishoeck answered: This is an excellent point. There are indeed types of interstellar clouds where parts of the chemistry can be tested in detail. The simplest case is that of diffuse and translucent molecular clouds, where one can see through the clouds even at optical wavelengths and where UV photons dominate the chemistry, resulting in very fast time scales to reach chemical equilibrium of only a few hundred years (with the notable exception of H_2 which is formed much more slowly on grain surfaces). In these clouds one can do "precision astrochemistry" at the 30% level, see discussion in van Dishoeck[1] (and references cited therein) and Gerin *et al.*[2] for successes and issues. The next step in complexity is similar benchmark studies of denser PDRs, like the Orion Bar and Horsehead nebula, where UV radiation still controls much of the chemistry and temperature structure.

For cold dark clouds, the TMC-1 core has long been used as a testing ground for basic ion–molecule reactions, see the papers by Eric Herbst, Tom Millar and collaborators as well as the UMIST database papers.[3–5] Here UV radiation is of minor importance, except for that what is produced internally in the cloud through cosmic rays interacting with H_2. As for diffuse clouds, there are successes but also puzzles. The agreement is usually considered "good" if the models are within an order of magnitude of observations. This particular dark cloud is now known to be particularly rich in carbon-bearing molecules (likely because there is considerable free atomic carbon) and not representative of the bulk of the cold dark clouds. Gas–grain reactions and ice formation is taken into account as well, some of which can be tested against observations of ices. The ice-mapping projects now being carried out with JWST toward background stars will be very helpful in further constraining models for the formation of simple ices.

The final step is going to more complex regions such as protostars and protoplanetary disks where there are large gradients in temperatures, densities and UV irradiation, and where a combination of all these processes plays a role. See the van Dishoeck review (and references cited therein) for a discussion.[1]

1 E. F. van Dishoeck, *Faraday Discuss.*, 2014, **168**, 9.
2 M. Gerin, *et al.*, *Annu. Rev. Astron. Astrophys.*, 2016, **54**, 181.
3 E. Herbst, N. G. Adams and D. Smith, *Astrophys. J.*, 1984, **285**, 618.
4 R. Terzieva and E. Herbst, *Astrophys. J.*, 1998, **501**, 207.
5 T. J. Millar, P. R. A. Farquhar and K. Willacy, *Astron. Astrophys., Suppl. Ser.*, 1997, **121**, 139.

Serena Viti opened discussion of the paper by Nadia Balucani: One of your conclusions is that it would be good to know if a reaction is important (under ISM conditions) before investing large amounts of time investigating it experimentally. While traditionally sensitivity analyses may have provided some help there, I wanted to point out that in the past few years we have investigated and published various proofs of concepts based on Bayesian, MCMC and ML techniques to determine how important a reaction is, or rather, how sensitive abundances may be to the rate of that reaction.[1–7] These techniques are very promising and ultimately may be the best way to optimise experimental efforts.

1 A. Makrymallis and S. Viti, *Astrophys. J.*, 2014, **794**, 45, DOI: 10.1088/0004-637X/794/1/45.
2 J. Holdship, N. Jeffrey, A. Makrymallis, S. Viti and J. Yates, *Astrophys. J.*, 2018, **866**, 116, DOI: 10.3847/1538-4357/aae1fa.
3 D. de Mijolla, S. Viti, J. Holdship, I. Manolopoulou and J. Yates, *Astron. Astrophys.*, 2019, **630**, A117, DOI: 10.1051/0004-6361/201935973.
4 J. Heyl, S. Viti, J. Holdship and S. M. Feeney, *Astrophys. J.*, 2020, **904**, 197, DOI: 10.3847/1538-4357/abbeed.
5 J. Holdship, S. Viti, T. J. Haworth and J. D. Ilee, *Astron. Astrophys.*, 2021, **653**, A76, DOI: 10.1051/0004-6361/202140357.
6 J. Heyl, J. Holdship and S. Viti, *Astrophys. J.*, 2022, **931**, 26, DOI: 10.3847/1538-4357/ac6606.
7 J. Heyl, E. Sellentin, J. Holdship and S. Viti, *Mon. Not. R. Astron. Soc.*, 2022, **517**, 38, DOI: 10.1093/mnras/stac2652.

Nadia Balucani responded: These new approaches will be certainly of great help.

Olivia H. Wilkins opened discussion of the paper by Divita Gupta: Given it is unlikely to detect $C_3H_2D^+$ in space, are there any plans to measure the doubly-deuterated isotopologue of this cation? What other ions you are targeting next, and can you please comment on whether it is likely they will be detected in space?

Divita Gupta answered: It is in our potential future plans. However, given that it has an even lower dipole moment than the singly deuterated version and that a second deuteration step suggests lower abundance in space, it will be a difficult candidate to search for in the near future, at least based on its rotational spectrum.

Adam J. Fleisher asked: In leak-out spectroscopy (LOS), you mention the neutral-to-ion mass ratio as enabling vibrational energy transfer from vibrationally excited species into kinetic energy. Over what range of mass ratios can you perform LOS? What experimental "knobs" can you adjust to explore different mass ratios (*e.g.*, changing buffer gas or floating potential of the trap)?

Divita Gupta replied: Currently, we are testing the limitations of LOS in our laboratories. It seems that it is generally applicable. Also, higher ion masses are accessible by simply using N_2, Ar, Kr, or even Xe as collision partners (we tested already N_2 and Ar). One of the heaviest molecules studied so far is $HCCCO^+$,[1] where we operate with a mass ratio of $53:20 = 2.65$ ($HCCCO^+$: Ne). With Xe, we can thus imagine going to masses as high as $132 \times 2.65 = 350$ or even beyond.

1 O. Asvany, *et al.*, *Phys. Chem. Chem. Phys.*, 2023, **25**, 19740–19749, DOI: 10.1039/d3cp01976d.

Nikku Madhusudhan addressed Nadia Balucani and Cristina Puzzarini: New observations with JWST are expected to probe exoplanetary atmospheres with lower temperatures than observed before, but mostly H_2-rich atmospheres which are more observable. One of the key challenges in the interpretation of such spectra is that for many molecules which we expect to detect we don't have good absorption line lists. The line lists we do have available, *e.g.* from HITRAN, are applicable for terrestrial-like applications, but not necessarily for more diverse conditions. So, my broad question is what is the range of macroscopic parameters (P, T, spectral range) over which experiments can measure accurate line lists for new molecules? For example, if we want a new absorption line list for a molecule, *e.g.* like benzene, between 0.9 and 20 micron that is applicable over a broad range of temperatures and pressures, what are the challenges we might face in obtaining such a line list?

Nadia Balucani responded: This is not my field of expertise. In general, at high pressures there is a significant collision broadening while at high temperatures there is Doppler broadening as well as contributions from hot bands. I am aware that some effort to measure these spectra is being made in several research groups.

Cristina Puzzarini replied: To obtain an accurate line list for a given ro-vibrational band in T, P conditions very different from those on Earth, its spectrum needs to be recorded at a good resolution at different pressure values as well as different temperatures. This allows for determining the pressure-broadening and pressure-shift parameters and their temperature dependence. In fact, while the transition frequencies do not depend on temperature and pressure, the actual spectrum is made by features that are broadened by different effects, one of them being collisions (pressure). This can also cause a frequency shift of the maximum of the broadened feature with respect to the transition recorded without broadening effects (or them reduced to the minimum). Therefore, once we obtain from laboratory measurements the line catalog together with the pressure-broadening parameters and their temperature dependence, we can simulated the ro-vibrational band in any type of environment. However, getting all this information is a huge amount of work. Even if we consider only fundamental bands, the investigation described above has to be repeated for all them, which are $3N - 6$ ($3N - 5$ for linear molecules), where N is the number of atoms. The situation is further complicated by the fact that overtones and combination bands might need to be studied. The good news is that the frequency line catalogs obtained from the investigation of the ro-vibrational spectrum at a given pressure and temperature is often sufficient to guide the assignment of astronomically recorded spectra. The effect of the temperature can be modelled by adjusting the temperature-dependent parameters. Given the line intensities, the pressure effects can be simulated by convoluting the stick spectrum (*i.e.* line positions and their corresponding intensities) with a Gaussian profile with a reasonable linewidth (where reasonable refers to the conditions of interest combined with the experience on pressure-broadening effects).

Susanna Widicus Weaver opened discussion of the paper by Cristina Puzzarini: Even collection of and assignment of high resolution rovibrational spectra at one temperature is a time-consuming process. Extending this to the range of temperature and density conditions that we need to fully interpret JWST spectra is a huge challenge for the field. Can we implement the assignment-free spectral simulation approaches explored by Frank DeLucia[1] for rotational spectra to rovibrational spectra?

1 I. R. Medvedev and F. C. De Lucia, An Experimental Approach to the Prediction of Complete Millimeter and Submillimeter Spectra at Astrophysical Temperatures: Applications to Confusion-limited Astrophysical Observations, *Astrophys. J.*, 2007, **656**, 621.

Cristina Puzzarini answered: I do not think this is feasible. Ro-vibrational spectra combine together the complexity of rotational and vibrational spectra. However, in this respect, simulation tools might significantly help the analysis.

Arthur G. Suits continued discussion of the paper by Nadia Balucani: My question has two parts. First, does your beam contain any triplet N_2(A $^3\Sigma_u^+$)? Second, do you know whether this species is considered in models of Titan's upper atmosphere and whether it plays a significant role?

Nadia Balucani responded: We characterized our beam *via* Stern–Gerlach magnetic analysis and verified that some of the undissociated molecular nitrogen is present in the A $^3\Sigma_u^+$ excited state.[1] Unfortunately, since the ground singlet state is diamagnetic, the employed technique did not allow us to infer the relative abundance of N_2 in the two states. Nevertheless, in all the reactions that we have investigated with our experimental approach (ranging from reactions with simple molecules like H_2 or H_2O to more complex species like pyridine, benzene, ethene, propadiene or propyne, see ref. 2–7 and the current paper https://doi.org/10.1039/d3fd00057e) no interference from possible reactive N_2(A $^3\Sigma_u^+$) channels leading to products containing a novel C–N or N–H bond has ever been observed. This is in line with what is known from other experiments where the collisions of N_2(A $^3\Sigma_u^+$) with various molecular targets were not seen to induce the N_2 bond breakage (with subsequent formation of N-containing products). Rather, simple physical quenching or dissociation of the target molecule *via* dissociative energy transfer have always been observed (for a recent review see ref. 8).

Concerning a possible role of N_2(A $^3\Sigma_u^+$) in the upper atmosphere of Titan, the Vegard–Kaplan system was seen to contribute to the Titan limb airglow spectrum recorded by the Cassini Ultraviolet Imaging Spectrograph (UVIS) in the FUV region.[9,10] Therefore, the N_2(A $^3\Sigma_u^+$) abundance reaches an appreciable value *via* electron impact or cascade from other triplet states (the transition from the ground state upon the absorption of UV photons is forbidden). To model the N_2 state population in the upper atmosphere of Titan, since the A $^3\Sigma_u^+$ state has a lifetime of 2.37 s, Lavvas *et al.*[11] have recently considered the effects of the collisions involving excited N_2 and abundant hydrocarbons (this is because the quenching by N_2 itself is quite inefficient, see ref. 8 for a critical review). Bimolecular collisions with CH_4, C_2H_6, C_2H_2, and C_2H_4 were included with the rate coefficients recommended by Dutuit *et al.*[8] While in the case of CH_4, physical quenching is considered the only significant process, in the case of the reactions with C_2H_2 (or C_2H_4) the main channels are those leading to N_2 in the ground state in a dissociative energy transfer process producing either C_2H (or C_2H_3) + H or C_2 (or C_2H_2) + H_2 as demonstrated by the experimental work of Umemoto *et al.*[12] In no case has there been any evidence that product species containing a novel C–N bond are formed.

1 M. Alagia, V. Aquilanti, D. Ascenzi, N. Balucani, D. Cappelletti, L. Cartechini, P. Casavecchia, F. Pirani, G. Sanchini and G. G. Volpi, *Israel J. Chem.*, 1997, **37**, 329–342.
2 N. Balucani, M. Alagia, L. Cartechini, P. Casavecchia, G. G. Volpi, L. A. Pederson and G. C. Schatz, *J. Phys. Chem. A*, 2001, **105**, 2414–2422.
3 Z. Homayoon, J. M. Bowman, N. Balucani and P. Casavecchia, *J. Phys. Chem. Lett.*, 2014, **5**, 3508–3513.
4 P. Recio, D. Marchione, A. Caracciolo, V. J. Murray, L. Mancini, M. Rosi, P. Casavecchia and N. Balucani, *Chem. Phys. Lett.*, 2021, **779**, 138852.
5 N. Balucani, D. Skouteris, F. Leonori, R. Petrucci, M. Hamberg, W. D. Geppert, P. Casavecchia and M. Rosi, *J. Phys. Chem. A*, 2012, **116**, 10467–10479.
6 G. Vanuzzo, L. Mancini, G. Pannacci, P. Liang, D. Marchione, P. Recio, Y. Tan, M. Rosi, D. Skouteris, P. Casavecchia, N. Balucani, K. M. Hickson, J.-C. Loison and M. Dobrijevic, *ACS Earth Space Chem.*, 2022, **6**, 2305–2321.
7 L. Mancini, G. Vanuzzo, D. Marchione, G. Pannacci, P. Liang, P. Recio, M. Rosi, D. Skouteris, P. Casavecchia and N. Balucani, *J. Phys. Chem. A*, 2021, **125**, 8846–8859.
8 O. Dutuit, *et al.*, *Astrophys. J., Suppl. Ser.*, 2013, **204**, 20.
9 M. H. Stevens, J. Gustin, J. M. Ajello, J. S. Evans, R. R. Meier, A. J. Kochenash, A. W. Stephan, A. I. F. Stewart, L. W. Esposito, W. E. McClintock, G. Holsclaw, E. T. Bradley, B. R. Lewis and A. N. Heays, *J. Geophys. Res.*, 2011, **116**, A05304.
10 A. Bhardwaj and S. K. Jain, *Icarus*, 2012, **218**, 989–1005.

11 P. Lavvas, R. V. Yelle, A. N. Heays, L. Campbell, M. J. Brunger, M. Galand, V. Vuitton, *Icarus*, 2015, **260**, 29–59.
12 H. Umemoto, *J. Chem. Phys.*, 2007, **127**, 014304.

Ugo Jacovella continued discussion of the paper by Divita Gupta: Can the leak out spectroscopy technique be used to record electronic spectra?

Divita Gupta answered: In principle, yes it should be possible. But systematic studies will be needed to understand the efficiency of electronic-to-kinetic energy transfer.

Sandra Brünken added: I would assume that LOS for electronic spectroscopy is not as efficient as for vibrational spectroscopy due to a lower quenching rate for electronic transitions, *i.e.*, less energy transfer from the internal excitation to kinetic, translational energy. For systems with fast internal conversion and internal vibrational redistribution such as PAHs it might however work, and it would be interesting to test this. However, LOS has so far not yet been applied to such large systems even for vibrational spectroscopy; likely heavier collision partners are needed in this case for efficient energy transfer to the ionic collision partner.

Murthy S. Gudipati addressed Ugo Jacovella: How does your method differ from ZEKE (Zero Kinetic Energy) Electronic Spectroscopy?

Ugo Jacovella replied: I think this question is about the possibility of using the Schlemmer group's leakage spectroscopy (LOS) technique for electronic spectroscopy of molecular ions in the context of the diffuse interstellar bands (DIBs).

DIBs are absorption characteristics, most likely corresponding to transitions from the electronic ground state to an excited electronic state. The laboratory measurement of such a direct absorption spectrum can therefore be directly compared with astronomical observations.

In the case of PFI-ZEKE, this is photoelectron spectroscopy. The transitions measured correspond to transitions between the electronic ground state of the neutral molecule and the electronic states of the cation. The spectra cannot be compared directly, but the band origins and vibrational spacings can be extracted from the PFI-ZEKE spectra. However, the selection rules are different and the direct absorption of the cation is often due to the excitation of non-Koopman electronic states. In photoelectron spectroscopy, these states are generally not accessible. PFI-ZEKE spectroscopy is a fantastic tool, but it is not very well suited to solving the DIB puzzle.

Murthy S. Gudipati continued discussion of the paper by Divita Gupta: How do you cool the ions in your ion trap? By controlling the flow of cold buffer gas?

Divita Gupta answered: A helium (He) pulse enters through a piezoelectric valve after the ions are loaded into the trap. Sufficient collisions of ions with the He pulse ensure the thermalization of ions to the cryogenic temperatures.

Murthy S. Gudipati asked: The ion leak spectroscopy is an excellent and sensitive method – congratulations. My question is: as ions are vibrationally excited and subsequently undergo collisions with buffer gas, gain translational energy and escape the trap's threshold potential, after each resonant vibrational excitation, the number of ions trapped decreases in the ion-trap. Do you continuously supply ions into the trap to keep a constant ion population in the trap? Or, do you load ions after each resonant excitation? If the latter, this would be a time-consuming and painstaking procedure. Is it theoretically possible to continuously buffer ions into the trap, making measurement of a full ro-vibrational spectrum much easier?

Divita Gupta replied: The ions are loaded (at around 1 Hz) for each cycle and we optimize such that the total ions loading into the trap are similar for every cycle. But you are correct that leak-out is essentially a "destructive" method as the ions leak out of the trap. But since the repetition rate is fast enough and we use a very low number of ions (around 10 000 for each cycle), the amount of precursor used is very low. So, we are not too worried even when using the less-abundant isotopologues. But in principle, yes we can use the same sample multiple times, and that is one of the future plans. This way we can kick out ions of one kind from our trap and use the remaining for kinetics or even spectroscopy.

Murthy S. Gudipati continued discussion of the paper by Cristina Puzzarini: How do you model temperature gradients across your furnace tube to get accurate representation of temperature distribution (using a single rotational transition)? Or, do you assume one single uniform temperature in your model?

Cristina Puzzarini answered: When recording the rotational spectrum of molecules produced on-the-fly using flash vacuum pyrolysis, we are not interested in what the temperature of the gas inside the cell is or if there is a temperature gradient inside the cell. The reason is that we aim at measuring the transition frequencies which are not at all affected by temperature. Once the rotational line positions are known, their intensity can be easily and accurately evaluated at the temperature of interest.

Murthy S. Gudipati asked: When you focus on one rotational transition only, wouldn't it change (its intensity) with temperature? If yes, how do you calibrate to temperature or how do you derive the temperature of that molecule from that single rotational transition?

Cristina Puzzarini responded: Temperature has indeed a relevant impact on the intensity of rotational transitions, the main reason being the population of rotational levels (which affects the partition function). However, the intensity of rotational transitions at a given temperature can be easily and accurately obtained once the components of the electric dipole moments are known.[1,2] These can be experimentally (*e.g.* Stark spectroscopy) or theoretically (*ab initio* calculations) derived to the required accuracy.

1 C. Puzzarini, *Phys. Chem. Chem. Phys.*, 2013, **15**, 6595, DOI: 10.1039/c3cp44301a.

2 C. Puzzarini, J. Bloino, N. Tasinato and V. Barone, *Chem. Rev.*, 2019, **119**, 8131, DOI: 10.1021/acs.chemrev.9b00007.

Murthy S. Gudipati continued discussion of the paper by Nadia Balucani: Are there atomic nitrogen isotope enrichment observational data available for Titan's upper atmosphere?

Nadia Balucani answered: Yes, there are indeed several interesting studies about that. The $^{14}N/^{15}N$ ratio has been determined for molecular nitrogen and also for several nitriles.[1–5] These data are extremely useful to understand the evolution and the chemistry of Titan. For instance, mass-dependent fractionation during atmospheric escape has been modeled by Mandt *et al.*[6] and led to the conclusion that N_2 must have originated from the ammonia ice formed in the protosolar nebula. Other studies have focused on nitriles. The $^{14}N/^{15}N$ ratio for N_2 in the atmosphere of Titan is a factor of 2 higher than the corresponding ratio for HCN.[1–3] This difference was explained primarily by the photolytic fractionation of $^{14}N^{14}N$ and $^{14}N^{15}N$.[7]

A similar $^{14}N/^{15}N$ ratio is expected for species like HC_3N and C_2H_3CN that are supposed to be formed after dissociation of HCN, while a different $^{14}N/^{15}N$ ratio was predicted for other nitriles like CH_3CN by a recent photochemical model.[8] According to that model, CH_3CN and C_2H_5CN are formed by a mechanism that does not depend on the UV photodissociation of N_2, being rather dominated by the formation of N atoms induced by galactic cosmic rays at lower altitude. The recent $^{14}N/^{15}N$ isotopic ratio in CH_3CN measured with ALMA[5] seems to support this scenario.

1 H. B. Niemann, S. K. Atreya, S. J. Bauer, G. R. Carignan, J. E. Demick, R. L. Frost, D. Gautier, J. A. Haberman, D. N. Harpold, D. M. Hunten, G. Israel, J. I. Lunine, W. T. Kasprzak, T. C. Owen, M. Paulkovich, F. Raulin, E. Raaen and S. H. Way, *Nature*, 2005, **438**, 779–784.
2 A. Marten, T. Hidayat, Y. Biraud and R. Moreno, *Icarus*, 2002, **158**, 532–544.
3 E. M. Molter, C. A. Nixon, M. A. Cordiner, J. Serigano, P. G. J. Irwin, N. A. Teanby, S. B. Charnley and J. E. Lindberg, *Astron. J.*, 2016, **152**, 42.
4 M. A. Cordiner, C. A. Nixon, S. B. Charnley, *et al.*, *Astrophys. J.*, 2018, **859**, L15.
5 T. Iino, H. Sagawa and T. Tsukagoshi, *Astrophys. J.*, 2020, **890**, 95.
6 K. E. Mandt, O. Mousis, J. Lunine and D. Gautier, *Astrophys. J. Lett.*, 2014, **788**, L24.
7 M. C. Liang, A. N. Heays, B. R. Lewis, S. T. Gibson and Y. L. Yung, *Astrophys. J.*, 2007, **664**, L115.
8 M. Dobrijevic and J. C. Loison, *Icarus*, 2018, **307**, 371–379.

Nicholas Walker continued discussion of the paper by Divita Gupta: You mentioned that the measurement of the spectrum is quite fast, and now the bottleneck is assignment. Your paper (https://doi.org/10.1039/d3fd00068k) describes using PGOPHER for assignment and fitting. Have you experimented with the "Autofit" capability that PGOPHER has? These "Autofit" algorithms were originally developed by Pate and co-workers at the University of Virginia for broadband microwave spectroscopy as "open source" – and were later integrated into PGOPHER by Colin Western. Some microwavers have employed Autofit for accelerated assignment. Autofit for PGOPHER is briefly described here: https://pgopher.chm.bris.ac.uk/Help/autofitintro.htm and more fully described in the publication of Western and Billinghurst.[1] The paper describes application of the Autofit capability of PGOPHER to the rotational analysis of the ν_{11} band of *cis*-1,2-dichloroethene.

1 C. M. Western and B. E. Billinghurst, Automatic assignment and fitting of spectra with PGOPHER, *Phys. Chem. Chem. Phys.*, 2017, **19**, 10222, DOI: 10.1039/c7cp00266a.

Divita Gupta replied: We have not yet played with the "Autofit" feature of PGOPHER. This will be especially helpful in cases where we have an overlap of lines from other bands/combination bands/hot bands. We will certainly explore this in the near future.

Susanna Widicus Weaver commented: Nick Walker suggested you use Autofit to try and make assignments to other species or vibrational states in your spectrum. The Autofit program performs a triples fitting algorithm to sets of 3 lines in the spectrum and then gives an approximation of the rotational constants for all triples. One then matches the possible rotational constants to species that could be present in the mixture. It is not currently possible to do this with rovibrational spectra. But it is worth asking Steven Shipman if this could be implemented.

Adam J. Fleisher noted: This is a further comment on the AUTOFIT[1] discussion. There exists genetic algorithm (GA) routines for fitting rotationally resolved electronic spectra which may be applied to ro-vibrational spectra as well. These include the GA applied by Leo Meerts[2,3] and those integrated into the JB95 program by David Plusquellic.[4,5]

1 N. A. Seifert, I. A. Finneran, C. Perez, D. P. Zaleski, J. L. Neill, A. L. Steber, R. D. Suenram, A. Lesarri, S. T. Shipman and B. H. Pate, *J. Mol. Spectrosc.*, 2015, **312**, 13–21.
2 J. A. Hageman, R. Wehrens, R. de Gelder, W. Leo Meerts and L. M. C. Buydens, *J. Chem. Phys.*, 2000, **113**, 7955–7962.
3 M. Schmitt and W. L. Meerts, in *Handbook of High-resolution Spectroscopy*, ed. M. Quack and F. Merkts, 2011, pp. 1345–1371.
4 J. C. Keske, W. Lin, W. C. Pringle, S. E. Novick, T. A. Blake and D. F. Plusquellic, *J. Chem. Phys.*, 2006, **124**, 074309.
5 D. F. Plusquellic, JB95 (version v6.02.00), National Institute of Standards and Technology, Boulder, CO, 2016.

Divita Gupta responded: We will look to explore this in the near future.

Martin Cordiner continued discussion of the paper by Nadia Balucani: 1-Cyano-1,3-cyclopentadiene (C_6H_5N) has molecular mass 91 and the protonated ion has mass 92. Interestingly, a mass peak at 92 amu q^{-1} was measured by the Cassini INMS instrument during its flyby through Titan's atmosphere, so the protonated version of this molecule may have already been detected. C_6H_5N obviously has several isomers, so I'm wondering if high-frequency (millimeter-wave) rotational spectroscopy might be available for the various isomers, to enable a spectroscopic search using ALMA?

Nadia Balucani replied: There are three low lying cyanocyclopentadiene isomers: 1-, 2-, and 5-cyanocyclopentadiene.[1] Rotational spectroscopy information is available.[1–3] Indeed, 1- and 2-cyanocyclopentadiene have been recently identified in the dense cloud TMC-1 using radio observations with the Green Bank Telescope.[4,5] According to our photochemical model in the paper (https://doi.org/10.1039/d3fd00057e) and considering its large dipole moment, 1-

cyanocyclopendadiene should be detectable using a sensitive remote radiotelescope.

1 M. C. McCarthy, K. L. K. Lee, P. B. Carroll, J. P. Porterfield, P. B. Changala, J. H. Thorpe and J. F. Stanton, *J. Phys. Chem. A*, 2020, **124**, 5170–5181.
2 R. G. Ford and H. A. Seitzman, *J. Mol. Spectrosc.*, 1978, **69**, 326–329.
3 T. Sakaizumi, H. Kikuchi, O. Ohashi and I. Yamaguchi, *Bull. Chem. Soc. Jpn.*, 1987, **60**, 3903.
4 M. C. McCarthy, K. L. K. Lee, R. A. Loomis, A. M. Burkhardt, C. N. Shingledecker, S. B. Charnley, M. A. Cordiner, E. Herbst, S. Kalenskii, E. R. Willis and C. Xue, *Nat. Astron.*, 2021, **5**, 176–180.
5 K. L. K. Lee, P. B. Changala, R. A. Loomis, A. M. Burkhardt, C. Xue, M. A. Cordiner, S. B. Charnley, M. C. McCarthy and B. A. McGuire, *Astrophys. J., Lett.*, 2021, **910**, L2.

Ugo Jacovella continued discussion of the paper by Cristina Puzzarini: Step 1 of your approach (https://doi.org/10.1039/d3fd00052d) is to compute accurate spectroscopic parameters. Is this usually used as a tool to select the molecular targets to be measured in the lab. Or the target species are rather selected from astrochemical intuition?

Cristina Puzzarini answered: In my lab, we follow two strategies to select the molecules to be characterized using rotational spectroscopy with the final aim to search for them in space. First, as you mentioned, we use astrochemical intuition. These are often molecules that come out as suggestions from discussions with astronomers. The other strategy relies on the studies of interstellar gas-phase reactivity we carry out computationally. The investigation of gas-phase reactions often suggests new species that might be present in the ISM.

Ugo Jacovella asked: Few detections of linear molecular have been successfully made on the basis of theoretical predictions. To what extent can small non-linear (rigid) molecules be detected solely on the basis of quantum chemical calculations?

Cristina Puzzarini responded: Quantum-chemical calculations even when pushed to the limit cannot reach the accuracy usually required to assign astronomical spectra. These are often very complex and full of lines, thus requiring a frequency accuracy by far better than 1 MHz (usually 100 kHz). Such an accuracy can be obtained only experimentally. However, in a few cases, when dealing with linear molecules (thus showing very simple rotational spectra) and very sparse astronomical spectra, very accurate quantum-chemical predictions can be then sufficient to unequivocally detect a new species.

Susanna Widicus Weaver commented: Computational studies might be good enough to enable a molecular detection *via* direct comparison to observations in a source with a very low line density. But for hotter star-forming cores, the spectral line density is too great. Experimental measurements are essential.

Ugo Jacovella continued discussion of the paper by Divita Gupta: Would it be possible to replace the detection system with one that is sensitive to the kinetic energy of the ions (VMI?), so that the constant background noise of ions that are ejected without laser excitation could be reduced?

Divita Gupta replied: We currently do not have a velocity map imaging (VMI) detector available in our lab. But that is one of our future plans as that will really allow us to understand the kinetic energy distribution and its change on laser excitation. The background ions we have right now can be further limited by increasing the exit barrier under our current setup conditions too but a constant (small) background allows us to ensure that we do not miss any weak peak (where the kinetic energy change might be too low so that the barrier despite the increase can't be overcome).

Eric M. Crump asked: In leak out spectroscopy, is there a relationship between specific rovibrational states and the efficiency of kinetic energy transfer that would have an effect on measured counts?

Divita Gupta answered: Based on preliminary experimental results, we don't see any clear proof as of yet. But it would be fair to assume that the efficiency of vibrational to kinetic energy transfer depends on the vibrational mode. But this needs future systematic studies.

Arthur G. Suits enquired: I am interested in the hot bands you describe. How do you know when you have hot bands? Might it be possible to use them to study reactivity of vibrationally excited species?

Divita Gupta responded: Since we have predicted/computed band centers for our species of interest, we can tell if some features could be coming from hot bands. We have seen so for a couple of species measured around 20–40 K. Since we only have mass selectivity, we will have both the ground vibrational state and excited state in our trap. So, we would also need to know the quantitative measurement of the amount of excited *vs.* the ground state as a reactant. Comparing the intensities of features from each corresponding species could help with this or by varying the temperature (which will alter the ratio of ground/excited) it should be possible to study the reactivity of the two states.

Dwayne E. Heard continued discussion of the paper by Nadia Balucani: Your ability to measure both a rate coefficient together with branching ratios is a very nice feature of your experiments, and a significant advantage over other methods that typically only measure the total rate coefficient, or perhaps detect a product but cannot quantify the yield of it. The theory for your reaction predicts that the branching ratio is not changing much with lowering temperature, so I wondered if you could vary your collision energy, as it would be nice to have an experimental verification of this, although I appreciate that these would be difficult experiments.

Nadia Balucani replied: As a matter of fact, with the crossed molecular beam technique coupled to mass spectrometric detection and time-of-flight analysis we measure the product branching fractions (BFs) at specific collision energies rather than temperatures. This is why it is so important to combine our experimental results with the predictions of a theoretical treatment. The idea is that if the theory is good enough to reproduce our detailed experimental values, the potential energy surface and the statistical approach based on RRKM calculations (for indirect reactions) are reliable also for other values of the collision energy (in the microcanonical approach) or temperature (in the canonical approach). Concerning the possibility of varying our experimental conditions, indeed we can explore a limited range of collision energies either by resorting to different seeding gases (see for instance ref. 1–4) or by changing the crossing angle (we have three different experimental setups that allow us to cross the two beams at 45°, 90° and 135°).[5–9] In general, we can say that the effect of the temperature (or collision energy) on the branching fractions of indirect reactions is not expected to be large if the reaction channels are quite exothermic and all the involved transition states are well below the energy of the reactants. Important effects, instead, can be seen when the reaction channels are nearly thermoneutral or one or more transition states are close to the energy of the reactants asymptote or above.[5,6] Another case where the collision energy (or temperature) can significantly affect the product BFs is when the reaction is governed by processes like intersystem crossing as it is in the case of $O(^3P)$ reactions.[7–11]

1 N. Balucani, D. Stranges, P. Casavecchia and G. G. Volpi, *J. Chem. Phys.*, 2004, **120**, 9571.
2 N. Balucani, A. Bergeat, L. Cartechini, G. G. Volpi, P. Casavecchia, D. Skouteris and M. Rosi, *J. Phys. Chem. A*, 2009, **113**, 11138–11152.
3 M. Alagia, N. Balucani, L. Cartechini, P. Casavecchia, M. van Beek, G. G. Volpi, L. Bonnet and J. C. Rayez, *Faraday Discuss.*, 1999, **113**, 133.
4 Z. Homayoon, J. M. Bowman, N. Balucani and P. Casavecchia, *J. Phys. Chem. Lett.*, 2014, **5**, 3508–3513.
5 F. Leonori, R. Petrucci, E. Segoloni, A. Bergeat, K. H. Hickson, N. Balucani and P. Casavecchia, *J. Phys. Chem. A*, 2008, **112**, 1363–1379.
6 M. Costes, N. Daugey, C. Naulin, A. Bergeat, F. Leonori, E. Segoloni, N. Balucani and P. Casavecchia, *Faraday Discuss.*, 2008, **133**, 157–176.
7 N. Balucani, F. Leonori, P. Casavecchia, B. Fu and J. M. Bowman, *J. Phys. Chem. A*, 2015, **119**, 12498–12511.
8 B. Fu, Y.-C. Han, J. M. Bowman, F. Leonori, N. Balucani, L. Angelucci, A. Occhiogrosso, R. Petrucci and P. Casavecchia, *J. Chem. Phys.*, 2012, **137**, 22A532.
9 P. Casavecchia, F. Leonori and N. Balucani, *Int. Rev. Phys. Chem.*, 2015, **34**, 161–204.
10 A. Caracciolo, G. Vanuzzo, N. Balucani, D. Stranges, S. Tanteri, C. Cavallotti and P. Casavecchia, *Chin. J. Chem. Phys.*, 2019, **32**, 113–122.
11 C. Cavallotti, A. Della Libera, C.-W. Zhou, P. Recio, A. Caracciolo, N. Balucani and P. Casavecchia, *Faraday Discuss.*, 2022, **238**, 161–182.

Dwayne E. Heard questioned: In your paper (https://doi.org/10.1039/d3fd00057e) you compare the results of the current reaction of $N(^2D)$ + benzene, with the results you previously obtained for the reaction of $N(^2D)$ atoms plus pyridine. The results are quite different in terms of the branching ratios between the different modes of attack onto the aromatic ring and the subsequent product yields. Do you plan to perform any theory on the N plus pyridine reaction, as that would be very useful to understand the different behaviour for the 2 reactions? Do you have some ideas for the reason(s) for the different behaviour in the measured product branching ratios between N plus benzene and N plus pyridine?

Nadia Balucani answered: The calculations of the potential energy surface (PES) for reactions involving aromatic molecules are quite challenging *per se*. In addition, the reactions involving $N(^2D)$ start very high in energy and that means that there are many possible rearrangements that are energetically possible. The complexity of the PES for the $N(^2D)$ + benzene reaction clearly demonstrates that (see ref. 1 and the current paper: https://doi.org/10.1039/d3fd00057e). The PES for the reaction N(^{2}D) + pyridine will be much more complex because of the symmetry loss and the presence of different routes starting from the possible attack in the *ortho*, *meta*, *para* and also *ipso* positions. In this respect, it is quite illustrative to compare the reactions O + benzene (ref. 2 and 3) and O + pyridine (ref. 4). We are also calculating the PES for N(^{2}D) + toluene,[5] where again the symmetry of benzene is lost, and it is definitely much more complex being possible, in that case as well, to start with the attack in the *ortho*, *meta*, *para* and *ipso* positions.

By inspecting the PES for the N(^{2}D) + benzene reaction, we can note that the product BFs are regulated by (1) the competition between the formation of intermediate i2 and intermediate i4 (both formed by the isomerization of the initial intermediate i1) and (2) the fate of the seven-member ring *N*-cycloheptatriene intermediate (i12) that can decompose *via* three main competitive pathways: the main one leading to C_5H_5 + HCN (*via* a series of four isomerizations), the second one leading to *ortho-/para-N*-cycloheptatriene radical + H and the third one leading to C_4H_4N (pyrrolyl radical) + C_2H_2 (acetylene). Clearly, in the case of the N(^{2}D) pyridine reaction the characteristics of the analogous intermediates and transition state will favour one or more of the H-displacement channels for the dominant initial approach(es) thus explaining the experimental yields.[6] We note that the attacks in the *ortho*, *meta*, *para* and *ipso* positions of pyridine in the O(^{1}D) + pyridine reaction give rise to quite different PESs as far as the local minima and transition state are concerned (see Fig. 1 of ref. 4) even though the reaction channels are the same.

1 C.-H. Chin, T. Zhu and J. Z. H. Zhang, *Phys. Chem. Chem. Phys.*, 2021, **23**, 12408–12420.
2 C. Cavallotti, C. De Falco, L. Pratali Maffei, A. Caracciolo, G. Vanuzzo, N. Balucani and P. Casavecchia, *J. Phys. Chem. Lett.*, 2020, **11**, 9621–9628.
3 G. Vanuzzo, A. Caracciolo, T. K. Minton, N. Balucani, P. Casavecchia, C. de Falco, A. Baggioli and C. Cavallotti, *J. Phys. Chem. A*, 2021, **125**, 8434–8453.
4 P. Recio, S. Alessandrini, G. Vanuzzo, G. Pannacci, A. Baggioli, D. Marchione, A. Caracciolo, V. J. Murray, P. Casavecchia, N. Balucani, C. Cavallotti, C. Puzzarini and V. Barone, *Nat. Chem.*, 2022, **14**, 1405.
5 M. Rosi, N. Balucani, P. Casavecchia, N. Faginas-Lago, L. Mancini, D. Skouteris and G. Vanuzzo, in *Computational Science and Its Applications – ICCSA 2023 Workshops. ICCSA 2023. Lecture Notes in Computer Science*, ed. O. Gervasi *et al.*, Springer, Cham, 2023, vol. 14105, DOI: 10.1007/978-3-031-37108-0_46.
6 P. Recio, D. Marchione, A. Caracciolo, V. J. Murray, L. Mancini, M. Rosi, P. Casavecchia and N. Balucani, *Chem. Phys. Lett.*, 2021, **779**, 138852.

Dwayne E. Heard asked: Has anybody studied the reaction of hydrogen atoms with either benzene or pyridine?

Nadia Balucani responded: Concerning the reaction H + C_6H_6, there are several theoretical and experimental investigations. A recent theoretical study was able to reproduce most of the fragmentary experimental information.[1] According to those calculations, there is competition between the H-abstraction and H-addition mechanisms. H-abstraction can only make a contribution at very high temperatures since it is endothermic by *ca.* 10 kcal mol^{-1} and has an entrance TS of *ca.* 17 kcal mol^{-1}. H-addition with the formation of the C_6H_7 adduct is instead an exothermic channel with an entrance TS of around 4.5 kcal mol^{-1}. The information on the H + C_5H_5N (pyridine) reaction is very scarce. To the best of my knowledge there is only a theoretical estimate on the H-abstraction channel made to model pyridine combustion.[2] H-addition has not been considered even though

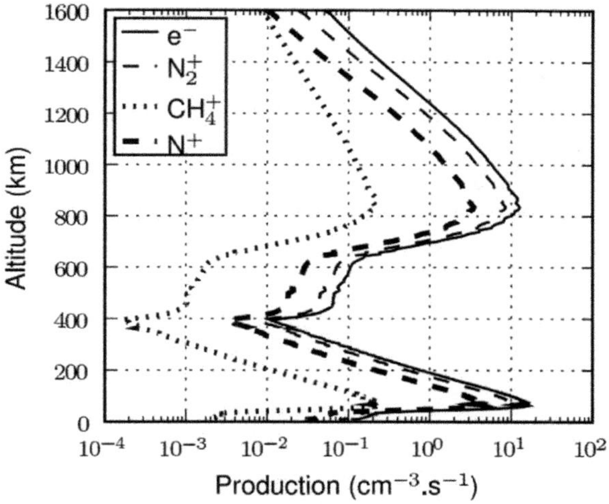

Fig. 4 Total ionization computed for the nightside T5 flyby conditions. Reprinted with permission from ref. 16.

it should be a competitive reaction pathway as in the case of the reaction with benzene.

1 H. Zhang, X. Zhang, D. G. Truhlar and X. Xu, *J. Phys. Chem. A*, 2017, **121**, 9033–9044 and references therein.
2 L.-N. Wu, Z.-Y. Tian, J.-J. Weng, D. Yu, Y.-X. Liu, D.-X. Tian, C.-C. Cao, J.-B. Zou, Y. Zhang and J.-Z. Yang, *Combust. Flame*, 2019, **202**, 394–404.

Ugo Jacovella enquired: I suppose it depends on altitude, but in Titan's atmosphere, how could we compare the predominance of neutral/neutral reactions *versus* ion–molecule reactions?

Nadia Balucani answered: Photochemical models of the atmosphere of Titan indicate that most of the chemistry is controlled by neutral–neutral reactions (see, for instance, Vuitton *et al.*[1] and Hörst[2]). However, some essential processes are governed by ion chemistry. For instance, the formation of two important species like benzene[3] and ammonia[4] in the upper atmosphere of Titan, while nitriles synthesis is governed by neutral–neutral reactions starting with the production of HCN.[5] As a matter of fact, nobody was expecting a rich ion chemistry and before the Cassini–Huygens mission molecular growth was believed to occur in the stratosphere *via* neutral–neutral reactions.[6–8] Titan's ionosphere, instead, is very rich in complex positive and negative ions. The Ion and Neutral Mass Spectrometer onboard Cassini identified *ca.* 50 positive ions with mass-to-charge ratio (m/z) values up to 100.[9] The Ion Beam Spectrometer of the Cassini Plasma Spectrometer indicated the presence of much heavier positive ions (m/z up to 350).[10] Another striking result came from the Electron Spectrometer (ELS) of the Cassini Plasma Spectrometer, which provided evidence of negatively charged ions with a mass up to 10 000 Da (and more) and accounting for *ca.* 10% of the negatively charged species.[11] Another interesting aspect is that, even though the ionization peak is around 1200 km (where a diurnal electron density of 2500–3500

cm^{-3} and a nocturnal electron density of 400–1000 cm^{-3} were measured[12]), the ionization level remains significant throughout the atmosphere[13] because of the different energy sources inducing ionization.[14–19] The Cassini–Huygens instruments demonstrated that solar photons dominate dayside ionization, while Saturnian magnetospheric electrons dominate nightside ionization (this contribution, however, is always much lower than solar radiation on the dayside).[12,20] Ionization from 500 to 1000 km is caused by energetic protons and oxygen ions coming from the magnetosphere of Saturn[21] while micrometeorite ablation causes ion production from 500 to 700 km.[22] Around 60–70 km a second peak in the electron distribution, almost as intense as the one at 1200 km, has been identified and associated with the deposition of galactic cosmic rays.[13] Unfortunately, the nature of the ions present at this altitude is unknown. In Fig. 4 here (taken from ref. 16) the total ionization computed for the nightside T5 flyby conditions is reported as a function of the altitude. All the above-mentioned contributions have been considered to derive the global production rates of the electrons, N_2^+, CH_4^+ and N^+. As clearly visible, some ion production occurs from the surface up to the exosphere.

In conclusion, neutral–neutral reactions and ion chemistry are mutually dependent and ion chemistry can play a role also in the stratosphere and below because of the contribution made by penetrating cosmic rays.

1 V. Vuitton, O. Dutuit, M. A. Smith and N. Balucani, Chemistry of Titan's atmosphere, in *Titan: Interior, Surface, Atmosphere, and Space Environment*, ed. I. Müller-Wodarg, C. Griffith, E. Lellouch and T. Cravens, Cambridge Planetary Science, Cambridge University Press, 2014, ch. 7, pp. 224–284, DOI: 10.1017/cbo9780511667398.010.
2 S. M. Hörst, *J. Geophys. Res.: Planets*, 2017, **122**, 432–482.
3 V. Vuitton, R. V. Yelle and J. Cui, *J. Geophys. Res.*, 2008, **113**, E05007.
4 R. V. Yelle, V. Vuitton, P. Lavvas, S. J. Klippenstein, M. A. Smith, S. M. Hörst and J. Cui, *Faraday Discuss.*, 2010, **147**, 31–49.
5 J. C. Loison, E. Hébrard, M. Dobrijevic, K. M. Hickson, F. Caralp, V. Hue, G. Gronoff, O. Venot and Y. Bénilan, *Icarus*, 2015, **247**, 218–247.
6 E. H. Wilson and S. K. Atreya, *J. Geophys. Res.*, 2004, **109**, E06002.
7 L. M. Lara, E. Lellouch, J. J. López-Moreno and R. Rodrigo, *J. Geophys. Res.*, 1996, **101**, 23261–23283.
8 Y. L. Yung, M. Allen and J. P. Pinto, *Astrophys. J.*, 1984, **55**, 465–506.
9 J. H. Waite, H. Niemann, R. V. Yelle, W. T. Kasprzak, T. E. Cravens, J. G. Luhmann, R. L. McNutt, W.-H. Ip, D. Gell, V. De La Haye, I. Müller-Wordag, B. Magee, N. Borggren, S. Ledvina, G. Fletcher, E. Walter, R. Miller, S. Scherer, R. Thorpe, J. Xu, B. Block and K. Arnett, *Science*, 2005, **308**, 982–986.
10 F. J. Crary, B. A. Magee, K. Mandt, J. H. Waite, J. Westlake and D. T. Young, *Planet. Space Sci.*, 2009, **57**, 1847–1856.
11 A. J. Coates, F. J. Crary, G. R. Lewis, D. T. Young, J. H. Waite and E. C. Sittler, *Geophys. Res. Lett.*, 2007, **34**, L22103.
12 K. Agren, J.-E. Wahlund, P. Garnier, R. Modolo, J. Cui, M. Galand, I. Muller-Wodarg, *Planet. Space Sci.*, 2009, **57**, 1821–1827.
13 J. J. Lopez-Moreno, G. J. Molina-Cuberos, M. Hamelin, R. Grard, F. Simoes, R. Godard, K. Schwingenschuh, C. Beghin, J. J. Berthelier, V. J. G. Brown, P. Falkner, F. Ferri, M. Fulchignoni, I. Jernej, J. M. Jeronimo, R. Rodrigo and R. Trautner, *Geophys. Res. Lett.*, 2008, **35**, L22104.
14 P. Lavvas, M. Galand, R. V. Yelle, A. N. Heays, B. R. Lewis, G. R. Lewis and A. J. Coates, *Icarus*, 2011, **213**, 233–251.
15 P. Lavvas, R. V. Yelle, A. N. Heays, L. Campbell, M. J. Brunger, M. Galand and V. Vuitton, *Icarus*, 2015, **260**, 29–59.
16 G. Gronoff, J. Lilensten and R. Modolo, *Astron. Astrophys.*, 2009, **506**, 965–970.
17 K. Willacy, S. Chen, D. J. Adams and Y. L. Yung, *Astrophys. J.*, 2022, **933**, 230.
18 V. Vuitton, R. V. Yelle, S. J. Klippenstein, S. M. Hörst and P. Lavvas, *Icarus*, 2019, **324**, 120–197.
19 M. Dobrijevic and J. C. Loison, *Icarus*, 2018, **307**, 371–379.

20 T. E. Cravens, I. P. Robertson, J. Clark, J.-E. Wahlund, J. H. Waite Jr, S. A. Ledvina, H. B. Niemann, R. V. Yelle, W. T. Kasprzak, J. G. Luhmann, R. L. McNutt, W.-H. Ip, V. De La Haye, I. Muller-Wodarg, D. T. Young and A. J. Coates, *Geophys. Res. Lett.*, 2005, **32**, L12108.
21 T. E. Cravens, I. P. Robertson, S. A. Ledvina, D. Mitchell, S. M. Krimigis and J. H. Waite, *Geophys. Res. Lett.*, 2008, **35**, L03103.
22 G. J. Molina-Cuberos, H. Lammer, W. Stumptner, K. Schwingenschuh, H. O. Rucker, J. J. López-Moreno, R. Rodrigo and T. Tokano, *Planet. Space Sci.*, 2001, **49**, 143–153.

Michelle Brann requested: Can you comment on role of electron chemistry? I'm also curious about whether you think secondary electrons play a role in the chemistry in Titan's upper atmosphere?

Nadia Balucani responded: Electrons play an important role in the atmosphere of Titan as they contribute to dissociate and ionize N_2 and CH_4 (the two main atmospheric species) and this is the starting point of all Titan chemistry.[1–9] There are essentially two main sources of electrons in the upper atmosphere of Titan: electrons coming from the magnetosphere of Saturn (and, indirectly, other charged species with the same origin, such as protons or O^+) and photoelectrons, that is, free electrons originating from the photoionization of N_2 and CH_4 caused by solar radiation in the extreme ultraviolet or soft X-rays. Depending on the photon wavelength, photoelectrons can be emitted with enough energy to induce secondary ionization or dissociation by colliding with other neutral molecules. Before the Cassini mission, there was some uncertainty on the role played by the two sources. Now, it has been well established that the role of magnetospheric electrons and charge particles is minor, because of the lack of UV emissions during the night and the large variation of thermal electron density between night and day. For instance, by analyzing the data of the instruments onboard Cassini (the Langmuir probe on the Cassini Radio and Plasma Wave Experiment and the Electron Spectrometer of the Cassini Plasma Spectrometer), Cravens *et al.*[6] were able to infer a dominant contribution from the solar radiation *vs.* magnetospheric particles while Agren *et al.*[7] found typical dayside electron densities of around 2500–3500 cm^{-3} to be compared with nightside densities of around 400–1000 cm^{-3}.

Photoionization is the main source of electrons down to 700 km.[9] Other ionisation sources (cosmic rays, meteors and energetic ions) are instead important in the lower layers of the atmosphere.[9] For instance, protons and other charged species from the magnetosphere of Saturn make a contribution in the middle atmosphere (500 km), while cosmic rays cause an ionization peak around 65 km.[9] Electron impact can induce numerous processes, including excitation, ionization, dissociation, dissociative ionization.[10,11] All of them are accounted for in the photochemical models of Titan.[1–9]

1 P. Lavvas, M. Galand, R. V. Yelle, A. N. Heays, B. R. Lewis, G. R. Lewis and A. J. Coates, *Icarus*, 2011, **213**, 233–251.
2 P. Lavvas, R. V. Yelle, A. N. Heays, L. Campbell, M. J. Brunger, M. Galand and V. Vuitton, *Icarus*, 2015, **260**, 29–59.
3 K. Willacy, S. Chen, D. J. Adams and Y. L. Yung, *Astrophys. J.*, 2022, **933**, 230.
4 V. Vuitton, R. V. Yelle, S. J. Klippenstein, S. M. Hörst and P. Lavvas, *Icarus*, 2019, **324**, 120–197.
5 M. Dobrijevic and J. C. Loison, *Icarus*, 2018, **307**, 371–379.

6 T. E. Cravens, I. P. Robertson, J. Clark, J.-E. Wahlund, J. H. Waite Jr, S. A. Ledvina, H. B. Niemann, R. V. Yelle, W. T. Kasprzak, J. G. Luhmann, R. L. McNutt, W.-H. Ip, V. De La Haye, I. Muller-Wodarg, D. T. Young, and A. J. Coates, *Geophys. Res. Lett.*, 2005, **32**, L12108.
7 K. Agren, J.-E. Wahlund, P. Garnier, R. Modolo, J. Cui, M. Galand, I. Muller-Wodarg, *Planet. Space Sci.*, 2009, **57**, 1821–1827.
8 T. E. Cravens, I. P. Robertson, J. H. Waite Jr, R. V. Yelle, V. Vuitton, A. J. Coates, J.-E. Wahlund, K. Agren, M. S. Richard, V. De La Haye, A. Wellbrock and F. M. Neubauer, *Icarus*, 2009, **199**, 174–188.
9 G. Gronoff, J. Lilensten and R. Modolo, *Astron. Astrophys.*, 2009, **506**, 965–970.
10 M.-Y. Song, H. Cho, G. P. Karwasz, V. Kokooline and J. Tennyson, *J. Phys. Chem. Ref. Data*, 2023, **52**, 023104.
11 O. Dutuit, *et al.*, *Astrophys. J. Suppl. Ser.*, 2013, **204**, 20.

Athena Flint asked: As you discuss the chemistry of Titan, you note the presence of higher-energy forms of nitrogen, notably that of electronically excited states of N. N(^{2}D) is capable of reacting with N_2 to form the azide radical (N_3) through a relatively low transition state.[1] Do you expect that N_3 might form in Titan's atmosphere, and that it may be an important part of chemical modelling of Titan?

1 C. S. Jamieson and R. I. Kaiser, *Chem. Phys. Lett.*, 2007, **440**, 98–104.

Nadia Balucani responded: I am not aware of anybody having considered the presence of the azide radical in the atmosphere of Titan, while the possible role of the N_3^+ ion in the stratosphere has been considered and its chemistry studied in laboratory experiments.[1] Indeed, at the altitude where N(^{2}D) is significantly formed (above 1000 km), the atmosphere is already quite rarefied ($P \sim 10^{-6}$ hPa), and this makes three-body collisions rare. The possible role of N_3^+ was invoked by Anicich *et al.*[1] for the ion chemistry of the stratosphere where penetrating cosmic rays can still induce ionization[2] and the pressure is around 1 hPa.

In addition to the rarefied condition of the thermosphere, the low T (less than 200 K also at that altitude) might also represent a problem for the reactions forming N_3 (the barrier for the N(^{2}D) + N_2 to form linear N_3 is *ca.* 13 kJ mol^{-1} while the one to form the cyclic N_3 is *ca.* 29 kJ mol^{-1}).[3] Since it is well established that some N(^{2}D) can also be formed by the impact of cosmic rays in the stratosphere, that is probably the best place where to search for N_3.

1 V. G. Anicich, D. B. Milligan, D. A. Fairley and M. J. McEwan, *Icarus*, 2000, **146**, 118–124.
2 G. Gronoff, J. Lilensten, L. Desorgher and E. Flückiger, *Astronon. Astrophys.*, 2009, **506**, 955–964.
3 C. S. Jamieson and R. I. Kaiser, *Chem. Phys. Lett.*, 2007, **440**, 98–104.

Cristina Puzzarini enquired: Can your molecular beam apparatus be use for ionic chemistry? What is your suggestion to complement neutral–neutral reaction studies from an experimental point of view?

Nadia Balucani answered: In principle, our apparatus could be adapted by using ion beam sources and a system of ion guides. In the context of astrochemistry, the main advantage of crossed molecular beam experiments is that they allow reproducing the low number density environments of most extraterrestrial gaseous media.[1,2] However, it is quite difficult to reproduce the low temperature conditions of several planetary atmosphere or cold regions of the interstellar medium, even though there are several promising techniques (see, for instance, ref. 3–6). The best way to complement the results of crossed molecular beam experiments is to measure either the total reactive cross sections as a function of the collision energy or (more appropriate for astrochemical models) the reaction rate coefficients as a function of the temperature. Our contribution at this *Faraday Discussions* meeting, where the crossed molecular beam approach is combined with the investigation of the chemical kinetics in a continuous supersonic flow reactor at very low temperatures (https://doi.org/10.1039/d3fd00057e), is an example of this combined approach. There are other notable examples (see, for instance, ref. 7–10).

1 P. Casavecchia, F. Leonori, N. Balucani, R. Petrucci, G. Capozza and E. Segoloni, *Phys. Chem. Chem. Phys.*, 2009, **11**, 46–65.
2 P. Casavecchia, F. Leonori and N. Balucani, *Int. Rev. Phys. Chem.*, 2015, **34**, 161–204.
3 F. Leonori, R. Petrucci, E. Segoloni, A. Bergeat, K. M. Hickson, N. Balucani and P. Casavecchia, *J. Phys. Chem. A*, 2008, **112**, 1363–1379.
4 M. Lara, S. Chefdeville, P. Larregaray, L. Bonnet, J.-M. Launay, M. Costes, C. Naulin and A. Bergeat, *J. Phys. Chem. A*, 2016, **120**, 5274–5281.
5 V. Plomp, J. Onvlee, F. Lique and S. Y. T. van de Meerakker, *J. Phys. Chem. A*, 2023, **127**, 2306–2313.
6 S. Tanteri, S. D. S. Gordon, J. Zou and A. Osterwalder, *J. Phys. Chem. A*, 2021, **125**, 10021–10034.
7 G. Vanuzzo, L. Mancini, G. Pannacci, P. Liang, D. Marchione, P. Recio, Y. Tan, M. Rosi, D. Skouteris, P. Casavecchia, N. Balucani, K. M. Hickson, J.-C. Loison and M. Dobrijevic, *ACS Earth Space Chem.*, 2022, **6**, 2305–2321.
8 F. Leonori, R. Petrucci, N. Balucani, P. Casavecchia, M. Rosi, D. Skouteris, C. Berteloite, S. D. Le Picard, A. Canosa and I. R Sims, *J. Phys. Chem. A*, 2009, **113**, 15328–15345.
9 C. J. Bennett, S. B. Morales, S. D. Le Picard, A. Canosa, I. R. Sims, Y. H. Shih, A. H. H. Chang, X. Gu, F. Zhang and R. I. Kaiser, *Phys. Chem. Chem. Phys.*, 2010, **12**, 8737–8749.
10 S. B. Morales, C. J. Bennett, S. D. Le Picard, A. Canosa, I. R. Sims, B. J. Sun, P. H. Chen, A. H. H. Chang, V. V. Kislov, A. M. Mebel, X. Gu, F. Zhang, P. Maksyutenko and R. I. Kaiser, *Astrophys. J.*, 2011, **742**, 26.

Murthy S. Gudipati asked: How does the chemical activation of nitrogen happen in Titan's lower atmosphere? We know that energetic photons and ions (solar UV and solar wind) as well as Saturnian magnetospheric radiation dissociate and ionize N_2 in Titan's upper atmosphere. However, this energetic radiation doesn't penetrate deeper into Titan's lower atmosphere. Are there processes that could activate N_2 in Titan's lower atmosphere to undergo chemistry with organic molecules?

Nadia Balucani replied: Cosmic rays are quite penetrating and they cause the presence of a peak of ionization around 70–80 km as revealed by the instruments onboard Huygens.[1]

Cosmic rays can ionize or induce dissociation. Indeed, in the most recent models[2,3] the interaction of N_2 with galactic cosmic rays (GCR) is considered to produce N_2^+ + e with a yield of *ca.* 50%, N^+ + N + e with a yield of *ca.* 12% and $N(^4S)$ + $N(^2D)$ with a yield of 38%. This is extremely interesting because, in the conditions of the stratosphere, three body collisions are possible and this opens up different reaction pathways. The chemistry occurring in the stratosphere after the ionization or dissociation of nitrogen is considered in the models and it has been found to influence the abundance of nitriles and their isotopic fractionations.

1 J. J. Lopez-Moreno, G. J. Molina-Cuberos, M. Hamelin, R. Grard, F. Simoes, R. Godard, K. Schwingenschuh, C. Beghin, J. J. Berthelier, V. J. G. Brown, P. Falkner, F. Ferri, M.

Fulchignoni, I. Jernej, J. M. Jeronimo, R. Rodrigo and R. Trautner, *Geophys. Res. Lett.*, 2008, **35**, L22104.
2 V. Vuitton, R. V. Yelle, S. J. Klippenstein, S. M. Hörst and P. Lavvas, *Icarus*, 2019, **324**, 120–197.
3 M. Dobrijevic and J. C. Loison, *Icarus*, 2018, **307**, 371–379.

Ardjan Sturm enquired: In an earlier session we saw that many low-mass protoplanetary disks show a very carbon rich chemistry, indicating high C/O ratios. Since the chemistry in those regions looks similar to the chemistry in Titan's atmosphere, can we extrapolate what we know about Titan to learn about the impact of this chemistry on protoplanetary disk chemistry and planet formation?

Nadia Balucani answered: Of course, Titan can be a good proxy of largely oxygen-depleted chemistry at a planetary scale.

Nikku Madhusudhan opened discussion of the paper by Adam J. Fleisher: What is the resolving power you can achieve with frequency comb spectroscopy for generating line lists? For well-known molecules such as HCN you can calibrate the experimental line list against existing ones. But, for new molecules where no good line lists exist what would be the big challenges?

Adam J. Fleisher replied: Near a wavelength of 1.6 microns we can achieve a resolving power $>10^6$. Near 4.6 μm, our resolving power is approximately 65 000. For new molecules, our goal is always to resolve the individual frequency comb teeth of our laser (as is done near 1.6 microns and demonstrated in the paper; https://doi.org/10.1039/d3fd00019b). Then we can assign a frequency axis based on the known optical frequencies of our laser, often with a relative precision of better than 10^{-9}. The next big challenges include efficient spectral assignment and analysis, particularly for congested spectra which may not be highly resolved due to closely spaced ro-vibrational lines.

Nikku Madhusudhan asked: What are the wavelength ranges you are speaking about here?

Adam J. Fleisher responded: We are currently working in the 1.5 μm to 5.0 μm wavelength range.

Ian R. Sims said: Your article (https://doi.org/10.1039/d3fd00019b) details beautiful high resolution measurements of a stable molecule, HCN, with high accuracy on the frequency and absorption axes. Would you care to comment on the application of these new developments taking into account the time axis, in particular the study of unstable species such as HNC which you mention in your perspectives. We recently made a very low resolution time domain study of collisional broadening in the HCN/HNC + He system in the 9–70 K temperature range, which is of interest in the interpretation of astronomical spectra.[1]

1 B. M. Hays, D. Gupta, T. Guillaume, O. Abdelkader Khedaoui, I. R. Cooke, F. Thibault, F. Lique and I. R. Sims, *Nat. Chem.*, 2022, **14**, 811–815, DOI: 10.1038/s41557-022-00936-x.

Adam J. Fleisher replied: I enjoyed the impressive work described in your collisional broadening study – and its implications for the interpretation of astronomical spectra. Generally, virtually imaged phased array (VIPA) spectrometers can achieved μs time resolution for the study of transient species. Importantly, the potential for achieving high signal-to-noise ratios using dispersive spectrometers potentially enables the study of non-reproducible dynamic events, or for the efficient deep averaging of spectra obtained from reproducible events. As an example, we first studied a photolysis reaction to form DOCO (deuterated HOCO) at room temperature.[1]

1 A. J. Fleisher, B. J. Bjork, T. Q. Bui, K. C. Cossel, M. Okumura and J. Ye, *J. Chem. Phys. Lett.*, 2014, **5**, 2241–2246.

Ian R. Sims requested: In studies of radical kinetics, sensitivity is a key issue – would you care to comment on this and if possible, estimate the detection limit with, say, a 5 μs time resolution and an interaction length of 1 cm (in terms of concentration in cm^{-3})?

Adam J. Fleisher responded: For HCN in the near-infrared ($\lambda \approx 1.53$ μm), the strongest ro-vibrational transitions have a HITRAN line strength $S \approx 7 \times 10^{-21}$ cm per molecule. Assuming a Doppler-broadened line at a temperature of 296 K and a baseline absorbance noise of 0.007 – estimated for a single spectrometer acquisition image with an integration time of 5 μs and calculated from the noise analysis equations in the paper (https://doi.org/10.1039/d3fd00019b), we estimate a noise-equivalent number density for HCN at a 1 cm pathlength of 2×10^{16} cm^{-3}. To enhance this estimated sensitivity, we could imagine coupling our frequency comb laser source to an enhancement cavity. With a modest finesse of $F \approx 3000$, the pathlength enhancement is about 1000-fold, resulting in an estimated noise-equivalent number density for HCN of 2×10^{13} cm^{-3} for cavity-enhanced direct frequency comb spectroscopy using a VIPA spectrometer in the $\lambda = 1.53$ μm region.

Laurent Wiesenfeld asked: How sensitive is this method to the lineshape, and especially so, to the far wings? Are you capable of having dynamics wide enough to be quantitative all the way and detect a line shape that could be a reliable signature of the gas conditions?

Adam J. Fleisher answered: Retrieving information from the far wings of ro-vibrational absorption lines using direct frequency comb spectroscopy requires precision treatment of the spectral baseline (background). Here we measure baseline by rapidly switching between a comb signal with and without sample, using either synchronized optical choppers or fast fiber-optic switches. This allows for baseline normalization at a rate of 10 Hz to 1 kHz, which is fast compared to laboratory drifts. Therefore, we anticipate exploring far wings to study phenomena like line mixing and non-Markovian collisions.

Inga Kamp opened discussion of the paper by Mark H. Stockett: If the PAHs cool efficiently by rapid radiative decay (so recurrent fluorescence), lowering their destruction in highly irradiated environments, we should observe this cooling radiation directly. You mention in your paper (https://doi.org/10.1039/d3fd00005b) the Extended Red Emission (ERE) as possibly being such a tracer. So, can such observations then be used to tell us how efficient this stabilising process is?

Mark H. Stockett responded: There have only been 2–3 direct observations of recurrent fluorescence (RF) photons in the laboratory, and so far no real spectroscopy. The RF emission spectra from PAHs and other nanocarbons is expected to be very broad, 100s of nm. The ERE is also broad, and the maximum wavelength shifts a bit looking at different environments, probably tracing a size range of the emitters. The situation is similar to the AIBs (Aromatic Infrared Bands), where the observations are not very specific and we need radiative models for many species benchmarked to lab data.

Ugo Jacovella asked: As the second excited state (D2) is very close in energy (about 1.7 eV above the ground state) and has a stronger oscillator force than the transition to the D1 state, what is the recurrent fluorescence rate of the D2 state?

Mark H. Stockett answered: We calculated it and found it to be much lower than for the D1 state. The RF rate depends exponentially on the transition energy but only linearly on the oscillator strength. This is a contrast to regular PAHs where the D1 → D0 transition is forbidden and RF occurs *via* the D2 state. There the dissociation energies are also much higher (5 eV *vs.* 3 eV) so RF is still competitive.

Ugo Jacovella opened discussion of the paper by Alexander K. Lemmens: Why have you chosen not to take into account PAHs containing heteroatoms when data exists on these species and they should have different characteristics to those of bare PAHs?

Alexander K. Lemmens responded: The presence of heteroatoms within the PAH skeleton in larger PAHs is still debated (ref. 1 and references therein discuss the reactivity of PAHs with nitrogen and stability of such NPAHs), however PAH derivatives containing cyano functional groups have now been detected in the ISM. A commonly cited article that concludes about the N-content in PAHs is ref. 2. Here, DFT calculations in the NASA Ames database – performed employing a relatively small basis set – are used to evaluate the band position of one peak corresponding to the cationic CC stretch. From the peak shift, the nitrogen content is deduced. This, however, is concluded for cationic endoskeletal NPAHs, that are closed shell, contrary to their neutral counterpart on which this work focuses. If a size of 50 C atoms is assumed, this would correspond to a substitution of about two nitrogen atoms, which is unlikely to have a large effect on the conclusions in this work.

1 A. G. G. M. Tielens, *EAS Publ. Ser.*, 2011, **46**, 3–10, DOI: 10.1051/eas/1146001.
2 D. M. Hudgins, *et al.*, *Astrophys. J.*, 2005, **632**, 316, DOI: 10.1086/432495.

Alexander K. Lemmens continued discussion of the paper by Mark H. Stockett: The recurrent fluorescence (RF) process relies on an active low-lying electronic state. For neutral PAHs these are typically higher than for cations, but for larger PAHs they tend to be lower. How likely is the RF process in neutral PAHs, in the astronomical perspective? And do you have ideas about RF measurements in neutrals?

Mark H. Stockett answered: As far as I know there is only one (theoretical) work on RF from neutral carbon clusters.[1] RF will not be important for small neutral PAHs. It could become competitive for very large neutrals or those with substituents which have low-lying states. RF experiments with neutrals would be challenging as they would need to be isolated in a low-pressure environment and excited to very high internal energies. It would be more difficult to separate emission by intact neutrals from that by decay products.

1 O. Lacinbala, *et al.*, *J. Chem. Phys.*, 2022, **156**, 144305.

Alexander K. Lemmens asked: Related to the previous question, what is the size limit of ions you could store in the storage ring? What would be the constraints?

Mark H. Stockett replied: One advantage of electrostatic ion beam storage rings compared to magnetic ones is that there is in principal no upper mass limit. We have stored fullerenes and other cluster ions with masses up to 3000 amu. At high masses we lose some mass resolution on the bending magnet used to select the ions prior to storage, we probably store multiple isotopologues.

Divita Gupta continued discussion of the paper by Alexander K. Lemmens: Can you implement a similar idea for nitrogen-substituted PAHs? Are there any two bands that could be compared to get an estimation of the sizes of such species in the ISM?

Alexander K. Lemmens responded: For PAH cations, the endoskeletal nitrogen content has been deduced using the CC stretch band at 6.2 μm. Estimates are about 2 nitrogen atoms for a PAH containing 50 carbon atoms.[1] While the endoskeletal nitrogen content is still highly debated, this amount of endoskeletal nitrogen is unlikely to affect the IR emission ratio to a large extent. Nitrogen containing functional groups are more likely to play a role, as evidenced by the observation of several cyano-substituted species by radio astronomy.[2] In this study, however, bare naphthalene is typically about two orders of magnitude higher in abundance than the substituted species. The bands used in the analysis in the current paper (https://doi.org/10.1039/d2fd00180b) are thus unlikely to be affected significantly by the nitrogen-substituted species. For deducing sizes for cyano-substituted PAHs specifically, one would need to identify IR bands that originate solely from such species, ideally at two well-separated wavelengths. The longer wavelength band would be hard to distinguish from bands originating from "pure" PAHs.

1 D. M. Hudgins, *et al.*, *Astrophys. J.*, 2005, **632**, 316, DOI: 10.1086/432495.
2 B. A. McGuire, *et al.*, *Science*, 2021, **371**, 1265–1269, DOI: 10.1126/science.abb7535.

Ugo Jacovella enquired: In your answer to my previous question, you said that you were looking at the most stable PAHs. In my opinion, one of the main reasons why we are struggling with this aromatic infrared bands is that in the laboratory we can only measure the spectra of very stable PAHs and not those that may exist in the harsh conditions of the interstellar medium. I was just wondering what you think would be the next step in resolving this aromatic infrared bands puzzle?

Alexander K. Lemmens answered: One aspect of the grandPAH hypothesis is the higher radiation resistance of larger compact PAHs.[1] It is assumed that radiation weeds out the less stable (smaller, non-compact and/or irregular structure) species. This has been investigated both from a laboratory perspective[2] and by using observations.[3] This would result in mostly the stable species being responsible for the observed IR emission. Nevertheless, it is possible to investigate the IR spectra of radical or intermediate species in collisionless conditions in the laboratory as demonstrated by our group;[4] however it is unlikely that they significantly contribute to the overall IR emission. Using only the mid-IR emission it is in my opinion unlikely to completely “solve” the aromatic infrared band puzzle, or in other words teasing out the exact molecular composition, because of the spectral overlap in emission from most species of the PAH family. Higher spectral resolution and more accurate predicted IR spectra *will* enable us to put more accurate and precise constraints on the size, charge state and general structure of PAHs.

1 A. G. G. M. Tielens, *Annu. Rev. Astron. Astrophys.*, 2008, **46**, 289–337, DOI: 10.1146/annurev.astro.46.060407.145211.
2 J. Bouwman, *et al.*, *Astron. Astrophys.*, 2019, **621**, A80, DOI: 10.1051/0004-6361/201834130.
3 H. Andrews, *et al.*, *Astrophys. J.*, 2015, **807**, 99, DOI: 10.1088/0004-637X/807/1/99.
4 A. K. Lemmens, *et al.*, *Nat. Commun.*, 2020, **11**, 269, DOI: 10.1038/s41467-019-14092-3.

Dwayne E. Heard commented: In response to your answer, we do now have better spectral resolution with JWST.

Alexander K. Lemmens responded: Having a better resolution will aid greatly in putting more accurate and precise constraints on the properties of the PAH population in space. The improved resolution will not improve the analysis performed in this work immediately, since the CH-stretch and CH-out-of-plane bands used here are spectrally well separated.

Dwayne E. Heard asked: Your paper (https://doi.org/10.1039/d2fd00180b) has provided an improved understanding of how the ratio of the intensity of the two lines changes with the carbon number of the PAH. Have you applied this to any new observations? How much has the size distribution of PAHs changed when the new understanding (parameterisation of the functional form of how the intensity ratio changes with size) presented in the paper is applied? Is the size distribution particularly sensitive to the ratio once you reach a certain size?

Alexander K. Lemmens replied: We have applied our findings to two observed intensity ratio distributions. The horizontal black lines in Fig. 4 in the paper (https://doi.org/10.1039/d2fd00180b) – that contains the improved relation between intensity ratio and size – indicate lower and higher limits of the intensity ratios in Orion Bar and NGC 7023. Without correction, one would deduce a range between 50 and 70 carbon atoms. Using the corrected model, however, this range would be reduced to 40–55 carbon atoms for NGC 7023 and 37–53 carbon atoms for the Orion Bar region. The heat capacity, and hence temperature of PAHs, which in turn mainly influences the emission ratio is well reflected in the ratio since the bands are spectrally well separated and are therefore a robust tracer of PAH size.

Ewine F. van Dishoeck enquired: It is interesting to see that your new analysis of the 11.3/3.3 micron flux ratio shifts the inferred size distribution of PAHs to lower numbers of carbon atoms, typically less than 50 carbon atoms. Have you checked whether this also holds for PAHs in protoplanetary disks around intermediate mass Herbig stars? In those high UV environments, the intense UV radiation is thought to destroy PAH molecules with typically less than 100 carbon atoms. Would this stability argument get into conflict with the smaller sizes derived from the 11.3/3.3 micron ratio?

Alexander K. Lemmens responded: Indeed it seems the case that the 11.2/3.3 μm ratio is relatively high around Herbig stars,[1] which would at first glance point to the presence of larger sized PAHs. However, care must be taken when applying the models from the current study to interpret these observations. First and foremost, because they assume a single UV photon process. Secondly, upon photodestruction, PAH fragments may emit from yet undefined internal energies. Much more elaborate models, that include fragmentation and ionization processes would be needed to accurately determine the size distribution in such sources.

1 B. Acke and M. E. van den Ancker, *Astron. Astrophys.*, 2004, **426**, 151–170, DOI: 10.1051/0004-6361:20040400.

Daniël B. Rap asked: Some wavelength regions in the interstellar aromatic infrared bands spectra are thought to show signatures of cationic PAHs. Do we need to re-assess the intensity of these cationic modes, as well in comparison to other vibrational modes to correctly establish good numbers of neutral and cationic PAH (size) distributions?

Alexander K. Lemmens responded: The results of this study indeed call for similar analysis to be performed on the size distribution of cationic PAHs, although tracers for this tend to be less accurate as the used bands are spectrally not as well separated as for neutral PAHs. Benchmarking the models used to establish the ratio of neutral *versus* cationic PAHs would be similarly important as the benchmark performed in the current work (https://doi.org/10.1039/d2fd00180b).

Michelle Brann continued discussion of the paper by Adam J. Fleisher: I was wondering what the sources and approximate background levels are in your system and how they influence the frequency location and shapes.

Adam J. Fleisher replied: To date, we have identified the following sources of spectral noise: detector noise, photon shot noise and quantization noise due to finite camera bit depth. A quantitative discussion of our estimated limits for each component noise source can be found in the paper (https://doi.org/10.1039/d3fd00019b). The signal-to-noise ratio in our 2-D dispersed spectral images will limit our ability to identify and assign resolved frequency comb modes. So this may be a limitation of useable optical bandwidth when we approach the edges of the power spectrum of our laser frequency comb sources. More so than the spectrometer noise, we anticipate that challenges in power spectral normalization are most likely to influence the retrieved instrument line shape – if frequency comb modes are not resolved. For an analysis example, please see ref. 1. Otherwise, our ability to fit molecular line shape functions beyond the Voigt profile should also be limited by achievable signal-to-noise ratios.

1 D. M. Bailey, G. Zhao and A. J. Fleisher, *Anal. Chem.*, 2020, **92**, 13759–13766.

Sandra Brünken continued discussion of the paper by Mark H. Stockett: For your measurements of the spontaneous and laser-induced dissociation rates you monitor the neutral fragments as a function of time. Can you distinguish between different mass channels? In the paper (https://doi.org/10.1039/d3fd00005b) you only mention the HCN fragment that you monitor, but should there not be other dissociation channels open depending on the internal energy of the ion?

Mark H. Stockett answered: Indeed we cannot uniquely determine the mass of the neutral products. For cyanonaphthalene it was previously reported that the main fragmentation channels are HCN- and H-loss. Because of their low mass, H atoms are spread out over a wide area and also detected with much lower efficiency, contributing to the background term in the kinetic energy release (KER) distribution. C_2H_2-loss opens only at significantly higher energy (5 eV *vs.* 3 eV) and is not expected to contribute here.

Sandra Brünken questioned: You say you assume a constant ratio between the HCN and other fragments, which you cannot detect. But is this really true? If the branching ratio changes depending on internal energy, *e.g.*, if some dissociation channels close off at lower internal energy, should this then not have an influence on the observed time-dependent dissociation curves?

Mark H. Stockett answered: From the breakdown curves of West *et al.*,[1] the branching between H- and HCN-loss only varies by a few percent across the energy range to which we are sensitive. But in principle yes such effects could be included in more detailed modelling.

1 B. J. West, *et al.*, *J. Phys. Chem. A*, 2019, **123**, 3569.

Daniël B. Rap asked: Could one use a spectroscopic technique (*i.e.* laser-induced fluorescence (LIF)) to characterize the structure of the neutral fragment (HCN/HNC) or is this experimentally challenging due to the outline of the storage ring setup.

Mark H. Stockett replied: The neutral products leave the ring very quickly after emission and it would be difficult to do experiments with them. A new system will be installed this summer for collecting photons emitted from neutrals formed in ion–ion recombination experiments. This sounds crazy but we have already seen some photons without any collection optics. Even there though the neutrals are formed in an excited state that emits a photon, LIF experiments would be another layer of difficulty.

Inga Kamp continued discussion of the paper by Adam J. Fleisher: I know we ask a lot when we like to obtain full molecular spectra from 5–28 μm for our JWST/MIRI spectral analysis. The new technique you present is very powerful and I was curious why you focus on two specific bands, 1.56 and 4.7 μm. What are the technical issues/limitations – if there are any – to go to longer wavelengths?

Adam J. Fleisher responded: We focus in this paper (https://doi.org/10.1039/d3fd00019b) on the bands near $\lambda \approx 1.56$ μm and $\lambda \approx 4.7$ μm because they are accessible with our existing frequency comb laser sources. Generally speaking, the 1 μm to 5 μm range is accessible using commercial (but expensive) laser technology. Optical frequency combs and components to build dispersive spectrometers from 5 μm to 15 μm and beyond are an active area of research, with several groups around the world having demonstrated laboratory operations capable of performing spectroscopy at high resolution for these longer wavelengths using other read-out techniques like FTS or dual-comb spectroscopy. The largest technical challenges include low noise camera/detector arrays at wavelengths greater than approximately 8 μm, as well as generating high enough optical frequency comb power to perform dispersive spectroscopy at high signal-to-noise ratios. I believe that the 1 μm to 5 μm spectral region is well suited for laboratory spectroscopy to support astrophysical and astrochemical observations right now.

Mark H. Stockett asked: Do you have a detector array that works at 11 microns?

Adam J. Fleisher replied: Detector arrays and long-wave infrared cameras using strained layer superlattice (SLS) detector technology are available at wavelengths of 11 microns.

Dwayne E. Heard enquired: Do you also use an FTIR to analyse the frequency comb absorption spectrum at longer IR wavelengths, rather than a VIPA array/camera? There would be less limitation on the wavelengths you could analyse, and perhaps achieve a higher resolution?

Adam J. Fleisher answered: Although not a capability that is currently built into our laboratory, several research groups use FTS instruments to analyze broadband frequency comb spectra in the infrared (wavelengths of about 1.5 microns to 10 microns). While broader bandwidth detectors used in FTS are an advantage, acquisition speed is slower for FTS when compared to dispersive spectroscopy techniques like our VIPA spectrometer. In either technique, once comb modes are resolved, the resolution can be limited by only the optical frequency comb itself.

For additional details on FTS with frequency combs, see ref. 1 and references therein.

1 M. Germann, A. Hjalten, V. Boudon, C. Richard, K. Krzempek, A. Hudzikowski, A. Gluszek, G. Sobon and A. Foltynowicz, *J. Quant. Spectrosc. Radiat. Transfer*, 2022, **288**, 108252.

Dwayne E. Heard asked: You have an amazing machine that can do line surveys of chosen molecules. In order to decide on what to study, have you engaged with the observational astronomers or perhaps the astrochemical modellers about the best species to prioritize in your programme of research?

Adam J. Fleisher replied: We are at the early stages of engaging with observational astronomers and astrochemical modelers, and are open to discussing collaborations regarding specific molecular targets.

Sandra Brünken continued discussion of the paper by Alexander K. Lemmens: In your PAH emission model you assume a certain heat capacity and temperature, based on the energy the PAH gains upon absorption of a single UV photon. Would the inclusion of recurrent fluorescence, as shown to be an important process for cationic PAHs in Mark Stockett's work (https://doi.org/10.1039/d3fd00005b), impact the emission intensity, both overall and also the distribution over different vibrational modes?

Alexander K. Lemmens answered: Recurrent fluorescence (RF) would be an additional process through which the PAH could lose vibrational energy. Since quantitative data on RF is not yet available for neutral PAHs, it is hard to estimate the extent of the effect. The efficiency of the inverse internal conversion of vibrational energy to an excited electronic state strongly decreases with increasing energy of the involved electronic excited state.[1] Hence, it would be larger in larger PAHs (and cationic PAHs for that matter) since they possess optically active electronic states at lower energy. If such process would compete with the emission of IR photons, it is likely to decrease the emission at high vibrational energy more than at low vibrational energy, because the molecule has an additional pathway by which it can cool down more quickly especially when it possesses larger amounts of vibrational energy.

1 K. Peeters, *et al.*, *Phys. Rev. Res.*, 2021, **3**, 033225, DOI: 10.1103/PhysRevResearch.3.033225.

Mark H. Stockett added: Inclusion of RF in emission models would reduce the overall IR intensity as a significant fraction of the internal energy would be stolen by the RF photons. The IR spectrum may be slightly affected as the emitting ions would be a bit cooler than when RF is neglected, but still very warm.

Adam J. Fleisher asked: Can you apply information from fluorescence excitation spectroscopy (FES) of neutral species?

Alexander K. Lemmens responded: Much spectroscopic information has been obtained by laser induced fluorescence (LIF). Not only locating electronic energy levels, but also information on the redistribution of the electronic energy. For example, efficient intersystem crossing in PAHs has been established and this information should definitely be included in future models that describe PAH IR emission.

Adam J. Fleisher enquired: Are these measurements performed in collisional environments, or in collisionless environments (*e.g.*, a molecular beam)?

Alexander K. Lemmens replied: The measurements used to benchmark the theoretical calculations in this study are extracted from the NIST database and performed in a static gas cell at elevated temperatures, hence absorption lines are expected to be broadened. We do not expect that this would influence the integrated intensities. In the discussion there was mention of IR emission within the triplet manifold after the PAHs undergo intersystem crossing (ISC) which is not taken into account in this paper. The time scale and significant quantum yield – of up to 50% for *e.g.* coronene – of the ISC process has been investigated by us in molecular beam experiments for a number of PAHs.

Murthy S. Gudipati addressed Alexander K. Lemmens and Mark H. Stockett: The average size of PAHs found in the ISM is thought to be ~50 carbons. We have shown that this size of PAHs can easily be multiply ionized.[1] Have you considered multiple ionization of PAHs in your models?

1 M. S. Gudipati and L. J. Allamandola, *J. Phys. Chem. A*, 2006, **110**, 9020–9024.

Alexander K. Lemmens answered: In this work we deduce a size range with an upper limit of about 55 carbon atoms. Besides that, we indeed should consider the possibility of (multiply) ionized species. The two prominent infrared emission bands at 3.3 μm and 11.2 μm, however, are mostly the result of CH modes in neutral, not charged PAHs.[1,2] Charged PAHs tend to emit more in the 6–9 μm region. Therefore, the 11.2/3.3 μm ratio is a good measure of the size of neutral PAHs and rather unaffected by the charged PAH population. The 6.2 and 7.7 μm peaks are prominent in ionized PAHs and the 6.2/11.2 μm ratio is a good measure of the ionization degree of a population.[3]

1 S. R. Langhoff, *J. Phys. Chem.*, 1996, **100**, 2819–2841, DOI: 10.1021/jp952074g.
2 L. J. Allamandola, *et al.*, *Astrophys. J.*, 1999, **511**, L115, DOI: 10.1086/311843.
3 B. A. Croiset, *et al.*, *Astron. Astrophys.*, 2016, **590**, A26, DOI: 10.1051/0004-6361/201527714.

Mark H. Stockett responded: Indeed we do! Multiply charged PAHs have slightly different energy barriers towards dissociation processes like H_2 emission.[1] Multiple ionization also opens up new dissociation pathways like fission and Coulomb explosion. For large PAHs in the ISM, the most relevant "new" pathways upon multiple ionization are probably the emission of small charged fragments like H^+ and $C_2H_2^+$.[2]

1 T. Chen, *et al.*, *J. Chem. Phys.*, 2015, **142**, 144305.
2 A. I. S. Holm, *et al.*, *J. Chem. Phys.*, 2011, **134**, 044301.

Sandra Brünken said: I have a comment in response to Murthy Gudipati's question about what is known spectroscopically about PAH dications. There are only a handful of spectroscopic infrared studies on PAH dications, *e.g.*, on the anthracene dication,[1] the naphthalene, anthracene and phenanthrene dications,[2] and hexa-*peri*-hexabenzocoronene dications.[3] One of the outcomes of these studies is that also for dications the role of anharmonicity should not be underestimated. Experimentally observed band intensities differ from those calculated at the harmonic level, which is what is present in the PAH databases, due to the presence of overtone and combination modes, and resonances. It would be interesting to use then (anharmonically) calculated vibrational emission spectra of dications and compare them with a large sample of observational data to infer the presence of these dications in the interstellar PAH family.

1 J. Roithová, *et al.*, *Faraday Discuss.*, 2019, **217**, 98–113, DOI: 10.1039/c8fd00196k.
2 S. Banhatti, *et al.*, *Astron. Astrophys.*, 2021, **648**, A61, DOI: 10.1051/0004-6361/202039744.
3 J. Zhen, *et al.*, *Astrophys. J.*, 2017, **836**, 28, DOI: 10.3847/1538-4357/836/1/28.

Murthy S. Gudipati remarked: In a model that does not take all possible components into account, oftentimes errors are smeared out for the lack of a full set of components. For example, here, if you are only considering neutral PAHs or other molecules, but leave out ionized PAHs and other ionized stable species (in the given environment), the analysis may give results that could be misinterpreted. Perhaps ionized molecular species could be included in the analysis.

Sandra Brünken noted: I would like to mention that many astronomical retrieval codes used to decompose the PAH family (for size, charge, *etc.*), like PAHFIT[1] and PAHTAT,[2] use decomposition models[3] or astronomical PAH template spectra,[4,5] respectively, and not calculated PAH vibrational spectra, or even experimental spectra. This highlights the need for more, and more direct, communication between laboratory astrophysicists, theoretical chemists, and observers.

1 J. D. T. Smith, *et al.*, *Astrophys. J.*, 2007, **656**, 770, DOI: 10.1086/510549.
2 P. Pilleri, *et al.*, *Astron. Astrophys.*, 2012, **542**, A69, DOI: 10.1051/0004-6361/201015915.
3 F. Boulanger, *et al.*, *Astron. Astrophys.*, 1998, **339**, 194–200, https://ui.adsabs.harvard.edu/abs/1998A&A...339..194B.
4 C. Joblin, *et al.*, *Astron. Astrophys.*, 2008, **490**, 189–196, DOI: 10.1051/0004-6361:20079061.
5 S. Foschino, *et al.*, *Astron. Astrophys.*, 2019, **632**, A84, DOI: 10.1051/0004-6361/201935085.

Alexander K. Lemmens responded: The results of this work indeed exemplify the importance of benchmarking theoretical predictions with suitable experiments, before applying them to interpret observations. On the one hand, as theory is often used to interpret observations, one should be aware of the shortcomings of theory in those instances. On the other hand, when benchmarking theory, theorists and experimentalists should communicate clearly about the shortcomings of experiments.

Ugo Jacovella continued discussion of the paper by Mark H. Stockett: What do you think is the influence of the CN substitution on PAHs for recurrent fluorescence efficiency? What should be the next step, changing the carbon skeleton with the cyano group attached or changing the substituent on the naphthalene?

Mark H. Stockett answered: The CN-substituent breaks the symmetry and makes the lowest energy D1 → D0 transition weakly allowed. This transition is strictly forbidden for regular naphthalene. On top of that, Herzberg–Teller coupling increases the oscillator strength of this transition by two orders of magnitude. We have looked at some other substituted PAHs[1] and each has needed to be treated on a case-by-case basis. The electronic structure calculations are extremely challenging. More experiments are needed to identify patterns.

1 B. Zhu, *et al.*, *J. Chem. Phys.*, 2022, **157**, 044303.

Dwayne E. Heard asked: Is recurrent fluorescence (RF) unique to PAHs? Is this type of fluorescence useful for anything else in this community?

Mark H. Stockett replied: RF has also been experimentally shown to be the dominant cooling mechanism for carbon chain anions (even numbers)[1] and cations (odd numbers),[2] and $C_{60}{}^{-/0/+}$ (*e.g.* ref. 3).

1 G. Ito, *et al.*, *Phys. Rev. Lett.*, 2014, **112**, 183001.
2 S. Iida, *et al.*, *Mon. Not. R. Astron. Soc.*, 2022, **514**, 844–851.
3 K. Hansen and E. E. B. Campbell, *J. Chem. Phys.*, 1996, **104**, 5012–5018.

Conflicts of interest

There are no conflicts to declare.

Faraday Discussions

PAPER

Infrared photodesorption of CO from astrophysically relevant ices studied with a free-electron laser

Emily R. Ingman,[a] Domantas Laurinavicius,[b] Jin Zhang,[c] Johanna G. M. Schrauwen,[d] Britta Redlich,[d] Jennifer A. Noble,[ef] Sergio Ioppolo,[cg] Martin R. S. McCoustra[b] and Wendy A. Brown[*a]

Received 31st January 2023, Accepted 22nd March 2023
DOI: 10.1039/d3fd00024a

The infrared excitation and photodesorption of carbon monoxide (CO) and water-containing ices have been investigated using the FEL-2 free-electron laser light source at the FELIX laboratory, Radboud University, The Netherlands. CO–water mixed ices grown on a gold-coated copper substrate at 18 K were investigated. No CO photodesorption was observed, within our detection limits, following irradiation with light resonant with the C–O vibration (4.67 μm). CO photodesorption was seen as a result of irradiation with infrared light resonant with water vibrational modes at 2.9 μm and 12 μm. Changes to the structure of the water ice, which modifies the environment of the CO in the mixed ice, were also seen subsequent to irradiation at these wavelengths. No water desorption was observed at any wavelength of irradiation. Photodesorption at both wavelengths is due to a single-photon process. Photodesorption arises due to a combination of fast and slow processes of indirect resonant photodesorption (fast), and photon-induced desorption resulting from energy accumulation in the librational heat bath of the solid water (slow) and metal-substrate-mediated laser-induced thermal desorption (slow). Estimated cross-sections for the slow processes at 2.9 μm and 12 μm were found to be $\sim 7.5 \times 10^{-18}$ cm^2 and $\sim 4.5 \times 10^{-19}$ cm^2, respectively.

[a] *Department of Chemistry, University of Sussex, Falmer, Brighton, BN1 9QJ, UK. E-mail: w.a.brown@sussex.ac.uk*
[b] *Institute of Chemical Sciences, Heriot-Watt University, Edinburgh, EH14 4AS, UK*
[c] *School of Electronic Engineering and Computer Science, Queen Mary University of London, London E1 4NS, UK*
[d] *FELIX Laboratory, Radboud University, Nijmegen 6525 ED, The Netherlands*
[e] *Physique des Interactions Ioniques et Moléculaires (PIIM), CNRS, Aix-Marseille Université, Marseille, France*
[f] *School of Physical Sciences, University of Kent, Canterbury, CT2 7NH, UK*
[g] *Department of Physics and Astronomy, University of Aarhus, Ny Munkegade 120, 8000 Aarhus C, Denmark*

1. Introduction

To date, more than 240 different species have been identified in space in the gas phase.[1] As well as gas-phase species, molecules are also found frozen out on dust-grain surfaces, forming icy mantles.[2–4] The composition of these icy mantles depends on the region of space, but includes water ice, CO, CO_2 and methanol ice[4–7] and several organic species.[8] Molecular ices frozen out on dust grains act as an important reservoir for molecules in astrophysical environments, such as the interstellar medium (ISM),[9,10] comets[11] and circumstellar disks.[12] In many regions of space, water and CO dominate the solid state.[6,7] Water accretes reactively on grain surfaces from early in the cooling phase of cold, dense environments,[13] initially forming solid water clusters due to diffusion over the grain surface.[14] However, as temperatures fall below ~30 K, CO deposition also occurs from the gas phase. In parallel with continued reactive accretion of water, this then forms a mixed film of CO and water ice that coats the grain surface. At the very lowest temperatures, a CO-rich layer is formed on the water-rich layer to also give rise to layered ices.[15]

Processing of these ice-covered grains by energetic particles including vacuum ultra-violet (VUV),[16–18] X-rays,[19] electrons[20–22] and cosmic rays[19,20,23] leads to desorption into the gas phase. Processing of mixed ices can also lead to the formation of complex organic molecules (COMs),[24] many of which have an important role in forming prebiotic species.[25] In addition to these more energetic particles, there is also a large amount of infrared radiation in astrophysical environments.[26] To date, however, there have been very few investigations of the effects of infrared radiation on the physics and chemistry of astrophysically relevant ices and this source of radiation is currently neglected in models of star- and planet-forming regions, other than through radiative transport. This contrasts markedly with the extensive literature associated with the impact of VUV and cosmic-ray interactions on the physics and chemistry of ices. As well as understanding the effect of the resonant excitation of ice components by light at different wavelengths, it is also important to understand how any energy deposited into the icy mantle of a dust grain is dissipated within the grain.

Solid CO absorbs in the infrared around 4.67 μm [27] and solid water exhibits infrared bands around 3 μm (O–H stretch), 4.5 μm (a combination of the H–O–H bend and librational motions), 6.10 μm (the H–O–H bend) and 12.8 μm (librational motions).[28] Numerous studies report on the infrared spectra of mixtures of CO and water ice in the solid state.[29–34] As the prototypical model of interstellar ices, temperature-programmed desorption (TPD) of CO from solid water surfaces has also been extensively studied,[34–36] and binding energies for CO on solid water typically range from 6–12 kJ mol^{-1}, corresponding to photon wavelengths of 20 and 10 μm, respectively. Hence, it might be expected that irradiation of CO on solid water at any wavelength shorter than 20 μm should result in CO desorption. In contrast to CO desorption promoted by VUV photons, which has been extensively studied,[37,38] observations of infrared-promoted desorption are few and far between, even though the infrared flux in the interstellar radiation field in molecular clouds is comparable to the VUV flux in the same region (10^4 photons cm^{-2} s^{-1}).[39] Given that infrared radiation penetrates more deeply than VUV radiation into dense objects, the presence of such radiation in cold, dense cores may help, in part, to explain observations of excess gas-phase densities of species like CO that would be expected to be condensed out on grains in dense, low-temperature environments.

Previous investigations of the effect of selective infrared irradiation of water ice have shown that desorption[40,41] and restructuring[42,43] of the water ice can take place, with the exact effects depending on which vibrational modes of the water are excited. A more recent study of the infrared free-electron laser irradiation of CO_2 ice[44] under astrophysically relevant conditions has also shown that restructuring of the ice, as well as photodesorption, occurs following irradiation with infrared light. Infrared free-electron laser studies of CO irradiation in the presence of methanol ice have also been undertaken very recently and show the photodesorption of CO.[45]

To investigate the role of infrared absorption and excitation in processing astrophysically relevant ices, we have studied the infrared excitation and photodesorption of model astrophysical ices containing water and CO. A free-electron laser source was used to selectively excite vibrational modes of water and CO in CO–water mixed ices. As well as reporting observations of infrared-induced photodesorption of CO and commenting on the mechanisms that could lead to such photodesorption, we also present estimates of the cross-section, and photodesorption yield, for infrared-induced photodesorption of CO from water-containing ices.

2. Experimental methods

Experiments were undertaken using the laboratory ice surface astrophysics (LISA) ultrahigh vacuum (UHV) end-station at the HFML-FELIX laboratory, Radboud University, The Netherlands. Details of the experimental set-up have been reported previously.[42,44] CO and water ices were grown on an Au-coated Cu substrate cooled to a temperature of between 18–25 K using a closed-cycle He cryostat (SHI Sumitomo Cryogenics). In this substrate temperature range, water forms a well-characterised porous structure known as porous amorphous solid water p-(ASW).[46] Solid CO on the other hand exhibits amorphous behaviour at the lower temperature limit of the range and is known to be crystalline at the higher temperature limit.[47,48] The sample was cleaned before each experiment by heating to 200 K to desorb any remaining water and CO ice. The irradiation of mixtures of CO and water (deionised) ranging from 6–42% of CO with respect to water was investigated. CO was used as received from Hoekloos/Praxair (99.9% purity). Ices were grown *via* background deposition of molecules through an all-metal leak valve. Water doses are given in L_m where 1 L_m corresponds to deposition at a pressure of 1 $\times 10^{-6}$ mbar for 1 s. Mixed ices were grown by co-depositing 780 L_m water and the appropriate amount of CO to give the required ice composition.

The FEL-2 light source at FELIX, consisting of 10 μs macropulses at a 5 Hz repetition rate, was used to irradiate the ices at base temperature for a total of 5 minutes. FEL-2 allowed the irradiation of CO and water in the mid-infrared range, due to its tunability over the 3–45 μm wavelength region. The energy of the laser pulses depended on the wavelength of light used, and the attenuation of the beam, but generally varied between 1–140 mJ. Power-dependent studies were undertaken by applying up to 38 dB of attenuation to the laser source to vary the average energy of the beam. The spectral FWHM of the FEL-2 beam is estimated to be ~0.8% $\delta\lambda/\lambda$ for the entire wavelength range. THE FEL-2 beam impinges on the Au-plated Cu substrate at an angle of 45° with respect to the surface. The optical configuration of the FEL-2 source was optimised to ensure that the infrared light used to irradiate the sample was p-polarised, so that the absorption of infrared light at the metal surface was maximal.[49] The FEL-2 laser spot on the sample was around 2 mm in height. Since the substrate was 38 × 57 mm in size, this allowed multiple clean spots on the ice to be irradiated without needing to completely clean the sample and re-dose the ice mixture between irradiations. Although this increased the number of experiments that could be undertaken without cleaning and re-dosing the ice, it meant that some re-adsorption of CO onto the ice occurred during the long time of an experiment. Ices were irradiated with a range of different wavelengths corresponding to the water and CO infrared bands shown in Fig. 1.

Irradiation of the sample was undertaken at 2.90 ± 0.02 μm (OH stretch mode), 3.10 ± 0.02 μm (OH stretch mode), 3.47 ± 0.03 μm (water off-resonance), 4.50 ± 0.04 μm (CO off-resonance), 4.67 ± 0.04 μm (CO stretch), 6.00 ± 0.04 μm (water bending mode) and 12.00 ± 0.09 μm (water librational mode). The upper and lower limits on the irradiation wavelengths indicate the FWHM of the FEL-2 beam. These wavelengths were chosen to selectively irradiate the ices at wavelengths resonant with the different vibrational modes of water and CO. Off-resonant wavelengths were also investigated to confirm that any observations occurred because of direct absorption by the water and/or CO.

A combination of reflection–absorption infrared spectroscopy (RAIRS) and mass spectrometry measurements of photon-induced photo-induced desorption (PID) was used to quantify the results of the infrared irradiation. A Fourier transform infrared spectrometer (Vertex 80v, Bruker) coupled to a liquid-nitrogen-cooled mercury cadmium telluride detector was used for the RAIRS experiments. The beam angle of the infrared light onto the substrate in the RAIRS experiments was 13°. RAIR spectra were recorded prior to irradiation, and after irradiation of the ice at 18 K. The pre-irradiation spectrum consisted of the co-addition of 512 scans, and the post-irradiation spectra were the result of the co-addition of 256 scans. All spectra were recorded with a resolution of 0.5 cm^{-1}. Post-irradiation

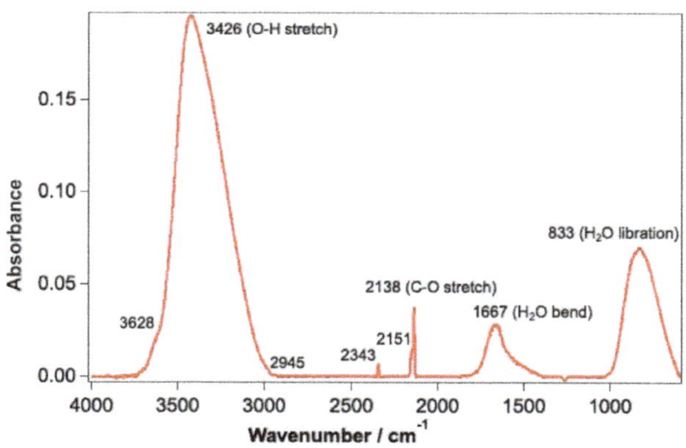

Fig. 1 Infrared spectrum of a mixed CO and water ice dosed on the Au-coated Cu substrate at 18 K. The ice consists of 780 L_m of ASW and 13.3% CO.

infrared spectra are reported as difference spectra, which are obtained by subtracting the pre-irradiation spectrum from the post-irradiation spectrum to show any spectral changes that occur because of the FEL-2 irradiation. PID data were recorded using a quadrupole mass spectrometer (QMS) (Hiden Analytical HAL 201). Mass spectrometer fragments of $m/z = 18$ u (water) and $m/z = 28$ u (CO) were recorded. These are referred to as mass 18 and mass 28, respectively, throughout the discussion. Hereafter we discuss RAIR spectra in wavenumbers and FEL-2 irradiation in wavelength to highlight the higher spectral resolution of the RAIRS data compared to the transform-limited bandwidth of the FEL-2 beam.

3. Results and discussion

3.1 Infrared spectra of unirradiated ices

RAIR spectra were initially recorded for the unirradiated ices to allow a comparison to be made with the irradiated ices. RAIR spectra of pure CO and CO in the presence of ASW have been well characterised in the literature. Fig. 2 shows RAIR spectra for unirradiated ices consisting of mixed CO and water ice at various compositions. Only the spectral region from 2100–2200 cm^{-1} is shown, to highlight the C–O stretching mode. The RAIR spectra in Fig. 2 show two main bands, at 2137 and 2151 cm^{-1}, which have previously been assigned to CO adsorbed in a water matrix and dispersed as a monolayer in the p-ASW structure.[34,50] For the 41.7% mixed ice, a strong feature is seen at 2143 cm^{-1}, which overlays the 2137 cm^{-1} feature. This band is associated with the presence of pure CO adsorbed on the surface[50–54] and suggests that there is also a layer of CO adsorbed on top of the CO–water mixed ice due to re-adsorption from the background in the UHV chamber in this case. Hence, at low mixing fraction, the spectra are consistent with CO isolated in, and monolayer CO adsorbed on, the matrix of p-ASW. As the mixing ratio increases, the appearance of a feature at 2143 cm^{-1} due to solid CO points to the presence of small clusters of CO isolated in, and multilayer CO adsorbed on, the p-ASW matrix.

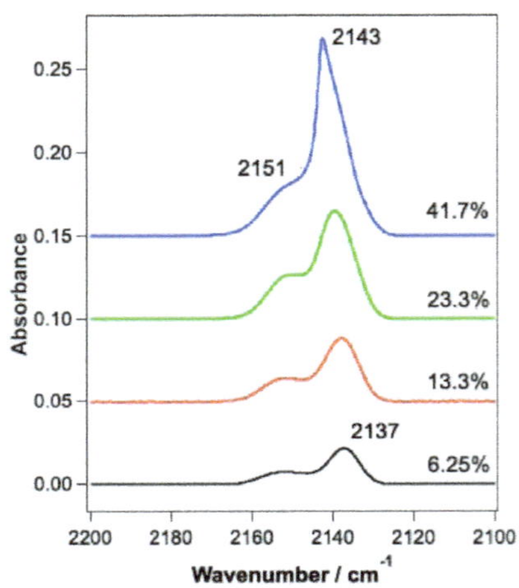

Fig. 2 RAIR spectra of mixtures of CO and water ranging from 6.25–41.7% CO with respect to 780 L_m ASW. Ices were grown on the Au-plated Cu substrate at 18 K. Traces are offset vertically for clarity of presentation.

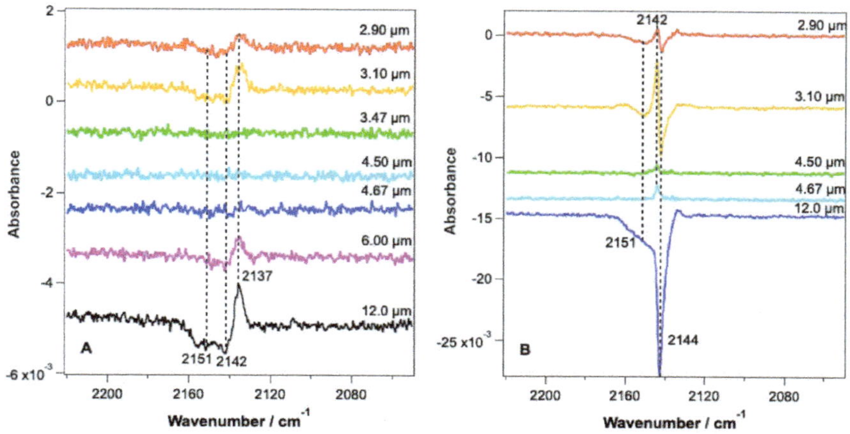

Fig. 3 RAIR difference spectra following the irradiation of CO and water mixtures consisting of (A) 6.25% and (B) 41.7% CO in 780 L_m water. Ices were deposited on Au-coated Cu at 18 K. Irradiation was carried out at a range of wavelengths, as marked on the figure. Traces are offset vertically for clarity of presentation.

3.2 Infrared irradiation of CO and water-containing ices

Fig. 3 shows RAIR difference spectra for CO–water mixed ices at a range of irradiation wavelengths. Fig. 3A shows spectra for an ice consisting of 6.25% CO with respect to water ice and Fig. 3B shows spectra recorded for an ice consisting of 41.7% CO with respect to water ice. Due to the tunability of the FEL-2 light source, different energies were available at different wavelengths, as shown in Table 1.

The RAIR spectra seen in Fig. 3 show clear wavelength-dependent spectral changes following irradiation at 2.9, 3.1, 6 and 12 μm. The largest effects are seen following irradiation of the ices at 12 μm. Fig. 3A shows that for all wavelengths where an effect is observed for the 6.25% CO–water ices, the change in the CO spectrum is similar. There is a loss of infrared signal intensity between 2141 and 2155 cm^{-1} and an increase in signal intensity at 2137 cm^{-1}. No effect is seen following irradiation at 3.47, 4.5 and 4.67 μm, at least within the detection limits

Table 1 Wavelengths used to irradiate the 6.25% and 41.7% mixed CO–water ices shown in Fig. 3, with the corresponding energies of irradiation at that wavelength

	Average irradiation energy/mJ	
Wavelength of irradiation/μm	6.25% CO–water mixture	41.7% CO–water mixture
2.90	1.6	5
3.10	12	7
3.47	12	—
4.50	4	1.5
4.67	20	1
6.00	50	—
12.0	110	100

of our experiment. The observed spectral changes can be assigned either to a loss of CO from the ice, or to a change in the environment of CO within the mixed ice promoted by the irradiation. The observed increase in the CO band at 2137 cm^{-1} suggests an increase in the CO–water interactions in the ice due to irradiation.

Fig. 3B shows spectral changes that occur as the result of irradiation of the 41.7% mixed ice at various wavelengths. Irradiation at 2.9 and 3.1 μm (the O–H stretching mode) results in a loss of the CO signal at 2144 cm^{-1} and 2151 cm^{-1} and a small gain in signal intensity at 2142 cm^{-1}. Irradiation at 12 μm shows the largest spectral changes, with a large loss in CO band intensity at 2144 and 2151 cm^{-1} being observed. The observed spectral changes in Fig. 3B are larger than those in Fig. 3A due to the larger proportion of CO in the ice. The increase in the band at 2142 cm^{-1} seen for all wavelengths of irradiation can be assigned to re-adsorption of CO during the time of the experiment (data for the irradiation of a 41.7% CO mixed ice were recorded several hours after the initial dosing of the sample), and not to radiation-induced changes. This is confirmed by the fact that this effect is seen following irradiation off-resonance (4.5 μm), when no other spectral changes are seen. It is therefore likely that the CO–water mixtures in these experiments also have a thin, dispersed, CO layer on top of the mixed ice due to re-adsorption during the time between the initial ice dose and the irradiation experiments. The decreases in the CO band at 2144 and 2151 cm^{-1} are assigned to the loss of CO from the ice that occurs following irradiation at 2.9, 3.1 and 12 μm. These bands are the same as those seen for the unirradiated ice in Fig. 2 and hence represent loss of CO from the ice.

Fig. 4 shows PID traces recorded with the QMS during the irradiation shown in Fig. 3. Although an overall loss of solid CO signal intensity due to irradiation at 2.90, 3.1, 6 and 12 μm is observed for the ice consisting of 6.25% CO (Fig. 3A), no

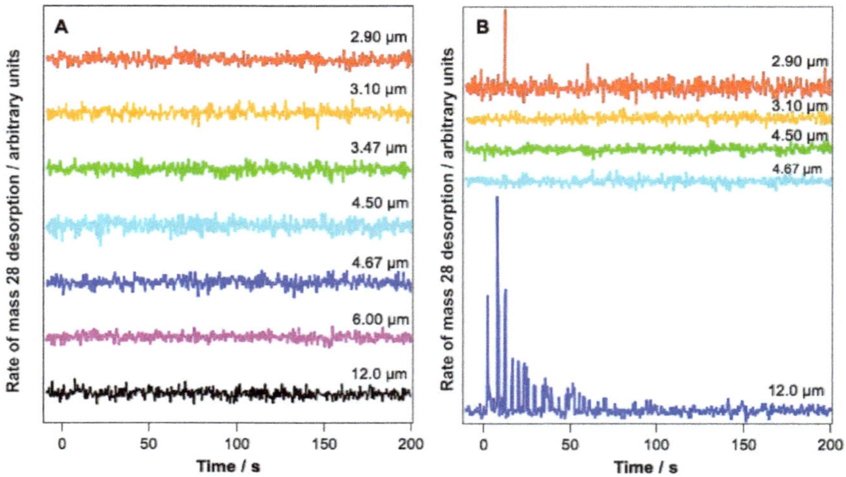

Fig. 4 PID traces recorded during irradiation of (A) a 6.25% CO–water mixed ice and (B) a 41.7% CO–water mixed ice with light of various wavelengths for a total of 5 minutes. Ices were dosed onto Au-coated Cu at 18 K. Time 0 s represents the time at which the irradiation began.

photodesorption is observed at any wavelength for this ice within the detection limits of our QMS (Fig. 4A).

In contrast, Fig. 4B shows that for the 41.7% CO–water mixture, photodesorption of CO is seen when the ice is irradiated at wavelengths of 2.9 μm and 12 μm. This observation agrees with the infrared difference spectra in Fig. 3B, which clearly show a loss of CO signal at 2144 and 2151 cm^{-1} resulting from irradiation at 2.9 and 12 μm. No CO PID signal was observed following irradiation at 4.67 μm, at least within the detection limits of the QMS. No PID signal for water was observed at any irradiation wavelength. Photodesorption was also not observed for either ice composition at an off-resonance wavelength of 4.5 μm. Off-resonance irradiation was therefore not investigated further.

Although no previous work has investigated the resonant infrared irradiation of mixed CO and water-containing ices to the best of our knowledge, Santos *et al.*[45] have studied the irradiation of mixed CO and methanol-containing ices. They observed the loss of CO signal in RAIRS following irradiation at both the methanol and CO (4.67 μm) wavelengths. The lack of any detectable effect due to irradiation at 4.67 μm in our experiments is therefore in contrast to observations made by Santos *et al.*[45] However, the ices used here are considerably thinner than those studied by Santos *et al.*[45] Hence it is expected that the small effects seen in RAIRS for their thicker ices, following irradiation at 4.67 μm, would not be detectable for our much thinner ices. We also note that Santos *et al.*[45] did not observe a CO PID signal during irradiation at 4.67 μm, in agreement with our observations. It is also likely that CO–water ices show different behaviour when compared to CO–methanol ices due to the different interactions between CO and methanol compared to those between CO and water.

Noble *et al.*[42] also studied the irradiation of a pure ASW ice with the FEL-2 beam using wavelengths like those used in this work. RAIR spectra showed a change from amorphous to crystalline water following infrared irradiation with light corresponding to the water vibrational modes.[42] Hence it is expected that the observed changes in the CO bands shown in Fig. 3 may arise due to a combination of CO loss from the ice, and a change in the environment of the CO caused by structural changes in the water ice. This is confirmed by RAIR spectra (shown later) that show that the water bands also change upon infrared irradiation.

3.3 Detailed studies of irradiation at 2.90 and 12.0 μm

As shown by the data in Fig. 3 and 4, the largest changes occur in the mixed ice due to irradiation at 2.9 μm and 12 μm, corresponding to the O–H stretch and librational modes of water, respectively. Hence a more detailed study of the irradiation of mixed CO–water ices at these wavelengths was undertaken. Fig. 5 shows RAIR difference spectra for the OH and CO stretching regions for various mixed CO–water ices irradiated at 2.9 μm with beam energies between 1 and 5 mJ.

As seen in Fig. 5A, the percentage of CO in the ice does not affect the features seen in the O–H stretching region. The O–H band (Fig. 5A) shows a loss of signal intensity at around 3530 cm^{-1} and 3658 cm^{-1} and an increase in signal intensity at $\sim$3364 cm^{-1}. Previous studies have reported the same changes in the RAIR spectrum following irradiation of pure ASW ice at 2.7 μm.[42,55–57] These spectral changes can therefore be assigned to a loss of OH dangling bonds as the water ice undergoes a structural change from an amorphous structure to a more crystalline structure,

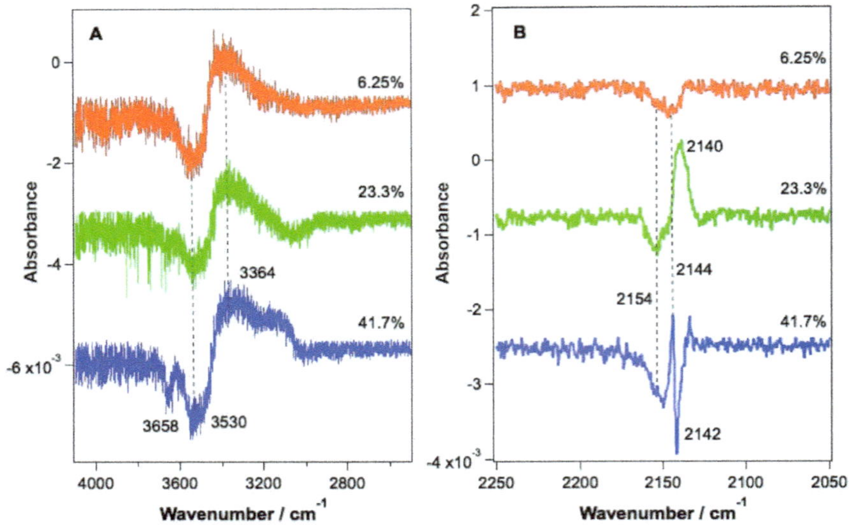

Fig. 5 RAIR difference spectra following the irradiation of various mixed ices at 2.9 μm for 5 minutes. Irradiation energies were 1.6, 4 and 5 mJ for ices with compositions of 6.25%, 23.3% and 41.7%, respectively. (A) shows the O–H stretching region and (B) shows the C–O stretching region. Ice mixtures were dosed on Au-coated Cu at 18 K. Traces are offset vertically for clarity of presentation.

due to irradiation. The CO infrared band shows the same changes on irradiation as already described in Fig. 3, with the changes becoming more pronounced for increasing percentages of CO within the ice. We note that the increase in the CO band at ~2140 cm^{-1} for the 23.3% and 41.7% ice suggests the re-adsorption of a layer of CO on top of the mixed CO–water ice, as described earlier.

PID traces (not shown) were also recorded with the QMS during the irradiation of mixed CO–water ices at 2.9 μm. Data showed that photon-induced desorption of CO was observed at energies above around 9 mJ. It is unlikely that this value represents a threshold energy for observation of PID, and it is most likely that PID is not seen below this energy due to the relatively low sensitivity of the QMS and the high residual CO signal in some experiments. We note that the presence of a re-adsorbed layer of CO ice on top of the mixed CO–water ice suggests that it is most likely CO from this "capping" layer that desorbs into the gas phase following ice irradiation. This is discussed in more detail below with respect to irradiation of CO–water ices at 12 μm.

Fig. 6 shows RAIR difference spectra that result from the irradiation of a 13.3% CO–water ice at various energies at a wavelength of 12 μm. Fig. 6A shows changes observed for the water ice in the O–H stretching region and Fig. 6B shows changes observed for the CO ice. As already shown in Fig. 3, the observed spectral changes are larger than those observed following irradiation at 2.9 μm. Similar spectral changes were observed for all ice mixtures investigated over the range from 6.25% CO to 41.7% CO.

The spectral changes observed in the O–H stretching region in Fig. 6 are the same as those observed following irradiation at 2.9 μm (Fig. 5) and can hence be

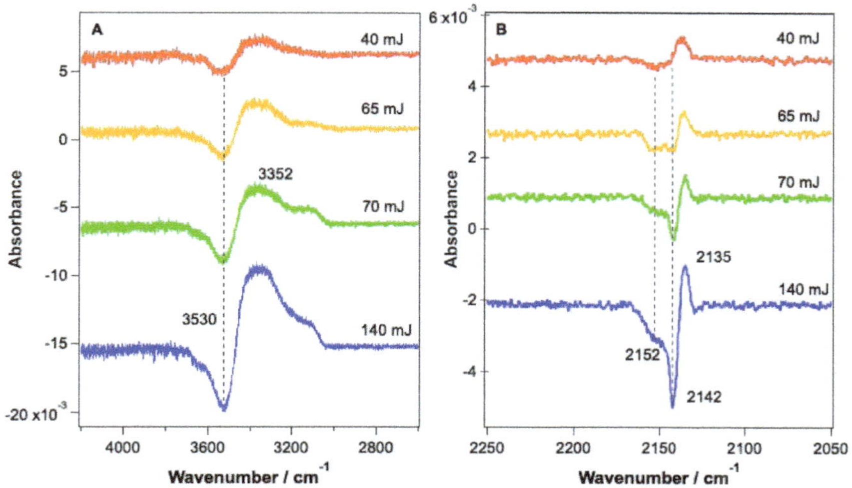

Fig. 6 RAIR difference spectra following the irradiation of a CO–water mixed ice, consisting of 13.3% CO, dosed on Au-coated Cu at 18 K, at 12 μm for 5 minutes. (A) shows the O–H stretching region and (B) shows the C–O stretching region. Traces are offset vertically for clarity of presentation.

assigned to the same structural changes in the water ice, observed previously by Noble *et al.*[42,58] As the energy of the infrared irradiation increases, the intensity of the observed spectral changes increases. Fig. 6B shows changes in the CO spectral region that occur following irradiation at 12 μm. The changes seen in Fig. 6B are like those observed in Fig. 5B, which arise from irradiation at 2.9 μm, with a loss of the bands at 2142 cm^{-1} and 2152 cm^{-1} and an increase in the C–O mode at 2135 cm^{-1}. The loss of CO bands at 2152 and 2142 cm^{-1} increases as the laser power increases.

To determine whether the observed changes in the C–O vibrational bands seen in Fig. 6B arise as a result of a change in environment of the water ice due to structural rearrangement, or due to loss of CO from the ice, PID traces were also recorded for a range of mixtures during the infrared irradiation. Fig. 7 shows example PID traces for a 13.3% CO–water mixed ice at a range of irradiation energies. Photodesorption was not observed for a 6.25% CO–water ice at any energy, even though infrared difference spectra do show spectral changes. This is most likely due to the low sensitivity of the QMS used for the PID measurements. For mixtures containing 13.3%, 23.2% and 41.7% CO, no photodesorption was observed at irradiation energies below 32 mJ. However, above these energies, photodesorption was observed for all ice percentages investigated. It is therefore likely that the changes in the CO spectral region observed for these ice percentages (and illustrated for a 13.3% ice in Fig. 6) arise due to a combination of photodesorption and a change in the CO environment due to the structural changes that occur in the water ice.

Since the infrared data suggest the re-adsorption of CO during the relatively lengthy experiments described here, studies were also undertaken where pure CO was deliberately dosed on top of the CO–water mixed ices. The resulting ice was

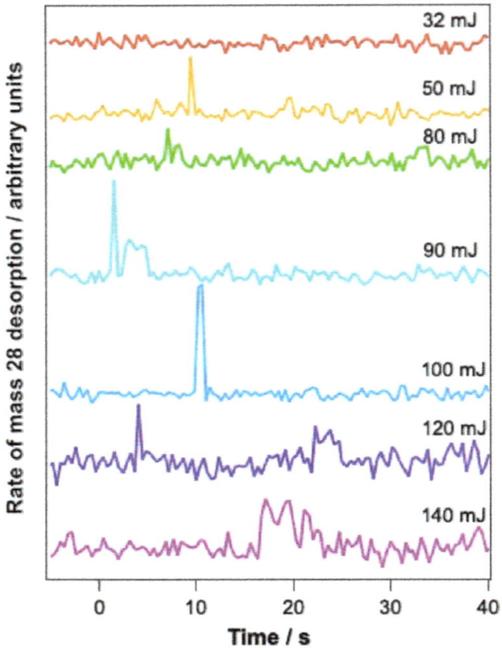

Fig. 7 PID traces recorded during irradiation of a 13.3% CO–water mixed ice, dosed on Au-coated Cu at 18 K, at 12 μm for a total of 5 minutes at different energies of irradiation. Ices were dosed onto Au-coated Cu at 18 K. Time 0 s represents the time at which the irradiation began.

then irradiated and PID traces were recorded during the irradiation. Fig. 8 shows the results of these investigations for 13.3% and 23.3% ice mixtures onto which pure CO was deposited.

Fig. 8 clearly shows a strong PID signal for CO due to irradiation at 12 μm. This suggests that CO is being kicked out from the surface because of irradiation of the ice at a wavelength resonant with the water vibrational modes. These data therefore suggest that it is likely that the CO PID that is observed arises from the CO re-adsorbed on top of the ice, as opposed to CO PID occurring due to CO molecules escaping from the mixed ice.

3.4 Mechanism of photodesorption

As a first step towards gaining a more detailed insight into the mechanism of CO photodesorption from CO and water-containing ices, the laser power dependence for the process was investigated. Power-dependent studies were undertaken at 2.9 μm and 12 μm using the integrated attenuator in the FEL-2 system to reduce the laser power incident on the sample. Typically, in photodesorption processes, the desorption signal scales with laser power, P, in terms of the number of photons involved in the photodesorption process, n, *i.e.*, P^n. Fig. 7 gives an example of the data recorded in these measurements. Given that the QMS response time is slower than ideally desirable, the initial peak intensity for CO photodesorption was not always recorded accurately. Consequently, the experimental QMS signals

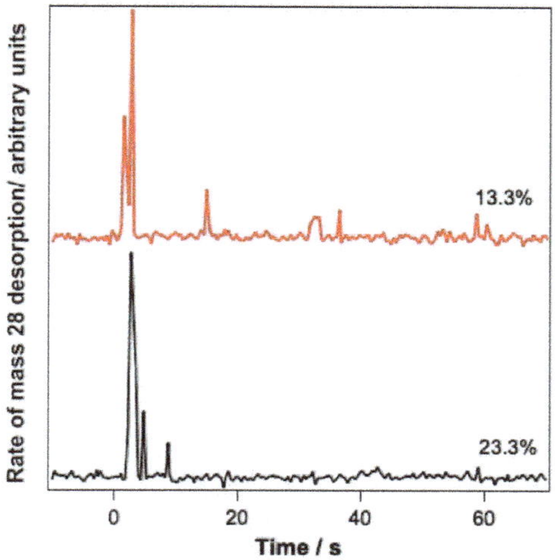

Fig. 8 PID spectra resulting from the irradiat on of 13.3% and 23.3% CO–water mixed ices, onto which a layer of pure CO had been deliberately deposited. Irradiation took place at 12 μm for 5 minutes. Time = 0 corresponds to the point at which irradiation started.

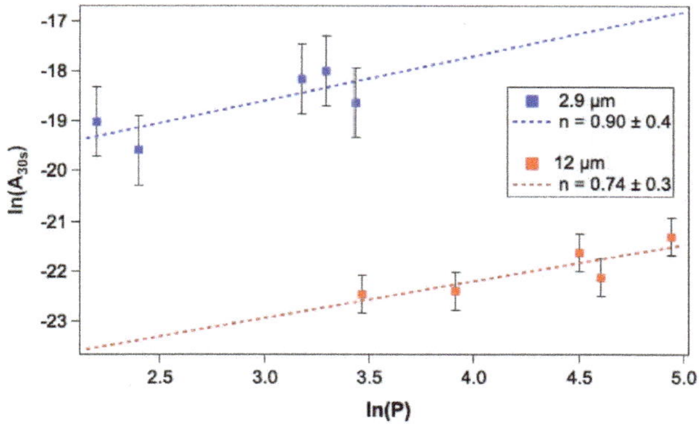

Fig. 9 Plot of ln($A_{30\ s}$) as a function of ln(P) to determine the laser power dependence of irradiation of mixed CO–water ices, comprising 13.3% CO with respect to 780 L_m water ice dosed on Au-coated Cu at 18 K, at 2.9 μm and 12 μm for 5 minutes. Within the errors, excitation at both wavelengths is consistent with a single-photon process.

were integrated for up to 30 s, after which no further photodesorption was observed. Fig. 9 shows a plot of the natural logarithm of the integrated area of the photodesorption traces up to 30 s (A_{30s}) *versus* the natural logarithm of the laser power, P, for a range of irradiation energies at 2.9 μm and 12 μm. The gradient of this plot gives n, the photon order. It is clear from Fig. 9 that photodesorption following irradiation at 2.9 μm and 12 μm occurs *via* a single-photon process ($n \approx 1$). The observation that infrared photon-induced photodesorption is a single-photon process means that it is potentially astrophysically relevant as multi-photon processes are unlikely at either stellar or interstellar radiation field fluxes.

There are four possible mechanisms that could lead to the observed CO photodesorption from thin films of dispersed CO on porous, amorphous water–CO ices on a metallic substrate:

(i) Photodesorption *via* resonant excitation of the C–O stretching vibration of CO adsorbed at a surface water molecule, resulting in dissociation along the weak intermolecular interaction coordinate. This is *direct resonant photodesorption*.

(ii) Photodesorption *via* resonant excitation of the vibrational modes of water in the surface of the ice to which CO is attached, resulting in dissociation along the weak intermolecular interaction coordinate. This is *indirect resonant photodesorption*.

(iii) Photodesorption resulting from energy accumulation in the librational heat bath of the bulk solid water substrate following internal vibrational-energy redistribution (IVR). This is *photon-induced desorption* or, given that in this instance it is driven specifically by a laser, *laser-induced desorption* (LID).

(iv) *Laser-induced thermal desorption* (LITD)[59–61] in which the metal substrate is heated sufficiently by the laser to raise the local temperature under the laser spot to above the desorption temperature of CO. It is worthwhile noting that this process is dependent on the peak power of the laser and not the average power.[62,63]

The wavelength dependence of the photodesorption discussed above allows us to begin to explore these mechanisms, as previously demonstrated in studies of benzene photodesorption following excitation at 250 nm of benzene adsorbed on a compact solid water surface.[64–66]

Excitation of the ice mixtures at 4.67 μm, in resonance with the observed C–O stretching frequency, does not lead to desorption of CO from any ice configuration, at least within the detection limits of our experiment. We can hence conclude that infrared-induced CO photodesorption from dispersed monolayer CO in porous mixtures of CO and water is unlikely to occur *via* mechanism (i). Why might this be the case? The infrared absorption of adsorbed CO overlays the broad pseudo-continuum associated with the combination band of the solid water libration and H–O–H bend. Classically, as the CO is coupled to the solid water surface *via* one or more weak hydrogen bonds or van der Waals interactions, this therefore provides a pathway for IVR into modes strongly coupled to the librational heat sink, resulting in relaxation of the CO stretch on the timescale of a few ps.[67] An alternative view is that quantum mechanically the narrow CO resonance is coupling to the broad underlying continuum *via* Fano coupling,[68] which provides an efficient relaxation pathway consistent with the lack of photodesorption.

Let us now consider CO photodesorption resulting from excitation of the solid water vibrational modes. Two mechanisms are possible in this instance: (ii) and (iii) above. With reference to the work of Thrower *et al.*,[66] QMS detection of desorbates is biased toward detection of fast processes, while RAIRS is more effective at observing slower processes. The results of both sets of observations must therefore be considered if the photodesorption mechanism is to be identified. The discussion above has focussed on the RAIRS data. To complement these data, photodesorption transients were also recorded using the QMS. Transients recorded during the irradiation of equivalent mixtures at the same wavelength and same average beam energy were co-added to improve the quality of the data. Fig. 10 shows an example of the co-added data recorded by the QMS

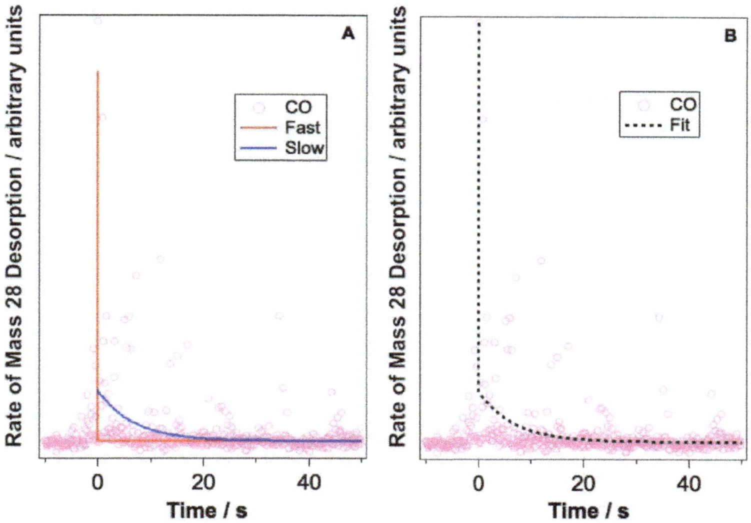

Fig. 10 Example of a bi-exponential fit to the photodesorption transient recorded by the QMS during irradiation of a 13.3% CO–water mixed ice, dosed onto Au-coated Cu at 18 K, with 2.9 μm light (average energy 8.9–11 mJ) for 5 minutes. (A) shows the fast and slow components of the bi-exponential fit and (B) shows the total function, showing a good fit to the recorded data.

during the irradiation of a 13.3% CO–water mixture at 2.9 μm with an average beam energy of 8.9–11 mJ. The observed CO photodesorption transient in Fig. 10 is multicomponent in nature, with an apparent "fast" (larger cross-section) and "slow" (small cross-section) desorption process occurring simultaneously.

This type of multicomponent behaviour has previously been observed in studies of benzene desorption from compact ASW following either electron[69] or ultraviolet irradiation.[66] We will follow the analysis developed in these references given the apparent two-component nature of the observed desorption, and use a bi-exponential function to describe the photodesorption transients:

$$I(t) = I_1 \exp\left(\frac{-t}{\tau_1}\right) + I_2 \exp\left(\frac{-t}{\tau_2}\right) + I_\infty$$

where $I(t)$ is the intensity of the QMS signal as a function of time t, I_1 and I_2 are the initial intensities at $t = 0$ for each of the fast and slow components, respectively, τ_1 and τ_2 are the time constants for the fast and slow processes, respectively, and I_∞ is the residual baseline QMS signal intensity. The time constants τ_i for the fast and slow desorption processes are given by:

$$\tau_i = \frac{1}{\sigma_i \phi_i}$$

where σ_i is the cross-section for the desorption process at the relevant wavelength and ϕ_i is the laser fluence at the relevant wavelength and energy.

Fig. 10 shows an example of the photodesorption transient data, and the bi-exponential fit to that data, recorded during the irradiation of a 13.3% CO–water ice mixture at 2.9 μm with an average energy of between 8.9–11 mJ. As shown in Fig. 10, the bi-exponential function provides a good fit to the data, suggesting the existence of two components of desorption. Additionally, single-exponential decay and a three-component decay–appearance–decay system were also explored in the fitting of these data. For both alternatives, the fits as described by the correlation coefficient were poorer than the bi-exponential model suggesting that the latter is a better description for the transient decay kinetics. Co-added photodesorption traces for irradiation of 13.3% CO–water mixed ices at 2.9 μm with an average energy of 29 mJ, and for a 13.3% CO–water ice irradiated at 12 μm with an average beam energy of 97.5 mJ, were also fitted with a bi-exponential decay function. The time constants, τ_i for the two components in all fitted systems are given in Table 2.

Table 2 also lists the desorption cross-sections that result from the time constants, derived using the calculated photon fluxes given in Table 2. Cross-sections were not calculated for the fast process because of the large error associated with the time constants for the fast process. This arises because the response time of the QMS is relatively slow and hence it is not able to properly capture faster transient processes. Despite this, we are confident that, at the different wavelengths and energies reported, there is evidence of a fast and a slow desorption process. This is consistent with a combination of process (ii), which is a faster process, and process (iii), which is a slower process, occurring due to infrared irradiation of the CO and water-containing ice.

The cross-sections reported in Table 2 for the infrared photodesorption of CO are for photodesorption driven by the absorption of infrared light at wavelengths corresponding to the O–H stretch and the libration of the solid water substrate. To estimate the efficiency of photodesorption, the quantum yield, it is necessary to compare the CO photodesorption cross-section values with relevant water absorption cross-sections. To date, no infrared absorption cross-sections have been reported in the literature for solid water ice at 2.9 μm and 12 μm. However, an infrared absorption cross-section for dangling O–H bonds in water ice has been reported using transmission spectroscopy to be $1.0 \pm 0.2 \times 10^{-18}$ cm^2.[70] Although no values exist for absorption cross-sections at other wavelengths, it is possible to estimate these (in a "back of an envelope" calculation) by comparing the strength of the infrared absorption of the dangling O–H bonds with the strength of the water O–H stretching mode and the librational mode. Noble *et al.*[55] show that the peak intensity of the O–H stretching mode at 2.9 μm is some 27 times that of the O–H dangling bond. This gives an absorption cross-section at 2.9

Table 2 Table showing the time constants for the fast and slow processes (as illustrated in Fig. 10) for irradiation of a 13.3% CO–water ice at different wavelengths and different average energies. Error limits represent one standard deviation on the fitted value. The photodesorption cross-sections for the slow process, σ_2, and their estimated uncertainties are also shown

Irradiation wavelength/μm	Average irradiation energy/mJ	ϕ/cm^{-2} s^{-1}	τ_1/s	τ_2/s	σ_2/cm^2
12	97.5	6.3×10^{17}	0.002 ± 7	3.5 ± 1.2	$4.5 \pm 1.6 \times 10^{-19}$
2.9	9.6	1.5×10^{16}	$3.7 \times 10^{-5} \pm 0.07$	6.7 ± 0.1	$9.9 \pm 0.1 \times 10^{-18}$
2.9	29	4.5×10^{16}	$2.5 \times 10^{-4} \pm 0.07$	4.3 ± 0.5	$5.2 \pm 0.6 \times 10^{-18}$

μm of ~2.7×10^{-17} cm^2. Data from Mifsud *et al.*[71] show that the libration band at 12 μm has a peak height roughly 10 times smaller than that of the O–H stretching mode, giving an approximate absorption cross-section at 12 μm of 2.7×10^{-18} cm^2. Finally, to give an estimate of the infrared absorption cross-sections at these two wavelengths for ice adsorbed on a metal surface, we need to multiply these two values by the metal surface enhancement factor[49] of around 50, since the value reported for the dangling O–H modes was measured using transmission spectroscopy. This gives estimated absorption cross-sections for solid water of 1.4×10^{-16} cm^2 and 1.4×10^{-15} cm^2 at 12 μm and 2.9 μm, respectively.

Having estimated approximate values for the absorption cross-section of water at 12 μm and 2.9 μm and having shown that both infrared wavelengths result in a single-photon photodesorption process, it is possible to obtain an estimate of the quantum yield for CO photodesorption that occurs following infrared irradiation at these wavelengths. Considering only the slow channel at both wavelengths, given the high uncertainty in the time constants of the fast process, this gives values of the CO photodesorption yields of ~3×10^{-3} per photon at 12 μm and ~5×10^{-3} per photon at 2.9 μm. These values are in good agreement with those previously determined by Öberg and co-workers[72] and by Paardekooper *et al.*[37] for VUV photodesorption of CO from solid CO ices.

What of process (iv), LITD? With a typical micropulse peak energy of 40 μJ and pulse width of around 1 ps, the peak power delivered to the metal substrate by the FEL-2 beam is of the order of 40 MW. Even with the relatively gentle focussing employed in the current work, the laser fluence on the metal substrate surface will exceed 100 MW per cm^2 per micropulse. This is certainly sufficient for LITD. The question is then whether this process is observed. Ultimately, this is determined by whether the local temperature rise under the laser spot is sufficient to desorb CO. However, it is also determined by the bias in our chosen experimental probes: QMS measurements of transients is biased in favour of sampling fast processes while RAIRS favours detection of slow processes.[66] LITD hence may be occurring, but it is likely occurring in parallel with, and hence indistinguishable from, process (iii). Only with future experiments focussing on measuring desorbate time-of-flight distributions will it be possible to disentangle the mechanistic complexities of this system.

4. Conclusions and astrophysical implications

The data reported here show that infrared irradiation of mixed CO–water ices, at wavelengths corresponding to the water O–H stretching and librational modes, leads to changes in the ice that occur because of both rearrangement of the water ice structure and photodesorption of the CO. No effects are seen for irradiation at wavelengths resonant with the C–O stretching mode, at least within our detection limits. No water desorption is observed at any wavelength of irradiation.

The observed CO photodesorption arises because of a single-photon excitation process at both wavelengths of irradiation (2.9 μm and 12 μm). Mechanistically, photodesorption transient measurements suggest photodesorption occurs by a combination of fast and slow processes *via* indirect resonant photodesorption (fast); *via* photon-induced desorption resulting from energy accumulation in the librational heat bath of the bulk solid water substrate following IVR (slow); and *via* metal-substrate-mediated laser-induced thermal desorption (slow).

The estimated infrared photodesorption yields at 2.9 μm and 12 μm (respectively ~5×10^{-3} per photon and ~3×10^{-3} per photon) are similar in magnitude to those previously reported[72] for VUV photodesorption of CO. Given that VUV and infrared fluxes from the interstellar radiation field are comparable, this would imply that infrared photodesorption may be equivalent to, and perhaps exceed, the impact of VUV radiation in returning CO to the gas phase in cold, dense environments. This effect could slow the accumulation of CO on solid water ice surfaces during gas-cloud collapse from the diffuse toward the dense medium. This, in turn, might delay the onset of the accumulation of simple organics associated with hydrogenation of CO. Opportunities exist today, given the power of the James Webb Space Telescope, to explore this phenomenon.

Author contributions

Emily R. Ingman – experiments, data analysis, writing – original draft, data interpretation and commented on the paper. Domantas Laurinavicius – experiments, data analysis, data interpretation and commented on the paper. Jin Zhang – experiments, data interpretation and commented on the paper. Johanna G. M. Schrauwen – experiments and commented on the paper. Britta Redlich – initiation and management of the project, data interpretation and commented on the paper. Jennifer A. Noble – initiation and management of the project, experiments, data interpretation and commented on the paper. Sergio Ioppolo – initiation and management of the project, experiments, data interpretation and commented on the paper. Martin R. S. McCoustra – initiation and management of the project, experiments, data analysis, writing – original draft, data interpretation and commented on the paper. Wendy A Brown – initiation and management of the project, experiments, data analysis, writing – original draft, data interpretation and commented on the paper.

Data Availability

Raw experimental data are available and can be found at http://dx.doi.org/10.25377/sussex.22347769.

Conflicts of interest

The authors declare that there are no conflicts of interest that influence the work reported in this paper.

Acknowledgements

The authors thank the HFML-FELIX Laboratory team at Radboud University, Nijmegen in the Netherlands, for their experimental assistance and scientific support. The LISA end-station was designed and constructed at the HFML-FELIX Laboratory by the groups of S. Ioppolo and B. Redlich, who also manage the end-station. LISA is funded through the Royal Society University Research Fellowships Renewals 2019 (URF/R/191018), the Royal Society University Research Fellowship (UF130409), the Royal Society Research Fellow Enhancement Award (RGF/EA/180306), and the Royal Society Research Grant (RSG/R1/180418). Travel support is acknowledged from the

UK Engineering and Physical Sciences Research Council (UK EPSRC Grant EP/R007926/1 – FLUENCE: Felix Light for the UK: Exploiting Novel Characteristics and Expertise). ERI and DL thank the EPSRC for their provision of funding to enable their PhD studies. SI gratefully acknowledges the Royal Society and the Danish National Research foundation through the Centre of Excellence "Intercat" (grant agreement number DNRF150) for financial support. JAN acknowledges funding from the Agence Nationale de la Recherche (ANR, HYDRAE project ANR-21-CE30-0004-01) and additional support from the French Programme National "Physique et Chimie du Milieu Interstellaire" (PCMI) of the CNRS/INSU with the INC/INP, co-funded by the CEA and the CNES. A. Traspas Muiña is thanked for help with initial experiments on the LISA end-station.

References

1 B. A. McGuire, *Astrophys. J., Suppl. Ser.*, 2022, **259**, 30.

2 D. J. Burke and W. A. Brown, *Phys. Chem. Chem. Phys.*, 2010, **12**, 5947.

3 E. Herbst, *Phys. Chem. Chem. Phys.*, 2014, **16**, 3344–3359.

4 E. L. Gibb, D. C. B. Whittet, A. C. A. Boogert and A. G. G. M. Tielens, *Astrophys. J., Suppl. Ser.*, 2004, **151**, 35–73.

5 E. L. Gibb, D. C. B. Whittet, W. A. Schutte, A. C. A. Boogert, J. E. Chiar, P. Ehrenfreund, P. A. Gerakines, J. v. Keane, A. G. G. M. Tielens, E. F. van Dishoeck and O. Kerkhof, *Astrophys. J.*, 2000, **536**, 347–356.

6 K. I. Öberg, A. C. A. Boogert, K. M. Pontoppidan, S. van den Broek, E. F. van Dishoeck, S. Bottinelli, G. A. Blake and N. J. Evans, *Astrophys. J.*, 2011, **740**, 109.

7 E. Dartois, *Space Sci. Rev.*, 2005, **119**, 293–310.

8 M. K. McClure, W. R. M. Rocha, K. M. Pontoppidan, N. Crouzet, L. E. U. Chu, E. Dartois, T. Lamberts, J. A. Noble, Y. J. Pendleton, G. Perotti, D. Qasim, M. G. Rachid, Z. L. Smith, F. Sun, T. L. Beck, A. C. A. Boogert, W. A. Brown, P. Caselli, S. B. Charnley, H. M. Cuppen, H. Dickinson, M. N. Drozdovskaya, E. Egami, J. Erkal, H. Fraser, R. T. Garrod, D. Harsono, S. Ioppolo, I. Jiménez-Serra, M. Jin, J. K. Jørgensen, L. E. Kristensen, D. C. Lis, M. R. S. McCoustra, B. A. McGuire, G. J. Melnick, K. I. Öberg, M. E. Palumbo, T. Shimonishi, J. A. Sturm, E. F. van Dishoeck and H. Linnartz, *Nat. Astron.*, 2023, **7**, 431–443.

9 J. M. Greenberg, *Astrophys. Space Sci.*, 1986, **128**, 17–31.

10 D. A. Williams and E. Herbst, *Surf. Sci.*, 2002, **500**, 823–837.

11 D. Bockelée-Morvan and N. Biver, *Philos. Trans. R. Soc., A*, 2017, **375**, 20160252.

12 R. Visser, E. F. van Dishoeck, S. D. Doty and C. P. Dullemond, *Astron. Astrophys.*, 2009, **495**, 881–897.

13 F. Dulieu, L. Amiaud, E. Congiu, J.-H. Fillion, E. Matar, A. Momeni, V. Pirronello and J. L. Lemaire, *Astron. Astrophys.*, 2010, **512**, A30.

14 A. Rosu-Finsen, D. Marchione, T. L. Salter, J. W. Stubbing, W. A. Brown and M. R. S. McCoustra, *Phys. Chem. Chem. Phys.*, 2016, **18**, 31930–31935.

15 K. M. Pontoppidan, A. C. A. Boogert, H. J. Fraser, E. F. van Dishoeck, G. A. Blake, F. Lahuis, K. I. Öberg, N. J. Evans II and C. Salyk, *Astrophys. J.*, 2008, **678**, 1005–1031.

16 J. M. Greenberg, *Astrophys. Space Sci.*, 1976, **39**, 9–18.

17 S. S. Prasad and S. P. Tarafdar, *Astrophys. J.*, 1983, **267**, 603.

18 L. J. Allamandola, M. P. Bernstein, S. A. Sandford and R. L. Walker, *Space Sci. Rev.*, 1999, **90**, 219–232.
19 M. H. Moore, B. Donn, R. Khanna and M. F. A'Hearn, *Icarus*, 1983, **54**, 388–405.
20 C. J. Bennett, C. S. Jamieson, Y. Osamura and R. I. Kaiser, *Astrophys. J.*, 2006, **653**, 792–811.
21 A. G. M. Abdulgalil, A. Rosu-Finsen, D. Marchione, J. D. Thrower, M. P. Collings and M. R. S. McCoustra, *ACS Earth Space Chem.*, 2017, **1**, 209–215.
22 A. G. M. Abdulgalil, D. Marchione, J. D. Thrower, M. P. Collings, M. R. S. McCoustra, F. Islam, M. E. Palumbo, E. Congiu and F. Dulieu, *Philos. Trans. R. Soc., A*, 2013, **371**, 20110586.
23 R. I. Kaiser, G. Eich, A. Gabrysch and K. Roessler, *Astrophys. J.*, 1997, **484**, 487–498.
24 E. Herbst and E. F. van Dishoeck, *Annu. Rev. Astron. Astrophys.*, 2009, **47**, 427–480.
25 M. Guélin and J. Cernicharo, *Front. Astron. Space Sci.*, 2022, **9**, 787567.
26 J. S. Mathis, P. G. Mezger and N. Panagia, *Astron. Astrophys.*, 1983, **128**, 212–229.
27 G. E. Ewing and G. C. Pimentel, *J. Chem. Phys.*, 1961, **35**, 925–930.
28 W. Hagen, A. G. G. M. Tielens and J. M. Greenberg, *Chem. Phys.*, 1981, **56**, 367–379.
29 E. Seperuelo Duarte, A. L. F. de Barros, E. F. da Silveira, A. Domaracka, P. Boduch and H. Rothard, *Mon. Not. R. Astron. Soc.*, 2021, **508**, 4297–4309.
30 S. A. Sandford, L. J. Allamandola, A. G. G. M. Tielens and G. J. Valero, *Astrophys. J.*, 1988, **329**, 498.
31 B. Schmitt, J. M. Greenberg and R. J. A. Grim, *Astrophys. J.*, 1989, **340**, L33.
32 M. E. Palumbo, *J. Phys. Chem. A*, 1997, **101**, 4298–4301.
33 C. Manca, C. Martin, A. Allouche and P. Roubin, *J. Phys. Chem. B*, 2001, **105**, 12861–12869.
34 M. P. Collings, J. W. Dever, H. J. Fraser, M. R. S. McCoustra and D. A. Williams, *Astrophys. J.*, 2003, **583**, 1058–1062.
35 M. Minissale, Y. Aikawa, E. Bergin, M. Bertin, W. A. Brown, S. Cazaux, S. B. Charnley, A. Coutens, H. M. Cuppen, V. Guzman, H. Linnartz, M. R. S. McCoustra, A. Rimola, J. G. M. Schrauwen, C. Toubin, P. Ugliengo, N. Watanabe, V. Wakelam and F. Dulieu, *ACS Earth Space Chem.*, 2022, **6**, 597–630.
36 R. S. Smith, R. A. May and B. D. Kay, *J. Phys. Chem. B*, 2016, **120**, 1979–1987.
37 D. M. Paardekooper, G. Fedoseev, A. Riedo and H. Linnartz, *Astron. Astrophys.*, 2016, **596**, A72.
38 M. Bertin, E. C. Fayolle, C. Romanzin, K. I. Öberg, X. Michaut, A. Moudens, L. Philippe, P. Jeseck, H. Linnartz and J.-H. Fillion, *Phys. Chem. Chem. Phys.*, 2012, **14**, 9929.
39 C. Cecchi-Pestellini and S. Aiello, *Mon. Not. R. Astron. Soc.*, 1992, **258**, 125–133.
40 A. Krasnopoler and S. M. George, *J. Phys. Chem. B*, 1998, **102**, 788–794.
41 C. Focsa, B. Chazallon and J. L. Destombes, *Surf. Sci.*, 2003, **528**, 189–195.
42 J. A. Noble, H. M. Cuppen, S. Coussan, B. Redlich and S. Ioppolo, *J. Phys. Chem. C*, 2020, **124**, 20864–20873.
43 H. M. Cuppen, J. A. Noble, S. Coussan, B. Redlich and S. Ioppolo, *J. Phys. Chem. A*, 2022, **126**, 8859–8870.

44 S. Ioppolo, J. A. Noble, A. Traspas Muiña, H. M. Cuppen, S. Coussan and B. Redlich, *J. Mol. Spectrosc.*, 2022, **385**, 111601.

45 J. C. Santos, K.-J. Chuang, J. G. M. Schrauwen, A. Traspas Muiña, J. Zhang, H. M. Cuppen, B. Redlich, H. Linnartz and S. Ioppolo, *Astron. Astrophys.*, 2023, **672**, 112.

46 G. A. Kimmel, K. P. Stevenson, Z. Dohnálek, R. S. Smith and B. D. Kay, *J. Chem. Phys.*, 2001, **114**, 5284–5294.

47 J. Lasne, A. Rosu-Finsen, A. Cassidy, M. R. S. McCoustra and D. Field, *Phys. Chem. Chem. Phys.*, 2015, **17**, 30177–30187.

48 A. Rosu-Finsen, J. Lasne, A. Cassidy, M. R. S. McCoustra and D. Field, *Phys. Chem. Chem. Phys.*, 2016, **18**, 5159–5171.

49 R. G. Greenler, *J. Vac. Sci. Technol.*, 1975, **12**, 1410–1417.

50 M. P. Collings, J. Dever, H. J. Fraser and M. R. S. McCoustra, *Astrophys. Space Sci.*, 2003, **285**, 633–659.

51 G. E. Ewing and G. C. Pimentel, *J. Chem. Phys.*, 1961, **35**, 925–930.

52 G. E. Ewing, *J. Chem. Phys.*, 1962, **37**, 2250–2256.

53 H. Vu, M. R. Atwood and B. Vodar, *J. Chem. Phys.*, 1963, **38**, 2671–2677.

54 L. Marchese, S. Coluccia, G. Martra and A. Zecchina, *Surf. Sci.*, 1992, **269–270**, 135–140.

55 J. A. Noble, C. Martin, H. J. Fraser, P. Roubin and S. Coussan, *J. Phys. Chem. Lett.*, 2014, **5**, 826–829.

56 J. A. Noble, C. Martin, H. J. Fraser, P. Roubin and S. Coussan, *J. Phys. Chem. C*, 2014, **118**, 20488–20495.

57 S. Coussan, P. Roubin and J. A. Noble, *Phys. Chem. Chem. Phys.*, 2015, **17**, 9429–9435.

58 S. Coussan, J. A. Noble, H. M. Cuppen, B. Redlich and S. Ioppolo, *J. Phys. Chem. A*, 2022, **126**, 2262–2269.

59 D. P. Land, C. L. Pettiette-Hall, J. C. Hemminger and R. T. McIver, *Acc. Chem. Res.*, 1991, **24**, 42–47.

60 D. P. Land, T. L. Tai, J. M. Lindquist, J. C. Hemminger and R. T. McIver, *Anal. Chem.*, 1987, **59**, 2924–2927.

61 S. M. George, ed. B. W. Rossiter and R. C. Baetzold, *Investigations of surfaces and interfaces - part A*, Wiley, 1993, 2nd edn, pp. 452–498.

62 J. L. Brand and S. M. George, *Surf. Sci.*, 1986, **167**, 341–362.

63 A. A. Deckert and S. M. George, *Surf. Sci.*, 1987, **182**, L215–L220.

64 J. D. Thrower, A. G. M. Abdulgalil, M. P. Collings, M. R. S. McCoustra, D. J. Burke, W. A. Brown, A. Dawes, P. J. Holtom, P. Kendall, N. J. Mason, F. Jamme, H. J. Fraser and F. J. M. Rutten, *J. Vac. Sci. Technol., A*, 2010, **28**, 799.

65 J. D. Thrower, M. P. Collings, M. R. S. McCoustra, D. J. Burke, W. A. Brown, A. Dawes, P. D. Holtom, P. Kendall, N. J. Mason, F. Jamme, H. J. Fraser, I. P. Clark and A. W. Parker, *J. Vac. Sci. Technol., A*, 2008, **26**, 919.

66 J. D. Thrower, D. J. Burke, M. P. Collings, A. Dawes, P. D. Holtom, F. Jamme, P. Kendall, W. A. Brown, I. P. Clark, H. J. Fraser, M. R. S. McCoustra, N. J. Mason and A. W. Parker, *Astrophys. J.*, 2008, **673**, 1233–1239.

67 S. Taj, D. Baird, A. Rosu-Finsen and M. R. S. McCoustra, *Phys. Chem. Chem. Phys.*, 2017, **19**, 7990–7995.

68 A. E. Miroshnichenko, S. Flach and Y. S. Kivshar, *Rev. Mod. Phys.*, 2010, **82**, 2257–2298.

69 J. D. Thrower, M. P. Collings, F. J. M. Rutten and M. R. S. McCoustra, *Chem. Phys. Lett.*, 2011, **505**, 106–111.
70 T. Nagasawa, R. Sato, T. Hasegawa, N. Numadate, N. Shioya, T. Shimoaka, T. Hasegawa and T. Hama, *Astrophys. J., Lett.*, 2021, **923**, L3.
71 D. V. Mifsud, P. A. Hailey, P. Herczku, Z. Juhász, S. T. S. Kovács, B. Sulik, S. Ioppolo, Z. Kaňuchová, R. W. McCullough, B. Paripás and N. J. Mason, *Eur. Phys. J. D*, 2022, **76**, 87.
72 K. I. Öberg, E. F. van Dishoeck and H. Linnartz, *Astron. Astrophys.*, 2009, **496**, 281–293.

Faraday Discussions

PAPER

Thermal behavior of astrophysical amorphous molecular ices

Murthy S. Gudipati, *[a] Benjamin Fleury, [b] Robert Wagner, [c] Bryana L. Henderson, [a] Kathrin Altwegg [d] and Martin Rubin [d]

Received 20th February 2023, Accepted 7th March 2023
DOI: 10.1039/d3fd00048f

Ice is a major component of astrophysical environments – from interstellar molecular clouds through protoplanetary disks to evolved solar systems. Ice and complex organic matter coexist in these environments as well, and it is thought primordial ice brought the molecules of life to Earth four billion years ago, which could have kickstarted the origin of life on Earth. To understand the journey of ice and organics from their origins to becoming a part of evolved planetary systems, it is important to complement high spatial and spectral resolution telescopes such as JWST with laboratory experimental studies that provide deeper insight into the processes that occur in these astrophysical environments. Our laboratory studies are aimed at providing this knowledge. In this article we present simultaneous mass spectrometric and infrared spectroscopic investigation on how molecular ice mixtures behave at different temperatures and how this information is critical to interpret observational data from protoplanetary disks as well as comets. We find that amorphous to crystalline water ice transformation is the most critical phenomenon that differentiates between outgassing of trapped volatiles such as CO_2 *vs.* outgassing of pure molecular ice domains of the same in a mixed molecular ice. Crystalline water ice is found to trap only a small fraction of other volatiles (<5%), indicating ice grain composition in astrophysical and planetary environments must be different depending on whether the ice is in amorphous phase or transformed into crystalline phase, even if the crystalline ice undergoes radiation-induced amorphization subsequently. Crystallization of water ice is a key differentiator for many ices in astronomical environments as well as in our Solar System.

Introduction

Astrochemistry is aimed at simulating astrophysical environments to predict chemical processes as well as providing their underlying mechanisms in these

[a]*Science Division, Jet Propulsion Laboratory, California Institute of Technology, 4800 Oak Grove Drive, Pasadena, CA 91109, USA. E-mail: murthy.gudipati@jpl.nasa.gov*
[b]*Univ Paris Est Creteil and Université Paris Cité, CNRS, LISA, F-94010 Créteil, France*
[c]*Institute of Meteorology and Climate Research, Karlsruhe Institute of Technology, Karlsruhe, Germany*
[d]*Physikalisches Institut, University of Bern, Sidlerstrasse 5, CH-3012 Bern, Switzerland*

environments. As JWST continues to provide outstanding data at better spatial and spectral resolution and sensitivity than the previous telescopic measurements,[1–4] the need to obtain laboratory data at higher resolution is now more important than ever before. In addition to higher spectral resolution that enables better separation of individual absorption/emission bands and higher spatial resolution of observations that provides better contrast of the observed objects (whether interstellar molecular clouds or protoplanetary disks or other solar system objects), we need a better understanding of the effect of temperature, pressure, and local radiation in these environments that are often spatially heterogeneous. Small temperature variations could have a significant effect on the ice *vs.* gas-phase composition in these environments.[5,6] Temperature measurements in astrophysical environments are still averaged through the line of sight and often need to be approximated. For this reason, if laboratory studies could provide better handle on the effect of temperature on the ice/gas partitions, we should be able to use these data to better constrain temperatures of local astrophysical environments. Observations with higher spatial resolution can then be better interpreted using laboratory data.

In protoplanetary environments, ice-lines are drawn at various radial and lateral distances from the protostar.[6–8] However, these lines are dictated by not only ice composition and its phase (amorphous or crystalline), but also the exact temperature of the local environment. For this reason, we focused in our study on outgassing of mixed molecular ices and their composition at various temperatures. In protoplanetary disks, though temperature decreases with radial distance, volatile and supervolatile molecular species (with respect to water) would form pure ices or mixed ices trapped in water and other ices based on their local abundances. It is still not clear whether ice particles form in the protoplanetary disks freshly or retain their interstellar ice grain properties as they accrete on to a protoplanetary disk during its formation. Though radial transportation is used in many models, how far the grains and gas are transported and whether transport occurs primarily in one direction or both directions (high to low and low to high temperatures) are still open questions,[9,10] particularly, when gas-phase observations are used as a proxy for ice composition.[11] It is also assumed that the icy grains in the protoplanetary disks become a part of cometary nuclei retaining their chemical, structural, and compositional integrity.[12–14] Thus, there is an effort to correlate cometary outgassing (reflecting thermal evolution of a comet during its travel around the Sun) with the ice grain composition of protoplanetary disks. In order to gain better constraints on interpreting observational data from icy environments – be they interstellar, protoplanetary, or cometary nuclei – we need a wide range of laboratory data. We conducted temperature programmed desorption (TPD) studies of mixed molecular ices by means of simultaneous mass spectrometry of outgassed molecules and infrared spectroscopy of mixed molecular ices. We also carried out calorimetric study of macroscopic water ice. Our experiments provide some of these critically needed data and understanding of the processes involved, complement existing experimental data as discussed in the following, and provide a deeper understanding of the role of local temperature on the ice physics.

Experimental

All the experiments were conducted at the Ice Spectroscopy Laboratory (ISL) of the Jet Propulsion Laboratory. We used standard Temperature Programmed Desorption (TPD) methods to determine sublimation of mixed molecular ices at various starting mixing ratios. The experimental station (named Himalaya) used for this work is shown in Fig. 1.

TPD of mixed molecular ices

Molecules used here are CO, O_2, CO_2, and H_2O as a realistic representative of supervolatiles (CO and O_2), and volatiles (CO_2) in water ice. We chose much higher mixing ratios such that the volatiles are in "excess" compared to what have been observed in astrophysical or cometary ices.[15–18] This was deliberately done to

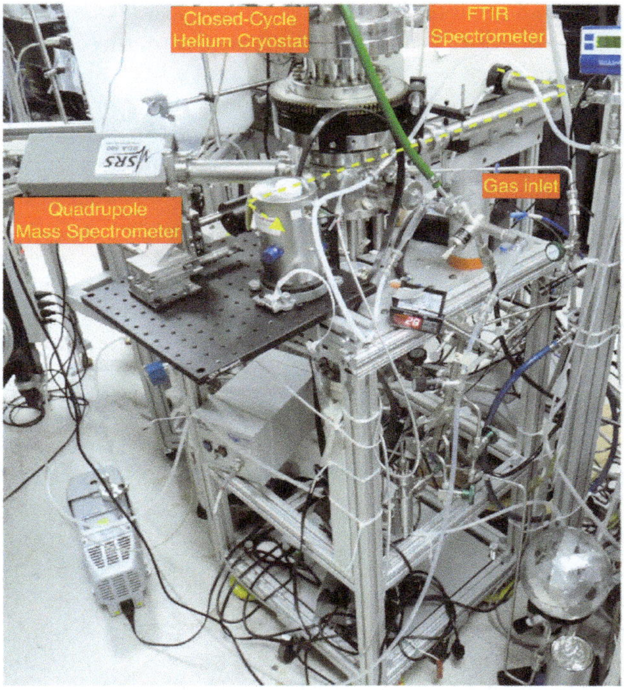

Fig. 1 Experimental station (named Himalaya) that is used to conduct TPD-MS, TPD-IR, and calorimetry experiments at the Ice Spectroscopy Laboratory (ISL) of the Jet Propulsion Laboratory (JPL). FTIR (Thermo-Nicolet 6700) beam is guided through Ni-coated tubes with high-throughput reflecting mirror and passed through the sapphire optical window mounted on a sample holder at the end of the closed-cycle helium cryostat in the vacuum chamber. The transmitted IR beam is focused with a parabolic gold-coated mirror on to the MCT-A detector. Once ice is deposited, the sample is then rotated to be perpendicular to the IR beam and held in that position for the entire TPD experiment. Also mounted on the vacuum chamber is the quadrupole mass spectrometer (Stanford Research Instruments, SRS 300) which is equipped with an electron multiplier. In this configuration we have been able to simultaneously record FTIR and mass spectra at a given temperature during the TPD experiments.

ensure that there are pure volatile and supervolatile ice domains in the mixed molecular ices. This method differs slightly from previous methods,[19,20] where either observed mixing ratios were used for mixed molecular ices or layered ices,[21–33] or pure ices were studied through TPD.[34,35] In astrophysical environments it is unlikely that pure supervolatile ice grains exist where no other molecule is present in the ice grains, such as pure CO ice, though pure CO ice domains could exist in an ice grain that contains other volatiles as well. Properties of such mixed-molecular ices or layered ices also depend upon whether these ice grains travelled from cold to warmer regions of the protoplanetary disk or *vice versa*. It is easy to imagine grains moving from the inner protoplanetary disk to the outer, which would lead to less volatile molecules to condense first forming layered ice grains with supervolatiles condensing on the top of the ice grain. The situation is more complex if the movement is from colder to warmer regions as the supervolatiles that are trapped within the less volatile molecular ices only survive in that ice grain. For this reason, we think our molecular mixing ratio provides a more realistic representative of pure ice domains in mixed molecular ice grains and complements data so far available on other ice compositions mentioned earlier.

All the ice mixtures were deposited at 10 K on a sapphire substrate at normal incidence for a duration of 3–6 min at a background pressure of $\sim 1 \times 10^{-6}$ mbar. The end of the vapor deposition tube (1/4-inch outer diameter stainless steel) was kept ~2 inches away from the sapphire optical window that was cooled with a Sumitomo (SHI-4-15, Janis Research) closed-cycle helium cryostat that reaches a minimum of 4.2 K. Calibration studies done on water ice under these conditions show the ice films are of ~250 nm (about 1000 monolayers) in thickness. Initial mixing ratios were controlled based on pressure in the ice chamber determined by an inverted magnetron pressure gauge (Agilent Varian IMG-300). We did not use the mass spectrometer to adjust the mixing ratios because of fragmentation of CO_2 to CO that interferes with the exact ratios. Final mixing ratios are determined from the FTIR spectral integrated intensities normalized to the respective band strengths (see Results section) of the individual molecules.

The temperature ramp was held at 0.5 K min^{-1}. Our Lakeshore (Model 340) temperature controller allows for a ramp rate as slow as 0.1 K min^{-1}, but the long duration experiments introduce artefacts such as background deposition that may lead to erroneous results. We experimented with faster ramp rates (1 K min^{-1}) and found that the rate did not show significant effect on the TPD-mass spectrometry (TPD-MS) of these ices. Most of the experiments were done with 0.5 K min^{-1} heating rate to ensure better equilibration of ice at any given temperature and better applicability of TPD data to typical astrophysical environments.

Desorption of molecular volatiles during TPD was quantified through quadrupole ion mass spectrometry (MS) using a Stanford Research Systems (SRS RGA-300) quadrupole mass spectrometer and infrared (IR) spectroscopy using a Thermo-Nicolet 6700 FTIR spectrometer equipped with a KBr beam splitter and an external liquid-nitrogen cooled MCT-A detector. IR spectroscopy provides independent and complementary data on the desorption process in terms of molecules remaining in the ice phase, whereas mass spectrometry provides composition of desorbed molecules in vacuum. We found that this combination of techniques was particularly useful in cases where species are less-efficiently pumped by the turbomolecular pump (*i.e.*, CO molecules) and thus can be over-represented in the mass spectra, or for those that are unable to be quantified by the infrared spectra alone (*i.e.*, N_2 and O_2

molecules). With IR spectroscopy, we were also able to track a slight increase in the IR absorption of the H_2O ice band as the temperature is ramped due to sublimation from other parts of the cryostat and recondensation on the sapphire window.

Ionization of gases in the mass spectrometer was done at 70 eV. Electron-impact ionization cross-sections at 70 eV for these molecules are taken from NIST data (O_2 = 2.441 Å^2; CO = 2.516 Å^2; CO_2 = 3.521 Å^2; H_2O = 2.275 Å^2). Integrated mass spectral data were then normalized with respect to the CO_2 ionization cross-section to obtain molecular mixing ratios of deposited gas mixture (which in turn represents closely the ice composition determined through infrared spectroscopy). The scaling factors used are: O_2 (1.44); CO (1.4); CO_2 (1); H_2O (1.55). We note here that the CO mass peak at 28 *m*/*z* has contribution from N_2 (which can be estimated to be about 8 times the partial pressure of N detected at 14 *m*/*z*) as well as ionization fragmentation of CO_2 to CO, which is about 10% of CO_2 partial pressure. These and slow pumping of CO by turbomolecular pumps cause difficulties to quantitatively interpret the CO mixing ratio in the outgassing under our experimental conditions.

Using the integrated IR spectral absorbances, the relative molecular abundance was calculated using band strengths from the literature[36–39] as follows: CO (1.1×10^{-17} cm per molecule); CO_2 (7.7×10^{-17} cm per molecule); H_2O (1.6×10^{-16} cm per molecule). Please note that O_2 is IR inactive, and we rely upon mass spectrometry for O_2 mixing ratios. Further, we conducted calorimetry using the same conventional closed-cycle cryostat that is also used for the above mentioned TPD studies.

Calorimetry of pure H_2O ice

For the calorimetry experiments, we deposited macroscopic amounts of pure water at 10 K to form amorphous ice. Our goal was to capture the amorphous to crystalline phase transition of pure H_2O ice and determine if there were differences in temperature range or any dependence on ice thickness on the crystallization of amorphous ice during TPD. For these experiments, we made an internal calibration between two temperature-measuring silicon diodes, the first diode (SD1) is mounted at the end of the heating tape where the sample holder is connected to the closed-cycle cryostat and the second silicon diode (SD2) is mounted at the bottom of the sample holder (see Fig. 2). Amorphous ice was directly deposited on the copper sample holder where the SD2 is located for different durations. The water vapor inlet tube was brought about 5 mm closer to the SD2 location of the sample holder to avoid ice formation on the entire sample holder, but to limit it at the SD2 location. First, we ran a pre-programmed 2 K min^{-1} reference ramp prior to ice deposition using the Lakeshore temperature controller's temperature ramping with PID (proportional, integral, derivative) control. To obtain a smooth heating ramp devoid of the oscillations inherent in the PID control, we fitted the recorded time-dependent heater wattage values with a three-exponential fit and used these values to prescribe the heating ramp for the ice experiments. If there were no changes in the heat absorbed or released by the ice, at each pre-programmed ramping stage, the temperatures of both the diodes SD1 and SD2 would have read the same as without the sample. However, if there were to be heat released due to the exothermic amorphous to crystalline phase transition, the SD2 would read a higher temperature than SD1 and the ratio SD2/SD1 would increase. Similarly, for

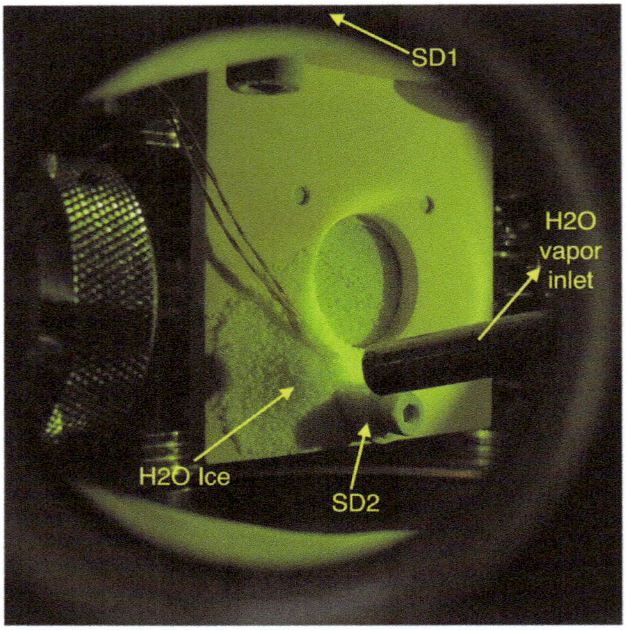

Fig. 2 Configuration of calorimetry on conventional closed-cycle helium cryostat used for ice TPD-MS and TPD-IR experiments. The locations of the Silicon Diode (SD) temperature sensors are shown. SD1, which cannot be seen in this photo, is mounted at the top of the ice sample holder that is connected to the end of the closed-cycle helium cryostat. SD2 is mounted at the bottom of the sample holder, which is made of oxygen-free copper. Also seen is the optical window mounted between the copper sample holder, on which normally ice is deposited for TPD experiments. The water vapor inlet is mounted closer to the sample holder and directed to the bottom part to ensure ice deposition is predominantly localized close to SD2.

an endothermic phase transition, this ratio would decrease. Indeed, subtle changes were measured corresponding to exothermic ice crystallization around 140 K and endothermic ice sublimation around 190 K. The ice samples deposited were estimated to be a few milligrams and the heating ramp was conducted at the maximum power setting on the heater controller at 100 W, starting at 35% of the output at 10 K and ending at ~90% at 200 K. Despite the high heater wattage needed to work against the cooling power, we were able to measure extremely small exothermic and endothermic peaks because we have used the two temperature sensor diodes located apart from each other that would otherwise have exactly the same conditions except for instantaneous minuscule differences in temperature. This is almost like weighing an elephant with and without a sparrow on it! But it indeed worked quantitatively as will be presented in the results section.

Results

TPD of mixed molecular ices

We studied several mixed molecular ices, including two component $CO:CO_2$, three component $CO:O_2:CO_2$, and four component $CO:O_2:CO_2:H_2O$ ices. We also varied the mixing ratios, except the CO to O_2 ratio was kept constant close to 1. In the TPD-MS, we followed gas-phase N (proxy of N_2 that comes at the same *m*/*z* as CO at 28; N : N_2 peak ratio from NIST Webbook is ~12 : 100), CO, O_2, CO_2, CO_3 (that could be an impurity in CO gas cylinders through reaction of CO with O_2), and H_2O. Some of these desorption mass spectral plots are shown in Fig. 3–7 along with their counterpart from integration of the FTIR spectra. In Fig. 8, integrated mass spectral data, of outgassed volatiles trapped in amorphous water ice (>90 K) that are released during amorphous to crystalline ice transition ~140 K (black bars) and trapped in crystalline water ice (>145 K) that are released along with subliming water (orange bars), are shown as bar graphs. Full range mass spectra and infrared spectra for four component mixed molecular ice are shown in Fig. 9 and 10. Our TPD-MS profiles are qualitatively in agreement with earlier work, especially pure and mixed-molecular ices containing H_2O, CO, CO_2, CH_3OH and NH_3,[19] and H_2O, CO, CO_2 and CH_4.[40] In all the experiments we conducted, most of the CO either exists as a pure CO ice domain or CO is outgassed prior to the start of CO_2 outgassing, even if CO is trapped in CO_2 ice, except with CO : O_2 : CO_2 : H_2O mixed molecular ices. During the formation of water-dominant amorphous ice, all the other volatiles are trapped to some extent within the amorphous ice. We also observed some entrapment of CO_2 in pure CO or O_2 ice domains evident from CO_2 outgassing along with CO and O_2, both of which have similar sublimation temperatures. N_2 sublimation starts after CO sublimation

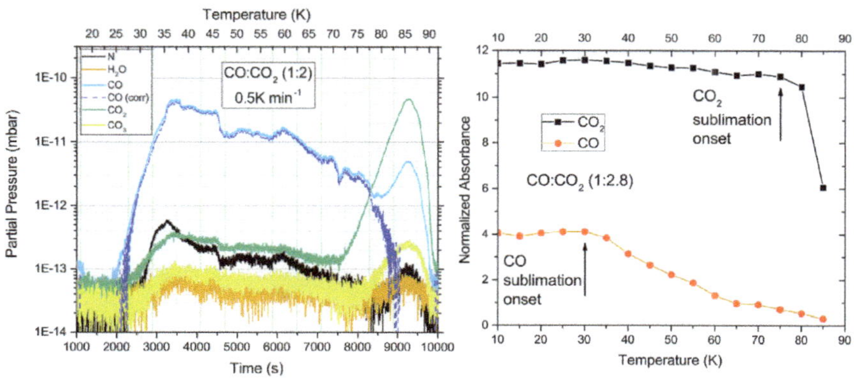

Fig. 3 TPD-MS profiles (left) and integrated TPD-IR absorption intensities that are normalized to respective band-strengths (right) of a two-component CO : CO_2 ice deposited at 10 K and heated at 0.5 K min^{-1}. While the TPD-MS data show an onset of CO sublimation at ~25 K, FTIR spectra show observable sublimation starting after 30 K. N_2 sublimation (with N as the proxy) is offset by ~5 K. A small amount of CO_2 that is trapped in CO ice is also sublimed along with CO between 30 K and 70 K. Only a small fraction of CO (~10%) remains trapped in CO_2 ice at 70 K, the onset of CO_2 sublimation, as indicated by the IR data. Please note that CO_2 fragmentation in the mass spectrometer results in around 10% of CO, which makes it difficult to quantify the mass spectra, as can be seen in the curve obtained after CO partial pressure correcting for CO_2 fragmentation and N_2 contribution. We skip this procedure in the rest of the TPD-MS profiles to keep clarity. Here, FTIR spectra are useful for quantification. For mass spectra the initial deposition chamber pressure ratios are given. For IR spectral data, the ratios obtained from integrated and normalized (to their band strengths) relative absorbances are given. Both are from the same ice and are simultaneously measured.

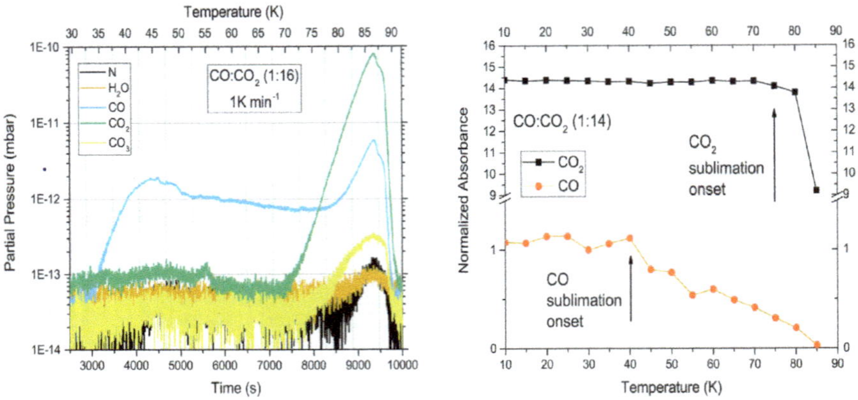

Fig. 4 TPD-MS profiles (left) and integrated TPD-IR absorption intensities (right) of two-component CO : CO_2 ice deposited at 10 K and heated at 1 K min^{-1}. Initial mixing ratio of CO is about 6–7% in CO_2 and most of the CO molecules are trapped in CO_2 ice. As a result, CO outgassing is delayed by about 10 K from the pure CO ice outgassing (refer to Fig. 3). About 3% of CO remained trapped in CO_2 ice at 70 K, the onset of CO_2 sublimation. For the rest of the information on the figure details, please refer to Fig. 3.

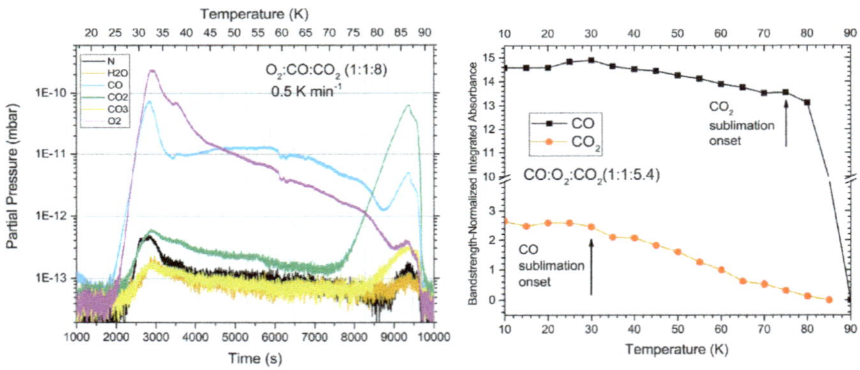

Fig. 5 TPD-MS profiles (left) and integrated TPD-IR absorption intensities (right) of three-component CO : O_2 : CO_2 ice deposited at 10 K and heated at 0.5 K min^{-1}. In the left figure, it can be seen that the O_2 partial pressure is slightly higher between 30 and 35 K and then drops rapidly while CO continues to remain in the vacuum chamber. Though the initial mixing ratio of CO : CO_2 is 1 : 5.4 (O_2 is IR inactive and assumed to be about the same mixing ratio as CO), most of the CO sublimes off between 30 K and 70 K, in agreement with the two-component experiments (Fig. 3 and 4), leaving only about 4% CO in CO_2 ice at the onset of CO_2 sublimation. For the rest of the information on the figure details, please refer to Fig. 3.

and along with O_2 sublimation, as inferred from tracing N in the TPD-MS profiles. Also, minor entrapment of CO and O_2 was observed in CO_2 ice domains as reflected in the CO_2 sublimation region of the TPD-MS profiles. Above 100 K, when all the excess pure ices of CO, O_2, and CO_2 are sublimed, the remaining ice on the substrate must be amorphous water ice and any residual volatiles/supervolatiles trapped in it. Indeed at ~130 K, we start observing another desorption event that

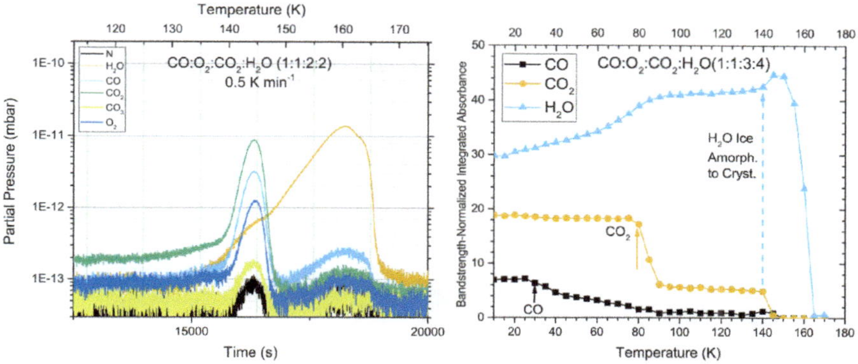

Fig. 6 Zoomed-in part of the TPD-MS profiles (left) and integrated TPD-IR absorption intensities (right) of four-component $CO : O_2 : CO_2 : H_2O$ ice deposited at 10 K and heated at 0.5 K min^{-1}. In the left figure, we see between ~137 K and ~147 K a significant outgassing of trapped volatiles in amorphous water ice. In the right figure, we observe similar outgassing onset of CO (hence O_2 as well) at 30 K and gradual outgassing until CO_2 outgassing, which occurs between 70 K and 90 K. Above 90 K, the amount of CO trapped in the amorphous ice is far less than CO_2. The data confirm outgassing of almost all CO, O_2, and CO_2 during crystallization of amorphous ice. During water–ice sublimation only a few percent of these volatiles are still trapped. This experiment demonstrates that if crystalline water ice in astrophysical environments were to be formed, it could accommodate a maximum of 5% of other volatiles in its lattice. If crystalline water ice is observed along with other ices, they must have been formed separately after amorphous water ice has been crystallized (*i.e.*, subjected to temperatures above 120 K for a given amount of time, which is dependent on the maximum exposed temperature below 160 K under vacuum conditions). Please note that water ice integrated absorbance increases constantly and reaches its maximum around 150 K. This is less likely to be due to the increased band strength of water ice with temperature, but rather due to sublimation of a small amount of water from warmer parts of the cryostat and redeposition on the sapphire optical window. For the rest of the information on the figure details, please refer to Fig. 3.

maximizes at 145 K and completes at 150 K. During this period, a significant amount of all three volatiles CO, O_2, and CO_2 are released simultaneously into the gas-phase, along with a small amount of H_2O sublimation. We see consistently that more CO_2 is trapped in amorphous water ice than CO or O_2. From the integrated TPD-MS data we find that the sum of all trapped species in amorphous water ice is about 30%. Our data indicates that indeed amorphous water ice can trap significant amounts of other volatile species even at temperatures as high as 130 K. Interestingly, as soon as the temperature reaches 150 K all the trapped volatiles are desorbed and the amorphous to crystalline ice transformation is then complete as well. Above 150 K, water ice starts sublimation alone. At this point only an extremely small portion of other volatiles are trapped in the crystalline water ice, as evident from the desorption maxima around 160 K, where only about 2–3% of CO is seen to desorb along with H_2O. The rest of the molecules have even lower mixing ratios. We can use the TPD-MS and IR data of our experiments to backtrack ice composition at different temperatures.

Infrared spectra have also been simultaneously obtained and the discrepancies between IR and mass spectral data reveal values obtained for CO from

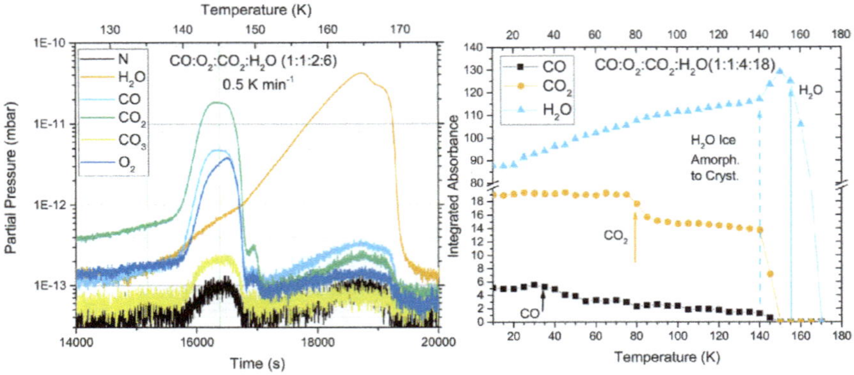

Fig. 7 Zoomed-in part of the TPD-MS profiles (left) and integrated TPD-IR absorption intensities (right) of a four-component $CO : O_2 : CO_2 : H_2O$ ice deposited at 10 K and heated at 0.5 K min^{-1}. In this experiment, we increased the mixing ratio of water to test whether there would be any change in the thermal or trapping behavior. We found no changes and, as expected, significant amounts of CO (O_2) and CO_2 were trapped in amorphous ice and released during crystallization. Otherwise, both the two four-component ice types behaved very similarly. Please refer to Fig. 6 and 3 for additional details.

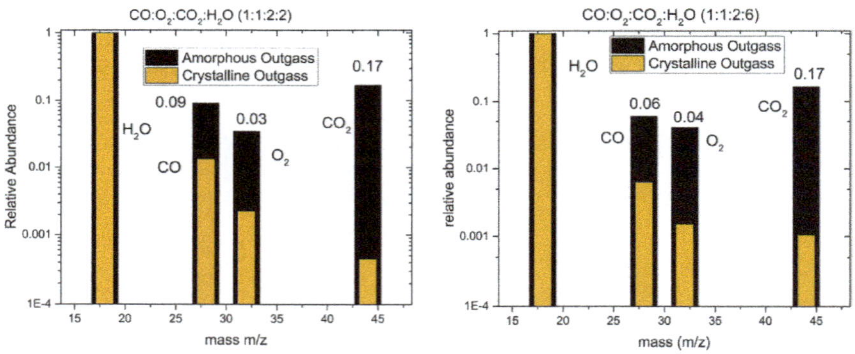

Fig. 8 Relative abundances with respect to H_2O calculated from integrated TPD-MS spectra (shown in Fig. 6 and 7) of four-component $CO : O_2 : CO_2 : H_2O$ ice deposited at 10 K and heated at 0.5 K min^{-1}. Volatiles CO, O_2, and CO_2 outgassed during crystallization of amorphous water ice between 130 K and 150 K normalized to the total water content are shown in black bars and the same for crystalline water ice outgassing between 150 K and 190 K shown in orange bars. In spite of different mixing ratios, in both experiments above 90 K, CO_2 is trapped in amorphous ice at an abundance of 17%, while CO and O_2 are much lower in abundance (6–9% and 3–4%, respectively). After amorphous ice crystallizes, the total amount of volatiles trapped is reduced to around 1%. Again, our experiments demonstrate that crystalline water ice does not trap significant amounts of volatiles.

mass spectral data alone cannot be quantitatively used due to the longer residence time of CO in the vacuum chamber. For example, total integrated TPD-MS partial pressure ratio of $CO : CO_2$ is ~3 : 1, whereas IR spectral data indicate a $CO : CO_2$ ratio of 1 : 2.8 (Fig. 3). It is well known that conventional turbomolecular

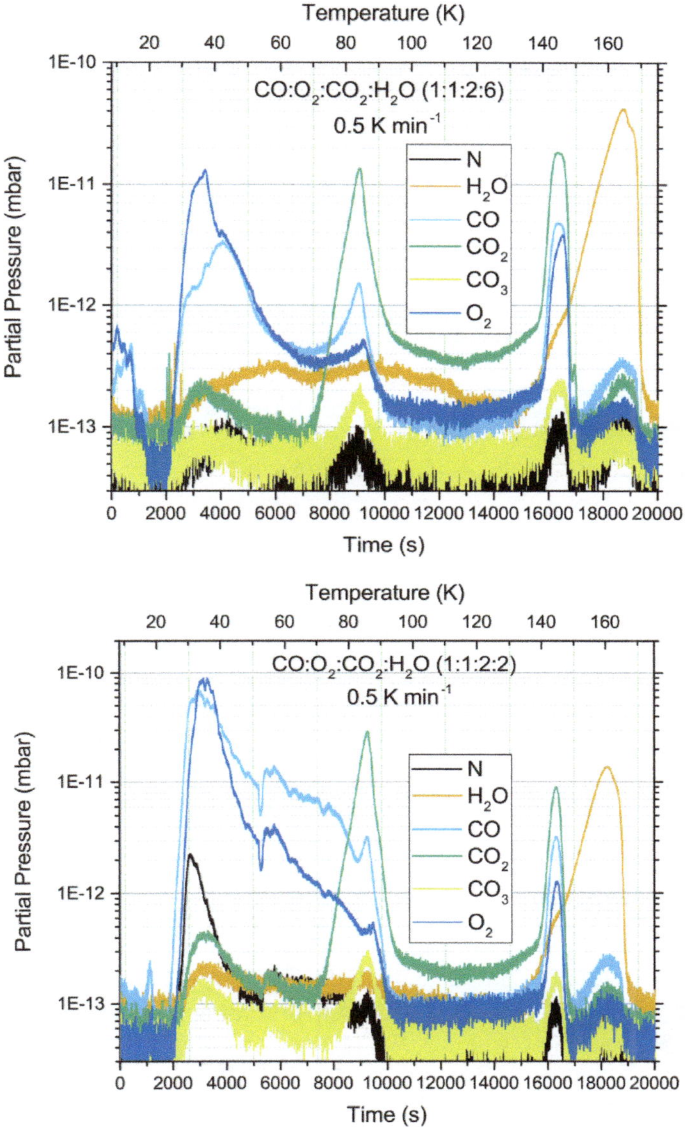

Fig. 9 Mass spectral profiles during TPD from 10 K to 190 K of four component mixed molecular ices CO : O_2 : CO_2 : H_2O.

pumps even at 10^{-9} mbar vacuum are less efficient at pumping CO. As a result, integrated mass spectra over-estimate CO outgassing. This is further complicated by continuous CO release from the ice into the vacuum that is clearly seen from the IR spectra of the ices (Fig. 3–7). In the astrochemistry community, the majority of the cryogenic systems are equipped with these turbomolecular pumps, at $\sim 10^{-9}$ mbar working pressures. As discussed below, under those conditions caution must be exercised in the interpretation of TPD-MS data alone.

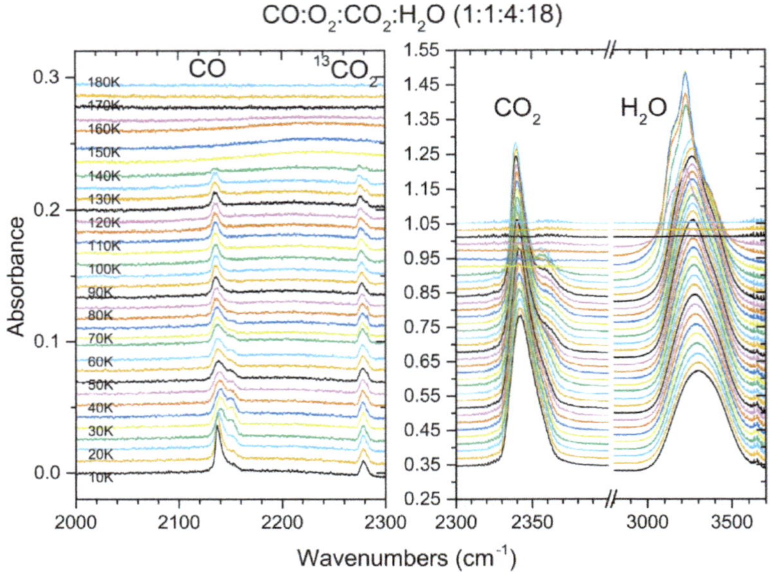

Fig. 10 FTIR absorbance spectra of four component CO : O_2 : CO_2 : H_2O mixed molecular ice during TPD. Spectra are stacked to provide clarity. CO absorption around 2140 cm^{-1} quickly splits into two components, the higher energy one disappearing by 90 K, indicating the lower energy part should be due to CO trapped in amorphous water–ice. CO_2 absorption at 2340 cm^{-1} starts as a single broad band and as the temperature increases a shoulder develops on the higher energy side, which becomes a clear band beyond CO_2 ice sublimation. Around 145 K all CO_2 sublimes and the amorphous water ice band centered around 3250 cm^{-1} develops into a three-band structure, typical of crystalline water ice. Around 185 K, all the water ice sublimes, leaving no residue behind.

Calorimetry of pure H_2O ice

In order to quantify the exothermicity of the amorphous to crystalline ice transition, we have relied on an internal standard, namely the heat of sublimation of crystalline ice. In our first experiment, where ice was deposited at 2–3 $\times$ 10^{-5} mbar for 900 min at a rate of 1 $\times$ 10^{-5} g min^{-1} (black line curve in Fig. 11), the endothermic sublimation as well as the exothermic crystallization are well behaved. Subsequently, we increased the time of deposition as shown in Fig. 11 and both the exotherm and endotherm peaks increased, with the sublimation endotherm out of bound with thicker ice sample. At this point, ice sublimation is so fast that the vacuum in the system raised significantly (to $\sim$1 $\times$ 10^{-3} mbar) and the data was no longer reliable. We excluded such data from calculating the sublimation enthalpy. From the literature, we took the sublimation endotherm to be 2830 kJ kg^{-1},[41] and multiplied it by the ratio of the integrated crystallization exotherm to the integrated sublimation endotherm. By this procedure, we obtained the heat of crystallization exotherm for amorphous ice to be 109 kJ kg^{-1}, 109 kJ kg^{-1}, 116 kJ kg^{-1}, and 138 kJ kg^{-1} for the four different experiments we conducted at different ice deposition durations ranging from 900 min to 5160 min, with an arithmetic mean and standard deviation of 118 $\pm$ 12 kJ kg^{-1}. This value is in very good agreement with the literature reports of $\sim$100 kJ

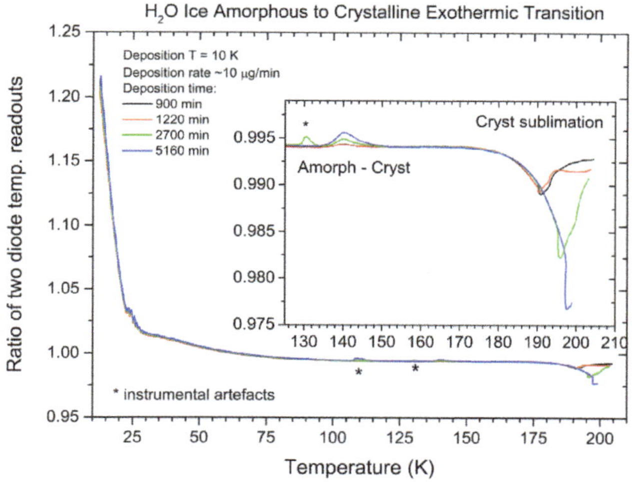

Fig. 11 Differential calorimetry during 2 K min^{-1} pure water ice annealing from amorphous to crystalline to sublimation transitions. Ratio of SD2/SD1 (SD = silicon diode for temperature readout) at various temperatures is plotted for four different ice experiments. SD2 is directly in contact with the ice, whereas SD1 acts as an internal reference at the top of the sample holder. For an exothermic phase transition, this ratio is higher and for an endothermic phase transition it is smaller with respect to a common baseline ratio. We consistently see an exothermic transition between 130 and 150 K with maximum at 140 K due to the amorphous to crystalline phase transition. This peak position remains independent of ice thickness, but the transition intensifies with thickness as expected (more ice deposition mass induces a greater heat release and larger differences in the temperatures read at SD1 and SD2). We also observe the endothermic sublimation transition with an onset around 170 K, which also intensifies with thickness. The sublimation onset is the same in all cases, but the sublimation maximum shifts depending on thickness, since it is a surface-driven phenomenon and it takes longer for thicker ices to complete the sublimation process. For thinner ice it maximizes at ~190 K, and moves to higher temperatures for thicker ice.

kg^{-1}.[42–45] Our method is simple and easy to implement for a wide range of equipment and ices.

Discussion

There are a few major take aways from our experiments presented here. The first one is that super volatiles CO, O_2, and N_2 (proxy N) can only exist in significant mixing ratios in interstellar, protoplanetary, and cometary ices that are extremely cold – at or less than 30 K. The second take away is that the exothermicity of amorphous to crystalline pure water ice transition is confirmed again. At the same time, we also found that irrespective of the ice thickness – whether a few hundred nanometers thin ice films or few hundred micrometers to millimeters thick ice grains – the amorphous to crystalline ice transition occurs within a narrow temperature range of 130 K to 150 K, with a maximum at ~140 K. Another experimental take away is that quantification of CO in the gas-phase under vacuum should be carefully conducted and should be verified by complementary experiments. Finally, conventional cryogenic systems with no additional instrument modification can be used to carryout calorimetry of ices and we plan to extend our calorimetric studies to mixed-molecular ices in the future. In the following we discuss more in detail temperature ranges and their effect on the ice/gas composition.

Below 30 K

Most interstellar ices, protoplanetary ices at greater radial distances, and Oort Cloud comet interiors are expected to be at or below 30 K. Under these conditions, based on our data presented here, we can expect no outgassing of any supervolatiles such as CO, O_2, N_2 (CH_4, Ar *etc.*, by inference) either as pure ices or trapped in mixed molecular ices. So, supervolatile-rich ices should be present at temperatures below 30 K. We can approximately refer to this the CO ice line.

30–75 K

This is the temperature range one expects in protoplanetary disks at closer distances than the CO ice line mentioned above. These temperatures are also present as equilibrium temperatures of solar system icy bodies such as TNOs (Trans Neptune Objects) and Kuiper Belt Objects (KBOs), and Centaurs.[46,47] In these environments, the ice grains are expected to be depleted of significant amounts of supervolatiles CO, O_2, and N_2 (CH_4, Ar *etc.*, by inference), but still contain significant amounts of CO_2. For this reason, Jupiter family short-period comets that originate from KBOs and Centaurs should be significantly depleted of CO and other supervolatiles, but still retain high amounts of CO_2. In the protoplanetary disks, the ice grains accreting from the interstellar environment and moving radially into the inner disk would have significant amounts of CO_2, but otherwise be depleted in supervolatiles. From the TPD-MS and TPD-IR data presented here, we find that though significant amount of CO is outgassed early (below 50 K), CO continues to leave the ice-phase gradually until CO_2 starts sublimation. This indicates that some of the CO molecules, either trapped in CO_2 or existing as a CO ice domain covered by CO_2 ice, outgas at higher temperature. We should be cautious to over interpret this data, as this outgassing is expected to be both temperature and time dependent.

75–120 K

This temperature range leads to the loss of any excess pure CO_2 ice and any volatiles left behind are now completely trapped within amorphous water ice cages in the 4-component ices. In three and two component ices, where CO_2 is the dominant species, we observe very small amounts (<3%) of supervolatiles (including N_2, O_2, and CO) to outgas along with CO_2 (Fig. 3–5), indicating CO_2 ice has crystallized expelling other impurities well before its sublimation onset.[48] Centaurs, the precursors of Jupiter-family comets, could have their equilibrium temperatures in this range.[49,50]

120–150 K

This is the most critical temperature in astrophysical environments, and it is a point of no return, when amorphous water ices irreversibly undergo phase transition into crystalline water ice, during which process all the volatiles trapped within the amorphous ice are released. Our experiments show that about 17% CO_2, ~9% CO, and ~4% O_2 are trapped in warm amorphous ices beyond 90 K – and these are released during ice crystallization between 130 K and 150 K. These values are close to the modeling work done recently to reproduce comet 67P/Churyumov–Gerasimenko nucleus properties.[51] It is expected that crystallization of amorphous water ice is slower at lower temperatures and faster as the temperatures approach 150 K.[52] This temperature region in protoplanetary disks should be narrow but clearly marked by crystalline water–ice features in the IR spectral observations. JWST should be able to detect these regions and constrain the temperatures. Comets should also show unusual CO_2-rich outgassing and H_2O inactivity in these temperature ranges. Indeed, such dichotomy is observed in the outgassing of many comets, including 67P/CG,[53–56] whereby subsolar regions with higher temperatures show H_2O outgassing (crystalline water ice sublimation – see below), whereas other regions with lower temperatures show enhanced CO_2 outgassing. It is very likely that amorphous water ice loaded with CO_2 and other volatiles, at depths where thermal equilibrium reaches ~130 K, eject these volatiles while undergoing crystallization. As a result, a significant amount of CO_2, CO, and other volatiles are observed from the regions where the surface temperatures are lower. Once crystallized, water ice from the same region, when temperatures are higher due to direct exposure to Sun, results in predominantly H_2O sublimation (and depleted in CO_2). Based on this laboratory observation alone, we can confidently conclude that cometary nuclei are predominantly made of amorphous water ice, at least deep inside the nucleus. We also note that presence of silicate dust and refractory organic mantle on short-period comets could further effect outgassing from comets.[53,54,57]

150–190 K

Above 150 K crystalline water ice starts to sublimate rapidly. Based on the thickness of the crystalline water ice film, sublimation could be complete at 170 K or above. The sublimation temperature also depends on the ambient pressure of the ice, with higher pressures shifting sublimation to higher temperatures. Such a scenario does not exist in low-pressure protoplanetary environments but could be present in the near sub-surface ices of the outer Solar System icy bodies and cometary near sub-surface ices, as a comet gets closer to its perihelion. The location of crystalline water ice sublimation in the protoplanetary disks is also designated as H_2O ice line that is closer to 150 K. However, based on the above-mentioned variables, it is possible that water ice could persist at higher temperatures.

If an accreted interstellar amorphous ice grain makes its journey through each of the ice lines in a protoplanetary disk, shedding away volatiles, the remaining material on these grains would be highly refractory (silicates and complex organics). If these grains or silicate dust from the inner protoplanetary disk were subsequently to move radially outwards, then the ice condensation on these grains will be layered (onion-like), with the non-volatiles such as H_2O condensing first, followed by other volatiles such as CO_2, and finally the supervolatiles such as CO. Thus, the composition of an ice grain – layered *vs.* intimately mixed – depends on whether it was formed within the protoplanetary disk or retained from interstellar grains.

The structural composition of these grains (layered *vs.* intimately mixed) plays an important role in the subsequent chemistry of these ices. Because highly energetic particles can deposit energy in ionization tracks as they penetrate through matter, reactions tend to occur with nearby species. As a result, any radiation-induced chemistry of protoplanetary-disk-formed ice grains should be different from interstellar ice grain chemistry.

Cometary nuclei have been puzzling researchers for decades. The recent Rosetta mission provided more insight into the outgassing properties of comet 67P/CG.[53] High porosity of cometary nuclei and large thermal gradients make predictions of cometary nucleus composition based on outgassing difficult. However, based on our data and earlier experimental work, we can provide some insight into cometary nuclei: (a) if the entire nucleus of a comet reaches its equilibrium temperature above 70 K, it would be unlikely that these comets contain regions of pure molecular ices of super volatiles such as CO, N_2, CH_4, *etc.* We observe continuous outgassing of CO from the mixed molecular ice above 30 K until CO_2 sublimation starts at 70 K. More laboratory work is needed to understand the long-term effect of temperature between 30 K and 70 K on the outgassing of these supervolatiles to determine if any of their pure ice domains would survive these temperatures. CO_2 ice domains are more likely to survive in the interiors of a comet nucleus with equilibrium temperature at or below 70 K, followed by trapped volatiles in amorphous H_2O ice below ~120 K. At closer heliocentric distances, where surface temperatures raise beyond H_2O ice sublimation temperatures below the dust mantle, outgassing properties of a comet would be most complex due to the thermal wave propagating deeper that results in not only sublimation of crystalline water ice, but also outgassing from amorphous to crystalline phase transition at deeper regions releasing significant amounts of trapped volatiles (also known as molecular volcanos on a comet).[33,58,59] However, at even greater depths CO_2 ice domains could start sublimation reaching the surface. However, the outgassing dichotomy of comets indicates that CO_2 is predominantly from relatively lower temperature regions on the surface compared to H_2O, indicating CO_2 outgassing could have contribution from all these processes. At larger heliocentric distances, however, H_2O amorphous-to-crystalline transition and sublimation are significantly suppressed, and it is expected that CO_2 ice domains dominate the outgassing in these regions. The Rosetta mission also found strong correlation between H_2O and O_2 outgassing, up to 5% mixing ratio of O_2.[60,61] In our experiments, we see a much smaller fraction of O_2 outgassing along with H_2O (see Fig. 8). One possibility is that cometary nuclei and their precursors may have been irradiated by cosmic rays over the past 4.6 billion years, resulting in radiation-induced trapping of O_2 in H_2O ice as has been seen on the surface of Galilean moons Europa, Ganymede, and Callisto[62] and experimentally demonstrated during ion irradiation of ice.[63] More laboratory studies are needed to understand complex thermal properties of cometary nuclei and their outgassing properties during their journey around the Sun.

In the outer Solar System, on icy satellites such as Europa and other Jovian and Saturnian moons, surface temperatures can vary between 150 K and 60 K depending on latitude and sub-solar regions.[64–67] Sub-solar equatorial regions sustain the highest possible temperatures, whereas polar regions remain cooler. Similarly, day and night temperatures also vary significantly. Under these conditions, water ice sublimation occurs from the crystalline ice phase into vapor, forming a tenuous atmosphere. When the water vapor condenses on the surface, amorphous water ice could form in cold higher latitude regions or at lower latitudes on the night side. At temperatures below 70 K, this refrosting process could also lead to trapping of CO_2 if it is present in the atmosphere.

Conclusions

Temperature plays an important role in the compositional properties of astrophysical ices. Laboratory data under various conditions will provide better constraints on the interpretation of observational data, whether it is protoplanetary disks, comets, or other Solar System icy bodies. H_2O can co-condense with significant amounts of other volatiles and supervolatiles at very low temperatures (<30 K), whereas at warmer – above the CO_2 sublimation – temperature (>90 K), amorphous water ice has only a limited ability to trap volatiles, of which CO_2 has the highest mixing ratio. Crystallization of amorphous ice expels all the trapped volatiles around 130–150 K. This temperature region is the same, whether the ice grains are thin (a few hundred nanometers) or thick (a few hundred microns), based on our differential calorimetry experiments. Such a process could lead to outgassing of significant amounts of CO_2, whereas H_2O is still in the solid form. This important process likely explains the outgassing dichotomy observed in comets where CO_2 outgassing occurs from a different region of the comet than H_2O outgassing, which is predominantly from the subsolar regions. We also found that crystalline water ice traps only a small fraction of other volatiles (<5%), indicating ice grain composition in astrophysical and planetary environments must be different depending on whether the ice is in amorphous phase or transformed into crystalline phase. Once irreversibly transformed into crystalline phase, water ice is expected not to trap anymore volatiles, even if the crystalline ice undergoes radiation-induced amorphization subsequently. As for the protoplanetary disks, any radial mixing over long distances from the protostar could lead to entirely different ice compositions, namely layered ices, instead of the mixed molecular ices formed in the interstellar molecular clouds. The morphology of these layered ices could also lead to different radiation chemistry compared to mixed molecular ices. JWST observations should provide better insight and constraints on the formation and mobility of ice grains in protoplanetary environments.

Conflicts of interest

There are no conflicts to declare.

Acknowledgements

This research work was carried out at the Jet Propulsion Laboratory, California Institute of Technology, under a contract with the National Aeronautics and Space Administration. This research was enabled through funding to MSG from NASA SSW and DDAP Programs.

References

1 M. Jin, K. H. Lam, M. K. McClure, J. T. van Scheltinga, Z. Y. Li, A. Boogert, E. Herbst, S. W. Davis and R. T. Garrod, *Astrophys. J.*, 2022, **935**, 133.

2 B. Muller, B. M. Giuliano, A. Vasyunin, G. Fedoseev and P. Caselli, *Astron. Astrophys.*, 2022, **668**, A46.
3 S. Hinkley, A. L. Carter, S. Ray, A. Skemer, B. Biller, E. Choquet, M. A. Millar-Blanchaer, S. Sallum, B. Miles, N. Whiteford, P. Patapis, M. Perrin, L. Pueyo, G. Schneider, K. Stapelfeldt, J. Wang, K. Ward-Duong, B. P. Bowler, A. Boccaletti, J. H. Girard, D. Hines, P. Kalas, J. Kammerer, P. Kervella, J. Leisenring, E. Pantin, Y. F. Zhou, M. Meyer, M. C. Liu, M. Bonnefoy, T. Currie, M. McElwain, S. Metchev, M. Wyatt, O. Absil, J. Adams, T. Barman, I. Baraffe, M. Bonavita, M. Booth, M. Bryan, G. Chauvin, C. Chen, C. Danielski, M. De Furio, S. M. Factor, M. P. Fitzgerald, J. J. Fortney, C. Grady, A. Greenbaum, T. Henning, K. K. W. Hoch, M. Janson, G. Kennedy, M. Kenworthy, A. Kraus, M. Kuzuhara, P. O. Lagage, A. M. Lagrange, R. Launhardt, C. Lazzoni, J. Lloyd, S. Marino, M. Marley, R. Martinez, C. Marois, B. Matthews, E. C. Matthews, D. Mawet, J. Mazoyer, M. Phillips, S. Petrus, S. P. Quanz, A. Quirrenbach, J. Rameau, I. Rebollido, E. Rickman, M. Samland, B. Sargent, J. E. Schlieder, A. Sivaramakrishnan, J. M. Stone, M. Tamura, P. Tremblin, T. Uyama, M. Vasist, A. Vigan, K. Wagner and M. Ygouf, *Publ. Astron. Soc. Pac.*, 2022, **134**, 095003.
4 N. P. Ballering, L. I. Cleeves and D. E. Anderson, *Astrophys. J.*, 2021, **920**, 115.
5 M. Lecar, M. Podolak, D. Sasselov and E. Chiang, *Astrophys. J.*, 2006, **640**, 1115–1118.
6 K. Baillie, S. Charnoz and E. Pantin, *Astron. Astrophys.*, 2015, **577**, A65.
7 J. E. Owen, *Astrophys. J., Lett.*, 2014, **790**, L7.
8 N. van der Marel, R. B. Dong, J. di Francesco, J. P. Williams and J. Tobin, *Astrophys. J.*, 2019, **872**, 112.
9 A. D. Bosman, A. G. G. M. Tielens and E. F. van Dishoeck, *Astron. Astrophys.*, 2018, **611**, A80.
10 Y. Aikawa, in *Highlights of Astronomy*, ed. O. Engvold, Astronomical Soc Pacific, San Francisco, 2005, vol. 13, pp. 515–517.
11 C. Walsh, T. J. Millar, H. Nomura, E. Herbst, S. W. Weaver, Y. Aikawa, J. C. Laas and A. I. Vasyunin, *Astron. Astrophys.*, 2014, **563**, A33.
12 C. M. O. Alexander, L. R. Nittler, J. Davidson and F. J. Ciesla, *Meteorit. Planet. Sci.*, 2017, **52**, 1797–1821.
13 K. I. Oberg, V. V. Guzman, K. Furuya, C. H. Qi, Y. Aikawa, S. M. Andrews, R. Loomis and D. J. Wilner, *Nature*, 2015, **520**, 198–201.
14 M. K. McClure, C. Espaillat, N. Calvet, E. Bergin, P. D'Alessio, D. M. Watson, P. Manoj, B. Sargent and L. I. Cleeves, *Astrophys. J.*, 2015, **799**, 162.
15 O. Mousis, A. Aguichine, A. Bouquet, J. I. Lunine, G. Danger, K. E. Mandt and A. Luspay-Kuti, *Planet. Sci. J.*, 2021, **2**, 72.
16 A. M. A. Piso, J. Pegues and K. I. Oberg, *Astrophys. J.*, 2016, **833**, 203.
17 K. I. Oberg, A. C. A. Boogert, K. M. Pontoppidan, S. van den Broek, E. F. van Dishoeck, S. Bottinelli, G. A. Blake and N. J. Evans, *Astrophys. J.*, 2011, **740**, 109.
18 M. Rubin, C. Engrand, C. Snodgrass, P. Weissman, K. Altwegg, H. Busemann, A. Morbidelli and M. Mumma, *Space Sci. Rev.*, 2020, **216**, 102.
19 R. Martin-Domenech, G. M. Munoz Caro, J. Bueno and F. Goesmann, *Astron. Astrophys.*, 2014, **564**, A8.
20 S. Malyk, G. Kumi, H. Reisler and C. Wittig, *J. Phys. Chem. A*, 2007, **111**, 13365–13370.

21 K. Acharyya, G. W. Fuchs, H. J. Fraser, E. F. van Dishoeck and H. Linnartz, *Astron. Astrophys.*, 2007, **466**, 1005–1169.

22 G. W. Fuchs, K. Acharyya, S. E. Bisschop, K. I. Oberg, F. A. van Broekhuizen, H. J. Fraser, S. Schlemmer, E. F. van Dishoeck and H. Linnartz, *Faraday Discuss.*, 2006, **133**, 331–345.

23 J. L. Edridge, K. Freimann, D. J. Burke and W. A. Brown, *Philos. Trans. R. Soc., A*, 2013, **371**, 20110578.

24 A. Bar-Nun, D. Laufer, O. Rebolledo, S. Malyk, H. Reisler and C. Wittig, in *The Science of Solar System Ices*, ed. M. S. Gudipati and J. Castillo-Rogez, Springer, New York, 2013, vol. 356, ch. 14, pp. 487–499.

25 A. N. Greenberg, D. Laufer and A. Bar-Nun, *Mon. Not. R. Astron. Soc.*, 2017, **469**, S517–S521.

26 B. Muller, B. M. Giuliano, L. Bizzocchi, A. I. Vasyunin and P. Caselli, *Astron. Astrophys.*, 2018, **620**, A46.

27 S. Taj and M. R. S. McCoustra, *Mon. Not. R. Astron. Soc.*, 2020, **498**, 1693–1699.

28 *Faraday Discuss.*, 2006, **133**, 347–374.

29 M. P. Collings, J. W. Dever, H. J. Fraser, M. R. S. McCoustra and D. A. Williams, *Astrophys. J.*, 2003, **583**, 1058–1062.

30 E. C. Fayolle, K. I. Oberg, H. M. Cuppen, R. Visser and H. Linnartz, *Astron. Astrophys.*, 2011, **529**, A74.

31 M. P. Collings, J. W. Dever and M. R. S. McCoustra, *Chem. Phys. Lett.*, 2005, **415**, 40–45.

32 H. J. Fraser, M. P. Collings, J. W. Dever and M. R. S. McCoustra, *Mon. Not. R. Astron. Soc.*, 2004, **353**, 59–68.

33 M. P. Collings, M. A. Anderson, R. Chen, J. W. Dever, S. Viti, D. A. Williams and M. R. S. McCoustra, *Mon. Not. R. Astron. Soc.*, 2004, **354**, 1133–1140.

34 K. Isokoski, J. B. Bossa, T. Triemstra and H. Linnartz, *Phys. Chem. Chem. Phys.*, 2014, **16**, 3456–3465.

35 B. Mate, O. Galvez, B. Martin-Llorente, M. A. Moreno, V. J. Herrero, R. Escribano and E. Artacho, *J. Phys. Chem. A*, 2008, **112**, 457–465.

36 D. M. Hudgins, S. A. Sandford, L. J. Allamandola and A. Tielens, *Astrophys. J., Suppl. Ser.*, 1993, **86**, 713–870.

37 P. A. Gerakines, W. A. Schutte, J. M. Greenberg and E. F. Vandishoeck, *Astron. Astrophys.*, 1995, **296**, 810–818.

38 P. A. Gerakines and R. L. Hudson, *Astrophys. J., Lett.*, 2015, **808**, L40.

39 B. M. Giuliano, R. M. Escribano, R. Martin-Domenech, E. Dartois and G. M. Munoz Caro, *Astron. Astrophys.*, 2014, **565**, A108.

40 A. Kouchi and T. Yamamoto, *Prog. Cryst. Growth Charact. Mater.*, 1995, **30**, 83–108.

41 R. Feistel and W. Wagner, *Geochim. Cosmochim. Acta*, 2007, **71**, 36–45.

42 A. Kouchi and S. Sirono, *Geophys. Res. Lett.*, 2001, **28**, 827–830.

43 R. S. Smith, J. Matthiesen, J. Knox and B. D. Kay, *J. Phys. Chem. A*, 2011, **115**, 5908–5917.

44 M. Chonde, M. Brindza and V. Sadtchenko, *J. Chem. Phys.*, 2006, **125**, 094501.

45 Y. P. Handa, O. Mishima and E. Whalley, *J. Chem. Phys.*, 1986, **84**, 2766–2770.

46 A. Guilbert-Lepoutre, S. Besse, O. Mousis, M. Ali-Dib, S. Hoefner, D. Koschny and P. Hager, *Space Sci. Rev.*, 2015, **197**, 271–296.

47 A. Guilbert-Lepoutre, A. Gkotsinas, S. N. Raymond and D. Nesvorny, *Astrophys. J.*, 2023, **942**, 92.

48 R. M. Escribano, G. M. M. Caro, G. A. Cruz-Diaz, Y. Rodriguez-Lazcano and B. Mate, *Proc. Natl. Acad. Sci. U. S. A.*, 2013, **110**, 12899–12904.

49 M. Drahus, B. Yang, D. C. Lis and D. Jewitt, *Mon. Not. R. Astron. Soc.*, 2017, **468**, 2897–2909.

50 D. Jewitt, *Astron. J.*, 2009, **137**, 4296–4312.

51 B. J. R. Davidsson, N. H. Samarasinha, D. Farnocchia and P. J. Gutierrez, *Mon. Not. R. Astron. Soc.*, 2022, **509**, 3065–3085.

52 R. E. Mastrapa, W. Grundy and M. Gudipati, in *The Science of Solar System Ices*, ed. M. S. Gudipati and J. Castillo-Rogez, Springer, New York, 2013, vol. 356, ch. 11, pp. 371–408.

53 C. Herny, O. Mousis, R. Marschall, N. Thomas, M. Rubin, O. Pinzón-Rodríguez and I. P. Wright, *Planet. Space Sci.*, 2021, **200**, 105194.

54 M. Hoang, P. Garnier, H. Gourlaouen, J. Lasue, H. Reme, K. Altwegg, H. Balsiger, A. Beth, U. Calmonte, B. Fiethe, A. Galli, S. Gasc, A. Jackel, A. Korth, L. Le Roy, U. Mall, M. Rubin, T. Semon, C. Y. Tzou, J. H. Waite and P. Wurz, *Astron. Astrophys.*, 2019, **630**, A33.

55 C. Tubiana, G. Rinaldi, C. Guttler, C. Snodgrass, X. Shi, X. Hu, R. Marschall, M. Fulle, D. Bockeele-Morvan, G. Naletto, F. Capaccioni, H. Sierks, G. Arnold, M. A. Barucci, J. L. Bertaux, I. Bertini, D. Bodewits, M. T. Capria, M. Ciarniello, G. Cremonese, J. Crovisier, V. Da Deppo, S. Debei, M. De Cecco, J. Deller, M. C. De Sanctis, B. Davidsson, L. Doose, S. Erard, G. Filacchione, U. Fink, M. Formisano, S. Fornasier, P. J. Gutierrez, W. H. Ip, S. Ivanovski, D. Kappel, H. U. Keller, L. Kolokolova, D. Koschny, H. Krueger, F. La Forgia, P. L. Lamy, L. M. Lara, M. Lazzarin, A. C. Levasseur-Regourd, Z. Y. Lin, A. Longobardo, J. J. Lopez-Moreno, F. Marzari, A. Migliorini, S. Mottola, R. Rodrigo, F. Taylor, I. Toth and V. Zakharov, *Astron. Astrophys.*, 2019, **630**, A23.

56 U. Fink, L. Doose, G. Rinaldi, A. Bieler, F. Capaccioni, D. Bockelee-Morvan, G. Filacchione, S. Erard, C. Leyrat, M. Blecka, M. T. Capria, M. Combi, J. Crovisier, M. C. De Sanctis, N. Fougere, F. Taylor, A. Migliorini and G. Piccioni, *Icarus*, 2016, **277**, 78–97.

57 M. Hoang, P. Garnier, J. Lasue, H. Rème, M. T. Capria, K. Altwegg, M. Läuter, T. Kramer and M. Rubin, *Astron. Astrophys.*, 2020, **638**, A106.

58 R. S. Smith, C. Huang, E. K. L. Wong and B. D. Kay, *Phys. Rev. Lett.*, 1997, **79**, 909.

59 R. S. Smith, C. Huang and B. D. Kay, *J. Phys. Chem. B*, 1997, **101**, 6123–6126.

60 K. L. Heritier, K. Altwegg, J. J. Berthelier, A. Beth, C. M. Carr, J. De Keyser, A. I. Eriksson, S. A. Fuselier, M. Galand, T. I. Gombosi, P. Henri, F. L. Johansson, H. Nilsson, M. Rubin, C. S. Wedlund, M. G. G. T. Taylor and E. Vigren, *Nat. Commun.*, 2018, **9**, 2580.

61 A. Bieler, K. Altwegg, H. Balsiger, A. Bar-Nun, J. J. Berthelier, P. Bochsler, C. Briois, U. Calmonte, M. Combi, J. De Keyser, E. F. van Dishoeck, B. Fiethe, S. A. Fuselier, S. Gasc, T. I. Gombosi, K. C. Hansen, M. Hassig, A. Jackel, E. Kopp, A. Korth, L. Le Roy, U. Mall, R. Maggiolo, B. Marty, O. Mousis, T. Owen, H. Reme, M. Rubin, T. Semon, C. Y. Tzou, J. H. Waite, C. Walsh and P. Wurz, *Nature*, 2015, **526**, 678–681.

62 J. R. Spencer, W. M. Calvin and M. J. Person, *J. Geophys. Res.: Planets*, 1995, **100**, 19049–19056.

63 B. D. Teolis, J. Shi and R. A. Baragiola, *J. Chem. Phys.*, 2009, **130**, 134704.

64 J. R. Berdis, M. S. Gudipati, J. R. Murphy and N. J. Chanover, *Icarus*, 2020, **341**, 113660.

65 A. V. Oza, F. Leblanc, R. E. Johnson, C. Schmidt, L. Leclercq, T. A. Cassidy and J. Y. Chaufray, *Planet. Space Sci.*, 2019, **167**, 23–32.

66 Y. Ashkenazy, *Heliyon*, 2019, **5**, e01908.

67 J. A. Rathbun, N. J. Rodriguez and J. R. Spencer, *Icarus*, 2010, **210**, 763–769.

Faraday Discussions

ROYAL SOCIETY OF CHEMISTRY

PAPER

Vacuum UV photodesorption of organics in the interstellar medium: an experimental study of formic acid HCOOH and methyl formate $HCOOCH_3$-containing ices

Mathieu Bertin, *[a] Romain Basalgète, [a] Antonio J. Ocaña,[a] Géraldine Féraud, [a] Claire Romanzin,[b] Laurent Philippe, [a] Xavier Michaut [a] and Jean-Hugues Fillion [a]

Received 10th January 2023, Accepted 20th February 2023
DOI: 10.1039/d3fd00004d

Being a potential process that could explain gas phase abundances of so-called complex organic molecules (COMs) in the cold interstellar medium (ISM), the UV photon-induced desorption from organics-containing molecular ices has been experimentally studied. In this work, we focused on the observation of the photodesorbed products and the measurement of the associated photodesorption yields from pure and mixed molecular ices, each containing organic molecules whose detection has been achieved in the gas phase of the cold ISM, namely formic acid HCOOH and methyl formate $HCOOCH_3$. Each molecule, in pure ice or in ice mixed with CO or water, was irradiated at 15 K with monochromatic vacuum UV photons in the 7–14 eV range using synchrotron radiation from the SOLEIL synchrotron facility, DESIRS beamline. Photodesorption yields of the intact molecules and of the photoproducts were derived as a function of the incident photon energy. Experiments have revealed that the desorbing species match the photodissociation pattern of each isolated molecule, with little influence of the kind of ice (pure or mixed in CO or H_2O-rich environment). For both species, the photodesorption of the intact organics is found to be negligible in our experimental conditions, resulting in yields typically below 10^{-5} ejected molecules per incident photon. The results obtained on HCOOH and $HCOOCH_3$-containing ices are similar to what has already been found for methanol-containing ices, but contrast with the case of another complex molecule, CH_3CN, photodesorption of which has been recently studied. Such experimental results may be linked to the observation of COMs in protoplanetary disks, in which CH_3CN is commonly observed whereas HCOOH or methanol are detected only in some sources, $HCOOCH_3$ not being detected at all.

[a]*Sorbonne Université, Observatoire de Paris, PSL University, CNRS, LERMA, F-75005, Paris, France. E-mail: mathieu.bertin@sorbonne-universite.fr*
[b]*Univ. Paris-Saclay, CNRS UMR 8000, ICP, F-91405, Orsay, France*

1 Introduction

This last decade, the presence of small organic molecules – usually referred to as complex organic molecules (COMs) – was evidenced in the gas phase of the coldest regions of the interstellar medium (ISM). Among COMs, gaseous methyl formate $HCOOCH_3$ is detected in several prestellar cores,[1–3] while gaseous formic acid HCOOH is observed in prestellar cores,[3,4] in low UV flux photon-dominated regions (PDR)[5] and in protoplanetary disks.[6] In any case, the observations of such COMs in these cold regions ($T \sim$ 10–30 K) is still puzzling, since both their formation and their detection in the gas phase have to be explained without available thermal energy.

The exact formation mechanisms of these species are still unclear. Gas phase chemical routes are possible at low temperature, but rather unlikely – they involve the presence of gaseous ethanol for the formation of HCOOH, or a several step chemical process initiated from methanol for the formation of $HCOOCH_3$.[7–9] Other potential origins of these molecules may involve the solid phase chemistry in the icy mantles coating the interstellar dust grains in the cold regions. For instance, the formation of HCOOH and $HCOOCH_3$ can proceed through hydrogenation and hydroxylation of the CO-rich phase of the ice mantle,[10,11] or be triggered by cosmic rays (CR)-induced or UV photon-induced chemistry in pure methanol ices[12] or more realistic $H_2O:CO$ and $H_2O:CH_3OH$ icy mixtures.[13]

In any case, due to the very low temperatures of the media in which they are detected, the presence of HCOOH and $HCOOCH_3$ in the gas phase has to involve a non-thermal mechanism which either counterbalances their depletion on the dust grains if they are formed in the gas, or ejects the molecules from the icy mantles if they are formed in the solid phase. Several non-thermal desorption mechanisms may fulfill this role, such as the sputtering of the ice mantle by shocks between grains, the CR-induced sputtering of the icy mantles, the exothermicity of the formation reaction on the grain surface which releases a part of the product in the gas phase – a process usually referred to as chemical desorption – and the desorption induced by VUV or X-ray photons absorption in the ice mantle, also called VUV or X-ray photodesorption. Therefore, proper modelling of the chemistry of COMs in the cold ISM requires to quantitatively constrain both the chemical formation of organics, and their non-thermal desorption processes.[14] This in return implies it is necessary to perform quantitative studies that provide desorption yields for these organics in physical–chemical conditions relevant to the cold ISM. However, to date, very little experimental data dealing with non-thermal desorption efficiencies of COMs exist, with the notable exception of the methanol molecule. VUV photodesorption,[15,16] X-ray photodesorption[17,18] and CR sputtering[19,20] yields of methanol have been experimentally measured from methanol-containing ices, which concluded on the preponderant role played by CR in dense cores, compared to the secondary UV photons. Chemical desorption of methanol has also been experimentally studied, although only upper limits could be proposed because of the lack of experimental evidence.[21] The non-thermal desorption of the other COMs is left mainly unexplored, and usually the results obtained in the case of methanol are used as representative of a general trend followed by all the other organics. However, a recent study of the UV photodesorption of acetonitrile CH_3CN has revealed a different behaviour as compared to methanol, hinting that experimental results obtained from methanol-containing ice cannot *a priori* be extrapolated to the case of any other organics for the UV-induced processes.[22] In the case of both HCOOH and $HCOOCH_3$, there exists to our knowledge no experimental data for non-thermal desorption in general, and for UV photodesorption in particular.

Here, we present an experimental study dealing with the UV photodesorption from HCOOH and $HCOOCH_3$ containing ices, with the double aim of shedding light on the photodesorption mechanisms and properties, and of providing photodesorption yields for any desorbing species from these ices applicable to different regions of the ISM. Each molecule is studied as a pure ice, and in physisorbed binary mixtures organics : CO and organics : H_2O, believed to be more representative of the composition of the outer layers of the interstellar ices on either side of the CO freeze-out region.[23,24] The tunable output of the SOLEIL synchrotron facility, DESIRS beamline, is used to irradiate the ices at several energies in the 7–14 eV range, allowing derivation of photon energy-resolved photodesorption yields of each desorbing species, and finally astrophysical desorption yields which take into account the spectral structure of the UV fields in different regions of the interstellar medium. The obtained values are then discussed in regards to observations of these COMs in the cold ISM, and of already-characterized photodesorption from other COM-containing ices, such as methanol and acetonitrile.

2 Experimental methods

The studies are realized in the Ultra High Vacuum (UHV, base pressure $\sim 2 \times 10^{-10}$ torr) setup SPICES (Surface Processes & ICES) installed at the LERMA (Sorbonne University, Paris). The molecular ices are grown on a polycrystalline High Conductivity, Oxygen-Free (OFHC) copper substrate, mounted on the tip of the rotatable cold head of a closed-cycle helium cryostat. Its temperature can be varied from 13 K to 300 K, with an absolute precision better than 0.5 K, and a relative stability within 0.1 K, even during UV irradiation. The ices are prepared *in situ* by exposing the cold substrate to a partial pressure of the desired molecules, through a dosing tube positioned a few millimeters away in front of the surface – thus allowing deposition of relatively thick ices ($\sim$100 monolayers ML) with a limited increase in the base pressure in the chamber (typically only a few 10^{-9} torr during the ice growth). A calibration method based on the use of the Temperature Programmed Desorption (TPD) technique allows preparation of ice of chosen thickness, expressed in monolayers ML – one ML being equivalent to the surface density of a saturated molecular layer on the surface, roughly 10^{15} molecules per cm^2 – with a relative precision of about 10% (see *e.g.* ref. 25 or ref. 26).

In this study, ices of formic acid HCOOH and methyl formate $HCOOCH_3$ were prepared, pure and mixed with either CO or H_2O. Pure liquids of HCOOH (Sigma-Aldrich, >98% purity), $HCOOCH_3$ (Sigma-Aldrich, 99% purity) and H_2O (Fluka, liquid chromatography standard) were further purified using freeze–pump–thaw cycles. CO (Air Liquide, >99.9% purity) was used with no further purification. Each species' purity was checked *in situ* using mass spectrometry. Pure organic ices and ices mixed with CO were grown at 15 K, while ices containing H_2O were prepared at 90 K, to ensure the resulting water ice is in its compact amorphous phase (also referred to as c-ASW, standing for compact Amorphous Solid Water), and then cooled down to 15 K for the irradiation experiments. Icy mixtures were grown by first preparing a few mbar of gaseous mixture of the two constituents, composition of which is controlled by two sets of capacitive pressure gauges in the gas introduction system. The gaseous mixture was then introduced in the UHV chamber and deposited onto the cold copper substrate. The resulting ice composition was first calibrated using TPD technique before the experiments, and further checked by systematically performing TPD on each sample after irradiation.

To study the photodesorption from the ices, the Vacuum UV (VUV) output of the DESIRS beamline from the synchrotron SOLEIL facility (St Aubin, France) was used. The SPICES setup was directly connected to the end of the beamline, without any window, thus preventing cut-off at high photon energies. The DESIRS beamline provides VUV photons, the energy of which is tunable by means of an undulator, coupled to a gas filter to suppress high energy harmonics.[27] In these experiments, a typical photon energy range of 7–14 eV was used. Prior to its introduction in the chamber, the photon beam is reflected onto a grating. In the presented results, the 0 diffraction order was used, resulting in a higher flux but lower resolution mode. The photon flux is measured using a calibrated photodiode, and varies between 6×10^{14} photons per s at 7 eV and 6×10^{13} photons per s at 14 eV. The energy resolution in this mode is of $\sim$1 eV.

The VUV beam is sent with a 47° incidence angle on the samples, and irradiates a surface of $\sim$1 cm^2. During irradiation, the photodesorption of neutrals is monitored as a function of the photon energy by means of a Quadripolar Mass Spectrometer (QMS) using electron-impact ionization at 70 eV. The photon energy scan is made as follows: the ice is irradiated at a fixed energy using the 0 order of the grating during about 30 s, the beam is shut and the energy changed to the next desired value. During each irradiation step, the desorbing species are probed using the mass channel corresponding to their intact cation. An example of the acquired data is given in Fig. 1 in the case of photodesorption from pure formic acid at 15 K. At each irradiation energy, the amplitude of the mass signals – corrected from the background – is defined as the desorption signal, and is converted to photodesorption yields using the following protocol. The energy resolved desorption yield $\Gamma_X(E)$ of a given species X is derived from the desorption signal $I_{X^+}(E)$ of the cation X^+, corrected from the photon flux $\Phi(E)$ at the photon energy E following:

$$\Gamma_X(E) = A_X \frac{I_{X^+}(E)}{\Phi(E)}$$

where A_X is a calibration factor for the species X. The factors are determined experimentally by first recording a calibration photodesorption spectrum of CO from a 50 ML CO ice, and comparing it to the already well-constrained CO photodesorption yields in the literature.[28] From this, the calibration factor A_{CO} for the CO molecule is determined. All the other calibration factors are then extrapolated from A_{CO} considering different electron-impact ionization efficiencies in the mass spectrometer and different sensitivity of the mass spectrometer depending on the mass of X, also referred to as the apparatus function AF:

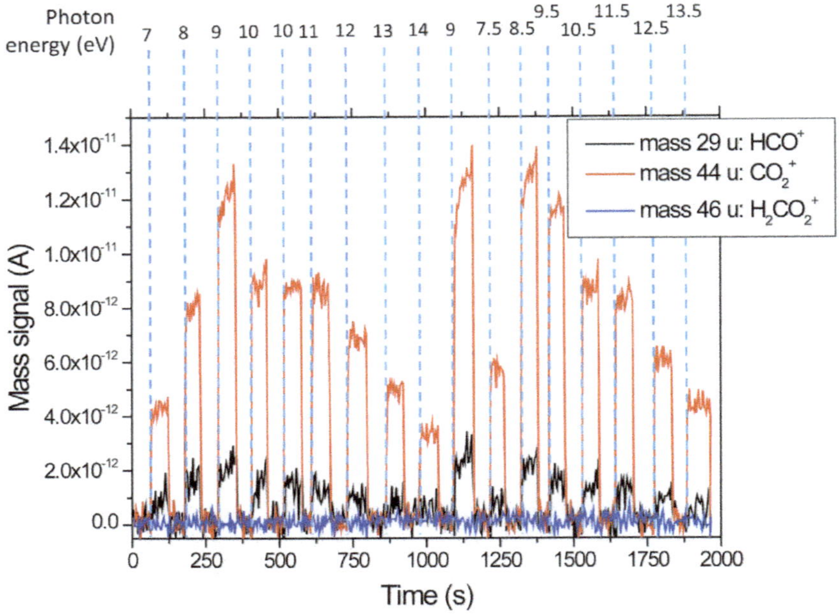

Fig. 1 Mass signals of several desorbing species during irradiation at several energies of pure 60 ML of formic acid at 15 K. Vertical lines indicate the moment where the VUV photons are introduced in the experiment. The mass signals are corrected from the continuous background signal due to the residual vacuum in the chamber.

$$A_{\mathrm{X}} = A_{\mathrm{CO}} \frac{\mathrm{AF(X)}}{\mathrm{AF(CO)}} \frac{\sigma(\mathrm{CO}^+/\mathrm{CO})}{\sigma(\mathrm{X}^+/\mathrm{X})}$$

with $\sigma(X^+/X)$ being the partial electron-impact ionization cross section at 70 eV of X giving X^+ and AF(X) the apparatus function of our spectrometer determined by comparing mass spectra of known species to reference data from the NIST chemical webbook.[29] Values from the literature were used for the ionization of CH_3, OH, H_2O, CO, H_2CO, O_2, CO_2, HCOOH and $HCOOCH_3$.[30–35] Values for HCO and H_3CO were not available, and the ionization cross section for H_2CO was used for these two species. Finally, since the dissociative ionization of heavier species may participate in the signal of X^+, each desorption signal was corrected from the dissociation of the higher mass signals by taking into account the dissociation branching ratio at 70 eV and the mass-dependent apparatus functions.

Independently from the signal-to-noise ratio, this calibration procedure comes with a systematic uncertainty which comprises uncertainties on (i) the published values of the ionization cross sections, (ii) on the CO molecule desorption yields, and (iii) on the apparatus function determination, but also uncertainties associated with the method itself, which in particular neglects the effects of different kinetic energies and spatial dispersion the desorbing molecules may have as compared to the CO molecules. Overall, we estimate this systematic uncertainty to be in the order of 50% of the derived photodesorption yields.

It should be noted here that only desorption of neutrals has been measured, and searched for, in this study. However, we do not expect the positive ion photodesorption to be an important desorption pathway in this energy domain. Indeed, it is suggested in past studies, *e.g.* by Philippe *et al.*[36] in the case of a CO ice, that the threshold for ion desorption exceeds 15 eV, and proceeds through the excitation of high-energy lying autoionized Rydberg states, well above the ionization potential – and also out of our probed energy range. Furthermore, the ion desorption yield becomes important only at even higher energies (>25 eV) when the excited state of the ion can be populated. Accordingly, previous attempts to measure photodesorption of ions in the 7–14 eV range from more simple ices (pure H_2O, CO and CO_2) were not successful in our case. We expect that ion desorption from HCOOH and $HCOOCH_3$ is also very weak in our energy range, and probably negligible as compared to neutral desorption.

Finally, the ices are probed during and after irradiation by two complementary techniques: Reflection Absorption InfraRed Spectroscopy (RAIRS), performed by means of a Brucker vec22 Fourier-transform infrared spectrometer used with a resolution of 1 cm^{-1}, coupled to the main chamber by in-vacuum optics and KBr windows, and the TPD technique, which consists of applying a constant heating ramp to the sample, and monitoring the desorbed species as a function of the sample's temperature. Both methods are used in order to identify any photochemistry which may have occurred during the irradiation.

3 Results and discussion

3.1 Photon-induced modifications in the HCOOH- and $HCOOCH_3$-containing ices probed by infrared spectroscopy

Fourier Transform-Reflection Absorption InfraRed Spectroscopy (FT-RAIRS) was systematically performed on the studied ices, after their growth and their irradiation at several energies in the 7–14 eV range. The obtained infrared spectra are shown in Fig. 2, on HCOOH-containing ices and $HCOOCH_3$-containing ices, after deposition, and after irradiation at 15 K. Each ice has been irradiated by the equivalent of a photon energy scan in the 7–14 eV energy range, representing a total photon fluence of about 10^{17} cm^{-2}. The IR spectra of the pure organics in the solid phase were attributed following works by Bertin *et al.*[37] for $HCOOCH_3$ and Bennett *et al.*[38] for HCOOH. IR spectra of mixed ices consist mainly of the superimposition of the organics-related vibrations with the most intense vibrations of solid water ice (broad OH-stretching features around 3400 cm^{-1} and libration mode features around 850 cm^{-1}, the bending mode overlapping with the C=O stretching vibration of the organics) or the solid CO vibration (sharp peak at $\sim$2140 cm^{-1}). Slight shifts in the band positions or modifications in the width of the vibrational features are observed depending on the composition of the ice; an effect being associated with different intermolecular interactions which modify the involved bond-strengths. After VUV irradiation, the only clearly observed modification of the IR spectra is the appearance of two bands at 2345 and 2140 cm^{-1} which are associated with condensed CO_2 and CO respectively, and a general slight decrease in the organics-related vibrational signatures. In the case of organics mixed with CO, a very small increase in the intensity of the CO band is observed. Our experiments do not highlight the appearance of more complex molecules after irradiation – that would be observed *via* the appearance of new vibrational features – indicating that either more complex products are formed in too small quantity for us to detect, or that their vibrational signatures overlap with

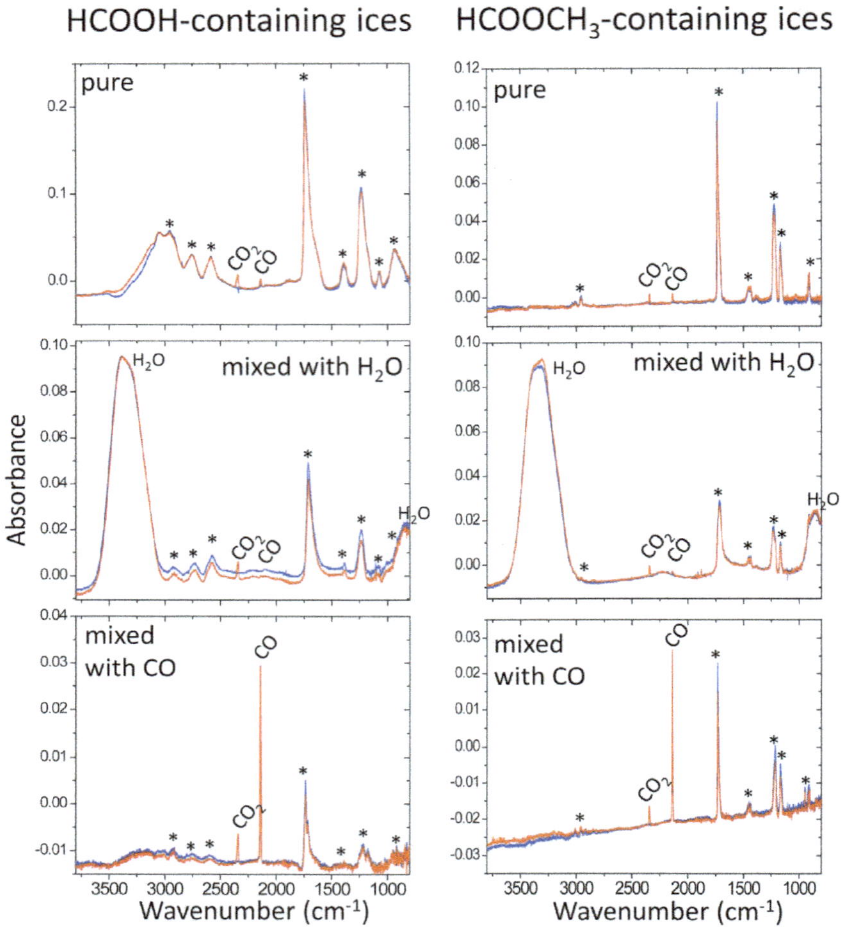

Fig. 2 IR spectra of HCOOH (left) and $HCOOCH_3$ (right) -containing ices after deposition (blue spectra) and after irradiation (red spectra) at 15 K. Each organic has been deposited in 60 ML of pure ices (upper line), in condensed mixture organics : H_2O (1 : 10 for HCOOH : H_2O and 1 : 3 for $HCOOCH_3$: H_2O – central line), and in mixture organics : CO (1 : 10 – lower line). Positions of the vibrational features associated with each organic molecule are indicated with *. Each ice was irradiated with varying incident photon energy between 7 and 14 eV with a total fluence of about 10^{17} cm^{-2}.

those of the non-irradiated ices. Except for the CO and CO_2 formation in the solids, the pure and mixed ices remain mostly intact after the irradiation in our flux and fluence conditions, as shown by their IR spectra which are only weakly altered, and still dominated by the organics and either water or CO vibrational features.

3.2 Desorption from pure HCOOH and $HCOOCH_3$ ices

Experiments realized on pure ices of formic acid and methyl formate kept at 15 K have revealed the desorption of a collection of neutral species. The associated photodesorption yields as a function of the photon energy are presented in

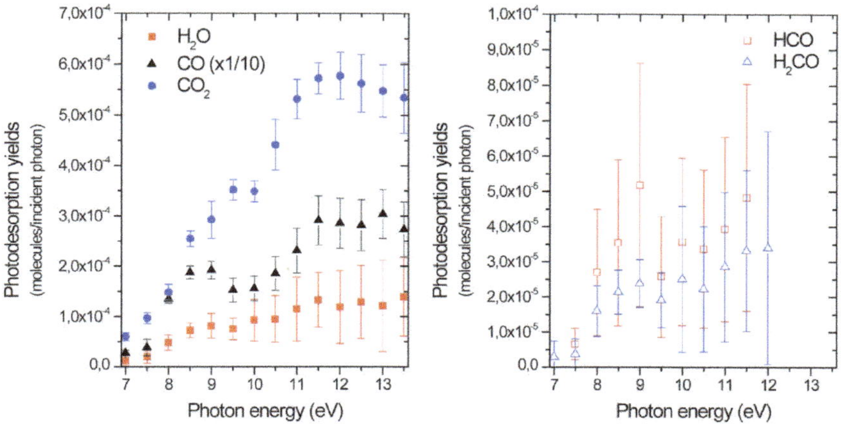

Fig. 3 Photodesorption yields of all the detected neutral species from 60 ML of pure ice of formic acid HCOOH at 15 K. Yields for the mass channels 29 and 30 *u* (associated with desorption of HCO and H_2CO respectively) could not be confidently isolated from the noise for photon energies above 12 eV, and are thus not shown. For clarity, the photodesorption yield of CO was divided by 10.

Fig. 3 and 4. In these figures, the mass channels 18 *u*, 28 *u*, and 44 *u* were attributed to the desorption of H_2O, CO and CO_2 respectively, desorption of which are observed both from HCOOH and $HCOOCH_3$. The origin of the mass 32 *u* signal measured in the case of photodesorption from $HCOOCH_3$ (Fig. 4) has been mainly attributed to O_2 desorption. Mass 32 *u* could also correspond to the CH_3OH^+ ion resulting from the ionization of some desorbing methanol CH_3OH or dissociative ionization of $HCOOCH_3$ in the QMS. The latter has been ruled out since no $HCOOCH_3^+$ ion was detected during the experiments. Methanol dissociative ionization in the QMS is also expected to produce an important signal on the 31 *u* mass channel – actually bigger than the 32 *u* signal – associated with the CH_3O^+ ion.[29] In our case, we indeed detect a small signal on the mass channel 31 *u*, but almost three times weaker than the signal on the 32 *u* channel, thus ruling out the fact that 32 *u* solely corresponds to methanol. Moreover, if one considers that the signal at 32 *u* is only O_2, then the correction of its dissociative ionization into the QMS brings the level of a small signal detected on the mass 16 *u* channel exactly to 0 (within error bars, and after correction from the water dissociative ionization). This further motivated our attribution of mass 32 *u* as O_2 and mass 31 *u* as the H_3CO radical ionization, even if the desorption of a very small quantity of methanol, participating as a small amount to these two mass signals, cannot be totally ruled out. Finally, the signals at 15, 29 and 30 *u* were attributed to the desorption of radicals CH_3, HCO, and H_2CO respectively.

In the case of condensed HCOOH, the desorption yields of each species follow the same energy dependence, depicting a constant increase over the whole energy range, with a small local depth at around 10 eV. We thus expect this structure to be representative of the VUV absorption spectrum of solid HCOOH, even if, such data being not available in the literature, a direct comparison was not possible. The same observation can be made from the desorption yields from solid $HCOOCH_3$, for which most of the energy-dependences are identical (within error

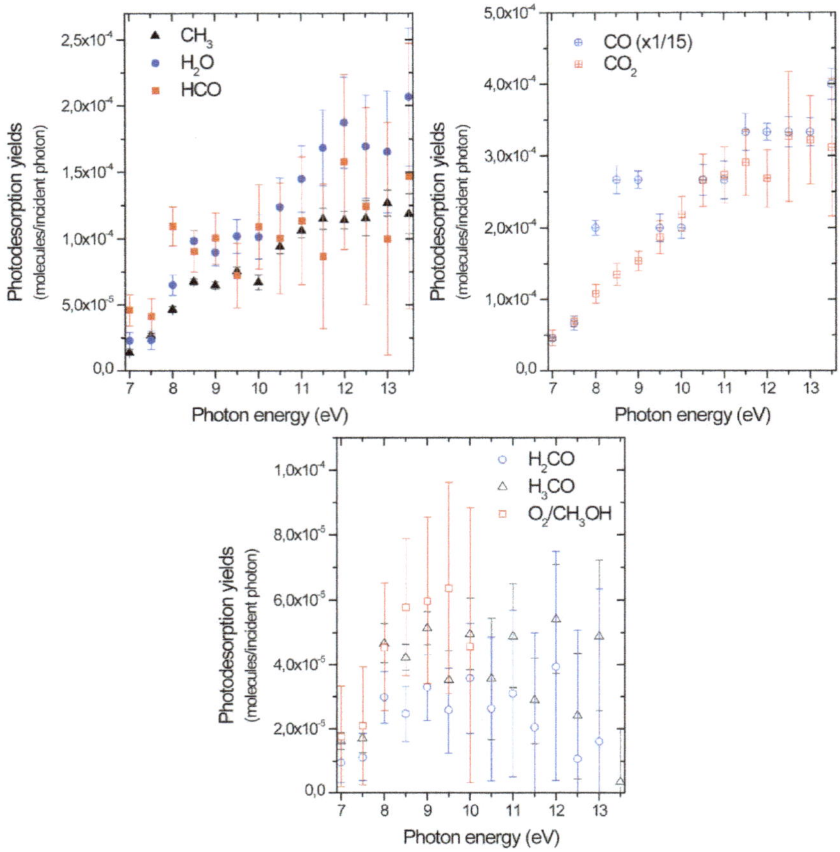

Fig. 4 Photodesorption yields of all the detected neutral species from 60 ML of pure ice of methyl formate $HCOOCH_3$ at 15 K. Yields for the mass channel 32 *u* (associated with desorption of mainly O_2, and probably some CH_3OH) could not be isolated from the noise for photon energies above 10 eV, and are thus not shown. For clarity, the photodesorption yield of CO was divided by 15.

bars), but the VUV absorption spectrum of which is not available to our knowledge. However, a notable exception are the CO photodesorption yields, which present a clear local maximum at 8.5 ± 0.5 eV. This structure can be associated with the solid CO absorption feature related to the $A(^1\Pi)–X(^1\Sigma^+)$ transition, which presents a structured vibronic band in this energy range[39,40] strongly coupled to the desorption.[28] Indeed, during irradiation, some of the formed CO can remain in the condensed phase, in return available to absorb another photon and desorb. This phenomenon, due to the accumulation of photoproduced CO in the solid, has already been observed in the case of pure CO_2 ices,[41] and infrared spectroscopy has confirmed the formation and accumulation of CO in the irradiated $HCOOCH_3$ ice after irradiation.

Pure ices of both HCOOH and $HCOOCH_3$ behave similarly under VUV irradiation. Probably the most striking finding of these experiments is that no photodesorption of the intact molecule could be measured. For both species, the photodesorption is dominated by the desorption of the CO molecule, followed by the desorption of CO_2 and H_2O. O_2, CH_3 and H_3CO could only be detected from pure $HCOOCH_3$, while HCO and H_2CO were detected from both HCOOH and $HCOOCH_3$ ices. In fact, those species match the direct photodissociation pattern of both HCOOH and $HCOOCH_3$ as studied in the gas phase. Photodissociation of gas phase HCOOH in the UV and VUV range results in the following fragmentations:[42–46]

$$HCOOH + h\nu \rightarrow CO + H_2O \tag{1}$$

$$HCOOH + h\nu \rightarrow CO + H + OH \tag{2}$$

$$HCOOH + h\nu \rightarrow CO_2 + H_2 \tag{3}$$

$$HCOOH + h\nu \rightarrow HCOO + H \tag{4}$$

$$HCOOH + h\nu \rightarrow HCO + OH \tag{5}$$

Except for the channel (4), each of these dissociation pathways can explain the detected photodesorbed species. H and H_2 cannot be confidently detected by our QMS, and OH may actually photodesorb, but could not be separated from background noise after correction of H_2O dissociative ionization on the 17 *u* mass signal. Desorption of some H_2CO is also detected, but is not explained by photodissociation of the HCOOH as known from gas phase experiments. This could be due either to a dissociation pathway specific to the condensed phase, or to the result of subsequent chemistry in the ice, *e.g.* triggered by hydrogen atoms released from channel (4), leading to the H_2CO formation and desorption.

Gas phase $HCOOCH_3$ photodissociation in the UV has also been studied, *e.g.* by Schwell *et al.*,[45] Lee[47] and Nakamura *et al.*[48] The main identified dissociations channels are:

$$HCOOCH_3 + h\nu \rightarrow CH_3O + CO + H \tag{6}$$

$$HCOOCH_3 + h\nu \rightarrow CH_3O + HCO \tag{7}$$

$$HCOOCH_3 + h\nu \rightarrow CH_3OH + CO \tag{8}$$

$$HCOOCH_3 + h\nu \rightarrow CH_3 + CO_2 + H \tag{9}$$

If one considers similar fragmentation patterns in the solid phase, detected desorbing CO, CO_2, CH_3, HCO and H_3CO from pure solid $HCOOCH_3$ can all be attributed to direct photodissociation of the condensed methyl formate molecules. However, gas phase studies do not highlight H_2O, O_2 or H_2CO as results of photodissociation of methyl formate. Similarly as the case of H_2CO from HCOOH, those species may originate from different dissociation pathways or from photochemistry in the solid $HCOOCH_3$ coupled to desorption.

3.3 Desorption from HCOOH and $HCOOCH_3$ in H_2O- and CO-rich ices

In order to estimate the role played by the molecular environment on the photodesorption, model ices of HCOOH and $HCOOCH_3$ embedded in a CO- or H_2O-

rich ice were studied. For each organic, four kinds of mixed ices were investigated: icy mixture of CO : organics and H_2O : organics, in which either CO or H_2O were in excess, or layered ices consisting of 1 ML of organics deposited onto a pure CO or pure compact amorphous H_2O ice. Fig. 5 presents an overview of the desorbing species, together with their photodesorption yields, for an irradiation at the photon energy of 10 $\pm$ 0.5 eV. As in the case of pure ices, no signal associated with the desorption of the intact HCOOH or $HCOOCH_3$ molecule could be detected. In these cases, only an upper limit, defined as the noise amplitude on the mass channel, can be derived from our data. For each ice, all the desorbing species are identical to the case of pure organics ices, with the notable exception of OH and O_2 desorptions which are detected from HCOOH-containing ices only when formic acid is coadsorbed with H_2O. This indicates the weak role played, in our flux and fluence conditions, by the photochemistry involving both the organics and the surrounding water or CO, or at least that potential photochemistry is not correlated to the desorption phenomenon. This is further corroborated by the fact that most of the desorbing species (HCO, H_2CO and to some extent CO_2 for HCOOH; CH_3, HCO, H_2CO, H_3CO, O_2 and CO_2 for $HCOOCH_3$) present very similar photodesorption yields, most of the time identical within error bars for each system, pure or mixed. This stresses that these species most likely originate from the dissociation of the organics alone, and that their desorption does not depend on the local molecular environment.

CO desorption is slightly enhanced when the organics are co-deposited with CO molecules – especially in the case of HCOOH-containing ices. Indeed, CO weakly absorbs, and photodesorbs, in the 9.5–10.5 eV region,[28] and thus some of the detected CO from mixed ices originates from the excitation and desorption of the co-deposited CO molecules. The H_2O photodesorption yields are also found to be poorly dependent on the ice, and almost not varying when the organics are mixed with water. This is surprising at first since some codeposited H_2O is expected to photodesorb when irradiated, and thus should, together with the

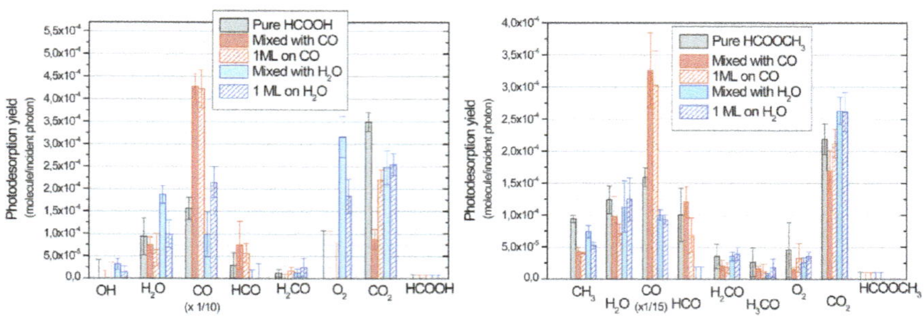

Fig. 5 Photodesorption yields of neutral species from HCOOH- and $HCOOCH_3$-containing ices at 15 K, during irradiation with 10 $\pm$ 0.5 eV photons and a flux of 2.3 $\times$ 10^{14} photons per s. Left: photodesorption from 60 ML of pure HCOOH, HCOOH mixed in 60 ML of CO (1 : 10), HCOOH mixed in 60 ML of H_2O (1 : 10), 1 ML HCOOH on top of 30 ML of CO, and 1 ML HCOOH on top of 30 ML of H_2O. Right: photodesorption from 60 ML of pure $HCOOCH_3$, $HCOOCH_3$ mixed in 60 ML of CO (1 : 10), $HCOOCH_3$ mixed in 60 ML of H_2O (1 : 3), 1 ML $HCOOCH_3$ on top of 30 ML of CO, and 1 ML $HCOOCH_3$ on top of 30 ML of H_2O. The photodesorption yields of CO were artificially divided by 10 (left panel) and 15 (right panel) for clarity.

dissociation of the organics, contribute to the total H_2O desorption yield. VUV photodesorption yield of H_2O from pure water ices at ~10 eV are experimentally estimated between 5×10^{-4} and 1×10^{-3} molecules per photon, the latter value being derived from an amorphous porous ice and not a compact one.[49–51] Here, H_2O desorption yields are found at best more than two times lower, even from ices mixed with water. An explanation would be that coadsorption with organics prevents the desorption from the condensed water. Indeed, Fillion *et al.*[51] propose that the water photodesorption process from pure water ice, under our flux and fluence conditions, is due to a large extent to the photochemistry at the ice surface, reforming H_2O molecules from H and OH radicals, which then desorb. Adjunction of organics in the vicinity may perturbate this chemistry and hinder the water desorption by recombination, leaving mainly the organics fragmentation as a source of gas phase H_2O. OH and O_2 desorption are observed during irradiation of HCOOH : H_2O mixed ices, but are not detected from pure HCOOH ice or in HCOOH : CO mixtures. These desorptions may originate from the desorption from the solid water itself, since the measured yields are comparable to the O_2 and OH photodesorption yields from pure water ice.[50–52] However, O_2 desorption is not modified in the case of $HCOOCH_3$ mixed with water, and no OH signal could be measured for these systems. However, contrary to the case of methyl formate, OH is expected to be formed *via* the dissociation of HCOOH, even if its desorption from pure formic acid could not be detected. Since OH is also expected to be produced from water itself, the irradiation of the mixture HCOOH : H_2O may increase its production to the point where it exceeds our detection limit, explaining thus its observation in this case. The origin of O_2 desorption from HCOOH : H_2O mixed ice is still unclear, but it may also be linked to the increased production of OH radicals, as compared to the other systems. Indeed, OH radicals are at the origin of the O_2 production in irradiated water ice as demonstrated by UV photon or electron irradiation experiments.[51,53] Such process in our case may trigger some surface chemistry releasing some O_2 in the gas phase. A different HCOOH dissociation branching ratio at the surface of the water ice, leading to O_2 formation, cannot not however be ruled-out. Finally, desorption of HCO is not detected anymore when $HCOOCH_3$ is mixed in a H_2O-rich environment. HCO is believed to originate from the photodissociation of methyl formate also leading to some H_3CO production, desorption of which is not affected by the coadsorption with water molecules. The absence of HCO desorption is still unclear, but may be associated to some chemistry between the radical HCO and H_2O molecules or other photoproducts. A similar effect is also observed in the HCOOH : H_2O ices, for which the HCO desorption is also found below the detection limit. Our studies do not bring evidence, however, of this chemistry, which may lead *e.g.* to some CO and CO_2 production, as it cannot be disentangled from the much higher quantities already detected directly from the dissociation of the organics.

Photon energy-dependent photodesorption yields from the mixed ices were also measured. Some examples of photodesorption yields from mixed ices are shown in Fig. 6 and 7. Except for the CO desorption from CO-containing ices, the global shapes of the desorption spectra reflect the ones of the pure HCOOH or $HCOOCH_3$. There again, the strong similarities between the pure and the mixed organics suggests that most of the observed desorption is due to the organics absorption and subsequent dissociation, and that the matrix plays here a minor role. The low energy resolution employed here, the error bars and the broadness

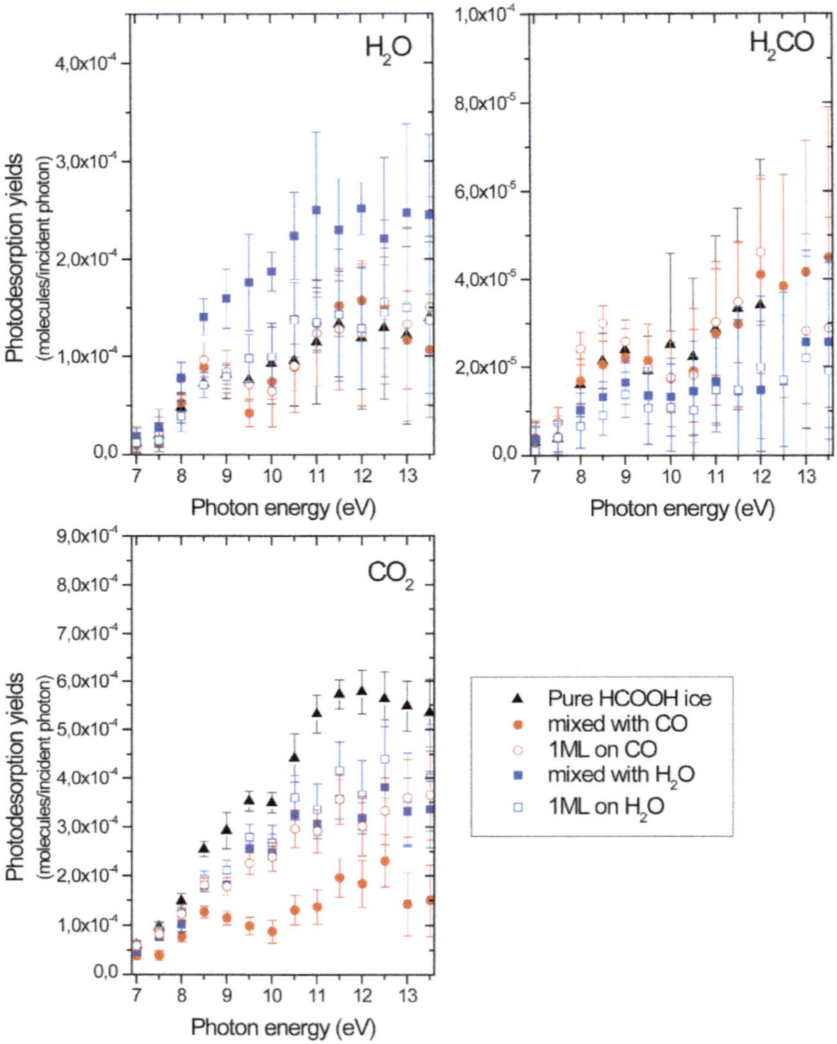

Fig. 6 Photodesorption yields of some neutral species (H_2O, H_2CO and CO_2) from 60 ML formic acid-containing ices at 15 K: pure ice of HCOOH, 1 ML of HCOOH onto pure ices of CO and H_2O, HCOOH : CO and HCOOH : H_2O mixed ice in proportion 1 : 10.

of the spectra prevent however totally ruling out any contribution of the H_2O or CO absorption in the photodesorption yields of water and CO-containing ice. It is clear however that those contributions should be minimal.

3.4 Discussions and astrophysical implications

Our experiments have revealed that the UV photodesorption from HCOOH- and $HCOOCH_3$-containing ices is dominated by the photodissociation of the organics and the subsequent desorption of the photofragments in the gas phase. Whatever is the composition of the studied ices, the desorption of the intact organics could

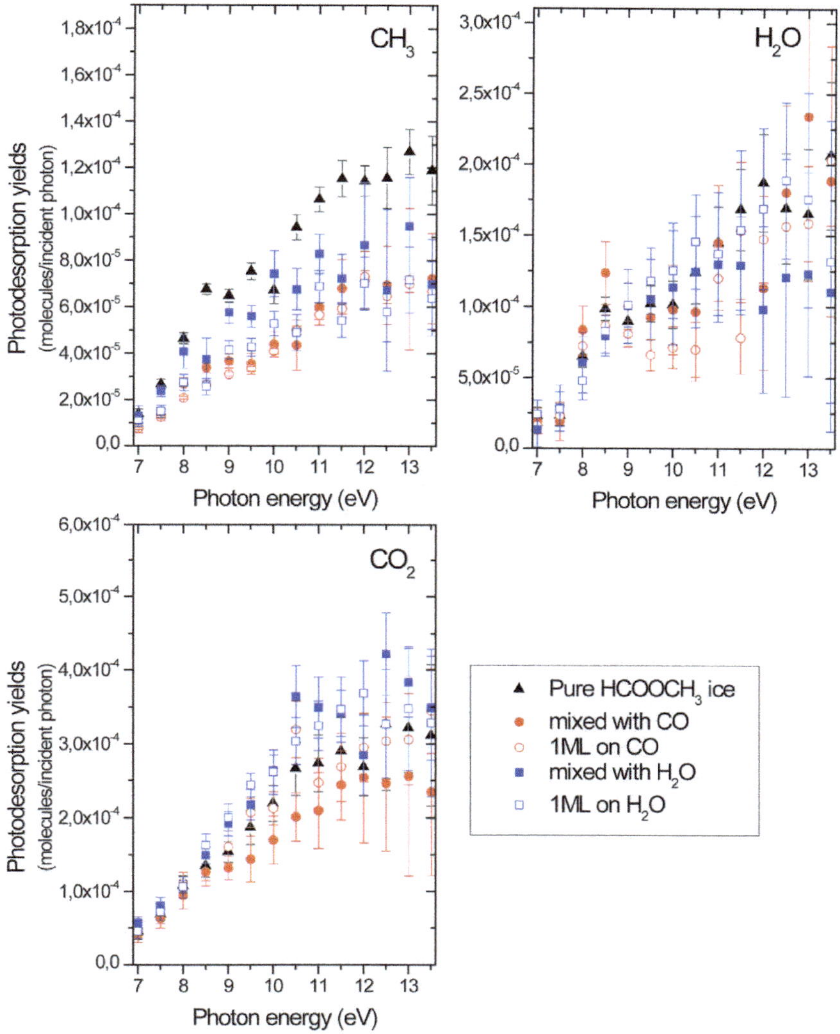

Fig. 7 Photodesorption yields of some neutral species (CH_3, H_2O and CO_2) from methyl formate-containing ices at 15 K: pure ice of $HCOOCH_3$, 1 ML of $HCOOCH_3$ onto pure ices of CO and H_2O, $HCOOCH_3$: CO and $HCOOCH_3$: H_2O mixed ice in proportions 1 : 10 and 1 : 3 respectively.

not be detected. Only an upper limit for the desorption yields can be derived from our measurements. The experimental photodesorption yields as a function of the photon energy can be used to calculate averaged photodesorption yields for each detected species, taking into account the spectral structure of the UV fields in the ISM. Table 1 presents these astronomical photodesorption yields for several interstellar UV spectra, that may be of use in different regions of the ISM, such as the InterStellar Radiation Field (ISRF) for PDR-like border regions,[54] secondary photons from cosmic-ray ionization for dense cores[55] and inner regions of disks.[56] The latter two are mostly dominated by a strong H Ly-α component at 10.2 eV, and

Table 1 Astronomical photodesorption yields ($\times 10^{-5}$ desorbed molecule per incident photon) derived from the experimental photon-energy resolved yields of HCOOH- and $HCOOCH_3$-containing ices and UV spectra relevant to PDR and dense cores or inner regions of disks. Associated uncertainties take into account signal-over-noise ratio and the systematic relative uncertainty of ~50% related to the quantification method

Photodesorbed species	ISRF[a]	Dense clouds[b] and disks[c]
From HCOOH-containing ices		
HCOOH	<0.5	<0.5
CO_2	14 ± 7	36 ± 18
H_2CO	0.7 ± 0.4	2.1 ± 1.1
HCO[d]	1.0 ± 0.5	3.1 ± 1.6
CO	73 ± 36	170 ± 90
H_2O	3.4 ± 1.7	8.7 ± 4.4
From $HCOOCH_3$-containing ices		
$HCOOCH_3$	<1	<1
CO_2	7.9 ± 4	22 ± 11
H_3CO	1.0 ± 0.5	3.0 ± 1.5
H_2CO	1.6 ± 0.8	4.1 ± 2.1
O_2	1.0 ± 0.7	1.2 ± 0.9
HCO[d]	3.8 ± 1.9	10 ± 5
CO	140 ± 70	340 ± 170
H_2O	4.3 ± 2.2	11 ± 6
CH_3	3.1 ± 1.6	7.5 ± 3.8

[a] Using UV fields from Mathis *et al.*[54] [b] Using UV fields from Gredel *et al.*[55] [c] Using UV fields from France *et al.*[56] [d] When mixed in H_2O-rich ices, the photodesorption yields of HCO are hindered and drop below our detection limit ($<2 \times 10^{-5}$).

they both lead to very similar values – equal within our error bars – for the averaged photodesorption yields. As discussed previously, we found that the VUV photodesorption from the COMs does not strongly depend on the presence of either CO or H_2O molecules in the ices (an exception however is the case of HCO, photodesorption of which being hindered in the case of mixed water : HCOOH and water : $HCOOCH_3$ ices). Accordingly, we took the photodesorption spectra from pure ices of COMs to derive the yields of Table 1, which we assume to be applicable in the case of any mixed ices containing each of the COMs and CO or H_2O.

The fact that the desorption yields of the intact molecules are found below our detection limits, independently from the ice (pure, mixed with H_2O or CO), is an important and interesting finding. First, it points out that the photodissociation of the molecules is a more efficient decay pathway subsequent to the absorption of a photon by the organics, thus competing with the desorption which becomes comparatively negligible. Potential photochemistry in the ice does not participate in the reformation and desorption of the initial organics, as it is for instance in the case of H_2O desorption from pure water ices.[51] Second, it also points out that the surrounding molecules, in our case CO and H_2O, do not promote the desorption of the organics, as is the case for more simple coadsorbed molecules. Indeed, some studies on UV photodesorption from simple molecules (N_2, CH_4, CO, CO_2) or rare gas atoms adsorbed on a CO or H_2O rich ice have shown that the absorption of a photon by the dominant species, *i.e.* CO or H_2O, could trigger the desorption of the coadsorbed species. This indirect desorption process is known to proceed either *via* a kick-out mechanism in the case of species coadsorbed with water – during which a hydrogen atom from H_2O photodissociation collides with a surface molecule or atom, giving it enough momentum to overcome the adsorption barrier and desorb[57] – or *via* another kind of energy transfer, which remains to be clarified, in the case of CO.[58,59] Such indirect processes may have a strong influence when it comes to model UV photodesorption from interstellar ices since the photodesorption yields of such simple species actually depend on the ice composition. Obviously, in the case of HCOOH and $HCOOCH_3$, such indirect desorption, promoted by the photon absorption by the main constituents CO or H_2O, is either not operative, or too weak to produce enough gas phase organics for us to detect. There can be several explanations for this. One of those could be that HCOOH and $HCOOCH_3$ present many internal degrees of freedom as compared to simple molecules such as N_2, and thus any external energy brought by the matrix to the molecule can be dissipated into internal energy, lowering the amount of energy left in the translation for the molecules to desorb. Another could be due to the higher strength of intermolecular interactions in the solids when organics are involved, *via* hydrogen bonding implying the hydroxyl –OH and carboxyl –C=O groups for instance, which are not encountered in the case of simpler species for which only weaker van der Waals interactions are at play.

In many respects, the UV photodesorption from HCOOH- and $HCOOCH_3$-containing ices is very similar to the one from methanol-containing ice. Indeed, UV photodesorption yields of CH_3OH are close to our detection limit from pure methanol ices, and are not detected in the case of mixed ices methanol : CO (yields below 3×10^{-6} molecules per photon). Instead, the photodesorption from methanol is dominated by the desorption of its photodissociation products.[15,16] HCOOH and $HCOOCH_3$ seem to follow the same tendencies, *e.g.* a weak photodesorption of the intact molecules – in these cases not even detected – not promoted by the molecular environment, and a desorption mostly reflecting the photodissociation of the species. This does not, however, mean that it is a general trend for all the organic species that can be encountered in the interstellar ices. Indeed, recent studies on the photodesorption of acetonitrile CH_3CN containing ices have shown an efficient photodesorption pathway of the intact molecule, reaching yields of several 10^{-5} molecules per photons, and which is comparable to the dissociative channels.[22,60] These different trends in the UV photodesorption of organics species may be related to the observations of gaseous organics in the protoplanetary disks: molecules which dissociate efficiently, and whose photodesorption of intact molecules has been experimentally found to be very low, such as HCOOH or CH_3OH, are very difficult to detect, and observed only in one source,[6,61] or even not seen at all in the case of $HCOOCH_3$. On the other hand, acetonitrile, whose intact photodesorption has been observed and measured experimentally, is seen in many sources (*e.g.* Bergner *et al.*[62]). Such correlation between experimental findings and observation is a strong argument in favor of the key role played by UV photodesorption to explain observed gas phase molecular abundances, including those of organics, in protoplanetary disks and in PDR regions – a hypothesis already formulated in Bergner *et al.*[62] for the CH_3OH and CH_3CN abundances in disks, and in Guzmán *et al.*[5] in the Horsehead Nebula. Generalizing the role of UV photodesorption of organics to any cold region of the ISM is however not straightforward, since many other non-thermal desorption processes can be at play, such as chemical desorption or cosmic-ray sputtering of the ices.

4 Conclusions

In this study, we have measured the UV photodesorption yields of neutral species from formic acid HCOOH and methyl formate $HCOOCH_3$ ices – pure and mixed with CO or H_2O – in the 7–14 eV range. Our experiments show that the photodesorption process is mostly dominated by the absorption and photodissociation of the solid organics, resulting in the desorption of photofragments or products. Photodesorption yields of H_2O, CO, HCO, H_2CO, and CO_2 were measured from HCOOH-containing ice, and of CH_3, H_2O, CO, HCO, H_2CO, H_3CO, O_2 and CO_2 from $HCOOCH_3$-containing ices. In both cases, the photodesorption yields of the intact HCOOH or $HCOOCH_3$ could not be measured, and only higher limits for the desorption, of 0.5×10^{-5} and 1×10^{-5} ejected molecules per incident photon respectively, could be estimated. The yields are found equal, within our experimental error bars, for pure and mixed ices with CO or H_2O, allowing proposition of astrophysical UV photodesorption yields from these two species independent from the ices composition.

In many respects, the UV photodesorption from HCOOH- and $HCOOCH_3$-containing ices is similar to UV photodesorption from methanol ices, *i.e.* a process dominated by the photodissociation of the organics, and in which the desorption of the intact molecule is a minority pathway. This is in contrast to the case of other COMs such as acetonitrile, molecular desorption of which has been shown to compete with its dissociation. This study thus brings new hints that the photodesorption of organics from interstellar ices may actually follow different trends depending on the nature of the considered COM, and that some of them present a methanol-like behaviour, such as HCOOH or $HCOOCH_3$, while others may behave differently.

Many authors tentatively attribute the origin of gas phase HCOOH or $HCOOCH_3$ in the cold media either to the photodesorption (*e.g.* ref. 5 and 6) or to other processes (*e.g.* ref. 3), but the lack of quantitative data on the efficiency of each process prevented actually being able to conclude. In addition, even if the desorption of the intact organics is rather weak as compared to the dissociative channels, the photodesorption of radicals from organics-containing ices is also a potential route to bring reactive material in the gas phase and trigger chemistry leading to the formation of organics. Only a proper astrochemical modelling including the gas–grain exchanges can give a good idea of the relative role played by each process, which in return implies knowing the desorption yields associated with each species. Even if the present study could only provide upper limits to the UV photodesorption of HCOOH and $HCOOCH_3$, it also brings the desorption yields for the desorption of radicals and chemical products from the ices, and it may participate in a better quantification of the role of UV photodesorption as compared to other non-thermal desorption mechanisms in the ISM.

Conflicts of interest

There are no conflicts to declare.

Acknowledgements

This work was funded by (i) the Region Ile-de-France DIM-ACAV+ program, (ii) the Sorbonne Université "Emergence" program, (iii) the ANR PIXyES project, grant ANR-20-CE30-0018 of the French "Agence Nationale de la Recherche" and (iv) the Programme National "Physique et Chimie du Milieu Interstellaire" (PCMI) of CNRS/INSU with INC/INP cofunded by CEA and CNES. We would like to acknowledge SOLEIL for provision of synchrotron radiation facilities under Project No. 20191298, and we gratefully thank L. Nahon for his help on the DESIRS beamline.

Notes and references

1 A. Bacmann, V. Taquet, A. Faure, C. Kahane and C. Ceccarelli, *Astron. Astrophys.*, 2012, **541**, L12.
2 I. Jiménez-Serra, A. I. Vasyunin, P. Caselli, N. Marcelino, N. Billot, S. Viti, L. Testi, C. Vastel, B. Lefloch and R. Bachiller, *Astrophys. J.*, 2016, **830**, L6.
3 V. Taquet, E. S. Wirström, S. B. Charnley, A. Faure, A. López-Sepulcre and C. M. Persson, *Astron. Astrophys.*, 2017, **607**, A20.
4 C. Vastel, C. Ceccarelli, B. Lefloch and R. Bachiller, *Astrophys. J.*, 2014, **795**, L2.
5 V. V. Guzmán, J. Pety, P. Gratier, J. R. Goicoechea, M. Gerin, E. Roueff, F. L. Petit and J. L. Bourlot, *Faraday Discuss.*, 2014, **168**, 103–127.
6 C. Favre, D. Fedele, D. Semenov, S. Parfenov, C. Codella, C. Ceccarelli, E. A. Bergin, E. Chapillon, L. Testi, F. Hersant, B. Lefloch, F. Fontani, G. A. Blake, L. I. Cleeves, C. Qi, K. R. Schwarz and V. Taquet, *Astrophys. J., Lett.*, 2018, **862**, L2.
7 D. Skouteris, N. Balucani, C. Ceccarelli, F. Vazart, C. Puzzarini, V. Barone, C. Codella and B. Lefloch, *Astrophys. J.*, 2018, **854**, 135.
8 N. Balucani, C. Ceccarelli and V. Taquet, *Mon. Not. R. Astron. Soc.: Lett.*, 2015, **449**, L16–L20.
9 A. J. Ocaña, S. Blázquez, D. González, A. Potapov, B. Ballesteros, A. Canosa, M. Antiñolo, J. Albaladejo and E. Jiménez, *Proc. Int. Astron. Union*, 2019, **15**, 365–367.
10 D. Qasim, T. Lamberts, J. He, K.-J. Chuang, G. Fedoseev, S. Ioppolo, A. C. A. Boogert and H. Linnartz, *Astron. Astrophys.*, 2019, **626**, A118.
11 C. Walsh, E. Herbst, H. Nomura, T. J. Millar and S. W. Weaver, *Faraday Discuss.*, 2014, **168**, 389.
12 K. I. Öberg, R. T. Garrod, E. F. van Dishoeck and H. Linnartz, *Astron. Astrophys.*, 2009, **504**, 891.
13 A. Ishibashi, H. Hidaka, Y. Oba, A. Kouchi and N. Watanabe, *Astrophys. J., Lett.*, 2021, **921**, L13.
14 V. Wakelam, E. Dartois, M. Chabot, S. Spezzano, D. Navarro-Almaida, J.-C. Loison and A. Fuente, *Astron. Astrophys.*, 2021, **652**, A63.
15 M. Bertin, C. Romanzin, M. Doronin, L. Philippe, P. Jeseck, N. Ligterink, H. Linnartz, X. Michaut and J.-H. Fillion, *Astrophys. J.*, 2016, **817**, L12.
16 G. A. Cruz-Diaz, R. Martín-Doménech, G. M. Muñoz Caro and Y.-J. Chen, *Astron. Astrophys.*, 2016, **592**, A68.
17 R. Basalgète, R. Dupuy, G. Féraud, C. Romanzin, L. Philippe, X. Michaut, J. Michoud, L. Amiaud, A. Lafosse, J.-H. Fillion and M. Bertin, *Astron. Astrophys.*, 2021, **647**, A35.

18 R. Basalgète, R. Dupuy, G. Féraud, C. Romanzin, L. Philippe, X. Michaut, J. Michoud, L. Amiaud, A. Lafosse, J.-H. Fillion and M. Bertin, *Astron. Astrophys.*, 2021, **647**, A36.

19 E. Dartois, M. Chabot, T. Barkach, H. Rothard, B. Augé, A. Agnihotri, A. Domaracka and P. Boduch, *Astron. Astrophys.*, 2019, **627**, A55.

20 E. Dartois, M. Chabot, A. Bacmann, P. Boduch, A. Domaracka and H. Rothard, *Astron. Astrophys.*, 2020, **634**, A103.

21 M. Minissale, F. Dulieu, S. Cazaux and S. Hocuk, *Astron. Astrophys.*, 2016, **585**, A24.

22 R. Basalgète, A. J. Ocaña, G. Féraud, C. Romanzin, L. Philippe, X. Michaut, J.-H. Fillion and M. Bertin, *Astrophys. J.*, 2021, **922**, 213.

23 K. M. Pontoppidan, H. J. Fraser, E. Dartois, W. F. Thi, E. F. van Dishoeck, A. C. A. Boogert, L. d'Hendecourt, A. Tielens and S. E. Bisschop, *Astron. Astrophys.*, 2003, **408**, 981–1007.

24 A. A. Boogert, P. A. Gerakines and D. C. Whittet, *Annu. Rev. Astron. Astrophys.*, 2015, **53**, 541–581.

25 M. Doronin, M. Bertin, X. Michaut, L. Philippe and J.-H. Fillion, *J. Chem. Phys.*, 2015, **143**, 084703.

26 M. Bertin, M. Doronin, J.-H. Fillion, X. Michaut, L. Philippe, M. Lattelais, A. Markovits, F. Pauzat, Y. Ellinger and J.-C. Guillemin, *Astron. Astrophys.*, 2017, **598**, A18.

27 L. Nahon, N. de Oliveira, G. Garcia, J.-F. Gil, B. Pilette, O. Marcouillé, B. Lagarde and F. Polack, *J. Synchrotron Radiat.*, 2012, **19**, 508.

28 E. C. Fayolle, M. Bertin, C. Romanzin, X. Michaut, K. I. Öberg, H. Linnartz and J.-H. Fillion, *Astrophys. J., Lett.*, 2011, **739**, L36.

29 NIST Mass Spectrometry Data Center and W. E. Wallace, in *NIST chemical webbook, NIST standard database no. 69*, ed. P. Linstrom and W. Mallard, National Institute of Standard and Technology, 2022.

30 F. A. Baiocchi, R. C. Wetzel and R. S. Freund, *Phys. Rev. Lett.*, 1984, **53**, 771.

31 K. N. Joshipura, M. Vinodkumar and U. M. Patel, *J. Phys. B: At., Mol. Opt. Phys.*, 2001, **34**, 509.

32 J. R. Vacher, F. Jorand, N. Blin-Simiand and S. Pasquiers, *Chem. Phys. Lett.*, 2009, **476**, 178–181.

33 R. S. Freund, R. C. Wetzel and R. J. Shul, *Phys. Rev. A: At., Mol., Opt. Phys.*, 1990, **41**, 5861.

34 H. C. Straub, P. Renault, B. G. Lindsay, K. A. Smith and R. F. Stebbings, *Phys. Rev. A: At., Mol., Opt. Phys.*, 1996, **54**, 2146.

35 P. Możejko, *Eur. Phys. J.: Spec. Top.*, 2007, **144**, 233–237.

36 L. Philippe, T. Hirayama, M. J. Ramage, G. Comtet, M. Rose, L. Hellner and G. Dujardin, *J. Chem. Phys.*, 1997, **106**, 7072.

37 M. Bertin, C. Romanzin, X. Michaut, P. Jeseck and J.-H. Fillion, *J. Phys. Chem. C*, 2011, **115**, 12920–12928.

38 C. J. Bennett, T. Hama, Y. S. Kim, M. Kawasaki and R. I. Kaiser, *Astrophys. J.*, 2011, **727**, 27.

39 H.-C. Lu, H.-K. Chen, B.-M. Cheng, Y.-P. Kuo and J. F. Ogilvie, *J. Phys. B: At., Mol. Opt. Phys.*, 2005, **38**, 3693–3704.

40 G. A. Cruz-Diaz, G. M. Muñoz Caro, Y.-J. Chen and T.-S. Yih, *Astron. Astrophys.*, 2014, **562**, A119.

41 J.-H. Fillion, E. C. Fayolle, X. Michaut, M. Doronin, L. Philippe, J. Rakovsky, C. Romanzin, N. Champion, K. I. Oberg, H. Linnartz and M. Bertin, *Faraday Discuss.*, 2014, **168**, 533–552.

42 K. Tabayashi, J.-i. Aoyama, M. Matsui, T. Hino and K. Saito, *J. Chem. Phys.*, 1999, **110**, 9547–9554.

43 H. Su, Y. He, F. Kong, W. Fang and R. Liu, *J. Chem. Phys.*, 2000, **113**, 1891–1897.

44 M. Hashinokuchi, R. Koumura, D.-C. Che, H. Ohoyama and T. Kasai, *J. Mass Spectrom. Soc. Jpn.*, 2002, **50**, 7–10.

45 M. Schwell, H.-W. Jochims, H. Baumgärtel, F. Dulieu and S. Leach, *Planet. Space Sci.*, 2006, **54**, 1073–1085.

46 C.-J. Tso, T. Kasai and K.-C. Lin, *Sci. Rep.*, 2020, **10**, 4769.

47 S.-H. Lee, *J. Chem. Phys.*, 2008, **129**, 194304.

48 M. Nakamura, P.-Y. Tsai, T. Kasai, K.-C. Lin, F. Palazzetti, A. Lombardi and V. Aquilanti, *Faraday Discuss.*, 2015, **177**, 77–98.

49 K. I. Öberg, H. Linnartz, R. Visser and E. F. van Dishoeck, *Astrophys. J.*, 2009, **693**, 1209–1218.

50 G. A. Cruz-Diaz, R. Martín-Doménech, E. Moreno, G. M. Muñoz Caro and Y.-J. Chen, *Mon. Not. R. Astron. Soc.*, 2018, **474**, 3080–3089.

51 J.-H. Fillion, R. Dupuy, G. Féraud, C. Romanzin, L. Philippe, T. Putaud, V. Baglin, R. Cimino, P. Marie-Jeanne, P. Jeseck, X. Michaut and M. Bertin, *ACS Earth Space Chem.*, 2022, **6**, 100–115.

52 M. Bulak, D. M. Paardekooper, G. Fedoseev, K.-J. Chuang, J. Terwisscha van Scheltinga, C. Eistrup and H. Linnartz, *Astron. Astrophys.*, 2022, **657**, A120.

53 N. G. Petrik, A. G. Kavetsky and G. A. Kimmel, *J. Phys. Chem. B*, 2006, **110**, 2723–2731.

54 J. S. Mathis, P. G. Mezger and N. Panagia, *Astron. Astrophys.*, 1983, **128**, 212.

55 R. Gredel, S. Lepp and A. Dalgarno, *Astrophys. J.*, 1987, **323**, L137.

56 K. France, E. Schindhelm, E. A. Bergin, E. Roueff and H. Abgrall, *Astrophys. J.*, 2014, **784**, 127.

57 R. Dupuy, M. Bertin, G. Féraud, X. Michaut, P. Marie-Jeanne, P. Jeseck, L. Philippe, V. Baglin, R. Cimino, C. Romanzin and J.-H. Fillion, *Phys. Rev. Lett.*, 2021, **126**, 156001.

58 M. Bertin, E. C. Fayolle, C. Romanzin, H. A. M. Poderoso, X. Michaut, L. Philippe, P. Jeseck, K. I. Öberg, H. Linnartz and J.-H. Fillion, *Astrophys. J.*, 2013, **779**, 120.

59 R. Dupuy, M. Bertin, G. Féraud, X. Michaut, P. Jeseck, M. Doronin, L. Philippe, C. Romanzin and J.-H. Fillion, *Astron. Astrophys.*, 2017, **603**, A61.

60 M. Bulak, D. M. Paardekooper, G. Fedoseev and H. Linnartz, *Astron. Astrophys.*, 2021, **647**, A82.

61 C. Walsh, R. A. Loomis, K. I. Öberg, M. Kama, M. L. R. van 't Hoff, T. J. Millar, Y. Aikawa, E. Herbst, S. L. Widicus Weaver and H. Nomura, *Astrophys. J., Lett.*, 2016, **823**, L10.

62 J. B. Bergner, V. G. Guzmán, K. I. Öberg, R. A. Loomis and J. Pegues, *Astrophys. J.*, 2018, **857**, 69.

Faraday Discussions

PAPER

A systematic mechanistic survey on the reactions between OH radical and CH_3OH on ice†

W. M. C. Sameera, *[ab] Avon P. Jayaweera, [b] Atsuki Ishibashi, [a] Hiroshi Hidaka, [a] Yasuhiro Oba [a] and Naoki Watanabe [a]

Received 8th February 2023, Accepted 17th March 2023
DOI: 10.1039/d3fd00033h

A systematic mechanistic survey was performed for the CH_3OH + OH reaction on ice. ONIOM(ωB97X-D/Def2-TZVP:AMOEBA09) calculations suggested a range of binding energies for the CH_2OH radical (0.29–0.69 eV) and CH_3OH (0.15–0.72 eV) molecule on hexagonal water ice (I_h) and amorphous solid water (ASW). Computed average binding energies of CH_2OH radical (0.49 eV) and CH_3OH (0.41 eV) are relatively stronger compared to the CH_3O radical binding energies (0.32 eV, Sameera *et al.*, *J. Phy. Chem. A*, 2021, **125**, 387–393). Thus, the CH_3OH molecule, CH_2OH and CH_3O radicals can adsorb on ice, where the binding energies follow the order $CH_2OH > CH_3OH > CH_3O$. The multi-component artificial force-induced reaction (MC-AFIR) method systematically determined the reaction mechanisms for the CH_3OH + OH reaction on ice, where two reaction paths, giving rise to CH_2OH and CH_3O radicals, were confirmed. A range of reaction barriers, employing the ωB97X-D/Def2-TZVP level of theory, was found for each reaction (0.03–0.11 eV for CH_2OH radical formation, and 0.03–0.44 eV for CH_3O radical formation). Based on the lowest energy reaction paths, we suspect that both reactions operate on ice. The computed data in this study evidence that the nature of the binding site or the reaction site has a significant effect on the computed binding energies or reaction barriers. Thus, the outcomes of the present study will be very useful for the computational astrochemistry community to determine reliable binding energies and reaction barriers on ice.

Introduction

Radical species in the interstellar medium (ISM) play a key role in forming complex organic molecules (COMs).[1–8] The radicals in the ISM, such as the primary radicals (H, OH, CO, HCO, CH_3O, CH_2OH, CH_3, NH, and NH_2)[9,10] can be

[a]*Institute of Low Temperature Science, Hokkaido University, Sapporo, Hokkaido 060-0819, Japan. E-mail: wmcsameera@lowtem.hokudai.ac.jp*

[b]*Department of Chemistry, University of Colombo, Colombo 00300, Sri Lanka*

† Electronic supplementary information (ESI) available. See DOI: https://doi.org/10.1039/d3fd00033h

formed through surface reactions and the photodissociation of small molecules in interstellar ice. At very low temperatures (*e.g.*, 10 K), radical species adsorb on ice grain surfaces. Among the primary radicals, only H can diffuse on the ice at 10 K. The other primary radicals may start migrating on the ice at relatively high temperatures, the so-called warming-up stage. When a radical species encounters another radical or a molecule on ice, relatively large radicals or molecules can be formed. Therefore, quantitative details of the radical adsorption, radical diffusion, and radical reactions on ice are essential to understand the mechanisms of COMs formation. The radical processes on ice are challenging to characterize from experimental methods. Thus, computational studies are indispensable.[11]

Quantum mechanics (QM) and quantum mechanics/molecular mechanics (QM/MM) methods, typically employing density functional theory (DFT), allow fast and accurate determination of potential energy surfaces (PES) of complex molecular systems.[11–13] Then, the properties of the molecular systems, radical binding energy, reaction barriers, diffusion barriers, and reaction mechanisms can be calculated. We have used QM and QM/MM methods to calculate the binding energy of radical species on ice. The radical species interact with the dangling hydrogen (d-H) or dangling oxygen (d-O) on ice. Depending on the combination of dangling atoms at the binding site on ice, a range of binding energies was found for OH (0.06–0.74 eV),[14] HCO (0.12–0.42 eV),[14] CH_3 (0.11–0.26 eV),[14] CH_3O (0.10–0.50 eV),[15] PH_2 (0.16–0.21 eV),[16] PH (0.12–0.16 eV),[16] P (0.07–0.15 eV),[16] OCSH (0.19–0.46 eV).[17] We have also calculated radical reactions on ices; PH_3 + D,[16] and OCS + H,[17] and CH_3SH + H,[18] where the computed reaction mechanisms explained the experimental results.

CH_3OH is one of the molecules in the ISM that plays a crucial role in COMs formation. Thus, the formation of CH_3OH in the ISM and the reactions between CH_3OH and the primary radicals are very important. The successive hydrogenation of CO, an abundant molecule in the ISM, gives rise to CH_3OH (Scheme 1).[19,20] The first hydrogenation yields the HCO radical, and the second hydrogenation gives rise to the H_2CO molecule. CH_3O or CH_2OH radicals are the products of the third hydrogenation. Final hydrogenation gives CH_3OH. Among the three intermediate radicals in the mechanism, only the CHO and CH_3O radicals were found in the ISM.[21–23]

Reactions between the CH_3OH and OH radical, an abundant primary radical in the ISM, is a major interest to the astrochemistry field. Two reaction mechanisms have been proposed for the CH_3OH + OH reaction in the gas phase or on ice (Scheme 2). When CH_3OH and an OH radical come closer, OH···CH_3OH complex, a spectroscopically characterized complex,[24] is formed. According to *ab initio* computations on the gas phase reaction, CH_2OH radical formation [*i.e.*, reaction (i)] has a relatively low energy barrier, where quantum tunnelling plays a key role.[25–29] Another theoretical survey suggested that the cross-section yielding the CH_3O radical is higher than that for the CH_2OH radical.[27] The computed branching ratios at low temperatures and low pressure suggested that the CH_2OH

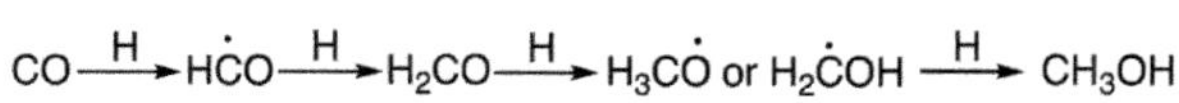

Scheme 1 Mechanism of CH_3OH formation through successive hydrogenation of CO.

$$CH_3OH + {}^{\bullet}OH \longrightarrow {}^{\bullet}CH_2OH + H_2O \text{ ------ (i)}$$

$$CH_3OH + {}^{\bullet}OH \longrightarrow CH_3O^{\bullet} + H_2O \text{ ------ (ii)}$$

Scheme 2 Reaction mechanisms for the reaction between CH_3OH and the OH radical.

radical remains,[28] while below 40 K, the computed branching ratio was uncertain and sensitive to the computational methodology. Thus, outcomes of the reaction (*i.e.*, CH_2OH or CH_3O) at low temperatures are currently under active discussion. Recently, the importance of this reaction on ice dust was also proposed.[30]

In the present study, we calculated the binding energy of the CH_2OH radical and CH_2OH molecule on ice, employing a number of binding sites, and compared it with the binding energy of the CH_3O radical on ice.[15] Then, reaction mechanisms between CH_3OH and an OH radical on the ice were systematically determined.

Computational methods and models

Binding energy

Ice cluster models were taken from our previous study[15] to replicate amorphous solid water (ASW) and crystalline water ice (I_h) (Fig. 1). The ASW model has 162 H_2O molecules, where 50 H_2O molecules are in the QM region, and 112 H_2O molecules are in the MM region. The I_h models B and C, have 162 and 168 H_2O molecules, respectively. The QM region of model B has 48 H_2O molecules, and the MM region has 114 H_2O molecules. In model C, the QM and MM regions have 156 and 112 H_2O molecules, respectively.

Geometry optimizations were performed using the two-layer ONIOM method as implemented in the PyQM/MM interface.[31] The ωB97X-D,[32] functional and def2-TZVP,[33] basis sets were used for the ONIOM,[34,35] high-layer. The AMOEB09 [36–38] polarizable force field was used for the ONIOM low-layer. To avoid structure deformations, atoms in the ONIOM-low layer were frozen during the structure optimization. Vibrational frequency calculations were performed for the optimized structures to calculate the zero-point energy and to confirm the local

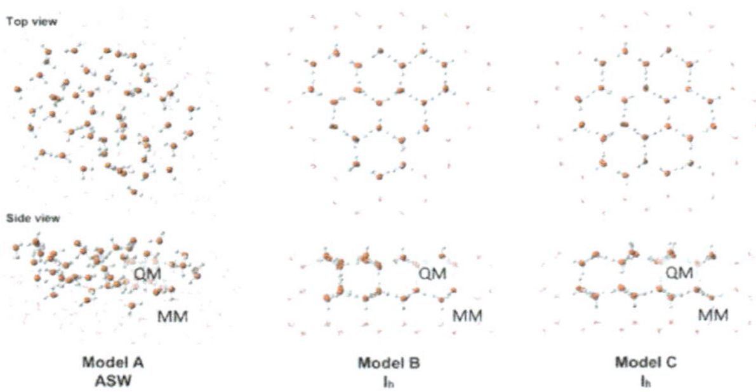

Fig. 1 Top and side views of the ice cluster models. The QM region is shown with "ball and stick" illustration, and the MM region is shown in "wireframe" illustration.

minima (*i.e.*, no imaginary frequencies). Binding energies were calculated using the following formula;

$$\text{Binding energy} = |E_{(\text{ice-molecule/radical})} - E_{\text{ice}} - E_{\text{molecule/radical}}|$$

where $E_{(\text{ice-molecule/radical})}$ is the total energy of the optimized CH_3OH-ice or CH_2OH-ice complexes. The E_{ice} term is the total energy of an optimized ice model system. The $E_{\text{molecule/radical}}$ is the energy of optimized CH_3OH or CH_2OH.

Reaction mechanisms

Reaction mechanisms between a CH_3OH molecule and an OH radical on ASW were determined by using the multi-component artificial force-induced reaction (MC-AFIR)[13,39] method as implemented in the GRRM17 program.[40] For this purpose, an ice cluster model, consisting of 76 H_2O molecules, was employed. In this model, 28 H_2O molecules on the side and bottom walls were frozen during the structure optimization, as highlighted in green in Fig. 2a and b, and the remaining 48 H_2O molecules were left flexible. The ωB97X-D method and the 6-31G* basis sets,[41–44] were used for the AFIR calculations. The CH_3OH molecule and OH radical were placed randomly on the ice surface. Then, an artificial force parameter of 100 kJ mol^{-1} was added between the CH_3OH and OH radical (Fig. 2c). After that, the resulting reaction paths (*i.e.*, AFIR paths), were inspected and approximate transition states (TSs) were identified, and fully optimized using the ωB97X-D/6-31G* level of theory, as implemented in the Gaussian16 program.[45]

Starting from the TS optimized structures, Pseudo intrinsic reaction coordinate calculations (IRC)[46,47] were performed for 20 steps forward and 20 steps backward from the optimized TS structures. Then, the local minima (LM) connecting the TSs (*i.e.*, the reactant complex and product) were calculated. Vibrational frequency calculations were performed to confirm that the optimized LM have no imaginary frequency and the optimized TSs have one imaginary frequency; and to calculate zero-point energies. Potential energies of the optimized LM or TSs were calculated as the single-point energy using the ωB97X-D/def2-TZVP level of theory.

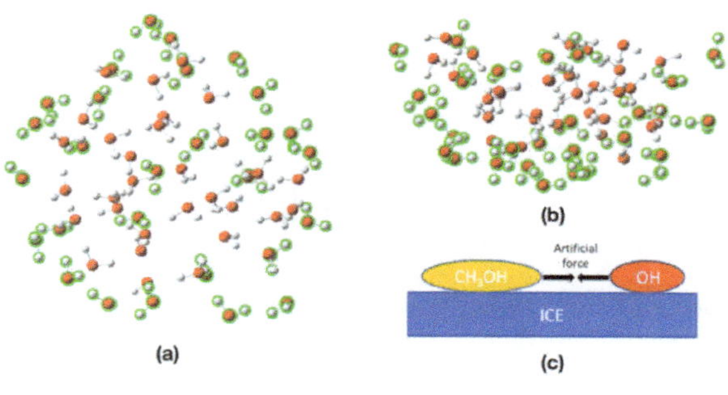

Fig. 2 (a) Top and (b) side views of the ice cluster model employed in this study. The water molecules highlighted in green were frozen during the structure optimization. (c) Adding artificial force between the CH_3OH and OH radical on ASW.

Results and discussion

Binding energy of CH_2OH and CH_3OH on ice

We have chosen 16 binding sites, eight from I_h (A–H) and eight from ASW (I–P) ice cluster models, to calculate binding energies. The computed binding energies are shown in Fig. 3. The molecular structures of the optimized structures are shown in Fig. S1 and S2.† The computed binding energies of the CH_2OH radical are in the range 0.29–0.69 eV, and the average binding energy is 0.49 eV (Fig. 3a). In the case of CH_3OH, the computed binding energies are in the range 0.15–0.72 eV, and the computed average binding energy is 0.41 eV (Fig. 3b).

Based on the average binding energies, we concluded that CH_2OH radical binding energy on ice is larger than that of the CH_3OH radical. From our previous study, using the same computational methods, we found that the average binding energy of the CH_3O radical is 0.32 eV.[15] Thus, the binding energy of the CH_3OH molecule, CH_2OH radical, and CH_3O radical follows the order CH_3O (0.32 eV) < CH_3OH (0.41 eV) < CH_2OH (0.49 eV). Therefore, if CH_3OH, CH_2OH, and CH_3O are formed on ice, the thermal desorption probability of these species would follow the opposite order, CH_3O < CH_3OH < CH_2OH.

In our ice models, a relatively large MM region is chosen. To check whether the MM region has a significant effect on the computed binding energies, we have compared binding energies (without ZPE) from ONIOM(ωB97X-D/Def2-TZVP:AMOEBA09) and ωB97X-D/Def2-TZVP methods (Fig. 4). In the case of the CH_2OH radical, the R^2 value of the plot is 0.99, indicating good agreement

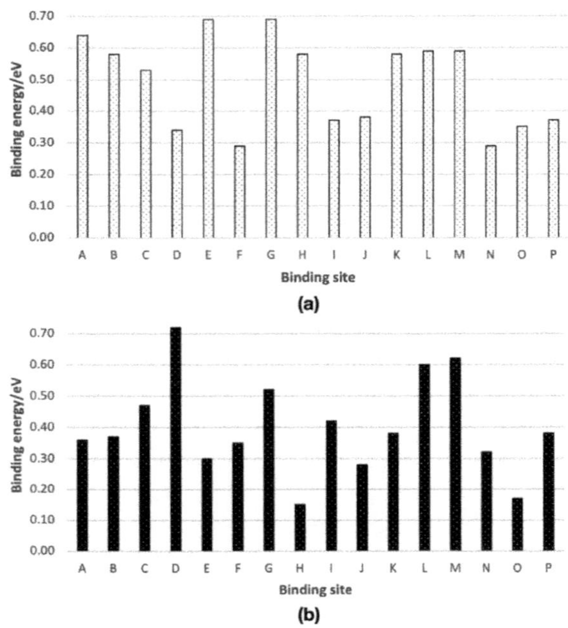

Fig. 3 Calculated binding energies of (a) a CH_2OH radical and (b) a CH_3OH on ice, from ONIOM(ωB97X-D/Def2-TZVP:AMOEBA09) calculations. Optimized molecular structures are shown in Fig. S1 and S2.† Binding energies are summarized in Table S1.†

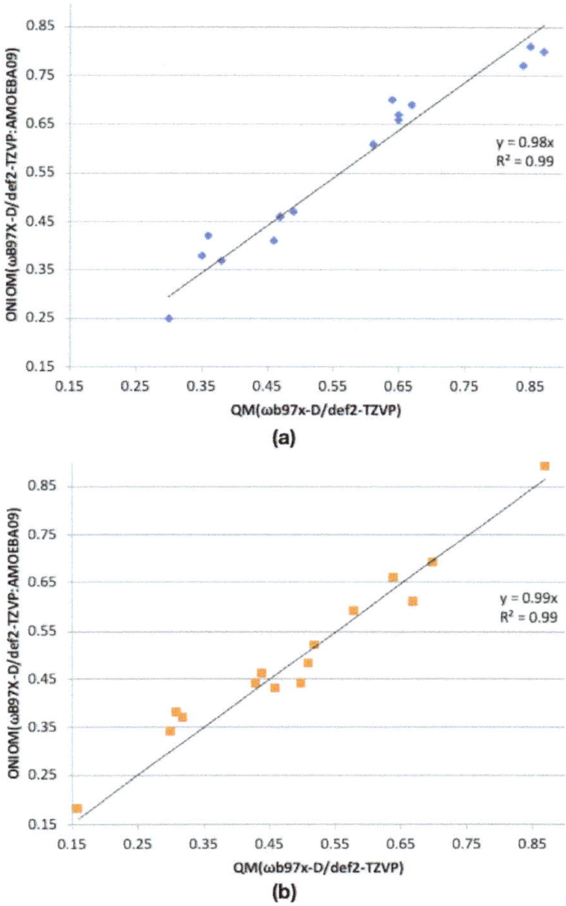

Fig. 4 ONIOM(ωB97X-D/def2-TZVP:AMOEBA09) binding energy (without ZPE) *vs.* ωB97X-D/def2-TZVP binding energy (without ZPE) of (a) CH_2OH radical, and (b) CH_3OH. Intercept of the linear trendlines were set to 0.0.

between the ONIOM(QM:MM) and QM binding energies. Similarly, the R^2 value of the plot of CH_3OH of 0.99, shows the good agreement between the computed binding energies from the ONIOM(ωB97X-D/Def2-TZVP:AMOEBA09) and ωB97X-D/Def2-TZVP methods.

The maximum discrepancy of the CH_2OH radical binding energies (without ZPE, Table S2†) of 0.07 eV of binding site A, is very small compared the computed binding energy of A, 0.77 eV. When we consider the lowest binding energy (0.25 eV), the discrepancy between the computed binding energy from ONIOM(ωB97X-D/Def2-TZVP:AMOEBA09) and ωB97X-D/Def2-TZVP methods is also small (0.05 eV). Thus, in general, binding energies from ONIOM(QM:MM) and QM methods are in agreement. Similarly, the maximum discrepancy of the CH_3OH radical binding energies (without ZPE, Table S2†) is 0.06 eV, belonging to binding sites E and N with the computed binding energy of 0.38 and 0.37 eV, respectively. If we consider the lowest binding energy of CH_3OH (0.18 eV, binding site H), the discrepancy between the computed binding energy from ONIOM(QM:MM) and QM methods is very small (0.02 eV). Therefore, the ONIOM(QM:MM) method gave reliable binding energies.

In the present study, it is important to note that we have chosen 16 binding sites to calculate binding energies. Thus, we were able to cover a number of arrangements of the dangling atoms at the binding sites, where the interactions between the radical and the binding site are diverse, allowing us to collect a range of binding energies.

CH_3OH + OH reaction on ice

This section focuses on the mechanisms for the reaction between CH_3OH and an OH radical on ice. An MC-AFIR search was performed to determine reaction paths systematically, and two reaction paths were found on ice; (a) $CH_3OH + OH \rightarrow CH_2OH + H_2O$ and (b) $CH_3OH + OH \rightarrow CH_3O + H_2O$. We have chosen four AFIR paths for each reaction mechanism and potential energy surfaces were calculated (Fig. 5). The lowest and highest computed potential energy barrier for the CH_2OH radical formation is 0.03 eV and 0.11 eV, respectively (Fig. 5a), and therefore the computed reaction barrier is in a range. Similarly, the computed potential energy barrier for the CH_3O radical formation is also in a range (Fig. 5b), where the lowest energy barrier is 0.03 eV, while the highest barrier is 0.44 eV. If the lowest energy

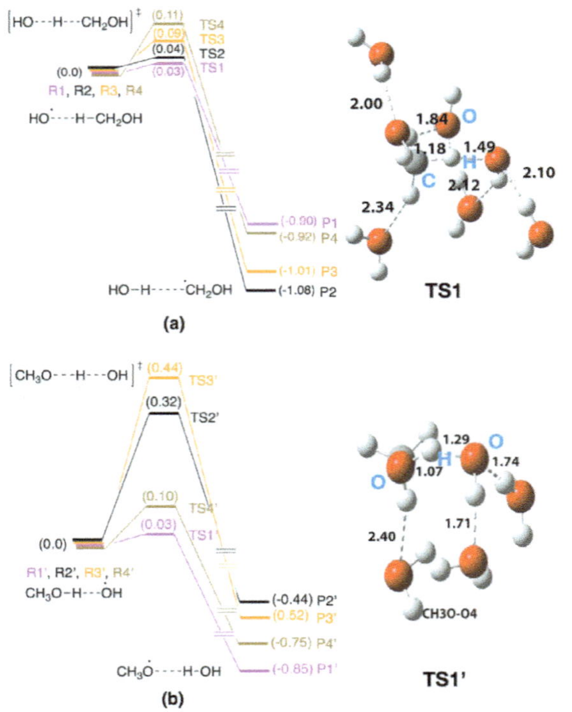

Fig. 5 Computed potential energy surfaces for (a) $CH_3OH + OH \rightarrow CH_2OH + H_2O$ and (b) $CH_3OH + OH \rightarrow CH_3O + H_2O$ reactions. Optimized transition state structures are shown in Fig. S3 and S4†).

reaction path is taken into account (*i.e.*, 0.03 eV for the formation of the CH_2OH radical and 0.03 eV for the formation of the CH_3O radical), both reaction mechanisms may operate on ice. In the case of the products of the reactions, the CH_2OH radical is thermodynamically more stable compared to the CH_3O radical in all cases. Further, if we take the lowest energy paths of the two mechanisms, the HO–H···CH_2OH complex is −1.08 eV more stable compared to the entry point of the potential energy profile, while the HO···CH_3O complex is −0.85 eV more stable compared to its entry point.

In this reaction path survey, we restricted the analysis to four reaction paths for each reaction (eight reaction paths in total). Even with these data, we can see a range of reaction barriers for each reaction. If we extend the analysis for collecting more reaction paths, we suspect that the reaction barrier range will be broadened. The lowest barrier limit is the key to the rate of the reaction. In this present case, both reactions operate at a very low reaction barrier of 0.03 eV. Therefore, both reaction mechanisms would operate on interstellar ice.

Conclusions

We have used the ONIOM(ωB97X-D/def2-TZVP:AMOEBA09) method to calculate CH_2OH and CH_3OH radical binding on ASW and I_h. A range of binding energies were found for each species; CH_3OH: 0.15–0.72 eV; CH_2OH: 0.29–0.69 eV. Thus, both CH_2OH and CH_3OH radicals would adsorb on interstellar ice. The calculated average binding energies, 0.41 eV of CH_3OH and 0.49 eV of CH_2OH, indicate the strong binding preference of CH_2OH, and are relatively stronger compared to the CH_3O radical binding energies (0.32 eV).[15] Reaction paths for the CH_3OH + OH were searched using the MC-AFIR method. Two reaction mechanisms were determined, giving rise to CH_2OH and CH_3O products. A range of reaction barriers was found for each reaction mechanism; 0.03–0.11 eV for the CH_2OH radical formation and 0.02–0.44 eV for the CH_3O radical formation. Both radicals can be formed on ice if the lowest energy paths are taken into account. As radical binding on ice and radical reactions on ice cover a range of numbers, a systematic sampling of binding energies or reaction barriers becomes critical, which is the key lesson we have learned from this study.

Author contributions

WMCS designed the project, performed calculations, and wrote the original manuscript with help from APJ. AI, HH, YO and NW performed the experiments relevant to this study. All co-authors have revised the manuscript and given approval to the final version of the manuscript.

Conflicts of interest

There are no conflicts to declare.

Acknowledgements

This work was partly supported by the JSPS KAKENHI grant numbers JP19K03940 (to WMCS), JP21H05416 (to WMCS), JP17H06087 (to NW), and JSPS grant number

JPJSBP 120229901 (to WMCS). Supercomputing resources at the Institute for Information Management and Communication at Kyoto University in Japan and the Institute for Molecular Science (IMS) in Japan are also acknowledged.

References

1 E. Herbst and E. F. Van Dishoeck, *Annu. Rev. Astron. Astrophys.*, 2009, **47**, 427–480.
2 S. Bottinelli, C. Ceccarelli, J. P. Williams and B. Lefloch, *Astron. Astrophys.*, 2007, **463**, 601–610.
3 E. Herbst and R. T. Garrod, *Front. Astron. Space Sci.*, 2022, **8**, 1–6.
4 A. Bacmann, V. Taquet, A. Faure, C. Kahane and C. Ceccarelli, *Astron. Astrophys.*, 2012, **541**, L12–5.
5 C. Vastel, C. Ceccarelli, B. Lefloch and R. Bachiller, *Astrophys. J.*, 2014, **795**, L2–11.
6 R. T. Garrod, S. L. W. Weaver and E. Herbst, *Astrophys. J.*, 2008, **682**, 283–302.
7 K. I. Öberg, E. F. Van Dishoeck, H. Linnartz and S. Andersson, *Astrophys. J.*, 2010, **718**, 832–840.
8 K. J. Chuang, G. Fedoseev, S. Ioppolo, E. F. van Dishoeck and H. Linnartz, *Mon. Not. R. Astron. Soc.*, 2016, **455**, 1702–1712.
9 R. T. Garrod, S. L. W. Weaver and E. Herbst, *Astrophys. J.*, 2008, **682**, 283–302.
10 K. Öberg, E. F. van Dishoeck, H. Linnartz and S. Andersson, *Astrophys. J.*, 2010, **718**, 832–840.
11 W. M. C. Sameera, B. Senevirathne, T. Nguyen, Y. Oba, A. Ishibashi, M. Tsuge, H. Hidaka and N. Watanabe, *Front. Astron. Space Sci.*, 2022, **9**, 890161.
12 W. M. C. Sameera and F. Maseras, *Wiley Interdiscip. Rev.: Comput. Mol. Sci.*, 2012, **2**, 375–385.
13 W. M. C. Sameera, S. Maeda and K. Morokuma, *Acc. Chem. Res.*, 2016, **49**, 763–773.
14 W. M. C. Sameera, B. Senevirathne, S. Andersson, F. Maseras and G. Nyman, *J. Phys. Chem. C*, 2017, **121**, 15223–15232.
15 W. M. C. Sameera, B. Senevirathne, S. Andersson, M. Al-Lbadi, H. Hidaka, A. Kouchi, G. Nyman and N. Watanabe, *J. Phys. Chem. A*, 2021, **125**, 387–393.
16 T. Nguyen, Y. Oba, W. M. C. Sameera, A. Kouchi and N. Watanabe, *Astrophys. J.*, 2021, **918**, 73.
17 T. Nguyen, Y. Oba, W. M. C. Sameera, A. Kouchi and N. Watanabe, *Astrophys. J.*, 2021, **922**, 146.
18 T. Nguyen, Y. Oba, W. M. C. Sameera, A. Kouchi and N. Watanabe, *Astrophys. J.*, 2023, **944**, 219.
19 A. G. G. M. Tielens and W. Hagen, *Astron. Astrophys.*, 1982, **114**, 245–260.
20 N. Watanabe and A. Kouchi, *Astrophys. J.*, 2002, **571**, L173–L176.
21 J. Cernicharo, N. Marcelino, E. Roueff, M. Gerin, A. Jiménez-Escobar and G. M. Caro, *Astrophys. J.*, 2012, **759**, L43.
22 L. E. Snyder, J. M. Hollis and B. L. Ulich, *Astrophys. J.*, 1976, **208**, L91–L94.
23 V. M. Rivilla, M. T. Beltrán, A. Vasyunin, P. Caselli, S. Viti, F. Fontani and R. Cesaroni, *Mon. Not. R. Astron. Soc.*, 2019, **483**, 806–823.
24 F. J. Hernandez, J. T. Brice, C. M. Leavitt, G. A. Pino and G. E. Douberly, *J. Phys. Chem. A*, 2015, **119**, 8125–8132.

25 S. Blázquez, D. González, A. García-Sáez, M. Antiñolo, A. Bergeat, F. Caralp, R. Mereau, A. Canosa and B. Ballesteros, *ACS Earth Space Chem.*, 2019, **3**, 1873–1883.
26 R. J. Shannon, M. A. Blitz, A. Goddard and D. E. Heard, *Nat. Chem.*, 2013, **5**, 745–749.
27 O. Roncero, A. Zanche and A. Aguado, *Phys. Chem. Chem. Phys.*, 2018, **20**, 25951–25958.
28 A. J. Ocaña, S. Blázquez, A. Potapov, B. Ballesteros, A. Canosa, M. Antiñolo, L. Vereecken, J. Albaladejo and E. Jiménez, *Phys. Chem. Chem. Phys.*, 2019, **21**, 6942–6957.
29 L. G. Gao, J. Zheng, A. Fernández-Ramo, D. G. Truhlar and X. Xu, *J. Am. Chem. Soc.*, 2018, **140**, 2906–2918.
30 A. Ishibashi, H. Hidaka, Y. Oba, A. Kouchi and N. Watanabe, *Astrophys. J.*, 2021, **921**, L13.
31 M. Rathnayake, W. M. C. Sameera and N. Watanabe, *ChemRxiv*, 2022, preprint, DOI: 10.26434/chemrxiv-2022-2c8mz.
32 J.-D. Chai and M. Head-Gordon, *Phys. Chem. Chem. Phys.*, 2008, **10**, 6615–6620.
33 F. Weigend and R. Ahlrichs, *Phys. Chem. Chem. Phys.*, 2005, **7**, 3297–3305.
34 W. M. C. Sameera and F. Maseras, *J. Chem. Inf. Model.*, 2018, **58**, 1828–1835.
35 L. W. Chung, W. M. C. Sameera, R. Ramozzi, A. J. Page, M. Hatanaka, G. P. Petrova, T. V. Harris, X. Li, Z. Ke, F. Liu, H.-B. Li, L. Ding and K. Morokuma, *Chem. Rev.*, 2015, **115**, 5678–5796.
36 J. W. Ponder and D. A. Case, Force Fields for Protein Simulations, *Adv. Protein Chem.*, 2003, **66**, 27–85.
37 P. Ren and J. W. Ponder, *J. Comput. Chem.*, 2002, **23**, 1497–1506.
38 P. Ren and J. W. Ponder, *J. Phys. Chem. B*, 2003, **107**, 5933–5947.
39 W. M. C. Sameera, A. K. Sharma, S. Maeda and K. Morokuma, *Chem. Rec.*, 2016, **16**, 2349–2363.
40 S. Maeda, Y. Harabuchi, M. Takagi, K. Saita, K. Suzuki, T. Ichino, Y. Sumiya, K. Sugiyama and Y. Ono, *J. Comput. Chem.*, 2018, **39**, 233–251.
41 R. Ditchfield, W. J. Hehre and J. A. Pople, *J. Chem. Phys.*, 1971, **54**, 724–728.
42 W. J. Hehre, R. Ditchfield and J. A. Pople, *J. Chem. Phys.*, 1972, **56**, 2257–2261.
43 P. C. Hariharan and J. A. Pople, *Theor. Chim. Acta*, 1973, **28**, 213–222.
44 M. M. Francl, W. J. Pietro, W. J. Hehre, J. S. Binkley, M. S. Gordon, D. J. DeFrees and J. A. Pople, *J. Chem. Phys.*, 1982, **77**, 3654–3665.
45 M. J. Frisch, G. W. Trucks, H. B. Schlegel, G. E. Scuseria, M. A. Robb, J. R. Cheeseman, G. Scalmani, V. Barone, G. A. Petersson, H. Nakatsuji, X. Li, M. Caricato, A. V. Marenich, J. Bloino, B. G. Janesko, R. Gomperts, B. Mennucci, H. P. Hratchian, J. V. Ortiz, A. F. Izmaylov, J. L. Sonnenberg, D. Williams-Young, F. Ding, F. Lipparini, F. Egidi, J. Goings, B. Peng, A. Petrone, T. Henderson, D. Ranasinghe, V. G. Zakrzewski, J. Gao, N. Rega, G. Zheng, W. Liang, M. Hada, M. Ehara, K. Toyota, R. Fukuda, J. Hasegawa, M. Ishida, T. Nakajima, Y. Honda, O. Kitao, H. Nakai, T. Vreven, K. Throssell, J. A. Montgomery Jr., J. E. Peralta, F. Ogliaro, M. J. Bearpark, J. J. Heyd, E. N. Brothers, K. N. Kudin, V. N. Staroverov, T. A. Keith, R. Kobayashi, J. Normand, K. Raghavachari, A. P. Rendell, J. C. Burant, S. S. Iyengar, J. Tomasi, M. Cossi, J. M. Millam, M. Klene, C. Adamo, R. Cammi, J. W. Ochterski, R. L. Martin, K. Morokuma, O. Farkas,

J. B. Foresman and D. J. Fox, GAUSSIAN 16 (Revision C.01), Gaussian, Inc., Wallingford, CT, 2016.

46 K. Fukui, *Acc. Chem. Res.*, 1981, **14**, 363–368.

47 H. P. Hratchian and H. B. Schlegel, *Theory and Applications of Computational Chemistry, The First Forty Years*, 2005, pp. 195–249.

Faraday Discussions

ROYAL SOCIETY OF CHEMISTRY

DISCUSSIONS

Laboratory astrochemistry of and on dust and ices: general discussion

Nadia Balucani, Mathieu Bertin, Michelle Brann, Wendy A. Brown, Cecilia Ceccarelli, Rafael Martín-Doménech, Jack Fulker, Robin T. Garrod, Joel Green, Murthy S. Gudipati, Dwayne E. Heard, Eric Herbst, Ugo Jacovella, Inga Kamp, Martin R. S. McCoustra, W. M. C. Sameera, Ian Sims, Ardjan Sturm, Serena Viti, Susanna Widicus Weaver, Laurent Wiesenfeld and Olivia H. Wilkins

DOI: 10.1039/d3fd90026f

Ugo Jacovella addressed all: What are the rigorous definitions and thus differences between ices, grains and dusts?

Murthy S. Gudipati answered: These definitions are a bit slippery. As per my understanding, here are the definitions and differences:

(a) Ice: any gas or liquid consisting of atoms and/or molecules of the same kind or mixed, at Earth's ambient temperature ($\sim$295 K), then when subjected to lower temperatures (295–5 K) results in the formation of a solid – such that the solid-phase chemical and physical properties can be measured distinctly. The ice should be at least a few tens of monolayers thick (in order to have bulk ice properties that dominate compared to surface binding that dominates at the heterogeneous interface).

(b) Grain: an ice or dust particle or a mix thereof, with a distinct boundary. A grain can exist as a single entity or an aggregate of grains that span from a few microns in size to a few meters or more, as long as gravity or temperature or both, do not lead to the centering of these grains forming larger grains.

(c) Dust: a dust particle is typically made of a non-volatile refractory material (at Earth's ambient temperature and pressure). Silicate grains (micron-sized), or large PAHs, or carbon-soot, or HAC (hydrogenated amorphous carbon), *etc.*, are dust grains.

Ugo Jacovella opened a general discussion of the paper by Mathieu Bertin: Have you checked the influence of sample thickness on the ratio between photodesorption and fragments – as well as the shape (round *vs.* flat) of the sample in your paper (https://doi.org/10.1039/d3fd00004d)?

Mathieu Bertin responded: Thickness dependence has not been studied for the specific ices that are presented here. However, past studies performed on other systems, such as condensed H_2O, CO, CO_2, or N_2, suggest that only the four to five outer molecular layers of the ices are involved in the desorption process. In the case of organic species, past experiments on condensed methanol did not evidence any thickness dependence for ices thicker than 6 monolayers (ML).[1] In our case, 60 ML thick ices were studied, and we thus expect our ice thickness to be large enough for reasonably considering the system to be in a semi-infinite regime. Since the photodesorption process involves only several molecular layers, it is a process involving energy decay over some nanometers at best. Thus, it is very improbable that the macroscopic shape of the solid-to-gas interface (*e.g.* flat *vs.* round) plays an important role on the photodesorption process. This may, however, not be true anymore in the case of surface structuration on the nanometer scale, such as porosity effects or surface clustering.

1 K. I. Öberg, R. T. Garrod, E. F. van Dishoeck and H. Linnartz, *Astron. Astrophys.*, 2009, **504**, 891–913.

Cecilia Ceccarelli asked: What is your opinion on the fact that CH_3CN is the only complex organic molecule (COM) that is not broken in pieces by photodesorption? What makes CH_3CN so special?

Mathieu Bertin responded: In fact, CH_3CN is also photodissociated upon vacuum UV photon absorption. The main difference with the other studied COMs is that the desorption of the intact molecule presents comparable yields with those of the desorption of the photofragments. This is not observed in the case of CH_3OH, HCOOH or $HCOOCH_3$, for which the desorption of the intact molecules is a minority pathway.

Several possibilities could explain why CH_3CN desorbs molecularly whereas the other studied COMs do not. One possible explanation would be that nitriles, such as CH_3CN, are known to be more resistant to the photodissociation as compared to other COMs, such as carboxylic acids (see ref. 1). The reason for this is not clearly known, but it might be due either to shorter living excited states of the molecule, or to more efficient geminate recombination of the fragments in the solid phase, constantly reforming CH_3CN.

1 M. P. Bernstein, S. F. M. Ashbourn, S. A. Sandford and L. J. Allamandola, *Astrophys. J.*, 2004, **601**, 365–370.

Nadia Balucani asked: What is the wavelength range of your experiments?

Mathieu Bertin answered: The photodesorption experiments were realized in the 7–13.6 eV range, that is a wavelength typically between 91 and 177 nm.

Nadia Balucani added: I have seen other experiments by your group where you used X-rays on ice to let molecules sublimate without dissociation. Can you comment on that?

Mathieu Bertin responded: The case of desorption induced by X-rays is very different to the case of UV photodesorption. In the case of X-ray irradiation, most of the excess energy is carried by the Auger electron, emitted in the ice after each X-ray photon absorption, being resonant or above the core electrons' ionization threshold. The Auger electrons thereafter thermalize in the ice over a large distance – typically several tens of layers, resulting in a cascade of secondary events eventually leading to the desorption of surface molecules. This process, referred to as X-ray induced electron stimulated desorption (XESD), has been recently shown to be able to desorb intact organics (*e.g.* ref. 1 and 2). Whereas UV irradiation of the condensed organics directly populates dissociative electronic states of the complex molecules, X-ray irradiation implies many secondary processes in the ice which in the end may induce the desorption of intact complex organic molecules.

1 R. Basalgète, R. Dupuy, G. Féraud, C. Romanzin, L. Philippe, X. Michaut, J. Michoud, L. Amiaud, A. Lafosse, J.-H. Fillion and M. Bertin, *Astron. Astrophys.*, 2021, **647**, A35.
2 R. Basalgète, D. Torres-Díaz, A. Lafosse, L. Amiaud, G. Féraud, P. Jeseck, L. Philippe, X. Michaut, J.-H. Fillion and M. Bertin, *Astron. Astrophys.*, 2023, **676**, A13.

Joel Green opened a general discussion of the paper by Murthy S. Gudipati: I have a definitional question – in your presentation, you describe the change from amorphous to crystalline as "bulk" (https://doi.org/10.1039/d3fd00048f). What does "bulk" mean in this context? Is that an experimental property that correlates to some properties in protoplanetary disks?

Murthy S. Gudipati answered: Bulk "ice" property is something that is characteristic of that molecular ice, irrespective of if the ice is a few tens of monolayers or a few hundred monolayers thick, or macroscopic – millimetres to centimetres. Bulk physical properties do not change with the size of the ice. This is the reason why, an amorphous to crystalline ice transition, which is an exothermic "bulk" property, occurs at the same temperature window irrespective of the thickness of the ice we make in the laboratory. All the H_2O molecules in the bulk ice behave very similar at a given temperature and pressure.

On the other hand, surface or interface properties are different from bulk ice properties. For example, surface layers of ice consist of molecules that are only partially bonded to other molecules. There is a significant heterogeneity in the surroundings of each of these molecule on the surface. When ice sublimes, the surface properties dominate the sublimation process. If the ice is thin, sublimation completes at a lower temperature or shorter time-scale, whereas thick ices take a longer time or continue sublimation at higher temperatures.

Finally, sub-monolayers of ice on a substrate (*e.g.*, silicate grain surface) have different binding energy with the substrate; and the ice structure for a few monolayers is dictated by the binding of molecules with the substrate at that heterogeneous interface. We discussed this elaborately when considering Cecilia Ceccarelli's paper (https://doi.org/10.1039/d3fd00018d). Under these conditions when ice is only a sub-monolayer or a couple of monolayers, phase transitions as well as sublimation are dictated by the type of interaction between the substrate and molecules that accrete on that substrate. These properties differ significantly from "bulk" ice properties. It is possible that under these conditions sublimation could occur at elevated temperatures due to strong binding with the substrate, or phase transitions may not follow bulk behavior due to the lack of long-range interactions in three dimensions. These systems could form dimensional materials that are different from the bulk.

Joel Green added: I study protoplanetary disks that are actively undergoing accretion outbursts, which cause a heatwave across the entire disk. The temperature rises dramatically over days to months, and recedes over years to decades or more. How do you think your measurement of the bulk transformation of amorphous *vs.* crystalline grains would apply to the surface layers of a disk in a protostellar system undergoing such a burst?

Murthy S. Gudipati responded: Thanks for asking this question and providing a context of your observations. Could you give more details on where (radially and vertically) the accretion outbursts occur? Interesting that the temperature rises dramatically over a short-period of time. There could be two possible reasons:

(a) Coulomb explosion, whereby charge builds up in an ice grain due to radiation processing through ionization and other mechanisms (see ref. 1–3). This kind of Coulomb explosion is expected at small grain-sizes and may not be observable as large outbursts, unless such charge accumulation happens on larger pebbles or planetesimals/cometesimals.

(b) Predominantly pure amorphous water ice grains accrete at temperatures much below 145 K, and either through shockwave transient heating or through turbulence, these amorphous ices are heated to the amorphous-to-crystalline phase transition temperature, which results in a run-away exothermic process of crystallization, outgassing of trapped volatiles, and even to some extent water ice sublimation, leading to temperatures of $\sim$170 K. Is this in line with your observations?

There is one caveat though – as discussed in our paper and our discussion – so far, there is only one experimental work published that infers exothermic amorphous to crystalline phase transitions occur in pure water ice only (which we confirmed in our lab as well) and that any trapped impurity ($\sim$10%) such as CO or CO_2 reverts this exothermic phase transition into an endothermic phase transition.[4] We are working on validating this publication. If this is the scenario and if the outbursts are due to pure water ice exothermic phase transition, then the ice accretion on to the protoplanetary disks you are studying must be due to predominantly pure water ice domains.

1 D. V. Bekaert, M. S. Gudipati, B. Henderson and B. Marty, *Geochem. J.*, 2019, **53**, 69–81.
2 A. Lignell, L. I. Tenelanda-Osorio and M. S. Gudipati, *Chem. Phys. Lett.*, 2021, **778**, 138814.
3 M. S. Gudipati and L. J. Allamandola, *J. Phys. Chem. A*, 2006, **110**, 9020–9024.
4 A. Kouchi and S. Sirono, *Geophys. Res. Lett.*, 2001, **28**, 827–830.

Robin T. Garrod commented: Firstly, I'd like to clarify what the microscopic structure of the ice is at the beginning of the experiment, as that may be a function of the method of deposition. As many as four molecules are being deposited, through background deposition. You state early on in your paper that you induce an overabundance of supervolatiles in the ices, specifically to ensure that there exist regions of pure supervolatiles. So this would presumably have the effect of creating a highly porous water ice, filled with seams of supervolatiles. So, is it possible to say anything about the connectivity within that structure? One could imagine long seams of porous water structure, filled with supervolatiles, connecting throughout the ice and thus affecting the desorption characteristics. Such a structure could produce different results from what we might think of as a more typical interstellar ice, which could be relatively well-mixed and lacking in these large, open, interconnected porous structures containing highly segregated supervolatiles. The situation produced in your experiments may be applicable to cometary ices, which we understand to be highly porous, but it may not be an appropriate analogue for interstellar ices.

Murthy S. Gudipati replied: The co-deposition of vapour molecules on a cold surface would essentially freeze those molecules as they arrive on the cold surface. When we take an excess supervolatile mixing ratio, those molecules form pure ice domains that could be a few tens of monolayers at a time, while trapping other molecules in those pure ice domains. This is essentially a statistical phenomenon and there are structures of every kind, water porous structure trapping supervolatiles and *vice versa* as well. As the excess supervolatiles sublimate, water molecules and their extended hydrogen bonding structures collapse, until ~145 K when amorphous water ice undergoes irreversible transition to crystalline ice, expelling the majority of impurities from its cages.

Robin T. Garrod added: Just to follow up on this point: It is important to consider how the ices are formed. The majority of interstellar ice is formed through surface chemistry, not through deposition. Deposition could be more relevant for protoplanetary disk ices, in which the molecules incorporated into the ices may already be present in the gas phase and are simply re-deposited. But in molecular clouds, *etc.*, the ices are mostly formed through a slow process of atomic surface diffusion and reaction. This would allow, for example, an oxygen atom to find an optimal site on an existing ice surface, prior to its conversion to OH and thence water (see ref. 1), resulting in a much less porous (water) ice, less capable of storing highly segregated volatiles and with lower interconnectivity between porous structures. The release of the energy of reaction may also allow the position of the reaction product(s) on the surface to be further optimized as the ice slowly builds. These effects would not be seen in the case of direct deposition on laboratory timescales.

1 R. T. Garrod, *Astrophys. J.*, 2013, **778**, 158.

Murthy S. Gudipati responded: Thank you for the explanation about the formation of H_2O ice in the interstellar medium (ISM). How about the other components such as CO, CO_2, *etc.*? Are these molecules also formed on a dust grain? We do have evidence that interstellar ices are mixed (or at least co-exist). How do you explain the entire infrared spectral signatures of ices in the ISM? We are simulating the final composition of the ISM ices (as per their IR spectral signatures), not so much on how they are formed (which I have no problem in accepting your explanation).

Further, do we have any idea, what is the average thickness of H_2O ice in the ISM? If it is not a couple of monolayers at the max, then the bulk ice properties start dominating (please see my earlier response to Cecilia Ceccarelli's comments on binding energies).

I think we need to have a more detailed discussion on the ices in the ISM, their formation and radiation processing before becoming a part of a protoplanetary disk.

Robin T. Garrod further responded: CO_2 does indeed appear to be formed on the grains. CO can be formed efficiently in the gas phase, but some fraction is likely produced on the grain surfaces as well. The thickness of the ice depends on the size of the grains; in the models, a grain of radius 0.1 micron produces a final ice thicknesses in the order of 100+ monolayers, but of course this is a mixture of all the different ice components, not just water. However, my point is that a surface-chemical formation mechanism for water, rather than direct deposition, should lead to a far more compact ice in the ISM, and where porosity exists within which volatiles such as CO might initially be trapped, those pores would be expected to be rather smaller and less well-connected than would be produced through direct deposition in the lab.

Jack Fulker said: The calorimetry experiments of pure water ices are really interesting, and having more concordant values for the heat of crystallisation exotherm for amorphous water ice will surely be extremely helpful for astrochemical modelling. I was wondering if you have (or have plans to) run the calorimetry experiments on the mixed ice systems? If so, how would you expect the crystallisation exotherm to change as a function of mixed ice composition?

Murthy S. Gudipati replied: To the best of my knowledge, there is one publication,[1] which has shown that the exothermic amorphous to crystalline phase transition in pure water ice becomes endothermic as soon as impurities are trapped in the ice (~10%). We were able to reproduce the exothermic amorphous to crystalline ice transition ~145 K and endothermic ice sublimation ~180 K. We couldn't observe an exothermic peak ~100 K in pure water ice. We are working on mixed ice calorimetry to verify Kouchi and Sirono's observations.

1 A. Kouchi and S. Sirono, *Geophys. Res. Lett.*, 2001, **28**, 827–830.

Ardjan Sturm highlighted: With the advent of JWST, we have the possibility to resolve ice features spatially in protoplanetary disks. It is very hard to say anything about variations in ice abundance radially because of the radiative transfer, but vertically we should trace the height of ices in the disk directly. In the first JWST observations of an edge-on disk, we see that CO ice is seen as high up in the disk as CO_2 and water. Is this consistent with the entrapment of CO in water ice as you presented in your presentation, or should it be another process (*e.g.* turbulence) that results in this vertical profile?

Murthy S. Gudipati responded: This is an exciting observation that JWST is able to resolve the vertical profile of ice in an edge-on disk. Can you provide some more information such as distance from the protostar and approximate local temperatures (estimated or observed) where CO ice is found and where only CO_2/H_2O ice (with no CO) is found and what the mixing ratio (CO/H_2O) is in the ice?

There are two different possible mechanisms: (a) as the interstellar ice grains (containing CO, CO_2, and H_2O) accrete onto the protoplanetary disk, their local temperature could be low enough to retain pure CO, CO_2, and H_2O ice domains as well as entrapped CO and CO_2 in amorphous H_2O ice (as formed in the interstellar environment). Based on the local equilibrium temperature all these phases could co-exist; (b) H_2O/CO_2 ice radially moves from the inner to outer disk, whereby CO accretes on the ice grains and diffuses into the amorphous H_2O ice structure, which would then be stable in an ice form up to ~140 K. I think this possibility is less likely than retaining an interstellar ice composition.

Ardjan Sturm replied: The CO ice is observed up to a height of 100 au or z/r equal to 0.7, which corresponds to a temperature of at least 50 K in our models. The mixing ratio of CO with H_2O or CO_2 is still uncertain, because the shape of the ice absorption features is dominated by radiative transfer effects. We are working on that to constrain the environment using radiative transfer models.

Murthy S. Gudipati added: Thank you Ardjan. The CO ice detection is in the expected region where CO can still be trapped in H_2O and/or CO_2 ice and is consistent with our experiments.

Inga Kamp opened a general discussion of the paper by Wendy A. Brown: We know that UV photodesorption occurs by absorbing the photon within the top surface layers of the ice. Do we have any idea about the depth in the ice that the IR photodesorption process happens, specifically? Are the ices thick enough in the experiments or do the photons interact with the Au-coated substrate on which the ice is adsorbed? So, in other words do we expect this IR desorption to be relevant in space where we may have thin ice mantles and different 'underlying' grain surface properties?

Wendy A. Brown answered: Inga Kamp poses an interesting question. While the penetration depths of IR and vacuum ultra-violet (VUV) photons are likely very different, the depth from which photodesorbed species may escape into the gas phase are likely very similar, at a few molecular layers from the surface. Simple inelastic scattering within the solid state will rapidly reduce the kinetic energy of a nascent photodesorbate to the level that insufficient energy remains to escape the solid state and hence it is most likely that any desorbing species come from the top few layers of the ice.

The comment about the gold surface is important. In our experiment (https://doi.org/10.1039/d3fd00024a), the IR radiation from both the FTIR spectrometer and the FEL beam, interact with (in fact are reflected by) the underlying gold surface. In fact, it is for this reason that we ensure that the FEL beam is p-polarised, to maximise the effect of the metal surface selection rule that applies for IR light at metal surfaces. There are two well-known processes commonly observed in the interaction of lasers with metals and monolayers on those metals: (i) hot-electron induced desorption, and (ii) laser-induced thermal desorption. Laser excitation of the metal surface rapidly heats the surface electrons on the femto- to pico-second timescale.

In process (i), the hot electrons can attach to empty states on adsorbates producing a transient negative ion that decays by desorption and loss of the attached electron back into the metallic band structure. This process is very likely with UV-visible radiation, but is likely to have less of a contribution with IR radiation because of the lower energies of the radiation. Nonetheless it needs to be considered.

Alternatively, scattering of the hot electrons from the metallic lattice can also induce rapid heating of the surface on the nanosecond timescale at rates of more than 1 000 000 K s^{-1}. Desorption of monolayer adsorbates follows. This is process (ii). The presence of additional layers on top of the monolayer can significantly reduce the impact of these processes at deeply buried interfaces such that we only really observe processes occurring near the surface of the icy film, independent of the nature of the substrate. Clearly neither of these substrate-mediated processes would be relevant to astrophysical situations. However, as discussed in the paper, we also observed CO photodesorption that was mediated by the excitation of water vibrational modes and by the accumulation of energy in the water librational modes. These two photodesorption mechanisms could occur in astrophysical situations, in cases where the ice thickness is sufficient.

Ian Sims suggested: In your article you make "back of the envelope" estimations of CO photodesorption yields from your CO–water ice mixture of $\sim 3 \times 10^{-3}$ per photon at 12 μm and $\sim 5 \times 10^{-3}$ per photon at 2.9 μm. In order for these numbers to be useful for modellers, and also to give meaning to any comparison with other studies, it would be helpful if you would also estimate (even very approximately) confidence limits on these numbers.

Wendy A. Brown replied: Our fitting procedure provides the CO photodesorption cross-sections with relevant fitting errors (φ_2) as listed in Table 2 in our paper. We can use these as the foundation for a propagation of error analysis based on the procedure that we employed in our "back-of-the-envelope" calculation. The foundational quantity in that analysis is the absorption cross section for the O–H dangling bond, $1.0 \pm 0.2 \times 10^{-18}$ cm^2.[1] A realistic estimate of the scaling factor for the intensity of the 2.9 μm band from Noble *et al.*,[2] is 27 ± 3, yielding an absorption cross-section of $2.70 \pm 0.62 \times 10^{-17}$ cm^2. Mifsud *et al.*[3] yield a scaling factor for the 12 μm band to 2.7 ± 0.3. The final correction factor for the effect of the reflection at the metal surface, we estimate to be 50 ± 5. Hence, the CO photodesorption quantum yield at 12 μm is $3.3 \pm 1.4 \times 10^{-3}$ and $5.6 \pm 1.4 \times 10^{-3}$ at 2.9 μm.

1 T. Nagasawa, R. Sato, T. Hasegawa, N. Numadate, N. Shioya, T. Shimoaka, T. Hasegawa and T. Hama, *Astrophys. J., Lett.*, 2021, **923**, L3.
2 J. A. Noble, C. Martin, H. J. Fraser, P. Roubin and S. Coussan, *J. Phys. Chem. Lett.*, 2014, **5**, 826–829.
3 D. V. Mifsud, P. A. Hailey, P. Herczku, Z. Juhász, S. T. S. Kovács, B. Sulik, S. Ioppolo, Z. Kaňuchová, R. W. McCullough, B. Paripás and N. J. Mason, *Eur. Phys. J. D*, 2022, **76**, 87.

Cecilia Ceccarelli asked: Do you have an idea of the amount of IR radiation used in the experiment and how it compares with typical IR illumination received by a standard galactic molecular cloud? This is to understand how the mechanism you find would be important in cold clouds, when considering that 12 and 6 micron photons are generally emitted by relatively warm sources (>300 K).

Wendy A. Brown replied: The infrared light used in these experiments generally had an energy of between 1–140 mJ, focussed down into a spot around 2 mm in height. This corresponds to an irradiation rate of approximately 10^{16}–10^{17} photons cm^{-2} s^{-1} (see Table 2 in our paper for details). Whilst it is hard to make a direct comparison of this photon flux with that in astrophysical regions, we can make an estimate by looking at the interstellar radiation field (ISRF). Looking at the ISRF (*e.g.*, ref. 1), we observe that mid-IR intensities are within an order of magnitude comparable to VUV intensities, though significantly smaller than near-UV and visible intensities. This ignores attenuation of the VUV and UV wavelengths in the dark environment which will raise the relative proportion of IR. If the fluxes and quantum yields are comparable for the VUV and IR radiation, then the infrared-induced process should be considered in models of cold, dense environments since we do assume that VUV has an important effect.

1 J. S. Mathis, P. G. Mezger and N. Panagia, *Astron. Astrophys.*, 1983, **128**, 212–229.

Martin R. S. McCoustra added: In studying photodesorption in the laboratory, we will inevitably use light sources (lamps or lasers) that are significantly more intense in comparison to the ISRF. Therefore, the key measurement that we must make in studying these processes is the photon order. Is the process single or multiple photon? The former are feasible with the ISRF, while the latter are clearly not. Our data indicate that these processes are clearly single photon and should be considered as a contributing process in relevant environments within a reaction network, including competition from radiative cooling of the water librational modes.

Mathieu Bertin commented: It is stated in the article that the photodesorption of CO is a 1 photon process. However, when irradiating the ice with photons having a 12 μm wavelength, the energy brought to the system is 100 meV, that is extremely close to the adsorption energy of CO which ranges between 80 and 150 meV. If the process is at play, that means that the entirety of the available energy is given as translational energy to the adsorbed CO molecules. This is very surprising, since, at 12 μm, the photons populate libration modes of the water substrate, *i.e.* intermolecular modes which strongly couple the water molecules in the hydrogen-bonded network. It would then be expected that the excess energy given to the water is very quickly distributed and dispersed into the solid water ice, leaving only a small fraction of it available for the desorption phenomenon. Therefore, it appears unlikely that, at 12 μm, a single photon is indeed at play. Another possibility, which is also suggested by the authors, would be that the process is a thermal process, in which the accumulation rate of the energy exceeds the relaxation rate. In this regime, the process would then appear linear with the photon flux, despite that several photons are needed for the temperature to locally rise.

Wendy A. Brown answered: Mathieu Bertin is correct in suggesting that a photothermal process (*i.e.*, a process where after photon absorption, energy is found to accumulate in the heat bath of the substrate resulting in desorption) may be in operation in terms of excitation, especially at 12 μm. We should, however, recognise that a 12 μm photon carries around 9.96 kJ mol^{-1} of energy. The binding energy of CO on the amorphous solid water (ASW) surface has been reported to range from 7.98 to 10.89 kJ mol^{-1}.[1,2]

Clearly, 12 μm photons carry sufficient energy to overcome the binding energy of the most weakly bound CO on ASW. Indeed, one can imagine exciting the librational motion of a surface H_2O molecule weakly attached to a CO, such that the CO is freed from the surface. However, we accept that there are many more librational degrees of freedom in the bulk solid and this will likely make up the majority of the 12 μm absorption which might indeed provide a photothermal desorption pathway.

In terms of excitation at 2.9 μm (corresponding to a photon energy of 41.25 kJ mol^{-1}, which is more than sufficient to desorb CO), we agree that relaxation on the picosecond timescale of the photo-excited O–H stretch into the librational heat bath is possible. Indeed, McCoustra and Taj make this argument in proffering an explanation for the observed IR line-width of the CO adsorption.[3] The key point we must remember is that the presence of this rapid pathway for relaxation of the O–H does not necessarily mean that other channels are switched off. Rather they become competitive. Hence, for CO desorption to be efficient, it must occur on a shorter, or similar, timescale to O–H vibrational relaxation into the librational heat bath.

However, we should recognise that the only effective method of distinguishing thermal (photothermal) from non-thermal processes is to consider the dynamics of the desorption. Our tentative data (limited by the response of the analogue detection system of the quadrupole mass spectrometer that we employed) suggests a combination of a fast desorption process (direct photodesorption) and a slow process (probably indirect in nature). The former, we suggest, represents the non-thermal process, and the latter perhaps the thermal (or photothermal). We currently have a number of additional shifts on FELIX forthcoming. With recent upgrades to the quadrupole mass spectrometer on the LISA End Station, we hope to be able to reveal the dynamics of these two processes, and resolve this key question.

The key take away message from our work presented here, is that there is IR-photon induced desorption of CO from ASW surfaces, which shows a wavelength dependence resonant with the vibrational spectroscopy of water. The absence of CO following excitation resonant with the CO absorption, we can readily explain by recourse to previous work reported in ref. 3.

1 S. Taj and M. R. S. McCoustra, *Mon. Not. R. Astron. Soc.*, 2020, **498**, 1693–1699.
2 M. Minissale, Y. Aikawa, E. Bergin, M. Bertin, W. A. Brown, S. Cazaux, S. B. Charnley, A. Coutens, H. M. Cuppen, V. Guzman, H. Linnartz, M. R. S. McCoustra, A. Rimola, J. G. M. Schrauwen, C. Toubin, P. Ugliengo, N. Watanabe, V. Wakelam and F. Dulieu, *ACS Earth Space Chem.*, 2022, **6**, 597–630.
3 S. Taj, A. Rosu-Finsen and M. R. S. McCoustra, *Mon. Not. R. Astron. Soc.*, 2021, **504**, 5806–5812.

Laurent Wiesenfeld asked: Ices come in very many allotropic forms, amorphous or crystalline. Does this influence the analysis in a profound way, and in particular the coupling adsorbed molecules/phonons?

Martin R. S. McCoustra answered: This is an insightful question. The simple answer is yes; the structure of the solid state matrix will impact on the energy transfer within the solid. As an illustrative example, please consider CO adsorbed at the surface of water ice. If the water is deposited at 10 K, we know from the literature that the hydrogen bond network is highly fragmented and the resulting hydrogen bond order is much less than 4. In that situation, I would expect that relaxation of CO *via* vibrational energy redistribution into the ice will be slower and perhaps the system would show a larger propensity toward desorption. Compare this with a crystalline ice with it's fully integrated hydrogen bond network with a hydrogen bond order of 4. Here, we might expect minimal hindrance of the energy flow from the excited CO vibrational state into the phonon bath of the ice, and so less desorption.

Rafael Martín-Doménech commented: The presence of methanol and other complex organic molecules in the cold regions of the ISM is currently puzzling, since non-thermal desorption yields measured in the laboratory cannot usually account for the observed abundances of these molecules. Could the IR photon-induced desorption of CO molecules in H_2O-bearing ices presented in your work be applied to larger, more complex molecules too?

Wendy A. Brown answered: That's an interesting question. The key factors that we would need to consider are: (i) the binding energy of the species to the surface, and (ii) the presence of competitive relaxation pathways for the vibrationally excited system. Clearly, the lower the binding energy, the easier it will be to overcome that energy with a photon. We must also remember that generally physisorption binding energies increase with increasing molecular mass, which will likely place a limit on the likelihood of the desorption of a large organic molecule.

Furthermore, if there is efficient vibrational relaxation of the vibrational excited system then the desorption yield will be significantly reduced. As suggested in ref. 1, the water ice surface is perhaps the worst surface on which to explore IR photodesorption, as there are very efficient pathways to relaxation of the resonant modes of water into the librational heat bath. It may be that on silica and silicate surfaces, where those relaxation pathways do not exist, IR photodesorption may be possible for larger molecules with the constraints offered by point (i). We note that there have also been investigations of the irradiation of CO and methanol containing ices with the FEL-2 beam by Santos and co-workers,[2] and these investigations show that irradiation of an ice with wavelengths resonant with the methanol vibrations lead to CO photodesorption, but also to a change in the infrared spectrum of the methanol ice. Further investigations are clearly needed to determine whether larger organic molecules could be photodesorbed by IR light.

1 S. Taj, A. Rosu-Finsen and M. R. S. McCoustra, *Mon. Not. R. Astron. Soc.*, 2021, **504**, 5806–5812.
2 J. C. Santos, K.-J. Chuang, J. G. M. Schrauwen, A. Traspas Muiña, J. Zhang, H. M. Cuppen, B. Redlich, H. Linnartz and S. Ioppolo, *Astron. Astrophys.*, 2023, **672**, A112.

Mathieu Bertin complemented: On the efficiency of the photon-induced desorption process for different species: as Wendy A. Brown has perfectly pointed out, the binding energy is one of the main parameters *a priori* involved in the desorption efficiency, as it represents the energy barrier to overcome. Another one is the number of atoms in the molecule – more specifically the number of internal degrees of freedom – since a part of the total excess energy given to the species may dissipate into the excitation of vibrational modes of the molecule, thereby limiting the amount of available energy transferred to the translation, and thus decreasing the desorption probability.

Michelle Brann communicated: What is the penetration of the FEL-2 beam?

Wendy A. Brown communicated in reply: The ice film is transparent in the mid-IR as our RAIR data reveal (see Fig. 1 in our paper for an example, https://doi.org/10.1039/d3fd00024a). The peak absorbance (at around 2.9 μm) is some 0.20. This implies that around 60% of the FEL-2 beam will be transmitted through the entire film. Interaction of the radiation from the FEL beam is therefore with the whole ice film and not just with the surface.

Michelle Brann communicated: How does film thickness impact your results since the 42% CO film would have more layers than the 6% film? Can this be normalized out?

Wendy A. Brown communicated in reply: The CO and H_2O are pre-mixed and the total dose is fixed. The resulting films are therefore basically all the same thickness provided that the sticking coefficients of CO and H_2O are both unity at the temperature of the substrate. All that changes is the relative amounts of CO and H_2O in the ice. Hence no normalisation is required for thickness.

Mathieu Bertin addressed Wendy A. Brown and Martin R. S. McCoustra: The process of photodesorption has been highlighted – and quantified – in the study partly from the gas phase detection of desorbed CO species using mass spectrometry, which is measured as a function of the irradiation time. However, the IR photon source used here is a pulsed source, with macropulses of several microseconds at a 5 Hz repetition rate. That means that, most of the time, the surface does not receive any photon. A continuous detection of the desorption signal with a large integration time will lead to artifacts when it comes to detecting and quantifying the desorption signal as a function of irradiation time. Has this been taken into account when quantitative parameters were extracted from the curves?

Wendy A. Brown communicated in reply: The combination of a pulsed light source and continuous detection using a QMS in studies of photodesorption is well known in surface kinetics and dynamics. Indeed, we have some 15 years ago demonstrated the effectiveness of this combination in studies of UV photodesorption of benzene (C_6H_6) from cASW surfaces.[1–3] The key issue is not so much the pulsed nature of the light source (synchrotrons are pulsed and not continuous), but rather the timescale of the pulsed irradiation relative to the rate of the process being observed. Provided that the pulse is fast compared to the rate of the process being observed and the integration time of the QMS is matched to the desorption kinetics, then it is feasible to extract rate constants for desorption from QMS traces at fixed mass on the multi-second timescale. The discussion in ref. 3 is particularly pertinent here. In our case, the analogue detection system is best matched to the timescale of the long decay, and the integration time is too long to accurately reproduce the short decay. Of course, for dynamical studies, *i.e.*, measurement of time-of-flight, analogue electronics are entirely inappropriate and gated ion counting methods should be used.

1 J. D. Thrower, M. P. Collings, M. R. S. McCoustra, D. J. Burke, W. A. Brown, A. Dawes, P. D. Holtom, P. Kendall, N. J. Mason, F. Jamme, H. J. Fraser, I. P. Clark and A. W. Parker, *J. Vac. Sci. Technol., A*, 2008, **26**, 919–924.

2 J. D. Thrower, D. J. Burke, M. P. Collings, A. Dawes, P. D. Holtom, F. Jamme, P. Kendall, W. A. Brown, I. P. Clark, H. J. Fraser, M. R. S. McCoustra, N. J. Mason and A. W. Parker, *Astrophys. J.*, 2008, **673**, 1233–1239.
3 J. D. Thrower, A. G. M. Abdulgalil, M. P. Collings, M. R. S. McCoustra, D. J. Burke, W. A. Brown, A. Dawes, P. J. Holtom, P. Kendall, N. J. Mason, F. Jamme, H. J. Fraser and F. J. M. Rutten, *J. Vac. Sci. Technol., A*, 2010, 28, 799–806.

Martin R. S. McCoustra added: Mathieu Bertin is correct that the response time of the quadrupole mass spectrometer system is important. We were well aware of this in having conducted some of the earliest dynamical experiments in photodesorption relevant to laboratory astrochemistry some 15 years ago, as mentioned above by Wendy A. Brown.[1,2] Certainly, using an analogue detection system, rather than the pulsed system previously used,[1,2] will limit the time-resolution to the relatively long analogue integration time. It does not mean that the mass spectrometer will produce spurious artefacts, it simply broadens out the response and may consequently have reduced sensitivity. This simply means that we do not have sufficient time-resolution to accurately determine the kinetics of the fast transient compared to that of the slow component.

We've have several outstanding shifts planned on FELIX and with a new pulse counting mass spectrometer fitted to the LISA end station, we plan to repeat these experiments with the goal of more accurately describing the kinetics of IR photodesorption.

1 J. D. Thrower, D. J. Burke, M. P. Collings, A. Dawes, P. J. Holtom, F. Jamme, P. Kendall, W. A. Brown, I. P. Clark, H. J. Fraser, M. R. S. McCoustra, N. J. Mason and A. W. Parker, *J. Vac. Sci. Technol., A.*, 2008, **26**, 919–924.
2 J. D. Thrower, D. J. Burke, M. P. Collings, A. Dawes, P. J. Holtom, F. Jamme, P. Kendall, W. A. Brown, I. P. Clark, H. J. Fraser, M. R. S. McCoustra, N. J. Mason and A. W. Parker, *Astrophys. J.*, 2008, **673**, 1233–1239.

Olivia H. Wilkins opened a general discussion of the paper by W. M. C. Sameera: In earlier discussion, we heard about some of the challenges of building networks to accurately study astrochemical reactions, specifically that increasing the parameter space for calculations makes networks complicated and calculation time too long. W. M. C. Sameera has made the point that calculations often use a single binding energy when in fact such parameters are typically a range of values rather than a fixed value (https://doi.org/10.1039/d3fd00033h). Given the previous discussion about computational limitations, could you please provide some perspective for this problem? Does anyone have any thoughts related to a cost–benefit analysis of using single binding energies *versus* ranges of values in the context of computations?

W. M. C. Sameera answered: The interaction energy between the radical and water molecules on ice is sensitive to the number of dangling atoms (*i.e.*, dangling hydrogen or dangling oxygen) and their arrangements at the binding site. As a result, the binding energy of a radical on ice is a range. Therefore, all possible arrangements of the dangling atoms and the radical must be taken into account to achieve quantitative binding energy calculations of a radical on ice, employing quantum chemical calculations. However, performing computations for all possible arrangements of the radical and dangling atoms on ice is not practical. What we can do is prepare some ice structures (*e.g.*, 10–20) to cover possible arrangements of the dangling atoms and the radicals on ice. In this way, a range of binding energies can be obtained within a reasonable time frame. On the other hand, quantum chemical calculations on relatively large ice cluster models (*e.g.*, $50H_2O$) give reliable binding energies, where the computational cost of the quantum chemical calculations is relatively high. However, modern high-performance computers are capable of performing these calculations in a short time. However, significant human input is required to prepare calculations and analyze computed data, which is the challenging part.

In my view, collecting a range of binding energies, reaction barriers, and reaction rates can be done within a reasonable time frame. Thus, now we are in the position to use a range of numbers, instead of a single number, in astrochemical modeling.

Wendy A. Brown added: I think that a sensitivity analysis might be the way to go in terms of single *versus* a range of binding energies. Inevitably, some binding energies will not affect models very much. Others, however, will be very important and hence a range may well be relevant and needed – and indeed may provide useful information. I of course do not know how difficult this would be for modellers to include in their models.

Dwayne E. Heard commented: For the gas-phase reaction between OH and methanol – the motivation for studying this reaction was the observation of the methoxy radical (CH_3O) using radioastronomy by Cernicharo *et al.* (ref. 21 in the paper).[1] At the time, gas–grain models considered desorption from grains *etc.*, and didn't explain the abundance of methoxy, and so it was hypothesised that the gas phase reaction between OH and methanol may be an important source. Do the calculations of the binding energies you show in your paper suggest that both of the channels producing CH_3O and CH_2OH are important on the surface and lead to desorption of both products? Does this change the situation regarding model predictions? Do we have more methoxy radical generated from the surface now because of your calculations? Are you then suggesting that the gas-phase reaction is possibly less important?

Following on from this, what was the situation before those experiments? What was the yield of the two products for the surface reaction? The previous understanding was that CH_2OH was the main product rather than CH_3O? Has that situation changed now?

1 J. Cernicharo, N. Marcelino, E. Roueff, M. Gerin, A. Jiménez-Escobar and G. M. Caro, *Astrophys. J.*, 2012, **759**, L43.

W. M. C. Sameera answered: Previously, photochemical experiments on solid methanol and CO–methanol mixed ice, which is a dust analog, indicated that the yields of CH_3O-derived COMs (*e.g.*, methyl formate) were small and inconsistent with astronomical observations in the gas phase. In other words, it was thought that CH_2OH was mainly formed by the photodissociation of methanol on dust. Recently, we performed an experiment incorporating the effect of H_2O ice on the photochemistry of CH_3OH with reference to a new interstellar dust model (ref. 1), confirming the major production of CH_3O-derived COMs and reporting that this result is in good agreement with the abundance ratios from astronomical observations (ref. 2). This suggests that the reaction of CH_3OH with OH produced by photodissociation of H_2O ice plays an important role in supplying CH_3O radicals at the dust surface. However, information on the CH_3OH + OH reaction was only available for the gas phase. Therefore, we conducted detailed studies of the CH_3OH + OH reaction on ice surfaces, both experimentally and theoretically.

We cannot argue the importance of gas phase reactions. We suggest that the supply of CH_3O from the dust surface to the gas phase is more significant than previously thought because the CH_3O formation was found to be rather dominant on the dust surface, and the binding energy of CH_3O on ice is smaller than that of CH_2OH.[3]

1 A. Kouchi, M. Tsuge, T. Hama, H. Niinomi, N. Nakatani, T. Shimonishi, Y. Oba, Y. Kimura, S. Sirono, S. Okuzumi, M. Momose, K. Furuya and N. Watanabe, *Mon. Not. R. Astron. Soc.*, 2021, **505**, 1530–1542.
2 A. Ishibashi, H. Hidaka, Y. Oba, A. Kouchi and N. Watanabe, *Astrophys. J., Lett.*, 2021, **921**, L13.
3 A. Ishibashi, H. Hidaka, W. M. C. Sameera, Y. Oba and N. Watanabe, *Astrophys. J.*, 2023, submitted.

Robin T. Garrod added: In the case of the chemical kinetics models, the production of gas-phase CH_3O is unlikely to be strongly affected by the OH reaction with methanol on grain surfaces – this, at least, is what our own dark cloud models indicate.

Dwayne E. Heard addressed Eric Herbst and Robin T. Garrod: I would be interested to hear a comment from Eric Herbst or Robin T. Garrod on the impact within models of methoxy being a major product from the reaction of OH and methanol in ices rather than it just being a minor product. There surely must be a major impact of this change?

Eric Herbst responded: Regarding the question of proton exchange on ice, we have not studied proton reactions *per se* but have begun to look at reactions on ice surfaces involving the ions C^+ and HCO^+ with water ice. These reactions lead to initial products which subsequently react to include a number of COMs as products. Chemistry occurs mainly on the edges of so-called lattices of assorted sizes. The reactions are assumed to follow an Eley–Rideal mechanism, and are added to a large gas–grain simulation with mainly diffusive reactions. Our present results show that the abundances of a number of COMs in the gas and on the grain surface can undergo small increases or decreases as functions of time and of temperature (https://doi.org/10.1039/d3fd00014a). More research is planned.

Robin T. Garrod added: We do in fact already have these reactions in the networks. In terms of kinetic behaviour, the OH reaction goes with high efficiency upon contact, because the OH is immobile on the surface. However, H abstraction by atomic H tends to be much more important in producing either CH_3O or CH_2OH, due to the mobility of atomic H. The reaction of OH with CH_3OH on the surface occurs as a non-diffusive reaction, which is intrinsically slower.

Because we deal with kinetics in the chemical models of interstellar grain-surface chemistry, we must consider competition between reaction on the one hand and diffusion of OH away from methanol on the other. It's never just a one-sided equation: we have to know about the competition with other processes to determine the efficiency of a reaction. In our models, at low temperatures (at which the OH radical is immobile on the surface), any OH that happens to be contiguous to methanol will react with it at high efficiency, simply because it cannot do anything else. W. M. C. Sameera, what do your models tell us about that competition between reaction of OH with methanol, *versus* diffusion?

W. M. C. Sameera answered: We assumed that OH and methanol met on the ice, giving rise to a local minimum on the potential energy surface. Then, we systematically determined the possible reaction paths for the reaction between OH and methanol on ice, and computed the barrier for the reaction. Present computational results do not provide a barrier for OH diffusion. Thus, our computations do not give data for competition between the reaction of OH with methanol, *versus* diffusion. We can certainly focus on this problem in future work.

Cecilia Ceccarelli added: As I said in my presentation, the binding energy of a species is not a single value but rather a distribution of values, which reflects the different sites of the adsorbate and the orientation that the molecule can have on it. There are actually a number of theoretical studies now published.[1–3] In contrast, astrochemical models use one single value (and not necessarily the most up-to-date value) with the exception of ref. 2 and 4. In the latter, we also modelled the impact of the water binding energy distribution on the predicted water snowline in protoplanetary disks.

1 L. Tinacci, A. Germain, S. Pantaleone, S. Ferrero, C. Ceccarelli and P. Ugliengo, *ACS Earth Space Chem.*, 2022, **6**, 1514–1526.
2 L. Tinacci, A. Germain, S. Pantaleone, C. Ceccarelli, N. Balucani and P. Ugliengo, *Astrophys. J.*, 2023, **951**, 32.
3 G. M. Bovolenta, S. Vogt-Geisse, S. Bovino and T. Grassi, *Astrophys. J., Suppl. Ser.*, 2022, **262**, 17.
4 T. Grassi, S. Bovino, P. Caselli, G. Bovolenta, S. Vogt-Geisse and B. Ercolano, *Astron. Astrophys.*, 2020, **643**, A155.

Murthy S. Gudipati responded: I agree that binding energy can't be a single value, but perhaps at least three different values: heterogeneous surface binding energy, when sub-monolayer gas adsorbs on another surface; then homogeneous bulk binding energy, which is within a homogeneous ice; and then homogeneous surface binding energy, where molecules are bound on the surface. However, the significance of these binding energies depends upon how thick the ice layer is. For very thin (a few monolayers) ice, all three different binding energies are critical, whereas for several hundred monolayer thick ice, the bulk binding energy becomes more important. Heterogeneous mixed molecular ices complicate this situation further. Frequency factors add another dimension of complexity. For this reason, we should indeed take binding energies with a pinch of salt! Thank you for bringing this important aspect into discussion.

These are important points regarding binding energies and their corresponding frequency factors – that these are interdependent and strongly correlated; and that the binding energy cannot be used independently without the corresponding frequency factor derived in the same experimental work. We recently published work which stresses this exact point and warns the community not to use binding energies alone.[1]

1 L. R. Smith, M. S. Gudipati, R. L. Smith and R. D. Lewis, *Astron. Astrophys.*, 2021, **656**, A82.

Serena Viti added: Large grids of astrochemical model computations, where ranges of binding energies are varied to determine sensitivity, are done but can not cover the whole parameter space due to both CPU issues but also the fact that we lack enough constraints to get robust inferences. This is discussed further in Session 4 (https://doi.org/10.1039/d3fd90027d).

Susanna Widicus Weaver highlighted: Garrod, Herbst and my student Laas and I, wrote a paper investigating the impact of methanol photolysis branching fractions in 2011.[1] In this work, we changed the branching fractions for the methoxy, hydroxymethyl, and methyl radicals to the extremes and ran the models for each limiting case. The results indicate that the complex organic chemistry in ices changes completely depending on which radical is the dominant product. This is of fundamental importance to studying COMs in star-forming regions because methanol provides many of the organic functional groups that comprise larger molecules. If we can experimentally measure the branching fractions under the right conditions it's a huge step forward for the field.

1 J. C. Laas, R. T. Garrod, E. Herbst and S. L. Widicus Weaver, *Astrophys. J.*, 2011, **728**, 71.

Dwayne E. Heard responded: Many thanks for the further information on this paper. I agree entirely with the last statement. I sincerely hope that someone will be able to measure the branching fractions in ices under the right conditions for the reaction of OH with methanol.

Rafael Martín-Doménech continued the general discussion of the paper by Mathieu Bertin: The presence of methanol and other complex organic molecules in the cold regions of the ISM is currently puzzling, since non-thermal desorption yields measured in the laboratory cannot usually account for the observed abundances of these molecules. In particular, photoprocessing of COM-containing ices usually leads to the photon-induced desorption of COM fragments, but not intact COMs, as reported in your work. At the same time, previous works have reported the photon-induced formation and subsequent desorption of COM precursors such as H_2CO.[1] Considering both results, do you think there is potential in the photoprocessing of ices containing simpler species, to lead to the formation and subsequent desorption of larger COMs?

1 R. Martín-Doménech, G. M. Muñoz Caro and G. A. Cruz-Díaz, *Astron. Astrophys.*, 2016, **589**, A107.

Mathieu Bertin responded: You point-out an interesting question. Indeed, except for acetonitrile (CH_3CN), UV irradiation of most of the COM-containing ices studied so far lead mostly to the dissociation of the molecule and desorption of its fragments, the photodesorption of the intact species at best, is very minor in comparison. The exothermic production of organics from the UV irradiation of a composite ice, and its subsequent desorption, as Rafael Martín-Doménech refers, has only been evidenced so far for formaldehyde. The process may indeed also be at play for more complex species. However, it should also be noted that the more complex the species is, usually the more it is bound to the surface, and the more it may dissipate the excess energy into internal degrees of freedom, such as vibrational modes, making it the more difficult to desorb. Proper studies from mixed ices under UV irradiation are needed to answer this question.

Finally, one should also keep in mind that other non-thermal desorption processes may be at play to desorb intact COMs, including the cosmic rays sputtering which has been demonstrated to be an efficient process (*e.g.* ref. 1); or photodesorption with higher energy photons, such as soft X-rays, which are also a good vector of desorption of intact organics (*e.g.* ref. 2).

1 E. Dartois, M. Chabot, T. Id Barkach, H. Rothard, B. Augé, A. N. Agnihotri, A. Domaracka and P. Boduch, *Astron. Astrophys.*, 2019, **627**, A55.
2 R. Basalgète, R. Dupuy, G. Féraud, C. Romanzin, L. Philippe, X. Michaut, J. Michoud, L. Amiaud, A. Lafosse, J.-H. Fillion and M. Bertin, *Astron. Astrophys.*, 2021, **647**, A36.

Martin R. S. McCoustra added: Mathieu Bertin is correct in his assertion that we should be careful in looking to the gas phase as a guide to solid state photochemistry. The many-body effects present in the solid state can significantly alter the photochemistry of a species. For example, as we are all aware, the low lying electronic excited states of water in the gas phase are dissociative on the timescale of an O–H vibration. However, in small water clusters, there is computational evidence that these excited states are sufficiently bound to allow for electronic energy transfer between water molecules.[1] Indeed, we have offered this as an explanation for the large cross-section for low-energy electron-stimulated desorption of benzene from compact amorphous solid water (cASW) surfaces.[2]

1 A. Acocella, G. A. Jones and F. Zerbetto, *J. Phys. Chem. Lett.*, 2012, **3**, 3610–3615.
2 D. Marchione and M. R. S. McCoustra, *Phys. Chem. Chem. Phys.*, 2016, **18**, 29747–29755.

Robin T. Garrod commented: On the modelling perspective regarding the chemical desorption of methanol: the chemical models have changed in the past 20 years since we first introduced the mechanism (see ref. 1). We originally incorporated the mechanism into the models using an efficiency factor of 1%, meaning that surface reactions in general would lead to the desorption of the product in no more than 1% of cases. The formulation also included a statistical unimolecular treatment (RRK) to take account of the exothermicity of the reaction and the binding energy of the product, meaning that the efficiency of desorption could be yet less than that maximum of 1%. However, the latter value was rather arbitrary, and is something that has to be tuned to achieve an appropriate reproduction of observational gas-phase abundances (*e.g.* of methanol). In the most recent chemical models, we require that universal maximum value to be around 0.1% (nominally for a water ice surface). This is far below the current detection thresholds in the laboratory, while it is consistent with the experiments, which produced only upper limits for most molecules on water ice.

The reason the value required by the models has changed is that we now use a more complete methanol network, in which we consider the abstraction of hydrogen from molecules, including methanol, by H atoms on the surface. This is a barrier-mediated process that forms a radical, such as CH_2OH or CH_3O, which can then recombine with another H atom, producing methanol in this case. Because methanol and other species can be repetitively destroyed by H-abstraction and then reformed by H-addition, the overall production rate of methanol is raised, even though the total amount of methanol present in the ice is essentially unchanged. This process amplifies the efficiency of chemical desorption; thus the models do not need it to be as efficient. Incidentally, these H-abstraction processes are a widespread class of reactions that we include routinely in our grain-surface networks for the majority of species, including essentially all complex organic molecules, which tends to increase the efficiency of chemical desorption for those species as well.

Regarding Dwayne's earlier question to Eric Herbst and myself, about whether methoxy production on grains would be more significant if we included the reactions of OH with methanol in our models: we do in fact already have these reactions in the networks. In terms of kinetic behaviour, the OH reaction goes with high efficiency upon contact, because the OH is immobile on the surface. However, H abstraction by atomic H tends to be much more important in producing either CH_3O or CH_2OH, due to the mobility of atomic H. The reaction of OH with CH_3OH on the surface occurs as a non-diffusive reaction, which is intrinsically slower.

1 R. Garrod, I. H. Park, P. Caselli and E. Herbst, *Faraday Discuss.*, 2006, **133**, 15.

Eric Herbst addressed all: Is the OH + CH_3OH reaction a key reaction for molecule formation in the gas as well as in the solid phase, where both CH_3O and CH_2OH are products? The rate of the reaction in the gas phase has a most unusual temperature dependence with a peak reactivity at intermediate temperatures. The CH_3O product is a precursor of dimethyl ether and methyl formate. Theoretical studies indicate the need for a complex potential involving tunnelling.

The situation for the surface reaction on hexagonal water ice has been studied theoretically. Reaction routes involving both CH_3O and CH_2OH have been calculated, and also show barriers as well as binding energies. It should be useful to include this process in gas–grain simulations. It is likely that along with atoms and smaller radicals such as OH, the products can lead to complex species in the solid phase.

Dwayne E. Heard answered: In the gas-phase our group detected the CH_3O product of the gas-phase reaction at around 60 K or so,[1] but we were not able to experimentally measure the branching ratio to form this product. Using reaction rate theory (specifically the MESMER package) on the calculated PES we predicted that CH_3O is the dominant product at very low temperatures, whereas at 200–300 K the yields of the two become similar, and CH_2OH becomes dominant at high temperatures. Thank you for the information regarding the theoretical studies on hexagonal water ice, showing that both products can be formed. I agree that the latest calculated barrier heights for the two channels of the OH + CH_3O reaction must be included in gas–grain simulations, as indeed the products may go to form more complex species on the grain. What is also needed are experimental measurements of the absolute product branching ratios for the two channels following the reaction both in the gas phase and on the surface of icy grains.

1 R. J. Shannon, M. A. Blitz, A. Goddard and D. E. Heard, *Nat. Chem.*, 2013, **5**, 745–749.

Susanna Widicus Weaver added: There is a challenge in deciphering methanol photodissociation products. Experimentally, when one studies ice analogs, one uses infrared spectroscopy to examine the chemistry in the ice and mass spectrometry to examine the chemistry in the gas phase. Neither of these techniques offer a means to distinguish between methoxy and hydroxymethyl radicals. The best that experimentalists can do is to look at the products that form after warm-up of the ice and use the astrochemical models developed to interpret observations to infer the chemistry that must have occurred in the lab. This approach is fundamentally flawed because we are already making assumptions about the products of photolysis and their branching fractions in these models.

Dwayne E. Heard replied: Many thanks for that reply. I agree with your final statement.

Dwayne E. Heard continued the general discussion of the paper by W. M. C. Sameera: Did you look at desorption products in your mass spectrometer for the studies of the OH reaction with methanol in ices (which you referred to in your talk)? Can you please describe your experiment in further detail – you will still not be able to tell the difference between methoxy and hydroxymethyl products as they have the same mass, unless some other method was used also?

W. M. C. Sameera replied: The Cs^+ ion pickup method can detect the changes in chemical components adsorbed on the ice surface through the association of Cs^+ and adsorbates (for details of Cs^+ ion pickup, see ref. 1). Since this method is designed for highly sensitive mass detection of trace adsorbates on the ice surface, only the molecules that remain on the surface are measured. We also have a conventional quadrupole mass spectrometer (QMS) equipped on our apparatus for gas analysis, which allows us to observe desorbed molecules. However, no desorption products were detected. Please note that this does not prove that the chemical desorption did not occur. This is because the absolute amount of products formed in our experiments are so small that even if chemical desorption is actively occurring, the detection of desorption products using a conventional QMS would be difficult due to sensitivity.

The separation of CH_3O and CH_2OH, produced from methanol + OH, was performed using methanol isotopologues (CD_3OH, CH_3OD). In these cases, the product radicals have different masses and can be distinguished from each other. Interestingly, the isotope effects in the formation of methoxy and hydroxymethyl radicals through the abstraction reaction of hydrogen and deuterium atoms, were found to be quite small. Therefore, we conclude that the branching ratio obtained from the isotopologue experiments can be applied to the CH_3OH + OH branching ratio. This experimental paper has been submitted.[2]

1 A. Ishibashi, H. Hidaka, Y. Oba, A. Kouchi and N. Watanabe, *Astrophys. J., Lett.*, 2021, **921**, L13.

2 A. Ishibashi, H. Hidaka, W. M. C. Sameera, Y. Oba and N. Watanabe, *Astrophys. J.*, 2023, submitted.

Dwayne E. Heard addressed all: You said in your earlier response to me, that there is too much methanol in your models, suggesting the sources of methanol in models are overestimated, and the previous discussion could lead to even further methanol. How do you know though, that there is not a missing methanol sink which reduces levels and helps with model agreement? Is there likely to be an unknown methanol sink process that could help, do other measurements or models suggest that the methanol sink term is accurate?

Robin T. Garrod responded: It is perhaps more accurate to say that the efficiency of chemical desorption of methanol from water ice is still poorly constrained, and that if we use the earlier prescription of $\sim$1% of methanol formation on surfaces leading to chemical desorption, then the amount of gas-phase methanol achieved in the most recent models is too high. An efficiency of $\sim$0.1% seems to be sufficient in the new models, for reasons I have mentioned in answering other questions. The question of whether there is a sink for gas-phase methanol that we have not yet considered is of course hard to answer, but I suspect that we have all the main destruction mechanisms already in the model: ion-molecule proton-transfer reactions, followed by dissociative electronic recombination of protonated methanol, should be fairly efficient at destroying the molecule. For the solid-phase methanol, the models seem to be in the right ballpark in terms of the fraction of methanol with respect to water measured for dense-gas sight-lines using infrared instruments. So I think that the main area of uncertainty is indeed in the efficiency of methanol transfer from the grain/ice surface to the gas phase. We can at least tune this efficiency in the models without contradicting current experimental upper limits, as the efficiency required is very small.

Dwayne E. Heard commented further: Is it true then that the rate of desorption of methanol in astrochemical environments is not at all well constrained, and rather is more of an educated guess, meaning that the uncertainty in methanol abundances predicted from models is rather uncertain?

Robin T. Garrod responded: Yes, in the case of chemical desorption the efficiency is not a well constrained quantity, and has to be tuned to produce the right order-of-magnitude behaviour for methanol in cold, dark clouds (although obviously in hot regions, thermal desorption is in play). The actual rate of chemical desorption is also complicated by the fact that H-atom abstraction from methanol by diffusive H atoms from the gas phase, followed by re-addition of H to the product radical, provides additional chances to chemical desorb a surface methanol molecule, beyond its original moment of formation. This effect tends to amplify the efficiency of (methanol) chemical desorption.

Ian Sims commented: What are the prospects for directly detecting desorbing species from analogue interstellar ices by microwave spectroscopy? Would this require improved sensitivity? Is there any prospect of detecting desorbing radicals, which for example could help answer the question of the identity of the products from the ice-phase reaction of OH with methanol?

Susanna Widicus Weaver answered: We have attempted this using the SubLIME technique.[1] We are working to optimize spectral sensitivity to see neutral, closed-shell photodesorbed species starting with water and CO. This goal is within reach. Detection of desorbing radicals is much more difficult. As has been shown by Bertin *et al.* (https://doi.org/10.1039/d3fd00004d), the fraction of desorbed radicals in a methanol photolysis experiment is quite low (*i.e.*, 1×10^{-6} at most). Even for direct photodesorption of methanol, this is an incredible challenge in terms of spectral sensitivity.

1 K. M. Yocum, S. N. Milam, P. A. Gerakines and S. L. Widicus Weaver, *Astrophys. J.*, 2020, **913**, 61.

Nadia Balucani added: Concerning OH reactions in space, I would like to mention a recent investigation we have performed in our laboratory on the reaction OH + C_2H_4.[1] This reaction is not very efficient at low temperatures, but it could make a contribution to the formation of vinyl alcohol in the warm regions of the interstellar medium. In addition to that, we estimated a radiative association rate coefficient of *ca.* 1×10^{-12} cm^3 molec^{-1} s^{-1} at 10 K. Radiative association to the CH_2CH_2OH radical could explain the formation of vinyl alcohol in very cold objects as well as to contribute to the formation of glycolaldehyde of which CH_2CH_2OH is an important precursor.[2] Furthermore, this process could also be very efficient when occurring on the icy surface of interstellar grains.

1 P. Liang, E. Valença Ferreira de Aragão, L. Giani, L. Mancini, G. Pannacci, D. Marchione, G. Vanuzzo, N. Faginas-Lago, M. Rosi, D. Skouteris, P. Casavecchia and N Balucani, *J. Phys. Chem. A*, 2023, **127**, 4609–4623.
2 D. Skouteris, N. Balucani, C. Ceccarelli, F. Vazart, C. Puzzarini, V. Barone, C. Codella and B. Lefloch, *Astrophys. J.*, 2018, **854**, 135.

Conflicts of interest

There are no conflicts to declare.

Faraday Discussions

PAPER

Preferential destruction of NH_2-bearing complex interstellar molecules *via* gas-phase proton-transfer reactions

Robin T. Garrod * and Eric Herbst

Received 23rd January 2023, Accepted 24th February 2023
DOI: 10.1039/d3fd00014a

Complex, nitrogen-bearing interstellar molecules, especially amines, are targets of particular interest for detection in star- and planet-forming regions, due to their possible relevance to prebiotic chemistry. However, these NH_2-bearing molecules are not universally detected in sources where other, oxygen-bearing complex organic molecules (COMs) are often plentiful. Nevertheless, recent astrochemical models have often predicted large abundances for NH_2-bearing complex organics, based on their putative production on dust grains. Here we investigate a range of new gas-phase proton-transfer reactions and their influence on the destruction of COMs. As in past studies, reactions between protonated COMs and ammonia (NH_3) are found to be important in prolonging gas-phase COM lifetimes. However, for molecules with proton affinities (PA) greater than that of ammonia, proton-transfer reactions result in drastic reductions in abundances and lifetimes. Ammonia acts as a sink for proton transfer from low-PA COMs, while passing on protons to high-PA species; dissociative recombination with electrons then destroys the resulting ions. Species strongly affected include methylamine (CH_3NH_2), urea ($NH_2C(O)NH_2$) and others bearing the NH_2 group. The abundances of these species show a sharp time dependence, indicating that their detectability may rest on the precise chemical age of the source. Rapid gas-phase destruction of glycine (NH_2CH_2COOH) in the models suggests that its future detection may be yet more challenging than previously hoped.

1 Introduction

The presence of so-called complex organic molecules (COMs) in the interstellar medium (ISM) has been a topic of great interest and sustained discussion among astrochemists over the past several decades. COMs are typically defined as organic species composed of six or more atoms,[1] which conveniently include methanol (CH_3OH), one of the most pervasive molecules observed in dense regions of the ISM. COMs are now routinely observed toward both high-mass and low-mass star-

Depts. of Chemistry & Astronomy, University of Virginia, Charlottesville, VA 22904, USA. E-mail: rgarrod@virginia.edu forming regions, typically through the detection of their gas-phase microwave rotational spectra. Molecules identified include methyl formate[2] ($HCOOCH_3$), dimethyl ether[3] (CH_3OCH_3), several alcohols up to and including propanol,[4,5] as well as various nitrogen-bearing species such as methylamine (CH_3NH_2)[6–8] and recently urea ($NH_2C(O)NH_2$).[9]

The identification of new interstellar COMs has accelerated over the past few years, due partly to the availability of the ALMA interferometer and the superlative spatial resolution and spectral sensitivity that it offers. Meanwhile, many very recent COM detections have been driven by a dedicated, deep-integration line survey of the Galactic Center cloud G+0.693-0.027 using single-dish instruments. While G+0.693 is quiescent (*i.e.*, non-star forming), the plethora of COMs found toward this object are thought to result from the disruption of molecule-rich dust-grain ice mantles, caused by the passage of low-velocity shocks;[10,11] thermal desorption and/or sputtering of these molecules from the grains would allow them to be detected by rotational spectroscopy, as well as making them available for possible gas-phase chemistry. Recent new detections toward this source include a selection of nitrogen-bearing molecules such as hydroxylamine[12] (NH_2OH), vinylamine ($C_2H_3NH_2$) and ethylamine[13] ($C_2H_5NH_2$), and ethanolamine[14] ($NH_2CH_2CH_2OH$).

These latter species, along with several others, retain particular interest due to the importance of the amine ($-NH_2$) group in biochemical compounds, and thus the possibility that certain important prebiotic molecules may have had an interstellar origin. Perhaps chief among such targets for interstellar detection over the past decades – partly due to its totemic value – has been glycine (NH_2CH_2COOH), the simplest amino acid. Unfortunately, in spite of various searches,[15–18] interstellar glycine remains elusive, although it has now been detected in two comets within our own solar system.[19,20] Chemical kinetic simulations of star-forming regions have nevertheless predicted the presence of modest amounts of glycine on interstellar dust grains,[21–23] with the expectation that the heating caused by the star-formation process might allow it to be released, and thus detected in the gas phase given sufficient observing power. While other NH_2-bearing species are now beginning to be detected, in particular toward G+0.693, even simpler species like methylamine are by no means ubiquitous in the usual hot-core sources, while hydroxylamine has not yet been detected toward a star-forming source.[24]

COMs are most typically detected toward hot, star-forming cores, known as hot cores and hot corinos, in the case of high-mass and low-mass sources, respectively. The rich molecular emission spectra of these sources indicate various COMs reaching abundances as high as 10^{-7} with respect to total hydrogen, with excitation temperatures upward of 100 K (hence "hot"). While there continues to be much debate over the origins of specific molecules,[25] the general picture of the chemistry of hot cores involves firstly the formation of ice mantles on the surfaces of dust grains, fed by surface chemistry involving atoms and molecules adsorbed from the gas phase at temperatures of the order of 10 K. This is followed by a period of warming, driven by the evolution of the nascent protostar, that allows the thermal release of the molecules from the grain surfaces and thence their spectroscopic detection at temperatures reflective of the ambient hot conditions. This release also promotes further gas-phase chemistry. However, while undoubtedly many observed molecules are formed strongly in the gas-phase during this hot period (including some COMs, *e.g.*, dimethyl ether[26]), much research now points to a substantially earlier, cold grain-surface chemistry as the origin of perhaps the greater part of the inventory of detected COMs, particularly O-bearing species.[26–31]

Regardless of the originating formation mechanisms, gas-phase COMs are subject to destructive processes involving ion-driven chemistry. The dominant destruction pathways for COMs begin with proton-transfer (PT) reactions with abundant ions such as H_3^+, H_3O^+, and HCO^+; the latter two are formed strongly when abundant ice species H_2O and CO are released into the gas, undergoing proton transfer from H_3^+ (whose own origins lie in the cosmic ray ionization of gas-phase molecular hydrogen). Due to the large abundance of water in particular ($\sim 10^{-4} n_H$), following its desorption from the grains, the dominant PT reactions undergone by COMs tend to involve H_3O^+, which is the most abundant cation in the gas at that point. The protonated COMs may then undergo dissociative recombination (DR) with electrons, which typically breaks down the structure of the underlying molecule; only a small fraction of recombinations lead to the simple ejection of an H atom and the reformation of the COM,[27,32] with three or more fragments being typical. Thus, the gas-phase abundances of COMs would be expected to degrade gradually over time, especially if no efficient mechanism of gas-phase production is available following their desorption from the grains.

Proton-transfer reactions, when exothermic, typically occur at around the collisional rate,[35] which is usually of the order of 10^{-9} cm^3 s^{-1} at 300 K. For a given PT reaction, $AH^+ + B \rightarrow A + BH^+$, the enthalpy change associated with the reaction can be determined from the difference in proton affinities (PA) of the reactant and product neutrals; Table 1 shows a selection of such values. Astrochemical networks have typically included all energetically favourable (exothermic) PT processes between COMs and the abundant gas-phase ions H_3^+, H_3O^+, and HCO^+.[26] The low PAs of H_2, H_2O and CO ensure that proton transfer is usually effective in all three cases. For example, for a proton transfer from water (PA = 691.0 kJ mol^{-1}) to methanol (PA = 754.3 kJ mol^{-1}),

$$H_3O^+ + CH_3OH \rightarrow CH_3OH_2^+ + H_2O \tag{1}$$

the change in enthalpy is −63.3 kJ mol^{-1}, indicating a strongly exothermic process that proceeds unimpeded at the collisional rate. In cases where the difference in proton affinity is more extreme, such as for reactions between H_3^+ and COMs, corresponding to the transfer of a proton from H_2 (PA = 422.3 kJ mol^{-1}) to a COM, the exothermicity may be so great as to result in the break-up of the products, with simple proton transfer being only a minor branch, *e.g.*,[36]

$$\begin{aligned} H_3^+ + CH_3OH &\rightarrow CH_3^+ + H_2O + H_2 \ (46\%) \\ &\rightarrow CH_2OH^+ + 2H_2 \ (28\%) \\ &\rightarrow CH_3OH_2^+ + H_2 \ (26\%) \end{aligned}$$

where the change in enthalpy for the third branch, provided by the difference in proton affinities, is −332 kJ mol^{-1}. For reactions of H_3O^+ and HCO^+ with COMs, the destruction of the PT products typically does not occur,[35] as the higher PA of the underlying H_2O or CO molecule makes the exothermicity of the reaction less extreme. Thus, due to the position of H_3O^+ as the leading ionic reaction partner for COMs, direct PT dominates over any possible alternative destruction outcomes.

Table 1 Proton affinities for a selection of chemical species, used in the calculation of new proton-transfer rates[a]

Molecule	Proton affinity (kJ mol^{-1})	Reference
H_2	422.3	Hunter and Lias[33]
CO (at C atom)	594.0	Hunter and Lias[33]
H_2O	691.0	Hunter and Lias[33]
CH_3OH †	754.3	Hunter and Lias[33]
CH_3CN	779.2	Hunter and Lias[33]
$HCOOCH_3$	782.5	Hunter and Lias[33]
CH_3OCH_3	792.0	Hunter and Lias[33]
NH_2OH †	810.4	CCCBDB database[b]
NH_2CHO	822.2	Hunter and Lias[33]
CH_3NC †	839.1	Hunter and Lias[33]
H_2SiO †	841.0	Hunter and Lias[33]
NH_2OCH_3 †	844.8	Hunter and Lias[33]
NH_2CH_2OH †	844.8	Estimate
C_2H_5NC †	851.3	Hunter and Lias[33]
CH_3NHCHO †	851.3	Hunter and Lias[33]
CH_2NH †	852.9	Hunter and Lias[33]
N_2H_4 †	853.2	Hunter and Lias[33]
NH_3 †	853.6	Hunter and Lias[33]
NH_2CH_2CHO †	856.0	Estimate
$NH_2C(O)CHO$ †	860.0	Estimate
$NH_2C(O)OCH_3$ †	860.0	Estimate
$NH_2C(O)CH_2OH$ †	860.0	Estimate
$CH_3C(O)NH_2$ †	863.6	Hunter and Lias[33]
$NH_2C(O)NH_2$ †	868.4	Zheng and Cooks[34]
C_2S †	869.6	Hunter and Lias[33]
C_3O †	880.2	Hunter and Lias[33]
NH_2CH_2COOH †	886.5	Hunter and Lias[33]
CH_3NH_2 †	899.0	Hunter and Lias[33]
$C_2H_5NH_2$ †	912.0	Hunter and Lias[33]
C_3S †	933.0	Hunter and Lias[33]
NaOH †	1071.8	Hunter and Lias[33]

[a] Species marked † are included in the special set of additional reactions used in model M5, either as reactants or as products of reactions involving their protonated forms. [b] Computational Chemistry Comparison and Benchmark Database of the National Institute of Standards and Technology (NIST).

For a COM such as methanol, its main gas-phase destruction pathway following desorption would therefore proceed firstly through protonation *via* eqn (1), then through dissociative recombination of the resultant protonated methanol with electrons. The latter process has a number of branches, most of which irretrievably break down the original structure:[27]

$$\begin{aligned} CH_3OH_2^+ + e^- &\rightarrow CH_3 + OH + H\ (51 \pm 4\%) \\ &\rightarrow CH_2 + H_2O + H\ (21 \pm 2\%) \\ &\rightarrow H_2CO + H_2 + H\ (10 \pm 1\%) \\ &\rightarrow CH_3 + H_2O\ (9 \pm 2\%) \\ &\rightarrow CH_3O + H_2\ (6 \pm 2\%) \\ &\rightarrow CH_3OH + H\ (3 \pm 2\%) \end{aligned}$$

An important development in the astrochemical modelling of hot cores came with the work of Taquet *et al.*,[37] who included in their chemical network various additional PT reactions, involving a selection of COMs and abundant simple molecules, along with the protonated forms of each. This allowed for protonated COMs to react with other molecular species of higher PA, including the abundant methanol and ammonia released from the grains during the hot stage, thus introducing an alternative to electronic DR. Due to the higher PA of methanol and ammonia (*versus*, *e.g.*, H_2, CO or H_2O), their involvement in PT reactions with COMs, either as reactants or products, should typically involve smaller exothermicities, and thus a small likelihood of break-up rather than proton transfer; the somewhat limited experimental evidence available seems to confirm this expectation[35,38] (see also Sec. 2.3).

In their astrochemical models, Taquet *et al.* found that ammonia in particular, due to its large proton affinity and substantial abundance, would act as an important acceptor of protons from many species; for example, from protonated methanol:

$$CH_3OH_2^+ + NH_3 \rightarrow NH_4^+ + CH_3OH$$

which is exothermic, with an enthalpy change of -99.3 kJ mol^{-1}.

These proton-transfer reactions between ammonia and various protonated COMs would return the underlying molecule to its original state, without destruction. They are also highly competitive with electronic DR, thus diminishing the main process of destruction for those COMs and extending their gas-phase lifetimes considerably.

In the model of Taquet *et al.*, NH_3 was the species of highest PA, meaning that the cascade of proton transfer from various protonated molecules would terminate with ammonia. In spite of the resulting increase in conversion of NH_3 to NH_4^+, dissociative recombination of the latter typically leads only to the ejection of an H atom, allowing ammonia abundances to remain relatively stable.

A selection of proton-transfer reactions between protonated COMs and ammonia was incorporated into the recent hot-core models of Garrod *et al.* (2022);[23] these authors presented an advanced treatment of grain-surface and ice chemistry in which many COMs could be formed at low temperatures on the grains, as the result of so-called nondiffusive reactions (often *via* radical–radical recombination). As in the work of Taquet *et al.*, they found that the inclusion of the ammonia-related PT reactions produced a strong effect on the peak gas-phase abundance attained by methyl formate, one of the key COMs observed in hot cores, as well as an increase in the gas-phase lifetimes of COMs in general, particularly the O-bearing species, which have relatively low PAs.

The models of Garrod *et al.*[21,23,26] are fairly unique in simulating the gas and grain-surface chemistry of a large selection of COMs, including NH_2-bearing species such as methylamine, ethylamine, acetamide ($CH_3C(O)NH_2$), urea, and even glycine. However, as may be seen from Table 1, these species have substantially larger PAs than ammonia. This means firstly that proton transfer to ammonia from their protonated forms is not energetically favourable; thus, such processes were not considered in the above networks, and the underlying COMs did not experience enhanced abundances as a result. Secondly, it means that a full reckoning of their chemistry must make allowance for proton transfer from

NH_4^+ to those COMs, thus increasing their rates of destruction *via* protonation and dissociative recombination. PT reactions of those COMs with other protonated species, including those of other COMs (of lower PA), should also have a similar effect. In principle, the inclusion of PT reactions involving other species with even greater PA could also influence the behaviour of all the others, dependent on their abundances.

Some important gas-phase reaction mechanisms therefore appear to be missing from the existing networks, which may have a strong influence on the expected gas-phase abundances of NH_2-bearing species in particular. This effect would also be of interest due to the overproduction of certain nitrogen-bearing species in the models; these include hydroxylamine[21,24] and urea.[9] Given the large PA of glycine, the possibility of its future interstellar detection may also depend on the outcome of gas-phase proton-transfer reactions.

In the present study, we introduce new proton-transfer reactions incrementally into the network of Garrod *et al.* (2022).[23] The networks are implemented in the three-phase gas-grain astrochemical model *MAGICKAL*, with model runs applied to generic, hot-core physical treatments. Analysis of the results focuses especially on nitrogen-bearing species.

2 Methods

2.1 Chemical model

Chemical kinetic simulations of coupled gas-phase, grain-surface and bulk-ice (*i.e.*, three-phase) chemistry in hot, star-forming cores are carried out using the *MAGICKAL* code;[21] the latter has undergone significant updates since its inception, described in detail by Garrod *et al.* (2022,[23] hereafter G22). The most significant changes involve the introduction of nondiffusive reaction mechanisms on grain surfaces and in the ice mantles that build up on the grains during the early, cold stages of the models. Past hot-core models involving grain-surface chemistry have been based largely on diffusive chemistry, in which grain-surface chemical reactions are dominated by H diffusion at the lowest temperatures, leading to the production of hydrides such as water, methane and ammonia. At elevated temperatures (>20 K), the production of COMs in those models depended on surface diffusion of reactive radicals including CH_3, HCO and NH_2 to build up molecular structures group by group. While the present models still retain these diffusion-driven reaction processes, they no longer require grain-surface reactions to occur solely as the result of diffusive meetings; instead, a reactive species may be formed in a position close or contiguous to a potential reaction partner, leading to an instantaneous follow-on reaction without the need for diffusion across the surface. The driving rates for nondiffusive reactions are therefore determined either by the rate of some initiating reaction, or by the rate of photodissociation of some precursor species, either of which may result in the formation of a chemical species that in some fraction of cases happens to be formed next to its subsequent reaction partner. The formulation for the inclusion of these mechanisms in the models is described in detail by Jin & Garrod (2020).[31]

Within the framework of the physical conditions used to represent the hot-core environment (see Sec. 2.2), six sets of chemical models are run, each using a different chemical network setup (see Table 2 and Sec. 2.3). The number of chemical species in all the model runs remains the same: a total of 1385, which

Table 2 Description of chemical models run, based on the chemical reaction network adopted. For each model M0, M1, *etc.*, a single cold collapse simulation (Stage 1) is run, followed by a warm-up (Stage 2) using one of three characteristic timescales (fast, medium and slow)

Model	Description
M0	No proton-transfer (PT) reactions involving COMs (aside from the usual H_3^+, H_3O^+ and HCO^+ reactions, as per past models)
M1	Same as the final model of Garrod *et al.* (2022);[23] exothermic PT reactions from protonated molecules to NH_3 are allowed
M2	Exothermic PT reactions from protonated molecules to NH_3, and from NH_4^+ to neutral molecules, are included
M3	As M2, with exothermic PT reactions from protonated molecules to CH_3OH, and from $CH_3OH_2^+$ to neutral molecules are included
M4	Both exothermic and endothermic PT reactions from protonated molecules to NH_3 and CH_3OH, and from NH_4^+ and $CH_3OH_2^+$ to neutral molecules, are included
M5	As M4, with all PT reactions included for which either the neutral reactant or product molecule has a proton affinity close to or greater than PA(NH_3)

comprises 749 gas-phase species, along with 318 neutrals in the dust-grain surface layer and the same number in the bulk-ice component. The surface layer is restricted to a maximum of 1 monolayer thickness; net growth in the surface component leads to the transfer of material into the bulk beneath, while a net loss results in transfer from the bulk back into the surface layer,[39] such as during the warm-up stage when the ice mantles desorb into the gas phase. The chemical network used in the basic (M1) model presented here comprises a total of 22 401 reactions and processes, which include but are not limited to a range of gas-phase reactions, adsorption and desorption, transfer between surface and bulk ice, photodissociation of gas-phase, surface and bulk-ice molecules, and reactions on the surface and in the bulk ice. The models presented in this work use the same set of chemical initial conditions as in past models, such as G22, as listed in Table 3. The assumed gas : dust ratio by mass is 100, with a representative grain radius of 0.1 μm and 10^6 grain-surface binding sites.

The reaction-rate coefficients of all gas-phase reactions used in the networks presented here use the modified Arrhenius form:

$$k = \alpha\left(\frac{T}{300\ \mathrm{K}}\right)^{\beta}\exp\left(-\frac{\gamma}{k_\mathrm{b}T}\right) \tag{2}$$

2.2 Physical model

To replicate the time-dependent behaviour of the physical conditions within a hot core, a typical two-stage model is used.[40] The collapse stage, Stage 1, begins with a gas density $n_\mathrm{H} = 3000\ \mathrm{cm}^{-3}$ and visual extinction $A_\mathrm{V} = 3$. The gas temperature is fixed at 10 K during this stage. The density and extinction evolve according to an isothermal freefall collapse, reaching final values of $2 \times 10^8\ \mathrm{cm}^{-3}$ and 500 mag., respectively, over a period lasting $\sim 10^6$ years. Dust temperature is calculated as a function of the time-dependent visual extinction,[41] beginning at ∼14.6 K and

Table 3 Initial, gas-phase chemical abundances with respect to total hydrogen, applied at the beginning of the collapse stage (Stage 1)

Chemical species, i	n_i/n_H
H	2×10^{-3}
H_2	0.499
He	9×10^{-2}
C	1.4×10^{-4}
N	7.5×10^{-5}
O	3.2×10^{-4}
S^+	8×10^{-8}
Si^+	8×10^{-9}
Fe^+	3×10^{-9}
Na^+	2×10^{-8}
Mg^+	7×10^{-9}
P^+	3×10^{-9}
Cl^+	4×10^{-9}

falling to 8 K. The Stage 1 simulation begins from the chemical conditions given in Table 3, with all other species nominally at zero abundance, and the electron fraction set to the sum of the cation abundances.

The final chemical abundances calculated in Stage 1 are used as the starting point for the warm-up stage, Stage 2. During Stage 2, the gas density and visual extinction are held steady at the final Stage 1 values. Meanwhile, the gas and dust temperatures rise monotonically, from the $T_{gas} = 10$ K and $T_{dust} = 8$ K values achieved at the end of Stage 1 up to $T_{gas} = T_{dust} = 400$ K (T_{dust} rises independently until $T_{dust} = T_{gas}$, then they rise together). Following past treatments, three characteristic warm-up timescales are employed, resulting in three different Stage 2 runs for each Stage 1 collapse. The timescales correspond to the time taken to reach 200 K, and are labelled fast (5×10^4 years), medium (2×10^5 years) and slow (1×10^6 years). This warm-up is intended to reproduce the gradual warming experienced by the dust and gas as the central protostar evolves, over a plausible range of timescales; the precise warm-up timescale for individual sources is rather uncertain. Comparison of model results with observed molecular abundance ratios by G22 indicates that longer warm-up timescales are more consistent with hot-core abundances, while shorter timescales provide a better match with values observed in the lower-mass hot corinos. We refer the reader to Garrod & Herbst (2006)[40] and G22 for more detail and discussion of these points.

2.3 Implementation of new proton-transfer reactions

Using the above-described chemical and physical setups, six different chemical networks are tested, which vary only in the number of proton-transfer reactions included. The models (Stage 1 + Stage 2) using these different chemical networks are labelled M0–M5 (see Table 2). Descriptions of these models follow below. Table 4 shows a selection of the reactions that were added to the chemical network used in each model; the reactions shown are not comprehensive, but serve to illustrate the types of reactions added. Rate coefficients are also indicated for each reaction.

Models M0 and M1. The model M1 setup is identical to the final model setup of G22, and is used as a reference point; the associated chemical network includes exothermic PT reactions of protonated COMs (and other protonated species) with ammonia. In order to demonstrate the influence of those reactions alone, model M0 is also presented here, corresponding to the absence of those NH_3-related reactions in an otherwise identical setup (no such model was explicitly presented by G22).

In the M1 model network, in which reactions of the form $AH^+ + NH_3 \rightarrow A + NH_4^+$ are included, species "A" is selected from a list of ~300 neutral species in the network that have a known PA (or an approximate value), and that have a protonated form also included in the network; from these, only those species with a PA less than that of ammonia are included.

Model M2. Model M2 is the same as M1, except that its chemical network also includes exothermic PT reactions between NH_4^+ and neutral COMs chosen from the same basic set of ~300 potential reactants. For the reactions $NH_4^+ + A \rightarrow NH_3 + AH^+$, the set of species "A" used in the network includes only COM species with a PA greater than that of ammonia, of which there are nine; all such species in the network are shown in Table 1, listed beneath ammonia. There are few experimental examples of such reactions, perhaps partly due to the limited number of COMs with a sufficiently high PA. However, the reaction $NH_4^+ + CH_3NH_2 \rightarrow CH_3NH_3^+ + NH_3$ has been tested several times in the laboratory,[42–45] with proton transfer occurring in 100% of cases. The measured reaction rate ranges from $1.40–2.50 \times 10^{-9}$ cm^3 s^{-1} at 300 K. The calculated value used in our networks is $1.5 \times 10^{-9}(T/300\ \text{K})^{-1/2}$ cm^3 s^{-1} (see Sec. 2.3.1).

The non-COM species in the network with a PA greater than PA(NH_3), *i.e.*, C_2S, C_3O, C_3S and NaOH, do not have reactions added at this stage.

In a number of cases (relevant to all Models M1–M5) proton affinities were not available in the literature. For each species that was expected to have a PA value close to or greater than that of ammonia, a crude estimate was made for the purposes of the present work (as indicated in Table 1), based on relative values between similar species with some of the same functional groups, *e.g.*, PA($NH_2C(O)OCH_3$) $\simeq$ PA($NH_2C(O)NH_2$) − (PA(N_2H_4) − PA(NH_2OCH_3)). Since these estimated values happen to be close to PA(NH_3), they should be regarded with caution, and we do not linger on the results for those species; the associated PT reactions are included mainly for purposes of completeness.

In cases where two protonated forms of a COM exist in the network, branching is assumed to be statistical.

Model M3. Model M3 in addition includes a similar set of exothermic PT reactions involving reactions of CH_3OH and $CH_3OH_2^+$ with protonated species and neutrals, respectively. Whether the reaction included in the network corresponds to proton transfer to or from methanol is determined by the relative PAs of methanol and the other neutral. Due to the lower proton affinity of methanol with respect to ammonia, a larger set of neutrals is able to accept a proton from methanol than from ammonia. However, the resulting protonated species would in turn be able to transfer a proton exothermically to ammonia.

Model M4. Model M4 includes the same exothermic reactions involving NH_3, NH_4^+, CH_3OH and $CH_3OH_2^+$, but also introduces a selection of PT reactions that are moderately endothermic, defined here as $\Delta PA < 41$ kJ mol^{-1} (*i.e.*, ~5000 K).

Table 4 Selected proton-transfer reactions added to the chemical networks, and their reaction parameters as applied to eqn (2). The enthalpy change of the reaction, ΔH_r, is provided, based on the difference in proton affinities of the reactant and product neutrals. The final column indicates the first of the presented models in which the listed reaction first appears

Reaction	ΔH_r (kJ mol^{-1})	α (cm^3 s^{-1})	β	γ (K)	Model
$CH_3OH_2^+ + NH_3 \rightarrow NH_4^+ + CH_3OH$	−99.3	1.85×10^{-9}	−0.5	0	M1
$[HCOOCH_3]H^+ + NH_3 \rightarrow NH_4^+ + HCOOCH_3$	−71.1	1.70×10^{-9}	−0.5	0	M1
$CH_3OCH_4^+ + NH_3 \rightarrow NH_4^+ + CH_3OCH_3$	−61.6	1.76×10^{-9}	−0.5	0	M1
$NH_2OH_2^+ + NH_3 \rightarrow NH_4^+ + NH_2OH$	−43.2	1.84×10^{-9}	−0.5	0	M1
$NH_2CHOH^+ + NH_3 \rightarrow NH_4^+ + NH_2CHO$	−31.4	1.75×10^{-9}	−0.5	0	M1
$NH_4^+ + CH_3C(O)NH_2 \rightarrow CH_3C(O)NH_3^+ + NH_3$	−10.0	3.91×10^{-9}	−0.5	0	M2
$NH_4^+ + NH_2C(O)NH_2 \rightarrow NH_2C(O)NH_3^+ + NH_3$	−14.8	3.98×10^{-9}	−0.5	0	M2
$NH_4^+ + NH_2CH_2COOH \rightarrow NH_2CH_2COOH_2^+ + NH_3$	−32.9	1.22×10^{-9}	−0.5	0	M2
$NH_4^+ + CH_3NH_2 \rightarrow CH_3NH_3^+ + NH_3$	−45.4	1.51×10^{-9}	−0.5	0	M2
$NH_4^+ + C_2H_5NH_2 \rightarrow C_2H_5NH_3^+ + NH_3$	−58.4	1.51×10^{-9}	−0.5	0	M2
$CH_3OH_2^+ + HCOOCH_3 \rightarrow [HCOOCH_3]H^+ + CH_3OH$	−28.2	1.48×10^{-9}	−0.5	0	M3
$CH_3OH_2^+ + CH_3OCH_3 \rightarrow CH_3OCH_4^+ + CH_3OH$	−37.7	1.16×10^{-9}	−0.5	0	M3
$CH_3OH_2^+ + NH_2OH \rightarrow NH_2OH_2^+ + CH_3OH$	−56.1	5.61×10^{-10}	−0.5	0	M3
$CH_3OH_2^+ + NH_2CHO \rightarrow NH_2CHOH^+ + CH_3OH$	−67.9	3.30×10^{-9}	−0.5	0	M3
$CH_3OH_2^+ + CH_3C(O)NH_2 \rightarrow CH_3C(O)NH_3^+ + CH_3OH$	−109.3	3.16×10^{-9}	−0.5	0	M3
$CH_3OH_2^+ + NH_2C(O)NH_2 \rightarrow NH_2C(O)NH_3^+ + CH_3OH$	−114.1	3.21×10^{-9}	−0.5	0	M3
$CH_3OH_2^+ + NH_2CH_2COOH \rightarrow NH_2CH_2COOH_2^+ + CH_3OH$	−132.2	9.69×10^{-10}	−0.5	0	M3
$CH_3OH_2^+ + CH_3NH_2 \rightarrow CH_3NH_3^+ + CH_3OH$	−144.7	1.28×10^{-9}	−0.5	0	M3
$CH_3OH_2^+ + C_2H_5NH_2 \rightarrow C_2H_5NH_3^+ + CH_3OH$	−157.7	1.24×10^{-9}	−0.5	0	M3
$NH_4^+ + N_2H_4 \rightarrow N_2H_5^+ + NH_3$	0.4	1.99×10^{-9}	−0.5	50.3	M4
$NH_4^+ + NH_2CHO \rightarrow NH_2CHOH^+ + NH_3$	31.4	4.02×10^{-9}	−0.5	3770	M4
$NH_2C(O)NH_3^+ + NH_3 \rightarrow NH_4^+ + NH_2C(O)NH_2$	14.8	1.70×10^{-9}	−0.5	1810	M4
$NH_2CH_2COOH_2^+ + NH_3 \rightarrow NH_4^+ + NH_2CH_2COOH$	32.9	1.66×10^{-9}	−0.5	3980	M4
$[HCOOCH_3]H^+ + CH_3OH \rightarrow CH_3OH_2^+ + HCOOCH_3$	28.2	1.43×10^{-9}	−0.5	3370	M4
$CH_3CNH^+ + CH_3OH \rightarrow CH_3OH_2^+ + CH_3CN$	24.9	1.54×10^{-9}	−0.5	2970	M4

Table 4 (*Contd.*)

Reaction	ΔH_r (kJ mol^{-1})	α (cm^3 s^{-1})	β	γ (K)	Model
$CH_3OH_2^+ + NaOH \rightarrow NaOH_2^+ + CH_3OH$	−317.5	5.82×10^{-9}	−0.5	0	M5
$NH_4^+ + NaOH \rightarrow NaOH_2^+ + NH_3$	−218.2	7.02×10^{-9}	−0.5	0	M5
$NH_2CH_2COOH_2^+ + CH_3NH_2 \rightarrow CH_3NH_3^+ + NH_2CH_2COOH$	−12.5	1.10×10^{-9}	−0.5	0	M5
$CH_3NH_3^+ + NH_2CH_2COOH \rightarrow NH_2CH_2COOH_2^+ + CH_3NH_2$	12.5	1.00×10^{-9}	−0.5	1510	M5
$CH_3NH_3^+ + C_2H_5NH_2 \rightarrow C_2H_5NH_3^+ + CH_3NH_2$	−13.0	1.28×10^{-9}	−0.5	0	M5
$C_2H_5NH_3^+ + CH_3NH_2 \rightarrow CH_3NH_3^+ + C_2H_5NH_2$	13.0	1.19×10^{-9}	−0.5	1560	M5
$CH_3NH_3^+ + NaOH \rightarrow NaOH_2^+ + CH_3NH_2$	−72.8	5.98×10^{-9}	−0.5	0	M5

Thus, in cases where only a small difference in PA exists between reactant and product neutrals, the backward reaction is also included in the network along with its exothermic counterpart. In these cases, the rate coefficient of the backward PT reaction may become comparable to that of the forward reaction given sufficient temperature.

Model M5. In Model M5, a more comprehensive network is used, which allows exothermic and endothermic PT reactions to take place between a much broader range of species, based around those listed in Table 1. For all PT reactions $AH^+ + B \rightarrow A + BH^+$ and their reverse reactions, species "A" is drawn from the basic list of ∼300 viable neutrals in the network, while species "B" is drawn from the list shown in Table 1 (marked †) for which the PA is greater than or close to that of ammonia. This list also includes the four non-COM species with a PA greater than PA(NH_3). Also added to this list is hydroxylamine (NH_2OH), which is included due to its particular astronomical interest (see Sec. 4). Species set "B" is a subset of set "A", thus PT reactions amongst all species marked † in Table 1 are included, either in one direction or both. Species in set "A" that are not members of set "B" therefore are not, in general, allowed to undergo proton transfer amongst each other, except in the case of reactions with small molecular ions such as H_3^+, *etc.* The large gas-phase abundances of ammonia and methanol are assumed to render such reactions relatively unimportant.

2.3.1 Calculation of proton-transfer rate coefficients. For the few PT reactions involving apolar neutral species, collisional rate coefficients are calculated as straightforward Langevin values, with polarizabilities determined from the literature. For the majority of newly added reactions, the neutrals are polar; in these cases the Average Dipole Orientation (ADO) formulation is used,[46] with the assumption that the permanent-dipole term vastly outweighs the Langevin term, providing a simple $T^{-1/2}$ temperature dependence that is easily incorporated into a modified Arrhenius formula eqn (2).

For exothermic reactions, the above collisional rates are taken as the reaction rates. For endothermic reactions, the difference in proton affinities of the reactant and product neutrals is used as an activation energy (γ) in eqn (2), but the same collisional rate approach is otherwise retained, manifesting in the α and β coefficients; see Table 4 for selected examples.

3 Results

The description of model results presented here concentrates primarily on nitrogen-bearing species; these are the molecules most strongly affected by the changes made to the chemical networks. Further, since all of the additional reactions are gas-phase processes, a particular emphasis is placed on the Stage 2 evolution of the chemistry, during which the thermal desorption of dust-grain ice mantles occurs and gas-phase destruction processes for COMs become important. The chemistry of Stage 1 is described in detail by G22, but a basic description may be found immediately below. To provide a general picture of the behaviour of the different chemical network setups in Stage 2, the main focus is placed on the medium warm-up timescale models.

3.1 Stage 1 – collapse

During Stage 1, which simulates the gradual collapse from diffuse gas to a dense core, chemistry nevertheless occurs in both the gas phase and on the grains. Ice build-up on the grains occurs largely toward the end of Stage 1, when the gas density becomes highest. This coincides also with an increased visual extinction (diminution of the external UV field), resulting in low dust temperatures (minimum 8 K). The majority of the ice is composed of water, with substantial amounts of CO, CO_2, CH_4, NH_3, H_2CO and CH_3OH also present, as would be expected from interstellar ice observations. Much of this chemistry is driven by diffusive addition reactions of atomic H with atoms and simple radicals, all ultimately derived from the gas phase. Table 5 shows the final abundances of the

Table 5 Selected dust-grain ice abundances achieved at the end of the M1 Stage 1 model. Abundances for these species are identical among models M0–M5. Commonly observed ice species are shown in the upper half of the table; selected species/COMs of interest are shown in the lower half. Abundances are given as a fraction of total hydrogen in the simulation, and as a percentage of water ice. Notation $a(-b)$ indicates $a \times 10^{-b}$

Ice species, i	n_i/n_H	%/water ice
H_2O	1.63(−4)	100
CO	5.42(−5)	33.0
CO_2	3.36(−5)	20.6
CH_4	9.08(−6)	5.57
NH_3	2.81(−5)	17.2
H_2CO	2.98(−6)	1.83
CH_3OH	9.58(−6)	5.88
CH_3CN	8.97(−10)	5.50(−4)
NH_2CHO	1.15(−8)	7.24(−3)
NH_2OH	3.95(−6)	2.42
$HCOOCH_3$	1.12(−7)	6.87(−2)
CH_3OCH_3	8.07(−8)	4.95(−2)
CH_3NH_2	3.96(−7)	2.43(−1)
$C_2H_5NH_2$	2.46(−8)	1.15(−2)
$CH_3C(O)NH_2$	3.64(−9)	2.23(−3)
$NH_2C(O)NH_2$	7.46(−9)	4.58(−3)
NH_2CH_2COOH	2.93(−10)	1.80(−4)

main ice components and a selection of COMs on the grains at the end of Stage 1 for model M1. Since the chemical networks of models M0–M5 vary only in the proton-transfer reactions that become critical during Stage 2, there are no meaningful differences in ice composition achieved in Stage 1 between the different models.

Data for a selection of COMs and/or N-bearing species in the grain-surface ices are also shown in Table 5. The abundances of COMs in the gas phase during Stage 1 are relatively small and do not, in general, contribute significantly to the abundances studied in the "hot" Stage 2; however, a few COMs, such as dimethyl ether (CH_3OCH_3) have effective gas-phase production mechanisms in the model that can become dominant during Stage 2.

Various O-bearing COMs, including methyl formate ($HCOOCH_3$), are produced on the grains most strongly toward the end of Stage 1, through nondiffusive surface reactions between radicals related to the hydrogenation of CO to methanol. For methyl formate in particular, this occurs when the HCO radical is formed in proximity to the CH_3O radical (and *vice versa*), leading to immediate reaction. The production of the two radicals is, however, driven by H diffusion on the grain/ice surface.

Methylamine, on the other hand, is mostly produced very early on in Stage 1, when the ices are relatively thin but the visual extinction experienced is still low (around 3 mag.). This allows external UV photons to dissociate methane and ammonia in the bulk ice, allowing the resultant radicals to react to form CH_3NH_2. Ethylamine is formed in a similar way. Just such a mechanism also dominates the formation of glycine in the ice, based on the photodissociation of ammonia and acetic acid (CH_3COOH).

The very large solid-phase abundances of COMs produced on the grains during Stage 1 are retained through Stage 2, and may be enhanced at early periods in Stage 2 while still present on the grains, mainly through bulk-ice UV photolysis driven by the local UV field that is produced by cosmic ray collisions with gas-phase H_2. Table 22 of G22 indicates the periods in the model (as a function of temperature regime) when each molecule is formed. The N-bearing COMs of most interest here (those with a large PA) are produced on the grains, mainly during Stage 1, in the G22 models and in all those presented here.

3.2 Stage 2 – warm-up

Fig. 1 shows results for selected nitrogen-bearing molecules during Stage 2 (warm-up) for model setups M0–M5, using the medium warm-up timescale. In each figure panel, the chemical abundances of the molecules are presented with respect to total hydrogen, *i.e.*, $n(H) + 2n(H_2)$, and as a function of time in years. Time is plotted logarithmically on the lower axis, with the coupled gas and dust temperature shown along the top. Each plot begins a little way into the Stage 2 evolution, but no major change occurs before this time for the species shown. The gas-phase abundance of each molecule is indicated by a solid line, while the total amount of the same molecule on the grain surface (*i.e.*, surface layer + bulk ice) is shown with a dotted line of the same colour.

During this Stage 2 evolution, the dust gradually warms up, such that each of the five species shown, which originate primarily from grain-surface chemistry during Stage 1, are gradually desorbed. Due to the use of a three-phase approach

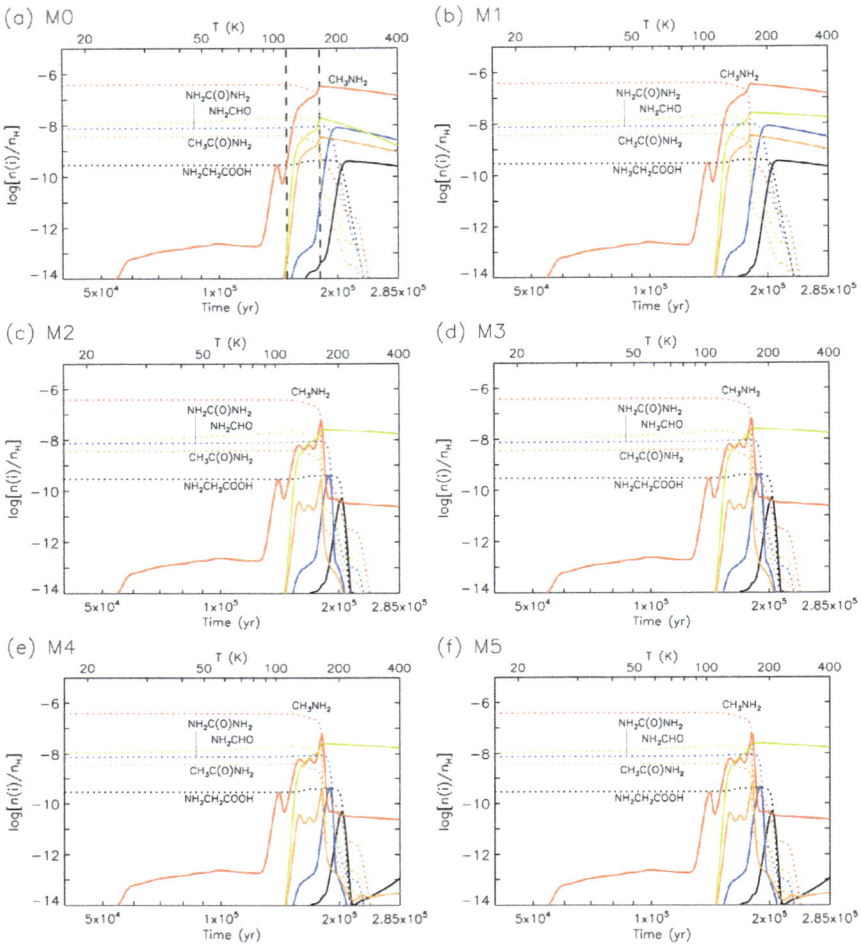

Fig. 1 Abundances with respect to total hydrogen of selected N-bearing species for the different proton-transfer model setups using the medium warm-up timescale. Solid lines indicate gas-phase abundances. Dotted lines of the same colour indicate the same species on the dust grains. Vertical dashed lines in panel (a) indicate the period when substantial thermal desorption of water is occurring.

in the *MAGICKAL* code, the surface layer of predominantly water ice may act to trap the more volatile species in the bulk ice until water itself begins to desorb strongly, although partial loss of volatiles can occur at lower temperatures. When water desorption begins in earnest (not shown directly in these figures – see G22), at ~114 K, molecules with surface binding energies comparable to or lower than that of water may also begin to desorb more strongly, usually reaching gas-phase peak abundances sometime between this initial release and the moment when most of the previously solid water has desorbed into the gas, which occurs by ~164 K (G22). These two key temperatures are marked with vertical dashed lines in Fig. 1, panel (a). In both panels (a) and (b), corresponding to models M0 and M1, this behaviour is observed for methylamine (CH_3NH_2), formamide (NH_2CHO)

and acetamide ($CH_3C(O)NH_2$). Each of these species has a slightly higher binding energy than water, but their own rapid release begins not long after water, which also encourages the faster release of water itself, so that water and all of these species tend to reach their gas-phase peak abundances at a similar point.

Urea ($NH_2C(O)NH_2$) and glycine (NH_2CH_2COOH) achieve their gas-phase peak values at later times, and thus higher temperatures; their stronger binding energies mean that they desorb primarily after most water has left the grains, although they may themselves act to trap a small amount of water (and other species of lower binding energy) beyond its own natural desorption temperature. Due to this trapping effect, involving multiple species each with somewhat different binding energies, the desorption behaviour of COMs during the warm-up stage is complex.

Panel (a) demonstrates the degradation of gas-phase abundances without the inclusion of any proton-transfer reactions from COMs directly to ammonia (model M0), once the molecules have been released from the grains; this gradual decay is caused mainly by PT from small ions and subsequent electronic dissociative recombination. Panel (b) shows the effect of the inclusion of PT from the protonated forms of lower-PA molecules to ammonia (model M1) to enhance the gas-phase lifetimes of certain species; the effect is seen only for NH_2CHO in panel (b), as the other molecules have proton affinities greater than PA(NH_3), meaning that no such reactions are included for them in M1.

Fig. 1, panel (c), shows the influence of the inclusion of PT reactions from protonated ammonia to COMs with very high proton affinities; while NH_2CHO is unaffected, the other four species, each with a large PA, readily accepts a proton from NH_4^+, whose own production rates are enhanced due to the transfer from protonated molecules with a PA lower than PA(NH_3). The effect is dramatic, reducing both the peak abundances achieved and the lifetimes of those high-PA COMs. Although less obvious in panel (c), each of the molecules shown does reach a somewhat stable post-desorption abundance (as in, *e.g.*, panel (a)), but at a much lower absolute value; CH_3NH_2, for example, falls rapidly by around 3 orders of magnitude from its peak down to a more stable value of a few $10^{-11}n_H$.

In Fig. 1, panel (c), the desorption behaviour of methylamine and acetamide in particular appears more complex than in panels (a) and (b), achieving multiple local peak abundances over the temperature range at which water is desorbing (shown as vertical dashed lines in panel (a)). This effect is caused by the combination of (i) multiple stages of faster/slower desorption related to the evolving ice-surface composition, and (ii) the more rapid destruction induced by the new PT reactions, which pulls down the peak gas-phase abundance achieved with each desorption spike. Note that ammonia itself is being released from the grains at this time, sufficient to have an immediate influence on proton transfer. The effect is an overall more stable abundance for these two COMs during the desorption period itself, although the final gas-phase peak is still achieved late in the desorption period. For methylamine, a fractional abundance of a little less than $10^{-8}n_H$ is maintained through this period, before the late peak and precipitous drop occur.

Panels (d)–(f) in Fig. 1 show the results for models M3–M5. Differences between these models are much smaller and are difficult to discern from the figures, aside from the slight upticks in the late-time abundances of glycine and acetamide in models M4 and M5. These are caused by the addition of the backward (endothermic) proton-transfer reactions from protonated glycine and acetamide to ammonia. Although the reaction rates are substantially less than the collisional rates (*e.g.*, a reaction efficiency of $\sim 10^{-6}$ for protonated glycine to react with NH_3 at 300 K), the abundance of ammonia is high enough to make these reactions competitive with electronic recombination, given high enough temperatures. The absolute abundances of these COMs are nevertheless very small when the backward reactions become more efficient.

In Fig. 2, panel (a), abundances are shown for a selection of other species, including the COMs methanol (CH_3OH), methyl formate ($HCOOCH_3$) and methyl cyanide (CH_3CN). Panel (b) in the same figure shows those same species for model M5, which is chosen as representative of the behaviour in all of models M1–M5 (models M2–M5 vary little from M1 for these species). The abundances of the three COMs, as well as hydroxylamine (NH_2OH), are seen to be enhanced in the gas phase to a greater or lesser extent; the most obvious effect is a slow-down in their decay. The behaviour of ammonia is seen to be little affected, reaching only a marginally lower abundance at the very end of the simulation in Fig. 2, panel (b). CH_3CN in the G22 models (generally) is formed mainly in the gas phase; the effect of the introduction of a PT reaction from protonated methyl cyanide to ammonia is to produce an even greater rise at late times.

Fig. 3 shows model M5 abundances for some of the remaining NH_2-bearing molecules shown in Table 1, some of which have a greater PA than ammonia. For ethylamine ($C_2H_5NH_2$), with one of the highest PA values, the gas-phase abundance peak is brief, due to reactions with NH_4^+. For methanolamine (NH_2CH_2OH), whose PA is estimated here to be less than PA(NH_3), a strong gas-phase abundance is maintained following desorption. For glycolamide ($NH_2C(O)CH_2OH$) and two other structures for which we have made estimates of the PA ($NH_2C(O)OCH_3$ and $NH_2C(O)CHO$), the backward reactions allowing PT back to ammonia are relatively efficient, lessening the effect of PT in the forward direction. The latter three species have not been detected in the ISM so far, and are included in the network mainly for completeness. The uncertain nature of their PA estimates leaves their gas-phase behaviour also rather unreliable.

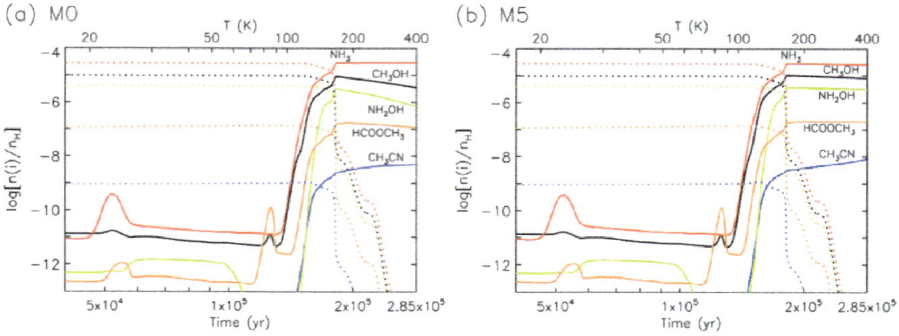

Fig. 2 Abundances with respect to total hydrogen of selected O- and N-bearing species, for the different proton-transfer model setups using the medium warm-up timescale. Solid lines indicate gas-phase abundances. Dotted lines of the same colour indicate the same species on the dust grains.

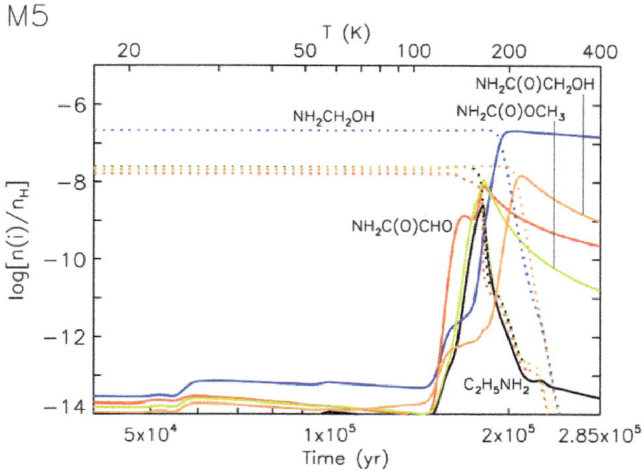

Fig. 3 Abundances with respect to total hydrogen of selected N-bearing species, for the M5 model setup using the medium warm-up timescale. Solid lines indicate gas-phase abundances. Dotted lines of the same colour indicate the same species on the dust grains.

The models were also run for the fast and slow warm-up timescales. Limited results for these two cases are shown in Fig. 4 and 5. Only the results for models M1 and M5 are shown; the behaviour of the M2–M4 models is largely in line with that described above for the medium-timescale models. With the shorter warm-up timescale, the effect of the enhanced gas-phase destruction mechanisms is somewhat more muted as a function of temperature, due to the shorter times spent in the gas phase. More specifically, the peak abundances achieved in the gas phase are somewhat higher in the fast model, due to the more rapid injection of material into the gas phase, *versus* the gas-phase destruction timescale. This is true throughout the main ice desorption period (114–164 K) for those species that undergo multiple desorption spikes. For the late-peaking glycine and urea, peak gas-phase values achieved in the fast model are crudely around an order of

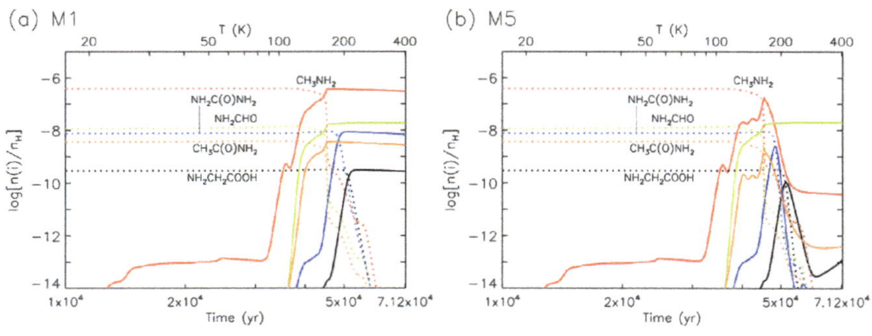

Fig. 4 Abundances with respect to total hydrogen of selected N-bearing species, for the different proton-transfer model setups using the fast warm-up timescale. Solid lines indicate gas-phase abundances. Dotted lines of the same colour indicate the same species on the dust grains.

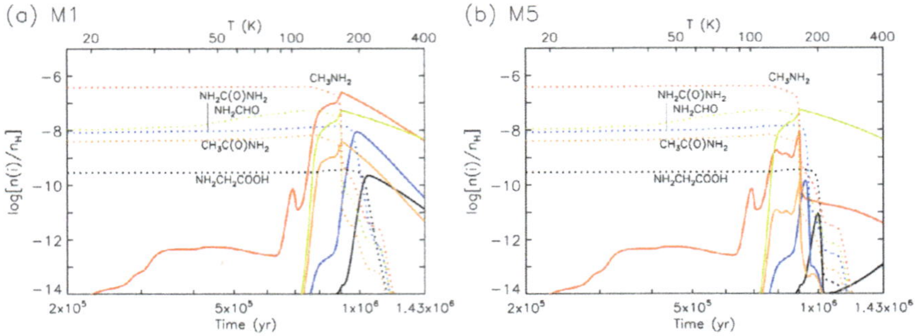

Fig. 5 Abundances with respect to total hydrogen of selected N-bearing species, for the different proton-transfer model setups using the slow warm-up timescale. Solid lines indicate gas-phase abundances. Dotted lines of the same colour indicate the same species on the dust grains.

magnitude higher than in the medium-timescale case, when comparing between model M5 runs. A similar relationship holds in reverse, when comparing the slow with the medium warm-up timescales for model M5.

Tables 6, 7 and 8 show instantaneous abundances obtained for a selection of molecules in models M0, M1, M3 and M5, corresponding to the medium, fast and slow warm-up timescales, respectively. Results for models M2 and M4 are omitted due to their similarity to those of models M3 and M5, respectively. In the top half of each table are shown molecules with proton affinities greater than PA(NH_3), for which some substantial change in results might be expected with the addition of the new PT reactions. Only model M5 includes new reactions directly involving production or destruction of NaOH, the species with the highest PA in the model. The lower half of each table shows molecules that are either commonly observed toward hot star-forming cores or those that are of particular interest, such as NH_2OH. The first column of abundance data in each table shows the peak ice abundance achieved for that species during Stage 2; in most cases, the majority of the gas-phase abundance of the molecule originates on the grains. This data therefore provides a reference point for the degree of destruction suffered by each molecule. This peak ice abundance is essentially the same in each of models M0–M5, as only the gas-phase reactions vary between models, although values vary between different warm-up timescales, due to the different amounts of thermal and/or UV processing experienced by the ice mantles prior to desorption. For each model and molecule, a peak gas-phase value is provided, along with the final gas-phase value achieved at the end of the model, corresponding to a temperature of 400 K.

As seen in the figures, the most drastic changes occur with the introduction of proton transfer from NH_4^+ to high-PA COMs in model M2 (whose results are very similar to the values shown for M3). Peak gas-phase abundances for those NH_2 group-bearing COMs, of which five are presented in the table, fall solidly by a factor of a few to more than an order of magnitude between M1 and M3, dependent on warm-up timescale, but the later gas-phase destruction is yet more drastic, as evidenced by the final abundances.

Table 6 Selected results from the medium warm-up timescale model runs. For each model, the peak abundance of the molecule achieved in the gas phase is provided along with the final value. For each molecule, the peak ice-mantle abundance value is also shown, which does not vary significantly between models M0–M5, assuming the same warm-up timescale is adopted. Chemical species in the upper half of the table are those that might be expected to demonstrate substantial variations in abundance between models M1–M5, and all have a proton affinity greater than PA(NH_3). The lower half of the table shows commonly observed or otherwise noteworthy species. Notation $a(-b)$ indicates $a \times 10^{-b}$

	Fractional abundance, $n(i)/n_H$								
		M0		M1		M3		M5	
Molecule	Peak ice	Peak gas	Final gas	Peak gas	Final gas	Peak gas	Final gas	Peak gas	Final gas
CH_3NH_2	3.89(−7)	3.39(−7)	1.40(−7)	3.39(−7)	1.48(−7)	6.33(−8)	2.39(−11)	6.16(−8)	2.40(−11)
$C_2H_5NH_2$	2.42(−8)	2.07(−08)	9.90(−9)	2.08(−8)	1.04(−8)	4.82(−9)	4.58(−14)	2.46(−9)	2.63(−14)
$CH_3C(O)NH_2$	3.91(−9)	3.48(−9)	8.78(−10)	3.45(−9)	9.78(−10)	2.98(−10)	8.70(−16)	6.40(−10)	3.05(−14)
$NH_2C(O)NH_2$	8.28(−9)	8.13(−9)	2.62(−9)	8.15(−9)	2.95(−9)	4.25(−10)	1.29(−19)	4.49(−10)	1.18(−18)
NH_2CH_2COOH	2.92(−10)	4.01(−10)	2.45(−10)	3.51(−10)	2.07(−10)	4.85(−11)	2.72(−15)	4.67(−11)	1.16(−13)
NaOH	1.65(−13)	2.47(−11)	2.47(−11)	2.48(−11)	2.48(−11)	2.47(−11)	2.47(−11)	7.26(−12)	3.66(−12)
CH_3OH	9.64(−6)	9.14(−6)	3.48(−6)	1.05(−5)	8.15(−6)	1.05(−5)	8.16(−6)	1.05(−6)	8.16(−6)
CH_3CN	9.53(−10)	4.94(−9)	4.94(−9)	8.20(−9)	8.20(−9)	8.53(−9)	8.53(−9)	8.51(−9)	8.51(−9)
NH_3	2.81(−5)	2.95(−5)	2.93(−5)	2.80(−5)	2.60(−5)	2.83(−5)	2.61(−5)	2.83(−5)	2.62(−5)
NH_2CHO	2.13(−8)	1.87(−8)	1.55(−9)	2.55(−8)	1.72(−8)	2.58(−8)	1.76(−8)	2.58(−8)	1.74(−8)
NH_2OH	3.98(−6)	3.15(−6)	6.62(−7)	3.58(−6)	3.15(−6)	3.56(−6)	3.13(−6)	3.57(−6)	3.10(−6)
$HCOOCH_3$	1.37(−7)	1.71(−7)	1.12(−7)	1.98(−7)	1.94(−7)	1.97(−7)	1.94(−7)	1.97(−7)	1.94(−7)
CH_3OCH_3	1.06(−7)	1.48(−7)	1.45(−7)	1.77(−7)	1.77(−7)	1.70(−7)	1.70(−7)	1.69(−7)	1.69(−7)

Table 7 Selected results from the fast warm-up timescale model runs; see Table 6 for details

Molecule	Fractional abundance, $n(i)/n_H$								
		M0		M1		M3		M5	
	Peak ice	Peak gas	Final gas	Peak gas	Final gas	Peak gas	Final gas	Peak gas	Final gas
CH_3NH_2	3.89(−7)	3.73(−7)	3.03(−7)	3.73(−7)	3.05(−7)	1.74(−7)	3.72(−11)	1.71(−7)	3.73(−11)
$C_2H_5NH_2$	2.42(−8)	1.48(−8)	1.28(−8)	1.48(−8)	1.28(−8)	8.51(−9)	4.04(−12)	5.39(−9)	3.35(−12)
$CH_3C(O)NH_2$	3.68(−9)	3.68(−9)	2.69(−9)	3.64(−9)	2.68(−9)	1.09(−9)	1.23(−14)	1.70(−9)	3.93(−13)
$NH_2C(O)NH_2$	7.62(−9)	8.72(−9)	6.83(−9)	8.70(−9)	6.86(−9)	1.95(−9)	1.98(−18)	2.41(−9)	1.96(−17)
NH_2CH_2COOH	2.92(−10)	3.25(−10)	3.06(−10)	3.09(−10)	2.73(−10)	1.16(−10)	2.93(−15)	1.07(−10)	1.24(−13)
NaOH	1.65(−13)	6.60(−12)	6.60(−12)	6.63(−12)	6.63(−12)	6.62(−12)	6.62(−12)	2.90(−12)	2.82(−12)
CH_3OH	9.61(−6)	1.04(−5)	8.29(−6)	1.08(−5)	1.02(−5)	1.08(−5)	1.02(−5)	1.08(−5)	1.02(−5)
CH_3CN	9.04(−10)	2.21(−9)	2.21(−9)	1.70(−9)	1.70(−9)	1.79(−9)	1.79(−9)	1.79(−9)	1.79(−9)
NH_3	2.81(−5)	2.88(−5)	2.88(−5)	2.82(−5)	2.76(−5)	2.85(−5)	2.79(−5)	2.85(−5)	2.79(−5)
NH_2CHO	1.39(−8)	1.51(−8)	1.08(−8)	1.88(−8)	1.85(−8)	2.02(−8)	2.00(−8)	1.97(−8)	1.95(−8)
NH_2OH	3.95(−6)	3.73(−6)	2.70(−6)	3.89(−6)	3.78(−6)	3.87(−6)	3.76(−6)	3.87(−6)	3.75(−6)
$HCOOCH_3$	1.20(−7)	1.75(−7)	1.66(−7)	1.88(−7)	1.88(−7)	1.89(−7)	1.88(−7)	1.88(−7)	1.88(−7)
CH_3OCH_3	8.77(−8)	1.16(−7)	1.16(−7)	1.12(−7)	1.12(−7)	1.10(−7)	1.10(−7)	1.10(−7)	1.10(−7)

Table 8 Selected results from the slow warm-up timescale model runs; see Table 6 for details

	Fractional abundance, $n(i)/n_H$								
		M0		M1		M3		M5	
Molecule	Peak ice	Peak gas	Final gas	Peak gas	Final gas	Peak gas	Final gas	Peak gas	Final gas
CH_3NH_2	4.00(−7)	2.48(−7)	2.60(−9)	2.57(−7)	4.30(−9)	9.22(−9)	4.91(−12)	9.01(−9)	3.98(−12)
$C_2H_5NH_2$	2.42(−8)	1.81(−8)	3.13(−10)	1.82(−8)	4.20(−10)	6.88(−10)	4.79(−19)	5.22(−10)	3.77(−19)
$CH_3C(O)NH_2$	6.51(−9)	4.03(−9)	2.10(−12)	4.16(−9)	4.85(−12)	7.30(−11)	8.68(−20)	1.30(−10)	1.50(−18)
$NH_2C(O)NH_2$	1.29(−8)	8.63(−9)	1.68(−11)	8.96(−9)	3.56(−11)	1.37(−10)	9.61(−22)	1.41(−10)	3.98(−21)
NH_2CH_2COOH	2.92(−10)	2.54(−10)	1.26(−11)	2.23(−10)	1.33(−11)	8.87(−12)	2.97(−15)	8.75(−12)	1.21(−13)
NaOH	1.65(−13)	1.14(−10)	1.14(−10)	1.13(−10)	1.13(−10)	1.12(−10)	1.12(−10)	7.55(−12)	2.01(−12)
CH_3OH	9.59(−6)	5.34(−6)	9.57(−8)	8.31(−6)	2.40(−6)	8.32(−6)	2.36(−6)	8.32(−6)	2.37(−6)
CH_3CN	1.68(−9)	1.05(−8)	7.55(−9)	7.04(−8)	7.04(−8)	7.08(−8)	7.08(−8)	7.06(−8)	7.06(−8)
NH_3	2.81(−5)	2.98(−5)	1.52(−5)	2.71(−5)	1.27(−5)	2.74(−5)	1.24(−5)	2.74(−5)	1.25(−5)
NH_2CHO	5.54(−8)	2.63(−8)	1.02(−13)	5.32(−8)	4.71(−9)	5.32(−8)	4.67(−9)	5.32(−8)	4.45(−9)
NH_2OH	4.13(−6)	1.87(−6)	2.36(−10)	2.92(−6)	1.51(−6)	2.89(−6)	1.49(−6)	2.90(−6)	1.37(−6)
$HCOOCH_3$	1.73(−7)	1.44(−7)	1.10(−8)	2.51(−7)	2.32(−7)	2.49(−7)	2.29(−7)	2.49(−7)	2.29(−7)
CH_3OCH_3	1.60(−7)	1.44(−7)	1.84(−8)	2.18(−7)	1.73(−7)	2.11(−7)	1.84(−7)	2.11(−7)	1.82(−7)

The effects on the important methanol molecule are noticeable but not very substantial for the purposes of astronomical comparison; it nevertheless undergoes a slight peak enhancement, and a longer survival time against gas-phase destruction, as with other COMs with a PA less than that of NH_3, such as the commonly observed O-bearing COMs methyl formate and dimethyl ether. The effects on NH_3 itself are also seen to be modest. Hydroxylamine shows only minor peak abundance increases due to the new reactions, while enjoying the same major increases in longevity, especially in the slow warm-up model. It also remains unaffected by its inclusion in the more comprehensive reaction set of model M5.

4 Discussion

While not every addition to the chemical networks implemented here shows an important effect, the inclusion of proton transfer from NH_4^+ to NH_2-bearing COMs has a drastic influence on the abundances of the latter when also combined with the enhanced proton transfer from low-PA COMs to NH_3. Peak abundances are lowered by around an order of magnitude in these models, when comparing between like warm-up timescales. This effect was suggested as a possible explanation for the overproduction of urea in earlier chemical models, when compared with its detected abundances toward the Galactic Center hot-core source Sgr B2(N1).[9] Although G22 included a far more comprehensive treatment for surface and ice chemistry than those models, the same overproduction was also found by G22. The observed abundance of urea toward position Sgr B2(N1S)[9] was determined to be 2.7×10^{16} cm^{-2}, yielding a ratio $NH_2CHO:NH_2CONH_2$ of ~107, while the M1 models (the same as G22) result in a value of 6, at best, when comparing peak formamide with peak urea values. Meanwhile, the new M5 models produce ratios 8.2, 58 and 380 for the fast, medium and slow models, respectively (using data from Tables 6–8). G22 showed that the slow warm-up timescales tend to produce the best overall match with higher-mass sources, so the new models may be within a factor of a few of reproducing the observed urea values, using this crude peak-to-peak comparison.

It is interesting also that the urea observations toward Sgr B2(N2), which is in the same collection of high-mass star-forming cores as core N1, yielded no detection of urea at all, with a lower limit for the ratio $NH_2CHO:NH_2CONH_2$ of ~1100. While this is not ideally consistent with the comparison of peak model abundance values, it would be far more consistent with the abundances reached either before urea is desorbed in the models, or after. The M5 abundance peak is relatively brief, indicating that the successful detection of urea may be dependent on the observed source or position being in just the right stage of evolution. At too low a temperature, urea would still be in the solid phase, while following desorption it would be short-lived.

The sharply peaked behaviour found for urea therefore makes a very meaningful comparison between these models and the observations rather tricky; the models are physically simple, containing no explicit spatial information. Direct comparison would require a chemical model with spatial structure. However, the peak-to-peak abundance ratios obtained from the models may provide at least a good best-case limit for observational detection of urea toward other sources.

Acetamide ($CH_3C(O)NH_2$), however, while still suffering from rapid gas-phase destruction, manages to maintain a substantial abundance for a more significant period of time. Like methylamine, its gradual desorption along with water over a range of temperatures means that it could potentially be more easily observed at hot-core temperatures up to around 160 K or so. Acetamide was indeed detected toward both sources Sgr B2(N1) and B2(N2).[9] However, it is less abundant in the models than would be expected from the observed ratio with NH_2CHO.

Methylamine (CH_3NH_2) shows a similar extended lifetime in the gas phase during its gradual desorption, followed by a brief peak, then a precipitous decline. Recent observations toward sources in the star-forming region NGC 6334I indicate[8] a $CH_3NH_2 : NH_2CHO$ ratio of at least 8.2, with some lines of sight producing upper limits greater than 60. Values as high as the latter are hard to reproduce with the models, although the ratio hovers around 10 in the M5 fast model during the extended period of desorption from 114–164 K (a ratio around unity is found for the medium model). The original M0/M1 models in fact reproduced CH_3NH_2 in NGC 6334I rather well, over a wider range of temperatures. But for a species with as high a proton affinity as methylamine, the existence of an effective protonation mechanism with ammonia and various other species seems hard to avoid. Based on the highest temperature (>200 K or so) abundances for model M5, the match with observations would appear very poor. This also presents a conundrum as the observational best-fit excitation temperatures for CH_3NH_2 toward NGC 6334I are greater than 200 K, and as high as 340 ± 60 K.

However, it may also be pertinent that methylamine has not been observed very frequently in the ISM until recently; furthermore, for the few other sources in which it has been observed, its abundance ratio with NH_2CHO is substantially lower than in NGC 6334I,[8] with a maximum around 3. It was not detected toward the well-known hot corino source IRAS 16293-2422, with an upper limit with respect to NH_2CHO of 0.053.[47,48] The abundances found toward NGC 6334I may therefore be exceptional.

The origins of methylamine in the model, along with several other NH_2-bearing species including ethylamine ($C_2H_5NH_2$) are also slightly unusual. CH_3NH_2 is formed mainly through photolysis in the early dust-grain ices, driven by external UV that penetrates the initially low visual extinction. It is unclear whether this is an accurate depiction of the chemical behaviour throughout a hot core, or whether such an effect would be highly localized to the more spatially extreme parts of the young core. If such a mechanism is indeed dominant, however, it would result in such molecules being most prevalent in the deepest layers of the ice. These might presumably be preserved to even higher temperatures than the present models are capable of reproducing, *via* trapping by species with yet higher binding energies; three-phase chemical models cannot distinguish between individual layers within the bulk ice. Such an effect would be perhaps more in agreement with the high excitation temperatures determined for methylamine toward NGC 6334I.

Glycine in these models is further diminished by the additional gas-phase destruction mechanisms, so that its already meagre fractional abundances appear to fall rapidly beneath any hope of detectability. It seems, then, that the new models would make it even less likely that the long-sought detection of glycine in the ISM might be forthcoming.

Hydroxylamine (NH_2OH) has been problematic for the models for a long time; it is strongly overproduced, while until very recently it had never been detected in the ISM at all.[24] Although it is present toward the quiescent Galactic Center cloud G+0.693,[12] it has not yet been detected toward a star-forming region. The models presented here indicate that whatever the cause of the overproduction – by several orders of magnitude – it is unlikely to be related to proton-transfer reactions. In spite of the NH_2-group borne by hydroxylamine, its PA is substantially lower than that of ammonia. The overproduction of NH_2OH in the models may well be due to inappropriate assumptions about the efficiency of its formation from NO and HNO on grain surfaces.

Toward the G+0.693 source, a wealth of NH_2-bearing molecules has been detected, including some molecules with very large proton affinities, such as ethylamine. It is natural to ask whether this source is unique in its molecular abundances (as compared with star-forming sources), or only in the means of the delivery of those molecules to the gas phase (*i.e.*, shock heating or sputtering, *versus* the assumed thermal desorption in the present hot-core models). It is notable that the gas number density in G+0.693 is estimated to be of the order of 10^4–10^5 cm^{-3},[49] *i.e.*, three to four orders of magnitude lower than in these models, meaning that the action of proton-transfer reactions would be correspondingly slower, producing greater peak abundances and much longer lifetimes for affected molecules. The rapid ejection of COMs from the grains following a shock would also help to keep abundances high, while potentially also preserving the relative abundances of the ice constituents more fully, assuming that the chemistry during the shock did not have a strong effect. It seems likely, then, that the PT mechanisms that are so destructive for certain species in hot cores may have a far less pronounced effect in a quiescent COM-rich source, and may provide an additional point of divergence between G+0.693 and the more typical environments in which COMs are detected.

The inclusion of endothermic reactions in the networks of models M4 and M5 is partly intended to avoid the situation where exothermic reactions of very modest reaction enthalpies would have an outsized influence on the chemistry even at high temperatures, at which the reverse process might be competitive. However, the overall effect of their inclusion is minor, except for a few cases (*e.g.*, Fig. 3), and the PAs of some of the species affected are only based on estimates rather than measured values, making these results less reliable. However, as seen with glycine, at high temperatures the endothermic reaction can become competitive with other processes, even when the endothermicity is very large. A more complete reaction network would have to consider additional branches with alternative reaction products, giving a reaction with smaller endothermicity, which could make proton transfer the less likely outcome.

For exothermic reactions involving PT to COMs with extremely high PA values, even a reaction with H_3O^+ might be sufficiently energetic to make destruction of the COM the more likely outcome, *versus* the simple proton transfer assumed here. However, since these species already suffer rapid destruction in the gas phase as the result of the new PT reactions with NH_4^+ (followed by dissociative recombination), the result may not be much worse for the abundances of those molecules.

We note finally that the proton-transfer reactions of protonated COMs with ammonia are important primarily because they are competitive with the electronic dissociative recombination that would otherwise lead to the destruction of the underlying molecules. Likewise, proton transfer from NH_4^+ to higher-PA COMs is critically destructive because the resulting protonated COMs have no alternative neutralization mechanism that is sufficiently competitive with DR. However, reactions of cations with negatively charged PAHs could provide a further competing mechanism, which might potentially mitigate the more extreme outcomes of the models presented here; modelling studies have indicated that charged PAHs can have an important influence on the chemistry of dark interstellar clouds.[50]

5 Conclusions

Much focus has necessarily been placed over the years on the origins of interstellar COMs, and thus on the particular formation mechanisms involved. However, destruction mechanisms, particularly in the gas phase, are equally important in determining their observed abundances. With the inclusion of a more comprehensive network of proton-transfer reactions for COMs, a clear distinction arises between those with and without an amine group. The former will tend to have higher proton affinities, greater even than ammonia, apparently dooming them to rapid destruction following their desorption into the high-density gas-phase conditions prevalent in hot, star-forming cores. This occurs due to the strong transfer of protons from various other protonated molecules to abundant ammonia, producing NH_4^+, which then passes the protons on to those COMs of even higher proton affinity. With no competing reaction process available, due to the relatively low abundances of all the other high-proton-affinity species, the resultant protonated, amine group-bearing COMs are destroyed by electronic dissociative recombination.

Methylamine (CH_3NH_2) has one of the larger proton affinities of molecules detected in such regions; this makes its substantial observed abundance in some sources difficult to explain, while being apparently consistent with its more general non-detection. Meanwhile, the recently detected hydroxylamine (NH_2OH), which is usually vastly overproduced in astrochemical models, remains unaffected by the newly added proton-transfer reactions, due to its lower proton affinity.

The destructive effect on NH_2-bearing molecules is helpful in more accurately reproducing the abundance of interstellar urea in the astrochemical models. However, the extreme gas-phase destruction found in the simulations for the amino acid glycine seems to further preclude its possible detection in the ISM.

While NH_2-bearing molecules are strongly affected by the reactions studied here, not all nitrogenated COMs are affected in this way. Nitriles, bearing the CN-group, are commonly observed in star-forming cores and are not subject to the rapid destruction described in this work. The new chemical networks explored here also included proton-transfer reactions for a number of other species that do not contain nitrogen, but which nevertheless have a very high proton affinity, *i.e.*, C_2S, C_3O, C_3S and NaOH. While these molecules were not the focus of this study, their abundances certainly decline with the inclusion of the new reactions. However, due to their generally much lower abundances, their presence in the suite of proton-transfer reactions has no major effect on the abundances of COMs.

Numerous new reactions have been tested here, especially in the M5 network. However, for the completion of existing astrochemical networks (that include high-PA COMs) used by other astrochemical modellers, it is reasonable to assume that only the exothermic reactions involving NH_3, NH_4^+, CH_3OH and $CH_3OH_2^+$ (model M3) are necessary to obtain a fairly accurate outcome.

In comparisons of hot-core models with observations, it has been convenient to use the peak abundances of various species as representative of the chemical composition of a hot core more broadly. The fidelity of this approach has relied on the gas-phase abundances of the various COMs remaining relatively stable with respect to each other over long periods of time, following their arrival in the gas phase (such as in models M0 and M1). Clearly, this requirement is not fulfilled for COMs with proton affinities greater than that of ammonia. Explaining the observed abundances of these species using simple models that do not account for dynamic physical conditions and molecular spatial distributions during the hot period will be challenging.

Author contributions

RTG calculated chemical reaction rates, ran astrochemical models, and contributed to the article text. EH advised on chemical reaction rates and contributed to the article text.

Conflicts of interest

There are no conflicts to declare.

Acknowledgements

We thank the Astronomy & Astrophysics program of the National Science Foundation (grant no. AST 19-06489) for funding this research.

Notes and references

1 E. Herbst and E. F. van Dishoeck, *Annu. Rev. Astron. Astrophys.*, 2009, **47**, 427–480.

2 R. D. Brown, J. G. Crofts, F. F. Gardner, P. D. Godfrey, B. J. Robinson and J. B. Whiteoak, *Astrophys. J., Lett.*, 1975, **197**, L29–L31.

3 L. E. Snyder, D. Buhl, P. R. Schwartz, F. O. Clark, D. R. Johnson, F. J. Lovas and P. T. Giguere, *Astrophys. J., Lett.*, 1974, **191**, L79.

4 I. Jiménez-Serra, L. F. Rodríguez-Almeida, J. Martín-Pintado, V. M. Rivilla, M. Melosso, S. Zeng, L. Colzi, Y. Kawashima, E. Hirota, C. Puzzarini, B. Tercero, P. de Vicente, F. Rico-Villas, M. A. Requena-Torres and S. Martín, *Astron. Astrophys.*, 2022, **663**, A181.

5 A. Belloche, R. T. Garrod, O. Zingsheim, H. S. P. Müller and K. M. Menten, *Astron. Astrophys.*, 2022, **662**, A110.

6 N. Kaifu, M. Morimoto, K. Nagane, K. Akabane, T. Iguchi and K. Takagi, *Astrophys. J., Lett.*, 1974, **191**, L135–L137.

7 N. Fourikis, K. Takagi and M. Morimoto, *Astrophys. J., Lett.*, 1974, **191**, L139.

8 E. G. Bøgelund, B. A. McGuire, M. R. Hogerheijde, E. F. van Dishoeck and N. F. W. Ligterink, *Astron. Astrophys.*, 2019, **624**, A82.
9 A. Belloche, R. T. Garrod, H. S. P. Müller, K. M. Menten, I. Medvedev, J. Thomas and Z. Kisiel, *Astron. Astrophys.*, 2019, **628**, A10.
10 S. Martín, M. A. Requena-Torres, J. Martín-Pintado and R. Mauersberger, *Astrophys. J.*, 2008, **678**, 245–254.
11 I. Jiménez-Serra, P. Caselli, J. Martín-Pintado and T. W. Hartquist, *Astron. Astrophys.*, 2008, **482**, 549–559.
12 V. M. Rivilla, J. Martín-Pintado, I. Jiménez-Serra, S. Martín, L. F. Rodríguez-Almeida, M. A. Requena-Torres, F. Rico-Villas, S. Zeng and C. Briones, *Astrophys. J., Lett.*, 2020, **899**, L28.
13 S. Zeng, I. Jiménez-Serra, V. M. Rivilla, J. Martín-Pintado, L. F. Rodríguez-Almeida, B. Tercero, P. de Vicente, F. Rico-Villas, L. Colzi, S. Martín and M. A. Requena-Torres, *Astrophys. J., Lett.*, 2021, **920**, L27.
14 V. M. Rivilla, I. Jiménez-Serra, J. Martín-Pintado, C. Briones, L. F. Rodríguez-Almeida, F. Rico-Villas, B. Tercero, S. Zeng, L. Colzi, P. de Vicente, S. Martín and M. A. Requena-Torres, *Proc. Natl. Acad. Sci.*, 2021, **118**, e2101314118.
15 Y.-J. Kuan, S. B. Charnley, H.-C. Huang, W.-L. Tseng and Z. Kisiel, *Astrophys. J.*, 2003, **593**, 848–867.
16 L. E. Snyder, F. J. Lovas, J. M. Hollis, D. N. Friedel, P. R. Jewell, A. Remijan, V. V. Ilyushin, E. A. Alekseev and S. F. Dyubko, *Astrophys. J.*, 2005, **619**, 914–930.
17 M. R. Cunningham, P. A. Jones, P. D. Godfrey, D. M. Cragg, I. Bains, M. G. Burton, P. Calisse, N. H. M. Crighton, S. J. Curran, T. M. Davis, J. T. Dempsey, B. Fulton, M. G. Hidas, T. Hill, L. Kedziora-Chudczer, V. Minier, M. B. Pracy, C. Purcell, J. Shobbrook and T. Travouillon, *Mon. Not. R. Astron. Soc.*, 2007, **376**, 1201–1210.
18 P. A. Jones, M. R. Cunningham, P. D. Godfrey and D. M. Cragg, *Mon. Not. R. Astron. Soc.*, 2007, **374**, 579–589.
19 J. E. Elsila, D. P. Glavin and J. P. Dworkin, *Meteorit. Planet. Sci.*, 2009, **44**, 1323–1330.
20 K. Altwegg, H. Balsiger, A. Bar-Nun, J. J. Berthelier, A. Bieler, P. Bochsler, C. Briois, U. Calmonte, M. R. Combi, H. Cottin, J. De Keyser, F. Dhooghe, B. Fiethe, S. A. Fuselier, S. Gasc, T. I. Gombosi, K. C. Hansen, M. Haessig, A. Ja ckel, E. Kopp, A. Korth, L. Le Roy, U. Mall, B. Marty, O. Mousis, T. Owen, H. Reme, M. Rubin, T. Semon, C. Y. Tzou, J. H. Waite and P. Wurz, *Sci. Adv.*, 2016, **2**, e1600285.
21 R. T. Garrod, *Astrophys. J.*, 2013, **765**, 60.
22 S. Ioppolo, G. Fedoseev, K. J. Chuang, H. M. Cuppen, A. R. Clements, M. Jin, R. T. Garrod, D. Qasim, V. Kofman, E. F. van Dishoeck and H. Linnartz, *Nat. Astron.*, 2021, **5**, 197–205.
23 R. T. Garrod, M. Jin, K. A. Matis, D. Jones, E. R. Willis and E. Herbst, *Astrophys. J., Suppl. Ser.*, 2022, **259**, 1.
24 R. L. Pulliam, B. A. McGuire and A. J. Remijan, *Astrophys. J.*, 2012, **751**, 1.
25 J. K. Jørgensen, A. Belloche and R. T. Garrod, *Annu. Rev. Astron. Astrophys.*, 2020, **58**, 727–778.
26 R. T. Garrod, S. L. Widicus Weaver and E. Herbst, *Astrophys. J.*, 2008, **682**, 283–302.
27 W. D. Geppert, M. Hamberg, R. D. Thomas, F. Österdahl, F. Hellberg, V. Zhaunerchyk, A. Ehlerding, T. J. Millar, H. Roberts, J. Semaniak,

M. A. Ugglas, A. Källberg, A. Simonsson, M. Kaminska and M. Larsson, *Faraday Discuss.*, 2006, **133**, 177.

28 A. Bacmann, V. Taquet, A. Faure, C. Kahane and C. Ceccarelli, *Astron. Astrophys.*, 2012, **541**, L12.

29 G. Fedoseev, H. M. Cuppen, S. Ioppolo, T. Lamberts and H. Linnartz, *Mon. Not. R. Astron. Soc.*, 2015, **448**, 1288–1297.

30 I. Jiménez-Serra, A. I. Vasyunin, P. Caselli, N. Marcelino, N. Billot, S. Viti, L. Testi, C. Vastel, B. Lefloch and R. Bachiller, *Astrophys. J., Lett.*, 2016, **830**, L6.

31 M. Jin and R. T. Garrod, *Astrophys. J., Suppl. Ser.*, 2020, **249**, 26.

32 M. Hamberg, V. Zhaunerchyk, E. Vigren, M. Kaminska, I. Kashperka, M. Zhang, S. Trippel, F. Österdahl, M. A. Ugglas, R. D. Thomas, A. Källberg, A. Simonsson, A. Paál, J. Semaniak, M. Larsson and W. D. Geppert, *Astron. Astrophys.*, 2010, **522**, A90.

33 E. P. Hunter and S. G. Lias, *J. Phys. Chem. Ref. Data*, 1998, **27**, 413–656.

34 X. Zheng and R. G. Cooks, *J. Phys. Chem. A*, 2002, **106**, 9939.

35 V. G. Anicich, *An Index of the Literature for Bimolecular Gas Phase Cation-Molecule Reaction Kinetics*, JPL publication, 2003, pp. 03–19.

36 V. G. Anicich, *J. Phys. Chem. Ref. Data*, 1993, **22**, 1469–1569.

37 V. Taquet, E. S. Wirström and S. B. Charnley, *Astrophys. J.*, 2016, **821**, 46.

38 W. Y. Feng and C. Lifshitz, *J. Phys. Chem.*, 1994, **98**, 3658–3663.

39 T. I. Hasegawa and E. Herbst, *Mon. Not. R. Astron. Soc.*, 1993, **263**, 589.

40 R. T. Garrod and E. Herbst, *Astron. Astrophys.*, 2006, **457**, 927–936.

41 R. T. Garrod and T. Pauly, *Astrophys. J.*, 2011, **735**, 15.

42 M. S. B. Munson, *J. Am. Chem. Soc.*, 1965, **87**, 2332–2336.

43 N. G. Adams, D. Smith and J. F. Paulson, *J. Chem. Phys.*, 1980, **72**, 288–297.

44 J. Barassin, A. Barassin and R. Thomas, *Int. J. Mass Spectrom. Ion Phys.*, 1983, **49**, 51–60.

45 N. G. Adams, L. M. Babcock, T. M. Mostefaouoi and M. S. Kerns, *Int. J. Mass Spectrom.*, 2003, **223–224**, 459–471.

46 E. Herbst and C. M. Leung, *Astrophys. J.*, 1986, **310**, 378.

47 A. Coutens, J. K. Jørgensen, M. H. D. van der Wiel, H. S. P. Müller, J. M. Lykke, P. Bjerkeli, T. L. Bourke, H. Calcutt, M. N. Drozdovskaya, C. Favre, E. C. Fayolle, R. T. Garrod, S. K. Jacobsen, N. F. W. Ligterink, K. I. Öberg, M. V. Persson, E. F. van Dishoeck and S. F. Wampfler, *Astron. Astrophys.*, 2016, **590**, L6.

48 N. F. W. Ligterink, H. Calcutt, A. Coutens, L. E. Kristensen, T. L. Bourke, M. N. Drozdovskaya, H. S. P. Müller, S. F. Wampfler, M. H. D. van der Wiel, E. F. van Dishoeck and J. K. Jørgensen, *Astron. Astrophys.*, 2018, **619**, A28.

49 S. Zeng, Q. Zhang, I. Jiménez-Serra, B. Tercero, X. Lu, J. Martín-Pintado, P. de Vicente, V. M. Rivilla and S. Li, *Mon. Not. R. Astron. Soc.*, 2020, **497**, 4896–4909.

50 V. Wakelam and E. Herbst, *Astrophys. J.*, 2008, **680**, 371–383.

Faraday Discussions

ROYAL SOCIETY OF CHEMISTRY

PAPER

A statistical and machine learning approach to the study of astrochemistry

Johannes Heyl, †[a] Serena Viti †*[ba] and Gijs Vermariën [b]

Received 15th January 2023, Accepted 8th February 2023
DOI: 10.1039/d3fd00008g

In order to obtain a good understanding of astrochemistry, it is crucial to better understand the key parameters that govern grain-surface chemistry. For many chemical networks, these crucial parameters are the binding energies of the species. However, there exists much disagreement regarding these values in the literature. In this work, a Bayesian inference approach is taken to estimate these values. It is found that this is difficult to do in the absence of enough data. The Massive Optimised Parameter Estimation and Data (MOPED) compression algorithm is then used to help determine which species should be prioritised for future detections in order to better constrain the values of binding energies. Finally, an interpretable machine learning approach is taken in order to better understand the non-linear relationship between binding energies and the final abundances of specific species of interest.

1 Introduction

Giant Molecular Clouds in our Milky Way as well as in other galaxies host gas which is almost entirely molecular, with densities above $\sim$100 cm^{-3} and temperature below $\sim$100 K. These denser, cooler regions contain a significant fraction of the non-stellar baryonic matter in a galaxy and they are usually much more massive than large tenuous ones. The importance of these regions lies in the fact that they are key for our understanding of how galaxies form and evolve because this denser, cooler gas is the reservoir of matter that forms stars and planets, as well as the gas that fuels the centres of galaxies.

From an astrochemical point of view, due to their high densities and low temperatures, these regions are great laboratories to study the interactions of gas and dust, with species from the gas phase 'freezing' onto the dust grains present, and forming icy mantles rich in hydrogenated as well as complex organic molecules (COMs), due to the many fast surface reactions that take place. As stars form in these clouds (or if any other energetic process takes place) then the dust

[a]Department of Physics and Astronomy, University College London, Gower Street, WC1E 6BT, London, UK. E-mail: johannes.heyl.19@ucl.ac.uk
[b]Leiden Observatory, Leiden University, PO Box 9513, 2300 RA Leiden, The Netherlands

† These authors contributed equally.

temperature may reach the mantle sublimation temperature ($\sim$100 K), and the molecules in the mantles are injected into the gas, where they react and form new, more complex, molecules. Associated with star formation, as well as with AGN (active galactic nucleus) activity, are highly supersonic collimated jets and molecular outflows. When the outflowing material encounters the quiescent gas of a molecular cloud, it creates shocks, where the grain mantles are (partially) sputtered and the refractory grains are shattered. Again, here, the interaction of gas and dust varies within very short timescales and the effects of chemistry and dynamics are interlocked in a complex non-linear fashion. In summary, the gas and dust surface compositions exhibit a complicated time dependent, non-linear chemistry that strongly depends on the physical environment. There are many open questions – still – about such interactions: what is the unprocessed ice composition? What are the efficiencies of the viable surface reactions? And how do the energetics of the interstellar medium (ISM) – cosmic rays, UV radiation, shocks – influence the processed ices? In order to determine accurate estimates of the abundances of molecular species as a function of all the parameters that influence their chemistry we need to be able to answer such questions. In other words, we need to understand the chemical pathways towards each molecule and its dependencies on the density, temperature and energetics of the gas and dust before molecules can be truly considered powerful tools.

In recent years, coupling chemical and radiative transfer models for the interpretation of molecular emission has been routinely done and the success of such techniques has varied to different degrees, depending on whether one wants to model the physical and chemical structure, or the hydrodynamical history of the gas.[1–4] However the shortcomings of such methods are two-fold: (i) understanding the physical conditions in molecular gas *via* a systematic and applicable to many galaxies methodology is an inverse problem subject to complicated chemistry that varies non-linearly with both time and the physical environment;[5] hence it may not have a solution, solutions might not be unique and/or might not depend continuously on the observational data. Traditionally astrochemistry has always been dominated by trial and error grid-based analysis combined with simple statistics,[6] an approach that becomes impossible or ineffective when datasets (*e.g.* from ALMA) and/or parameter space are large, complex, or heterogeneous; (ii) the knowledge of the micro-physics and chemistry of what occurs on the dust is well behind what is known for the gas-phase. While surface reactions and dynamics (including desorption and diffusion) can be experimentally investigated (but always within a constrained range of laboratory conditions), experimental data for interstellar ices are still limited. In order to make the best use of experimental resources, the chemical data that models require need to be prioritized according to what will have the most impact.

In recent years progress based on the use of Bayesian as well as Machine Learning (ML) techniques to deal with both the issues above has been made, from the creation of neural network based statistical emulators[7–9] in order to optimize the integration of chemical, radiative transfer and hydrodynamical models to the use of ML techniques to disentangle multiple gas components in unresolved beams.[10]

In this paper we will focus our attention to Bayesian and ML techniques applied to the study of chemical networks and the key parameters that govern their interactions. In recent years there has been a substantial body of work concentrating on reducing the cost of solving chemical networks' computations using various techniques from Monte Carlo approaches to constrain important reactions,[11] to automated reduction schemes,[12,13] to topological methods[14,15] to, finally, ML algorithms.[9,16] In parallel several studies have concentrated on the estimation of poorly known reaction rates, with particular emphasis on surface chemical networks: an initial approach considered a simple grain-surface network and applied a Bayesian inference method coupled with Markov Chain Monte Carlo sampling in order to infer reaction rates.[11] This was followed up with an approach that considered the topological structure of the network,[15] while another exploited the characteristics of the chemical reaction mechanism to significantly reduce the dimensionality of the problem under consideration by simply considering the binding energies and the role they play in the determination of grain-surface chemistry.[17] Subsequent work using the 'Massive Optimised Parameter Estimation and Data compression' (MOPED) algorithm, helped make predictions about which ice species needed to be detected to reduce the variance of binding energy estimates.[18]

Due to the significant role that binding energies play in grain-surface chemistry, we shall concentrate on the estimation of binding energies as well as on prioritization of the ice species that should be observed with instruments such as the JWST to better improve our understanding of their values. We will then use machine learning interpretability to consider the forward relationship between binding energies and the abundances of species of interest. Our methods are described in Sections 2. The results are presented in Section 3 and a brief conclusion is given in Section 4.

2 Methodology

In this section we first describe the chemical code we use and the chemical assumptions we make in our work (Section 2.1), followed by a description of the analytical approach we employ (Section 2.2).

2.1 The chemical code

All modelling in this work is done with the open-source astrochemical code UCLCHEM.‡[19] The chemistry of a collapsing dark cloud was modelled. The dark cloud collapsed isothermally at 10 K from 10^2 cm^{-3} to 10^6 cm^{-3} over 5 million years. The composition of the ices as a result of the ensuing chemistry was then compared to the recent ice observations with the James Webb Space Telescope (JWST).[20]

As this work focuses solely on grain-surface chemistry, it is pertinent to describe the details of the underlying reaction mechanisms we consider in this work. This will be used as justification to explain why binding energies are of such great importance in the context of this work.

In UCLCHEM, the main grain-surface reaction mechanism is the Langmuir–Hinshelwood mechanism.[21] The rate at which two species A and B react through diffusion is given by:

‡ https://uclchem.github.io/.

$$k_{\mathrm{AB}} = \kappa_{\mathrm{AB}} \frac{\left(k_{\mathrm{hop}}^{\mathrm{A}} + k_{\mathrm{hop}}^{\mathrm{B}}\right)}{N_{\mathrm{site}} n_{\mathrm{dust}}}, \tag{1}$$

where N_{site} is the number of sites on the grain surface and n_{dust} is the dust grain number density.

In eqn (1), $k_{\mathrm{hop}}^{\mathrm{X}}$ is the thermal hopping rate of species X on the grain surface which is given as:

$$k_{\mathrm{hop}}^{\mathrm{X}} = \nu_0 \exp\left(-\frac{E_{\mathrm{D}}}{T_{\mathrm{gr}}}\right), \tag{2}$$

where E_{D} is the diffusion energy of the species, T_{gr} is the grain temperature and ν_0 is the characteristic vibration frequency of species X. The diffusion energy is a fraction of the binding energy of the species, E_{b}.

The characteristic vibration frequency, ν_0, is defined as:

$$\nu_0 = \sqrt{\frac{2k_{\mathrm{b}} n_{\mathrm{s}} E_{\mathrm{b}}}{\pi^2 m}}, \tag{3}$$

where k_{b} is the Boltzmann constant, n_{s} is the grain site density and m is the mass of species.

The final term, κ_{AB}, which gives the reaction probability is:

$$\kappa_{\mathrm{AB}} = \max\left(\exp\left(-\frac{2a}{\hbar}\sqrt{2\mu k_{\mathrm{b}} E_{\mathrm{A}}}\right), \exp\left(-\frac{E_{\mathrm{A}}}{T_{\mathrm{gr}}}\right)\right), \tag{4}$$

where $\hbar$ is the reduced Planck constant, μ is the reduced mass, E_{A} is the reaction activation energy, k_{b} is Boltzmann's constant and a = 1.4 Angstrom is the thickness of a quantum mechanical barrier that is used as the default in UCLCHEM. The reaction probability encodes the competition between the quantum mechanical probability of a tunnelling through a rectangular barrier of thickness a, which is the first term, and the thermal reaction probability, which is the second term.

Species do not necessarily need to react with each other on the grains. It is also possible for them to diffuse away from a potential reactant or evaporate. As such, a modification needs to be made to the κ_{AB} term to take this into account. This is the reaction-diffusion competition.[22,23] The reaction probability is now defined as:

$$\kappa_{\mathrm{AB}}^{\mathrm{final}} = \frac{p_{\mathrm{reac}}}{p_{\mathrm{reac}} + p_{\mathrm{diff}} + p_{\mathrm{evap}}}, \tag{5}$$

where p_{reac}, p_{diff} and p_{evap} represent the probabilities of species A and B reacting, diffusing and evaporating per unit time, respectively. These quantities are defined as:

$$p_{\mathrm{reac}} = \max(\nu_0^{\mathrm{A}}, \nu_0^{\mathrm{B}})\kappa_{\mathrm{AB}}, \tag{6}$$

$$p_{\mathrm{diff}} = k_{\mathrm{hop}}^{\mathrm{A}} + k_{\mathrm{hop}}^{\mathrm{B}} \tag{7}$$

and

$$p_{\mathrm{evap}} = \nu_0^{\mathrm{A}} \exp\left(-\frac{E_{\mathrm{b}}^{\mathrm{A}}}{T_{\mathrm{gr}}}\right) + \nu_0^{\mathrm{B}} \exp\left(-\frac{E_{\mathrm{b}}^{\mathrm{B}}}{T_{\mathrm{gr}}}\right). \tag{8}$$

The term κ_{AB} in eqn (1) is replaced with κ_{AB}^{final}.

Eqn (1)–(8) show that the key quantities are ν_0, k_{hop}^{X}, E_b and E_A. The first two are functions of the binding energies of the reacting species. We assume that the activation energies in eqn (4) are well-known, so do not include these as parameters to be estimated.

If we wish to better understand grain-surface diffusion-based chemistry, we must have accurate values of the binding energies of species. For most cases, at 10 K, the reactant with the lower binding energy will dominate the total hopping rate, due to the exponential dependence of the hopping rate on the diffusion energy. Across the literature, there is often significant disagreement when it comes to the values of binding energies.[24–26] While there exist many different methods of estimating these values,[27–29] we utilise a Bayesian inference approach.

The chemical network used consists of a gas-phase and ice-phase network. The gas-phase network is the UMIST network.[24] The ice network used is the same as in previous work,[18] but augmented with a sulphur network based on work done to explain the sulphur depletion problem.[30] The inclusion of the sulphur network is important, since recently sulphur-bearing species have been confirmed in the ices.[20]

2.2 Analytical approach

2.2.1 Bayesian inference. One of the goals of this work is to estimate the binding energies of the most diffusive species in the network. These species were chosen based on a literature search that suggested they were amongst the species with the lowest values for their binding energies. The binding energy parameters are represented as a vector, $\boldsymbol{E} = (E_{b,H}, E_{b,H_2}, E_{b,C}, E_{b,CH}, E_{b,N}, E_{b,CH_3}, E_{b,NH}, E_{b,CH_4}, E_{b,O})$. UCLCHEM was rewritten so that it would take the vector as an input and output the abundances of species of interest. The mapping between the input and output can be summarised as $\boldsymbol{Y} = f(\boldsymbol{E})$, where f represents UCLCHEM. We are looking to estimate the binding energies that give us abundances that match our measurements best. This is an inverse problem, as we are trying to determine the best-fitting inputs that give an output of interest.

Bayes' Law was used to solve this inference problem. Given the data, $\boldsymbol{d}$, of abundance measurements of species, the probability distributions of the binding energies of interest are given by:

$$P(\boldsymbol{E}|\boldsymbol{d}) = \frac{P(\boldsymbol{d}|\boldsymbol{E})P(\boldsymbol{E})}{P(\boldsymbol{d})}, \tag{9}$$

where $P(\boldsymbol{E}|\boldsymbol{d})$ is the posterior probability distribution, $P(\boldsymbol{E})$ is the prior, $P(\boldsymbol{d}|\boldsymbol{E})$ is the likelihood and $P(\boldsymbol{d})$ is referred to as the evidence. The prior distribution encodes the initial understanding of the binding energy distribution. The likelihood gives the data's likelihood as a function of the binding energies. Within the likelihood function, the physical model is encoded. The evidence serves as a normalising factor and represents the marginalised likelihood. The posterior distribution represents the updated probability distribution of reaction rates based on the data, the prior distribution, and the physical model.

The prior for all binding energies was selected as a uniform distribution between 400 K and 2000 K. The abundance measurements, given in Table 1, were

Table 1 The abundances and uncertainties taken from McClure *et al.*[20] These abundances were taken from sources with an visual extinction, A_V, of 95

Species	Abundances relative to H
H_2O	$(8.8 \pm 1.1) \times 10^{-5}$
CO	$(2.2 \pm 0.3) \times 10^{-5}$
CO_2	$(1.1 \pm 0.2) \times 10^{-5}$
CH_3OH	$(3.1 \pm 0.7) \times 10^{-6}$
NH_3	$(8.8 \pm 1.6) \times 10^{-6}$
CH_4	$(1.8 \pm 0.1) \times 10^{-6}$
OCN	$\sim 2.0 \times 10^{-7}$
SO_2	$\sim 6.6 \times 10^{-8}$
OCS	$\sim 1.3 \times 10^{-7}$

assumed to be Gaussian. The species without associated uncertainty, OCN, SO_2 and OCS, were given a relative uncertainty of 50%. Assuming a Gaussian distribution, the likelihood function can be specified as:

$$P(\boldsymbol{d}|\boldsymbol{E}) = \prod_{i=1}^{n_d} \frac{1}{\sqrt{2\pi}\sigma_i} \exp\left(-\frac{(d_i - Y_i)^2}{2{\sigma_i}^2} \right), \tag{10}$$

where n_d is the number of observations and σ_i is the uncertainty of the ith observation. Only the species for which there are abundance measurements are indexed over.

The inference was implemented using the UltraNest Python package.[31] The package implements efficient methods to construct a neighbourhood to sample from, allowing for better convergence of the sampling of the likelihood.[32,33] The package conveniently also outputs the maximum likelihood-estimator, $\boldsymbol{E}_{\mathbf{ML}}$, which will be utilised later.

2.2.2 The MOPED algorithm. While our knowledge of the molecular inventory in the gas-phase is quite complete, we are still far from being confident about the ice composition as well as the ice chemistry. To this end, we employ the "Massive Optimised Parameter Estimation and Data compression" (MOPED) algorithm.[34–36]

The aim of the algorithm is to determine which of the M species in our chemical network would best constrain our knowledge for our p binding energy parameters. In this work, $p = 9$ and $M = 119$. Some binding energies will have a greater influence on certain species than others. The key is to determine the species that are most sensitive to the binding energies of interest. In doing so, we can then make recommendations for future ice observations as was done in a proof-of-concept work recently.[18]

There will be uncertainty associated with all of our potential future abundance measurements. It is likely that these uncertainties will vary by species. However, it is difficult to determine these species-dependent uncertainties *a priori*. As such, we assume the uncertainty on each abundance measurement is the same. This is summarised in a covariance matrix: $\boldsymbol{C} = \mathrm{diag}({\sigma_1}^2, {\sigma_2}^2, \ldots {\sigma_M}^2)$.

We apply a filtering technique developed by Heavens *et al.*,[34,35,36] who propose using a linear combination of the final abundances of a network, $\boldsymbol{Y}$, to compress data points into numbers. Such a compression takes the form:

$$c_\alpha = \boldsymbol{b}_\alpha^{\mathrm{T}} \boldsymbol{Y}, \tag{11}$$

where α ranges from 1 to p and $\boldsymbol{b}_\alpha$ is a set of orthonormal linear filters. Each filter vector is unique to each parameter and does not contain information contained in any of the other vectors. $\boldsymbol{Y}$ represents a vector containing the final, steady-state abundances for some value of $\boldsymbol{E}$, though we employ $\boldsymbol{E} = \boldsymbol{E}_{\mathbf{ML}}$, which can be obtained from the Bayesian inference, as this has been found to be sufficient as the fiducial model.[34,35] For each c_α, there will be greater dependence on some of the components of $\boldsymbol{b}_\alpha$ than others. As each component represents a different species, this implies that a component with a greater magnitude has more information about that parameter.

The vectors $\boldsymbol{b}_\alpha$ are given by

$$\boldsymbol{b}_1 = \frac{\boldsymbol{C}^{-1}\boldsymbol{Y}_{,1}}{\sqrt{\boldsymbol{Y}_{,1}^{\mathrm{T}}\boldsymbol{C}^{-1}\boldsymbol{Y}_{,1}}} \tag{12}$$

and

$$\boldsymbol{b}_\alpha = \frac{\boldsymbol{C}^{-1}\boldsymbol{Y}_{,\alpha} - \sum_{\beta=1}^{\alpha-1}\left(\boldsymbol{Y}_{,\alpha}^{\mathrm{T}}\boldsymbol{b}_{,\beta}\right)\boldsymbol{b}_{,\beta}}{\sqrt{\boldsymbol{Y}_{,\alpha}^{\mathrm{T}}\boldsymbol{C}^{-1}\boldsymbol{Y}_{,\alpha} - \sum_{\beta=1}^{\alpha-1}\left(\boldsymbol{Y}_{,\alpha}^{\mathrm{T}}\boldsymbol{b}_{,\beta}\right)^2}}, \tag{13}$$

where $\boldsymbol{Y}_{,\alpha}$ is the partial derivative of $\boldsymbol{Y}$ with respect to the parameter α around the point $\boldsymbol{Y} = f(\boldsymbol{E}_{\mathbf{ML}})$. The equations for $\boldsymbol{b}_\alpha$ were derived *via* a Lagrange multiplier procedure.[34] When it is said that all filters are orthonormal, this means that

$$\boldsymbol{b}_\alpha^{\mathrm{T}}\boldsymbol{C}\boldsymbol{b}_\beta = \delta_{\alpha\beta}, \tag{14}$$

which is another way of saying that all filter vectors are uncorrelated. Each component of $\boldsymbol{b}_\alpha$ is weighted:

• inversely by the size of the uncertainties associated with each species, as encoded by the covariance matrix.

• by the sensitivity of the species' abundance to the value of the binding energy, which is represented by the $\boldsymbol{Y}_{,\alpha}$.

If one wished to obtain a ranking of species in terms of their importance in helping constrain binding energies, one would need to come up with a 'score' for each species. Recall that as the magnitude of each component of $\boldsymbol{b}_\alpha$ is a weight for that species' influence on the parameter α, one would need to sum over the absolute values of the components of $\boldsymbol{b}_\alpha$ for species across all α. That is, we perform the sum over our linear filters

$$\sum_{\alpha=1}^{p}\left[\left|b_\alpha^1\right|, \left|b_\alpha^2\right| \ldots, \left|b_\alpha^M\right|\right]. \tag{15}$$

We now have a "filter sum" for each of the M species in our network, which serves as a means of comparing the importance of each species in helping us better constrain binding energy distributions. A species with a larger filter sum will have a larger influence in helping constrain the p binding energies.

2.2.3 Machine learning interpretability. The previous methods explore the influence of the abundances on the values of the binding energies. This is an inverse problem. In order to tackle the forward problem of assessing the impact of the binding energies on the abundances instead, one needs to use a different set of methods.

As UCLCHEM solves a system of coupled ordinary differential equations, it stands to reason that the relationship between the input parameters (the binding energies) and the output parameters (the abundances of species) is non-linear. As such, the relationship between the input and output is not necessarily intuitive and is likely to be different for various 'binding energy regimes'. We make use of machine learning interpretability to help uncover this relationship.

In order to better understand the relationships between the inputs and outputs, we utilise SHapley Additive exPlanations (SHAP).[37] SHAP approximates Shapley values: these are measures of the marginal contribution of a feature to the output value, relative to the mean value of all outputs in the dataset.[38] This is done by considering various coalitions of feature values. A coalition of features represents all subsets of the total set of features. The Shapley value of a feature represents the average change in the prediction when that feature is included in the coalition of features selected. This change is assessed by considering the change in the prediction when the feature is included, averaged over all coalitions.[39] However, this becomes computationally unfeasible as the number of features grows, as the number of subsets grows exponentially with the number of features. SHAP is particularly useful, as it approximates the Shapley values, greatly reducing the time taken to compute them. This is done through the use of the TreeSHAP algorithm.[40]

500 000 data points were created from UCLCHEM by using a Latin Hypercube sampling scheme[41] implemented with the help of the Python surrogate modelling toolbox.[42] We employ the XGBoost Python package§ to build an XGBoost regressor[43] that is made to fit the relationship between the input parameters and the output abundance for each species.

3 Results

3.1 Results of the Bayesian inference

At first, the Bayesian inference was run using the original dataset. However, it was found that despite running the inference in parallel using MPI over 128 cores, that there was no convergence, even after several days. This was attributed to the fact that the model struggled to match the constraints. Many of these constraints have very low relative error, compared to the data used in previous works which typically had relative errors of the order of 50%.[11,15,17] A nested sampler will move from areas of low likelihood to areas of high likelihood. However, if the model struggles to find combinations of parameters that lead to a higher likelihood, then it will inevitably take longer to perform the inference. To properly run the inference, a significantly larger computing cluster would be required. As an alternative, we decided to investigate how the relative error, ε, impacted the obtained posterior probability distributions. We used values of 0.5, 0.33 and 0.25

§ https://xgboost.readthedocs.io/en/stable/index.html.

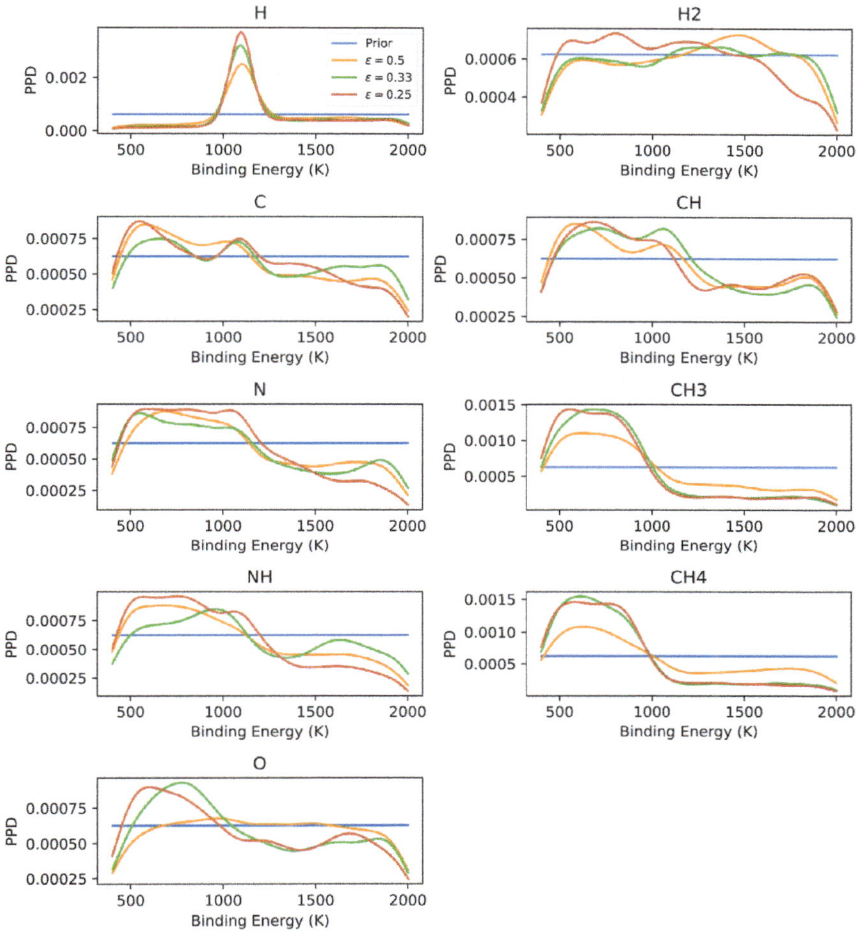

Fig. 1 Marginalised posterior distributions of the binding energies of the diffusive species we consider of interest in this work. We also plot the uniform prior distribution. Only H's binding energy marginalised posterior distribution differs significantly from the prior distribution. For the other binding energies, there is less difference. This is due to the lack of enough sufficiently constraining data. We also observe that decreasing the value of ε in general decreases the variance of the distribution. Both of these points motivate the need for further ice observations to reduce the variance of the distributions.

and ran the inference each time. Our results are displayed in Fig. 1. Also plotted are the prior distributions.

We observe that with the exception of hydrogen's binding energy, the binding energy posteriors are prior-dominated. However, it can also be seen that a decrease in the relative error of the data appears to be accompanied by a decrease in the variance of some of the posteriors, such as for CH_3, CH_4, NH and O. This is consistent with lower variance posteriors for H and O binding energies with the artificially reduced uncertainties for H_2O observation in prior work.[18] However, even in this scenario we are finding that our posteriors have a relatively large variance. The best way to address this is to figure out which other species we should observe to further constrain the distributions.

3.2 Using the MOPED algorithm

We now look to analyse the results of the MOPED algorithm. The fiducial model we use is the one with $\varepsilon = 0.25$. In Fig. 2 we plot the filter sums for each species to provide us with an initial ranking. We only consider species formed on the grains. As the UCLCHEM code models both the bulk and the surface abundances, we sum the abundances of each species on the surface as well as in the bulk to provide us with a total abundance on the ices.

However, in order to inform future ice observations, it would be useful to also consider the likely abundances of each species. Ideally, we would wish to observe

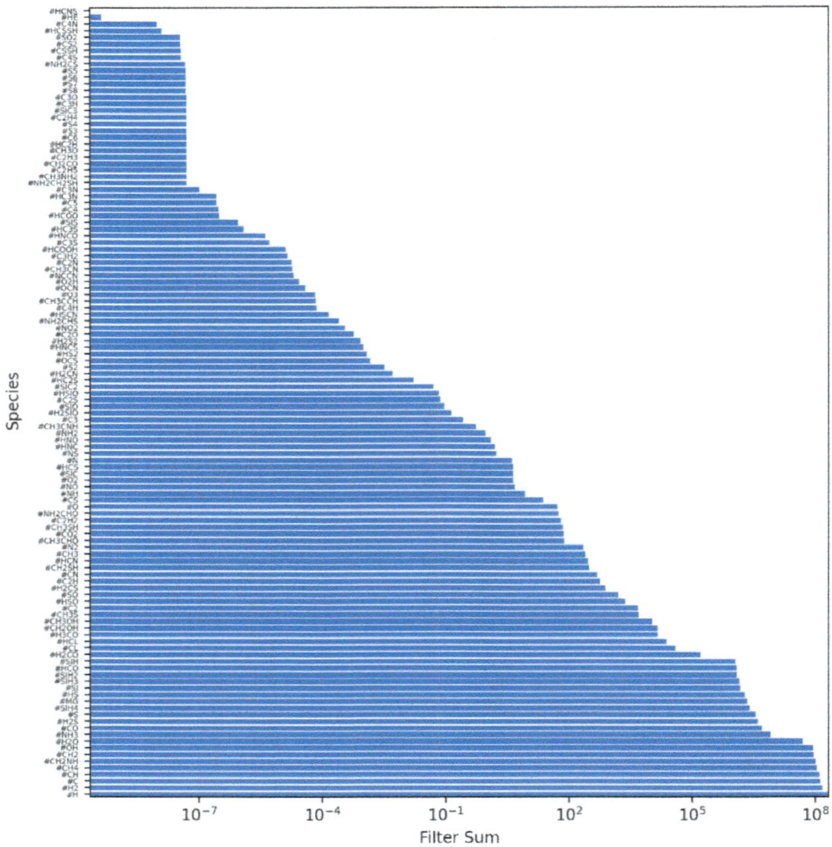

Fig. 2 Bar chart displaying the filter sums for all grain-surface species. Species with a larger filter sum are higher priority detection targets, as they are more affected by the binding energies of the species we consider. Some of the highest-ranked species have already been detected, which potentially implies that future observations should aim to improve the level of precision of these abundance measurements.

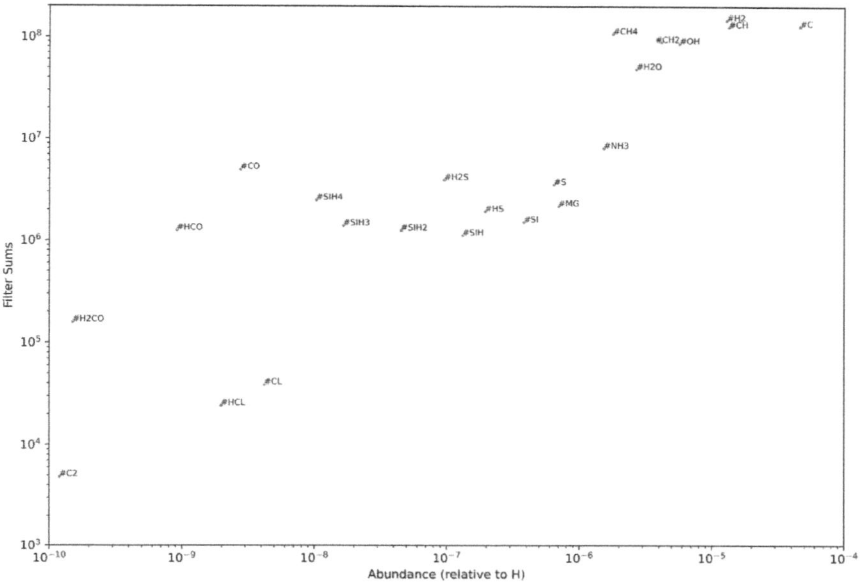

Fig. 3 Scatter plot depicting filter sum against the predicted abundances when the maximum-likelihood estimate for the binding energies is input into UCLCHEM. Given constraints on instrumental uncertainties, we should look to prioritise species that are not only important, as determined by their filter sums, but that can also be realistically detected. These include saturated species such as #CH_4, #NH_3, #SiH_4, #H_2S and #H_2O, as well as their precursors. We find that many of the species we observe are the intermediate species formed during the creation of the saturated species in Table 1. This indicates that understanding these intermediate products is essential to better constraining the binding energies of interest.

species that are highly abundant and that have large filter sums. The first requirement means it is easier to observe a species given a particular instrumental uncertainty, whilst the second ensures that we are observing species that are dependent on the binding energies and are therefore relevant to the chemistry we are considering. To do this, we plot the filter sum of each species against the abundance produced when we use binding energies equal to $\boldsymbol{E}_{\mathbf{ML}}$. The resulting plot is shown in Fig. 3. We only consider species with an abundance greater than 10^{-10} relative to H, as anything less abundant is unlikely to be detected in the ices. As in previous work,[18] we observe that the species H_2O, CH_4, NH_3, H_2S, SiH_4, CO and H_2CO are amongst the highest-ranked species with abundances that are predicted to be detectable. These species all have modes in the range considered by JWST. Unlike in previous work, however, CO_2, CH_3OH and HCN are not amongst the most significant species. This can be attributed to the fiducial model, as we used different constraints, which lead to the maximum-likelihood estimate being different.

3.3 Insights from the machine learning interpretability

Previously, we considered the impact of the data, *i.e.* the species abundances, on the binding energy values and their distributions. We now wish to consider the opposite situation, which is the impact of the binding energy values on the final steady-state abundances of molecules of interest. This is important to consider as the binding energy of a species can be dependent on the ice-composition as well as on the individual sites.[44,45]

In the interest of brevity, we consider a subset of the molecules so as to demonstrate the effectiveness of this approach as a proof-of-concept. We are interested in better understanding the importance of each of the features in predicting the final abundance of a species of interest, as well as the relative importances of the features. Fig. 4 and 5 are so-called beeswarm plots for H_2O and CO respectively. The features are listed from top to bottom in decreasing order of importance to the model output. Along the horizontal axis, individual predictions are plotted in terms of their SHAP value. Recall that the SHAP value states the difference in the value of the model output for that prediction relative to the global average. Furthermore, the points are colour-coded in terms of the size of the feature value. From this, we can attempt to better understand the directionality of each feature's relationship with the output.

From the beeswarm plots, we can make a number of comments about which binding energies are most relevant for that species. For example, H_2O is unsurprisingly dependent on the H and O. Others seems less intuitive, such as CO's strong dependence on the H binding energy or CO_2's dependence on nitrogen. These can typically be reasoned out by considering the chemical network used.

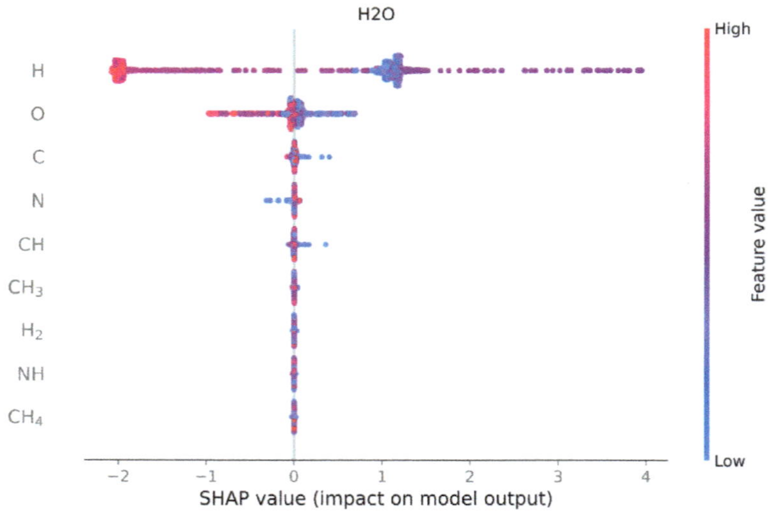

Fig. 4 A beeswarm plot for the statistical emulator trained to predict H_2O's abundance. The features are listed from top to bottom in decreasing order of importance to the model output. Along the horizontal axis, individual predictions are plotted in terms of their SHAP value, that is the change to the log-abundance relative to the average value in the dataset. Recall that the SHAP value states the difference in the value of the model output for that prediction relative to the global average. Furthermore, the points are colour-coded in terms of the size of the feature value. We observe that the binding energies of H and O are the most important features. This makes sense, as both species are necessary to form water *via* successive hydrogenations of an oxygen atom.

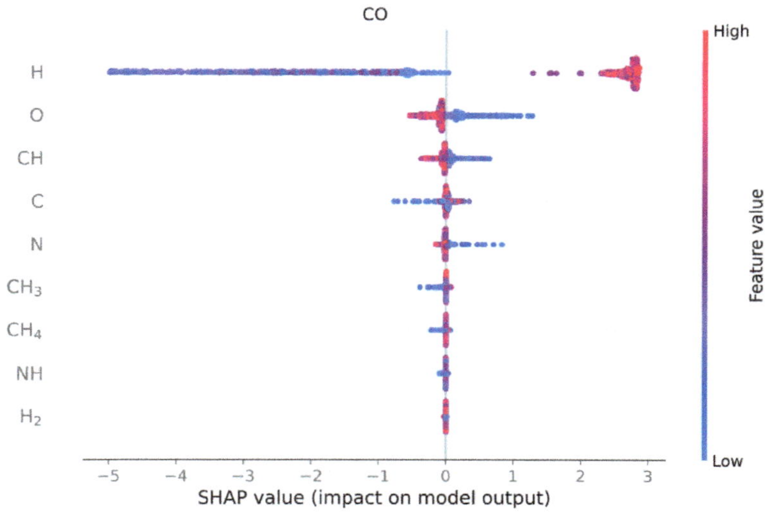

Fig. 5 A beeswarm plot for the statistical emulator trained to predict CO's abundance. The features are listed from top to bottom in decreasing order of importance to the model output. Along the horizontal axis, individual predictions are plotted in terms of their SHAP value, that is the change to the log-abundance relative to the average value in the dataset. Recall that the SHAP value states the difference in the value of the model output for that prediction relative to the global average. Furthermore, the points are colour-coded in terms of the size of the feature value. We observe that the binding energies of H and O are the most important features. Increasing H's binding energy appears to increase CO's abundance, which can be attributed to a decrease in the efficiency of the hydrogenation of CO.

We can also consider the exact nature of the relationship between the features and the final abundance. To do this, we consider the partial dependence of specific variables relative to the output variable. The partial dependence is defined as the marginal effect of one or several features on the output of a machine learning model.[39,46] To demonstrate the utility of the partial dependence, we consider H_2O and CO. Both of these molecules are largely dependent on two binding energies: that of H and O. We plot their 1-D and 2-D partial dependences in Fig. 6 and 7. Note that the y-axis of the 1-D plots are simply the log-abundance of the respective species.

We observe that for water, there is a small area of parameter space in which the abundance peaks. This roughly matches the maximum-posterior hydrogen binding energy value obtained in Fig. 1. Despite the oxygen's binding energy being the second most important feature, we observe that over the range of binding energies considered, it has far less impact in changing the obtained water abundance. Even so, the parameter favoring binding energies lower than ~1000 K for oxygen is consistent with the posterior for the inverse problem.

We can make a similar comment about carbon monoxide. The abundance peaks for hydrogen binding energy values greater than 1100 K. This makes sense, as having too low a binding energy for hydrogen would result in CO being hydrogenated efficiently. For binding energies above 600 K for oxygen, we notice a slight decrease in the abundance of carbon monoxide.

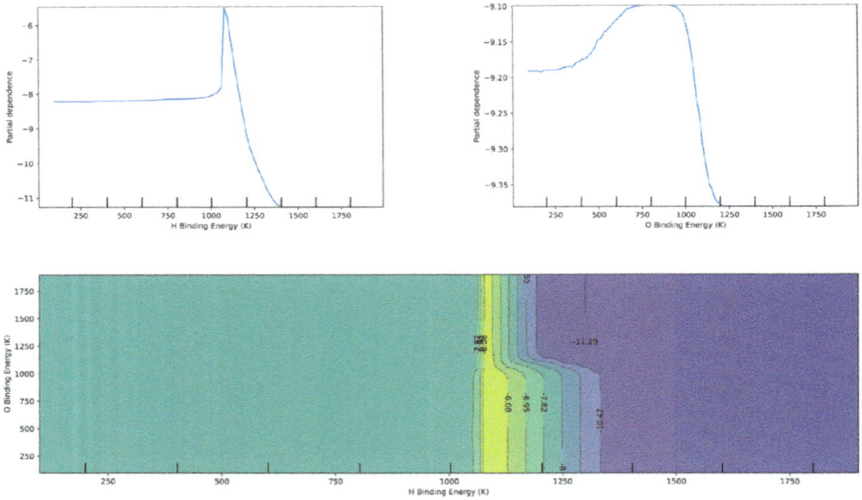

Fig. 6 Top: A plot of the 1-D partial dependence plots of the binding energies of H and O for water. The partial dependence represents the expected value of the log-abundance of water as a function of the variable in features, marginalised over all other features. We observe that for a narrow range of atomic hydrogen's binding energies at around 1100 K, there is a sharp increase in the abundance of water. This is roughly the point at which the marginalised posterior distribution for H's binding energy in Fig. 1 peaks. The dependence for O's binding energy shows a similar consistency with the posteriors, having a clear preference for energies smaller than ~1000 K. Bottom: A 2-D partial dependence plot for the binding energies of H and O. Yellow represents the region with the highest abundance of water.

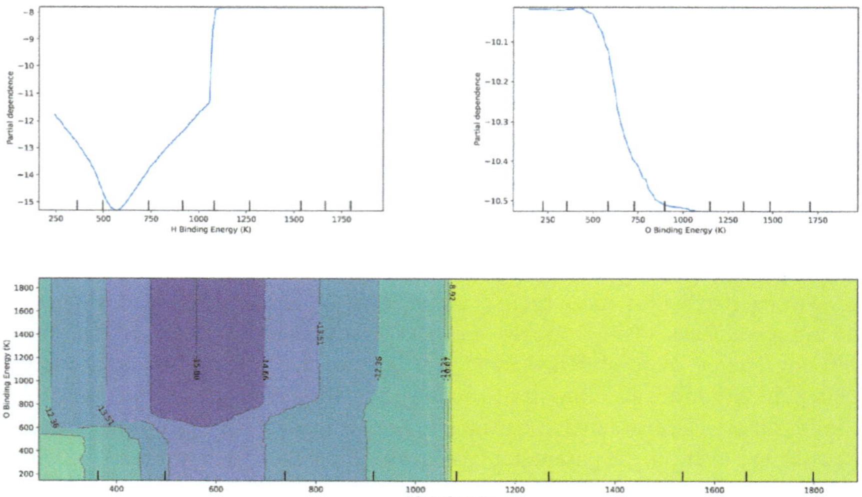

Fig. 7 Top: A plot of the 1-D partial dependence plots of the binding energies of H and O for CO. The partial dependence represents the expected value of the log-abundance of CO as a function of the variable in features, marginalised over all other features. Bottom: A 2-D partial dependence plot for the binding energies of H and O. Yellow represents the region with the highest abundance of CO.

4 Conclusions

In this article we focus our attention on the estimation of binding energies, key parameters in the interaction among surface reactions in ice. We use three statistical approaches to estimate binding energies, prioritise future ice species to be observed, and to understand better the non-linear relationship between binding energies and abundances of such species. Our conclusions can be summarized as followed:

• As in our previous work, we find that Bayesian inference can be a very useful tool to constrain binding energies. However further ice observations are needed in order to reduce the variance of the distributions.

• Indeed, the MOPED algorithm can help towards the prioritization of such observations. As in previous work, we find that solid H_2O, CH_4, NH_3, H_2S, SiH_4, CO and H_2CO are the most important species to observe; surprisingly ice observations of CO_2, CH_3OH and HCN are not amongst the most significant species.

• Using SHAP we establish the key relationships between binding energies and the abundances of the ice species. For example, we find that for water and CO the key parameter is the hydrogen binding energy, and to a much lesser extent the oxygen one. Prioritizing which binding energies are keys for the potentially observable species may be of use in prioritizing experiments and calculations of such energies to reduce their errors.

Probabilistic methodologies as well as Machine Learning methods have now started to be used to solve astrochemical problems. As larger chemical reaction networks and more complex models are being employed in astrochemistry, statistical methods and machine learning (ML) techniques will become ever more necessary in order to reduce the uncertainty in such networks.

Author contributions

Johannes Heyl: conceptualization, data curation, formal analysis, validation, writing – original draft, writing – review & editing. Serena Viti: conceptualization, data curation, formal analysis, writing – original draft, writing – review & editing. Gijs Vermariën: writing – review & editing.

Conflicts of interest

There are no conflicts to declare.

Acknowledgements

J. Heyl is funded by an STFC studentship in Data-Intensive Science (grant number ST/P006736/1). This work was also supported by European Research Council (ERC) Advanced Grant MOPPEX 833460.

Notes and references

1 T. G. Bisbas, T. A. Bell, S. Viti, M. J. Barlow, J. Yates and M. Vasta, *Mon. Not. R. Astron. Soc.*, 2014, **443**, 111–121.

2 S. Viti, S. García-Burillo, A. Fuente, L. K. Hunt, A. Usero, C. Henkel, A. Eckart, S. Martin, M. Spaans, S. Muller, F. Combes, M. Krips, E. Schinnerer, V. Casasola, F. Costagliola, I. Marquez, P. Planesas, P. P. van der Werf, S. Aalto, A. J. Baker, F. Boone and L. J. Tacconi, *Astron. Astrophys.*, 2014, **570**, A28.

3 M. V. Kazandjian, I. Pelupessy, R. Meijerink, F. P. Israel, C. M. Coppola, M. J. F. Rosenberg and M. Spaans, *Astron. Astrophys.*, 2016, **595**, A124.

4 K. Y. Huang, S. Viti, J. Holdship, S. García-Burillo, K. Kohno, A. Taniguchi, S. Martn, R. Aladro, A. Fuente and M. Sánchez-García, *Astron. Astrophys.*, 2022, **666**, A102.

5 A. Makrymallis and S. Viti, *Astrophys. J.*, 2014, **794**, 45.

6 C. Lefèvre, L. Pagani, M. Juvela, R. Paladini, R. Lallement, D. J. Marshall, M. Andersen, A. Bacmann, P. M. McGehee, L. Montier, A. Noriega-Crespo, V. M. Pelkonen, I. Ristorcelli and J. Steinacker, *Astron. Astrophys.*, 2014, **572**, A20.

7 D. de Mijolla, S. Viti, J. Holdship, I. Manolopoulou and J. Yates, *Astron. Astrophys.*, 2019, **630**, A117.

8 J. Holdship, S. Viti, T. J. Haworth and J. D. Ilee, *Astron. Astrophys.*, 2021, **653**, A76.

9 T. Grassi, F. Nauman, J. P. Ramsey, S. Bovino, G. Picogna and B. Ercolano, *Astron. Astrophys.*, 2022, **668**, A139.

10 D. de Mijolla, J. Holdship, S. Viti and J. Heyl, *Astrophys. J.*, submitted.

11 J. Holdship, N. Jeffrey, A. Makrymallis, S. Viti and J. Yates, *Astrophys. J.*, 2018, **866**, 116.

12 T. Grassi, S. Bovino, F. A. Gianturco, P. Baiocchi and E. Merlin, *Mon. Not. R. Astron. Soc.*, 2012, **425**, 1332–1340.

13 R. Xu, X.-N. Bai, K. Öberg and H. Zhang, *Astrophys. J.*, 2019, **872**, 107.

14 T. Grassi, S. Bovino, D. Schleicher and F. A. Gianturco, *Mon. Not. R. Astron. Soc.*, 2013, **431**, 1659–1668.

15 J. Heyl, S. Viti, J. Holdship and S. M. Feeney, *Astrophys. J.*, 2020, **904**, 197.

16 K. S. Tang and M. Turk, *arXiv e-prints*, 2022, preprint, arXiv:2207.07159, DOI: 10.48550/arXiv.2207.07159.

17 J. Heyl, J. Holdship and S. Viti, *Astrophys. J.*, 2022, **931**, 26.

18 J. Heyl, E. Sellentin, J. Holdship and S. Viti, *Mon. Not. R. Astron. Soc.*, 2022, **517**, 38–46.

19 J. Holdship, S. Viti, I. Jiménez-Serra, A. Makrymallis and F. Priestley, *Astron. J.*, 2017, **154**, 38.

20 M. K. McClure, W. R. M. Rocha, K. M. Pontoppidan, N. Crouzet, L. E. U. Chu, E. Dartois, T. Lamberts, J. A. Noble, Y. J. Pendleton, G. Perotti, D. Qasim, M. G. Rachid, Z. L. Smith, F. Sun, T. L. Beck, A. C. A. Boogert, W. A. Brown, P. Caselli, S. B. Charnley, H. M. Cuppen, H. Dickinson, M. N. Drozdovskaya, E. Egami, J. Erkal, H. Fraser, R. T. Garrod, D. Harsono, S. Ioppolo, I. Jiménez-Serra, M. Jin, J. K. Jørgensen, L. E. Kristensen, D. C. Lis, M. R. S. McCoustra, B. A. McGuire, G. J. Melnick, K. I. Ã-berg, M. E. Palumbo, T. Shimonishi, J. A. Sturm, E. F. van Dishoeck and H. Linnartz, *Nat. Astron.*, 2023, **7**, 431–443.

21 T. I. Hasegawa, E. Herbst and C. M. Leung, *Astrophys. J., Suppl. Ser.*, 1992, **82**, 167.

22 Q. Chang, H. M. Cuppen and E. Herbst, *Astron. Astrophys.*, 2007, **469**, 973–983.

23 R. T. Garrod and T. Pauly, *Astrophys. J.*, 2011, **735**, 15.

24 D. McElroy, C. Walsh, A. J. Markwick, M. A. Cordiner, K. Smith and T. J. Millar, *Astron. Astrophys.*, 2013, **550**, A36.

25 V. Wakelam, J. C. Loison, R. Mereau and M. Ruaud, *Mol. Astrophys.*, 2017, **6**, 22–35.

26 D. Quénard, I. Jiménez-Serra, S. Viti, J. Holdship and A. Coutens, *Mon. Not. R. Astron. Soc.*, 2018, **474**, 2796–2812.

27 J. He, K. Acharyya and G. Vidali, *Astrophys. J.*, 2016, **825**, 89.

28 S. Ferrero, L. Zamirri, C. Ceccarelli, A. Witzel, A. Rimola and P. Ugliengo, *Astrophys. J.*, 2020, **904**, 11.

29 T. Villadsen, N. F. W. Ligterink and M. Andersen, *Astron. Astrophys.*, 2022, **666**, A45.

30 J. C. Laas and P. Caselli, *Astron. Astrophys.*, 2019, **624**, A108.

31 J. Buchner, *J. Open Source Softw.*, 2021, **6**, 3001.

32 J. Buchner, *Stat. Comput.*, 2016, **26**, 383–392.

33 J. Buchner, *Publ. Astron. Soc. Pac.*, 2019, **131**, 108005.

34 A. F. Heavens, R. Jimenez and O. Lahav, *Mon. Not. R. Astron. Soc.*, 2000, **317**, 965–972.

35 A. F. Heavens, E. Sellentin, D. de Mijolla and A. Vianello, *Mon. Not. R. Astron. Soc.*, 2017, **472**, 4244–4250.

36 A. F. Heavens, E. Sellentin and A. H. Jaffe, *Mon. Not. R. Astron. Soc.*, 2020, **498**, 3440–3451.

37 S. M. Lundberg and S.-I. Lee, A Unified Approach to Interpreting Model Predictions, *Proceedings of the 31st International Conference on Neural Information Processing Systems*, Curran Associates Inc., Red Hook, NY, USA, 2017, pp. 4768–4777.

38 L. S. Shapley, in *17. A Value for n-Person Games*, ed. H. W. Kuhn and A. W. Tucker, Princeton University Press, Princeton, 2016, pp. 307–318.

39 C. Molnar, *Interpretable Machine Learning*, 2nd edn, 2022.

40 S. M. Lundberg, G. G. Erion and S.-I. Lee, *arXiv e-prints*, 2018, preprint, arXiv:1802.03888, DOI: 10.48550/arXiv.1802.03888.

41 M. D. McKay, R. J. Beckman and W. J. Conover, *Technometrics*, 1979, **21**, 239–245.

42 M. A. Bouhlel, J. T. Hwang, N. Bartoli, R. Lafage, J. Morlier and J. R. R. A. Martins, *Adv. Eng. Softw.*, 2019, 102662.

43 T. Chen and C. Guestrin, *arXiv e-prints*, 2016, preprint, arXiv:1603.02754, DOI: 10.48550/arXiv.1603.02754.

44 T. Grassi, S. Bovino, P. Caselli, G. Bovolenta, S. Vogt-Geisse and B. Ercolano, *Astron. Astrophys.*, 2020, **643**, A155.

45 A. Das, M. Sil, R. Ghosh, P. Gorai, S. Adak, S. Samanta and S. K. Chakrabarti, *Front. Astron. Space Sci.*, 2021, **8**, 78.

46 J. H. Friedman, *Ann. Stat.*, 2001, **29**, 1189–1232.

Faraday Discussions

ROYAL SOCIETY OF CHEMISTRY

PAPER

Disentangling physics and chemistry in AGB outflows: revealing degeneracies when adding complexity†

Marie Van de Sande, *[a] Catherine Walsh [a] and Tom J. Millar [b]

Received 15th February 2023, Accepted 14th April 2023

DOI: 10.1039/d3fd00039g

Observations of the outflows of asymptotic giant branch (AGB) stars continue to reveal their chemical and dynamical complexity. Spherical asymmetries, such as spirals and disks, are prevalent and thought to be caused by binary interaction with a (sub)stellar companion. Furthermore, high density outflows show evidence of dust–gas interactions. The classical chemical model of these outflows – a gas-phase only, spherically symmetric chemical kinetics model – is hence not appropriate for the majority of observed outflows. We have included several physical and chemical advancements step-by-step: a porous density distribution, dust–gas chemistry, and internal UV photons originating from a close-by stellar companion. Now, we combine these layers of complexity into the most chemically and physically advanced chemical kinetics model of AGB outflows to date. By varying over all model parameters, we obtain a holistic view of the outflow's composition and how it (inter)depends on the different complexities. A stellar companion has the largest influence, especially when combined with a porous outflow. We compile sets of gas-phase molecules that trace the importance of dust–gas chemistry and allow us to infer the presence of a companion and porosity of the outflow. This shows that our new chemical model can be used to infer physical and chemical properties of specific outflows, as long as a suitable range of molecules is observed.

1 Introduction

During the asymptotic giant branch (AGB) phase, stars with an initial mass up to 8 $M_{\odot}$ lose their outer layers by means of a stellar outflow or wind. This process of

[a]School of Physics and Astronomy, University of Leeds, Leeds LS2 9JT, UK. E-mail: m.vandesande@leeds.ac.uk

[b]Astrophysics Research Centre, School of Mathematics and Physics, Queen's University Belfast, University Road, Belfast BT7 1NN, UK

† Electronic supplementary information (ESI) available: Fractional abundance profiles of all parents and their daughters along with all species that show a significant change in abundance when including a stellar companion, together with figures showing the number of monolayers of dust and ice for all assumed GSDs and the ice and refractory coverage at the end of the outflow for the GSDs different to the MRN distribution. See DOI: https://doi.org/10.1039/d3fd00039g

mass-loss is efficient, with mass-loss rates between 10^{-8} and 10^{-4} $M_\odot$ y^{-1}, and determines the star's remaining lifetime rather than exhaustion of nuclear fuel in its core. The outflow is thought to be launched by a two-step mechanism: stellar pulsations facilitate the formation of dust grains, subsequently launching a dust-driven wind.[1,2] Thanks to their outflows, AGB stars are important contributors to the chemical enrichment of the interstellar medium (ISM), contributing about 80% of the stellar gas injection rate[3] and about 70% of the total stellar dust production rate.[4]

The chemical content of the outflow is determined by the elemental carbon-to-oxygen ratio (C/O) of the AGB star, linked to its initial mass and evolutionary stage. Stars with C/O < 1 give rise to an oxygen-rich (O-rich) outflow, while stars with C/O > 1 have a carbon-rich (C-rich) outflow. The outflows of AGB stars are rich astrochemical laboratories: close to 100 different molecules and some 15 types of newly formed dust species have been detected so far. Moreover, they are host to different types of chemistry thanks to the large gradients in density and temperature present throughout the outflow. Three main regions can be distinguished: non-equilibrium chemistry caused by shocks due to stellar pulsations close to the star, followed by dust condensation and dust–gas interaction, to photochemistry initiated by interstellar UV photons in the tenuous outer region.[5]

Chemical models of AGB outflows are divided into two main groups, dealing with the non-equilibrium chemistry close to the star[6–9] or with the photochemistry of the outer wind.[10–13] Both types of models include gas-phase chemistry only and assume a spherically symmetric outflow. Observations continue to reveal discrepancies with model predictions, putting an increasingly larger strain on these two assumptions. Spherically asymmetric outflows are common: both small-scale inhomogeneities or clumps[14–16] and large-scale structures such as spirals[17,18] and disks,[19,20] are widely observed. These large-scale structures are thought to be caused by binary interaction with a (sub)stellar companion. Most AGB stars with mass-loss rates larger than $\sim 10^{-7}$ $M_\odot$ y^{-1} will have at least one planetary and/or stellar companion.[21] The potential presence of another star complicates the chemical picture even further as it produces an internal UV radiation field.

Moreover, in both O-rich and C-rich outflows, the abundance of refractory species (*e.g.*, SiO, SiS, and CS) is observed to decrease with increasing outflow density, hinting towards depletion onto dust grains.[22,23] The abundance of SiO and SiS is observed to decrease in the intermediate wind of high density O-rich outflows, before the onset of dust formation,[24,25] and H_2O ice has been detected around OH/IR stars.[26]

To better explain observations, we have increased the physical and chemical complexity of earlier gas-phase only models through the inclusion of: a clumpy density distribution,[28] dust–gas chemistry,[29–31] and internal UV photons originating from a close-by stellar companion.[32,33] This was done step-by-step, adding separate layers of complexity to investigate their effects on the composition of the outflow in isolation. While clumpy outflows were considered when including internal UV photons, the interdependencies of the other complexities have not yet been explored. To this end, we have combined all previous developments in the most physically and chemically complex chemical model of AGB outflows to date. This increases the model's resolution by extending the chemical validity of the model from the outer wind into the dust interaction zone and increasing its applicability to spherically asymmetric outflows.

Exploring the interplay of the model developments enables us to determine how the chemistry of the outflow depends on its structure and the presence of a (sub)stellar companion. By maintaining a holistic view of the chemical composition throughout the outflow and the different chemical and physical processes included, we are able to determine a suite of molecules that indicate the presence of (a combination of) these complexities. This makes the chemical model a powerful tool to interpret observations and reveal the hidden cause of dynamical structures.

We describe the chemical model in Section 2, along with the different complexities and their limitations. We present the results for C-rich and O-rich outflows in Section 3. Discussion and conclusions follow in Section 4 and 5.

2 AGB outflow chemical model

The one-dimensional chemical kinetics model is based on the publicly available UDfA circumstellar envelope model.[‡,34] It describes a spherically symmetric outflow with constant mass-loss rate, $\dot{M}$, and expansion velocity, v_∞. The temperature profile of the gas is described by a power law[28]

$$T(r) = T_* \left(\frac{r}{R_*}\right)^{-\varepsilon}, \tag{1}$$

with T_* and R_* the stellar temperature and radius, respectively, and r the distance from the centre of the star. CO self-shielding is taken into account using a single-band approximation;[35] H_2 is assumed to be fully self-shielded.

Table 1 lists the different fixed and variable model parameters along with their values adopted in this work. For each configuration, we calculate an O-rich and a C-rich outflow. The parent species for the O-rich and C-rich outflows are listed in Table 2. These species are assumed to be present at the start of the model; they are taken from Agúndez *et al.*,[36] who compiled (ranges of) observed abundances in the inner region. We consider three outflow densities, determined by the mass-loss rate, $\dot{M}$ and the expansion velocity, v_∞. The value of v_∞ and the drift velocity v_{drift} for each mass-loss rate is estimated from observations.[22,37–40]

We outline important features and parameters of the physical and chemical developments along with our previous main findings in Section 2.1 and 2.2, respectively. For additional details, we refer to the original works. In Section 2.3, we elaborate on the assumptions made when implementing the developments and the limitations of the model. The different complexities included in the chemical model are illustrated in Fig. 1.

2.1 Physical complexity

Clumps are taken into account using a porosity formalism, which divides the outflow into a stochastic two-component medium of overdense clumps within a rarified interclump component.[41–44] The formalism allows for a one-dimensional approximation of a clumpy outflow by including the effects of the

‡ http://udfa.ajmarkwick.net/index.php?mode=downloads

Table 1 Physical parameters of the grid of chemical models

Fixed parameters	
Physical parameters	
Stellar radius, R_*	5×10^{13} cm
Stellar temperature, T_*	2330 K
Exponent $T(r)$, ε	0.7
Dust parameters	
O-rich dust composition	50/50 melilite and silica with Fe
C-rich dust composition	Amorphous carbon
Dust-to-gas mass ratio	2×10^{-3}
Surface density of binding sites	10^{15} cm^{-2}
Silicate dust bulk density[27]	3.5 g cm^{-3}
Carbonaceous dust bulk density[27]	2.24 g cm^{-3}
Variable parameters	
Physical parameters	
Initial radius	$1.025 \times R_{dust}$
Grid of outflow densities	
Mass-loss rate $\dot{M}$	10^{-5} M$_\odot$ y^{-1}; 15 km s^{-1}; 5 km s^{-1}
Expansion velocity, v_∞	10^{-6} M$_\odot$ y^{-1}; 10 km s^{-1}; 10 km s^{-1}
And drift velocity, v_{drift}	10^{-7} M$_\odot$ y^{-1}; 5 km s^{-1}; 15 km s^{-1}
Grid of density structures	
Smooth outflow	
Porous outflows, with	
Interclump density contrast	$f_{ic} = 0.0, 0.1, 0.3, 0.5$
Clump volume filling factor	$f_{vol} = 0.1, 0.3, 0.5$
Size of the clumps at R_*	$l_* = 5 \times 10^{12}, 10^{13}$ cm
Dust parameters	
Grain size distributions	
Minimum grain size, a_{min}	10^{-8} cm; 10^{-6} cm; -3.5
Maximum grain size, a_{max}	5×10^{-7} cm; 2.5×10^{-5} cm; -3.5
And power-law exponent, η	10^{-6} cm; 10^{-4} cm; -4.5
Companion parameters	
Onset of dust extinction, R_{dust}	2, 5 R_*
Companion temp., T_{comp}	4000 K; 1.53×10^{10} cm
And radius, R_{comp}	6000 K; 8.14×10^{10} cm
	10 000 K; 6.96×10^{8} cm

local overdensity within the clumps along with their effect on the outflow's optical depth.

The specific clumpiness of the outflow is described by three parameters: (i) the clump volume filling factor, f_{vol}, setting the fraction of the total volume of the outflow occupied by clumps, (ii) the interclump density contrast, f_{ic}, setting the fraction of the outflow density within the interclump component, and (iii) the size of the clumps at the stellar surface, l_*. Clump size and volume filling factor can be combined into the porosity length $h = l_*/f_{vol}$, which represents the local mean free path between clumps. Highly porous outflows have a large porosity length, due to a large l_* and/or small f_{vol}, combined with a low interclump density contrast.

Table 2 Parent species and their abundances relative to H_2 for the C-rich and O-rich outflows. Abundances are derived from observations, as compiled by Agúndez *et al.*[36]

Carbon-rich		Oxygen-rich	
Species	Abundance	Species	Abundance
He	0.17	He	0.17
CO	8.00×10^{-4}	CO	3.00×10^{-4}
N_2	4.00×10^{-5}	H_2O	2.15×10^{-4}
CH_4	3.50×10^{-6}	N_2	4.00×10^{-5}
H_2O	2.55×10^{-6}	SiO	2.71×10^{-5}
SiC_2	1.87×10^{-5}	H_2S	1.75×10^{-5}
CS	1.06×10^{-5}	SO_2	3.72×10^{-6}
C_2H_2	4.38×10^{-5}	SO	3.06×10^{-6}
HCN	4.09×10^{-5}	SiS	9.53×10^{-7}
SiS	5.98×10^{-6}	NH_3	6.25×10^{-7}
SiO	5.02×10^{-6}	CO_2	3.00×10^{-7}
HCl	3.25×10^{-7}	HCN	2.59×10^{-7}
C_2H_4	6.85×10^{-8}	PO	7.75×10^{-8}
NH_3	6.00×10^{-8}	CS	5.57×10^{-8}
HCP	2.50×10^{-8}	PN	1.50×10^{-8}
HF	1.70×10^{-8}	HCl	1.00×10^{-8}
H_2S	4.00×10^{-9}	HF	1.00×10^{-8}

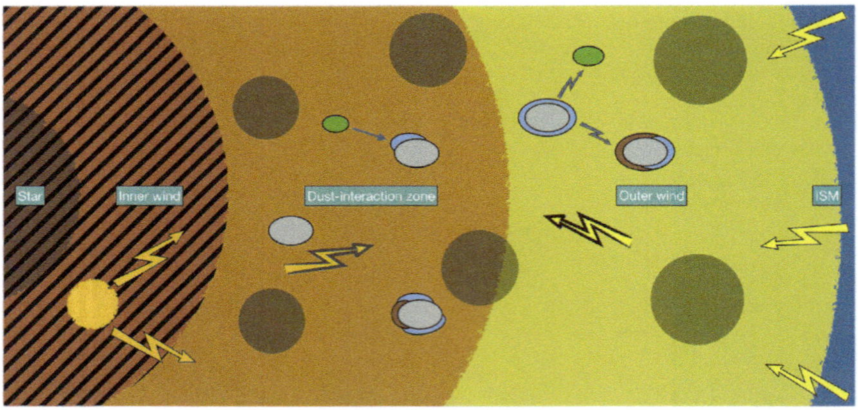

Fig. 1 Overview of an AGB star and its outflow with the different complexities included in the chemical model. The model does not describe the chemistry of the cross-hatched inner wind region. Interstellar UV radiation always irradiates the outer regions of the outflow (yellow lightning bolts). A companion star can be present within the inner wind region, resulting in an inner wind UV radiation field (orange lightning bolts). Including porosity distributes the outflow's material between overdense clumps (opaque black circles) and a rarified interclump medium. This affects extinction throughout the outflow, with interstellar photons reaching closer to the star and potential companion photons reaching further out in the outflow (lightning bolts with black outlines). Dust–gas chemistry leads to the accretion of gas-phase species (green ovals) onto dust grains (grey ovals), forming an ice mantle on top of the dust (blue). Ice species, which may be formed *via* grain-surface chemistry, are released into the gas phase by photodesorption. If the ice is complex, it can be photoprocessed into refractory organic material (red), both by interstellar and companion UV photons.

In our previous work,[28,45] which included gas-phase only chemistry and no internal UV photons, we found that the effects of the local overdensity of clumps is small, while those of the change in optical depth can be considerable. With increasing porosity, parent species are photodissociated closer to the star, though still in the outer region. This leads to larger abundances of daughter species as well as a shift in their peak abundances closer to the star. However, these changes are of the order of one magnitude or less.

2.2 Chemical complexity

The chemical reaction network used is a combination of the dust–gas chemical network used in Van de Sande *et al.*[31] and the inner UV photon reactions of Van de Sande and Millar.[33] We elaborate on these networks and their implementation below.

2.2.1 Dust–gas chemistry. The comprehensive dust–gas chemical reaction network is that of Walsh *et al.*,[46] which includes the original UDfA Rate12 gas-phase only chemical network[34] extended with grain-surface reactions and related reactions, such as needed gas-phase reactions from the 2008 version of the Ohio State University network.[47] Additionally, we included the photoprocessing of complex ices into refractory organic material as described in Van de Sande *et al.*[31] By including these dust–gas reactions, the chemical validity of the model is extended from the outer wind into the intermediate wind.

The network allows for gas-phase species to accrete onto dust grains, forming a physisorbed ice mantle which can be returned to the gas-phase *via* thermal desorption, photodesorption, and mechanical sputtering. Reactions in the ice mantle can occur *via* the diffusive Langmuir–Hinshelwood and stick-and-hit Eley–Rideal mechanisms, which both can give rise to reactive desorption. Icy species can be photodissociated, with the products either remaining on the dust or being returned to the gas-phase. The photoprocessing of ices is limited to complex species only, which we define as species containing at least three C atoms or two C atoms and another heavy atom.

As dust nucleation is not included in the network, dust grains are assumed to be present at the start of the model with a constant density. The dust has a drift velocity, $v_{\rm drift}$ relative to the gas, defined as $v_{\rm drift} = v_{\rm dust} - v_{\infty}$. The grain size distribution (GSD) is fixed and parametrised following the MRN grain-size distribution,[30,48] where the dust grain number density per unit volume is given by

$$n_{\rm dust} = \frac{C}{r^2}\int_{a_{\rm min}}^{a_{\rm max}} a^{\eta}{\rm d}a\ {\rm cm}^{-3}, \tag{2}$$

where $a_{\rm min}$ and $a_{\rm max}$ are the minimum and maximum grain size and η is the slope of the power law. We vary the GSD over the canonical MRN distribution, with $a_{\rm min} = 5 \times 10^{-7}$ cm, $a_{\rm max} = 2.5 \times 10^{-5}$, and $\eta = -3.5$,[48] and GSDs with smaller and larger grains than the MRN distribution (Table 1).

The dust's temperature is approximated by a power law

$$T_{\rm dust} = T_{\rm dust,*}\left(\frac{2r}{R_*}\right)^{2/(4+s)}, \tag{3}$$

where s and $T_{\rm dust,*}$ are free parameters. These parameters are constrained by fitting eqn (3) to the results of a continuum radiative transfer model for different outflow densities and dust compositions. We assume the same dust composition as in Van de Sande *et al.*:[31] a 50/50 mixture of melilite and silicate with Fe for O-rich outflows and amorphous carbon for C-rich outflows.

In our previous work,[29–31] we found that dust–gas chemistry can have a large impact on higher density outflows ($\dot{M} \geq 10^{-6}$ M$_\odot$ y^{-1}). Parent species with higher binding energies are depleted onto dust grain, while daughter species can be formed in the ice mantle and released by photodesorption in the outer regions.[29] The depletion level is influenced by the GSD, where GSDs with larger average surface areas (caused by a smaller $a_{\rm min}$ and/or steeper slopes) lead to larger depletion of parent species.[30] The formation of an ice mantle mainly depends on outflow density and the chemistry of the outflow, with up to 100 monolayers of H_2O-rich ice building up in very high density O-rich outflows and sub-monolayer coverage only (~0.1 monolayers) in C-rich outflows.[29] Refractory organic material is more readily formed in C-rich outflows thanks to a larger abundance of C-rich feedstock ices, forming up to 0.2 monolayers, compared to negligible coverage (~10^{-5} only) for O-rich outflows.

2.2.2 Internal UV photons. The binary fraction of AGB stars strongly depends on the their initial mass. For AGB stars with an initial mass smaller than 1.5 M$_\odot$, ~33% have a stellar companion and ~10% a planetary or brown dwarf companion capable of shaping the outflow. However, these stars only have a mass-loss rate of 10^{-7} M$_\odot$ y^{-1}, the smallest value in our model grid, for a short period. Therefore, observed AGB outflows with $\dot{M} \geq 10^{-7}$ M$_\odot$ y^{-1} will generally have an initial mass larger than 1.5 M$_\odot$, with ~56% having a stellar companion and ~40% a planetary or brown dwarf companion. Roughly 10% of all stellar companions are thought to be white dwarf companions. While these fractions are reduced by roughly 10–20% during the AGB phase due to tidal inspiral and stellar mass-loss, most stars with $\dot{M} \geq 10^{-7}$ M$_\odot$ y^{-1}, as considered here, will have at least one (sub)stellar companion.[21]

Unlike a planetary companion, a stellar companion's UV radiation field initiates UV-driven chemistry in this otherwise shielded inner wind. UV photons originating from either the AGB star itself[32] or a companion star[33] are included in the reaction network as additional photoreactions. UV fluxes are approximated by blackbody radiation, set by the temperature of the AGB star, T*, or the temperature of the companion, $T_{\rm comp}$, together with the radius of the companion, $R_{\rm comp}$. We consider three types of companion: a red dwarf ($T_{\rm comp} = 4000$ K, $R_{\rm comp} = 1.53 \times 10^{10}$ cm), a solar-like star ($T_{\rm comp} = 6000$ K, $R_{\rm comp} = 8.14 \times 10^{10}$ cm), and a white dwarf ($T_{\rm comp} = 10\,000$ K and $R_{\rm comp} = 6.96 \times 10^{8}$ cm). Unshielded photodissociation and photoionisation rates were calculated using cross sections where available, which were mainly taken from the Leiden Observatory Database.§ [49] If not available, the rates were estimated by scaling the unshielded interstellar rate by the ratio of the integrated fluxes of the stellar or companion photons in the 912–2150 Å range. The scaling factors and the corrections for overestimated photoionisation rates can be found in Van de Sande and Millar.[33]

Internal UV photons are diluted geometrically as well as extinguished by dust. The visual extinction caused by dust experienced by internal photons, determined by the outflow density, porosity, and $R_{\rm dust}$, and the intensity of the radiation,

§ https://home.strw.leidenuniv.nl/~ewine/photo/

determined by T_{comp} and R_{comp}. The companion star is assumed to lie within the dust free region before R_{dust}.

While the impact of AGB UV photons is limited because of their low blackbody temperatures,[32] that of companion UV photons can be substantial throughout the entire outflow. Solar-like companions have the largest influence, whilst white dwarf companions have a smaller effect. The impact of red dwarf companions is not significant. The result of the photochemistry induced by companion UV photons depends mostly on the extinction experienced by the photons and the intensity of the radiation. In low extinction outflows, characterised by a low outflow density and/or a high porosity, photoreactions occur faster than two-body reactions. This inhibits the reformation of parent species and chemistry among photoproducts, reducing the outflow to a mostly atomic and ionised state and making it appear apparently molecule-poor. In high extinction outflows, characterised by high outflow density and/or low porosity, two-body reactions occur faster than photoreactions, increasing the chemical complexity of the outflow. Daughter species that are formed in the outer regions can now be produced in the inner regions, resulting in a parent-like abundance profile which consists of a large inner region abundance followed by a Gaussian decline due to photodissociation by interstellar UV photons in the outer envelope.

2.3 Model limitations

Several assumptions are made when introducing the physical and chemical complexities, limiting the one-dimensional model. As the porosity formalism assumes a stochastic mixture of clumps within an interclump medium, the clumps' specific configuration is not needed as input to the model. While this is an advantage for model setup and exploration of the $[f_{vol}, f_{ic}, l_*]$ parameter space, reproducing specific (observed) outflows is not possible. Specific spherical asymmetries can only be approximated.

Dust nucleation is not included in the model: all dust is assumed to present at the start of the model with a fixed GSD. Hence, we assume that dust formation has taken place in the inner region before the start of the model and that GSD is kept constant throughout the calculations, as ices and refractory organic material are physisorbed rather than chemisorbed.

Since the same method of calculating photoreaction rates is used for companion and AGB UV photons, this implies that the companion is located at the centre of the star. However, misplacing the companion by up to 5 R_*, the largest value of R_{dust} (Table 1), is negligible compared to the scale of the outflow. Additionally, the companion's radiation field is continuously present in our one-dimensional model. In Van de Sande and Millar,[33] we argue that occultation of the close-by companion by the AGB star likely has a limited effect on the chemistry, making our model a reasonable first-order approximation of the effects of a companion within the dust forming region.

3 Results

Including different complexities in the chemical model can result in large differences compared to the fiducial gas-phase chemistry only, spherically symmetric outflow model. Both the gas-phase chemistry throughout the outflow and the ice and refractory coverage of its dust grains can be severely affected. We find that including companion UV photons has the largest effect, especially in combination with a porous outflow. The specific influence depends on the outflow's density and porosity, as well as the type of companion included.

The influence of different (combinations of) complexities on the gas-phase composition and the dust's ice and refractory coverage at the end of the outflow is discussed in Section 3.1 and 3.2 for O-rich and C-rich outflows, respectively. The results shown in this Section all assume the MRN distribution for the GSD. Any differences caused by changing the GSD are discussed in Section 4. All fractional abundances are expressed relative to H_2, which follows a r^{-2} distribution.

The ESI† shows the fractional abundance profiles of all parent species and their daughters, along with all species that show a significant change in abundance when including a companion. It also includes the number of monolayers of dust and ice for all assumed GSDs and the ice and refractory coverage at the end of the outflow for the GSDs different to the MRN distribution.

3.1 O-rich outflows

To illustrate the combined effects of the different complexities, we consider two different species: the parent SO, which can show a bumped abundance profile, and the daughter HS, whose abundance profile can change dramatically.

Fig. 2 shows the abundance of the parent SO in outflows with different densities, porosities, and companions. Including porosity and dust–gas chemistry, separately or combined, does not affect its abundance profile significantly. Including a companion only decreases the inner wind abundance in lower density outflows, creating a bumped abundance profile. The largest effects are seen when combining a companion and porosity. For a red dwarf companion, the SO profile becomes bumped, with a decreased inner wind abundance by up to four orders of magnitude followed by a peak in abundance, as outflow density decreases and porosity increases. Higher density outflows with a smooth or low porosity density distribution are not affected. Including a white dwarf leads to similar profiles, with lower porosities now also resulting in a bumped profile in the higher density outflows, with a decrease in the inner wind abundance of up to two orders of magnitude. For $\dot{M} = 10^{-7}\ M_{\odot}\ y^{-1}$, the peak of the bump decreases with increasing porosity. In higher density outflows, a highly porous outflow gives rise to a wider bump, ranging from the inner wind to where SO is photodissociated by interstellar UV photons.

The abundance of the daughter HS is shown in Fig. 3. In the fiducial model, HS is present in a shell in the outer wind, where it is formed *via* photodissociation of the parent H_2S. Including porosity alone leads to a widening of the shell, especially in lower density outflows. Dust–gas chemistry alone has some influence on the inner wind abundance, increasing it by a factor of a few in higher density outflows. Again, including a companion shows the largest influence. Including a red dwarf or white dwarf changes the abundance profile to a high inner wind abundance followed by a sharp decline of at least five orders of magnitude and a bump in abundance in the outer wind. The bump's abundance decreases with increasing porosity, flattening the abundance profile. In high density outflows with a low porosity, we find that the inner wind abundance of HS increases by up

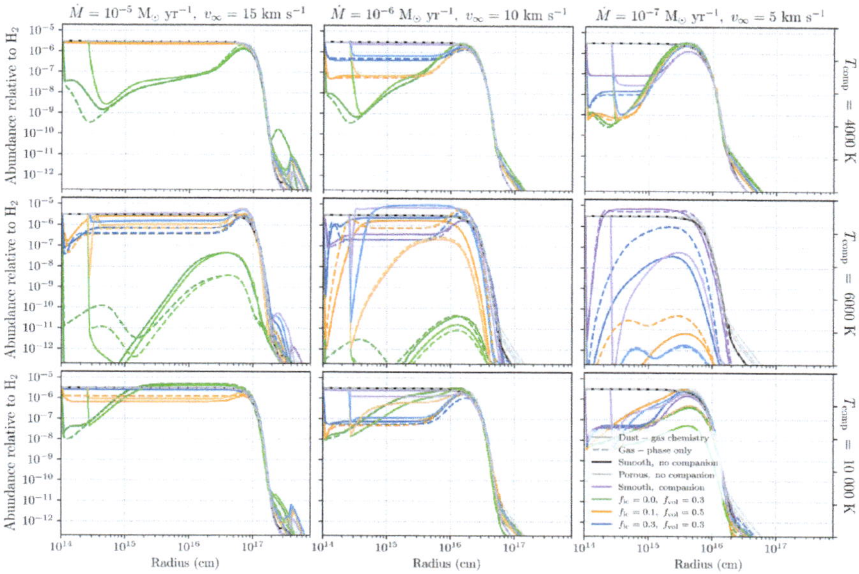

Fig. 2 Fractional abundance of the parent SO relative to H_2 for a selection of O-rich outflows with different outflow densities (columns) and companions (rows). Different line styles show different chemistries: dashed lines: gas-phase chemistry only, solid lines: including dust–gas chemistry. Different colours show different density structures and the presence of a companion: black: smooth outflow without a companion, gray: porous outflow without a companion; purple: smooth outflow with a companion, green: porous outflow with a companion, $f_{ic} = 0.0$, $f_{vol} = 0.3$, $l_* = 10^{13}$ cm, orange: porous outflow with a companion, $f_{ic} = 0.1$, $f_{vol} = 0.5$, $l_* = 10^{13}$ cm, blue: porous outflow with a companion, $f_{ic} = 0.3$, $f_{vol} = 0.3$, $l_* = 5 \times 10^{12}$ cm. When including a companion, different shades correspond to the value of R_{dust}: darker shade: $R_{dust} = 2\ R_*$, lighter shade: $R_{dust} = 5\ R_*$.

to two orders of magnitude, even making it appear parent-like in a smooth outflow with $\dot{M} = 10^{-6}$ M$_\odot$ y^{-1}. For a solar-like companion, HS is not efficiently reformed, lowering its abundance in the outer wind. In lower density, high porosity outflows, the abundance profile is flattened. Dust–gas chemistry leads to a large difference of more than five orders of magnitude compared to gas-phase only chemistry in porous, high density outflows, with HS severely decreased in the inner regions when including dust–gas chemistry.

The number of ice monolayers covering the dust at the end of the outflow is shown in Fig. 4. We take the end of the outflow where the H_2 number density is equal to 10 cm^{-3}. In outflows with $\dot{M} = 10^{-5}$ M$_\odot$ y^{-1}, porosity increases the ice coverage by up to an order of magnitude, from a coverage of ~1% to ~10%. In lower density outflows, this ice coverage increases by up to an order of magnitude, but remains low: we find an ice coverage of up to ~0.01% for $\dot{M} = 10^{-6}$ M$_\odot$ y^{-1} and ~0.001% for $\dot{M} = 10^{-7}$ M$_\odot$ y^{-1}. Including porosity, increases the coverage of the dust by up to a monolayer, with the largest increases seen for smaller values of f_{ic} as a smaller fraction of the envelope mass is in the interclump component for these outflows. When adding a companion, an increase of up to five orders of magnitude and a decrease of up to two orders of magnitude is possible. The final

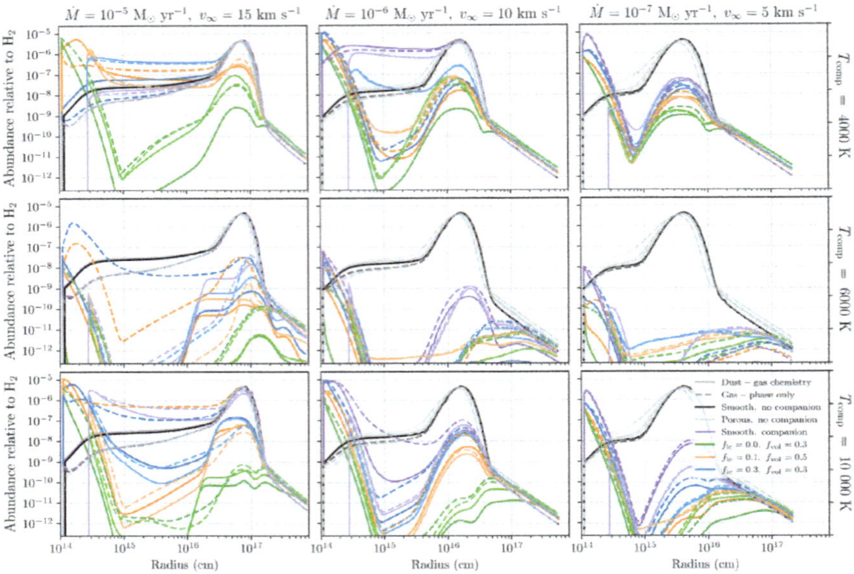

Fig. 3 Fractional abundance of HS relative to H_2 for a selection of O-rich outflows with different outflow densities (columns) and companions (rows). Different line styles show different chemistries: dashed lines: gas-phase chemistry only, solid lines: including dust–gas chemistry. Different colours show different density structures and the presence of a companion: black: smooth outflow without a companion, gray: porous outflow without a companion; purple: smooth outflow with a companion, green: porous outflow with a companion, $f_{ic} = 0.0$, $f_{vol} = 0.3$, $l_* = 10^{13}$ cm, orange: porous outflow with a companion, $f_{ic} = 0.1$, $f_{vol} = 0.5$, $l_* = 10^{13}$ cm, blue: porous outflow with a companion, $f_{ic} = 0.3$, $f_{vol} = 0.3$, $l_* = 5 \times 10^{12}$ cm. When including a companion, different shades correspond to the value of R_{dust}. Darker shade: $R_{dust} = 2\ R_*$, lighter shade: $R_{dust} = 5\ R_*$.

coverage can increase to several monolayers for $\dot{M} = 10^{-5}$ M$_\odot$ y^{-1}, and stays below 10% and 0.001% for $\dot{M} = 10^{-6}$ and 10^{-7} M$_\odot$ y^{-1}.

3.2 C-rich outflows

We consider again two species to showcase the combined effects of the different complexities: the parent SiO, whose abundance decreases when including a solar-like companion and can show a bumped profile for a white dwarf; and the daughter H_2CS, which can become observable for specific combinations of complexities.

Fig. 5 shows the abundance of the parent SiO in outflows with different densities, porosities, and companions. Porosity and dust–gas chemistry, combined or separate, do not have a significant impact on the chemistry, neither does including a red dwarf companion. Including a solar-like companion leads to decreases in abundance as porosity increases, with smooth outflows showing a decreases of 2 to 3 orders of magnitude in lower density outflows. When including a white dwarf companion, the profile changes to a bumped profile, similar to SO in O-rich outflows. The inner wind abundances decrease by 2 to 4 orders of magnitude, depending on the outflow's density and porosity, with

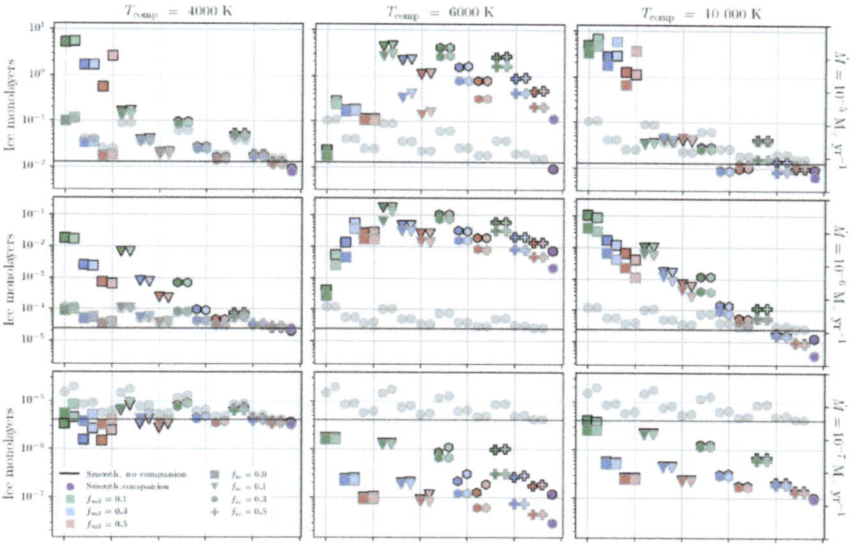

Fig. 4 Number of ice monolayers at the end of the outflow for all calculated O-rich outflows with different density structures and companions. The black solid line shows the number of monolayers in a smooth outflow without a companion. Different shapes and colours correspond to different density structures and the presence of a companion: purple circles: smooth outflow with companion, grey circles: porous outflows without a companion. The other shapes show the results for porous outflows with a companion. The shape signifies the value of f_{ic}: squares, triangles, hexagons, crosses correspond to $f_{ic} = 0.0, 0.1, 0.3, 0.5$, respectively. The colour signifies the value of f_{vol}: green, blue, red correspond to $f_{vol} = 0.1, 0.3, 0.5$, respectively. Different shades signify different values of l_*: lighter colour: $l_* = 5 \times 10^{12}$ cm, darker colour: $l_* = 10^{13}$ cm. The edge colour of the shapes shows different values of R_{dust}: black edge: $R_{dust} = 2\ R_*$, grey edge: $R_{dust} = 5\ R_*$.

a bump in abundance around 10^{-7} in the outer wind. The abundance of the bump decreases with increasing porosity in lower density outflows.

Fig. 6 shows the abundance of H_2CS. We find that including porosity increases the abundance peak up to a factor of five and widens the peak in the lower density outflows. Including a red dwarf leads to parent-like abundance profiles with increasing porosity. For a solar-like star, this occurs with decreasing porosity. In lower density outflows, the abundance profile flattens with increasing porosity. Including a white dwarf leads to a mix of the two previous effects: at high outflow densities, H_2CS becomes parent-like with increasing porosity; at low densities with decreasing porosity. In the outflow with $\dot{M} = 10^{-7}\ M_\odot\ y^{-1}$, a high porosity results in a large inner wind abundance followed by a steep decline.

The number of refractory monolayers covering the dust at the end of the outflow is shown in Fig. 7. The dust's refractory coverage in C-rich outflows with $\dot{M} = 10^{-5}\ M_\odot\ y^{-1}$ increases with porosity from ~1.5% in the smooth outflow up to two orders of magnitude to a complete grain coverage of 1. In lower density outflows, the refractory coverage is increased by up to three orders of magnitude with porosity reaching coverages of ~10%. In C-rich outflows, a companion can lead to an increase in refractory coverage of up to one order of magnitude and a decrease of up to two orders of magnitude at low outflow densities.

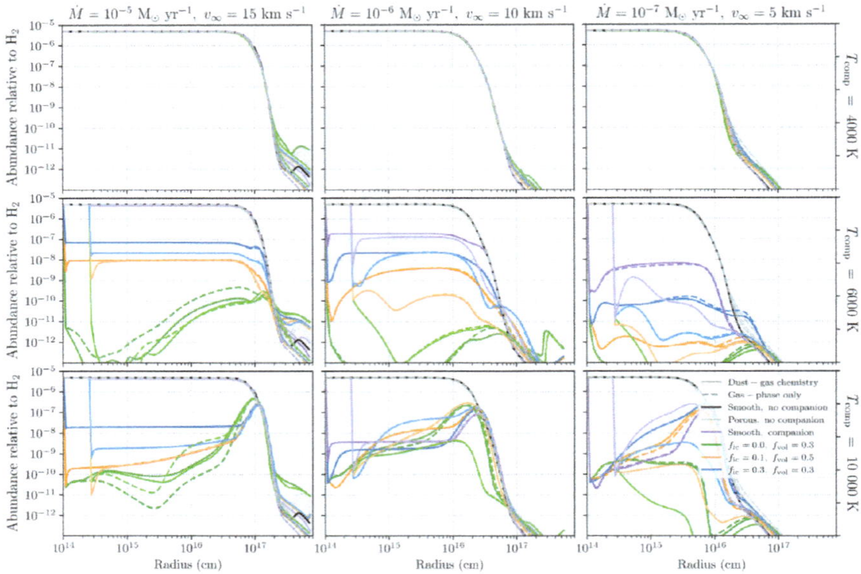

Fig. 5 Fractional abundance of the parent SiO relative to H_2 for a selection of C-rich outflows with different outflow densities (columns) and companions (rows). Different line styles show different chemistries: dashed lines: gas-phase chemistry only, solid lines: including dust–gas chemistry. Different colours show different density structures and the presence of a companion: black: smooth outflow without a companion, gray: porous outflow without a companion; purple: smooth outflow with a companion, green: porous outflow with a companion, $f_{ic} = 0.0$, $f_{vol} = 0.3$, $l_* = 10^{13}$ cm, orange: porous outflow with a companion, $f_{ic} = 0.1$, $f_{vol} = 0.5$, $l_* = 10^{13}$ cm, blue: porous outflow with a companion, $f_{ic} = 0.3$, $f_{vol} = 0.3$, $l_* = 5 \times 10^{12}$ cm. Including a companion, different shades correspond to the value of R_{dust}: darker shade: $R_{dust} = 2\ R_*$, lighter shade: $R_{dust} = 5\ R_*$.

Nevertheless, the coverage stays below 10%, 1%, and 0.01% in outflows with $\dot{M} = 10^{-5}$, 10^{-6}, and 10^{-7} M$_\odot$ y^{-1}, respectively.

4 Discussion

The chemical composition throughout the outflow can be strongly influenced when including different layers of complexity. While including a companion in a porous outflow has the largest effects, the other (combinations of) complexities show different signatures in the gas-phase and the ice and refractory coverage of the dust. In Section 4.1, we discuss the effects of the different complexities separately as well as the interplay between them. In Section 4.2, we identify sets of gas-phase molecules that trace the dust–gas chemistry as well as the presence of a companion in a (porous) outflow.

4.1 Interplay of complexities

Separately, porosity and dust–gas chemistry have a small effect on the overall chemistry of the outflow. When including only porosity, *e.g.* caused by a planetary companion that induces a large-scale structure in the outflow, the envelopes of parent species become mainly smaller with increasing outflow porosity, but the

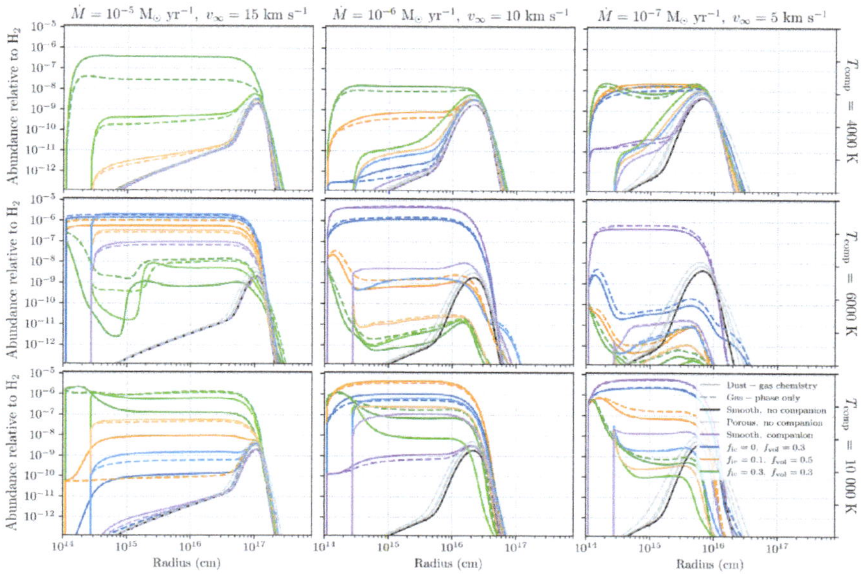

Fig. 6 Fractional abundance of the H_2CS relative to H_2 for a selection of C-rich outflows with different outflow densities (columns) and companions (rows). Different line styles show different chemistries: dashed lines: gas-phase chemistry only, solid lines: including dust–gas chemistry. Different colours show different density structures and the inclusion of a companion: black: smooth outflow without a companion, gray: porous outflow without a companion, purple: smooth outflow with a companion, green: porous outflow with a companion, $f_{ic} = 0.0$, $f_{vol} = 0.3$, $l_* = 10^{13}$ cm, orange: porous outflow with a companion, $f_{ic} = 0.1$, $f_{vol} = 0.5$, $l_* = 10^{13}$ cm, blue: porous outflow with a companion, $f_{ic} = 0.3$, $f_{vol} = 0.3$, $l_* = 5 \times 10^{12}$ cm. When including a companion, different shades correspond to the value of R_{dust}: darker shade: $R_{dust} = 2\ R_*$, lighter shade: $R_{dust} = 5\ R_*$.

reduction in size is too small to be observed. The abundance profile of certain parents changes with increased porosity and lower outflow density: SO in O-rich outflows shows a small bump before the onset of photodissociation, and the envelope of CS in O-rich outflows and C_2H_4 in C-rich outflows become smaller with the knee in the outer wind widening. As porosity increases, peak abundances of daughter species increase and shift closer to the star, but this effect is again likely too small to be observed.

Including dust–gas chemistry does not lead to significant depletion of parent species for dust grains following the canonical MRN GSD and the GSD with larger grains than MRN. Assuming silicates without Fe only, likely leads to more significant depletion in O-rich outflows.[29] However, most O-rich outflows host a mixture of dust grains, including warmer grains that have Fe inclusions,[50] leading to smaller depletion levels. Parent species are depleted onto dust assuming a GSD with smaller dust grains because of the larger number density of grain surface sites per unit volume. For dust grains following the GSD with smaller dust grains, parent species with high binding energies (*e.g.*, SiO, SiS, H_2O) are depleted onto dust grains in outflows with $\dot{M} = 10^{-5}$ M$_\odot$ y^{-1}. In O-rich outflows, the depletion takes place around 2×10^{16} cm, in C-rich outflows around 4×10^{16} cm. Certain species, such as SiH_4, are efficiently formed *via*

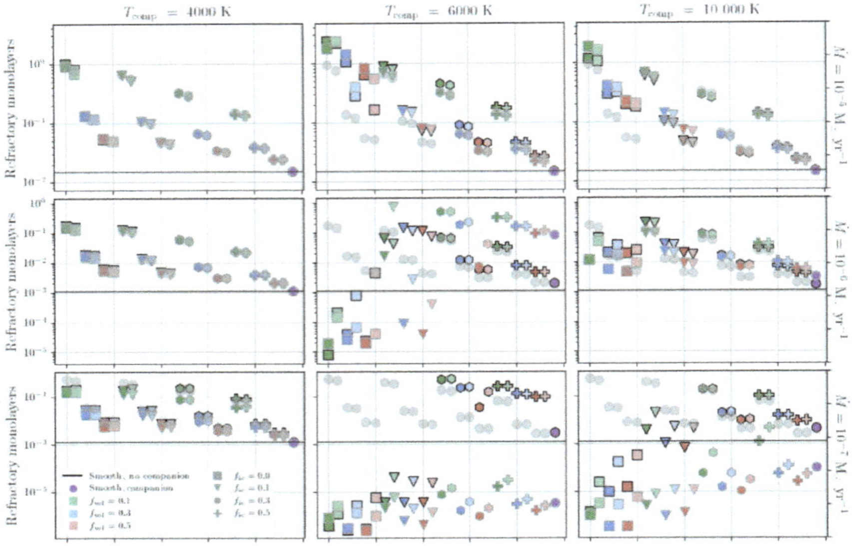

Fig. 7 Number of refractory monolayers at the end of the outflow for all calculated C-rich outflows with different density structures and companions. The black solid line shows the number of monolayers in a smooth outflow without a companion. Different shapes and colours correspond to different density structures and the presence of a companion: purple circles: smooth outflow with companion, grey circles: porous outflows without a companion. The other shapes show the results for porous outflows with a companion. The shape signifies the value of f_{ic}: squares, triangles, hexagons, crosses correspond to $f_{ic} = 0.0, 0.1, 0.3, 0.5$, respectively. The colour signifies the value of f_{vol}: green, blue, red correspond to $f_{vol} = 0.1, 0.3, 0.5$, respectively. Different shades signify different values of l_*: lighter colour: $l_* = 5 \times 10^{12}$ cm, darker colour: $l_* = 10^{13}$ cm. The edge colour of the shapes shows different values of R_{dust}: black edge: $R_{dust} = 2\ R_*$, grey edge: $R_{dust} = 5\ R_*$.

surface chemistry. These gas-phase tracers of dust–gas chemistry are released into the gas phase in the outer wind, forming a molecular shell (Section 4.2.1). Assuming a GSD with smaller grains increases the abundance of the gas-phase tracers by about an order of magnitude.

Including a stellar companion alone has the largest effect on the chemistry. For smooth outflows, its effect is limited to lower density outflows (corresponding to lower extinction outflows) with a solar-like or white dwarf companion. A red dwarf companion does not significantly impact the chemistry. Sub-stellar brown dwarfs are therefore not expected to have a significant impact.

When combining complexities, the impact on the dust- and gas-phase chemistry becomes more apparent. Porosity together with a stellar companion in a gas-phase only chemistry chemical model has been described in Van de Sande and Millar[33] (see Section 2.2.2).

4.1.1 Porosity and dust–gas chemistry. Combining porosity and dust–gas chemistry, we find that a larger porosity increases the abundance of the gas-phase tracers and the grain surface coverage. The ice coverage in higher density O-rich outflows, where H_2O is the main ice component, is expected to be observable with JWST considering the ISO detections.[51,52] The abundance of most gas-phase tracers can increase up to an order of magnitude (Section 4.2.1). Assuming a GSD with smaller grains than MRN leads to an additional increase by an order of magnitude.

The coverages are not significantly changed when assuming a GSD with larger dust grains than MRN. When assuming smaller dust grains, the ice and refractory coverage in both outflows decreases by a factor of a few, with porosity leading to a smaller possible increase. While gas-phase species are now depleted, leading to a larger total abundance of ice and refractory material, the larger number density of dust surface sites per unit volume leads to a smaller number of monolayers.

4.1.2 Porosity, dust–gas chemistry, and a stellar companion. When combining all three complexities, we find that including a stellar companion has the largest effects on the overall chemistry of the outflow, significantly impacting both the gas phase composition and grain surface coverage. The effects on the gas phase are similar to our previous results, which included gas-phase chemistry only:[33] outflows where companion UV photons experience a low extinction (low outflow density and/or high porosity) appear molecule poor, while outflows with a high extinction (high outflow density and/or low porosity) show an increase in chemical complexity, with daughter species showing a parent-like abundance profile.

4.1.2.1 Influence of dust–gas chemistry on the gas phase. Including dust–gas chemistry generally does not significantly affect the inner wind abundances in outflows with an increased chemical complexity. These newly formed daughters are not depleted onto dust. While some accretion does occur, most of the accreted ices are rapidly thermally desorbed or photoprocessed into refractory organic material if the ice is already complex. Any changes to the gas-phase are predominantly caused by the availability of feedstock material. As parents are depleted onto dust grains, this lowers the abundance of their photodissociation products in the gas phase and hinders the formation of more complex daughters in the inner wind. A notable example is HS in both C-rich and O-rich outflows which shows several orders of magnitude difference, rendering it effectively unobservable when including dust–gas chemistry. This is linked to depletion of the parent CS: photodissociation by internal photons of CS is an important source of S, removing it from the gas phase reduces the available S for HS formation, and for H_2S formation as well in C-rich outflows.

Including a companion generally increases the abundance of the gas-phase tracers of dust–gas chemistry, present in a shell in the outer wind. We find increases of about two orders of magnitude compared to a smooth outflow and of about one order of magnitude compared to clumpy outflows. However, certain species are destroyed by including a companion (see Section 4.2). Assuming a GSD with smaller dust grains than MRN leads to a further increase by about an order of magnitude.

4.1.2.2 Influence of a companion on the ice coverage. The influence of a companion on the ice coverage of the dust depends mainly on surface chemistry. When including a companion, the abundance of atoms in the gas phase and on the dust increases in the outer region. The atoms are highly reactive on the dust surface, efficiently (re)forming, *e.g.*, ice H_2O, H_2S, HCN, NH_3, and SO, and increasing the ice coverage. In both O-rich and C-rich outflows, the final ice coverage mainly depends on H_2O formation on the surface, linked to the availability of H. This leads to a larger coverage compared to the fiducial dust–gas chemistry model for $= 10^{-6}\ M_{\odot}\ y^{-1}$ with a solar-like or white dwarf companion.

In C-rich outflows, the reactivity of the surface chemistry is the main cause for an increase in final ice coverage. The increase mainly takes place in low extinction outflows, where surface chemistry efficiently forms H_2O ice. Surface chemistry also has an indirect effect on the ice coverage. In outflows with $\dot{M} = 10^{-5}\ M_\odot\ y^{-1}$, the peak abundance of SiS ice can increase by an order of magnitude with the peak broadening by half a magnitude in both O-rich and C-rich outflows. SiS ice is only formed *via* accretion from the gas phase. The gas-phase abundance of SiS increases in this region *via* the reaction $SiH_2^+ + S \rightarrow HSiS^+$, which dissociatively recombines with electrons to form SiS. SiH_2^+ is produced by photoionisation of SiH_2, formed on the surface and released into the gas phase by photodesorption.

4.1.2.3 Influence of a companion on the refractory coverage. In general, a companion does not lead to a larger range of possible coverages compared to including porosity alone. However, it can affect the refractory coverage of individual porous outflows. In high extinction C-rich outflows, the complexity of the gas phase and ices is reduced, leading to a smaller refractory coverage. In low extinction outflows, a solar-like companion can lead to an increase in coverage of up to two orders of magnitude, a white dwarf to up to one order of magnitude, and a red dwarf to a factor of a few. The observability of the refractory material with JWST depends on its composition. As refractory material is assumed to be chemically inert in our model, estimates are difficult to make. For O-rich outflows, a companion does not significantly increase the coverage of any specific outflow. A solar-like companion significantly reduces the coverage because of its large effect on the gas phase, rendering the dust effectively bare. Nonetheless, the refractory coverage of O-rich dust is low, with a negligible coverage for all outflow densities.

4.2 Chemical tracers of outflow structure

Specific (combinations of) species allow us to constrain the underlying physics and chemistry at play in the outflow. The effects of porosity only on the chemistry is too small to constrain. While dust–gas chemistry can lead to a significant ice and/or refractory coverage of the dust grains, these are not easily observable, although observations of AGB envelopes with JWST may reveal clues on the dust composition and presence or otherwise of ices. We identify gas-phase tracers of active dust–gas chemistry in O-rich and C-rich outflows in Section 4.2.1. The presence of a companion, along with whether the outflow is highly porous or not, can be constrained as well, allowing us to distinguish between planetary and stellar companions shaping the outflow. Section 4.2.1 discusses the combinations of species for O-rich and C-rich outflows.

4.2.1 Dust–gas chemistry tracers. When including dust–gas chemistry, SiH_2, SiH_3, SiH_4, and H_2SiO appear in a shell around 10^{17} cm in both O-rich and C-rich outflows with $\dot{M} = 10^{-5}\ M_\odot\ y^{-1}$. In O-rich outflows, SiH_2, SiH_3, and SiH_4 have peak abundances of around 10^{-9} and H_2SiO has one of $\sim 5 \times 10^{-10}$. In C-rich outflows, this decreases to 10^{-11} for the silicon hydrides and $\sim 5 \times 10^{-13}$ for H_2SiO.

Their peak abundances depend on the presence of other complexities. When porosity is included, they increase by roughly an order of magnitude. Including a companion leads to an increase of two orders of magnitude, which can increase by an additional order of magnitude when combining it with porosity.

O-rich outflows have two additional tracers: NH_2OH and H_2O_2. NH_2OH reaches abundances of 5×10^{-9} around 3×10^{17} cm. Its abundance decreases by an order of magnitude when including a solar-like companion, decreasing further when increasing the outflow's porosity. H_2O_2 is present in a broad shell from about 10^{16} to 10^{17} cm, with a peak abundance of a few times 10^{-7}. The shell becomes narrower when including a companion, shifting to 10^{17} cm. Highly porous outflows decrease its abundance by an order of magnitude; in the case of a solar-like companion, its abundance is drastically reduced to 10^{-11}.

4.2.2 Companion and porosity tracers. The presence of a companion star, especially if the outflow is porous, can be traced by a combination of parent species, daughter species, and uncommon species formed *via* inner wind photochemistry. The sets of species depend on the type of chemistry, the outflow's porosity, and the type of companion.

We find that the chemistry in the inner region is mainly driven by the competition between O and S for the newly liberated C in the inner wind. A large number of the tracer species therefore contain sulphur. Newly liberated N in the inner wind also leads to large abundances of NS, SiN, and NO.

4.2.2.1 C-rich outflows. Table 3 lists all molecular tracers for different types of companions, densities, and porosities in C-rich outflows. We distinguish different types of behaviour: for parents (i) a decrease or increase in parent

Table 3 Molecular tracers of different companions and porosities in C-rich outflows. If not specified, the tracer is valid for all densities and porosities

	Red dwarf	Solar-like	White dwarf
Parents			
(i) ↘ in abundance with ↗ porosity	C_2H_4, NH_3	All except HCN	All except HCN
			High porosity: C_2H_2, CH_4 CS, NH_3, SiS H_2O
↗ in abundance with ↗ porosity	Low porosity: H_2S		
(ii) Bump for ↗ porosity	High porosity: H_2S		Low density: SiO
Bump for ↘ porosity		High density: H_2S	High density: SiO
Daughters			
(iii) ↗ inner wind abundance with ↗ porosity	CH_3, HS, NH	Low density: HS	High density: CH_3
	NH_2, C_6H H_2CS, HCSi, NS	C_2H High density: HCSi	High density: C_6H
With ↘ porosity		High density: CH_3 CN, OH, HS	Low density: CH_3
			High density: NH, HS
		HC_3N, HC_5N, HC_7N	Low density: C_6H
(iv) Parent-like with ↗ porosity		High density: SiN, SiC, SiNC	SiNC
With ↘ porosity	HS	CH_3CN, H_2CS, NS	Low density: CH_3CN
		Low density: SiN, SiC, SiNC	H_2CS, NS, SiC, SiN

abundance with increasing porosity, (ii) the appearance of a bump in the abundance profile with increasing or decreasing porosity; for daughters, (iii) a larger inner wind abundance followed by a peak in abundance as porosity increases or decreases, and (iv) the abundance profile changing to a parent-like profile with a large inner wind abundance followed by a Gaussian decline as porosity increases or decreases.

We find that the parents CO, N_2, and HCN are unaffected by the presence of a companion. A solar-like companion leads only to a smaller envelope size of HCN in high porosity, lower density outflows. The parent H_2S is most strongly affected, its abundance profile changing to a steep decline followed by a bump, whose peak abundance decreases as the outflow density decreases, in all low density outflows and in high porosity, high density outflows. For SiO, including a white dwarf changes its abundance profile to a lower inner wind abundance followed by a bump in higher density outflows and to a shell-like profile in lower density outflows, with the abundance peak decreasing with increasing porosity.

The daughter HS follows the shape of the H_2S profile. The abundance profile of SO is sensitive to the type of companion and porosity of the outflow. For a red dwarf companion, the peak in abundance of SO in the outer wind widens with increasing porosity, creating a broader shell. For a solar-like and white dwarf companion, SO becomes parent-like with increasing porosity in high density outflows, and with decreasing porosity in lower density outflows.

The behaviour of the cyanopolyynes and hydrocarbon radicals depends on outflow density and porosity for a solar-like and white dwarf companion. In low density, high porosity outflows, they show dramatically smaller abundances. For a solar-like companion, we find that the inner wind abundance of HC_3N, HC_5N, and HC_7N increases with decreasing porosity. As porosity increases, their profiles change to a broad shell. The peak abundance drastically decreases in lower density outflows, rendering the longer chains unobservable in low density, high porosity outflows. The inner wind abundance of C_2H increases in high density, high porosity outflows. For lower densities, the profile is flattened with increasing porosity, where the inner wind abundance increases while the peak abundance decreases. The profiles of C_4H and C_6H change to a broad peak, again disappearing in low density, high porosity outflows. For a white dwarf companion, the inner wind abundance of HC_3N, HC_5N, HC_7N, C_2H, and C_4H increases with increasing porosity, with their profiles flattening in lower density outflows. The inner wind abundance of C_6H increases with increasing porosity in higher density outflows. In lower density outflows, the profile flattens with increasing porosity.

4.2.2.2 O-rich outflows. Table 4 lists all molecular tracers for different types of companions, densities, and porosities in O-rich outflows. We distinguish the same types of behaviour as for the C-rich outflow.

We find that the parents CO and N_2 are unaffected by the presence of a companion. H_2S is efficiently destroyed when including a solar-like companion, rendering it apparently absent in lower density outflows and strongly reducing its abundance in high density outflows. In high porosity outflows with a red dwarf or white dwarf companion, the abundance profile of H_2S changes to a high inner wind abundance followed by a bump. In low porosity outflows, it has a smaller abundance, but parent-like shape. The abundance profiles of SO and SO_2 changes to a lower initial abundance followed by a bump as porosity increases, becoming

Table 4 Molecular tracers of different companions and porosities in O-rich outflows. If not specified, the tracer is valid for all densities and porosities

	Red dwarf	Solar-like	White dwarf
Parents			
(i) ↘ in abundance with ↗ porosity	SiS, CO_2, CS	CS, HCN, NH_3	CS, HCN, NH_3
	NH_3, H_2S	SiO, SiS, H_2O CO2 (low porosity)	SiS, H_2S
↗ in abundance with ↗ porosity		CO_2 (high porosity)	
(ii) Bump for ↗ porosity	SO, SO_2	SO, SO_2	SO, SO_2, SiO
Daughters			
(iii) ↗ inner wind abundance with ↗ porosity	Low density: HS	CN	CN, OH
	NO, N_2O		NO, N_2O
With ↘ porosity			
Parent-like with ↗ porosity	High density: HS, NS	Low density: OH	High density: SiN, NS
With ↘ porosity		NS, SiN, N_2O	
(iv) Flattening with ↗ porosity			HS, SiC, SiN, NS

shell-like for lower densities. The envelope size of H_2O and HCN becomes smaller in low density, high porosity outflows when including a white dwarf companion.

5 Conclusions

We presented the results of the most physically and chemically complex AGB chemical kinetics model to date, including porosity, dust–gas chemistry, and companion UV photons. The chemical validity of the model is extended from the outer wind into the dust interaction zone, while being able to include the effects of spherically symmetric outflows, produced by a stellar or substellar companion. The effects of each of these complexities was studied separately in previous work. Here, we combined them all into one model and considered their interdependencies.

Porosity alone does not significantly impact the gas-phase composition of the outflow. It increases the ice and refractory coverage of the dust grains, with the specific increase depending on the chemistry of the outflow and the material considered.

Dust–gas chemistry does not significantly impact the gas-phase composition, except when a grain size distribution with large number of dust surface sites is assumed. However, certain gas-phase species, such as SiH_4, H_2SiO, and N_2O, are efficiently formed *via* surface chemistry and released into the gas phase *via* photodesorption by interstellar photons in the outer wind, forming a molecular shell. The abundance of this shell increases up to an order of magnitude when including porosity. Including a companion increases the abundance of SiH_4 and H_2SiO, but decreases that of N_2O.

We find that a companion star combined with a porous outflow, with or without dust–gas chemistry, shows the largest effect. The impact on the gas-phase chemistry depends on the outflow density, porosity, and the assumed onset of dust formation, as well as the intensity of the companion's radiation. Depending on the configuration, certain parent and daughter species are efficiently destroyed, with other daughters showing an increased inner wind abundance or even parent-like behaviour. We have identified molecular tracers that allow us to discern the type of companion and porosity of the outflow for both O-rich and C-rich outflows. Solar-like companions show the largest effects and can be inferred using our tracer molecules. The effect of white dwarfs is smaller, but their presence can be inferred as well. Red dwarf companions can be inferred only in high porosity outflows. In this case, specific outflow structures are needed for chemistry to be used as a tool to distinguish between stellar and substellar companions.

Our results confirm that chemistry can be used to distinguish between a stellar or substellar companion shaping the outflow and constraining the underlying density distribution. Even though the new model has more input parameters, it is possible to constrain them based on observations, as long as a suite of molecules is targeted. This makes the model an important tool to help interpret and guide observations.

While the chemical model is the most complex to date, it does not cover the chemistry of the entire outflow. In order to achieve such a model, dust formation and growth need to be included. Additionally, to accurately model the chemistry in outflows with specific observed density distributions, such as spirals and disks, appropriate parameterisations or hydrodynamical modelling output are required. Nevertheless, this model is an important first step in the development of such three-dimensional, full outflow chemical models.

Author contributions

Marie Van de Sande: conceptualisation, methodology, software, validation, investigation, visualisation, writing – original draft. Catherine Walsh: conceptualisation, software, investigation, writing – review & editing. Tom J. Millar: conceptualisation, software, investigation, writing – review & editing.

Conflicts of interest

There are no conflicts to declare.

Acknowledgements

MVdS acknowledges support from the European Union's Horizon 2020 research and innovation programme under the Marie Skłodowska-Curie grant agreement No 882991. CW acknowledges financial support from the University of Leeds and from the Science and Technology Facilities Council (STFC, grant numbers ST/T000287/1 and MR/T040726/1) TJM gratefully acknowledges the receipt of a Leverhulme Emeritus Fellowship and the STFC for support under grant numbers ST/P000312/1 and ST/T000198/1.

Notes and references

1 *Asymptotic Giant Branch Stars*, ed. H. J. Habing and H. Olofsson, Springer, 2003.

2 S. Höfner and H. Olofsson, *Astron. Astrophys. Rev.*, 2018, **26**, 1.
3 A. G. G. M. Tielens, *The Physics and Chemistry of the Interstellar Medium*, Cambridge University Press, 2005.
4 S. Zhukovska and T. Henning, *Astron. Astrophys.*, 2013, **555**, A99.
5 L. Decin, *Annu. Rev. Astron. Astrophys.*, 2021, **59**, 337–389.
6 T. M. Koehler, H.-P. Gail and E. Sedlmayr, *Astron. Astrophys.*, 1997, **320**, 553-567.
7 I. Cherchneff, *Astron. Astrophys.*, 2006, **456**, 1001–1012.
8 T. P. M. Goumans and S. T. Bromley, *Mon. Not. R. Astron. Soc.*, 2012, **420**, 3344-3349.
9 D. Gobrecht, I. Cherchneff, A. Sarangi, J. M. C. Plane and S. T. Bromley, *Astron. Astrophys.*, 2016, **585**, A6.
10 P. J. Huggins and A. E. Glassgold, *Astrophys. J.*, 1982, **252**, 201–207.
11 L. A. M. Nejad, T. J. Millar and A. Freeman, *Astron. Astrophys.*, 1984, **134**, 129-133.
12 T. J. Millar and E. Herbst, *Astron. Astrophys.*, 1994, **288**, 561–571.
13 X. Li, T. J. Millar, A. N. Heays, C. Walsh, E. F. van Dishoeck and I. Cherchneff, *Astron. Astrophys.*, 2016, **588**, A4.
14 I. C. Leão, P. de Laverny, D. Mékarnia, J. R. de Medeiros and B. Vandame, *Astron. Astrophys.*, 2006, **455**, 187–194.
15 T. Khouri, M. Maercker, L. B. F. M. Waters, W. H. T. Vlemmings, P. Kervella, A. de Koter, C. Ginski, E. De Beck, L. Decin, M. Min, C. Dominik, E. O'Gorman, H.-M. Schmid, R. Lombaert and E. Lagadec, *Astron. Astrophys.*, 2016, **591**, A70.
16 M. Agúndez, J. Cernicharo, G. Quintana-Lacaci, A. Castro-Carrizo, L. Velilla Prieto, N. Marcelino, M. Guélin, C. Joblin, J. A. Martín-Gago, C. A. Gottlieb, N. A. Patel and M. C. McCarthy, *Astron. Astrophys.*, 2017, **601**, A4.
17 N. Mauron and P. J. Huggins, *Astron. Astrophys.*, 2006, **452**, 257–268.
18 M. Maercker, T. Danilovich, H. Olofsson, E. De Beck, K. Justtanont, R. Lombaert and P. Royer, *Astron. Astrophys.*, 2016, **591**, A44.
19 P. Kervella, M. Montargès, S. T. Ridgway, G. Perrin, O. Chesneau, S. Lacour, A. Chiavassa, X. Haubois and A. Gallenne, *Astron. Astrophys.*, 2014, **564**, A88.
20 W. Homan, M. Montargès, B. Pimpanuwat, A. M. S. Richards, S. H. J. Wallström, P. Kervella, L. Decin, A. Zijlstra, T. Danilovich, A. de Koter, K. Menten, R. Sahai, J. Plane, K. Lee, R. Waters, A. Baudry, K. T. Wong, T. J. Millar, M. Van de Sande, E. Lagadec, D. Gobrecht, J. Yates, D. Price, E. Cannon, J. Bolte, F. De Ceuster, F. Herpin, J. Nuth, J. Philip Sindel, D. Kee, M. D. Grey, S. Etoka, M. Jeste, C. A. Gottlieb, E. Gottlieb, I. McDonald, I. El Mellah and H. S. P. Müller, *Astron. Astrophys.*, 2020, **644**, A61.
21 L. Decin, M. Montargès, A. M. S. Richards, C. A. Gottlieb, W. Homan, I. McDonald, I. El Mellah, T. Danilovich, S. H. J. Wallström, A. Zijlstra, A. Baudry, J. Bolte, E. Cannon, E. De Beck, F. De Ceuster, A. de Koter, J. De Ridder, S. Etoka, D. Gobrecht, M. Gray, F. Herpin, M. Jeste, E. Lagadec, P. Kervella, T. Khouri, K. Menten, T. J. Millar, H. S. P. Müller, J. M. C. Plane, R. Sahai, H. Sana, M. Van de Sande, L. B. F. M. Waters, K. T. Wong and J. Yates, *Science*, 2020, **369**, 1497–1500.
22 D. González Delgado, H. Olofsson, F. Kerschbaum, F. L. Schöier, M. Lindqvist and M. A. T. Groenewegen, *Astron. Astrophys.*, 2003, **411**, 123–147.
23 S. Massalkhi, M. Agúndez and J. Cernicharo, *Astron. Astrophys.*, 2019, **628**, A62.

24 V. Bujarrabal, J. Gomez-Gonzalez and P. Planesas, *Astron. Astrophys.*, 1989, **219**, 256–264.
25 L. Decin, E. De Beck, S. Brünken, H. S. P. Müller, K. M. Menten, H. Kim, K. Willacy, A. de Koter and F. Wyrowski, *Astron. Astrophys.*, 2010, **516**, A69.
26 R. J. Sylvester, F. Kemper, M. J. Barlow, T. de Jong, L. B. F. M. Waters, A. G. G. M. Tielens and A. Omont, *Astron. Astrophys.*, 1999, **352**, 587–599.
27 B. T. Draine, *Annu. Rev. Astron. Astrophys.*, 2003, **41**, 241–289.
28 M. Van de Sande, J. O. Sundqvist, T. J. Millar, D. Keller, W. Homan, A. de Koter, L. Decin and F. De Ceuster, *Astron. Astrophys.*, 2018, **616**, A106.
29 M. Van de Sande, C. Walsh, T. P. Mangan and L. Decin, *Mon. Not. R. Astron. Soc.*, 2019, 2325.
30 M. Van de Sande, C. Walsh and T. Danilovich, *Mon. Not. R. Astron. Soc.*, 2020, **495**, 1650–1665.
31 M. Van de Sande, C. Walsh and T. J. Millar, *Mon. Not. R. Astron. Soc.*, 2020, **501**, 491–506.
32 M. Van de Sande and T. J. Millar, *Astrophys. J.*, 2019, **873**, 36.
33 M. Van de Sande and T. J. Millar, *Mon. Not. R. Astron. Soc.*, 2021, **510**, 1204–1222.
34 D. McElroy, C. Walsh, A. J. Markwick, M. A. Cordiner, K. Smith and T. J. Millar, *Astron. Astrophys.*, 2013, **550**, A36.
35 M. Morris and M. Jura, *Astrophys. J.*, 1983, **264**, 546–553.
36 M. Agúndez, J. I. Martínez, P. L. de Andres, J. Cernicharo and J. A. Martín-Gago, *Astron. Astrophys.*, 2020, **637**, A59.
37 F. L. Schöier and H. Olofsson, *Astron. Astrophys.*, 2001, **368**, 969–993.
38 H. Olofsson, D. González Delgado, F. Kerschbaum and F. L. Schöier, *Astron. Astrophys.*, 2002, **391**, 1053–1067.
39 S. Ramstedt, F. L. Schöier, H. Olofsson and A. A. Lundgren, *Astron. Astrophys.*, 2008, **487**, 645–657.
40 T. Danilovich, D. Teyssier, K. Justtanont, H. Olofsson, L. Cerrigone, V. Bujarrabal, J. Alcolea, J. Cernicharo, A. Castro-Carrizo, P. García-Lario and A. Marston, *Astron. Astrophys.*, 2015, **581**, A60.
41 S. P. Owocki, K. G. Gayley and N. J. Shaviv, *Astrophys. J.*, 2004, **616**, 525–541.
42 S. P. Owocki and D. H. Cohen, *Astrophys. J.*, 2006, **648**, 565–571.
43 J. O. Sundqvist, S. P. Owocki, D. H. Cohen, M. A. Leutenegger and R. H. D. Townsend, *Mon. Not. R. Astron. Soc.*, 2012, **420**, 1553–1561.
44 J. O. Sundqvist, J. Puls and S. P. Owocki, *Astron. Astrophys.*, 2014, **568**, A59.
45 M. Van de Sande, J. O. Sundqvist, T. J. Millar, D. Keller, W. Homan, A. de Koter, L. Decin and F. De Ceuster, *Astron. Astrophys.*, 2020, **634**, C1.
46 C. Walsh, T. J. Millar, H. Nomura, E. Herbst, S. Widicus Weaver, Y. Aikawa, J. C. Laas and A. I. Vasyunin, *Astron. Astrophys.*, 2014, **563**, A33.
47 R. T. Garrod, S. L. Widicus Weaver and E. Herbst, *Astrophys. J.*, 2008, **682**, 283–302.
48 J. S. Mathis, W. Rumpl and K. H. Nordsieck, *Astrophys. J.*, 1977, **217**, 425–433.
49 A. N. Heays, A. D. Bosman and E. F. van Dishoeck, *Astron. Astrophys.*, 2017, **602**, A105.
50 A. M. Heras and S. Hony, *Astron. Astrophys.*, 2005, **439**, 171–182.
51 A. Omont, S. H. Moseley, T. Forveille, W. J. Glaccum, P. M. Harvey, L. Likkel, R. F. Loewenstein and C. M. Lisse, *Astrophys. J.*, 1990, **355**, L27.
52 K. Justtanont and A. G. G. M. Tielens, *Astrophys. J.*, 1992, **389**, 400.

Faraday Discussions

ROYAL SOCIETY OF CHEMISTRY

PAPER

Predicting observable infrared signatures of nanosilicates in the diffuse interstellar medium

Sascha T. Zeegers, [a] Joan Mariñoso Guiu, [b] Francisca Kemper, [cde] Jonathan P. Marshall [f] and Stefan T. Bromley *[bd]

Received 2nd March 2023, Accepted 27th March 2023
DOI: 10.1039/d3fd00055a

The destruction time scale of dust in the diffuse interstellar medium is estimated to be an order of magnitude shorter than its residence time. Nevertheless, dust is observed in the interstellar medium, leading to the conclusion that reformation and grain growth must take place. Direct observations of nanometre-sized silicate grains, the main constituent of interstellar dust, would provide a smoking gun for the occurrence of grain condensation in the diffuse interstellar medium. Here we employ quantum chemical calculations to obtain the mid-infrared (IR) optical properties of a library of Mg-end member silicate nanoparticles with olivine (Mg_2SiO_4) and pyroxene ($MgSiO_3$) stoichiometries. We use this library as an input for a foreground-screen model to predict the spectral appearance of the absorption profile due to mixtures of bulk and nanoparticle silicates towards bright background sources. The mid-IR spectrum observed towards an O8V star or a carbon-rich Wolf–Rayet star starts to change when ~3% of the silicate mass is in the form of nanosilicates. We predict that a 3–10% nanosilicate fraction can be detected with the James Webb Space Telescope (JWST) using the mid-IR instrument (MIRI). With our upcoming JWST observations using MIRI, we will be able to detect or place limits on the nanosilicate content in the diffuse interstellar medium, and thus potentially directly confirm interstellar dust formation.

[a]*European Space Agency (ESA), European Space Research and Technology Centre (ESTEC), Keplerlaan 1, 2201 AZ Noordwijk, The Netherlands*

[b]*Departament de Ciència de Materials i Química Física, Institut de Química Teòrica i Computacional (IQTCUB), Universitat de Barcelona, c/Martí i Franquès 1-11, E-08028 Barcelona, Spain. E-mail: s.bromley@ub.edu*

[c]*Institut de Ciències de l'Espai (ICE, CSIC), Can Magrans, s/n, E-08193 Cerdanyola del Vallès, Barcelona, Spain*

[d]*Institució Catalana de Recerca i Estudis Avançats (ICREA), Passeig Lluís Companys 23, E-08010 Barcelona, Spain*

[e]*Institut d'Estudis Espacials de Catalunya (IEEC), E-08034 Barcelona, Spain*

[f]*Academia Sinica Institute of Astronomy and Astrophysics (ASIAA), s, 11F of AS/NTU Astronomy-Mathematics Building, No. 1, Sect. 4, Roosevelt Rd, Taipei 10617, Taiwan*

Introduction

Observations of the nearby universe are significantly affected by the presence of dust in the interstellar medium (ISM). The formation of most dust is thought to occur during high-temperature nucleation processes around dying stars[1,2] after which it is ejected into the diffuse ISM as small dust grains. Dust grains tend to be either carbonaceous or silicaceous in nature, with the latter being more common.[3] In the diffuse ISM the nascent grains are subject to energetic processing by supernovae shockwaves, cosmic rays[4] and ultraviolet (UV) radiation which lead to sputtering, shattering and ionization.[5,6] Dust grain processing in the ISM is predicted to create a huge population of nano-sized silicate grains[7] which could be astrochemically relevant for formation/dissociation of H_2,[8,9] and water molecules,[10,11] and are a prime candidate for being the source of the anomalous microwave emission (AME).[12–14] Due to the dominance of photoelectric electron ejection over recombination in very small grains in the ISM, nanosilicates are also likely to be positively ionized[15] enhancing their role as seed species for dust (re)nucleation.[16] Ionised nanosilicates could also act as sinks for oxygen depletion – potentially contributing to the unresolved problem of the "missing" oxygen in the ISM.[17,18]

Silicate dust is mainly thought to be Mg-rich and of olivine (Mg_2SiO_4) or pyroxene ($MgSiO_3$) composition[19] and originally more olivinic when first formed.[20] However, sputtering in the diffuse ISM leads to depletion of Mg cations[21,22] tending to relatively increase the fraction of pyroxenic dust.[23] Generally, the chemical composition of interstellar silicate dust is thus the result of a combination of production and destruction processes and varies considerably with the type of astrophysical environment. It is estimated that the rate of dust destruction by supernova shocks in the ISM is far in excess of the stardust production rate.[6,24] As such, we do not yet understand the observed abundance of dust in the ISM.

Infrared (IR) spectroscopy is ideal for identifying and quantifying the composition and structure of interstellar dust. Diagnostic bands for silicates fall almost exclusively in the near- and mid-IR region between 8 and 25 μm where features due to Si–O stretching (~10 μm) and O–Si–O bending (~18–20 μm) modes are dominant.[19] Until recently, only sightlines with high column densities of dust against bright background sources, including towards the Galactic Center (GC), provided sufficient signal-to-noise to study the dust composition and mineralogy. Indeed, much of our current information about dust composition is obtained from such sightlines. However, our understanding of these sightlines is complicated because they include both the diffuse ISM and denser molecular regions, for instance the GC sightline includes the Galactic *Central Molecular Zone* and three intervening spiral arms. Dense molecular clouds typically shield the dust grains from dust destruction processes, and they also facilitate grain growth, so a larger typical grain size is expected in the denser regions of the ISM. The highly sensitive spectrometers of the mid-IR instrument (MIRI) on the James Webb Space Telescope (JWST) allow for a detailed study of the characteristics of dust in the diffuse ISM, including the opportunity to detect silicate nanoclusters for the first time. Thus far, interpretations of silicate dust grain observations have employed laboratory-based measurements on bulk silicate samples.[20]

Nanosilicates, however, are known to have specific non-bulk-like structures with associated distinct infrared features,[25] which should be taken into account when interpreting IR observations.

In this work we use a combination of accurate quantum chemical calculations and models of the observational performance of JWST to predict potential IR signatures that could be used to determine the presence, abundance and composition of nanosilicate dust in the ISM. Observations of dust along the line of sight of twelve O and B stars in the diffuse ISM with JWST MIRI are already planned.[26] Here we present our modelled predictions for the diffuse ISM using an O8V star and for the dense ISM on a longer sightline using a carbon-rich Wolf–Rayet star.

Methodology

Computational chemistry calculations

We populated our spectral library using calculated IR spectra from nanoclusters of olivine ($(Mg_2SiO_4)_N$) and pyroxene ($(MgSiO_3)_N$) stoichiometry, with sizes $N = 1$–10. The structures of the nanosilicates were obtained from systematic global optimization searches using the Monte Carlo Basin-Hopping (MCBH) algorithm[27] and a genetic algorithm.[28] Full details on this procedure can be found in previous work.[29] As an example, the three lowest energy structures used for $(Mg_2SiO_4)_{10}$ are shown in Fig. 1. We determined the energies and structures of the 50 most stable nanosilicates from the global optimisation searches for each stoichiometry and size employing density functional theory (DFT) calculations using the hybrid PBE0 exchange and correlation functional[30] with a tier1/light numerical atom-centered orbital (NAO) basis set. This basis set has a quality similar to a valence triple-ζ plus polarization Gaussian-type orbital basis set.[31] For the DFT optimizations we used the all-electron, full potential Fritz Haber Institute *Ab Initio* Molecular Simulation package (FHI-AIMS).[32] For both nanosilicate stoichiometries and for each N, we then calculated the IR spectra for the five lowest energy isomers, except for the monomers (*i.e.* $N = 1$) for which we only included the lowest energy isomer. The resulting nanosilicate spectral library thus contains 92 distinct IR spectra (*i.e.* 46 from nano-pyroxenes and 46 from nano-olivines).

The harmonic IR spectra of 96 nanosilicates were obtained using the same DFT specification described above. We note that DFT-calculated IR vibrational frequencies using this approach agree very well with those obtained from high level quantum chemical calculations[33] and experimental data.[34] By summing all the as-calculated spectra for each stoichiometry separately, we obtain two stoichiometrically pure spectra which we take to be characteristic of pyroxene

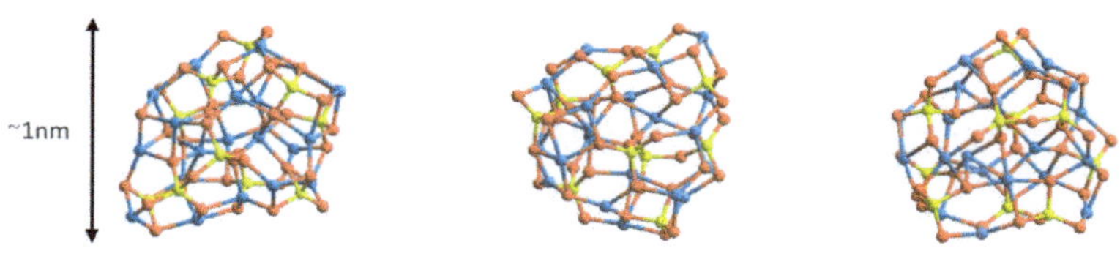

Fig. 1 Examples of optimized low energy nanosilicate structures with an olivinic ($(Mg_2SiO_4)_{10}$) chemical composition. Atom colour key: Si – yellow, Mg – blue, O – red.

nanosilicates and olivine nanosilicates (see Fig. 2a and b). Using these two pure spectra we can then create characteristic spectra for nanosilicate populations with mixed-stoichiometries by proportionally combining these pure spectra. In Fig. 2c–e we show the mixed stoichiometry nanosilicate spectra for three different pyroxene : olivine ratios. Note that we assume that there is no particular preference for size and/or energetic stability in producing these spectra, and all individual spectra have an equal weighting.

Observational modelling

In order to use the calculated nanosilicate spectra in astronomical extinction models, the intensities of nanosilicate IR spectra in kilometers per mole (km mol^{-1}) are converted to cross sections integrated over the frequency. The conversion is described in detail in paragraph 2.2 of Mattioda *et al.*,[35] who use the same conversion for polycyclic aromatic hydrocarbons (PAHs) from the NASA AMES PAH database.[36,37] Below, we summarize the conversion process.

The intensities of the spectral lines are designated as *A*-values (in km mol^{-1}), not to be confused with Einstein *A* coefficients. The cross section is related to the *A*-values in the following way:

$$\sigma_{\nu,\mathrm{int}} = \int_{\nu} \sigma(\nu)\mathrm{d}\nu = 10^5 \frac{c \cdot A\text{-value}}{N_\mathrm{A}}.$$

Here, the cross section is indicated by σ, which is expressed as integrated over the frequency ν (cm^{-1}), c is the speed of light and N_A is Avogadro's number.

For this study, we estimate that 3–10% of the silicates in the ISM have $\leq$1 nm diameter (as characterised by our set of nanosilicates – see Fig. 1). Although the amount of nanosilicates may vary for different sightlines depending on grain

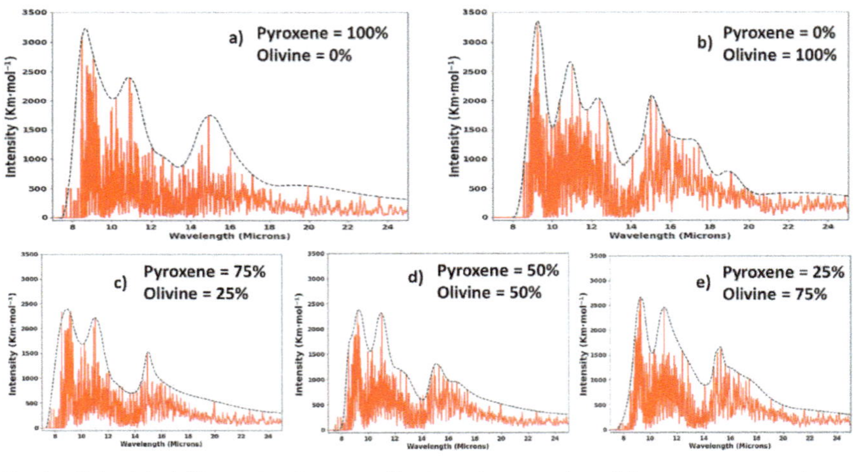

Fig. 2 Calculated IR spectra for nanosilicates where each red line corresponds to an individual vibrational mode. Top: pure stoichiometric nanosilicate IR spectra corresponding to: (a) 46 pyroxene nanosilicates and (b) 46 olivine nanosilicates. Bottom: mixed stoichiometric nanosilicate IR spectra for which the pyroxene : olivine ratio is varied as: (c) 3 : 1, (d) 1.1, (e) 1 : 3. The dashed lines indicate tentative approximations to the limiting overall spectral shape for more realistic larger nanosilicate populations.

destruction processes in the environment, we assume here that in both the dense and diffuse ISM this value may range between 3 and 10%. These percentages have been derived assuming an MRN size distribution,[38] which considers a size range 0.005–0.25 μm. We extended this distribution to the smallest nanosilicates (*i.e.* $N = 1$). Draine and Li[7,39] predict that similar quantities of nanosilicates could be present in the ISM. Based on IR emission spectra it is estimated that an upper limit mass fraction of ~15% of amorphous nanosilicates could have diameters of ≤1 nm. We note that all small nanosilicates in the size regime we consider are intrinsically non-crystalline (*i.e.* amorphous),[25] and so crystalline silicates are not relevant. The other 90–97% of the silicates modelled in this distribution consist of larger particles, which we refer to as bulk matter. For simplicity we assume that these grains are all 0.1 μm radii particles, which is the average size of grains in the ISM. These larger particles can have different shapes. We consider this using a continuous distribution of ellipsoids (CDE) model.[40]

The reddening by dust along the line of sight can be expressed in visual extinctions, A_V. On average there is about two magnitudes of visual extinction per kiloparsec (kpc).[41] The amount of extinction can be converted to an estimate of the column density of hydrogen particles along the line of sight toward a - background source: $\frac{N_H}{A_V} = 1.9 \times 10^{21}\ \text{cm}^{-2}\ \text{magn}^{-1}$.[42] From the column density N_H we can derive the number of silicon atoms along the line of sight using the abundance of this element. A fraction of the silicon atoms will be present in the gas phase, but most of them are depleted in dust. In this study we assume that 98% of the silicates are locked up in dust particles.[43,44] To calculate the extinction by dust along the line of sight we compute the optical depth from our modelling:

$$\frac{I}{I_0} = e^{-\tau} = e^{-N_d \cdot \sigma},$$

where I_0 is the intrinsic brightness of the source, I is the observed brightness, and τ is the optical depth. τ can be expressed as the column density of particles (N_d) times the extinction cross section (σ). We calculate the optical depth contribution for both the bulk silicates and the nanosilicate particles.

We model two scenarios:

(1) The nearby diffuse ISM

(2) The dense ISM in the Galactic Center area

Since we are observing the nanosilicates in extinction, we need to make use of a background source to observe the extinction by dust along the line of sight. For this study we selected sources that are featureless in the IR, because intrinsic features from the source may hinder the analysis of the nanosilicates. In the first scenario we use an O8V star. Main sequence O stars are bright in the IR and, most importantly, the stellar atmosphere can be easily modelled using stellar template models.[45,46] To model the nearby diffuse ISM, we use a star located at a distance of 2 kpc. In the case of scenario 1 we model the extinction along the line of sight with $A_V = 4$. This degree of extinction allows for sufficiently deep silicate features, as can be seen in Fig. 3, and is common in the nearby diffuse ISM. We model the bulk matter with an amorphous pyroxene ($MgSiO_3$) model consisting of particles of radius 0.1 μm as indicated above.[47] The medium-resolution spectroscopy (MRS) observing mode for MIRI on JWST, enables us to obtain detailed spectroscopic data between 4.9 to 27.9 μm with a resolving power of ~3700 and 1300.[48] The modelled spectra are shown with the

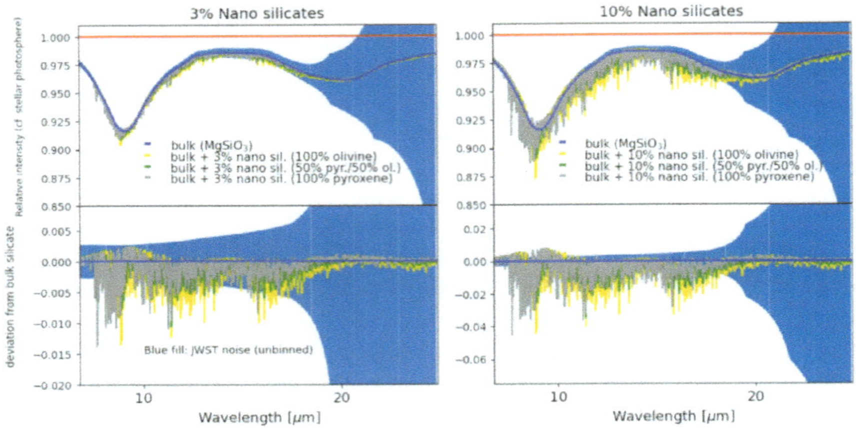

Fig. 3 Diffuse ISM: the upper two panels show absorption by dust relative to the photosphere of an O8V star (red line) for 3 and 10% nanosilicates along the line of sight. The blue line shows the contribution of 100% bulk matter modelled with $MgSiO_3$. Three models with nanosilicates are shown in the native (unbinned) resolution of JWST, containing different compositions: yellow 100% olivine composition, green 50% pyroxene and 50% olivine composition and grey 100% pyroxene composition. The models are shown with the uncertainties in the unbinned data (with blue fill). The two lower panels show the same models but highlight the deviation from the bulk matter.

uncertainties from MIRI MRS in the unbinned data (with blue fill). Signal-to-noise values were obtained using the JWST exposure time calculator (ETC).[49,50] The exposure time used in this modelling is 2220 seconds. We demonstrate the different nanosilicate compositions using three models from Fig. 2 – 100% olivine, 50% pyroxene + 50% olivine and 100% pyroxene. The silicate features are shown relative to the unabsorbed stellar continuum. The two different panels show the results for 3% and 10% of nanosilicates along the line of sight. In the case of the olivinic $MgSiO_3$ bulk silicate (blue line) we assume that 100% of the silicates are in the 0.1 μm particles.

In the second scenario we model the dense ISM along a more distant sightline of 8 kpc towards the Galactic Center area with an extinction of $A_V = 16$. In this case the source is a carbon-rich Wolf–Rayet star. This source is bright in the IR and we are able to observe these sources at high extinction.[23,51] As a template we use a WC7 star from the PoWR database with an effective temperature $T = 50$ K and radius $\log R_t = 1.0$ in solar radii.[52] Carbon rich Wolf–Rayet stars are assumed to be devoid of silicates. Therefore, we can assume that the silicate features around 10–20 μm belong entirely to interstellar dust. In this scenario, we model the dense ISM bulk matter using amorphous olivine (Mg_2SiO_4) with grain radii of 0.1 μm.[47] We also demonstrate the same three nanosilicate compositions as in scenario 1 in our modelling of this sightline. The modelling is shown in Fig. 4. The modelled exposure time is the same as in scenario 1.

Results and discussion

Fig. 3 and 4 show simulations of modelled JWST observations of two stars in different environments in the Galaxy. The difference in density between the two

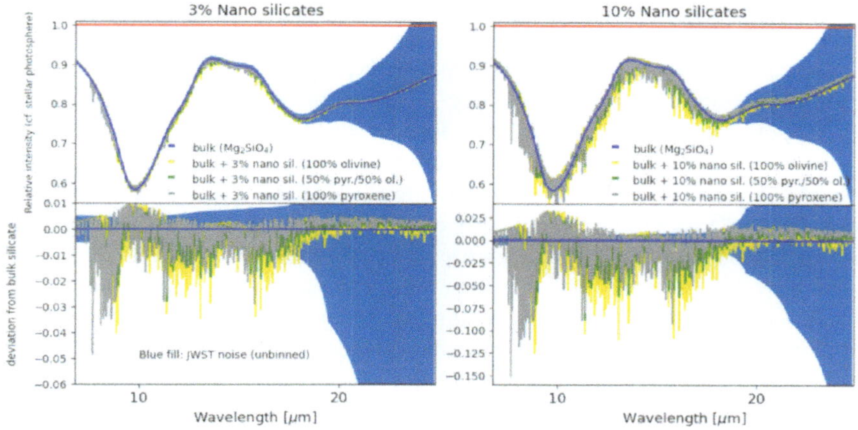

Fig. 4 Dense ISM: the upper two panels show absorption by dust relative to the photosphere of a Wolf–Rayet star (red line) for 3 and 10% nanosilicates along the line of sight. The blue line shows the contribution of 100% bulk matter modelled with Mg_2SiO_4. Three models with nanosilicates are shown in the native (unbinned) resolution of JWST, containing different compositions: yellow 100% olivine composition, green 50% pyroxene and 50% olivine composition and grey 100% pyroxene composition. The models are shown with the uncertainties in the unbinned data (with blue fill). The two lower panels show the same models but highlight the deviation from the bulk matter.

environments can be observed in the depth of the silicate features. We also made a distinction between dust species selected for the bulk matter. We assume that the diffuse ISM contains pyroxene silicates and the dense ISM contains olivine silicates. This distinction, although exaggerated for this analysis, is based on the prediction that pyroxene dust is more dominant in the diffuse ISM due to destruction events.[23] Recent findings from X-ray spectroscopy provide observational support for this effect, since the dense sightlines in these studies are dominated by olivine type,[53,54] whereas pyroxene is the dominant dust species along diffuse sightlines.[55]

From the modelling of both sightlines, we can distinguish different types of nanosilicates from the amorphous bulk silicates along the line of sight of two modelled sources. In the case of 3% of nanosilicates along the line of sight this is more challenging. However, we do note that the uncertainties (shown by the blue filled lines) are smaller than the silicate features and that this is the native (unbinned) resolution of MRS using MIRI on JWST. Binning the data will decrease the errors further and will highlight the general effect of the nanosilicates. In the case of 10% nanosilicates along the line of sight, we can clearly distinguish the effect of these particles in both scenarios: the deviation from the bulk matter as well as their different spectral features. The effect of the nanosilicates is most prominent in the spectra between 10 and 20 μm. From observations of dusty sightlines with the Spitzer Space Telescope it is already clear that this part of the spectrum cannot be fitted entirely with the existing silicate models and that excess extinction is observed between 10 and 20 μm.[56] Furthermore, the 10 μm feature itself also cannot be fitted well with bulk matter models. Van Breemen *et al.* concluded that although the best fitting dust to the feature around 10 μm consists of magnesium rich olivine, this does not explain the shift in the feature to shorter wavelength, which is exactly where the pyroxene nanosilicates dominate the spectrum.[57]

Our simulations of detection of nanosilicates open the possibility to study grain formation and destruction along these sightlines. The amount of nanosilicates gives an indication of the processes at work in these environments. We can try to piece together the cycle of shattering, destruction, formation, and grain growth. So far, grain formation has been invoked, as the residence time of silicates in the ISM is about ten times longer than the destruction time scale, but it has been assumed that this process occurs in molecular clouds, and it has not been directly observed.[58] By sampling different sightlines with a variety of distances and magnitudes of extinction, we can probe differences in various environments of the Galaxy. Of course, the distance to the clouds of gas and dust in the ISM is degenerate since we do not know exactly where along the line of sight these clouds are situated from just these observations. This problem can be partly solved by comparing the observations with Galaxy dust maps.[59] If the nanosilicates are abundantly present in the spectra we may observe areas along the line of sight with active grain formation in the diffuse ISM.

Conclusions

We have modelled the IR spectral features associated with Mg-rich olivine and pyroxene nanosilicate dust grains (and a 50 : 50 mixture) based on a library of nanosilicate IR spectra calculated using an accurate quantum chemical based approach. We consider the impact of intervening nanosilicates on mid-IR spectra of background stars in two distinct astrophysical environments, namely the nearby diffuse ISM ($A_V = 4$) and the dense galactic centre ($A_V = 16$). We compare the features imprinted by nanosilicates with bulk material of the same compositions, finding distinctive features and deviations at the few per cent level across mid-IR wavelengths. Differentiation between the species is easier toward denser intervening material, but still possible for the diffuse ISM at high signal-to-noise ratios. These models demonstrate the feasibility of the detection of nanosilicates in already planned JWST MIRI observations for a selection of nearby O stars. Furthermore, these observations will potentially allow us to directly trace grain destruction and grain growth in the ISM for the first time.

Conflicts of interest

There are no conflicts to declare.

Acknowledgements

STB acknowledges support from the MICINN funded project grants: PID2021-127957NB-I00 and TED2021-132550B-C21 and grant 2021SGR00354 funded by the Generalitat de Catalunya. This work was also partly supported by the Spanish program Unidad de Excelencia María de Maeztu *via* grants CEX2020-001058-M and CEX2021-001202-M, financed by MCINN. J. M. acknowledges the Generalitat de Catalunya for the pre-doctoral grant 2020 FI-B-00617. S. Z. acknowledges support from the European Space Agency (ESA) as an ESA Research Fellow. JPM acknowledges research support by the Ministry of Science and Technology of

Taiwan under grant MOST109-2112-M-001-036-MY3. We also acknowledge the Red Espanola de Supercomputación (RES) for the provision of supercomputing time.

References

1 H.-P. Gail and E. Sedlmayr, Inorganic dust formation in astrophysical environments, *Faraday Discuss.*, 1998, **109**, 303–319.

2 T. P. M. Goumans and S. T. Bromley, Efficient Nucleation of Stardust Silicates via Heteromolecular Homogeneous Condensation, *Mon. Not. R. Astron. Soc.*, 2012, **420**, 3344–3349.

3 A. G. G. M. Tielens, The Composition of Circumstellar and Interstellar Dust, *ASP Conf. Ser.*, 2001, **231**, 92.

4 E. M. Bringa, S. O. Kucheyev, R. A. Loeffler, *et al.*, Energetic Processing of Interstellar Silicate Grains by Cosmic Rays, *Adv. Pharm. J.*, 2007, **662**, 372.

5 A. P. Jones, A. G. G. M. Tielens and D. J. Hollenbach, Grain Shattering in Shocks: The Interstellar Grain Size Distribution, *Astrophys. J.*, 1996, **469**, 740.

6 A. P. Jones and J. A. Nuth III, Dust destruction in the ISM: a re-evaluation of dust lifetimes, *Astron. Astrophys.*, 2011, **530**, A44.

7 A. Li and B. T. Draine, On Ultrasmall Silicate Grains in the Diffuse Interstellar Medium, *Astrophys. J.*, 2001, **550**, L213–L217.

8 I. Oueslati, B. Kerkeni and S. T. Bromley, Trends in the adsorption and reactivity of hydrogen on magnesium silicate nanoclusters, *Phys. Chem. Chem. Phys.*, 2015, **17**, 8951–8963.

9 B. Kerkeni and S. T. Bromley, Competing mechanisms of catalytic H_2 formation and dissociation on ultrasmall silicate nanocluster dust grains, *Mon. Not. R. Astron. Soc.*, 2013, **435**(2), 1486–1492.

10 T. P. M. Goumans and S. T. Bromley, Hydrogen and oxygen adsorption on a nanosilicate – a quantum chemical study, *Mon. Not. R. Astron. Soc.*, 2011, **414**, 1285–1291.

11 B. Kerkeni, M.-C. Bacchus-Montabonel and S. T. Bromley, How hydroxylation affects hydrogen adsorption and formation on nanosilicates, *Mol. Astrophys.*, 2017, **7**, 1–8.

12 A. Macia Escatller and S. T. Bromley, Assessing the viability of silicate nanoclusters as carriers of the anomalous microwave emission: a quantum mechanical study, *Astron. Astrophys.*, 2020, **634**, A77.

13 B. S. Hensley and B. T. Draine, Modeling the Anomalous Microwave Emission with Spinning Nanoparticles: No PAHs Required, *Astrophys. J.*, 2017, **836**, 179–192.

14 T. Hoang, N.-A. Vinh and N. Q. Lan, Spinning Dust Emission from Ultra-Small Silicates: Emissivity and Polarization Spectrum, *Astrophys. J.*, 2016, **824**, 18–28.

15 J. C. Weingartner and B. T. Draine, Photoelectric Emission from Interstellar Dust: Grain Charging and Gas Heating, *Astrophys. J. Suppl.*, 2001, **134**, 263.

16 S. Zhukovska, C. Dobbs, E. B. Jenkins and R. S. Klessen, Modeling Dust Evolution in Galaxies with a Multiphase, Inhomogeneous ISM, *Astrophys. J.*, 2016, **831**, 147.

17 D. C. B. Whittet, Oxygen Depletion in the Interstellar Medium: Implications for Grain Models and the Distribution of Elemental Oxygen, *Astrophys. J.*, 2010, **710**, 1009–1016.

18 J. Mariñoso Guiu, B.-A. Ghejan, T. M. Bernhardt, J. M. Bakker, S. M. Lang and S. T. Bromley, Cluster Beam Study of $(MgSiO_3)^+$-Based Monomeric Silicate Species and Their Interaction with Oxygen: Implications for Interstellar Astrochemistry, *ACS Earth Space Chem.*, 2022, **6**(10), 2465–2470.
19 T. Henning, Cosmic Silicates, *Annu. Rev. Astron. Astrophys.*, 2010, **48**, 21.
20 J. Dorschner, B. Begemann, Th. Henning, C. Jäger and H. Mutschke, Steps toward interstellar silicate mineralogy. II. Study of Mg-Fe-silicate glasses of variable composition, *Astron. Astrophys.*, 1995, **300**, 503–520.
21 L. P. Keller and D. S. McKay, The nature and origin of rims on lunar soil grains, *Geochim. Cosmochim. Acta*, 1997, **61**, 2331.
22 J. P. Bradley, Chemically anomalous, preaccretionally irradiated grains in interplanetary dust from comets, *Science*, 1994, **265**, 925.
23 F. Kemper, W. J. Vriend and A. Tielens, The absence of crystalline silicates in the diffuse interstellar medium, *Astrophys.J.*, 2004, **609**, 826.
24 J. D. Slavin, E. Dwek and A. P. Jones, Destruction of Interstellar Dust in Evolving Supernova Remnant Shock Waves, *Astrophys. J.*, 2015, **803**, 7.
25 A. Macia Escatller, T. Lazaukas, S. M. Woodley and S. T. Bromley, Structure and Properties of Nanosilicates with Olivine $(Mg_2SiO_4)_N$ and Pyroxene $(MgSiO_3)_N$ Compositions, *ACS Earth Space Chem.*, 2019, **3**, 2390.
26 Illuminating the dust properties in the diffuse ISM with JWST, Zeegers *et al.*, 2021, JWST Proposal. Cycle 1, ID. #2183.
27 D. Wales and J. P. K. Doye, Global Optimization by Basin Hopping and the Lowest Energy Structures of Lennard-Jones Clusters Containing up to 110 Atoms, *J. Phys. Chem. A*, 1997, **101**(97), 5111–5116.
28 T. Lazauskas, A. A. Sokol and S. M. Woodley, An Efficient Genetic Algorithm for Structure Prediction at the Nanoscale, *Nanoscale*, 2017, **9**(11), 3850–3864.
29 A. Macià Escatllar, T. Lazaukas, S. M. Woodley and S. T. Bromley, Structure and Properties of Nanosilicates with Olivine $(Mg_2SiO_4)_N$ and Pyroxene $(MgSiO_3)_N$ Compositions, *ACS Earth Space Chem.*, 2019, **3**, 2390–2403.
30 C. Adamo and V. Barone, Toward Reliable Density Functional Methods without Adjustable Parameters: The PBE0 Model, *J. Chem. Phys.*, 1999, **110**, 6158–6170.
31 O. Lamiel-Garcia, K. C. Ko, J. Y. Lee, S. T. Bromley and F. Illas, When Anatase Nanoparticles Become Bulklike: Properties of Realistic TiO_2 Nanoparticles in the 1–6 nm Size Range from All Electron Relativistic Density Functional Theory Based Calculations, *J. Chem. Theory Comput.*, 2017, **13**, 1785–1793.
32 V. Blum, R. Gehrke, F. Hanke, P. Havu, V. Havu, X. Ren, K. Reuter and M. Scheffler, Ab Initio Molecular Simulations with Numeric Atom-Centered Orbitals, *Comput. Phys. Commun.*, 2009, **180**, 2175–2196.
33 J. Mariñoso Guiu, A. M. Escatllar and S. T. Bromley, How does temperature affect the infrared vibrational spectra of nanosized silicate dust?, *ACS Earth Space Chem.*, 2021, **5**, 812–823.
34 J. Mariñoso Guiu, B.-A. Ghejan, T. M. Bernhardt, J. M. Bakker, S. M. Lang and S. T. Bromley, *ACS Earth Space Chem.*, 2022, **6**(10), 2465–2470.
35 A. L. Mattioda, D. M. Hudgins, C. Boersma, A. Ricca, E. Peeters, J. Cami, F. Sánchez de Armas, G. Puerta Saborido, C. W. J. Bauschlicher and L. J. Allamandola, The Nasa Ames Pah Ir Spectroscopic Database: The Laboratory Spectra, *Astrophys. J., Suppl. Ser.*, 2020, **251**(22).

36 C. Boersma, C. W. Bauschlicher Jr, A. Ricca, A. L. Mattioda, J. Cami, E. Peeters, F. Sánchez de Armas, G. Puerta Saborido, D. M. Hudgins and L. J. Allamandola, The Nasa Ames Pah Ir Spectroscopic Database Version 2.00: Updated Content, Website And On/Offline Tools, *Astrophys. J. Suppl.*, 2014, **211**(1), 8.

37 https://www.astrochem.org/pahdb/.

38 J. S. Mathis, W. Rumpl and K. H. Nordsieck, The size distribution of interstellar grains, *Astrophys. J.*, 1977, **217**, 425.

39 B. T. Draine and A. Li, Infrared Emission from Interstellar Dust. I. Stochastic Heating of Small Grains, *Astrophys. J.*, 2001, **551**, 807–824.

40 C. F. Bohren and D. R. Huffman, *Absorption and Scattering of Light by Small Particles*, Wiley, New York, 1983.

41 A. Tielens, *The Physics and Chemistry of the Interstellar Medium*, Cambridge University Press, Cambridge, 2005.

42 R. C. Bohlin, B. D. Savage and J. F. Drake, A survey of interstellar H I from L-alpha absorption measurements. II, *Astrophys. J.*, 1978, **224**, 132–142.

43 J. Wilms, A. Allen and R. McCray, On the Absorption of X-Rays in the Interstellar Medium, *Astrophys. J.*, 2000, **542**, 914–924.

44 E. B. Jenkins, A Unified Representation of Gas-Phase Element Depletions in the Interstellar Medium, *Astrophys. J.*, 2009, **700**, 1299–1348.

45 *Spanish virtual observatory*, https://svo2.cab.inta-csic.es/theory/newov2/index.php.

46 F. Allard, D. Homeier and B. Freytag, *ASP Conference Series*, 2011, vol. 448, p. 91.

47 C. Jäger, J. Dorschner, H. Mutschke, T. Posch and T. Henning, Steps toward interstellar silicate mineralogy. VII. Spectral properties and crystallization behaviour of magnesium silicates produced by the sol-gel method, *Astron. Astrophys.*, 2003, **408**, 193–204.

48 https://jwst-docs.stsci.edu/jwst-mid-infrared-instrument/miri-observing-modes/miri-medium-resolution-spectroscopy.

49 K. M. Pontoppidan, T. E. Pickering, V. G. Laidler, *et al.*, *Proc. SPIE 9910, Observatory Operations: Strategies, Processes, and Systems VI*, 2016, p. 991016.

50 https://jwst.etc.stsci.edu/.

51 J. E. Chiar and A. G. G. M. Tielens, Pixie Dust: The Silicate Features in the Diffuse Interstellar Medium, *Astrophys. J.*, 2006, **637**, 774–785.

52 https://www.astro.physik.uni-potsdam.de/%7Ewrh/PoWR/wc/

53 S. T. Zeegers, Dust absorption and scattering in the silicon K-edge, *Astron. Astrophys.*, 2019, **627**, A16.

54 D. Rogantini, Magnesium and silicon in interstellar dust: X-ray overview, *Astron. Astrophys.*, 2020, **641**, A149.

55 I. Psaradaki, Oxygen and iron in interstellar dust: an X-ray investigation, *Astron. Astrophys.*, 2023, **670**, A30.

56 S. Fogerty, W. Forrest, D. M. Watson, B. A. Sargent and I. Koch, Silicate Composition of the Interstellar Medium, *Astrophys. J.*, 2016, **830**, 71.

57 J. M. van Breemen, The 9.7 and 18 μm silicate absorption profiles towards diffuse and molecular cloud lines-of-sight, *Astron. Astrophys.*, 2011, **526**, A152.

58 S. Zhukovska, C. Dobbs, E. B. Jenkins and R. S. Klessen, Modeling Dust Evolution in Galaxies with a Multiphase, Inhomogeneous ISM, *Astrophys. J.*, 2016, **831**, 147.

59 R. Lallement, Three-dimensional maps of interstellar dust in the Local Arm: using Gaia, 2MASS, and APOGEE-DR14, *Astron. Astrophys.*, 2018, **616**, A132.

Faraday Discussions

DISCUSSIONS

Computational astrochemistry: general discussion

Stefan T. Bromley, Sandra Brünken, Cecilia Ceccarelli, Martin Cordiner, Kristen Darnell, Gwenaëlle Dufour, Athena Flint, Robin T. Garrod, Murthy S. Gudipati, Joshua Halpern, Eric Herbst, Ko-Yun Huang, Inga Kamp, Nikku Madhusudhan, Martin R. S. McCoustra, Arthur G. Suits, Marie Van de Sande, Ewine F. van Dishoeck, Serena Viti and Laurent Wiesenfeld

DOI: 10.1039/d3fd90027d

Kristen Darnell opened a discussion of the paper by Serena Viti: You discussed using machine learning to determine the binding energies that are important to consider. Is your goal to use these binding energies in more traditional astrochemical models or are you envisioning a different approach?

Serena Viti replied: The motivation behind this study (https://doi.org/10.1039/d3fd00007a) was two fold: first of all, it is clear that binding energies are key parameters that govern the gas–grain "interactions"; secondly, there are disagreements on some of the values of the binding energies in the literature and, as we have seen at this Faraday Discussion (see also Table 4 in ref. 1), clear evidence that there is not just one value of binding energy per species. This particular study was an exploration of different methodologies to constrain binding energies. Indeed, at least from my side, once constrained, I would use the values (or the range of values) of the binding energies in more traditional astrochemical models.

1 J. Heyl, J. Holdship and S. Viti, *Astrophys. J.*, 2022, **931**, 26, DOI: 10.3847/1538-4357/ac6606.

Eric Herbst commented: While it is likely true, if I understand correctly, that machine learning cannot handle chemical and physical variations at the same time, it is true that they can handle chemistry and physics in the same sense that gas–grain kinetic simulations do. In other words, if we study cold cores such as TMC-1 and hot cores such as Orion, we will get different and possibly better chemical answers for the different physical conditions. For example, with the physical conditions in TMC-1, it has been shown that a deep learning approach does a slightly better job than kinetic models in reproducing the measured abundance of the anionic carbon chain $C_{10}H^-$.[1]

1 E. Bianchi, A. Remijan, C. Codella, C. Ceccarelli, F. Lique, S. Spezzano, N. Balucani, P. Caselli, E. Herbst, L. Podio, C. Vastel and B. McGuire, *Astrophys. J.*, 2023, **944**, 208.

Serena Viti answered: Yes, it all depends on the problem and its dimensionality. I do not think it's a matter of physics *vs.* chemistry.

Marie Van de Sande asked: What are the advantages of using machine learning over the more "classical" Markov chain Monte Carlo (MCMC) approach for the forward problem (the sensitivity analysis of the model)? Does it depend on the particular problem we want to address or is it just a question of speed?

Serena Viti responded: For the problems we have considered so far, the primary advantage is simply speed. It would be straightforward to perform the interpretability analysis using the original code instead of the emulator. However, the interpretability analysis requires many evaluations of the forward model (of the order of tens of thousands). With each evaluation taking 1–2 minutes, this becomes unfeasible. By using an emulator that approximates the forward model well, we can significantly reduce the time taken for this analysis.

Inga Kamp commented: I have a follow-up question on this. Your astrophysical environment contains physics and chemistry and these are included in the machine learning (ML) (and underlying astrophysical models) as free parameters. Should we deal with the chemical and physical parameters at the same level in these ML approaches, and so treat them on equal footing, or should we disentangle them?

Serena Viti replied: It really depends on what problems one is tackling. For example, our group has done some work on disentangling the physical and chemical parameters[1] in the context of disentangling multiple gas components in extra-galactic beams and trying to retrieve the physical parameters of each of these components; however, this is not something we have focused on for this paper, where we were concentrating on the details of the chemistry.

1 D. de Mijolla, S. Viti, J. Holdship, I. Manolopoulou and J. Yates, *Astron. Astrophys.*, 2019, **630**, A117.

Marie Van de Sande answered: Disentangling physics and chemistry is currently being done in the asymptotic giant branch (AGB) community. Maes *et al.*[1] have assessed the sensitivity of the predicted gas-phase chemistry to the input physics of the model (the density and temperature profiles). We are working on a sensitivity analysis of the chemical network, keeping the physics constant. Combining the two would be an interesting next step.

1 S. Maes, M. Van de Sande, T. Danilovich, F. De Ceuster and L. Decin, *Mon. Not. R. Astron. Soc.*, 2023, **522**, 4654.

Ko-Yun Huang remarked: This can be tricky sometimes. As we presented in our work (https://doi.org/10.1039/d3fd00007a), in "most" cases we find the physical modeling result is compatible with the chemical modeling result. However, there are very few cases where we found no match from these two sides; for example, in one case we could not find sufficient chemical abundance produced by the chemical modeling based on the gas physical condition, as derived from our physical modeling with radiative transfer analysis. This could arise from the incomplete chemical network involved, potentially some flaws from the physical modeling still out of our current radar, or other factors. One needs to be cautious with consistency/compatibility between the physical and chemical modeling before combining them for further analysis and interpretation.

Serena Viti opened the discussion of the paper by Robin T. Garrod: On the subject of charges of grains and how we deal with them in models, at the moment most models consider only an average charge per grain. How do you deal with the charges in your model?

Robin T. Garrod replied: The MAGICKAL model uses the same method as was implemented in the earlier Ohio State University (OSU) gas-phase and gas–grain codes; two populations of grains are modeled, corresponding to (i) a neutral grain and (ii) a grain with a single negative charge. The two populations are treated as additional chemical species that can interact with charged gas-phase species. For example, an electron may attach to the neutral grain; certain abundant atomic and molecular ions are then allowed to adsorb to the negatively charged grain, resulting in immediate electronic recombination and the desorption of the products. The intention is mainly to ensure that negatively charged grains can help neutralize positively charged gas-phase species, and this also allows the fraction of charged *versus* neutral grains to be calculated in the model in a simple way.

However, this treatment is entirely independent of the grain-surface chemical network in which we are typically most interested; the two treatments run in parallel. Thus, there is no explicit treatment of charge in the grain chemistry – it is all neutral. This is certainly something that will need further attention from all gas–grain modelers over the coming years.

Robin T. Garrod opened a general discussion: We should avoid thinking about whether the models are "wrong"; rather it should be "by how much they are wrong". The models aren't reality, and in a sense they are all "wrong" to some degree. The models are just a representation of our current understanding. But, at the same time, they are the best way of synthesizing all of what we think we know. We can no longer do back-of-an-envelope astrochemistry calculations; the systems are too complex, so complicated models are necessary. But there is the issue of using the correct prefactors for desorption rate calculations in the models – this is something that I would like to improve.

I would also point out that the context in which a kinetic model is applied is important to the question of which aspects of the models are more or less correct. For example, when treating grain-surface chemistry involving atomic hydrogen, one question is: how fast does H diffuse on the grain/ice surface? This question might appear at first glance to be of premier importance. But in regions of parameter space where H diffusion is fast, compared to other surface processes, the precise diffusion rate may be unimportant in determining the overall production rate for many reactions.[1] This is because, under such conditions, diffusion is so fast that it is no longer the rate limiting step. Instead, it is the rate of delivery of H atoms to the grain surface that determines the overall reaction rate (*i.e.* the system is in the "accretion limit"). This is just an example, but it shows that sometimes even certain fairly fundamental quantities may not matter at all to the results when modeling kinetics. The bigger danger may in fact be the failure to properly treat the kinetics of a particular type of process from a mathematical point of view, rather than just whether an inaccurate input value has been used. The above example would be a case in point, in which a completely different method of calculating the reaction rates is necessary when in the accretion limit.

1 R. T. Garrod, *Astron. Astrophys.*, 2008, **491**, 239.

Cecilia Ceccarelli responded: I agree that models are "just a representation of our current understanding", I couldn't have said it better! I, however, don't agree that the more models are complex, the better: sometimes, making them more complicated just puts our problems of ignorance under a thicker rug, actually making them more difficult to pin down. I add that there is also the danger of thinking that the more sophisticated the model, the more reliable it is, which obviously is not true. As also emphasized by Murthy's comment in an earlier session (Session 3), it is extremely important to use the couple binding energy AND prefactor correctly. This is another field where our ignorance is still very big.

Finally, I am not sure that I fully agree with the first part of the comment on the H diffusion. When a lot of uncertain values and processes are included in a model, we cannot be sure what matters when.

On the other hand, I fully agree on the second part of the comment, about how we model the kinetics of the processes being something we should also place attention on.

Murthy S. Gudipati addressed Robin T. Garrod and Serena Viti: Have you considered charged ice grains for grain-accretion or gas–grain reactions? Water clusters are known to stabilize electrons by up to 4 eV and similarly, if ice grains contain polycyclic aromatic hydrocarbons (PAHs), these PAHs can easily be ionized, leading to electron and ion entrapment in the ice grain (see our recent publication and references therein[1]). Further, it is known that the electronegativity of OH is very high, trapping electrons to form OH^-. Based on all these facts, it is highly likely that interstellar and even protoplanetary ice grains that receive moderate amounts of radiation tend to become charged (also see ref. 2 and 3).

1 A. Lignell, L. I. Tenelanda-Osorio and M. S. Gudipati, *Chem. Phys. Lett.*, 2021, **778**, 138814.
2 L. Matteini, T. S. Horbury, F. Pantellini, M. Velli and S. J. Schwartz, *Astrophys. J.*, 2015, **802**, 11.
3 S. F. Tigik, A. Vaivads, D. M. Malaspina and S. D. Bale, *Astrophys. J.*, 2022, **936**, 7.

Robin T. Garrod answered: As noted in my response to an earlier question, charged grains are included in a more or less trivial way in the case of accretion of charged species onto the grains. These are all very interesting and important points; however, I believe that none of the astrochemical models are presently configured to take such effects into account in a meaningful way, other than work that Eric Herbst and his group have been doing in considering specific ion–molecule surface reactions. I think that the consideration of charges on grains will be an important area for model development over the coming years.

Serena Viti replied: The models we used for our statistical and ML works do not include grain charges. In general, for this particular work so far, our chemical networks on the surface were simplified in order to reduce dimensionality. I agree that standard chemical models may need to consider charged ice grains when modelling specific interstellar medium (ISM) regions.

Arthur G. Suits returned to the discussion of the paper by Robin T. Garrod: Did you examine the effect of the overall ionization rate? I can imagine if it is reduced it will lower the overall impact on the subsequent reactions of course, but I am wondering if it would affect the relative impact on particular reactions.

Robin T. Garrod responded: We have not done an investigation into the effects of varying the cosmic-ray ionization rate (CRIR) specifically in this work (https://doi.org/10.1039/d3fd00014a). However, we do have a large model data grid available for an earlier version of the models (essentially Model M1 shown in this paper), which includes variation of the CRIR. Within a reasonable range of CRIR values (say an order of magnitude above and below the canonical rate of $\sim 10^{-17}\ s^{-1}$), the influence of specific reactions on the overall chemistry is similar.

Nikku Madhusudhan returned to the discussion of the paper by Serena Viti: The problem setup you described is close to what we have in atmospheric retrieval of exoplanets. We started with MCMC, but in the last decade we have moved away from MCMC and explored other Bayesian parameter estimation techniques like nested sampling,[1] which can handle large numbers of parameters and are highly parallelizable. We have also explored ML,[2] but so far the Bayesian approaches are still the bread-and-butter, because they are much more scalable and more efficient than ML. Have these approaches been explored in enough detail in your area before switching to ML?

1. N. Madhusudhan, in *Handbook of Exoplanets*, ed. H. J. Deeg and J. A. Belmonte, Springer International Publishing AG, 2018, pp. 2153–2182, DOI: 10.1007/978-3-319-55333-7.
2. M. C. Nixon and N. Madhusudhan, *MNRAS*, 2020, **496**, 269–281, DOI: 10.1093/mnras/staa1150.

Serena Viti answered: In fact, we only utilised machine learning in order to accelerate the machine learning interpretability process and did not use it to infer the parameters themselves. However, we have also moved from MCMC to nested sampling over the last few years (*e.g.* ref. 1), hence our use of the UltraNest algorithm. Despite its superior efficiency, it is not able to necessarily overcome the data sparsity that plagues the area of astrochemistry. We are in a situation where we have more parameters than data points, leading to great degeneracy. This is ultimately the limiting factor and we have found that the choice of inference algorithm is unable to bridge the gap.

1 J. Heyl, E. Sellentin, J. Holdship and S. Viti, *MNRAS*, 2022, **517**, 38–46, DOI: 10.1093/mnras/stac2652.

Nikku Madhusudhan asked: Can you throw some more light on the dimensionality of your problem, *e.g.*, how big is the parameter space and the computational demands involved?

Serena Viti replied: The dimensionality depends on the problem we tackled, *e.g.*, when dealing with inferring posterior distributions for rates of surface reactions then the number of dimensions is high. However for this particular paper (https://doi.org/10.1039/d3fd00007a), the dimensionality is 9, which corresponds to the number of binding energies we consider. However, performing the Bayesian inference using the original data in Table 1 in the paper was difficult to do, even with a parallelised UltraNest (a nested sampling algorithm) implementation spread over 128 cores and several days. It is for this reason that we ended up increasing the relative uncertainty of the constraints in order to allow the sampler to converge faster. Even with 128 cores, each run could take up to a day. The Massive Optimised Parameter Estimation and Data (MOPED) algorithm was not computationally expensive to run. The machine learning interpretability was not inherently expensive to run, but was dependent on running the UCLCHEM forward model tens of thousands of times. With each evaluation taking between 1–2 minutes, this would be unfeasible to run quickly, which is why we implemented a statistical emulator.

Sandra Brünken returned to the discussion of the paper by Robin T. Garrod: In your paper (https://doi.org/10.1039/d3fd00014a) you discuss cases where dissociative recombination with electrons leads to destruction of protonated complex organic molecules (COMs). However, as you discuss in your introduction, there are also cases where protonation itself, *e.g.*, proton transfer from H_3^+ or H_3O^+, leads directly to dissociation in cases where the proton affinity of the COM is very large. The isomers methyl formate, acetic acid, and glycolaldehyde have been shown, for example, to behave very differently upon protonation, with methyl formate and acetic acid being dissociated upon proton transfer from H_3^+ and other proton donors, whereas glycolaldehyde can be protonated by H_3^+.[1] Are these processes included in the models, and what influence do they have on the relative abundance of isomers?

1 P. A. Lawson, D. S. Osborne, Jr. and N. G. Adams, *J. Phys. Chem. A*, 2012, **116**, 2880.

Robin T. Garrod answered: The models presented here only include non-dissociative proton transfer for the reactions of H_3^+ with those COMs. However, calculations have been done for methyl formate,[1] in which, while dissociation dominates, there is a non-dissociative branch (~17%). Those authors claim that using these rates and branching ratios (along with some others that they calculated for other species) in astrochemical simulations, rather than the standard, entirely non-dissociative branches used in the existing networks, they find variations in the relevant species' abundances of up to ~40%.

For the reaction of methyl formate with H_3O^+ (the most likely reaction partner of methyl formate), proton transfer appears to be non-dissociative, based on experimental evidence from some decades ago.[2]

1 D. Ascenzi, A. Cernuto, N. Balucani, P. Tosi, C. Ceccarelli, L. M. Martini and F. Pirani, *Astron. Astrophys.*, 2019, **625**, A72.
2 A. C. Hopkinson, G. I. Mackay, and D. K. Bohme, *Can. J. Chem.*, 1979, **57**, 2996.

Eric Herbst responded: Isomers of COMs are very important in distinguishing the chemistry of different interstellar clouds, and differences in dissociative recombination and large proton affinities should also be looked at carefully. Dissociative recombination with electrons and COMs (*e.g.* $CH_3OH_2^+$ + e) have been studied experimentally using storage rings, and the different products measured quantitatively. We have included a number of product distributions to the best of my knowledge, and estimated others. The situation regarding proton transfer involving COMs is mentioned in our *Faraday Discussions* paper (https://doi.org/10.1039/d3fd00014a), which references Garrod *et al.* (2022), a paper in the *Astrophysical Journal, Supplement Series.*[1] More work is certainly needed, especially in COM isomers. We certainly cannot guarantee that all important cases of these two types of processes involving COMs have been studied carefully. We are certainly aware of their existence and hope to improve upon our current basis of knowledge. We welcome new experiments.

1 R. T. Garrod, M. Jin, K. A. Matis, D. Jones, E. R. Willis and E. Herbst, *Astrophys. J., Suppl. Ser.*, 2022, **259**, 1.

Sandra Brünken addressed Robin T. Garrod and Eric Herbst: As a follow-up to my previous question, and in light of earlier discussions throughout the days on which systems are important to study in the laboratory: would it be helpful to study the dissociative proton transfer to, *e.g.*, methyl formate and acetic acid in the laboratory? Are these rate coefficients and branching ratios important for the overall COM chemical network?

Robin T. Garrod replied: I would very much welcome measurements of this sort, particularly if branching ratios can be obtained. It is likely that the most important ion–molecule reactions leading to protonation of COMs in the hot-core scenario would be those involving transfer from protonated water. We assume, as a default position in the model (in the absence of other information), that proton transfer from H_3O^+ is non-dissociative, but this may vary depending on the proton affinities involved. Proton transfer from H_3^+ is more likely to be dissociative, although the abundance of this ion should be rather low relative to that of H_3O^+. But with a large enough fraction of reactions leading to dissociation, in the case of a particular COM, it is possible that proton transfer from H_3^+ could affect the peak gas-phase abundances and lifetimes for some molecules. However, we do not have much direct information from the laboratory for the individual branching ratios for H_3^+ reactions.

The sensitivity of the model results to the branching ratios (*e.g.* proton transfer *versus* dissociation) for particular reactions is something that we could test, if there is interest in pursuing such reactions in the lab.

Eric Herbst answered: Dissociative proton transfer could probably be measured in more systems than currently known, especially involving COMs, but we have attempted to include them when they are exothermic. You might look at the Anicich and Huntress collection (1986)[1] for a large listing of ion–neutral reactions, often with products. We would indeed like to see the results of experiments involving COMs.

1 V. G. Anicich and W. T. Huntress, Jr, *Astrophys. J., Suppl. Ser.*, 1986, **62**, 553.

Ko-Yun Huang commented: This article (https://doi.org/10.1039/d3fd00014a) investigated the proton transfer occurrence in the gas phase. Would proton transfer on ice be interesting to explore as well? Maybe you have considered it? What could the potential differences in these two scenarios be?

Martin R. S. McCoustra said: It is interesting to see the discussion come around to the issue of ion–molecule chemistry at surfaces. We have reported measurements on the interaction of H^+ and D^+ with H_2O and D_2O surfaces, in which we observed H_2^+ and D_2^+ formation mass spectrometrically.[1] This broad area remains a major omission in chemical networks, as highlighted by Woon.[2] Perhaps we need to explore this area more thoroughly both computationally and experimentally.

1 S. Bag, M. R. S. McCoustra and T. Pradeep, *J. Phys. Chem. C*, 2011, **115**, 13813.
2 D. E. Woon, *Acc. Chem. Res.*, 2021, **54**, 490.

Martin R. S. McCoustra opened a general discussion: My first foray into studying the chemistry and physics of icy solid surfaces focussed on the destruction of halogen reservoirs in the Antarctic stratosphere.[1] The reaction

$$ClONO_2(g) + HCl(g) \rightarrow Cl_2(g) + HNO_3(g)$$

is known to be very slow. However, the icy-surface-catalysed chemistry, proceeding *via* the corresponding adsorbed species

$$HCl(ads) + nH_2O(s) \rightarrow H_3O^+(ads) + Cl^-(ads)$$

$$ClONO_2(ads) + Cl^-(ads) \rightarrow Cl_2(g) + NO_3^-(ads)$$

$$H_3O^+(ads) + NO_3^-(ads) \rightarrow HNO_3(ads),$$

which undergo dissociative ionisation to yield the corresponding hydrated ionic species, is known to be very much faster. Perhaps it might be useful to reflect on this difference in rates of gas-phase *vs.* condensed-phase chemistry when looking at COM-forming organic chemistry. The formation of esters and amides, for example, is known to be proton-catalysed in the liquid phase and likely to be so in the solid state. Should we be looking towards investigating this type of chemistry and then including it in relevant models?

1 T. G. Koch, S. F. Banham, J. R. Sodeau, A. B. Horn, M. R. S. McCoustra and M. A. Chesters, *J. Geophys. Res. D*, 1997, **102**, 1513–1522.

Robin T. Garrod responded: This is an interesting point. I would certainly advocate for further work to be done in his area. The modeling framework for treatment of such reactions is not really existent at this point, so it would not be trivial to include them in the models in a systematic way. This is an area where I think the models will have to be improved in the coming years, but we will certainly require substantial guidance both from experimental and quantum chemists in putting together the kinetic networks and framework for introducing this chemistry into the models.

Eric Herbst commented: In trying to understand the chemistry involved in the topmost granular surface layers of water ice, there is still much to be learned. Whether the surface layers are composed mainly of ice, or whether they are mainly silicates or even a mixture of molecules such as CO and CO_2, the motions of molecules have to be treated so as to result in surface reactions. Until recently, simple diffusive motion was utilized, in which pairs of molecules would move quasi-randomly and react when they approached one another in a potential well. The motion, however, is slowed down *via* so-called diffusive barriers or by chemical activation energy between the two reactants. More recently, it was suggested that the rate of surface reactions can be faster in the absence of diffusive motion as long as one of the reactants has a high abundance and moves slowly, if at all, allowing a fast-moving light species to react with it. As an example, the fast-moving OH, formed by the rapid diffusion of H and O, can subsequently react with a nearby static CO to form CO_2 + H.[1] The purely diffusive motion involving CO and O is slow because of a larger barrier.

1 R. T. Garrod, M. Jin, K. A. Matis, D. Jones, E. R. Willis and E. Herbst, *Astrophys. J. Suppl.*, 2022, **259**, 1.

Martin R. S. McCoustra replied: Yes, indeed, it is possible to consider the classical Langmuir–Hinshelwood chemistry of reactant diffusion being modified to Kasemo–Harris chemistry, where one of the reactants carries sufficient energy (from a preceding radical recombination reaction or photodissociation process) to be placed in a weakly adsorbed state lying above the classical diffusion barrier but below the desorption barrier, such that it rapidly samples the adjacent surface without the influence of the diffusional barrier (akin to a precursor state to adsorption).

Ewine F. van Dishoeck returned to the discussion of the paper by Serena Viti: I appreciate the analysis in your paper (https://doi.org/10.1039/d3fd00007a) and what has been said in the discussion so far about machine learning. Can you elaborate some more on the outcomes of your analysis, in particular what we are now learning about molecular astrophysics? For example, your Fig. 6 shows the distribution of most probable binding energies for atomic H to a grain surface to explain the water ice observations. But the most probable value of about 1100 K seems to be quite large compared with what has been recommended based on laboratory experiments,[1] which give a range of 300–700 K with a recommended value of 450 K on amorphous water ice. So how should we now interpret the higher value that machine learning is telling us?

1 M. Minissale, Y. Aikawa, E. Bergin, M. Bertin, W. A. Brown, S. Cazaux, S. B. Charnley, A. Coutens, H. M. Cuppen, V. Guzman, H. Linnartz, M. R. S. McCoustra, A. Rimola, J. G.M. Schrauwen, C. Toubin, P. Ugliengo, N. Watanabe, V. Wakelam and F. Dulieu, *ACS Earth Space Chem.*, 2022, **6**, 597.

Serena Viti responded: In terms of the binding energy of atomic H, we can say that, given the chemical network UCLCHEM uses as well as the reaction mechanisms implemented, the value of 1100 K works best at reproducing ice observations. However, it is not a given that the network and the choice of reaction mechanisms are necessarily complete. Further work would need to be done to (a) implement additional reaction mechanisms and (b) infer binding energy parameters from codes that currently use these mechanisms. The inference algorithms operate under the assumption that our UCLCHEM forward model is complete. They are unable to identify any deficiencies in the forward modelling, which is why collaboration with chemists will be needed to improve the modelling. This should hopefully ensure better fidelity with laboratory binding energies.

Ewine F. van Dishoeck asked: Related to my previous question: suppose that you used a larger chemical network, could the most probable value for the H binding energy come down to a lower value that is more consistent with experiments? Or can you only force it to a low value by adjusting the prior?

Serena Viti answered: It is possible that the choice of network may play a role, but we do include a significant number of hydrogen addition reactions and our results are able to reproduce measured abundances reasonably well using only the Langmuir–Hinshelwood diffusion mechanism. This suggests that decreasing the H binding energy at present might overproduce some hydrogen species. To counter this, we would need to introduce something that would serve as competition to grain-surface hydrogen. It is possible that including the non-diffusive reaction mechanisms described in the literature[1,2] may help with this, as it would allow other reactions to be more efficient and therefore force H to need a lower binding energy to produce the required abundances.

1 R. T. Garrod and T. Pauly, *Astrophys. J.*, 2011, **735**, 15.
2 M. Jin and R. T. Garrod, *Astrophys. J., Suppl. Ser.*, 2020, **249**, 26.

Ewine F. van Dishoeck said: I appreciate the answers to my previous questions, and especially the comment that machine learning should be used as a tool to find out which reactions matter most in the production of certain molecules. It is certainly true that for a significant fraction of the reactions in a chemical network, the precise value does not matter much for the physical or chemical question that one is trying to answer.

Serena Viti replied: Indeed, and, in my opinion, within the context of this particular problem, ML should definitely never be used as a "substitute" for experimental or computational astrochemistry; however, it can help to "prune" the reactions to be investigated.

Laurent Wiesenfeld returned to the discussion of the paper by Robin T. Garrod: Quite some discussion in your paper (https://doi.org/10.1039/d3fd00014a) is devoted to the amines (R–NH_2) and their stability, if I am not mistaken.

However, in anhydrous media, the cyano functionality (R–C≡N) is known to be much more stable. Also, when dissolved in acidic water, it readily converts into an amide (R–C(O)NH_2). It is even true for polymers like the tholins, supposed to be prevalent in the Titan haze.

I am wondering if a route to complexity for nitrogen-bearing molecules would be the cyano route instead of the amino route, to bypass the proton affinity and dissociative recombination barrier.

Robin T. Garrod responded: Certainly we have begun to see the detection of nitriles of substantial complexity (by interstellar standards) both in star-forming regions (*e.g.*, normal- and iso-propyl cyanide[1]) and in dark clouds (*e.g.*, cyanobenzene[2]). You are also correct that the enhanced destruction described in our paper (https://doi.org/10.1039/d3fd00014a) does not affect the nitriles, due to their lower proton affinities *vs.* that of ammonia. In the gas phase, certainly, they should be longer-lived than amines.

But I would suggest also that the amines, nitriles, and other molecules that we detect in the gas phase in various star-forming sources may not themselves survive for long enough to influence later (planetary) chemistry, as their presence in the hot gas necessarily makes them subject to destruction mechanisms that are not expected to enhance chemical complexity. Rather, they are a detectable signpost of the chemical complexity present in the dust grain-surface ices. Complex species that might influence later chemical complexity instead may be those that remain in the solid phase during the star-formation process, simply because they exist in the more distant regions within a star-forming core or circumstellar disk that do not experience such high temperatures. These may be the complex organics, formed at early times, that ultimately manifest in comets and may contribute to other, cold planetary chemistry. I would expect both amines and nitriles to be present in these ices, along with a range of other interesting species.

1 A. Belloche, R. T. Garrod, H. S. P. Müller and K. M. Menten, *Science*, 2014, **345**, 1584.
2 B. A. McGuire, A. M. Burkhardt, S. Kalenskii, C. N. Shingledecker, A. J. Remijan, E. Herbst and M. C. McCarthy, *Science*, 2018, **359**, 202.

Serena Viti opened a discussion of the paper by Marie Van de Sande: Do your models, at the moment, consider the dynamics of the outflows? Or are they static?

Marie Van de Sande answered: The dynamics of the outflow are not considered; the model (https://doi.org/10.1039/d3fd00039g) is a static "snapshot" model. Combining chemistry and dynamics would be valuable to assess the interaction between large-scale structures within the outflow and its chemistry. However, it is also prohibitively computationally expensive. Silke Maes, a PhD student at KU Leuven, is currently working on a chemistry emulator that can be slotted into hydrodynamic models.

Kristen Darnell opened a discussion of the paper by Stefan T. Bromley: When silicates clump together in space to form dust grains, do you expect to see nanosilicates with varying structure, or will the grain be composed of nanosilicates with a uniform structure?

Stefan T. Bromley replied: The detailed structure of nanosilicates is very sensitive to size and chemical composition. Even when having the same size and composition, nanosilicates can possess quite different structures with only small differences in energetic stability. Thus, any nano-aggregated silicate grain would be composed of nanosilicates with non-uniform structure. We note that, although the detailed structures of nanosilicates are non-uniform, these structures tend to differ from those of larger silicate grains in a systematic way. This leads to a discernible overall infrared (IR) signature for nanosilicate grain populations.

Inga Kamp returned to the discussion of the paper by Marie Van de Sande: The Atacama Large Millimeter/submillimeter Array (ALMA) observations of the continuum and lines show shells and complex structures and this motivated you to develop more complex 'clumpy' astrochemical models. In the end, you may want to demonstrate that your new porosity/filling factor approach reproduces the observational data better than the previous homogeneous approach. Can you comment on how your new models quantitatively compare to the ALMA data? And which other quantitative benchmarking with observations do you foresee to validate your new approach?

Marie Van de Sande answered: So far, the porous model has been used to interpret observations on a case-by-case basis, when retrieved abundance profiles did not correspond to the output of smooth chemical models (*e.g.*, ref. 1). A more thorough, quantitative benchmarking to observations has not been done so far. This is largely because retrieving abundance profiles from observational data is time intensive and computationally expensive.

1 T. Danilovich, A. M. S. Richards, L. Decin, M. Van de Sande and C. A. Gottlieb, *Mon. Not. R. Astron. Soc.*, 2023, **494**, 1323.

Robin T. Garrod commented: I am interested in the chemical network you are using. Within the physical structure, there must be some radius where the gas density is high enough for three-body reactions to occur – that is, where a third body stabilises the reaction product. Are those reactions in your network, and how important are they to the overall response?

Marie Van de Sande responded: Three-body reactions are not included in the chemical network as the densities are not large enough for them to occur within our model. Such high densities are reached in the inner wind and three-body reactions are included in inner-wind models.[1,2]

1 I. Cherchneff, *Astron. Astrophys.*, 2006, **456**, 1001.
2 D. Gobrecht, I. Cherchneff, A. Sarangi, J. M. C. Plane and S. T. Bromley, *Astron. Astrophys.*, 2016, **585**, A6.

Robin T. Garrod asked: If you do not include the evolution of the chemistry in the innermost circumstellar regions, do you start the models from some sort of atomic conditions?

Marie Van de Sande replied: The chemistry in the inner-wind region is characterised by shocks and non-equilibrium chemistry. We start our model after this region and assume mostly molecular parent species. These can be thought of as results of the inner-wind chemistry; our parent species and their abundances are based on observations.[1] Additionally, we assume that dust formation has taken place within the inner wind as dust nucleation is not included.

1 M. Agúndez, J. I. Martínez, P. L. de Andres, J. Cernicharo and J. A. Martín-Gago, *Astron. Astrophys.*, 2020, **637**, A59.

Athena Flint returned to the discussion of the paper by Stefan T. Bromley: You discuss that multiple local-minimum structures of the silicates you consider, in addition to your global-minimum structure, are important for creating an accurate observational model. If we consider the silicates that exist post-sputtering, we expect some to originate from the reaction of smaller constituents. In such processes, what forms of Mg (Mg atoms, MgH_2, *etc.*) would be expected to be the major Mg sources in light of the depletion of Mg^+ that you mention? If multiple forms of Mg are expected to be participating in the chemical pathways of silicate formation, could the form of Mg used affect the type of silicate structure that results?

Stefan T. Bromley responded: Supernova shockwaves of ionised H atoms (*i.e.*, protons) preferentially sputter Mg^+ ions from silicate grains. The interaction of protons with silicates during and after sputtering and the subsequent chemistry of liberated Mg^+ in the ISM is not well known. Potentially, these sputtered Mg^+ ions could subsequently participate in different chemical pathways, which could influence silicate (re)formation in the ISM. It is known that Mg^+ can combine with other elements (*e.g.* H, C, N) in circumstellar environments to form a range of molecules. However, apart from perhaps the simplest of these species (*e.g.* MgH^+), most of these molecules are not thought to be very stable in the diffuse ISM. The range of different Mg-bearing molecular species in the ISM is thus likely to be very limited and simple Mg^+ ions are probably the most dominant magnesium source for any silicate (re)formation. We note that there is indirect evidence of Mg^+ addition to small pyroxene monomers in cluster beam experiments.[1]

1 J. Mariñoso Guiu, B.-A. Ghejan, T. M. Bernhardt, J. M. Bakker, S. M. Lang and S. T. Bromley, *ACS Earth Space Chem.*, 2022, **6**, 2465–2470 (ref. 34 in the paper).

Martin Cordiner returned to the discussion of the paper by Marie Van de Sande: Could you comment on the timescales for material to flow out from the inner to outermost regions of the circumstellar envelope? I'm interested to know whether it's long enough to build significant ice mantles on the dust.

Marie Van de Sande replied: In outflows of 10^{-5} M per year, the collision time scale is smaller than the expansion time scale up to roughly 10^{16} cm; in outflows with 10^{-6} M per year, it is up to roughly 10^{15} cm. This is a sufficient amount of time for accretion onto dust to occur in these dense outflows. Besides outflow density, the thickness of the ice mantle also depends on the composition of the outflow and the temperature of the dust. We find that several monolayers of water ice are efficiently formed in oxygen-rich outflows, and generally a factor of a few fewer monolayers of ice are built up in carbon-rich outflows.[1]

1 M. Van de Sande, C. Walsh, T. P. Mangan and L. Decin, *Mon. Not. R. Astron. Soc.*, 2019, **490**, 2023.

Martin Cordiner commented: After the dust becomes irradiated and materials return to the gas phase, is there any chance of observing the grain surface products?

Marie Van de Sande answered: We have included photoprocessing of volatile complex ices into refractory organic material in our chemical model. Complex ices are assumed to contain three or more carbon atoms or at least two carbon atoms and one heavy atom. The refractory organics are treated as chemically inert and do not react among themselves nor with ices or gas-phase species. In carbon-rich outflows with a mass-loss rate of at least 10^{-6} M per year, the dust grain can be covered in refractory organics. As mentioned in the paper (https://doi.org/10.1039/d3fd00039g), the surface coverage depends on several factors, such as the outflow's density and density distribution, the grain size distribution, and the presence of a companion. If the coverage is large enough, this refractory material could be observed.

Ewine F. van Dishoeck returned to the discussion of the paper by Stefan T. Bromley: Interesting calculations. What is the width of individual nanosilicate features? Would they be slightly spectrally resolved or unresolved in James Webb Space Telescope Mid-Infrared Instrument (JWST MIRI) spectra? If resolved, they are more easily distinguished from gas-phase lines or instrumental features.

Stefan T. Bromley responded: The individual lines in the simulated IR spectra each relate to a single vibrational mode of a specific nanosilicate grain. Although these individual lines are thus very sharp, and would not themselves be resolvable, it is expected that nanosilicates would always be observed as a (sub-)population of grains. As each nanosilicate gives rise to $3n-6$ lines (where n is the number of atoms in the grain) and a population of nanosilicates is expected to involve huge numbers of distinct structures, dependent on size, composition and energetic stability, only the overall spectral signature from the resulting dense array of lines would be observable. In our preliminary calculations we consider only <100 distinct nanosilicate structures. Although we attempt to be representative with respect to size, composition and energetic stability, the number of nanosilicates is very small and individual lines can still be discerned in the calculated average IR spectra. In the final version of the paper, we include dashed lines to indicate tentative approximations to the limiting overall resolvable spectral shape for more realistic larger nanosilicate populations.

Ewine F. van Dishoeck returned to the discussion of the paper by Marie Van de Sande: Have any JWST spectra of a carbon-rich AGB star been taken yet? If not, are they planned?

Marie Van de Sande replied: There are no available JWST spectra of AGB stars. To my knowledge, none have been planned so far. Observations of carbon-rich AGB stars would be very interesting, especially in combination with the already planned observations of planetary nebulae, the next stage in their evolution.

Ko-Yun Huang asked: Can you illustrate more what the physical and/or chemical drivers that change the porosity parameter near AGB stars could be?

In other words, what physical and/or chemical drivers are at play that could bring variation in the porosity near AGB stars?

Marie Van de Sande answered: Large-scale density structures in the outflow, such as spirals and disks, are thought to be formed by binary interaction with a (sub)stellar companion.[1] Clumpy deviations from spherical symmetry are also seen on much smaller scales and closer to the star, likely caused by convective cells in the outer layers of the AGB star.[2] The main driver for porosity is hence physical in nature, although more dust appears to be forming in the density-enhanced clumps close to the star, driving an anisotropic outflow.[3]

1 L. Decin, M. Montargès, A. M. S. Richards, C. A. Gottlieb, W. Homan, I. McDonald, I. El Mellah, T. Danilovich, S. H. J. Wallström, A. Zijlstra, A. Baudry, J. Bolte, E. Cannon, E. De Beck, F. De Ceuster, A. de Koter, J. De Ridder, S. Etoka, D. Gobrecht, M. Gray, F. Herpin, M. Jeste, E. Lagadec, P. Kervella, T. Khouri, K. Menten, T. J. Millar, H. S. P. Müller, J. M. C. Plane, R. Sahai, H. Sana, M. Van de Sande, L. B. F. M. Waters, K. T. Wong and J. Yates, *Science*, 2020, **369**, 1497.
2 T. Khouri, M. Maercker, L. B. F. M. Waters, W. H. T. Vlemmings, P. Kervella, A. de Koter, C. Ginski, E. De Beck, L. Decin, M. Min, C. Dominik, E. O'Gorman, H.-M. Schmid, R. Lombaert and E. Lagadec, *Astron. Astrophys.*, 2016, **591**, A70.
3 L. Velilla-Prieto, J. P. Fonfría, M. Agúndez, A. Castro-Carrizo, M. Guélin, G. Quintana-Lacaci, I. Cherchneff, C. Joblin, M. C. McCarthy, J. A. Martín-Gago and J. Cernicharo, *Nature*, 2023, **617**, 696.

Ko-Yun Huang commented: I am in need of clarification concerning the definition of "porosity". Is only the distance from clump to clump involved in this parameter, or are other factors, such as how much density contrast there is between the two parts of the gas components, involved as well? Are any other factors involved?

Marie Van de Sande responded: Clumpiness in the outflow is described using a mathematical porosity formalism, in which the outflow is divided into a stochastic mixture of overdense clumps and a rarefied interclump medium.[1,2] The clumps are assumed to conserve their mass. The specific clumpiness of the outflow is described by three parameters: the clump volume filling factor, which describes the fraction of the total volume occupied by clumps, the interclump density contrast, which describes how much material resides in the interclump components, and the size of the clumps at the stellar surface, which cannot be larger than the star. These three factors describe the density structure and the resulting extinction throughout the outflow. Note that the porosity formalism is only a one-dimensional approximation of a three-dimensional structure; the location of the clumps cannot be specified. While individual observed structures cannot be reproduced, the model can still provide some chemical insights.

1 S. P. Owocki, K. G. Gayley and N. J. Shaviv, *Astrophys. J.*, 2004, **616**, 525.
2 J. O. Sundqvist, S. P. Owocki, D. H. Cohen, M. A. Leutenegger and R. H. D. Townsend, 2012, **420**, 1553.

Gwenaëlle Dufour asked: Is the grain size the same throughout the entire model? Are you putting collisions in?

Marie Van de Sande answered: The dust is assumed to have a constant grain size distribution in the model. Ices are assumed to be physisorbed onto the surface and not change the grain size. Collisions between dust particles are not included; they are not expected to be important at the drift velocities seen in AGB outflows. Dust–gas collisions are included: besides accretion of gas-phase species, forming an ice mantle, we also include mechanical sputtering, where H_2, He, CO, and N_2 can sputter off ices.

Martin Cordiner remarked: I understand that the importance of envelope porosity in your model is mainly for determining the radiative transfer and therefore the radiation field strength. The chemistry that occurs in circumstellar envelopes can also be density dependent,[1] so variations in chemical production rates could occur as a result of shell-like structures, spirals, clumps, *etc.* Please could you comment on the possible importance of this effect in the objects you studied?

1 M. A. Cordiner and T. J. Millar, *Astrophys. J.*, 2009, **697**, 68.

Marie Van de Sande replied: The porosity formalism also includes the effect of clumpiness on density, calculating the chemistry in the overdense clumps and the rarefied interclump medium. We find that changes in extinction have a larger effect on the chemistry than the local overdensities within the clumps.[1] Because the porosity formalism is a statistical formalism, the location and size of the clumps cannot be specified. The density dependence can be approximated, as was done for the shells around IRC+10216.[1,2] Neither the porosity formalism nor the model of Cordiner and Millar[3] can fully explain the relative locations of the cyanopolyynes and hydrocarbon radicals, indicating that a three-dimensional model is perhaps needed.

1 M. Van de Sande, J. O. Sundqvist, T. J. Millar, D. Keller, W. Homan, A. de Koter, L. Decin and F. De Ceuster, *Astron. Astrophys.*, 2018, **616**, 34.
2 M. A. Siebert, M. Van de Sande, T. J. Millar and A. J. Remijan, *Astrophys. J.*, 2022, **941**, 90.
3. M. A. Cordiner and T. J. Millar, *arXiv*, 2009, preprint, arXiv:0903.0890 [astro-ph.GA], DOI: 10.1088/0004-637X/697/1/68.

Martin Cordiner commented: Tracing the evolution of the observed circumstellar shells back in time to the point they were produced, we can consider the role of stellar pulsations, and possible physical interactions between binary companions to produce such density-enhanced structures. I was wondering if the initial chemical conditions in your model could therefore be different in the clumps *vs.* inter-clump medium, as a result of differing temperatures, densities and radiation fields. This may be something to consider in future studies.

Marie Van de Sande responded: The current model follows the evolution of the chemistry in a parcel of material as it moves away from the star. In the porosity formalism, the chemistry in the overdense clump and rarefied interclump component are calculated separately; the final result is a combination of the two. Consequently, the clump and interclump components never mix in the model. Moreover, the volume filling factor of the clumps is assumed to be constant. These are (over)simplifications of an AGB outflow. It would certainly be interesting to consider different initial conditions, and to further extend this to different density and radiation histories. This could be done, for example, by following the chemistry in the particles/parcels of gas of a smooth particle hydrodynamics model.

Inga Kamp returned to the discussion of the paper by Stefan T. Bromley: In your paper (https://doi.org/10.1039/d3fd00055a), you mention that there is already a database with nanosilicate spectra, and thus predicted IR emission spectra for different individual species (sizes). Could we hunt for those nanosilicates already in the JWST/MIRI emission spectra of planet-forming disks based on that database? Apart from a distribution of sizes/species, are these spectra for example also temperature dependent?

Stefan T. Bromley answered: We believe that nanosilicates could be present in any astrophysical environment in which grain processing/(re)formation is occurring. As such, planet-forming disks could indeed be a viable place for nanosilicate populations to form. Compared to the diffuse ISM, such populations would be found at higher temperatures, where infrared emission could become important. We have started to look at the issue of how temperature affects the IR spectra of nanosilicates (see ref. 1), but such effects have not yet been included in the full database of nanosilicates. The database is currently being curated and will be published online as an openly accessible resource in the coming months.

1 J. Mariñoso Guiu, A. Macià Escatller and S. T. Bromley, *ACS Earth Space Chem.*, 2021, **5**, 812.

Inga Kamp commented: Thanks, I understand the point about not looking for individual features but combining the individual spectra. So, a broader asymmetry or unexpected shoulder appearing on the usual silicate dust feature is what we could start looking for in our JWST/MIRI observations.

Stefan T. Bromley replied: Exactly. This is the main prediction of our preliminary study.

Murthy S. Gudipati returned to the discussion of the paper by Marie Van de Sande: In your outflow models, would there be gas *vs.* dust (ice particles included) separation that could potentially happen during a supernovae? Is there a possibility for gas–dust collisions that could result in re-formation of ice grains during the outflows?

Marie Van de Sande responded: While the dust is generally faster than the gas, dragging the gas along and driving the outflow, dust and gas are still well-mixed throughout AGB outflows. These outflows are very different from supernovae, with much lower velocities: the outflow velocity ranges from 5 to 30 km s^{-1}, and the drift velocity between dust and gas ranges from 5 to 15 km s^{-1}.[1]

1 S. Ramstedt, F. L. Schöier, H. Olofsson and A. A. Lundgren, *Astron. Astrophys.*, 2008, **487**, 645.

Joshua Halpern returned to the discussion of the paper by Stefan T. Bromley: As you point out, dust grain processing on silicate grains could be important for basic chemical processes in the ISM, including formation/dissociation of H_2 and water molecules. Does one need to know the structure of the particles or just those of chemically active sites on the surfaces, and how can the structure library you and your colleagues have built be used to understand chemistry occurring on the grains.

Stefan T. Bromley answered: Although it is always preferable to have the full atomistic and electronic structure of a particle to accurately calculate reactivity at any site on its surface, it is likely to be sufficient to estimate reactivity *via* only the structures of particularly important surface sites (*e.g.*, defects, low-coordinated sites). Such sites can be identified *via* quantum chemical calculations of specific reactions on a selection of nanosilicates from the library. The full structure library would then provide a useful resource for understanding/predicting nanosilicate grain chemistry based on the propensity of such sites in nanosilicate grain populations. More extensive use of the library could be used to machine learn simple descriptors for chemically relevant properties (*e.g.* molecular binding energies).

Conflicts of interest

Ms Darnell could not be contacted regarding any possible conflict of interest before going to print, and there are no other conflicts to declare.

Faraday Discussions

ROYAL SOCIETY OF CHEMISTRY

PAPER

Concluding remarks: *Faraday Discussion* on astrochemistry at high resolution

T. J. Millar *

Received 30th June 2023, Accepted 7th July 2023
DOI: 10.1039/d3fd00131h

Fifty years on from the first detailed chemical kinetic modelling of astronomical sources, I provide some introductory comments on the history of astrochemistry, summarise some personal views on the topics covered in this discussion meeting, and conclude with some thoughts on its future development. I have left out the jokes.

1 The past

Before 1970 only seven molecules were observed in interstellar space. CH, CH^+ and CN were detected in the period 1937–1941 through their optical absorption lines seen against the background of bright stars. The next molecular detections did not occur until the advent of radio astronomy and the development of receiver technology. Maser emission from OH was detected in 1963 and in the following six years ammonia (NH_3), water (H_2O) – again a maser – and formaldehyde (H_2CO), were identified. In their 1969 discovery paper of the latter molecule, Snyder *et al.*[1] noted that 'H_2CO ⋯ the first organic molecule ever detected in the interstellar medium ⋯ indicates that the process of chemical evolution may be much more complex than previously assumed'. With the current identification of close to 300 molecules, including the fullerenes C_{60} and C_{70}, Snyder *et al.*[1] could not have imagined the range of species contained in interstellar and circumstellar environments.

The earliest attempt to describe the gas phase chemical processes that might operate in interstellar clouds was published by Bates and Spitzer[2] in an effort to understand the CH and CH^+ observations. With the exception of ion–neutral reactions, they identified most of the important chemical processes likely to operate in the diffuse environment in which these species were found. Indeed, finding that their calculated abundances did not reproduce the observations, they even considered the possibility that the thermal desorption of CH_4 from grains might be their source. Of course, one should recognise that a major reason for model failures was their lack of knowledge of the physical properties of the diffuse clouds in which these molecules were found.

Astrophysics Research Centre, Queen's University Belfast, University Road, Belfast BT7 1NN, UK. E-mail: tom.millar@qub.ac.uk

The first detailed numerical model of chemical kinetics in interstellar clouds was published by Herbst and Klemperer[3] and is rightly regarded as a classic in our field. The authors included many of the processes at work and worried about some that still concern us today – the elemental abundance of oxygen, the carbon-to-oxygen (C/O) elemental abundance ratio, the role of grains as a source and sink of interstellar molecules, and the extent to which a global chemical kinetic model can be applied to individual molecular clouds. Although some arguments concerning time scales are somewhat dated, this is, I would argue, the foundational paper for our subject and one which led directly to the plethora of experimental, theoretical, gas–grain modelling approaches to astrochemistry presented at this meeting.

Starting around 1985, spectral line surveys of high-mass star-formation systems began to detect in a large number of objects, a plethora of interstellar complex organic molecules (iCOMs), very loosely defined as containing at least six atoms including one element in addition to carbon and hydrogen. Many of the sources were rather hot and dense by interstellar standards, ~100–300 K, $n(H_2) \sim 10^6$–10^9 cm^{-3}. These species, typified by their smallest member CH_3OH, are often saturated and tend to have larger abundances, often by several orders of magnitude, in these sources compared to those in cold (10 K) clouds. Observations of their D-substituted analogues, however, showed a surprising result – molecular D/H abundance ratios that are much larger, by factors of 10^3–10^5, than the cosmic D/H ratio of a few 10^{-5}. Indeed, the largest enhancement I am aware of occurs for triply-deuterated ammonia, ND_3, which has an abundance relative to NH_3 of about 10^{-3}, some 10^{11} times larger than the statistical D/H ratio. Such large fractionation can only occur at very low temperatures and the accepted paradigm is that iCOMs and their deuterated versions are formed in molecular ices on cold dust grains before being desorbed into hot gas through the effect of stellar heating or shock waves. Such observations led to a number of models in which H atom collisions with a cold ice surface was assumed to convert atoms such as N, C, O and S to NH_3, CH_4, H_2O and H_2S in a water-rich ice mantle. Subsequent heating due to the presence of a newly-formed hot, massive star liberated large abundances of these simple hydrogenated molecules to the gas phase where subsequent gas-phase chemistry produced the iCOMs.[4–6] The lack of any detailed description of the actual chemistry on these icy grains and the cry of the modellers that 'all' abundance anomalies could be explained by surface chemistry led Malcolm Walmsley to note that grain chemistry 'was the last refuge of the scoundrel'.

2 The present

As indicated earlier, there are now 300 or so molecules detected in space with more than 75 containing 8 or more atoms (see Fig. 1).

In the last five years there has been a substantial increase in the number of detections of complex hydrocarbons mainly driven through dedicated spectral line surveys of the cold, dark Taurus Molecular Cloud 1 (TMC-1). McGuire and colleagues have concentrated on using sophisticated analysis of rotational emission lines as part of their GOTHAM survey to identify a number of cyclic hydrocarbons including benzonitrile,[7] 1- and 2-cyanocyclopentadiene, C_5H_5CN,[8,9] indene,[10] and 2-cyanoindene.[11] Cernicharo and co-workers, using the very

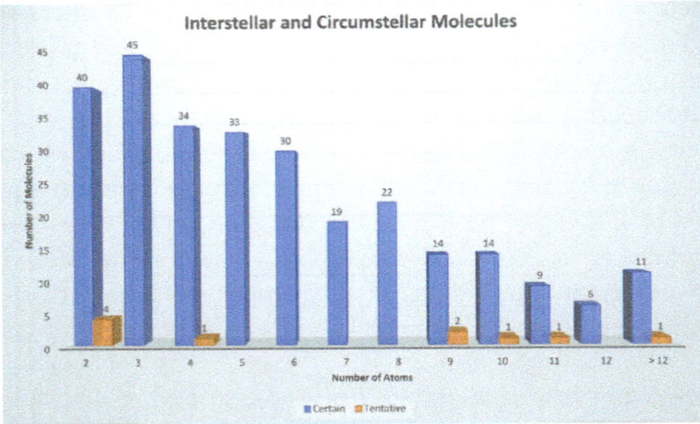

Fig. 1 Number of molecules detected in interstellar and circumstellar environments as a function of the number of atoms.

sensitive NANOCOSMOS receiver, have detected many more hydrocarbons in their QUIJOTE survey of TMC-1 over the past few years, including several 5- and 6-membered ring species and branched chains. Interesting species include 1- and 2-ethynyl cyclopentadiene (c-C_5H_5CCH), ethynyl benzene (C_6H_5CCH),[12] and ethynyl cyclopropenylidene (c-C_3HCCH).[13]

This discussion meeting was convened at a very appropriate time – and with a very timely agenda – given the recent development of very sensitive facilities such as the JWST (James Webb Space Telescope) with both higher sensitivity and spectral resolution than any other IR space observatory; as well as high spatial resolution instruments such as the ALMA (Atacama Large Millimeter/submillimeter Array). The early results from the JWST are breath-taking with an impact that affects every discipline in astronomy. The ALMA, which has been in operation for a decade, also continues to offer new views of the molecular universe and many of the observational papers at this meeting reflected the current dominant positions of these two facilities.

The meeting was arranged around four main themes: (1) observational astrochemistry; (2) laboratory gas-phase studies; (3) experimental studies of dust and ices; (4) computational astrochemistry. A number of poster papers were also on display during the meeting. My comments on the presentations are not meant to be comprehensive but reflect a number of personal factors including a lack of time to assimilate everything in the papers, presentations, extensive discussions and posters, a subjective bias toward certain topics, and a degree of ignorance in some areas. Nevertheless, I do commend the proceedings to the reader; they contain a wealth of information on both the rude health and the underlying concerns – the known unknowns – of our subject and its importance in answering fundamental questions in astrophysics, including the formation of stars and planetary systems. In particular, Cecilia Ceccarelli's Spiers Memorial Lecture (https://doi.org/10.1039/d3fd00106g), is a masterly introduction to the field, outlining recent successes, identifying complexities yet to be addressed in a comprehensive manner, and pointing to the future. For my part, the most important message of the meeting is that the foundation of astrochemistry is spectroscopy. Without the decades of advances in laboratory spectroscopy, and especially in astronomical instrumentation, our exploration of chemical processes in space would not be 'much more complex' than those considered 50 years ago.

2.1 Observational astrochemistry

Five papers discussed recent results from the JWST and ALMA. Van Dishoeck (https://doi.org/10.1039/d3fd00010a) and Kamp (https://doi.org/10.1039/d3fd00013c) presented a selection of spectacular results from the JWST mid-infrared (MIRI) spectral line surveys of protostars (JOYS) and of T Tauri and very low mass star (mostly brown dwarf) disks (MINDS). One clear advantage of the JWST observations over those from the ALMA, is that the former provides information on the abundances and excitation of non-polar molecules. The data presented were from a selection of the survey results but both papers show that there is a diversity within and between source types. Van Dishoeck's spectra show evidence of amorphous and crystalline silicates, ice bands, together with H_2 pure rotational and other molecular emission lines, most notably H_2O and CO_2. The disk around the very low-mass star J160532 shows perhaps the most surprising spectrum: line rich in strong hydrocarbon emissions from C_2H_2, C_4H_2 and C_6H_6. Indeed in some sources the hydrocarbons are so abundant that they form a 'quasi-continuum'. This source also contains very intense $^{13}C^{12}CH_2$ emission. The analyses presented here indicate the presence of hot (~500 K) acetylene with $N(C_2H_2) \sim 10^{21}$ cm^{-2}, an enormous value that indicates that either efficient 'bottom-up' combustion chemistry is occurring or that carbonaceous grains are being chemically or thermally destroyed in a 'top-down' chemistry.

Kamp derived similar results from fitting a combination of slab and 2D thermochemical disk models to her observed spectra. She noted that line lists for $^{13}C^{12}CH_2$ were incomplete, preventing a full analysis of the spectra and of the (likely large) $^{13}C/^{12}C$ ratio. Again differences within and between source types were evident. All T Tauri disks contain H_2O whereas it is only tentatively detected in the low mass disks. She also stressed the need to go beyond the slab retrieval models which treat only one molecule at a time and assume constant density and temperature. In addition, she noted that the underlying molecular line data and partition functions needed for such analyses are often missing. Her comparison between column densities derived with both slab and thermochemical models show that the latter typically predict values a factor of ten larger than the former. Clearly a more accurate approach is needed but again it was stressed that databases currently lack fundamental information – Einstein *A*-coefficients, partition functions at high *T* and *P* – as well as spectroscopy to provide hot band and isotopologue line lists for hydrocarbons. Finally, it is worth re-iterating that these are the early results from the JWST; we can expect a host of new species and new insights into the chemistry and physics of these planet-forming systems in the coming years. As always with new facilities, it will be a challenge for both laboratory and theoretical approaches to keep abreast of new observational developments.

Three papers presented high angular resolution ALMA observations on spatial scales ranging from the Solar System to 10^5 times larger. On the largest scale, Huang's observations (https://doi.org/10.1039/d3fd00007a) used molecular tracers of shock waves to probe the similarities and differences between molecular gas in two galaxies – NGC 253, a starburst galaxy, and NGC 1068, an active galactic nucleus harbouring a supermassive black hole which generates high X-ray and cosmic-ray fluxes. Detailed observations were interpreted by models of shock chemistry finding evidence for the presence of both slow shocks, *via* observations of HNCO, and fast shocks, *via* the detection of large abundances of SiO which arise from the sputtering of silicate dust grains. Wilkins and Blake (https://doi.org/10.1039/d3fd00003f) presented a pilot spectral line search in the 1.2 mm window toward 11 giant molecular clouds (GMCs) at a distance of 4–8 kpc from the Galactic Centre, while Bianchi (https://doi.org/10.1039/d3fd00018d) investigated molecular line emission in SVS13-A, proto-binary system, in particular from the streamers connecting the binary system to their larger scale molecular envelope. This source has a rich array of molecular lines generated by iCOMs in the hot gas in the innermost region. Her study of HDO and SO_2 found a large, warm (95–160 K), column density of the former, indicating that it has been released from dust grain mantles, most likely by shock processes generated in the infalling gas. Astrochemistry and the astrophysics that can be derived from it are likely to be dominated by the ALMA and the JWST for the next decade or two – an exciting prospect.

2.2 Gas-phase laboratory studies

One striking aspect, at least to me, of this discussion meeting was that it is probably the first time that there were no papers on ion–neutral reactions, which have historically been the dominant mechanism for molecular synthesis studied in astrochemistry. In the main part, this dominance was due to the fact that both reactants and products could be detected readily whereas for neutral–neutral reactions it was more common to follow only the decay of one reactant with no measured information on products or branching ratios. Experimental gas-phase kinetics of neutral–neutral systems were covered in the papers by Suits on the reaction of vibrationally excited CN with butadiene (https://doi.org/10.1039/d3fd00029j), by Douglas on the low temperature behaviour of NH_2 with acetaldehyde, CH_3CHO (https://doi.org/10.1039/d3fd00046j), and by Balucani (https://doi.org/10.1039/d3fd00057e) on the reaction of $N(^2D)$ with benzene. All three papers contain sophisticated, state-of-the-art experimental and theoretical techniques to detect products; the use of high-level quantum calculations to determine potential energy surfaces and the determination of total rate coefficients and branching ratios. The impetus for each study was to explain the formation of a specific interstellar molecule: Suits as a source of cyanobutadiene (C_4H_5CN), in TMC-1;[14] Douglas as a source of acetamide (CH_3CONH_2) in the interstellar medium;[15] and Balucani as a possible source of cyclopentadienyl (c-C_5H_5) in Titan. All required a significant investment of time and people. It is therefore rather disappointing for the authors, and for the wider astrochemical community, that none of the reported studies has had a significant impact on astrochemical models. I return to this point in Section 3.

The remainder of the session was devoted to spectroscopy and the study of transient species. This included papers by Brünken (https://doi.org/10.1039/d3fd00015j) on the fragmentation products of cyclic ions; Gupta who used a cryogenic ion trap at 4 K to record the rotational spectrum of c-C_3H_2D

(https://doi.org/10.1039/d3fd00068k); and Jacovella (https://doi.org/10.1039/d3fd00016h) who measured the rotational spectrum of the cross-branched molecule norbornadiene (C_8H_7), and its cyano-derivatives, which have much larger electric dipole moments, as a pre-cursor to an (unsuccessful) astronomical search. Puzzarini (https://doi.org/10.1039/d3fd00052d) presented a strategy for determining the rotational spectra of transient molecules using quantum calculations to predict properties and thereby guide laboratory searches and line identification.

2.3 Experimental studies of dust and ices

The third session, which concentrated on chemical reactions in and on dust grains, was marked by an animated and informed discussion session as perhaps befits a subject that is arguably the most important and the most uncertain aspect of understanding astrochemistry in molecular clouds and protoplanetary disks. Many astronomers could be forgiven for taking the view – to paraphrase the English essayist Charles Lamb's (1775–1834) comment about space and time:

'Nothing puzzles me more than molecules and ices yet nothing bothers me less as I never think about them'.

Nevertheless it is the job of the astrochemist to both worry and puzzle over the known and unknown unknowns of ice chemistry.

Four major questions on the cold (~10 K) ice chemistry that is thought to produce iCOMs, continue to exercise the community: what process removes these species from grains at temperatures of ~6–10 K to the cold gas phase in which they are detected?; do these processes return whole or fragmented species?; what are the binding energies of surface iCOMs since these determine the efficiency of thermal desorption processes?; and, given the fact that experiments and theory are limited to relatively few species, can we generalise results from the few to the many? A number of answers to these questions have been suggested in the literature and some were covered here.

Brown (https://doi.org/10.1039/d3fd00024a) discussed experiments which showed that infrared radiation could desorb CO from CO–water ice mixtures, noting that CO didn't desorb when irradiated with 4.6 μm photons but did so when the irradiation was at the wavelengths of the H_2O vibrational modes at 2.9 and 12 μm. If a single photon process and if the efficiency is high, this could be an important desorption mechanism in dark clouds which are much more transparent to IR radiation than to UV photons, which have long been thought to drive photodesorption. Brown identified two mechanisms: a fast process of indirect resonant desorption in which an excited H_2O molecule removes a neighbouring CO molecule, and a slow, photon-induced process in which energy is accumulated in the heat bath of water ice. It should be noted that the water ice layers are thin enough that the IR photons also heat the cold, metal substrate and that this may also contribute to desorption. Although poorly defined, Brown suggested yields of CO molecules removed per incident photon to be on the order of a few 10^{-5}.

Like Brown, Bertin (https://doi.org/10.1039/d3fd00004d) studied the photodesorption of ices, in his case with vacuum UV (VUV) photons and ices composed of either pure formic acid (HCOOH), or methyl formate ($HCOOCH_3$), or with these species embedded in CO or H_2O ices.

In addition to determining desorption yields, around 10^{-5} per incident VUV photon, another key issue is whether molecules are returned intact or as fragments. Bertin found similar results to previous experiments on methanol-containing ice, namely, that species detected in the gas phase are similar to the gas-phase photodesorption products of the molecule. Were this to be a universal rule it would be very convenient for modellers, but it is not – recent experiments on acetonitrile, CH_3CN, show that it photodesorbed intact.[16]

Gudipati (https://doi.org/10.1039/d3fd00048f) discussed an alternative desorption mechanism, one driven by the conversion of water ice from the amorphous to crystalline form. His experiments doped water ice with the volatiles CO, CO_2 and O_2 at 10 K and traced the outgassing species *via* mass spectrometry as the ice was heated to about 200 K. He showed that their abundance in the ice reduce by about a factor of 100 due to outgassing when the phase change occurs.

Finally, Sameera (https://doi.org/10.1039/d3fd00033h) discussed the important issues of binding energies and reaction barriers on ices. It has been known for some time, and certainly since the pioneering experimental work of Collings *et al.*,[17] that the binding energies of a molecule in water ice depends on the specific site in which it sits. More recently, Tinacci *et al.*[18] used theoretical calculations of NH_3 on a model cluster containing 100 H_2O molecules to show that binding energies followed a double peaked profile with that at higher energies well fit by experimental TPD measurements. Rather surprisingly, however, there was a second distribution with much lower binding energies comparable to those of CO.

Sameera concentrated on CH_3OH in water ice using a theoretical model to consider the reactions:

$$CH_3OH + OH \rightarrow CH_2OH + H_2O \tag{1}$$

$$CH_3OH + OH \rightarrow CH_3O + H_2O \tag{2}$$

His quantum mechanical calculations show that his model ice cluster has a number of binding sites and that the nature of these has a significant effect on the binding energies, on the computed reaction barriers, and on the detailed pathways that the reactants follow.

Many of the key issues raised here, including the effects of binding energies on grain chemistry, are discussed at length in Ceccarelli's Spiers Memorial Lecture (https://doi.org/10.1039/d3fd00106g).

2.4 Computational astrochemistry

This section contains a few papers which showed the complexity of computational approaches to chemical kinetics in the gas phase. I note that the focus of the meeting on astrochemistry meant that the complex issue of combining hydrodynamics with kinetics was not covered.

Garrod and Herbst (https://doi.org/10.1039/d3fd00014a) presented a very detailed time-dependent model of iCOM formation in molecular clouds based on their identification that several detected and candidate interstellar molecules containing the amine group, $-NH_2$, have proton affinities (PA) that are greater than that of NH_3, 853.6 kJ mol^{-1}. Because of its relatively large abundance in molecular clouds and its large PA, astrochemical models tend to include proton transfer reactions between ammonia and molecular ions as an important step in forming the neutral molecule. An example is:

$$H_2COH^+ + NH_3 \rightarrow NH_4^+ + H_2CO \tag{3}$$

The advantage of this proton transfer reaction is that it preserves the underlying neutral molecule, whereas the alternative loss of molecular ions, dissociative recombination with electrons, can destroy the molecule in a large fraction of cases. In the case of protonated formaldehyde, reaction with electrons has five product channels with that leading to formaldehyde occurring only around 30% of the time.[19] For protonated methanol, the situation is much worse with only 5% of recombinations re-cycling the neutral.[20] As a result, stable neutral molecules with PAs greater than that of ammonia are destroyed faster than those with PAs less than ammonia. Garrod's paper shows that this destruction process has a particular affect on the abundances of many amine species, including methylamine, ethylamine, urea and glycine. The latter molecule, yet to be identified in interstellar clouds, is a cornerstone in the argument that the precursors of life on Earth may be widely dispersed in the Galaxy – Garrod's results suggest that it may be more difficult to detect glycine than we currently envisage. One should note, however, that for reactions in which the difference between PAs is large, most importantly for reactions involving H_3^+, the excess reaction energy can lead to dissociative products rather than proton transfer, although not in every case.[21]

Viti (https://doi.org/10.1039/d3fd00008g) and Van de Sande (https://doi.org/10.1039/d3fd00039g) discussed novel ways of approaching the gas–grain modelling of molecular cloud and circumstellar chemistry, respectively. The former presented an alternative approach of using machine learning to model some of the uncertainties in complex chemical kinetic networks while the latter used a sophisticated physical model to discuss the effects of clumpy structures in the expanding envelopes around late-type stars. Viti's paper used both Bayesian and machine learning approaches to identify both those binding energies with the largest impact on calculated abundances and those species for which determining binding energies are a priority. Van de Sande's paper includes the words 'disentangling' and 'when adding complexity' in its title – a bold claim. As a co-author, I recuse myself at this point and leave it as an exercise for the reader as to whether the content lives up to its title!

Bromley (https://doi.org/10.1039/d3fd00055a) addressed the important topic of dust nucleation in the context of arguments – in essence that the dust destruction time scale is shorter than the dust formation time scale in evolved stars and supernovae – that a significant fraction of silicate dust must nucleate in the low density interstellar medium. Bromley concentrated on the creation of nanosilicates, combining quantum calculations on the lowest energy structures of small seed particles with their infrared absorption characteristics. He showed that the JWST could be sensitive enough to detect such particles if they composed 5–10% of the silicate grain population.

Finally, Madhusudhan (https://doi.org/10.1039/d3fd00075c) described the atmospheric physics and chemistry on a Hycean world, that is a sub-Neptunian exoplanet whose surface is covered by water oceans and with a H_2-rich atmosphere. Although no such planet has yet been detected, several candidate systems have been identified orbiting M dwarf stars. In addition to evolution time scales of several Gyr, Madhusudhan's paper showed one of the important differences between exoplanetary chemistry and astrochemistry, namely that the degree of physical complexity included in the former is of a much higher order than in the latter. In particular, the atmospheric models, internal planetary structures and delivery of exogenous materials are all informed by detailed observations and models of planet Earth and other solar-system bodies. Of course major uncertainties remain but he showed that the three essential ingredients for life – an energy source, liquid water and bio-essential elements – are all present in such worlds and that the JWST should be able to detect specific atmospheric molecules that are characteristic of Hycean worlds, an exciting prospect.

3 The future

The development of astrochemistry over the past 50 years has been astounding, driven forward by the innovation of thousands of scientists, engineers and technicians whose ingenuity has led to the availability of facilities such as the ALMA and JWST. With these, we are able to probe astrochemistry from galactic size scales at high redshift, to au scales in protoplanetary disks and to even smaller scales in exoplanetary atmospheres. Coupled with these advances in technology, we have seen concomitant developments in chemical physics. Experimental and theoretical spectroscopy has underpinned the identification and analysis of many of the thousands of interstellar lines that have been detected by new facilities and instruments. Such work will continue to be important, particularly for the identification of metal-containing molecules, small reactive radicals, and large iCOMs and hydrocarbons, isotopologues and excited states.

Advances in chemical kinetics have also been dramatic over the past 50 years. Many important low-temperature experimental studies are now routinely supplemented by theoretical calculations of potential energy surfaces and reaction rate coefficients and product branching ratios. Such research, built on advanced computational techniques and computing power, is now routine and, although expensive and time-consuming, it can answer important questions in astrochemistry.

In Section 2, I noted that in recent years, observational searches coupled with high resolution spectroscopy have identified more than 100 iCOMs and complex hydrocarbon species in interstellar clouds. Their observed abundances have led to innovative chemical models including both gas-phase and surface reactions. For example, Garrod's basic model, as described here, has over 22 000 reactions among 1385 species of which roughly half are in the gas phase with the rest equally divided between surface ice and bulk ice. Yet for many species, and as we heard in the discussions, abundance calculations rest on 'thin ice' (pardon the pun).

Let me use, as an example, some results from my own, as yet unpublished work,[22] discovered while updating the UMIST Database for Astrochemistry chemical network, Rate12.[23] Propene (CH_3CHCH_2), is a key intermediary in the synthesis of large hydrocarbons. It was detected in TMC-1 with a relatively large fractional abundance, 4×10^{-9}, by Marcelino *et al.*[24] and was thought to form *via* two sequential radiative association reactions:

$$C_3H_3^+ + H_2 \rightarrow C_3H_5^+ + h\nu \quad (4)$$

$$C_3H_5^+ + H_2 \rightarrow C_3H_7^+ + h\nu \quad (5)$$

followed by dissociative recombination of $C_3H_7^+$ with electrons. The Rate12 database included this synthesis and showed that the calculated abundance agreed reasonably well with that observed. Subsequently, however, Lin *et al.*[25] showed that both radiative association pathways had barriers in their entrance channels and could not operate in cold clouds such as TMC-1. The exclusion of these two reactions leads to a huge drop in the calculated abundance of propene as shown in the Rate22 curve in Fig. 2.

This fall in its calculated abundance by about 10 orders of magnitude leads to a 'propene catastrophe' in the models since it is known experimentally to be a reactive molecule even at very low temperatures. For example, it has fast measured rate coefficients in reactions with CH (15 K),[26] CN (23 K),[27] C_2 (77 K),[28] C_2H (15 K)[29] and C_4H (39 K),[30] where the value in parentheses refers to the lowest laboratory temperature at which the rate coefficient has been measured. The products of these reactions include many of the hydrocarbons and their cyano-derivatives that have been detected in TMC-1, including c-C_5H_6, CH_3C_4H, $H_2CCCHCCH$, $CH_2CHCHCH_2$, CH_3C_6H, $CH_3CHCHCN$, $CH_2C(CH_3)CN$ and CH_2CHCH_2CN. Their calculated abundances are now more than 10^4 times less than observed – and the misery goes on, since these species are also involved in the synthesis of other complex molecules including c-C_5H_5CN, c-$C_5H_4CCH_2$, c-C_5H_5CCH and H_2CCCHC_3N. Identifying efficient, low-temperature formation pathways to propene is one of the most important tasks for astrochemistry today. Such 'single-point failures' also exist for other species in the networks. Indeed it is likely that somewhere in the region of 70–100 detected molecules are not yet included in even the most complex networks because their synthesis – in the gas phase or in ices – is unknown.

To conclude, let me mention four generic areas that we will need to cultivate in the coming years if we are to ensure that the field of astrochemistry continues both its remarkable growth over the past 10–20 years and its relevance to astrophysics far and near. The first is **synergy**. To me, one of the central messages of

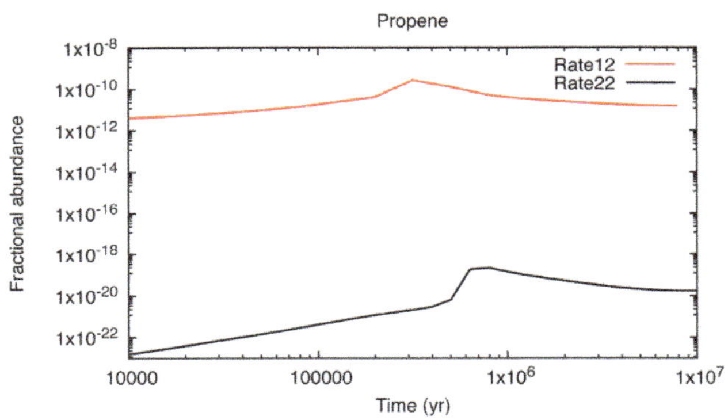

Fig. 2 The fractional abundance of propene relative to H_2 as a function of time for a model with properties similar to TMC-1 ($T = 10$ K, $n(H_2) = 10^4$ cm^{-3}).

this discussion meeting – funded by the Royal Society of Chemistry – is that research in core theoretical and experimental physical chemistry disciplines, including spectroscopy and gas-phase kinetics, plays a central role in astrochemistry. Collaborations between observers and modellers within the astronomy community were evident in the papers describing JWST and ALMA results and noticeably increase their impact.

Synergy across disciplines was also evident in and central to, many of the presentations and posters in this meeting and will increase in importance. For example, it is clear that many of the assumptions that underlie the grain-surface chemical kinetic models are too simplistic. Close co-operation between experimental and theoretical approaches on topics such as binding energies, surface and bulk diffusion, photodesorption yields, reaction products and desorption mechanisms, will be needed to ensure that theoretical models of interstellar chemistry become more predictive rather than reactive in nature. In a similar fashion, the astrochemical community needs to better identify critical reactions and processes for the chemical physics community to study.

The second is **strategy**. These synergistic links between chemistry and astronomy are often the result of individual actions, that is, where one researcher makes a measurement or calculation and reaches out to someone in the other field. There is an element of chance about these and indeed such interactions, while useful to a certain degree, may not have as much impact as one would like. We therefore need to find ways in which we can support research in a much more strategic manner. In the Netherlands, the Dutch Astrochemistry Network has been funded since 2010 and has supported many cross-disciplinary projects and, just as importantly, people. It not only brings together expertise in chemistry and astronomy but it also allows for the exchange of PhD students between institutions.

Such an approach is problematic in the UK where chemistry and astrophysics are funded by two separate research councils and where, in addition, many of the EU's cross-disciplinary programmes are ever more difficult to access. In the UK the joint RSC/RAS body, the Astrophysical Chemistry Group, provides a forum – but no funding – through which interested parties can interact.

There are major issues facing many of us in astrochemistry – can the astronomical community help shield experimental groups, often expensive to run, from the growing commodification of modern university operations, where the bottom line is, too often, profit? This is not a one-way support of course. Chemistry in its broadest sense is universal and as we move further into the era of the ALMA and the JWST, to more detailed explorations of exoplanet atmospheres and the planets, moons and minor bodies of our solar system, astronomers will need the chemical disciplines to help interpret and understand observational results.

The third area is **succession**. In my view, the paper by Herbst and Klemperer,[3] published 50 years ago, represents the birth of astrochemistry as a scientific discipline. Many of the scientists involved in the development of the foundations of the subject in its first 20 years, have retired or are close to retirement. They/we need to be active in planning their/our replacements. This means getting involved in departmental/university politics to ensure that the best people are recruited to, and promoted within, our discipline.

The final area is **students**. Our view of the molecular universe has improved enormously in the past 50 years. The development of high-resolution facilities, such as discussed here, provides us with an amazing panoply of images and spectra, coupled with a recognition of astrochemistry as central to answering many fundamental questions in astronomy, from the formation and evolution of galaxies to the formation of stars and planets. It was very noticeable at this meeting that the field has a wonderful array of talented, smart people who already possess a mature, informed, view of their research. The quality and impact of presentations and posters by PhD students and postdoctoral researchers was impressive as was their engagement in the discussion sessions. We need to ensure that, collectively, we harness those skills and encourage, excite, and enthuse our early career scientists. Where we can, let us support their intellectual growth, their exchange among institutions and across national boundaries, and let us find ways to employ them in full-time positions.

Conflicts of interest

There are no conflicts of interest to declare.

Acknowledgements

I am grateful to the Organising Committee for the invitation to present these closing remarks and to the Royal Society of Chemistry for the financial support that enabled me to do so in person. I am also grateful to the Leverhulme Trust for the award of an Emeritus Fellowship.

References

1 L. E. Snyder, D. Buhl, B. Zuckerman and P. Palmer, *Phys. Rev. Lett.*, 1969, **22**, 679–681.

2 D. R. Bates and L. Spitzer, Jr., *Astrophys. J.*, 1951, **113**, 441.

3 E. Herbst and W. Klemperer, *Astrophys. J.*, 1973, **185**, 505–534.

4 P. D. Brown, S. B. Charnley and T. J. Millar, *Mon. Not. R. Astron. Soc.*, 1988, **231**, 409–417.

5 S. B. Charnley, A. G. G. M. Tielens and T. J. Millar, *Astrophys. J.*, 1992, **399**, L71.

6 P. Caselli, T. I. Hasegawa and E. Herbst, *Astrophys. J.*, 1993, **408**, 548.

7 B. A. McGuire, A. M. Burkhardt, S. Kalenskii, C. N. Shingledecker, A. J. Remijan, E. Herbst and M. C. McCarthy, *Science*, 2018, **359**, 202–205.

8 M. C. McCarthy, K. L. K. Lee, R. A. Loomis, A. M. Burkhardt, C. N. Shingledecker, S. B. Charnley, M. A. Cordiner, E. Herbst, S. Kalenskii, E. R. Willis, C. Xue, A. J. Remijan and B. A. McGuire, *Nat. Astron.*, 2020, **5**, 176–180.

9 K. L. K. Lee, P. B. Changala, R. A. Loomis, A. M. Burkhardt, C. Xue, M. A. Cordiner, S. B. Charnley, M. C. McCarthy and B. A. McGuire, *Astrophys. J.*, 2021, **910**, 75.

10 A. M. Burkhardt, K. Long Kelvin Lee, P. Bryan Changala, C. N. Shingledecker, I. R. Cooke, R. A. Loomis, H. Wei, S. B. Charnley, E. Herbst, M. C. McCarthy and B. A. McGuire, *Astrophys. J.*, 2021, **913**, L18.

11 M. L. Sita, P. B. Changala, C. Xue, A. M. Burkhardt, C. N. Shingledecker, K. L. Kelvin Lee, R. A. Loomis, E. Momjian, M. A. Siebert, D. Gupta,

E. Herbst, A. J. Remijan, M. C. McCarthy, I. R. Cooke and B. A. McGuire, *Astrophys. J.*, 2022, **938**, L12.

12 J. Cernicharo, M. Agúndez, R. I. Kaiser, C. Cabezas, B. Tercero, N. Marcelino, J. R. Pardo and P. de Vicente, *Astron. Astrophys.*, 2021, **655**, L1.

13 J. Cernicharo, M. Agúndez, C. Cabezas, B. Tercero, N. Marcelino, J. R. Pardo and P. de Vicente, *Astron. Astrophys.*, 2021, **649**, L15.

14 I. R. Cooke, C. Xue, P. B. Changala, H. T. Shay, A. N. Byrne, Q. Y. Tang, Z. T. P. Fried, K. L. Kelvin Lee, R. A. Loomis, T. Lamberts, A. Remijan, A. M. Burkhardt, E. Herbst, M. C. McCarthy and B. A. McGuire, *Astrophys. J.*, 2023, **948**, 133.

15 S. Zeng, V. M. Rivilla, I. Jiménez-Serra, L. Colzi, J. Martín-Pintado, B. Tercero, P. de Vicente, S. Martín and M. A. Requena-Torres, *Mon. Not. R. Astron. Soc.*, 2023, **523**, 1448–1463.

16 R. Basalgète, A. J. Ocaña, G. Féraud, C. Romanzin, L. Philippe, X. Michaut, J.-H. Fillion and M. Bertin, *Astrophys. J.*, 2021, **922**, 213.

17 M. P. Collings, M. A. Anderson, R. Chen, J. W. Dever, S. Viti, D. A. Williams and M. R. S. McCoustra, *Mon. Not. R. Astron. Soc.*, 2004, **354**, 1133–1140.

18 L. Tinacci, A. Germain, S. Pantaleone, S. Ferrero, C. Ceccarelli and P. Ugliengo, *ACS Earth Space Chem.*, 2022, **6**, 1514–1526.

19 M. Hamberg, W. D. Geppert, R. D. Thomas, V. Zhaunerchyk, F. Österdahl, A. Ehlerding, M. Kaminska, J. Semaniak, M. af Ugglas, A. Källberg, A. Paal, A. Simonsson and M. Larsson, *Mol. Phys.*, 2007, **105**, 899–906.

20 W. D. Geppert, M. Hamberg, R. D. Thomas, F. Österdahl, F. Hellberg, V. Zhaunerchyk, A. Ehlerding, T. J. Millar, H. Roberts, J. Semaniak, M. A. Ugglas, A. Källberg, A. Simonsson, M. Kaminska and M. Larsson, *Faraday Discuss.*, 2006, **133**, 177.

21 D. Smith, P. Spanel and T. J. Millar, *Mon. Not. R. Astron. Soc.*, 1994, **266**, 31–34.

22 T. J. Millar, C. Walsh, M. Van de Sande and A. Markwick, *Astron. Astrophys.*, 2023, submitted for publication.

23 D. McElroy, C. Walsh, A. J. Markwick, M. A. Cordiner, K. Smith and T. J. Millar, *Astron. Astrophys.*, 2013, **550**, A36.

24 N. Marcelino, J. Cernicharo, M. Agúndez, E. Roueff, M. Gerin, J. Martín-Pintado, R. Mauersberger and C. Thum, *Astrophys. J.*, 2007, **665**, L127–L130.

25 Z. Lin, D. Talbi, E. Roueff, E. Herbst, N. Wehres, C. A. Cole, Z. Yang, T. P. Snow and V. M. Bierbaum, *Astrophys. J.*, 2013, **765**, 80.

26 J.-C. Loison and A. Bergeat, *Phys. Chem. Chem. Phys.*, 2009, **11**, 655–664.

27 S. B. Morales, S. D. Le Picard, A. Canosa and I. R. Sims, *Faraday Discuss.*, 2010, **147**, 155–171.

28 N. Daugey, P. Caubet, A. Bergeat, M. Costes and K. M. Hickson, *Phys. Chem. Chem. Phys.*, 2008, **10**, 729–737.

29 D. Chastaing, P. L. James, I. R. Sims and I. W. M. Smith, *Faraday Discuss.*, 1998, **109**, 165–181.

30 C. Berteloite, S. D. Le Picard, N. Balucani, A. Canosa and I. R. Sims, *Phys. Chem. Chem. Phys.*, 2010, **12**, 3677–3689.

Poster titles

The role of low-energy (<20 eV) electrons in astrochemistry, **Q. T. Wu, K. Barnes, M. Padovani, J. Battat and C. Arumainayagam,** *Wellesley College, USA*

Chemical complexity and role of condensed film structure: methane adsorption and propene oxidation, **M. Brann, R. Thompson and S. J. Sibener,** *The University of Chicago, USA*

Gas sources from the coma and nucleus of comet 46P/Wirtanen observed using ALMA, **M. A. Cordiner, N. X. Roth, S. N. Milam, G. L. Villanueva, D. Bockelee-Morvan, A. J. Remijan, S. B. Charnley, N. Biver, D. C. Lis, C. Qi, B. P. Bonev and J. Crovisier,** *NASA Goddard Space Flight Center, USA*

Discovery of an elevated $^{15}N/^{14}N$ ratio in the Jupiter-family comet 46P/Wirtanen using ALMA, **K. Darnell, M. A. Cordiner, S. Milam, N. Roth, S. Charnley, A. Remijan, N. Biver and D. Bockelee-Morvan,** *NASA Goddard Space Flight Center, USA*

Oscillations in astrochemical modelling, **G. Dufour, S. B. Charnley and J. Lindberg,** *NASA Goddard Space Flight Center, USA*

Anharmonic IR absorption spectra of prototypical neutral and cation astronomical PAHs, **V. J. Esposito, C. Boersma, L. J. Allamandola and R. C. Fortenberry,** *NASA Ames Research Center, USA*

Cyclopropenylidene chemistry in the interstellar medium, **A. R. Flint, C. Z. Palmer, A. G. Watrous, B. R. Westbrook, D. J. Patel and R. C. Fortenberry,** *University of Mississippi, USA*

The effects of simple alcohols on the crystallisation behaviour of ASW, **J. E. Fulker and W. A. Brown,** *University of Sussex, UK*

Enhanced star formation through the high-temperature formation of H_2 on carbonaceous dust grains, **F. Grieco, P. Theulé, I. De Looze and F. Dulieu,** *CY Cergy Paris University, France*

CPICE: A new instrument coupling broadband rotational spectroscopy astrochemical ices, **T. Hager, Q. Borengasser and B. Broderick,** *University of Missouri, USA*

Quantification of methanol photolysis branching ratios using multiplexed photoionization mass spectrometry, **E. K. Hockey, T. Howard, J. Palotás, D. L. Osborn and L. G. Dodson,** *University of Maryland, USA*

Photodissociation and photoionization of molecules of astronomical interest: Updates to the Leiden photodissociation & photoionization cross section database, **H. R. Hróðmarsson and E. F. van Dishoeck,** *Paris-East Créteil University, France*

Laboratory measurements of N_2 reacting with H_3^+ isotopologues, **D. Ivanov, C. Bu, P. -M. Hillenbrand, L. W. Isberner, D. Schury, X. Urbain and D. W. Savin,** *Columbia University, USA*

CO_2 formation on interstellar H_2O ice, **H. Kakkar and A. Rimola,** *Autonomous University of Barcelona, Spain*

Interpret mid-infrared gas-phase H_2O absorption ro-vibrational lines probed by JWST/MIRI spectroscopy, **J. Li, A. Boogert and A. Tielens,** *University of Maryland, USA*

Ice origins of OCS: preliminary results, **R. Martín-Doménech, K. I. Öberg and G. M. Muñoz Caro,** *Centre for Astrobiology (CSIC/INTA), Spain*

Measuring isotopic ratios in Titan's atmospheric nitriles, **J. Nosowitz, M. Cordiner, C. Nixon, A. Thelen, S. Charnley, N. Teanby and V. Vuitton,** *The Catholic University of America, USA*

Fingerprinting fragments of fragile interstellar molecules, **D. B. Rap, A. Simon, K. Steenbakkers, J. G. M. Schrauwen, B. Redlich, S. Brünken,** *Radboud University, Netherlands*

Measuring the icy content of protoplanetary disks with JWST, directly, **J. A. Sturm and M. K. McClure,** *Leiden University, Netherlands*

Low temperature reaction kinetics inside an extended Laval nozzle: REMPI characterization and detection by broadband rotational spectroscopy, **S. Thawoos, N. Suas-David, M. Edlin, R. Gurusinghe and A. G. Suits,** *University of Missouri, USA*

Quantifying non-thermal desorption from NH_3 ices: a comparative study of photon and electron irradiation in the valence- and core-shell energy ranges, **D. Torres-Díaz, R. Basalgète, L. Sala, A. Hacquard, D. Toulouse, L. Janin, J. A. Noble, S. Del Fré, X. Michaut, L. Philippe, G. Feraud, M. Bertin, J. -H. Fillion, L. Amiaud and A. Lafosse,** *Paris-Saclay University, France*

Low-energy rotational energy transfer in water-molecular hydrogen collisions: experiments and theory, **L. Wiesenfeld,** *Paris-Saclay University, France*

The affiliation of the presenting author is listed above.

List of participants

Miss Katie Ackermann, *Royal Society of Chemistry*, *United Kingdom*
Professor Nadia Balucani, *Università degli Studi di Perugia - DCBB*, *Italy*
Miss Kennedy Barnes, *Wellesley College*, *USA*
Dr Mathieu Bertin, *LERMA, Sorbonne Université*, *France*
Dr Eleonora Bianchi, *ORIGINS Excellence Cluster*, *Germany*
Dr Michelle Brann, *National Institute of Standards and Technology (NIST)*, *USA*
Professor Bernadette Broderick, *University of Missouri, Columbia*, *USA*
Professor Stefan Bromley, *University of Barcelona / ICREA*, *Spain*
Professor Wendy Brown, *University of Sussex*, *United Kingdom*
Dr Sandra Brünken, *FELIX Laboratory, Radboud University*, *Netherlands*
Professor Cecilia Ceccarelli, *IPAG, Universitè Grenoble Alpes*, *France*
Dr Martin Cordiner, *NASA Goddard Space Flight Center*, *USA*
Mr Eric Crump, *National Institute of Standards and Technology (NIST)*, *USA*
Ms Kristen Darnell, *NASA Goddard Space Flight Center*, *USA*
Dr Kevin Douglas, *University of Leeds*, *United Kingdom*
Dr Gwenaëlle Dufour, *Catholic University of America / NASA Goddard Space Flight Center*, *USA*
Dr Vincent Esposito, *NASA Ames Research Center*, *USA*
Dr Adam Fleisher, *National Institute of Standards and Technology (NIST)*, *USA*
Mx Athena Flint, *University of Mississippi*, *USA*
Mr Jack Fulker, *University of Sussex*, *United Kingdom*
Professor Robin Garrod, *University of Virginia*, *USA*
Dr Joel Green, *Space Telescope Science Institute*, *USA*
Mr Francesco Grieco, *Ghent University*, *Belgium*
Dr Murthy Gudipati, *NASA Jet Propulsion Laboratory / California Institute of Technology*, *USA*
Dr Divita Gupta, *University of Cologne*, *Germany*
Mr Travis Hager, *University of Missouri*, *USA*
Emer Prof Joshua Halpern, *LibreTexts*, *USA*
Dr Christine V. Hampton, *CV Hampton Consulting, LLC*, *USA*
Professor Dwayne Heard, *University of Leeds*, *United Kingdom*
Professor Eric Herbst, *University of Virginia*, *USA*
Miss Emily Hockey, *University of Maryland*, *USA*
Mr Samuel Howell, *Royal Society of Chemistry*, *United Kingdom*
Dr Helgi Hróðmarsson, *LISA Laboratory, Université Paris-Est Créteil*, *France*
Dr Ko-Yun Huang, *Leiden Observatory*, *Netherlands*
Dr Michael Hurst, *Royal Society of Chemistry*, *USA*
Dr Dmitry Ivanov, *Columbia University*, *USA*
Dr Ugo Jacovella, *Université Paris-Saclay / CNRS / ISMO*, *France*
Ms Harjasnoor Kakkar, *Universitat Autònoma de Barcelona*, *Spain*
Professor Inga Kamp, *Kapteyn Astronomical Institute, University of Groningen*, *Netherlands*

Dr Beatrice Kulterer, *Center for Space and Habitability, University of Bern, Switzerland*
Miss Sarah Latham, *Royal Society of Chemistry*, *United Kingdom*
Dr Alexander Lemmens, *Lawrence Berkeley Lab*, *USA*
Dr Jialu Li, *University of Maryland College Park*, *USA*
Professor Nikku Madhusudhan, *Institute of Astronomy, University of Cambridge*, *United Kingdom*
Dr Rafael Martin Doménech, *Centro de Astrobiologia*, *Spain*
Professor Martin McCoustra, *Heriot-Watt University*, *United Kingdom*
Professor Brett McGuire, *Massachusetts Institute of Technology / National Radio Astronomy Observatory*, *USA*
Professor Dr Anthony Meijer, *University of Sheffield*, *United Kingdom*
Dr Stefanie Milam, *NASA Goddard Space Flight Center*, *USA*
Professor Tom Millar, *Queen's University Belfast*, *United Kingdom*
Mr Jonathon Nosowitz, *Catholic University of America / NASA Goddard Space Flight Center*, *USA*
Dr Samuel Oldknow, *Royal Society of Chemistry*, *United Kingdom*
Mrs Karolina Plakitina, *Institute of Astronomy, Russian Academy of Sciences*, *Russian Federation*
Professor Dr Cristina Puzzarini, *University of Bologna*, *Italy*
Mr Daniel Rap, *FELIX Laboratory, Radboud University*, *Netherlands*
Dr Neill Reid, *Space Telescope Science Institute*, *USA*
Dr W. M. C. Sameera, *Hokkaido University*, *Japan*
Mrs Rachel Scott, *Royal Society of Chemistry*, *United Kingdom*
Professor Ian Sims, *University of Rennes*, *France*
Dr Mark Stockett, *Stockholm University*, *Sweden*
Mr Ardjan Sturm, *Leiden Observatory*, *Netherlands*
Professor Arthur Suits, *University of Missouri*, *USA*
Dr Ge Sun, *Dalian Institute of Chemical Physics, CAS*, *China*
Ms Daniela Torres Díaz, *Université Paris-Saclay / CNRS / ISMO*, *France*
Dr Takashi Toyao, *Hokkaido University*, *Japan*
Dr Marie Van de Sande, *University of Leeds*, *United Kingdom*
Professor Ewine van Dishoeck, *Leiden Observatory*, *Netherlands*
Professor Serena Viti, *Leiden University / University College London*, *Netherlands / United Kingdom*
Dr Nicholas Walker, *Newcastle University*, *United Kingdom*
Dr Susan Weatherby, *Royal Society of Chemistry*, *United Kingdom*
Dr Susanna Widicus Weaver, *University of Wisconsin-Madison*, *USA*
Dr Laurent Wiesenfeld, *Université Paris-Saclay / CNRS*, *France*
Dr Olivia Wilkins, *NASA Goddard Space Flight Center*, *USA*

